T0270912

Non-Hausdorff Topology and Domain Theory

This unique book on modern topology looks well beyond traditional treatises, and explores spaces that may, but need not, be Hausdorff. This is essential for domain theory, the cornerstone of semantics of computer languages, where the Scott topology is almost never Hausdorff. For the first time in a single volume, this book covers basic material on metric and topological spaces, advanced material on complete partial orders, Stone duality, stable compactness, quasi-metric spaces, and much more. An early chapter on metric spaces serves as an invitation to the topic (continuity, limits, compactness, completeness) and forms a complete introductory course by itself.

Graduate students and researchers alike will enjoy exploring this treasure trove of results. Full proofs are given, as well as motivating ideas, clear explanations, illuminating examples, application exercises, and some more challenging problems for advanced readers.

JEAN GOUBAULT-LARRECQ is Full Professor of Computer Science at Ecole Normale Supérieure de Cachan, France. He obtained his Ph.D. in 1993 from Ecole Polytechnique in the field of automated deduction, and since then he has had an active career in several fields of computer science: logic, computer security, semantics, domain theory, and probabilistic and non-deterministic systems. He is currently heading team SECSI (security of information systems) at INRIA, France's national institute for research in computer science and control. He is the recipient of the 2011 CNRS Silver Medal in the field of computer science and its interactions. This is the highest scientific distinction in computer science in France.

NEW MATHEMATICAL MONOGRAPHS

All the titles listed below can be obtained from good booksellers or from Cambridge University Press. For a complete series listing visit www.cambridge.org/mathematics.

1. M. Cabanes and M. Enguehard *Representation Theory of Finite Reductive Groups*
2. J. B. Garnett and D. E. Marshall *Harmonic Measure*
3. P. Cohn *Free Ideal Rings and Localization in General Rings*
4. E. Bombieri and W. Gubler *Heights in Diophantine Geometry*
5. Y. J. Ionin and M. S. Shrikhande *Combinatorics of Symmetric Designs*
6. S. Berhanu, P. D. Cordaro and J. Hounie *An Introduction to Involutive Structures*
7. A. Shlapentokh *Hilbert's Tenth Problem*
8. G. Michler *Theory of Finite Simple Groups I*
9. A. Baker and G. Wüstholz *Logarithmic Forms and Diophantine Geometry*
10. P. Kronheimer and T. Mrowka *Monopoles and Three-Manifolds*
11. B. Bekka, P. de la Harpe and A. Valette *Kazhdan's Property (T)*
12. J. Neisendorfer *Algebraic Methods in Unstable Homotopy Theory*
13. M. Grandis *Directed Algebraic Topology*
14. G. Michler *Theory of Finite Simple Groups II*
15. R. Schertz *Complex Multiplication*
16. S. Bloch *Lectures on Algebraic Cycles (2nd Edition)*
17. B. Conrad, O. Gabber and G. Prasad *Pseudo-reductive Groups*
18. T. Downarowicz *Entropy in Dynamical Systems*
19. C. Simpson *Homotopy Theory of Higher Categories*
20. E. Fricain and J. Mashreghi *The Theory of H(b) Spaces I*
21. E. Fricain and J. Mashreghi *The Theory of H(b) Spaces II*
22. J. Goubault-Larrecq *Non-Hausdorff Topology and Domain Theory*
23. J. Śniatycki *Differential Geometry of Singular Spaces and Reduction of Symmetry*

Non-Hausdorff Topology and Domain Theory

JEAN GOUBAULT-LARRECQ
Ecole Normale Supérieure de Cachan, France

Shaftesbury Road, Cambridge CB2 8EA, United Kingdom

One Liberty Plaza, 20th Floor, New York, NY 10006, USA

477 Williamstown Road, Port Melbourne, VIC 3207, Australia

314–321, 3rd Floor, Plot 3, Splendor Forum, Jasola District Centre, New Delhi – 110025, India

103 Penang Road, #05–06/07, Visioncrest Commercial, Singapore 238467

Cambridge University Press is part of Cambridge University Press & Assessment,
a department of the University of Cambridge.

We share the University's mission to contribute to society through the pursuit of education, learning and research at the highest international levels of excellence.

www.cambridge.org
Information on this title: www.cambridge.org/9781107034136

First published 2013

A catalogue record for this publication is available from the British Library

ISBN 978-1-107-03413-6 Hardback

Contents

1 Introduction 1

2 Elements of set theory 4
- 2.1 Foundations 4
- 2.2 Finiteness, countability 6
- 2.3 Order theory 8
- 2.4 The Axiom of Choice 15

3 A first tour of topology: metric spaces 18
- 3.1 Metric spaces 18
- 3.2 Convergence, limits 20
- 3.3 Compact subsets 25
- 3.4 Complete metric spaces 33
- 3.5 Continuous functions 37

4 Topology 46
- 4.1 Topology, topological spaces 46
- 4.2 Order and topology 53
- 4.3 Continuity 62
- 4.4 Compactness 67
- 4.5 Products 74
- 4.6 Coproducts 80
- 4.7 Convergence and limits 82
- 4.8 Local compactness 91
- 4.9 Subspaces 95
- 4.10 Homeomorphisms, embeddings, quotients, retracts 97
- 4.11 Connectedness 106
- 4.12 A bit of category theory I 111

5 Approximation and function spaces 120
- 5.1 The way-below relation 120
- 5.2 The lattice of open subsets of a space 140

5.3 Spaces of continuous maps 149
5.4 The exponential topology 151
5.5 A bit of category theory II 169
5.6 $\mathcal{C}$-generated spaces 180
5.7 bc-domains 194

6 Metrics, quasi-metrics, hemi-metrics 203
6.1 Metrics, hemi-metrics, and open balls 203
6.2 Continuous and Lipschitz maps 209
6.3 Topological equivalence, hemi-metrizability, metrizability 214
6.4 Coproducts, quotients 236
6.5 Products, subspaces 239
6.6 Function spaces 241
6.7 Compactness and symcompactness 251

7 Completeness 260
7.1 Limits, d-limits, and Cauchy nets 260
7.2 A strong form of completeness: Smyth-completeness 267
7.3 Formal balls 279
7.4 A weak form of completeness: Yoneda-completeness 288
7.5 The formal ball completion 311
7.6 Choquet-completeness 323
7.7 Polish spaces 334

8 Sober spaces 341
8.1 Frames and Stone duality 341
8.2 Sober spaces and sobrification 351
8.3 The Hofmann–Mislove Theorem 364
8.4 Colimits and limits of sober spaces 383

9 Stably compact spaces and compact pospaces 397
9.1 Stably locally compact spaces, stably compact spaces 397
9.2 Coproducts and retracts of stably compact spaces 411
9.3 Products and subspaces of stably compact spaces 413
9.4 Patch-continuous, perfect maps 419
9.5 Spectral spaces 423
9.6 Bifinite domains 440
9.7 Noetherian spaces 453

References 475
Notation index 480
Index 484

1

Introduction

Purpose This book is an introduction to some of the basic concepts of topology, especially of *non-Hausdorff* topology. I will of course explain what it means (Definition 4.1.12). The important point is that traditional topology textbooks assume the Hausdorff separation condition from the very start, and contain very little information on non-Hausdorff spaces. But the latter are important already in algebraic geometry, and crucial in fields such as domain theory.

Conversely, domain theory (Abramsky and Jung, 1994; Gierz *et al.*, 2003), which arose from logic and computer science, started as an outgrowth of theories of order. Progress in this domain rapidly required a lot of material on (non-Hausdorff) topologies.

After about 40 years of domain theory, one is forced to recognize that topology and domain theory have been beneficial to each other. I've already mentioned what domain theory owes to topology. Conversely, in several respects, domain theory, in a broad sense, is *topology done right*.

This book is an introduction to both fields, dealt with as one. This seems to fill a gap in the literature, while bringing them forth in a refreshing perspective.

Secondary purpose This book is self-contained. My main interest, though, as an author, was to produce a unique reference for the kind of results in topology and domain theory that I needed in research I started in 2004, on semantic models of mixed non-deterministic and probabilistic choice. The goal quickly grew out of proportion, and will therefore occupy several volumes. The current book can be seen as the preliminaries for other books on these other topics. Some colleagues of mine, starting with Professor Alain Finkel, have stressed that these preliminaries are worthy of interest, independently of any specific application.

Teaching This book is meant to be used as a reference. My hope is that it should be useful to researchers and to people who are curious to read about a modern view of point-set topology. It did not arise as lecture notes, and I don't think it can be used as such directly. If you plan on using this book as a basis for lectures, you should extract a few selected topics. Think of yourself as a script writer and this book as a novel, and pretend that your job is to produce a shorter script for a feature-length movie, and skip the less important subplots.

Exercises There are many exercises, spread over the whole text. Some of them are meant for training, i.e., to help you understand notions better, to make you more comfortable with definitions and theorems that we have just seen. Some others are there to help you understand notions in a deeper way, or to go further. This is traditional in mathematical textbooks. It is therefore profitable to read all exercises: those of the second kind in particular state additional theorems, which you can prove for yourself (sometimes with the help of hints), but do not need to. It will happen that not only solutions of exercises but also some proofs of theorems will depend on results we shall have seen in previous exercises. There is no pressure on you to actually do any exercise, and you can decide to take them as a mere source of additional information.

What is not covered Topology is an extremely rich topic, and I could not cover all subtopics in a book of reasonable size. (Notwithstanding the fact that I certainly do not know everything in topology.) I decided to make a selection among those subtopics that pleased me most. Some other topics were necessarily left out, despite them being equally interesting. For example, algebraic topology will not be touched upon at all. Uniform and quasi-uniform spaces, bitopological spaces, Lindelöf spaces, Souslin spaces, and topological group theory were left out as well. Topological convexity, topological measure theory, hyperspaces, and powerdomains will be treated in further volumes.

Dependencies Reading $u \leftarrow v$ as "v depends on u," the structure of the book is as follows. Dashed arrows mark weaker dependencies.

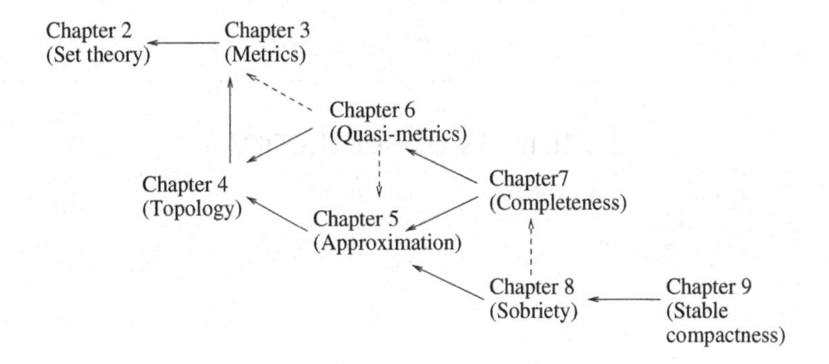

Acknowledgments There are many people without whom this book would never have become reality. I would like to address special thanks to Mike Mislove, Prakash Panangaden, Mai Gehrke, Achim Jung, Jean-Eric Pin, Klaus Keimel, Jimmie Lawson, Martín Escardó, Vincent Danos, Alain Finkel, Philippe Schnoebelen, Christophe Fouqueré, Ernst-Erich Doberkat, Alexander Kurz, Richard Lassaigne, Phil Scott, and Catuscia Palamidessi – some of them because they explained some proofs, or the importance of some notions, or because they expressed support at some time or another, in some way or another. I have probably forgotten some others, which I will regret not having cited later on; I hope they will forgive me.

This project was partially funded by the ACI nouvelles interfaces des mathématiques "GéoCal," and by the ARC "ProNobis" from INRIA, and was partially supported by a six-month leave (CRCT) by the author at Laboratoire PPS, Université Paris Diderot, from January to June 2010.

2

Elements of set theory

We recapitulate the axiomatic foundations on which this book is based in Section 2.1. We then recall a few points about finiteness and countability in Section 2.2 and some basics of order theory in Section 2.3, and discuss the Axiom of Choice and some of its consequences in Section 2.4. These points will be needed often in the rest of this book. If you prefer to read about topology right away, and feel confident enough, please proceed directly to Chapter 3.

2.1 Foundations

We shall rest on ordinary set theory. While the latter has been synonymous with Zermelo–Fraenkel (ZF) set theory with the Axiom of Choice (ZFC) for some time, we shall use von Neumann–Gödel–Bernays (VBG) set theory instead (Mendelson, 1997).

There is not much difference between these theories: VBG is a conservative extension of ZFC. That VBG is an extension means that any theorem of ZFC is also a theorem of VBG. That it is conservative means that any theorem of VBG that one can express in the language of ZFC is also provable in ZFC.

The main difference between VBG and ZFC is that the former allows one to talk about collections that are too big to be sets. This is required, in all rigor, in the definition of (big) graphs and categories of Section 4.12. VBG allows us to talk about, say, the collection V of all sets, although V cannot itself be a set. This is the essence of Russell's paradox: assume there were a set V of all sets. Then $A = \{x \in V \mid x \notin x\}$ is a set. The rather mind-boggling argument is that, first, $A \notin A$ since if A were in A, then A would be an x such that $x \notin x$ by definition of A. Since $A \notin A$, A is an x such that $x \notin A$, so A is in A, a contradiction.

VBG avoids this pitfall by not recognizing V as a set, but as something of a different nature called a collection (or a *class*). Here is an informal description of the axioms of VBG set theory.

VBG set theory is a theory of *collections*. The formulae of VBG are those obtained using the language of first-order logic, i.e., using true ($\top$), false ($\bot$), logical and ($\wedge$), or ($\vee$), negation ($\neg$), implication ($\Rightarrow$), equivalence ($\Leftrightarrow$), and universal ($\forall$) and existential ($\exists$) quantification, on a language whose predicate symbols are $\in$ ("belongs to," "is in," "is an element of"), $=$ (equality), and m ("being small"). The small collections, i.e., the collections A such that $m(A)$ holds, are called *sets*. Collections must obey the following axioms:

- (Elements are small) Any element of a collection is small: $\forall x, A \cdot x \in A \Rightarrow m(x)$.
- (Extensionality) Any two collections with the same elements are equal: $\forall A, B \cdot (\forall x \cdot x \in A \Leftrightarrow x \in B) \Rightarrow A = B$.
- (Comprehension) Say that a formula F is a set formula iff all the quantifications in F are over sets, not collections. That is, the syntax of set formulae is $F, G, \ldots ::= \top \mid \bot \mid m(x) \mid x \in y \mid x = y \mid F \wedge G \mid F \vee G \mid \neg F \mid F \Rightarrow G \mid F \Leftrightarrow G \mid \forall x \cdot m(x) \Rightarrow F \mid \exists x \cdot m(x) \wedge F$. The comprehension schema states that, for each set formula $F(x, \vec{y})$ (where we make explicit the set of collections variables $x, \vec{y} = y_1, \ldots, y_n$ that it may depend on), there is a collection $\{x \mid F(x, \vec{y})\}$ of all elements x such that $F(x, \vec{y})$ holds: $\forall \vec{y} \cdot \exists A \cdot \forall x \cdot x \in A \Leftrightarrow F(x, \vec{y})$.
- (Empty set) the empty collection $\emptyset = \{x \mid \bot\}$ is a set: $m(\emptyset)$.
- (Pairing) Given any two sets x, y, their pair $\{x, y\} = \{z \mid z = x \vee z = y\}$ is a set: $m(\{x, y\})$. The ordered pair (x, y) is encoded as $\{\{x, x\}, \{x, y\}\}$. It follows that $\{x\} = \{x, x\}$ is a set as well.
- (Union) Given any set x, the collection $\bigcup x = \bigcup_{y \in x} y = \{z \mid \exists y \cdot m(y) \wedge y \in x\}$ is a set; i.e., $m(\{z \mid \exists y \cdot m(y) \wedge y \in x\})$. We write $x \cup y$ for $\bigcup\{x, y\}$.
- (Powerset) Given any set x, the collection $\mathbb{P}(x)$ of all subsets of x is a set, i.e., $m(\{z \mid z \subseteq x\})$, where we write $z \subseteq x$ for $\forall y \cdot m(y) \Rightarrow (y \in z \Rightarrow y \in x)$, or equivalently $\forall y \cdot y \in z \Rightarrow y \in x$. For every set A, $\mathbb{P}(A)$ is called the *powerset* of A.
- (Infinity) Let $0 = \emptyset$, $x + 1 = x \cup \{x\}$. There is a collection $\mathbb{N}$ containing 0 and such that $x \in \mathbb{N}$ implies $x + 1 \in \mathbb{N}$. One may even define the smallest such collection, $\mathbb{N}$, as $\{x \mid \forall N \cdot m(N) \Rightarrow (0 \in N \wedge (\forall z \cdot m(z) \Rightarrow (z \in N \Rightarrow z + 1 \in N)) \Rightarrow x \in N)\}$. The axiom of infinity states that $\mathbb{N}$ is small: $m(\mathbb{N})$. $\mathbb{N}$ is the set of natural numbers $\{0, 1, 2, \ldots\}$.

- (Foundation) There is no infinite chain $\cdots \in x_k \in \cdots \in x_2 \in x_1$. This is usually formulated in the more cryptic, but equivalent form: Every non-empty class is disjoint from one of its elements. One may as well axiomatize this by an induction axiom, stating that we may induct along the relation $\in$: $\forall A \cdot (\forall x \cdot m(x) \wedge (\forall y \cdot y \in x \Rightarrow y \in A) \Rightarrow x \in A) \Rightarrow \forall x \cdot m(x) \Rightarrow x \in A$, i.e., any class A that contains every set x whenever it contains all smaller sets (i.e., all sets y such that $y \in x$) must in fact contain all sets.
- (Replacement) The image of a set under a function is a set. Functions f are encoded as their graphs, i.e., as the class of all pairs $(x, f(x))$: so a function is any class f such that $\forall x, y, z \cdot (x, y) \in f \wedge (x, z) \in f \Rightarrow y = z$. Replacement states that, for every function f and every set A, its *image* $f[A] = \{y \mid \exists x \cdot m(x) \wedge x \in A \wedge (x, y) \in f\}$ is small.
- (Axiom of Choice) Given any function f, call the *domain* of f the set of elements x such that $f(x)$ is defined, i.e., the set of elements x such that, for some y $(= f(x))$, (x, y) is in f. The Axiom of Choice states that, for every function F such that, for every x in its domain, $F(x)$ is a non-empty set, there is a function f whose domain is the same as F, and such that $f(x) \in F(x)$ for every x in the domain of F. In other words, f singles out one element $f(x)$ from each set $F(x)$. We discuss this axiom in Section 2.4.

We use standard abbreviations throughout; e.g., a set A *intersects*, or *meets*, a set B if and only if $A \cap B$ is non-empty, i.e., if and only if $A \cap B \neq \emptyset$.

2.2 Finiteness, countability

A set A is *finite* iff one can write it $\{x_1, \ldots, x_n\}$ for some $n \in \mathbb{N}$. We can assume that $x_1, \ldots, x_n$ are pairwise distinct. A formal definition may be: A is finite iff there is a bijection from A to some subset of the form $\downarrow n = \{m \in \mathbb{N} \mid m \leqslant n\}$, $n \in \mathbb{N}$.

A set is *infinite* iff it is not finite.

A set A is *countable* iff there is a bijection from A and some (arbitrary) subset of $\mathbb{N}$. It is equivalent to say that A is finite or *countably infinite*, where A is countably infinite iff there is a bijection from A to the whole of $\mathbb{N}$.

Every non-empty countable set A can be written $\{x_i \mid i \in \mathbb{N}\}$, i.e., there is a surjective map $i \mapsto x_i$ from $\mathbb{N}$ to A. (This is not true if A is empty.)

Every subset of a finite set is finite, and similarly every subset of a countable set is countable. In particular, any non-empty intersection of countable sets is countable.

We observe that $\mathbb{N} \times \mathbb{N}$ is countable. One possible bijection $m, n \mapsto \ulcorner m, n \urcorner$ is defined as follows. We can write any natural number in base 2, as an infinite word $\ldots a_i a_{i-1} \ldots a_2 a_1 a_0$, where each a_i is in $\{0, 1\}$, and all but finitely many equal 0: this word represents the natural number $\sum_{i=0}^{+\infty} a_i 2^i$. Now, when m is written in base 2 as $\ldots a_i a_{i-1} \ldots a_2 a_1 a_0$ and n as $\ldots b_i b_{i-1} \ldots b_2 b_1 b_0$, define $\ulcorner m, n \urcorner = \ldots a_i b_i a_{i-1} b_{i-1} \ldots a_2 b_2 a_1 b_1 a_0 b_0$.

It follows that the product $A \times B = \{(x, y) \mid x \in A, y \in B\}$ of two countable sets is countable.

Every countable union of countable sets is countable; i.e., (for non-empty sets) if A_k is the countable set $\{x_{ki} \mid i \in \mathbb{N}\}$ for each $k \in \mathbb{N}$, then $\bigcup_{k \in \mathbb{N}} A_k$ is the set $\{x_{ki} \mid \ulcorner k, i \urcorner \in \mathbb{N}\}$, which is countable.

We deduce that the set $\mathbb{Z}$ of all integers, negative, zero, or positive, is countable: indeed $\mathbb{Z}$ is in bijection with the union of $\mathbb{N}$ with $\{-n-1 \mid n \in \mathbb{N}\}$.

It also follows that the set $\mathbb{Q}$ of rational numbers is countable. Indeed, consider the set A of pairs $(m, n) \in \mathbb{Z} \times \mathbb{N}$ such that m and $n+1$ have no common divisor other than 1. A is countable, as a subset of the countable set $\mathbb{Z} \times \mathbb{N}$, and there is a bijection from A to $\mathbb{Q}$, which sends (m, n) to $\frac{m}{n+1}$.

The *finite powerset* $\mathbb{P}_{\text{fin}}(A)$ of a set A is the set of all finite subsets of A. When A is countable, $\mathbb{P}_{\text{fin}}(A)$ is countable, too. This is obvious when A is empty; otherwise it is enough to show that $\mathbb{P}_{\text{fin}}(\mathbb{N})$ is countable: the required bijection $E \in \mathbb{P}_{\text{fin}}(\mathbb{N}) \mapsto \ulcorner E \urcorner \in \mathbb{N}$ can be defined, for example, by $\ulcorner E \urcorner = \sum_{i \in E} 2^i$; its inverse maps every number m written in base 2 as $\ldots a_i a_{i-1} \ldots a_2 a_1 a_0$ to the finite set of all indices i such that $a_i = 1$.

However, the *powerset* $\mathbb{P}(A)$ of *all* subsets of A is not countable in general, even when A is countable. To wit, $\mathbb{P}(\mathbb{N})$ is not countable. This would indeed imply the existence of a surjective map from $\mathbb{N}$ to $\mathbb{P}(\mathbb{N})$, contradicting *Cantor's Theorem*:

Theorem 2.2.1 (Cantor) *For any set A, there is no surjective map from A to* $\mathbb{P}(A)$.

Proof Assume there was one, say f. Let $B = \{x \in A \mid x \notin f(x)\}$. Since f is surjective, there is an $x \in A$ such that $B = f(x)$. If $x \in B$, i.e., if $x \in f(x)$, by definition x is not in B: contradiction. So x is not in B. This implies that $x \notin f(x)$ since $B = f(x)$, so $x \in B$ by definition of B, again a contradiction. □

We call *uncountable* any set that is not countable.

The set $\{0, 1\}^\omega$ of all infinite words on the alphabet $\{0, 1\}$ is not countable either. An *infinite word* on an alphabet Σ is just a sequence of letters in Σ, i.e., a map from $\mathbb{N}$ to Σ. The map sending $w \in \{0, 1\}^\omega$ to the set $\{i \in \mathbb{N} \mid w(i) = 1\}$ is a bijection between $\{0, 1\}^\omega$ and $\mathbb{P}(\mathbb{N})$, so $\{0, 1\}^\omega$ cannot be countable.

It follows that the interval $[0, 1]$ of $\mathbb{R}$ is not countable either. Indeed, define the following function: for each $x \in [0, 1)$, write x in base 3 as $\sum_{k=0}^{+\infty} a_k/3^{k+1}$; if every a_k is in $\{0, 2\}$, then map x to the infinite word $\frac{a_0}{2}\frac{a_1}{2}\ldots\frac{a_k}{2}\ldots$; otherwise map x to an arbitrary word; finally, map 1 to the all one word $11\ldots1\ldots$ This is clearly surjective. If $[0, 1]$ were countable, there would also be a surjection from $\mathbb{N}$ to $[0, 1]$, which, composed with the above surjection, would yield a surjection from $\mathbb{N}$ to $\mathbb{P}(\mathbb{N})$, contradicting Cantor's Theorem.

From this, we deduce that $\mathbb{R}$ itself is not countable either. In fact, no subset A of $\mathbb{R}$ that contains a non-empty interval $[a-\epsilon, a+\epsilon]$, $\epsilon > 0$, can be countable. Indeed, this interval is in bijection with $[0, 1]$, by the map $t \mapsto (t-a+\epsilon)/(2\epsilon)$.

2.3 Order theory

2.3.1 Orderings, quasi-orderings

A *binary relation* R on X is any subset of $X \times X$. We write $x \mathrel{R} y$ instead of $(x, y) \in R$. R is *reflexive* iff $x \mathrel{R} x$ for every $x \in X$. R is *transitive* iff whenever $x \mathrel{R} y$ and $y \mathrel{R} z$, then $x \mathrel{R} z$. A reflexive and transitive relation is called a *quasi-ordering*, and is usually written $\leqslant$. A *quasi-ordered set* is any set equipped with a quasi-ordering. When $x \leqslant y$, we say that x is *below* y, or that y is *above* x.

R is *antisymmetric* iff whenever $x \mathrel{R} y$ and $y \mathrel{R} x$, then $x = y$. A *partial ordering*, or *ordering* for short, on X is any antisymmetric quasi-ordering on X. A set with an ordering is called a *partially ordered set* or a *poset*.

A binary relation R is *symmetric* iff whenever $x \mathrel{R} y$, then $y \mathrel{R} x$. An *equivalence relation* is any reflexive, symmetric, and transitive relation. We shall usually write $\equiv$ for equivalence relations. The *equivalence class* $q_\equiv(x)$ of $x \in X$ is the set of all $y \in X$ such that $x \equiv y$. The *quotient* $X/{\equiv}$ is the set of all equivalence classes of X.

Given any quasi-ordering $\leqslant$ on X, we can define an equivalence relation $\equiv$ by $x \equiv y$ iff $x \leqslant y$ and $y \leqslant x$. If $\leqslant$ is an ordering, then $\equiv$ is just the equality relation $=$. Otherwise, define $q_\equiv(x) \leqslant q_\equiv(y)$ iff $x \leqslant y$, and observe that this is well defined, i.e., it does not depend on the exact elements x and y that one picks from the equivalence classes $q_\equiv(x)$ and $q_\equiv(y)$. Then $\leqslant$ is not just a quasi-ordering, but is an ordering on $X/{\equiv}$. So, up to quotients, there is not much difference between quasi-ordered sets and posets.

We also write $<$ for the *strict part* of the quasi-ordering $\leqslant$, viz., $x < y$ iff $x \leqslant y$ and $x \not\equiv y$, iff $x \leqslant y$ and $y \not\leqslant x$.

Write $y \geqslant x$ iff $x \leqslant y$, and call $\geqslant$ the *opposite* of $\leqslant$. For every poset X, the set of all elements of X ordered by $\geqslant$ is the *opposite* of X, and is written X^{op}. We write $>$ for the strict part of $\geqslant$.

In any quasi-ordered set X, we say that a subset A is *upward closed* iff whenever $x \in A$ and $x \leqslant y$, then $y \in A$: going up from an element of A, we stay in A. The *upward closure* $\uparrow_X A$ of A in X is the set $\{y \in X \mid \exists x \in A \cdot x \leqslant y\}$ of all elements above some element of A. For short, we write $\uparrow A$ when X is clear from context, and $\uparrow x$ for $\uparrow\{x\}$ when $x \in X$. Note that A is upward closed iff $A = \uparrow A$. Conversely, we say that A is *downward closed* iff every element below some element of A is again in A. The *downward closure* $\downarrow_X A$, or $\downarrow A$ when X is clear, equals $\{y \in X \mid \exists x \in A \cdot y \leqslant x\}$. For short, we shall write $\downarrow x$ for $\downarrow\{x\}$ when $x \in X$.

2.3.2 Upper bounds, lower bounds

For any subset A of a poset X, we say that $x \in A$ is a *least element* of A iff $A \subseteq \uparrow x$, i.e., every element of A is above x. The least element, if it exists, is unique. This should not be confused with the notion of *minimal element*: $x \in A$ is *minimal* in A iff no element of A is strictly below x, i.e., iff whenever $y \leqslant x$ and $y \in A$, then $y = x$. The least element of A, if it exists, is minimal, but minimal elements may fail to be least. For example, in $\mathbb{N} \times \mathbb{N}$ with the product ordering, defined by $(m, n) \leqslant (m', n')$ iff $m \leqslant m'$ and $n \leqslant n'$, the set $\uparrow\{(1, 2), (3, 1)\}$ has two minimal elements, $(1, 2)$ and $(3, 1)$, but no least element.

The *greatest* element and the notion of *maximal* elements are defined similarly, using $\geqslant$ instead of $\leqslant$.

An *upper bound* x of a subset A of a poset X is an element such that $y \leqslant x$ for every $y \in A$. That is, an upper bound of A sits above every element of A. In $\mathbb{R} \times \mathbb{R}$, for example, $(3, 1)$, $(5, 5)$, and $(6, 1)$ are some of the upper bounds of the set $\{(x, y) \in \mathbb{R} \times \mathbb{R} \mid x < 3, y < 1\}$.

The *least upper bound* of a subset A of a poset X, if it exists, is the least element of the set of all upper bounds of A in X. For short, we shall also call it the *supremum* of A, and write it $\sup A$. For example, the subset $\{(x, y) \in \mathbb{R} \times \mathbb{R} \mid x < 3, y < 1\}$ of $\mathbb{R} \times \mathbb{R}$ admits $(3, 1)$ as least upper bound. When A is a family $(x_i)_{i \in I}$, we also write $\sup_{i \in I} x_i$ instead of $\sup A$.

If $\sup A$ exists and is in A, then $\sup A$ is necessarily the greatest element of A. But $\sup A$ may fail to be in A, as in the example just given.

The notions of *lower bound*, *greatest lower bound*, a.k.a. *infimum* $\inf A$ of a subset A of X, are defined similarly, replacing $\leqslant$ by its opposite $\geqslant$.

A *complete lattice* is a poset X such that every subset A of X has a least upper bound $\sup A$ and a greatest lower bound $\inf A$. It is equivalent to require the existence of least upper bounds only, since then the greatest lower bounds $\inf A$ exist as well, as least upper bounds of the set of lower bounds of A.

Similarly, any poset in which every subset has a greatest lower bound is also a complete lattice.

Every complete lattice has a least element $\bot$ (also called *bottom*), obtained as the least upper bound $\sup \emptyset$ of the empty family, and a greatest element $\top$ (also called *top*), obtained as $\inf \emptyset$.

For any set X, $\mathbb{P}(X)$ with the inclusion ordering $\subseteq$ is a complete lattice: the least upper bound of a family of subsets of X is their union, and the greatest lower bound is their intersection.

There are various weaker notions. For example, an *inf-semi-lattice* is a poset in which any two elements x, y have a greatest lower bound $x \wedge y$, and therefore any non-empty finite set of elements has a greatest lower bound. Symmetrically, a *sup-semi-lattice* is a poset in which any two elements x, y have a least upper bound $x \vee y$, and therefore any non-empty finite set of elements has a least upper bound.

We do not require inf-semi-lattices to have greatest lower bounds of all finite subsets, only the non-empty ones. That is, we do not require them to have a largest element $\top$. Similarly, we do not require sup-semi-lattices to have a least element $\bot$. Those that have one have least upper bounds of all finite subsets, and are called *pointed*.

A *lattice* is any poset that is both an inf-semi-lattice and a sup-semi-lattice. Those lattices that have a least element $\bot$ and a largest element $\top$ are called *bounded*.

2.3.3 Fixed point theorems

A map f from a poset X to a quasi-ordered set Y is *monotonic* iff, for every $x, x' \in X$ with $x \leqslant x'$, $f(x) \leqslant f(x')$. When X is a poset, an *order embedding* $f: X \to Y$ is a monotonic map such that, additionally, whenever $f(x) \leqslant f(x')$, then $x \leqslant x'$. In particular, every order embedding is injective, but there are injective monotonic maps that are not order embeddings, e.g., the identity map from $\{0, 1\}$ with equality as ordering to $\{0, 1\}$ with the ordering $0 \leqslant 1$. An *order isomorphism* $f: X \to Y$ is a bijective, monotonic map, whose inverse is also monotonic. Every order isomorphism is an order embedding, and every order embedding $f: X \to Y$ defines an order isomorphism from X to the image of X by f in Y, with the ordering induced by that of Y.

Two elements x and y of a quasi-ordered set X are *incomparable* iff $x \not\leqslant y$ and $y \not\leqslant x$. For example, $(5, 5)$ and $(6, 1)$ are incomparable in $\mathbb{R} \times \mathbb{R}$.

A subset A of a poset X is *totally ordered* iff it has no pair of incomparable elements, i.e., iff, for every pair of elements x, y of A, $x \leqslant y$ or $y \leqslant x$. The ordering $\leqslant$ itself is called *total*, or *linear*, iff X is totally ordered by $\leqslant$. A *chain* in X is any totally ordered, non-empty subset of X.

Call a poset X *chain-complete* iff every chain C in X has a least upper bound $\sup C$.

A *fixed point* of a map $f: X \to X$ is an element $x \in X$ such that $f(x) = x$. A map $f: X \to X$ is *inflationary* iff $x \leqslant f(x)$ for every $x \in X$. The following non-trivial result is our first fixed point existence theorem.

Theorem 2.3.1 (Bourbaki–Witt) *Let X be a non-empty chain-complete poset, and $f: X \to X$ be an inflationary map. Then f has a fixed point. More precisely, for every $x \in X$, f has a fixed point above x.*

If f is both inflationary and monotonic, then f even has a least *fixed point above x, for every $x \in X$.*

Proof Prologue. The intuitive idea is that we can produce an ascending sequence $x \leqslant f(x) \leqslant f^2(x) \leqslant \cdots \leqslant f^n(x) \leqslant \cdots$, and that if f did not have a fixed point above x, then all inequalities would be strict. If X is finite, then we would obtain an infinite set of elements $f^n(x)$, $n \in \mathbb{N}$, reaching a contradiction. In the general case, we need to index the sequence, not by natural numbers but by ordinals, which are generalizations of natural numbers with the additional property that any chain of ordinals has a least upper bound; e.g., $\mathbb{N}$ has a least upper bound, written ω, and which is the first infinite ordinal. Since X is chain-complete, if α is the least upper bound of a chain D of ordinals, we extend the definition of $f^n(x)$ to $f^\alpha(x)$ for ordinal α, letting $f^\alpha(x)$ be $\sup_{\beta \in D} f^\beta(x)$. Making the argument precise would lead us far astray. Instead, we give a proof from Lang (2005).

Call *admissible* any subset A of X such that: (i) $x \in A$, (ii) for every $y \in A$, $f(y) \in A$, and (iii) for every chain C in A, $\sup C$ is in A. It is easy to see that any non-empty intersection of admissible subsets is again admissible, and that X is itself admissible. Let A_0 be the intersection of all admissible subsets. (A_0 will be the collection of all $f^\alpha(x)$, α ordinal, from the above prologue.) It follows that A_0 is itself admissible, and must therefore be the least admissible subset.

1. We claim that x is the least element of A_0. First, $x \in A_0$ by (i). Next, $A_0 \cap \uparrow x$ is again admissible (in particular, for (ii), if $y \in A_0 \cap \uparrow x$, then $f(y) \in A_0$ and $f(y) \geqslant y \geqslant x$ since f is inflationary). Since A_0 is the least admissible subset, $A_0 \subseteq A_0 \cap \uparrow x$, so that every element of A_0 must be above x.

For every $y \in A_0$, we would like to show that f *jumps directly from y to $f(y)$ in A_0*, i.e., all elements of A_0 are either below y or above $f(y)$, and in particular none is in between. Let $A_y = \{z \in A_0 \mid z \leqslant y \text{ or } f(y) \leqslant z\}$. We would like to show that A_y is admissible; however this requires an additional assumption. We say that an element y of A_0 is *extreme* iff, for every $z < y$ in A_0, $f(z) \leqslant y$.

2. We claim that, for every extreme element y of A_0, A_y is admissible. Since x is the least element of A_0, $x \leqslant y$, so $x \in A_y$; this proves (i). For (ii), take any $z \in A_y$, and show that $f(z) \in A_y$. If $z \leqslant y$, then either $z < y$ and then $f(z) \leqslant y$ since y is extreme, or $z = y$ and $f(z) = f(y)$. If $z \geqslant f(y)$, then $f(z) \geqslant z \geqslant f(y)$ since f is inflationary. For (iii), consider a chain C in A_y. If y is an upper bound of C, then $\sup C \leqslant y$, so $\sup C$ is in A_y. Otherwise, there must be an element z of C above $f(y)$. So $\sup C \geqslant f(y)$. In any case $\sup C \in A_y$.

Since A_0 is the least admissible set again, $A_0 \subseteq A_y$, assuming y extreme. That is, for every extreme element y of A_0, f jumps directly from y to $f(y)$, i.e., all elements of A_0 are either below y or above $f(y)$.

3. We now claim that every element of A_0 is extreme. Consider indeed the set E of all extreme elements of A_0. Certainly, $x \in E$, since there is no element $z < x$ in A_0 (since x is the least element of A_0, x is certainly minimal in A_0). So E satisfies (i). For (ii), let y be any extreme element of A_0. Since f jumps directly from y to $f(y)$, every element $z < f(y)$ is such that $z \leqslant y$; if $z < y$, then $f(z) \leqslant y$ (y is extreme) $\leqslant f(y)$ (f is inflationary), and if $z = y$, then $f(z) = f(y)$. In any case, $f(z) \leqslant f(y)$. So $f(y) \in E$. For (iii), consider any chain C of extreme elements in A_0. We must show that, whenever $z < \sup C$ with $z \in A_0$, then $f(z) \leqslant \sup C$. If every element y of C were such that $f(y) \leqslant z$, then $y \leqslant z$ as well (f is inflationary), so z would be above $\sup C$, which is impossible. So there is an element y of C such that $f(y) \not\leqslant z$. But C is a chain of extreme elements, so y is extreme, and we have seen that, in this case, f jumped directly from y to $f(y)$, in particular $z \leqslant y$ or $f(y) \leqslant z$; since $f(y) \not\leqslant z$, we must have $z \leqslant y$. If $z < y$, then $f(z) \leqslant y \leqslant \sup C$, since y is extreme. If $z = y$, then, since y, hence z, is extreme, f must jump directly from z to $f(z)$, in particular $\sup C \leqslant z$ or $f(z) \leqslant \sup C$; however $z < \sup C$, so $f(z) \leqslant \sup C$. So $f(z) \leqslant \sup C$ in any case, whence (iii) holds.

We have just shown that E was admissible. So $A_0 \subseteq E$, that is, every element of A_0 is extreme.

4. It follows that A_0 is a chain. It is non-empty since $x \in A_0$, and for any two elements y, z of A_0, $y \leqslant z$ or $f(z) \leqslant y$ since z is extreme. Because f is inflationary, $f(z) \leqslant y$ implies $z \leqslant y$. So $y \leqslant z$ or $z \leqslant y$.

5. Since A_0 is a chain, it has a least upper bound x_0. By (iii) with $C = A_0$, $x_0 = \sup A_0$ is in A_0. By (ii), $f(x_0) \in A_0$, so $f(x_0) \leqslant x_0$. Since f is inflationary, it follows that $f(x_0) = x_0$. Moreover, since $x \in A_0$, $x \leqslant x_0$. So x_0 is a fixed point of f above x.

6. Finally, we claim that x_0 is least among the fixed points of f above x, provided that f is not only inflationary, but monotonic. Let x_1 be another one. Let $A_1 = A_0 \cap \downarrow x_1$. We claim that A_1 is admissible; (i) and (iii) are obvious,

and we show (ii) as follows: if $y \in A_1$, then $f(y) \in A_0$ since A_0 is admissible, and $f(y) \leqslant f(x_1) = x_1$ since f is monotonic, so $f(y) \in A_1$. Since A_0 is minimal, $A_0 \subseteq A_1$, and since $x_0 \in A_0$, $x_0 \leqslant x_1$. □

The following *Knaster–Tarski Fixed Point Theorem*, also known as *Tarski's Fixed Point Theorem*, follows easily, and will be used more routinely. Note that we are no longer assuming f inflationary, but monotonic.

Theorem 2.3.2 (Knaster–Tarski) *Let X be a chain-complete poset, and $f: X \to X$ be a monotonic map. Then, given any $x \in X$ such that $x \leqslant f(x)$, f has a least fixed point above x.*

In particular, if X is a chain-complete poset with a least element $\bot$, then any monotonic map from X to X has a least fixed point.

Proof Let $P = \{y \in X \mid x \leqslant y \leqslant f(y)\}$. P is non-empty, since $x \in P$. P is also chain-complete: if C is a chain in P, let $z = \sup C$, then $f(y) \leqslant f(z)$ for every $y \in C$ since $y \leqslant z$ by definition and f is monotonic; so $x \leqslant y \leqslant f(z)$ since $y \in P$, and since $y \in C$ is arbitrary, $x \leqslant \sup C = z \leqslant f(z)$; so $z \in P$.

For every $y \in P$, $f(y)$ is also in P: we must check that $x \leqslant f(y) \leqslant f(f(y))$, and the first inequality is by definition of $y \in P$, while the second inequality follows because f is monotonic. So f restricts to a monotonic map from P to P. On P, f is inflationary as well, by definition of P. By the Bourbaki–Witt Theorem 2.3.1, f has a least fixed point above x in P. This is clearly also a fixed point above x in X, and is least in X, not just in P, since every fixed point of f above x must be in P. The final claim, assuming X has a least element $\bot$, is clear. □

Given a property P of elements of X, we sometimes need to prove that $P(x_0)$ holds, where x_0 is the least fixed point of the monotonic map $f: X \to X$ above x given in Theorem 2.3.2. A standard way to achieve this is to apply the principle of *fixed point induction*:

Fact 2.3.3 (Fixed Point Induction) *Let X be a chain-complete poset, f be a monotonic map from X to X, $x \in X$ be such that $x \leqslant f(x)$, and x_0 be the least fixed point of f above x given by the Knaster–Tarski Theorem 2.3.2.*

Given any property P of elements of X, to show that $P(x_0)$ holds, it is enough to show that: (i) *$P(x)$ holds (base case),* (ii) *for every $y \in X$, if $P(y)$ holds then $P(f(y))$ holds (first inductive case), and* (iii) *for every chain C of elements y such that $P(y)$ holds, $P(\sup C)$ holds (second inductive case).*

Indeed, if (i)–(iii) hold, then $A = \{x \in X \mid P(x)\}$ is admissible in the sense of the proof of Theorem 2.3.1. Taking the notations from this proof, $x_0 \in A_0 \subseteq A$.

Exercise 2.3.4 The usual form of the Knaster–Tarski Fixed Point Theorem is: $(*)$ if X is a complete lattice and $f: X \to X$ is monotonic, then f has a least fixed point. This is of course a special case of Theorem 2.3.2, since every complete lattice is chain-complete and non-empty.

Give a simpler proof of $(*)$, as follows. Call $x \in X$ a *post-fixed point* of f iff $f(x) \leqslant x$, let A be the set of post-fixed points of f, and let $x_0 = \inf A$. Show that x_0 is a post-fixed point of f, hence is the least post-fixed point of f. Then show that $x_0 \leqslant f(x_0)$, hence that x_0 is a fixed point of f. Conclude that x_0 is also the least fixed point of f.

The additional assumption that X is a complete lattice gives us more: show that the set of all fixed points of f is also a complete lattice, with the order inherited from X, whose least element is x_0. You only need to show the existence of least upper bounds. Then greatest lower bounds will exist automatically; in particular, f will also have a greatest fixed point.

2.3.4 Well-founded orderings

A binary relation $\to$ on a set X is *Noetherian* iff there is no infinite *descending* chain $x_1 \to x_2 \to \cdots \to x_k \to \cdots$. In other words, when we follow any sequence of arrows, we must stop after finitely many steps. A quasi-ordering $\leqslant$ is *well-founded* if and only if the relation $>$ is Noetherian, i.e., iff one cannot find lower and lower elements indefinitely, iff there is no *strictly descending* chain $x_1 > x_2 > \cdots > x_k > \cdots$.

The *transitive closure* $\to^+$ of $\to$ is the smallest transitive binary relation containing $\to$. Concretely, $x \to^+ y$ iff there are finitely many elements x_0, $x_1, \ldots, x_n$ of X ($n \in \mathbb{N}, n \geqslant 1$) such that $x = x_0 \to x_1 \to \cdots \to x_n = y$. Note that $\to$ is Noetherian if and only if $\to^+$ is Noetherian.

The *reflexive-transitive closure* $\to^*$ of $\to$ is the smallest reflexive transitive relation, i.e., the smallest quasi-ordering, containing $\to$. Concretely, $x \to^* y$ iff there are finitely many elements $x_0, x_1, \ldots, x_n$ of X ($n \in \mathbb{N}$, but this time $n = 0$ is allowed) such that $x = x_0 \to x_1 \to \cdots \to x_n = y$.

Given any binary relation $\to$ on X, the *Noetherian part* of $\to$ is the set of all elements x such that there is no infinite descending chain $x = x_1 \to x_2 \to \cdots \to x_k \to \cdots$ starting from x. So $\to$ is Noetherian on X iff its Noetherian part is the whole of X.

Call $\to$-*successor* of $x \in X$ any $y \in X$ such that $x \to y$. The Noetherian part of $\to$ is the smallest subset W of X such that every element $x \in X$ whose $\to$-successors are all in W is itself in W. This allows one to prove theorems by *Noetherian induction* along $\to$: to show that some property $P(x)$ holds for every $x \in W$, it is enough to show it under the additional assumption that $P(y)$ already holds for every $\rightarrow$-successor y of x. This assumption is called the *induction hypothesis*. When $\rightarrow$ is the strict part $>$ of the opposite of a well-founded quasi-ordering $\leqslant$ on a set X, Noetherian induction is called *well-founded induction*: to show that $P(x)$ holds for every $x \in X$, it is enough to show it under the additional assumption that $P(y)$ already holds for every $y < x$.

A transitive relation $<$ that is additionally *irreflexive*, i.e., such that $x < x$ for no x, is called a *strict ordering*. The strict part $<$ of any quasi-ordering is a strict ordering. Conversely, given any strict ordering $<$, the relation defined by $x \leqslant y$ iff $x < y$ or $x = y$ is an ordering, whose strict part is $<$.

Notice that any transitive, Noetherian relation is automatically a strict ordering.

2.4 The Axiom of Choice

The Axiom of Choice (AC) is surrounded with an aura of mystery. There are objective reasons for this. First, it is incredibly obvious, to the point that one uses it without even realizing it. However, it is independent of ZF, as shown by Paul J. Cohen, following initial forays by Kurt Gödel (Cohen, 2008): this means that one cannot prove AC from the axioms of ordinary Zermelo–Fraenkel set theory (ZF), and that one cannot prove that AC is false either. Second, and despite its apparent vacuity, the Axiom of Choice has several deep consequences, sometimes even paradoxical.

We shall always assume the Axiom of Choice. Recall that AC states that every map $\varphi \colon X \to \mathbb{P}(Y)$ such that $\varphi(x)$ is non-empty for every $x \in X$ (a *multifunction* from X to Y) has a *selection* $f \colon X \to Y$, i.e., a map such that $f(x) \in \varphi(x)$ for each $x \in X$.

Using ZF only (or the axioms of VBG without AC), one can always pick an element from $\varphi(x)$, when x is fixed. What AC allows us is to pick an element $f(x)$ from $\varphi(x)$, simultaneously, for *all* $x \in X$. When X is infinite, this amounts to making infinitely many choices at once.

The Axiom of Choice has many interesting consequences. A first one is the *Hausdorff Maximality Theorem*:

Theorem 2.4.1 (Hausdorff Maximality Theorem) *Let X be any poset, C be a chain in X. Then C is included in some maximal chain in X.*

Proof By maximal, we of course mean that one cannot add any new element without destroying totality. Assume that there is no maximal chain containing C. In particular, for each chain C' containing C, there is an even larger chain,

C''. Using AC, pick such a C'' for each C', call it $next(C')$. Formally, let φ be the multifunction mapping each chain C' containing C to the set of all chains C'' such that $C' \subseteq C''$, $C' \neq C''$; then let $next$ be a selection for φ.

The collection of all chains C' of X that contain C, ordered by inclusion $\subseteq$, is chain-complete: for every chain $\mathcal{C} = \{C_i \mid i \in I\}$ of chains C_i of X, $\sup \mathcal{C} = \bigcup_{i \in I} C_i$ is indeed a chain. Clearly, $next$ is inflationary. By the Bourbaki–Witt Theorem 2.3.1, $next$ has a fixed point; but, by construction, $next(C') \neq C'$ for every chain C' containing C. Having reached a contradiction, we conclude. □

This entails what is probably the most well-known, and most useful, consequence of AC, namely *Zorn's Lemma*. Call a poset X *inductive* iff every chain in X has an upper bound. This is weaker than requiring X to be chain-complete: we are not requiring a *least* upper bound—although in most applications we use Zorn's Lemma in chain-complete posets.

Theorem 2.4.2 (Zorn's Lemma) *Every non-empty inductive poset has a maximal element. More precisely, if X is an inductive poset and $x \in X$, then there is a maximal element of X above x.*

Proof By the Hausdorff Maximality Theorem 2.4.1, there is a maximal chain C containing the chain $\{x\}$. Since X is inductive, C has an upper bound x_0. Clearly, $x \leqslant x_0$. Also, x_0 is maximal in X; otherwise there would be an even greater element x_1, and $C \cup \{x_1\}$ would then be an even larger chain, contradicting the fact that C is maximal. □

This also entails the following, apparently obvious result. Given any family $(X_i)_{i \in I}$ of sets, the *product* $\prod_{i \in I} X_i$ is the collection of tuples of elements $(x_i)_{i \in I}$ where, for each $i \in I$, x_i is in X_i. If one looks for a formal definition of a tuple, the easiest one is that such a tuple is a function from I to $\bigcup_{i \in I} X_i$, mapping each $i \in I$ to an element of X_i.

Theorem 2.4.3 *Every product of non-empty sets is non-empty.*

Proof Let $(X_i)_{i \in I}$ be a family of non-empty sets. Consider the multifunction from I to $\bigcup_{i \in I} X_i$ mapping each $i \in I$ to X_i. By AC, it has a selection $(x_i)_{i \in I}$, which is an element of $\prod_{i \in I} X_i$ by construction. □

One can define a quasi-ordering on sets by: *A has cardinality smaller than or equal to B* if and only if there is an injective map f from A into B. If A is non-empty, this is equivalent to requiring that there is a surjective map g from B to A: given f, define $g(b)$ to be the unique $a \in A$ such that $f(a) = b$

if such an a exists, otherwise some fixed element of A; conversely, given g, define $f(a)$ as being any $b \in B$ such $g(b) = a$, using the Axiom of Choice.

It is clear that A has cardinality smaller than or equal to $\mathbb{P}(A)$ for any set A. Cantor's Theorem 2.2.1 then implies that A has cardinality *strictly smaller than* $\mathbb{P}(A)$ (even when A is empty: there is no function at all from $\mathbb{P}(\emptyset)$ to $\emptyset$).

3

A first tour of topology: metric spaces

The most natural way one can introduce topology, and what topology is about, is to explain what a metric space is. A *metric space* is a set, X, equipped with a *metric* d, which serves to measure distances between points.

The purpose of this chapter is to introduce the basic notions that we shall explore throughout this book, in this more familiar setting.

However, we must face a conundrum. Imagine we explained what an *open* subset is in a metric space. In the more general settings we shall explore in later chapters, we would have to redefine opens. Redefinitions are bad mathematical practice, and for a good reason: we would never know which definition we would mean later on; i.e., are the opens we shall use meant to be those introduced in the metric case here, or the more general kind introduced in later chapters? So we shall only talk about *sequential opens*, not opens, here. Sequential opens will provide particular examples of opens, which we hope should be illuminating. We proceed similarly for sequentially closed subsets. The adjective "sequential" is itself justified by the fact that these notions will be defined through convergence of *sequences* – and convergence is arguably one of the central concepts in topology.

3.1 Metric spaces

Let us describe what a metric space, i.e., a set X with a metric d, is, in intuitive terms first.

The following is a prototypical example of a metric space; whenever one looks for intuitions about metric spaces, we should probably first imagine what is happening there. This example is the *real plane*. Think of it as a blank sheet of paper, extending indefinitely left, right, up, and down. On this sheet of paper, one can draw points. Some of them are close together, some of them are far

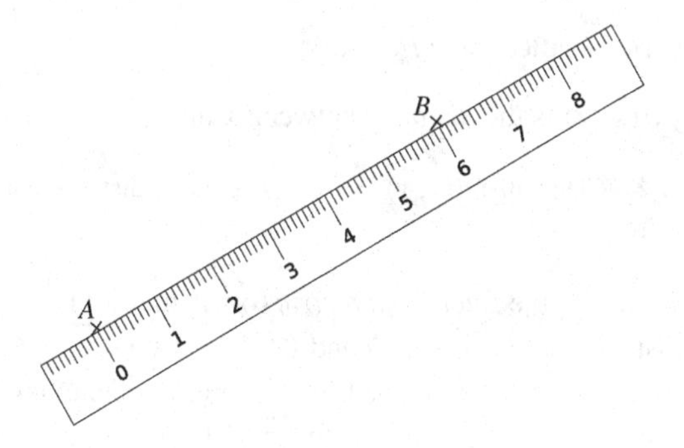

Figure 3.1 Measuring distance.

apart. The metric d on the real plane will serve to quantify how far two points are from one another: for any two given points A, B, evaluate their distance $d(A, B)$ by using a ruler, aligned so that its 0 mark lies precisely on the point A, and so that the ruler also passes through B: the mark at B on the ruler is the distance $d(A, B)$: see Figure 3.1, for example, where $d(A, B) = 6$.

Mathematically, the real plane can be thought of as $\mathbb{R}^2 = \mathbb{R} \times \mathbb{R}$, the set of all pairs (x, y) of reals: x and y are the coordinates of the point. One can then check that $d(A, B)$ is given by the formula $\sqrt{(x' - x)^2 + (y' - y)^2}$, when $A = (x, y)$ and $B = (x', y')$.

One can check easily that the distance $d(A, B)$ from A to B is the same as $d(B, A)$; that every point is at distance 0 from itself and that, conversely, the only point at distance 0 from a point A is A itself; and, finally, that the so-called *triangular inequality* holds: $d(A, B) \leqslant d(A, C) + d(C, B)$, meaning that taking a detour through a third point C only makes you travel a longer overall distance.

We now abstract away from the real plane and define a metric as *any* map d satisfying the above properties. We also allow for infinite distances, because there is no reason to disallow them. This means distances will be values in the set $\overline{\mathbb{R}^+}$, obtained by adjoining a fresh element $+\infty$ to $\mathbb{R}^+$, assumed to be above all reals. Addition extends from $\mathbb{R}^+$ to $\overline{\mathbb{R}^+}$, and even from $\mathbb{R}$ to $\mathbb{R} \cup \{+\infty\}$, by letting $t + (+\infty) = (+\infty) + t = +\infty$.

Definition 3.1.1 (Metric) Let X be a set. A *metric* d on X is a map from $X \times X$ to $\overline{\mathbb{R}^+}$ such that:

- for all $x, y \in X$, $d(x, y) = 0$ if and only if $x = y$;
- (Symmetry) $d(x, y) = d(y, x)$ for all $x, y \in X$;
- (Triangular inequality) $d(x, y) \leqslant d(x, z) + d(z, y)$ for all $x, y, z \in X$.

A set with a metric is called a *metric space*.

The quantity $d(x, y)$ is the *distance* between x and y.

Exercise 3.1.2 Check that $d(x, y) = |x - y|$ defines a distance between two real numbers x and y.

Example 3.1.3 In $\mathbb{R}^2$, the *Euclidean metric* (or L^2 metric) is the one we have just seen: the distance between (x, y) and (x', y') is $\sqrt{(x - x')^2 + (y - y')^2}$. There are others, e.g., $|x - x'| + |y - y'|$ (the L^1 metric) and $\max(|x - x'|, |y - y'|)$ (the L^∞ metric) also define metrics on $\mathbb{R}^2$. For every $p \geqslant 1$, the L^p metric is defined by $[|x - x'|^p + |y - y'|^p]^{\frac{1}{p}}$.

Even more generally, the L^p *metric* on $\mathbb{R}^n$, the set of n-tuples $(x_i)_{i=1}^n$ of real numbers, is defined by $d((x_i)_{i=1}^n, (y_i)_{i=1}^n) = \left(\sum_{i=1}^n |y_i - x_i|^p\right)^{\frac{1}{p}}$.

Example 3.1.4 We shall often refer to the L^1 *metric* on $\mathbb{R}$: this is defined by $d(x, y) = |y - x|$. Note that this is also the L^p metric on $\mathbb{R}$, whatever $p \geqslant 1$ is chosen.

Exercise 3.1.5 The purpose of this exercise is to show that the L^p metric is indeed a metric.

Call a function $f : \mathbb{R} \to \mathbb{R}^+$ *convex* if and only if $f(\lambda x + (1 - \lambda)y) \leqslant \lambda f(x) + (1 - \lambda) f(y)$ for all $x, y \in \mathbb{R}^+$ and $\lambda \in [0, 1]$. The function $x \mapsto |x|^p$ is convex whenever $p \geqslant 1$, a fact that you do not need to prove.

Use this to show that the formula $||(a_i)_{i=1}^n||_p = \left(\sum_{i=1}^n |a_i|^p\right)^{\frac{1}{p}}$ defines a *norm* on $\mathbb{R}^n$, called the L^p *norm*, for every $p \geqslant 1$: namely, $||(a_i)_{i=1}^n||_p = 0$ iff $a_i = 0$ for every i, $1 \leqslant i \leqslant n$, $||(a.a_i)_{i=1}^n||_p = |a|.||(a_i)_{i=1}^n||_p$ for every $a \in \mathbb{R}$, and $||(a_i + b_i)_{i=1}^n||_p \leqslant ||(a_i)_{i=1}^n||_p + ||(b_i)_{i=1}^n||_p$. To show the latter inequality, observe that $\left|\lambda \frac{a_i}{\lambda} + (1 - \lambda)\frac{b_i}{1-\lambda}\right|^p \leqslant \lambda \frac{|a_i|^p}{\lambda^p} + (1 - \lambda)\frac{|b_i|^p}{1-\lambda}$ for some well-chosen $\lambda \in (0, 1)$.

Now use the above facts to conclude that the L^p metric is indeed a metric.

3.2 Convergence, limits

One of the fundamental notions of topology is *convergence*. Consider the sequence of points $x_0, x_1, \ldots, x_n$, depicted in Figure 3.2. As n grows, x_n comes closer and closer to the point $(2, 1)$: we say that the sequence $(x_n)_{n \in \mathbb{N}}$ *converges to* $(2, 1)$, or that $(2, 1)$ is a *limit* of the sequence $(x_n)_{n \in \mathbb{N}}$.

The formal definition is as follows. We agree that the formulation "for every $\epsilon > 0$" means "for every real number ϵ such that $\epsilon > 0$."

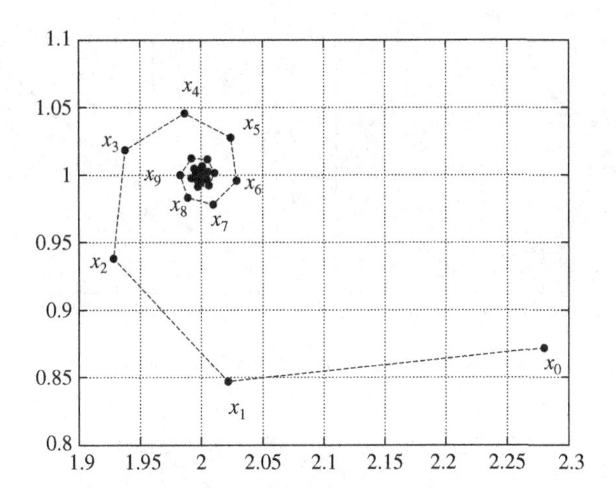

Figure 3.2 A sequence of points that converges.

Definition 3.2.1 (Convergence, metric case) A sequence $(x_n)_{n \in \mathbb{N}}$ of points in a metric space X, d *converges to* $x \in X$, alternatively x is a *limit* of $(x_n)_{n \in \mathbb{N}}$, if and only if, for every $\epsilon > 0$, there is an $n_0 \in \mathbb{N}$ such that, for every $n \geqslant n_0$, $d(x, x_n) < \epsilon$.

We shall abbreviate the formal statement "there is an $n_0 \in \mathbb{N}$ such that, for every $n \geqslant n_0$, $d(x, x_n) < \epsilon$" as "for n large enough, $d(x, x_n) < \epsilon$". In other words, for every $\epsilon > 0$, however small you choose it, the distance from x to x_n will eventually become smaller than ϵ, for n large enough.

For example, in Figure 3.2, one can check that, if we take $\epsilon = 0.1$, then $d(x, x_n) < \epsilon$ whenever $n \geqslant 3$. The same inequality holds for the smaller $\epsilon = 0.05$ whenever $n \geqslant 4$, or for $\epsilon = 0.02$ whenever $n \geqslant 9$.

In this example, it is not the case that x_n eventually equals x: while x_n gets closer and closer to x, it never actually reaches the limit x.

There are sequences $(x_n)_{n \in \mathbb{N}}$ that have no limit. In the real plane, one may take $x_n = (\cos n, \sin n)$: this is a point at distance 1 from the origin $(0, 0)$, rotated one radian from x_{n-1}; see Figure 3.3, left. Or one may take $x_n = (n, n)$, or $x_n = (\sqrt{n}, \sqrt{n})$ (Figure 3.3, right), for example, which cannot converge to any point since they eventually get infinitely far away from any given point.

On the other hand, if a sequence $(x_n)_{n \in \mathbb{N}}$ has a limit, then it is *unique*: indeed, if x and y are two limits, then, for every $\epsilon > 0$, we have $d(x, x_n) < \epsilon$ and $d(y, y_n) < \epsilon$ for n large enough; from this, we obtain $d(x, y) \leqslant d(x, x_n) + d(x_n, y)$ (by the triangular inequality) $= d(x, x_n) + d(y, x_n)$ (by symmetry) $< 2\epsilon$. Since $\epsilon > 0$ is arbitrary, $d(x, y) \leqslant \inf_{\epsilon > 0} 2\epsilon = 0$, therefore $x = y$.

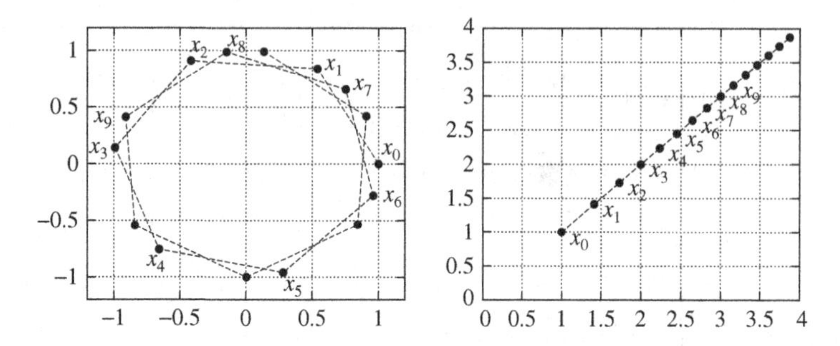

Figure 3.3 Sequences of points that do not converge.

Figure 3.4 Closed, non-closed-non-open, and open subsets of the plane.

We then write $\lim_{n \in \mathbb{N}} x_n$ for the unique limit of $(x_n)_{n \in \mathbb{N}}$.

A sequence that has a limit is called *convergent*. Since limits are unique in a metric space, we can therefore talk about *the* limit of a convergent sequence in a metric space.

That limits are unique is specific to so-called T_2, or Hausdorff, spaces, of which metric spaces are a special kind. In general, for non-Hausdorff spaces, which are the focus of this book, limits will be far from being unique.

A subset F of a metric space X, d is *sequentially closed* if and only if it is stable under the operation of taking limits, i.e., if and only if the limit of every convergent sequence $(x_n)_{n \in \mathbb{N}}$ with $x_n \in F$ for every $n \in \mathbb{N}$ is again in F.

The closed square $[0, 1] \times [0, 1] = \{(x, y) \mid 0 \leqslant x \leqslant 1, 0 \leqslant y \leqslant 1\}$ is sequentially closed. A pictorial representation is the square shown on the left of Figure 3.4. The border is shown in black, a convention by which we mean it to be part of the intended shape. On the other hand, we use dotted lines to represent borders that should not be part of the shape: the square shown on the right contains no point of its border, and is meant to depict $(0, 1) \times (0, 1) = \{(x, y) \mid 0 < x < 1, 0 < y < 1\}$, while the middle square is meant to represent $\{(x, y) \mid 0 \leqslant x < 1, 0 \leqslant y \leqslant 1\}$.

In a metric space, one can define sequentially closed sets in a number of equivalent ways. For every point x of a metric point X, d, let $B^d_{x,<\epsilon}$, the *open*

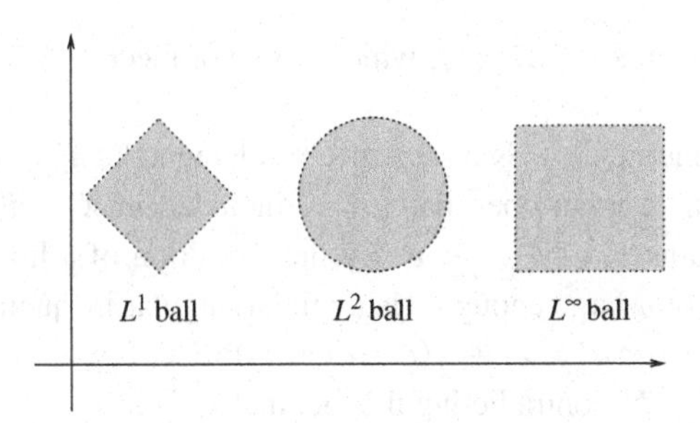

Figure 3.5 Open balls in $\mathbb{R}^2$.

ball with *center* x and *radius* $\epsilon > 0$, be the set of all points $y \in X$ such that $d(x, y) < \epsilon$.

Example 3.2.2 In $\mathbb{R}$ with distance $|x-y|$, the open ball of center x and radius r is the open interval $(x-r, x+r)$. In $\mathbb{R}^2$, the open balls look as in Figure 3.5, for various metrics (the gray areas, without the boundaries).

Proposition 3.2.3 (Sequentially closed) *Let X, d be a metric space. For every subset F of X, the following are equivalent:*

1. *F is sequentially closed;*
2. *for every point $x \in X$ that is not in F, there is an open ball $B^d_{x,<\epsilon}$ centered at x that does not meet F;*
3. *the complement of F is a union of open balls.*

Proof Instead of proving the three equivalences $1 \Leftrightarrow 2$, $1 \Leftrightarrow 3$, $2 \Leftrightarrow 3$, which would amount to proving the six implications $1 \Rightarrow 2$, $2 \Rightarrow 1$, $1 \Rightarrow 3$, $3 \Rightarrow 1$, $2 \Rightarrow 3$ and $3 \Rightarrow 2$, we only prove the circular list of implications $1 \Rightarrow 2 \Rightarrow 3 \Rightarrow 1$. The others are consequences, e.g., $2 \Rightarrow 3$ follows from $2 \Rightarrow 3$ and $3 \Rightarrow 1$. We shall use this trick often.

$1 \Rightarrow 2$. Let $x \in X \smallsetminus F$, and assume that every open ball centered at x meets F. In particular, for every $n \in \mathbb{N}$, $B^d_{x,<\frac{1}{2^n}}$ meets F, say at x_n. It is easy to see that the sequence $(x_n)_{n\in\mathbb{N}}$ converges to x: for every $\epsilon > 0$, $d(x, x_n) < \epsilon$ for every n such that $\frac{1}{2^n} \leqslant \epsilon$, which happens whenever n is larger than or equal to the number of leading zeroes in the base 2 representation of ϵ, plus two. Since $x_n \in F$ for every $n \in \mathbb{N}$ and F is sequentially closed, x is in F, too: contradiction.

$2 \Rightarrow 3$. For every $x \in X \smallsetminus F$, pick some $\epsilon_x > 0$ such that $B^d_{x,<\epsilon_x}$ does not meet F, using the Axiom of Choice. Then $X \smallsetminus F = \bigcup_{x\in X\smallsetminus F} B^d_{x,<\epsilon_x}$: every point x in the left-hand side is in $B^d_{x,<\epsilon_x}$, and conversely every point in the right-hand side is in some $B^d_{x,<\epsilon_x}$, which does not meet F, hence is included in $X \smallsetminus F$.

$3 \Rightarrow 1$. Assume $(x_n)_{n\in\mathbb{N}}$ is a sequence of elements of F that converges to $x \in X \smallsetminus F$. There is an open ball $B^d_{y,<\epsilon}$ included in $X \smallsetminus F$ that contains x. Since it contains x, $d(y,x) < \epsilon$. By the definition of a limit, $d(x, x_n) < \frac{1}{2}(\epsilon - d(y,x))$ for n large enough. Using the triangular inequality, $d(y, x_n) \leqslant d(y,x) + d(x,x_n) < d(y,x) + \frac{1}{2}(\epsilon - d(y,x)) = \frac{1}{2}(d(y,x)+\epsilon) < \epsilon$. So x_n is in $B^d_{y,<\epsilon} \subseteq X \smallsetminus F$, contradicting the fact that x_n is in F. □

Replacing the $<$ sign by $\leqslant$ in the definition of an open ball, we obtain the *closed ball* $B^d_{x,\leqslant\epsilon} = \{y \in X \mid d(x,y) \leqslant \epsilon\}$. As the name suggests, closed balls are sequentially closed: for any sequence $(y_n)_{n\in\mathbb{N}}$ of points of $B^d_{x,\leqslant\epsilon}$—namely, $d(x, y_n) \leqslant \epsilon$ for every $n \in \mathbb{N}$—that converges to y, for every $\eta > 0$, we must have $d(y, y_n) < \eta$ for n large enough, whence $d(x,y) \leqslant d(x,y_n) + d(y_n, y)$ (by the triangular inequality) $= d(x,y_n) + d(y,y_n)$ (by symmetry) $< \epsilon + \eta$; so $d(x,y) \leqslant \inf_{\eta>0}(\epsilon+\eta) = \epsilon$.

In non-Hausdorff *quasi*-metric spaces, where the axiom of symmetry will be dropped, closed balls will in general fail to be (sequentially) closed.

Items 2 and 3 of Proposition 3.2.3 suggest that the complements of closed sets are interesting in their own right. We call *sequentially open* any complement of a sequentially closed set. We obtain the following characterization.

Proposition 3.2.4 (Sequentially open) *Let X, d be a metric space. For every subset U of X, the following are equivalent:*

1. *U is sequentially open;*
2. *for every point $x \in U$, there is an open ball $B^d_{x,<\epsilon}$ centered at x that is included in U;*
3. *U is a union of open balls;*
4. *for every convergent sequence $(x_n)_{n\in\mathbb{N}}$ in X, if $\lim_{n\in\mathbb{N}} x_n$ is in U then $x_n \in U$ for n large enough.*

Proof Items 1, 2, 3 are rephrasings of the corresponding items in Proposition 3.2.3. Item 4 is a rephrasing of the fact that $X \smallsetminus U$ is sequentially closed. □

(Sequentially) open subsets are the key to an understanding of topology beyond the case of metric spaces. As we shall see in Chapter 4, a topology is purely described as a collection of open subsets satisfying some elementary properties. Here they are.

Proposition 3.2.5 *Let X, d be a metric space. Every intersection of sequentially closed subsets is sequentially closed (including the empty intersection, taken as X itself). Every finite union of sequentially closed subsets is sequentially closed (including the empty union, $\emptyset$).*

Proof Let $(F_i)_{i \in I}$ be an arbitrary family of sequentially closed subsets of X, and let $(x_n)_{n \in \mathbb{N}}$ be a sequence in $F = \bigcap_{i \in I} F_i$ with a limit x in X. For each $i \in I$, x is in F_i since F_i is sequentially closed. So x is in F, whence F is sequentially closed.

Let $(F_i)_{i \in I}$ be a family of sequentially closed subsets of X, with I finite. Let $(x_n)_{n \in \mathbb{N}}$ be a sequence in $F = \bigcap_{i \in I} F_i$ with a limit x in X. Since I is finite, there is an $i \in I$ such that F_i contains x_n for infinitely many $n \in \mathbb{N}$: otherwise, for every $i \in I$, F_i would contain x_n for finitely many ns, hence F would also contain finitely many x_ns, which is absurd. Let n_k, $k \in \mathbb{N}$, enumerate those infinitely many values of $n \in \mathbb{N}$ such that x_n is in F_i. The sequence $(x_{n_k})_{k \in \mathbb{N}}$ is a sequence of points in F_i, and converges to x: for every $\epsilon > 0$, $d(x, x_n) < \epsilon$ for n large enough, so $d(x, x_{n_k}) < \epsilon$ for k large enough. Since F_i is closed, x is in F_i. So x is in F, whence F is closed. □

It follows in particular that, for every subset A of X, there is a smallest sequentially closed subset $cl(A)$ of X that contains A, namely the intersection of all sequentially closed subsets of X that contain A. The subset $cl(A)$ is called the *closure* of A in X.

Exercise 3.2.6 Show that $cl(A)$ is, more simply, just the set of all limits of convergent sequences $(x_n)_{n \in \mathbb{N}}$ contained in A. In the hard direction, first show that any open ball centered at a point in $cl(A)$ must meet A.

Taking complements, we obtain:

Proposition 3.2.7 *Let X, d be a metric space. Every union of sequentially open subsets is sequentially open (including the empty union $\emptyset$). Every finite intersection of sequentially open subsets is sequentially open (including the empty intersection X).*

3.3 Compact subsets

A third, fundamental class of subsets is given by the *compact* subsets. This is harder to explain, but compact subsets are pervasive in topology, and will play a role in almost every part of this book.

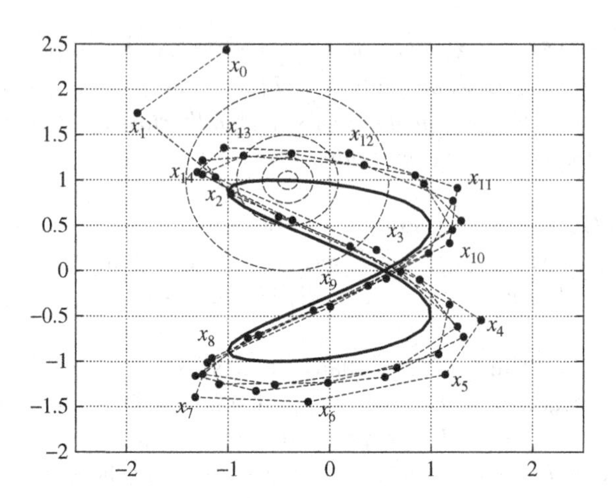

Figure 3.6 A sequence of points, with its cluster points.

One way of introducing them is to say that the compact subsets K are those that are so small that any sequence $(x_n)_{n \in \mathbb{N}}$ in K must coil up, accumulating in some places. The places where it accumulates will be called cluster points.

Formally, call a *cluster point* of a sequence $(x_n)_{n \in \mathbb{N}}$ in a metric space X, d any point $x \in X$ such that, for every $\epsilon > 0$, the open ball $B^d_{x,<\epsilon}$ contains x_n for infinitely many values of $n \in \mathbb{N}$. Look at Figure 3.6. (This was obtained by letting $x_n = \left((1 + \frac{1}{\ln(x+2)}) \cos(x + 2), (1 + \frac{1}{\ln(x+2)}) \sin(\frac{1}{2}x + \frac{3}{2})\right)$, a formula that we picked so as to make an interesting picture—not because it has any use. The numerical estimates given below are accurate up to 10^{-6}.)

While the sequence $(x_n)_{n \in \mathbb{N}}$ shown there has no limit point, it zeroes in on the 8-shaped solid curve. Given any fixed point of this curve, draw a small open ball around it: whatever the radius $\epsilon > 0$ of this open ball, there will be infinitely many values of $n \in \mathbb{N}$ such that x_n falls into this ball. We have shown four of these open balls around a point x on the solid curve, where $x = (-0.416147\ldots, 0.997495\ldots)$, with ϵ equal to 1, 0.5, 0.25, 0.1, respectively. In each case, there are infinitely many values of x_n in the open ball:

ϵ	Values that fall in $B^d_{x,<\epsilon}$				
1	x_2 =(−1.12515, 1.03018)	x_{12} =(0.18855, 1.29343)	x_{13} =(−1.04022, 1.3547)	x_{14} =(−1.30306, 1.08648)	...
0.5	x_{15} =(−0.372284, 0.557578)	x_{25} =(−0.380778, 1.29117)	x_{40} =(−0.507, 0.597824)	x_{50} =(−0.204241, 1.22722)	...
0.25	x_{113} =(−0.394475, 1.20212)	x_{176} =(−0.571963, 1.19224)	x_{201} =(−0.426654, 1.18206)	x_{226} =(−0.275191, 1.16749)	...
0.1	$x_{17\,995}$ =(−0.415405, 1.09739)	$x_{19\,038}$ =(−0.406231, 1.09635)	$x_{19\,415}$ =(−0.415232, 1.09661)	$x_{19\,792}$ =(−0.424199, 1.09685)	...

A *subsequence* of $(x_n)_{n\in\mathbb{N}}$ is any sequence of the form $(x_{n_k})_{k\in\mathbb{N}}$, where $n_0 < n_1 < \cdots < n_k < \cdots$. For example, $x_2, x_{12}, x_{13}, x_{14}, x_{15}, \ldots$ forms a subsequence of the sequence depicted on Figure 3.6, as well as $x_{17\,995}, x_{19\,038}, x_{19\,415}, x_{19\,792}, x_{20\,081}, \ldots$ We observe:

Lemma 3.3.1 *Let X, d be a metric space, and $(x_n)_{n\in\mathbb{N}}$ be a sequence in X. For every $x \in X$, x is a cluster point of $(x_n)_{n\in\mathbb{N}}$ if and only if x is the limit of some subsequence of $(x_n)_{n\in\mathbb{N}}$.*

Proof If x is the limit of some subsequence $(x_{n_k})_{k\in\mathbb{N}}$, where $n_0 < n_1 < \cdots < n_k < \cdots$, then, for every $\epsilon > 0$, $d(x, x_{n_k}) < \epsilon$ for k large enough. Since there are infinitely many values of n_k, $k \in \mathbb{N}$, x is indeed a cluster point of $(x_n)_{n\in\mathbb{N}}$.

Conversely, assume x is a cluster point of $(x_n)_{n\in\mathbb{N}}$. For each $k \in \mathbb{N}$, picking $\epsilon = \frac{1}{2^k}$, there are infinitely many points x_n such that $d(x, x_n) < \frac{1}{2^k}$, i.e., that lie in the open ball $B^d_{x,<\frac{1}{2^k}}$. For $k = 0$, pick one such point $x_{n_0} \in B^d_{x,<\frac{1}{2^0}}$. For $k = 1$, pick one point $x_{n_1} \in B^d_{x,<\frac{1}{2^1}}$, with $n_1 > n_0$: this is possible since there are infinitely many points x_n in $B^d_{x<\frac{1}{2^1}}$. Iterate this, picking one point $x_{n_2} \in B^d_{x,<\frac{1}{2^2}}$ with $n_2 > n_1, \ldots, x_{n_k} \in B^d_{x,<\frac{1}{2^k}}$ with $n_k > n_{k-1}, \ldots$ For every $\epsilon > 0$, there is a $k_0 \in \mathbb{N}$ such that $\frac{1}{2^{k_0}} < \epsilon$. For every $k \geqslant k_0$, x_{n_k} is in $B^d_{x,<\frac{1}{2^k}}$, so $d(x, x_{n_k}) < \frac{1}{2^k} \leqslant \frac{1}{2^{k_0}} < \epsilon$. So x is a limit of $(x_{n_k})_{j\in\mathbb{N}}$. □

We then define the sequentially compact subsets as those that are so cramped that any sequence must have cluster points.

Proposition 3.3.2 (Sequentially compact) *Let X, d be a metric space, and K be a subset of X. The following two properties are equivalent:*

1. *every sequence $(x_n)_{n\in\mathbb{N}}$ of points of K has a cluster point in K;*
2. *every sequence $(x_n)_{n\in\mathbb{N}}$ of points of K has a subsequence with a limit in K.*

We say that K is sequentially compact *if and only if either of these two properties is satisfied.*

Proof Direct consequence of Lemma 3.3.1. □

Sequentially compact and sequentially closed subsets are related. One can indeed observe that:

- In a metric space X, d, every sequentially closed subset F of a sequentially compact subset K is sequentially compact. Indeed, any sequence of points in F has a subsequence with a limit in K. Since the subsequence is in F and F is closed, this limit is in F, so F is sequentially compact.

- In a metric space X, d, every sequentially compact subset K is closed. Indeed, let $(x_n)_{n \in \mathbb{N}}$ be a sequence in K that has a limit $x \in X$. Since K is sequentially compact, there is a subsequence $(x_{n_k})_{k \in \mathbb{N}}$ that has a limit $x' \in K$. Using the definition of limits, x is also a limit of $(x_{n_k})_{k \in \mathbb{N}}$: for every $\epsilon > 0$, since $d(x, x_n) < \epsilon$ for n large enough, we also have $d(x, x_{n_k}) < \epsilon$ for k large enough. But limits are unique in metric spaces, so $x = x'$. It follows that x is in K, so K is closed.

The first property, that closed subsets of compact subsets are closed, will hold in any topological space, not just metric spaces. Our argument for the second property rests on limits being unique. We have seen that this was specific to Hausdorff (T_2) spaces, of which metric spaces are a special case. It will therefore be no surprise that:

In general, in non-Hausdorff spaces, which are the main topic of this book, compact subsets will *not* be closed.

Exercise 3.3.3 Let X, d be a metric space. Show that sequentially compact subsets have properties of closure under intersection and union that are similar to closed subsets. Namely, show that: (a) any finite union of sequentially compact subsets of X is sequentially compact; (b) any intersection of a non-empty family of sequentially compact subsets of a sequentially compact subset K of X is sequentially compact; (c) any finite non-empty intersection of sequentially compact subsets of X is sequentially compact. (Property (a) will continue to hold in non-Hausdorff spaces. Properties (b) and (c) will fail in general.)

In the real plane, and more generally in $\mathbb{R}^n$ with the L^p metric (Example 3.1.3), the following Borel–Lebesgue Theorem characterizes the compact subsets.

Proposition 3.3.4 (Borel–Lebesgue) *Call a subset K of $\mathbb{R}^n$* bounded *if and only if it is included in a subset of the form $[-a, a]^n$, for some $a \in \mathbb{R}^+$.*

The sequentially compact subsets of $\mathbb{R}^n$ with the L^p metric (whatever $p \geqslant 1$) are exactly the sequentially closed, bounded subsets of $\mathbb{R}^n$.

Proof Write $\vec{x}$ for the point $(x_i)_{i=1}^n$ of $\mathbb{R}^n$. Let d denote the L^p metric. Note that any open ball $B^d_{\vec{x},<\epsilon}$ is bounded: for any point $\vec{y}$ in this open ball, $d(\vec{x}, \vec{y}) < \epsilon$, so $|x_i - y_i|^p < \epsilon^p$ for every i, $1 \leqslant i \leqslant n$, whence $\vec{y}$ is in $[-a, a]^n$ with $a = \max_{i=1}^n |x_i| + \epsilon$.

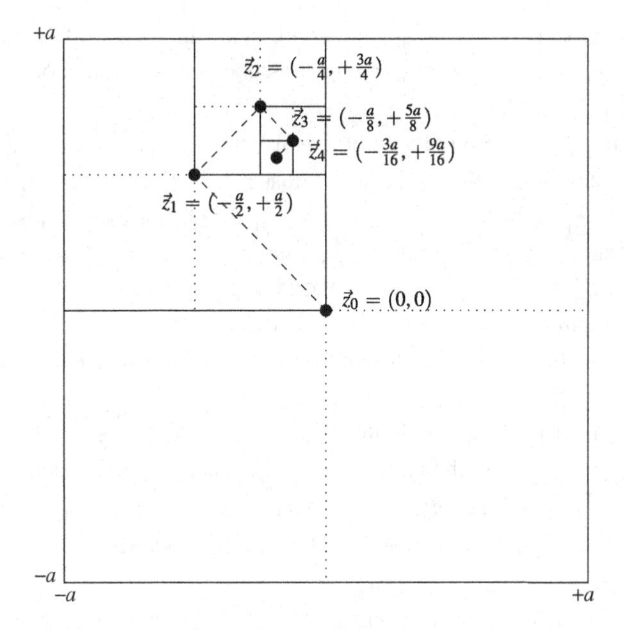

Figure 3.7 Cutting up cubes in the proof of the Borel–Lebesgue Theorem.

Let K be sequentially compact in $\mathbb{R}^n$. Then K is sequentially closed, as we have just seen. It must also be bounded; otherwise for each $k \in \mathbb{N}$ it would contain a point $\vec{x}_k$ that is not in $[-k, k]^n$. But the sequence $(\vec{x}_k)_{k \in \mathbb{N}}$ has no cluster point: if it had one, call it $\vec{x}$, then $\vec{x}_k$ would be in $B^d_{\vec{x},<1}$ for infinitely many values of k, contradicting the fact that $B^d_{\vec{x},<1}$ is bounded.

Conversely, let K be a sequentially closed and bounded subset of $\mathbb{R}^n$, and consider a sequence $(\vec{x}_j)_{j \in \mathbb{N}}$ in K. Since K is bounded, K is included in $[-a, a]^n$ for some $a \in \mathbb{R}^+$. By induction on $k \in \mathbb{N}$, we show that there is a point $\vec{z}_k$ such that $K \cap \prod_{i=1}^n [z_{ki} - \frac{a}{2^k}, z_{ki} + \frac{a}{2^k}]$ contains x_j for infinitely many values of j. (According to our writing convention, $\vec{z}_k$ is the point $(z_{ki})_{i=1}^n$.) The idea of the construction is depicted in Figure 3.7.

The claim is clear when $k = 0$: take $z_{0i} = 0$, then $K \cap \prod_{i=1}^n [z_{ki} - \frac{a}{2^k}, z_{ki} + \frac{a}{2^k}] = K \cap [-a, a]^n = K$ contains all the points x_j, $j \in \mathbb{N}$. Assume that we have constructed $\vec{z}_k$ already. We can cut up the cube $\prod_{i=1}^n [z_{ki} - \frac{a}{2^k}, z_{ki} + \frac{a}{2^k}]$ in 2^n sub-cubes, by cutting each of the n sides in two. Formally, $\prod_{i=1}^n [z_{ki} - \frac{a}{2^k}, z_{ki} + \frac{a}{2^k}]$ is included in the union of the 2^n sub-cubes $\prod_{i=1}^n [(z_{ki} \pm_i \frac{a}{2^{k+1}}) - \frac{a}{2^{k+1}}, (z_{ki} \pm_i \frac{a}{2^{k+1}}) + \frac{a}{2^{k+1}}]$, where the signs $\pm_1, \ldots, \pm_n$ are independently chosen from $\{-, +\}$. Since these sub-cubes are finitely many, one of them (at least) must contain infinitely many of the infinitely many x_js that lie in the containing cube $\prod_{i=1}^{n}[z_{ki} - \frac{a}{2^k}, z_{ki} + \frac{a}{2^k}]$. With the choice of signs $\pm_i$ we have made to define the corresponding sub-cube, we can then choose $\vec{z}_{k+1}$ as $(z_{ki} \pm_i \frac{a}{2^{k+1}})_{i=1}^{n}$.

This being done, we build a point $\vec{z}$ as follows. Let $z_{ki}^- = z_{ki} - \frac{a}{2^k}$, $z_{ki}^+ = z_{ki} + \frac{a}{2^k}$. The construction of $\vec{z}_k$ shows that $z_{0i}^- \leqslant z_{1i}^- \leqslant \cdots \leqslant z_{ki}^- \leqslant \cdots$, while $z_{0i}^+ \geqslant z_{1i}^+ \geqslant \cdots \geqslant z_{ki}^+ \geqslant \cdots$, and $z_{ki}^- \leqslant z_{ki}^+$, for each i, $1 \leqslant i \leqslant n$. Let $z_i^- = \sup_{k\in\mathbb{N}} z_{ki}^-$, $z_i^+ = \inf_{k\in\mathbb{N}} z_{ki}^+$. Since $z_{ki}^- \leqslant z_{ki}^+$, $z_i^- \leqslant z_i^+$. And since $z_i^+ - z_i^- \leqslant z_{ki}^+ - z_{ki}^- = \frac{2a}{2^k}$, for every $k \in \mathbb{N}$, $z_i^+ - z_i^- \leqslant 0$, so $z_i^+ = z_i^-$. Write z_i for the common value: $z_i = z_i^- = z_i^+$, and $\vec{z} = (z_i)_{i=1}^{n}$.

We observe that $\vec{z}$ is a cluster point of $(\vec{x}_j)_{j\in\mathbb{N}}$. For every $\epsilon > 0$, let k be so large that $\frac{2na}{2^k} < \epsilon^p$. Recall that the cube $\prod_{i=1}^{n}[z_{ki}^-, z_{ki}^+]$ was designed so that it contained x_j for infinitely many values of j. Also, every point $\vec{y}$ of the cube $\prod_{i=1}^{n}[z_{ki}^-, z_{ki}^+]$ is such that $|y_i - z_i| \leqslant \frac{2a}{2^k}$ (since $\vec{z}$ is in the same cube), so $d(\vec{z}, \vec{y})^p \leqslant \frac{2na}{2^k} < \epsilon^p$, i.e., $d(\vec{z}, \vec{y}) < \epsilon$: so the cube $\prod_{i=1}^{n}[z_{ki}^-, z_{ki}^+]$ is included in $B^d_{\vec{z},<\epsilon}$. Since $\vec{x}_j$ is in the cube for infinitely many values of j, it is also in $B^d_{\vec{z},<\epsilon}$ for infinitely many values of j.

So every sequence of points $(\vec{x}_j)_{j\in\mathbb{N}}$ in K has a cluster point $\vec{z}$. By appealing to Lemma 3.3.1, $\vec{z}$ is the limit of a subsequence, which must then consist of points in K as well. We use the fact that K is sequentially closed to note that $\vec{z}$ must be in K, and we conclude. ☐

Exercise 3.3.5 In the proof of Proposition 3.3.4, show that the constructed point $\vec{z}$ is the limit of $(\vec{z}_k)_{k\in\mathbb{N}}$.

Sequentially compact subsets have many pleasing properties. The following will be used in the proof of Proposition 3.3.7 below.

Proposition 3.3.6 (Lebesgue number) *Let X, d be a metric space, K be a sequentially compact subset of X, and $(U_i)_{i\in I}$ be a family of sequentially open subsets of X whose union contains K. Then there is an $\epsilon > 0$ such that, for every point x of K, the open ball $B^d_{x,<\epsilon}$ is included in some U_i, $i \in I$. The number $\epsilon > 0$ is called a* Lebesgue number *of $(U_i)_{i\in I}$.*

Proof By contradiction, assume no Lebesgue number exists. For every $n \in \mathbb{N}$, taking $\epsilon = \frac{1}{2^n}$, there is a point $x_n \in K$ such that the open ball $B^d_{x_n,<\frac{1}{2^n}}$ is not included in any U_i, $i \in I$. Since K is sequentially compact, $(x_n)_{n\in\mathbb{N}}$ has a subsequence $(x_{n_k})_{k\in\mathbb{N}}$ that has a limit x in K. By assumption, x is in some U_i, $i \in I$. Using Proposition 3.2.4, Item 2, there is an $\epsilon > 0$ such that $B^d_{x,<\epsilon} \subseteq U_i$. By the definition of a limit, $d(x, x_{n_k}) < \frac{\epsilon}{2}$ for k large enough. Pick k so large that, additionally, $\frac{1}{2^{n_k}} < \epsilon/2$. For every $z \in B^d_{x_{n_k},<\frac{1}{2^{n_k}}}$, $d(x, z) \leqslant$

$d(x, x_{n_k}) + d(x_{n_k}, z) < \frac{\epsilon}{2} + \frac{1}{2^{n_k}} < \epsilon$, so $z \in B^d_{x,<\epsilon} \subseteq U_i$. It follows that $B^d_{x_{n_k},<\frac{1}{2^{n_k}}}$ is included in U_i, a contradiction. □

Sequentially compact subsets can also be characterized using open subsets only, or closed subsets only, with no reference to points whatsoever. This will be how we understand compactness beyond the case of metric spaces: Item 2 below is sometimes called the *Heine–Borel Property*, or the *Borel–Lebesgue Property*.

Proposition 3.3.7 *Let X, d be a metric space, and K be a subset of X. The following are equivalent:*

1. *K is sequentially compact;*
2. *for every family $(U_i)_{i\in I}$ of sequentially open subsets of X such that $K \subseteq \bigcup_{i\in I} U_i$, there is a finite subset $J \subseteq I$ such that $K \subseteq \bigcup_{i\in J} U_i$;*
3. *for every ascending sequence $U_0 \subseteq U_1 \subseteq \cdots \subseteq U_n \subseteq \cdots$ of sequentially open subsets of X such that $K \subseteq \bigcup_{n\in\mathbb{N}} U_n$, there is an $n \in \mathbb{N}$ such that $K \subseteq U_n$;*
4. *for every family $(F_i)_{i\in I}$ of sequentially closed subsets of X such that K meets any finite intersection $\bigcap_{i\in J} F_i$ (J a finite subset of I), K also meets $\bigcap_{i\in I} F_i$;*
5. *for every descending sequence $F_0 \supseteq F_1 \supseteq \cdots \supseteq F_n \supseteq \cdots$ of sequentially closed subsets that meet K (i.e., $F_n \cap K \neq \emptyset$), $\bigcap_{n\in\mathbb{N}} F_n$ meets K.*

The key point in this Proposition is that J is required to be *finite* in Items 2 and 4.

Proof 1 ⇒ 2. Let $(U_i)_{i\in I}$ be a family of sequentially open subsets of X such that $K \subseteq \bigcup_{i\in I} U_i$, and assume that K is included in no union $\bigcup_{i\in J} U_i$ with J a finite subset of I. Also let $\epsilon > 0$ be a Lebesgue number of $(U_i)_{i\in I}$ (Proposition 3.3.6).

Build a sequence of indices $(i_n)_{n\in\mathbb{N}}$ in I, and a sequence of points $(x_n)_{n\in\mathbb{N}}$ of K, as follows. Since K is not included in $\bigcup_{i\in J} U_i$ with J empty, it contains some point x_0. By definition of the Lebesgue number $\epsilon > 0$, $B^d_{x_0,<\epsilon}$ is included in U_{i_0} for some $i_0 \in I$. Since K is not included in $\bigcup_{i\in J} U_i$ with $J = \{i_0\}$, it contains some point x_1 that is not in U_{i_0}. Again, $B^d_{x_1,<\epsilon}$ is included in U_{i_1} for some $i_1 \in I$. Repeating the construction, we build $x_n \in K$ such that $x_n \notin \bigcup_{k=0}^{n-1} U_{i_k}$, and i_n such that $B^d_{x_n,<\epsilon} \subseteq U_{i_n}$, by induction on $n \in \mathbb{N}$.

Note that $x_n \notin \bigcup_{k=0}^{n-1} U_{i_k}$ implies that x_n is not in the smaller set $\bigcup_{k=0}^{n-1} B^d_{x_k,<\epsilon}$, i.e., that $d(x_k, x_n) \geqslant \epsilon$ whenever $k < n$.

By assumption, $(x_n)_{n\in\mathbb{N}}$ has a cluster point x in K. By definition, $d(x, x_n) < \epsilon/2$ for infinitely many values of $n \in \mathbb{N}$. Pick two such values, call them

n and k, so that $k < n$. Then $d(x_k, x_n) \leqslant d(x_k, x) + d(x, x_n)$ (by the triangular inequality) $\leqslant d(x, x_k) + d(x, x_n)$ (by symmetry) $< \epsilon$, contradicting $d(x_k, x_n) \geqslant \epsilon$. So K must, in fact, be included in some finite union $\bigcup_{i \in J} U_i$.

2 $\Rightarrow$ 3. Immediate, since every union of finitely many sets $U_{n_0}, U_{n_1}, \ldots, U_{n_k}$ in an ascending sequence is included in U_n with $n = \max(n_0, n_1, \ldots, n_k)$ (or $n = 0$ if $k = 0$).

2 $\Leftrightarrow$ 4, 3 $\Leftrightarrow$ 5. Take complements. 4 $\Rightarrow$ 5 follows.

5 $\Rightarrow$ 1. Let $(x_n)_{n \in \mathbb{N}}$ be a sequence of points in K. For each $n \in \mathbb{N}$, let $A_n = \{x_k \mid k \geqslant n\}$, and F_n be the closure of A_n. Note that $F_{n-1} \supseteq F_n$ for every $n \geqslant 1$, since F_{n-1} is sequentially closed, contains A_{n-1} hence also A_n, and F_n is the smallest sequentially closed subset containing A_n.

By construction, F_n meets K for each $n \in \mathbb{N}$, namely at x_n. So, by 5, $\bigcap_{n \in \mathbb{N}} F_n$ meets K, say at x.

It remains to show that x is a cluster point of $(x_n)_{n \in \mathbb{N}}$. Fix $\epsilon > 0$. Assume by contradiction that there are only finitely many values of $k \in \mathbb{N}$ such that $x_k \in B^d_{x,<\epsilon}$. There is an $n \in \mathbb{N}$ such that, for every $k \geqslant n$, $x_k \notin B^d_{x,<\epsilon}$. Let F be the complement of $B^d_{x,<\epsilon}$: we have just shown that $A_n \subseteq F$. By Proposition 3.2.3, Item 3, F is sequentially closed. Since F_n is the smallest sequentially closed subset containing A_n, we obtain $F_n \subseteq F$. But $x \in \bigcap_{n \in \mathbb{N}} F_n$, so $x \in F$, contradicting the fact that F is the complement of $B^d_{x,<\epsilon}$. □

We conclude by the following result, which is a forerunner of one of the fundamental results of topology, Tychonoff's Theorem (which we shall state as Theorem 4.5.12).

Proposition 3.3.8 (Tychonoff, metric case, finite products) *Let X_i, d_i be finitely many metric spaces, $1 \leqslant i \leqslant m$. Equip the product space $X = X_1 \times \cdots \times X_m$ with the metric d defined by $d((x_1, \ldots, x_m), (y_1, \ldots, y_m)) = \max_{i=1}^m d_i(x_i, y_i)$.*

Every product $K_1 \times \cdots \times K_m$ of sequentially compact subsets K_i of X_i, $1 \leqslant i \leqslant m$, is sequentially compact in X.

Proof Write m-tuples $\vec{x}$ as $(x_1, \ldots, x_m)$. Given any sequence $(\vec{x}_n)_{n \in \mathbb{N}}$ in K – so that $\vec{x}_n = (x_{n1}, \ldots, x_{nm})$ – one can extract a sequence $(x_{n_k 1})_{k \in \mathbb{N}}$ from the sequence of first coordinates $(x_{n1})_{n \in \mathbb{N}}$ that converges to some point $x_1 \in K_1$. For purposes of readability, write $(\vec{x}^1_n)_{n \in \mathbb{N}}$ for the subsequence $(\vec{x}_{n_k})_{k \in \mathbb{N}}$: this is a subsequence of $(\vec{x}_n)_{n \in \mathbb{N}}$ whose first coordinates converges to $x_1 \in K_1$. Now extract a subsequence $(x^1_{n_k 2})_{k \in \mathbb{N}}$ from the sequence of second coordinates $(x^1_{n2})_{n \in \mathbb{N}}$, and write $(\vec{x}^2_n)_{n \in \mathbb{N}}$ for the subsequence $(\vec{x}^1_{n_k})_{k \in \mathbb{N}}$: this is a subsequence of $(\vec{x}_n)_{n \in \mathbb{N}}$ whose first coordinates converges to $x_1 \in K_1$ and whose second coordinates converges to $x_2 \in K_2$. Repeat the process, obtaining further subsequences $(\vec{x}_n^j)_{n\in\mathbb{N}}$ whose first j coordinates converge to points $x_1 \in K_1, \ldots, x_j \in K_j$, by induction on j, $1 \leqslant j \leqslant m$. Now look at the final subsequence $(\vec{x}_n^m)_{n\in\mathbb{N}}$. For every $\epsilon > 0$, for each i, $1 \leqslant i \leqslant m$, there is an $n_i \in \mathbb{N}$ such that, for every $n \geqslant n_i$, $d_i(x_i, x_i^m) < \epsilon$. So for n large enough, namely above $n_1, \ldots, n_m$, $d(\vec{x}, x^{\vec{m}}) < \epsilon$. So $(\vec{x}_n^m)_{n\in\mathbb{N}}$ converges to $\vec{x} \in K$. □

3.4 Complete metric spaces

In trying to determine whether a sequence $(x_n)_{n\in\mathbb{N}}$ in a metric space X, d has a limit x, one difficulty is to guess the value of the limit x itself.

Call a sequence $(x_n)_{n\in\mathbb{N}}$ of points in a metric space X, d a *Cauchy sequence* if and only if:

for every $\epsilon > 0$, there is an $n_0 \in \mathbb{N}$ such that, for all $m, n \geqslant n_0$, $d(x_m, x_n) < \epsilon$.

That is, for every $\epsilon > 0$, for m, n large enough, $d(x_m, x_n) < \epsilon$. If one wishes to check whether $(x_n)_{n\in\mathbb{N}}$ converges, one could start by checking whether it is Cauchy, because of the following.

Lemma 3.4.1 *Every convergent sequence in a metric space is Cauchy.*

Proof Let $(x_n)_{n\in\mathbb{N}}$ converge to x in the metric space X, d. For every $\epsilon > 0$, there is an $n_0 \in \mathbb{N}$ such that, for every $n \geqslant n_0$, $d(x, x_n) < \epsilon/2$. So, for any pair of natural numbers $m, n \geqslant n_0$, $d(x_m, x_n) \leqslant d(x_m, x) + d(x, x_n)$ (by the triangular inequality) $= d(x, x_m) + d(x, x_n)$ (by symmetry) $< \epsilon$. □

One may wonder whether this is enough. In $\mathbb{R}$, this is certainly true:

Proposition 3.4.2 *Every Cauchy sequence in $\mathbb{R}$ with the L^1 metric has a limit.*

Proof Let $(x_n)_{n\in\mathbb{N}}$ be a Cauchy sequence of reals. For every $\epsilon > 0$, there is an $n_0 \in \mathbb{N}$ such that, for all $m, n \geqslant n_0$, $|x_m - x_n| < \epsilon$. In particular, for $\epsilon = 1$, there is an $n_1 \in \mathbb{N}$ such that (taking $m = n_1$ itself) every x_n with $n \neq n_1$ lies in the interval $[x_{n_1} - 1, x_{n_1} + 1]$.

For $n \geqslant n_1$, define $x_n^- = \inf_{m\geqslant n} x_m$, $x_n^+ = \sup_{m\geqslant n} x_m$. These are values that lie in $[x_{n_1} - 1, x_{n_1} + 1]$, and are, in particular, well-defined reals. Let $x^- = \sup_{n\geqslant n_1} x_n^-$, $x^+ = \inf_{n\geqslant n_1} x_n^+$. It is easy to see that $x_{n_1}^- \leqslant x_{n_1+1}^- \leqslant \cdots \leqslant x_n^- \leqslant \cdots \leqslant x^- \leqslant x^+ \leqslant \cdots \leqslant x_n^+ \leqslant \cdots \leqslant x_{n_1+1}^+ \leqslant x_{n_1}^+$. We claim that $x^- = x^+$. For every $\epsilon > 0$, find $n_0 \in \mathbb{N}$ such that, for all $m, m' \geqslant n_0$, $|x_m - x_{m'}| < \epsilon$. So, for every $n \geqslant n_0$, $x_n^+ - x_n^- = \sup_{m,m' \geqslant n}(x_m - x'_m) \leqslant \epsilon$. As $\epsilon > 0$ is arbitrary, $x_n^+ - x_n^- \leqslant \inf_{\epsilon > 0} \epsilon = 0$. It follows $x^+ - x^- \leqslant x_n^+ - x_n^- \leqslant 0$, whence $x^- = x^+$.

Let $x = x^- = x^+$. We must finally show that x is the limit of $(x_n)_{n \in \mathbb{N}}$. For every $\epsilon > 0$, find n_0 such that, for all $m, m' \geqslant n_0$, $|x_m - x_{m'}| < \epsilon/2$. Then $x_{n_0}^- - x_{n_0}^+ \leqslant \epsilon/2$, and $x_{n_0}^- \leqslant x \leqslant x_{n_0}^+$. It follows that $x - x_{n_0}^- \leqslant \epsilon/2$, so $x - x_m \leqslant \epsilon/2 < \epsilon$ for every $m \geqslant n_0$, and $x_{n_0}^+ - x \leqslant \epsilon/2$, so $x_m - x \leqslant \epsilon/2 < \epsilon$ for every $m \geqslant n_0$. Therefore $|x_m - x| < \epsilon$, and we conclude. □

But this will not hold in other spaces. In particular, there are Cauchy sequences in $\mathbb{Q}$ (still with the L^1 norm) that have no limit.

This can be shown as follows. Let r be any irrational number, e.g., $\sqrt{2}$. Consider the sequence of numbers $r_n = \frac{1}{2^n}\lfloor 2^n r \rfloor$, $n \in \mathbb{N}$, i.e., r_n is obtained by writing r in binary and truncating after the nth bit after the decimal point. This is Cauchy in $\mathbb{R}$: when $m_0 \leqslant m \leqslant n$, $|r_m - r_n| \leqslant \frac{2}{2^{m_0}}$, which can be made arbitrarily small. The sequence $(r_n)_{n \in \mathbb{N}}$ converges to r in $\mathbb{R}$. But it does not converge in $\mathbb{Q}$, since its limit would also be one in $\mathbb{R}$; and, by uniqueness of limits, this limit would be r, which is not in $\mathbb{Q}$.

So we make this property into a definition.

Definition 3.4.3 (Complete metric space) A metric space X, d is *complete* if and only if every Cauchy sequence converges in X.

Summing up the above discussion:

Fact 3.4.4 *With the L^1 norm, $\mathbb{R}$ is complete, while $\mathbb{Q}$ is not.*

One of the key theorems of complete metric spaces is the Banach Fixed Point Theorem. For any $c \in \mathbb{R}^+$, a map f from a metric space X, d to a metric space Y, d' is called *c-Lipschitz* if and only if $d'(f(x), f(x')) \leqslant c.d(x, x')$ for all $x, x' \in X$.

Theorem 3.4.5 (Banach Fixed Point Theorem) *Let X, d be a non-empty complete metric space such that, for all $x, x' \in X$, $d(x, x') \neq +\infty$. Let $c \in [0, 1)$.*

Every c-Lipschitz map f from X to X has a unique fixed point $x = f(x)$.

This fixed point is obtained as the limit of the sequence of so-called iterates *$(f^n(x_0))_{n \in \mathbb{N}}$, where x_0 is any point from X.*

Proof Given any $x_0 \in X$, let $r_n = \frac{c^n}{1-c} d(x_0, f(x_0))$. Note that, since $c \in [0, 1)$, $\inf_{n \in \mathbb{N}} r_n = 0$. Indeed, letting $r = \inf_{n \in \mathbb{N}} r_n$, we first have $r \geqslant 0$ since

$r_n \geqslant 0$ for every $n \in \mathbb{N}$; second, $r \leqslant \inf_{n \geqslant 1} r_n = \inf_{n \in \mathbb{N}} r_{n+1} = \inf_{n \in \mathbb{N}} c.r_n = c.r$, and this implies that $(1-c).r \leqslant 0$, hence $r \leqslant 0$. For every $\epsilon > 0$, it follows that $r_n < \epsilon$ for n large enough.

Since f is c-Lipschitz, $d(f^n(x_0), f^{n+1}(x_0)) \leqslant c^n d(x_0, f(x_0)) = r_n - r_{n+1}$. So, if $n_0 \leqslant n \leqslant m$ then $d(f^n(x_0), f^m(x_0)) \leqslant d(f^n(x_0), f^{n+1}(x_0)) + d(f^{n+1}(x_0), f^{n+2}(x_0)) + \cdots + d(f^{m-1}(x_0), f^m(x_0)) \leqslant r_n - r_m \leqslant r_n \leqslant r_{n_0}$. The inequality also holds for $n_0 \leqslant m \leqslant n$, by symmetry. Choosing n_0 so large that $r_{n_0} < \epsilon$, we obtain that, for all $m, n \geqslant n_0$, $d(f^n(x_0), f^m(x_0)) < \epsilon$. So $(f^n(x_0))_{n \in \mathbb{N}}$ is a Cauchy sequence. Since X, d is complete, it has a limit x.

Given any $\epsilon > 0$, $d(f(x), f^n(x_0)) \leqslant c.d(x, f^{n-1}(x_0)) < c.\epsilon < \epsilon$ for n large enough, so $f(x)$ is also a limit of the same sequence. Since limits are unique in metric spaces, $f(x) = x$, i.e., x is indeed a fixed point of f.

Finally, if y is another fixed point of f, then $d(x, y) = d(f(x), f(y)) \leqslant c.d(x, y)$, which implies $(1-c).d(x, y) \leqslant 0$, hence $d(x, y) = 0$, i.e., $x = y$. So f has only one fixed point. □

The Banach Fixed Point Theorem has numerous applications in all areas of mathematics, e.g., the Cauchy–Lipschitz (a.k.a. Picard–Lindelöf) Theorem on local existence and uniqueness of solutions of differential equations.

Completeness and compactness are related, too. In a sense, compactness is a more demanding notion. We say that a metric space X, d is sequentially compact if and only if X, as a subset of the space, is sequentially compact.

Proposition 3.4.6 *Every Cauchy sequence in a sequentially compact subset K of a metric space converges to some point of K.*

In particular, every sequentially compact metric space is complete.

Proof Let $(x_n)_{n \in \mathbb{N}}$ be a Cauchy sequence in the compact subset K of the metric space X, d. Let $(x_{n_k})_{k \in \mathbb{N}}$ be a convergent subsequence, and let x be its limit, which is in K. For every $\epsilon > 0$, there is a $k_0 \in \mathbb{N}$ such that, for every $k \geqslant k_0$, $d(x, x_{n_k}) < \epsilon/2$. Since $(x_n)_{n \in \mathbb{N}}$ is Cauchy, there is an $m_0 \in \mathbb{N}$ such that, for all $m, n \geqslant m_0$, $d(x_m, x_n) < \epsilon/2$. Let $k \geqslant k_0$ be such that $n_{k_0} \geqslant m_0$, and let $m = n_k$: then, for every $n \geqslant m_0$, $d(x, x_n) \leqslant d(x, x_{n_k}) + d(x_m, x_n) < \epsilon$. So x is a limit of $(x_n)_{n \in \mathbb{N}}$. □

Exercise 3.4.7 One can generalize completeness to *subsets* of metric spaces. Call a subset C of a metric space X, d complete if and only if every Cauchy sequence of elements of C converges to an element of C. Equivalently, C is complete as a subset of X if and only if the metric space C, d is complete.

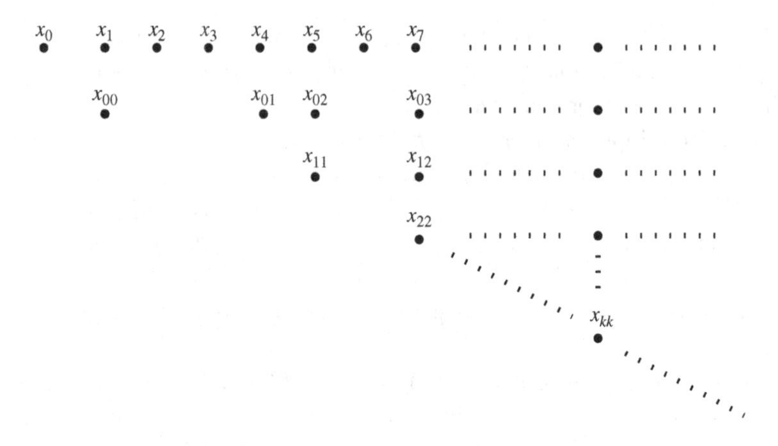

Figure 3.8 Extracting a convergent subsequence.

> Show that: (a) every sequentially compact subset of a metric space is complete in this new sense; (b) the complete subsets of a complete metric space are its sequentially closed subsets.

We can improve on this. Let X, d be a metric space. Call a subset K of X *precompact* if and only if, for every $\epsilon > 0$, K is included in a union of finitely many open balls $B^d_{x_i, <\epsilon}$ of radius ϵ, $1 \leqslant i \leqslant n$, where $x_1, \ldots, x_n \in K$. By Proposition 3.3.7, Item 2, every sequentially compact subset of a metric space is precompact.

Theorem 3.4.8 *The sequentially compact metric spaces are exactly the precompact, complete metric spaces.*

Proof It remains to show that if X, d is a precompact and complete metric space, then it is compact. Let $(x_n)_{n \in \mathbb{N}}$ be a sequence in X. We extract successive subsequences $(x_{kn})_{n \in \mathbb{N}}$ of $(x_n)_{n \in \mathbb{N}}$ by induction on $k \in \mathbb{N}$. See Figure 3.8. By extension, let $x_{(-1)n} = x_n$ for every $n \in \mathbb{N}$. Letting $\epsilon = \frac{1}{2^k}$ and using precompactness, X is included in a finite union of open balls of radius $\frac{1}{2^k}$. So there is one that contains $x_{(k-1)n}$ for infinitely many values of n: pick one such ball $B^d_{y_k, <1/2^k}$, and let $(x_{kn})_{n \in \mathbb{N}}$ be the subsequence of those elements of the previous subsequence $(x_{(k-1)n})_{n \in \mathbb{N}}$ that lie in $B^d_{y_k, <1/2^k}$.

Now consider the sequence $(x_{kk})_{k \in \mathbb{N}}$. This is a subsequence of the original sequence $(x_n)_{n \in \mathbb{N}}$. For all $m, n \geqslant k_0$, x_m and x_n both lie in $B^d_{y_{k_0}, <1/2^{k_0}}$, so $d(x_m, x_n) \leqslant d(x_m, y_{k_0}) + d(y_{k_0}, x_n)$ (by the triangular inequality) $= d(y_{k_0}, x_m) + d(y_{k_0}, x_n)$ (by symmetry) $< \frac{2}{2^{k_0}}$. It follows that $(x_{kk})_{k \in \mathbb{N}}$ is Cauchy. Since X, d is complete, it has a limit x. □

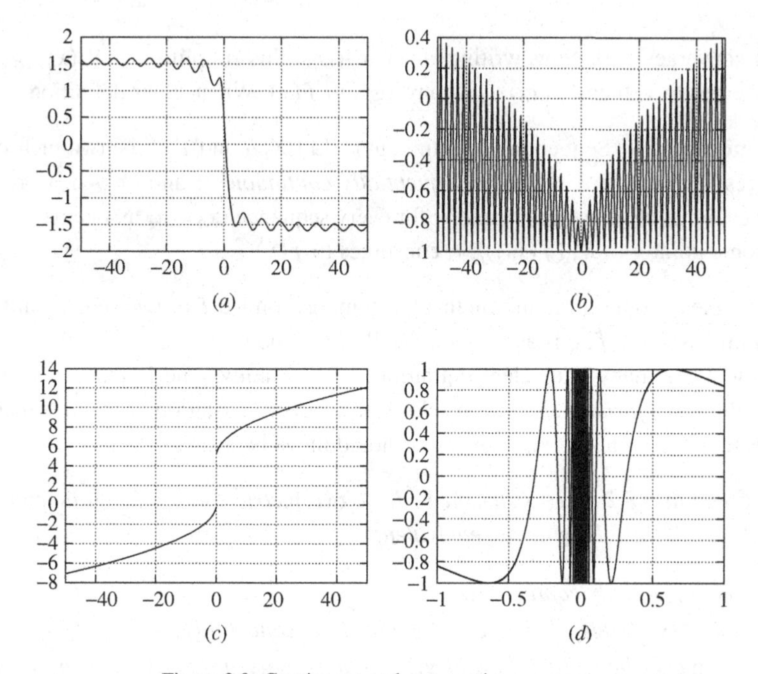

Figure 3.9 Continuous and non-continuous maps.

3.5 Continuous functions

Topology is about convergence, and also about continuity. A map $f: \mathbb{R} \to \mathbb{R}$ defines a curve in the real plane, as the set of points $(x, f(x))$. See Figure 3.9 for a few such plane curves. One would like to consider curves (a) and (b) as curves of continuous maps. Note in particular that (a) and (b) are in one piece, contrarily to (c), which exhibits a jump from $(0, 0)$ to $(0, 5)$. Accordingly, a map with a jump such as (c) should *not* be continuous.

The idea behind continuity is that if we move x only slightly, then $f(x)$ should move by a small amount as well. The less we move x around, the less $f(x)$ should change. This indeed excludes maps with jumps, but also maps with infinite oscillations such as (d)—the map that sends $x \neq 0$ to $\sin(1/x)$, and 0 to 0: if we move from 0 to an arbitrary small $\epsilon \neq 0$, $\sin(1/\epsilon)$ will assume values that are not small in general, and span the whole interval $[-1, 1]$ instead.

So, a continuous map should be such that if we replace x by $x + \eta$ for some small η (negative or positive), then $f(x + \eta)$ should equal $f(x) + \epsilon$ for some small ϵ (negative or positive). To make the idea of "small" precise, we imagine that we replace η by a sequence of values η_n that converge to 0—i.e., that are arbitrarily small. Then $f(x + \eta_n)$ should be of the form $f(x) + \epsilon_n$ for some ϵ_n

that converges to 0. Now, writing x_n for $x + \eta_n$, this simplifies to: if $(x_n)_{n \in \mathbb{N}}$ converges to x, then $(f(x_n))_{n \in \mathbb{N}}$ converges to $f(x)$. Whence our definition.

Definition 3.5.1 (Sequentially continuous) Let X, d and Y, d' be two metric spaces. A map $f : X \to Y$ is *sequentially continuous* if and only if it preserves limits of sequences, namely: for every sequence $(x_n)_{n \in \mathbb{N}}$ that converges to some point x in X, $(f(x_n))_{n \in \mathbb{N}}$ converges to $f(x)$.

It is clear from this definition that the composition $g \circ f$ of two sequentially continuous maps f, g is again sequentially continuous.

There are many equivalent definitions of continuity. The last one below, called the *ϵ-η formula*, is perhaps closer to a direct translation of the idea that if we move x by a small η, then $f(x)$ should move by a small ϵ.

Proposition 3.5.2 *Let X, d and Y, d' be two metric spaces. For a function $f : X \to Y$, the following are equivalent:*

1. *f is sequentially continuous;*
2. *the inverse image of every open ball by f is sequentially open;*
3. *the inverse image $f^{-1}(V)$ of every sequentially open subset V of Y is sequentially open in X;*
4. *the inverse image $f^{-1}(F)$ of every sequentially closed subset F of Y is sequentially closed in X;*
5. *(ϵ-η formula) for every $x \in X$, for every $\epsilon > 0$, there is an $\eta > 0$ such that, for every $x' \in X$ such that $d(x, x') < \eta$, $d'(f(x), f(x')) < \epsilon$.*

Proof 1 $\Rightarrow$ 2. Let $y \in Y$, and $\epsilon > 0$. We claim that $f^{-1}(B^{d'}_{y,<\epsilon})$ is sequentially open in X. Let $(x_n)_{n \in \mathbb{N}}$ be any sequence in X with a limit x in $f^{-1}(B^{d'}_{y,<\epsilon})$. So $d'(y, f(x)) < \epsilon$. By 1, $f(x)$ is a limit of $(f(x_n))_{n \in \mathbb{N}}$, so $d'(f(x), f(x_n)) < \frac{1}{2}(\epsilon - d'(y, f(x)))$ for n large enough. It follows that $d'(y, f(x_n)) \leqslant d'(y, f(x)) + d'(f(x), f(x_n)) < \frac{1}{2}(\epsilon + d'(y, f(x))) < \epsilon$ for n large enough. In other words, x_n is in $f^{-1}(B^{d'}_{y,<\epsilon})$ for n large enough. So $f^{-1}(B^{d'}_{y,<\epsilon})$ is sequentially open, by Proposition 3.2.4, Item 4.

2 $\Rightarrow$ 3. By Proposition 3.2.4, Item 3, V is a union of open balls, so $f^{-1}(V)$ is a union of inverse images of open balls, namely a union of sequentially open subsets of X by 2. It follows that $f^{-1}(V)$ is sequentially open, using Proposition 3.2.7.

3 $\Leftrightarrow$ 4. Take complements.

3 $\Rightarrow$ 2. Take $V = B^{d'}_{y,<\epsilon}$.

2 $\Rightarrow$ 5. Let $x \in X$, $\epsilon > 0$. By 2, $f^{-1}(B^{d'}_{f(x),<\epsilon})$ is sequentially open in X, and clearly contains x. By Proposition 3.2.4, Item 2, there is an open ball

$B^d_{x,<\eta}$ centered at x that is included in $f^{-1}(B^{d'}_{f(x),<\epsilon})$: this is exactly what Item 3 means.

$5 \Rightarrow 1$. Let $(x_n)_{n\in\mathbb{N}}$ be a sequence in X that converges to $x \in X$. For every $\epsilon > 0$, find $\eta > 0$ so that, for every x' with $d(x, x') < \eta$, $d'(f(x), f(x')) < \epsilon$. For n large enough, $d(x, x_n) < \eta$, by the definition of limits. So, for n large enough, $d'(f(x), f(x_n)) < \epsilon$. This means that $f(x)$ is a limit of $(f(x_n))_{n\in\mathbb{N}}$. □

Item 5 of Proposition 3.5.2 does not refer to sequences any longer: we shall simply call *continuous* the sequentially continuous functions between two metric spaces.

Exercise 3.5.3 Let X, d and Y, d' be metric spaces. Show that every c-Lipschitz map from X to Y is continuous, but that there are continuous maps that are not c-Lipschitz for any $c \in \mathbb{R}^+$ in general.

Exercise 3.5.4 Extend the L^1 metric to $\mathbb{R} \cup \{+\infty\}$ by declaring $+\infty$ to be at distance 0 from itself but $+\infty$ from all reals. Let X, d be a metric space. Show that, for each fixed $x_0 \in X$, the map $x \mapsto d(x, x_0)$ is 1-Lipschitz, hence continuous from X, d to $\overline{\mathbb{R}^+}$ with the L^1 metric.

Now continuity and compactness blend well together. We write $f[A]$ for the image of the subset A of X by the map f, i.e., the set $\{f(x) \mid x \in A\}$. A more customary notation is $f(A)$. However, if X is itself a set of subsets, the latter notation would be ambiguous.

Proposition 3.5.5 *Let X, d and Y, d' be metric spaces, and $f: X \to Y$ be continuous. The image $f[K]$ of any sequentially compact subset K of X is sequentially compact in Y.*

Proof Let $(f(x_n))_{n\in\mathbb{N}}$ be any sequence in $f[K]$, with $x_n \in K$ for every $n \in \mathbb{N}$. Extract a sequence $(x_{n_k})_{k\in\mathbb{N}}$ that converges to some point x of K. Then $(f(x_{n_k}))_{k\in\mathbb{N}}$ converges to $f(x)$ by sequential continuity, and $f(x)$ is in the $f[K]$. □

An immediate corollary is the following. Recall that the L^1 metric on $\mathbb{R}$ is defined by $d(x, y) = |y - x|$.

Corollary 3.5.6 *Let f be a continuous map from a metric space X, d to $\mathbb{R}$ with its L^1 metric. Let K be a sequentially compact subset of X. Then $f[K]$ is bounded, i.e., included in some interval $[a, b]$, with $a, b \in \mathbb{R}$, $a \leqslant b$.*

Moreover, f reaches its maximum and its minimum on K: there is a point $x_{\max} \in K$ such that $f(x_{\max}) = \max_{x\in K} f(x)$, and a point $x_{\min}$ in K such that $f(x_{\min}) = \min_{x\in K} f(x)$.

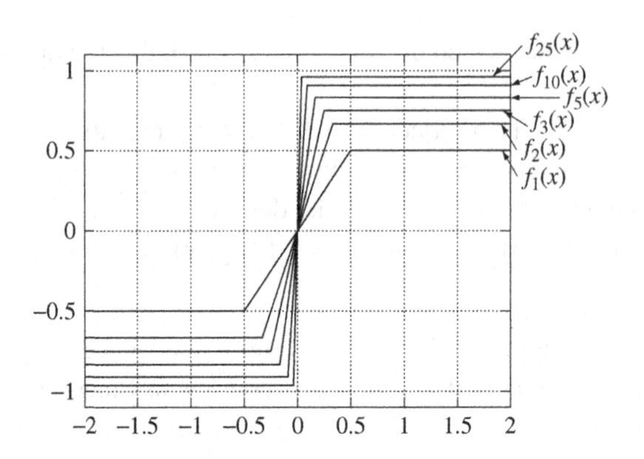

Figure 3.10 Functions that converge pointwise, not uniformly.

Proof By Proposition 3.5.5, $f[K]$ is sequentially compact, hence bounded by the Borel–Lebesgue Theorem (Proposition 3.3.4, with $n = 1$).

Let $a = \sup_{x \in K} f(x)$. For every $n \in \mathbb{N}$, there is a point $x_n \in K$ such that $f(x_n) > a - \frac{1}{2^n}$ (and, necessarily, $f(x_n) \leqslant a$). Extract a subsequence $(x_{n_k})_{k \in \mathbb{N}}$ that converges to some $x \in K$. Then $(f(x_{n_k}))_{k \in \mathbb{N}}$ converges to $f(x)$. Since $a - \frac{1}{2^n} < f(x_n) \leqslant a$ for every $n \in \mathbb{N}$, a is also a limit of $(f(x_{n_k}))_{k \in \mathbb{N}}$. But limits are unique in metric spaces, so $f(x) = a$, and we take $x_{\max} = x$. The argument for $x_{\min}$ is similar. □

There are several notions of convergence on spaces of functions. The simplest is *pointwise convergence*: let X be a set and Y, d' be a metric space, a sequence $(f_n)_{n \in \mathbb{N}}$ of functions from X to Y converges to $f : X \to Y$ if and only if $(f_n(x))_{n \in \mathbb{N}}$ converges to $f(x)$ for every $x \in X$.

While simple and intuitive, this notion of convergence has certain shortcomings. For example, the limit of a pointwise convergent sequence of continuous maps may fail to be continuous. See Figure 3.10 for an illustration: we take $X = Y = \mathbb{R}$ with the L^1 metric, $f_n(x) = -1 + \frac{1}{n+1}$ if $x \leqslant -\frac{1}{n+1}$, $f_n(x) = 1 - \frac{1}{n+1}$ if $x \geqslant \frac{1}{n+1}$, and $f_n(x) = n.x$ if $-\frac{1}{n+1} < x < \frac{1}{n+1}$. Each function f_n is continuous. However, $(f_n(x))_{n \in \mathbb{N}}$ converges to -1 if $x < 0$, 0 if $x = 0$, 1 if $x > 0$, and the latter defines a non-continuous map—with two jumps at $x = 0$.

Note that the fact that $(f_n)_{n \in \mathbb{N}}$ converges to f pointwise is described by the following formula:

$$\text{for every } x \in X, \text{ for every } \epsilon > 0, \text{ there exists } n_0 \in \mathbb{N} \text{ such that,}$$
$$\text{for every } n \geqslant n_0, d'(f(x), f_n(x)) < \epsilon.$$

We say that $(f_n)_{n\in\mathbb{N}}$ converges to f *uniformly* if and only if n_0 can be chosen the same for all points x; in other words, *independently* of $x \in X$. In other words, $(f_n)_{n\in\mathbb{N}}$ converges to f uniformly if and only if:

$$\text{for every } \epsilon > 0, \text{ there exists } n_0 \in \mathbb{N} \text{ such that, for every } n \geqslant n_0,\\ \text{for every } x \in X, d'(f(x), f_n(x)) < \epsilon.$$

Now, uniform convergence is just convergence with respect to the following metric.

Proposition 3.5.7 *Let X be a set, and Y, d' be a metric space. The* sup metric $\sup_X d'$ *is defined on the space of all maps from X to Y by* $\sup_X d(f, g) = \sup_{x\in X} d(f(x), g(x))$.

The sequence of maps $(f_n)_{n\in\mathbb{N}}$ from X to Y converges uniformly to f if and only if it converges to f in the sup metric.

Proof If: for every $\epsilon > 0$, there exists $n_0 \in \mathbb{N}$ such that, for every $n \geqslant n_0$, $\sup_X d'(f, f_n) < \epsilon$. In particular, for every $x \in X$, $d'(f(x), f_n(x)) < \epsilon$.

Only if: for every $\epsilon > 0$, there exists $n_0 \in \mathbb{N}$ such that, for every $n \geqslant n_0$, for every $x \in X$, $d'(f(x), f_n(x)) < \epsilon/2$. (The formula for uniform convergence must hold for any ϵ, in particular for $\epsilon/2$.) So $\sup_X d'(f, f_n) \leqslant \epsilon/2 < \epsilon$. □

Exercise 3.5.8 Let X, d and Y, d' be two metric spaces. We say that a sequence of continuous maps $(f_n)_{n\in\mathbb{N}}$ *converges continuously* to the continuous map f if and only if, for every convergent sequence $(x_n)_{n\in\mathbb{N}}$ of points of X, with limit x, $f(x) = \lim_{n\in\mathbb{N}} f_n(x_n)$. Show that: (a) if $(f_n)_{n\in\mathbb{N}}$ converges uniformly to f, then it converges continuously to f; (b) if $(f_n)_{n\in\mathbb{N}}$ converges continuously to f, then it converges pointwise to f; (c) in general, there are sequences $(f_n)_{n\in\mathbb{N}}$ of continuous maps that converge pointwise to f, but not continuously; (d) if X is sequentially compact, then if $(f_n)_{n\in\mathbb{N}}$ converges continuously to f, then it also converges uniformly to f. For (d), reason by contradiction.

The problem with pointwise limits of continuous maps is repaired:

Proposition 3.5.9 *Let X, d and Y, d' be metric spaces, and $(f_n)_{n\in\mathbb{N}}$ be a sequence of continuous maps from X to Y. If it converges uniformly, then its limit is continuous as well.*

Proof For every $\epsilon > 0$, let $n_0 \in \mathbb{N}$ be such that, for every $n \geqslant n_0$, for every $x \in X$, $d'(f(x), f_n(x)) < \epsilon/3$. Since f_{n_0} is continuous, there is an $\eta > 0$ such that, for every $x' \in X$, if $d(x, x') < \eta$ then $d'(f_{n_0}(x), f_{n_0}(x')) < \epsilon/3$. So $d'(f(x), f(x')) \leqslant d'(f(x), f_{n_0}(x)) +$

$d'(f_{n_0}(x), f_{n_0}(x')) + d'(f_{n_0}(x'), f(x'))$ (by the triangular inequality, applied twice) $= d'(f(x), f_{n_0}(x)) + d'(f_{n_0}(x), f_{n_0}(x')) + d'(f(x'), f_{n_0}(x'))$ (by symmetry) $< \epsilon$. □

Finally, the so-called Arzelà–Ascoli Theorem below characterizes compactness in spaces of continuous maps between metric spaces X and Y, provided X is compact as well. We need yet another notion.

Note that a family E of functions from X, d to Y, d' consists of continuous maps if and only if:

for every $\epsilon > 0$, for every $x \in X$, for every $f \in E$, there is an $\eta > 0$ such that, for every $x' \in X$ such that $d(x, x') < \eta$, $d'(f(x), f(x')) < \epsilon$.

If η is chosen independently of $x \in X$, then E consists of so-called *uniformly continuous* maps. (We won't say more about these here, except to notice that every c-Lipschitz map is uniformly continuous, and every uniformly continuous maps is continuous.) If on the other hand η is chosen independently of $f \in E$, then we say that E is an *equicontinuous* family of maps. In other words, E is an equicontinuous family of maps from X to Y if and only if:

for every $\epsilon > 0$, for every $x \in X$, there is an $\eta > 0$ such that, for every $f \in E$, for every $x' \in X$ such that $d(x, x') < \eta$, $d'(f(x), f(x')) < \epsilon$.

We also say that a sequence $(f_n)_{n \in \mathbb{N}}$ is equicontinuous if and only if the set $\{f_n \mid n \in \mathbb{N}\}$ is equicontinuous. Equicontinuity allows us to fill the gap between pointwise and uniform convergence, under a compactness assumption:

Proposition 3.5.10 *Let X, d be a compact metric space, and Y, d' be a metric space. Every equicontinuous sequence $(f_n)_{n \in \mathbb{N}}$ of maps from X to Y that converges pointwise to some map f actually converges uniformly to f.*

Proof Fix $\epsilon > 0$. For every $x \in X$, there is an $n_x \in \mathbb{N}$ such that, for every $n \geqslant n_x$, $d'(f(x), f_n(x)) < \epsilon/3$, by pointwise convergence. Equicontinuity gives us an $\eta_x > 0$ such that, for every $n \in \mathbb{N}$, for every $x' \in X$ such that $d(x, x') < \eta_x$, $d'(f_n(x), f_n(x')) < \epsilon/3$, in particular for $n \geqslant n_x$. Using the triangular inequality and the axiom of symmetry, $d'(f(x'), f_n(x')) \leqslant d'(f(x'), f(x)) + d'(f(x), f_n(x)) + d'(f_n(x), f_n(x')) < \epsilon$, for every $n \geqslant n_x$ and every x' in the open ball $B^d_{x, <\eta_x}$. Each open ball is sequentially open, and the union of the latter clearly contains the whole of X. Since X is compact, by the Heine-Borel Property (Proposition 3.3.7, Item 2), there is a finite subset E of X such that every point x' of X is in some ball B^d_{x, η_x} with $x \in E$. Let

$n_0 = \max_{x \in E} n_x$. By construction, for every $n \geqslant n_0$, n is greater than or equal to n_x, so $d'(f(x'), f_n(x')) < \epsilon$. □

Exercise 3.5.11 Let X, d and Y, d' be metric spaces, and $c \in \mathbb{R}^+$ be a constant. Show that every family E of c-Lipschitz maps from X to Y is equicontinuous—in fact, uniformly equicontinuous, in the sense that η can be chosen independently, not just of $f \in E$, but also of $x \in X$.

Theorem 3.5.12 (Arzelà–Ascoli, original form) *Let X, d be a compact metric space, Y, d' be a metric space, and $\mathcal{K}$ be a family of maps from X to Y that satisfy:*

1. *$\mathcal{K}$ is equicontinuous;*
2. *for every $x \in X$, there is a sequentially compact subset K_x of Y such that, for every $f \in \mathcal{K}$, $f(x) \in K_x$;*
3. *$\mathcal{K}$ is sequentially closed with respect to the sup metric $\sup_X d'$.*

Then $\mathcal{K}$ is a sequentially compact subset of continuous maps, with respect to the sup metric.

In Item 2, one may have the false impression that we should guess a sequentially compact subset K_x. However, if there is one, then $cl\{f(x) \mid f \in \mathcal{K}\}$ is one, too. So Item 2 can be stated equivalently as: $\{f(x) \mid f \in \mathcal{K}\}$ has sequentially compact closure.

Proof We first show that $\mathcal{K}$ is precompact.

Let $\epsilon > 0$. For every $x \in X$, there is an $\eta_x > 0$ such that, for every $f \in \mathcal{K}$, for every $x' \in X$ such that $d(x, x') < \eta_x$, $d'(f(x), f(x')) < \epsilon/3$. Clearly, $X \subseteq \bigcup_{x \in X} B^d_{x, <\eta_x}$. Since X is compact, there are finitely many points $x_1, \ldots, x_m$ in X such that $X \subseteq \bigcup_{i=1}^m B^d_{x_i, <\eta_{x_i}}$. This is the Heine–Borel Property (Proposition 3.3.7, Item 2). So, for every $x \in X$, there is an i, $1 \leqslant i \leqslant m$, such that $d(x_i, x) < \eta_{x_i}$, in particular $d'(f(x_i), f(x)) < \epsilon/3$. Using the Axiom of Choice, one can write such an i as $g(x)$, for some selection function g.

Since K_{x_i} is sequentially compact, it is precompact by Proposition 3.3.7, Item 2. So there is a finite subset E_i of Y such that $K \subseteq \bigcup_{y \in E_i} B^{d'}_{y, <\epsilon/3}$.

For every $f \in \mathcal{K}$, for every i, $1 \leqslant i \leqslant m$, since $f(x_i) \in K_{x_i}$, there is a point $y_i \in E_i$ such that $f(x_i) \in B^{d'}_{y_i, <\epsilon/3}$. In other words, there is an m-tuple $\vec{y} = (y_1, \ldots, y_m)$ in $E_1 \times \cdots \times E_m$ such that, for every i, $1 \leqslant i \leqslant m$, $d'(y_i, f(x_i)) < \epsilon/3$.

For every $x \in X$, recall that $d'(f(x_i), f(x)) < \epsilon/3$, where $i = g(x)$. Using the triangular inequality, $d'(y_{g(x)}, f(x)) < 2\epsilon/3$. Let $f_{\vec{y}}$ be the function that maps each $x \in X$ to $y_{g(x)}$. So $\sup_X d'(f_{\vec{y}}, f) \leqslant 2\epsilon/3 < \epsilon$, whence $f \in B^{\sup_X d'}_{f_{\vec{y}}, <\epsilon}$.

Since f is arbitrary in $\mathcal{K}$, we have shown that $\mathcal{K}$ is included in the finite union of open balls $B^{\sup_X d'}_{f_{\vec{y}}, <\epsilon}$, $\vec{y} \in E_1 \times \cdots \times E_m$. So $\mathcal{K}$ is precompact.

We now claim that $\mathcal{K}$ is complete in the sup metric $\sup_X d'$.

Any Cauchy sequence $(f_n)_{n \in \mathbb{N}}$ in $\mathcal{K}$ (with respect to the sup metric) is such that $(f_n(x))_{n \in \mathbb{N}}$ is Cauchy in Y, d' for every $x \in X$. Moreover, $f_n(x)$ is in K_x for every $n \in \mathbb{N}$. Since K_x is sequentially compact, $(f_n(x))_{n \in \mathbb{N}}$ converges to some point $f(x)$ of K_x by Proposition 3.4.6. (We collect all limits $f(x)$ in a single function f by using the Axiom of Choice again.) Now $(f_n)_{n \in \mathbb{N}}$ converges to f pointwise, hence uniformly since $\mathcal{K}$ is equicontinuous, using Proposition 3.5.10.

Since $\mathcal{K}$, equipped with the sup metric, is both precompact and complete, it is sequentially compact by Theorem 3.4.8. □

The most useful aspect of the Arzelà–Ascoli Theorem is the following corollary.

Corollary 3.5.13 *Let X, d be a compact metric space and Y, d' be a metric space. Every sequence $(f_n)_{n \in \mathbb{N}}$ of maps from X to Y:*

1. *that is equicontinuous,*
2. *and such that, for every $x \in X$, there is a compact subset K_x of Y such that, for every $n \in \mathbb{N}$, $f_n(x) \in K_x$,*

has a uniformly convergent subsequence.

Again, the second condition is equivalent to the fact that $\{f_n(x) \mid n \in \mathbb{N}\}$ has a sequentially compact closure.

Proof Let $\mathcal{K}$ be the closure of $\{f_n \mid n \in \mathbb{N}\}$ in the spaces of all maps from X to Y with the sup metric $\sup_X d'$.

We first check that $\mathcal{K}$ is equicontinuous. By Assumption 1, for every $\epsilon > 0$, for every $x \in X$, there is an $\eta > 0$ such that, for every $n \in \mathbb{N}$, for every x' such that $d(x, x') < \eta$, $d'(f_n(x), f_n(x')) < \epsilon/3$. For every $f \in \mathcal{K}$, $B^{\sup_X d'}_{f, <\epsilon/3}$ meets $\{f_n \mid n \in \mathbb{N}\}$ by Exercise 3.2.6: let n be such that $\sup_X d'(f, f_n) < \epsilon/3$. Then, for every x' such that $d(x, x') < \eta$, $d'(f(x), f(x')) \leqslant d'(f(x), f_n(x)) + d'(f_n(x), f_n(x')) + d'(f_n(x'), f(x'))$ (by the triangular inequality) $= d'(f(x), f_n(x)) + d'(f_n(x), f_n(x')) + d'(f(x'), f_n(x'))$ (by symmetry) $< \epsilon$, proving the claim.

For every $f \in \mathcal{K}$ and every $x \in X$, f is the uniform limit of some sequence of maps in $\{f_n \mid n \in \mathbb{N}\}$ (not necessarily in the same order as $(f_n)_{n \in \mathbb{N}}$). In particular, $f(x)$ is the limit of a sequence of points of the form $f_n(x)$. Since the latter are all in K_x and K_x, being sequentially compact, is sequentially closed, $f(x)$ is in K_x as well.

Finally, $\mathcal{K}$ is sequentially closed by construction. So Corollary 3.5.13 applies: $\mathcal{K}$ is sequentially compact in the sup metric. The result follows. □

Exercise 3.5.14 The assumptions of Theorem 3.5.12 may seen demanding. Show that they are all necessary. Namely, let X, d be a compact metric space, Y, d' be a metric space, and $\mathcal{K}$ be a sequentially compact subset of the set of continuous maps from X to Y with the sup metric $\sup_X d'$: show that (1) $\mathcal{K}$ must be equicontinuous, that (2) for every x, there is a sequentially compact subset of Y such that $f(x) \in K_x$ for every $f \in \mathcal{K}$ (one can even take $K_x = \{f(x) \mid f \in \mathcal{K}\}$), and that (3) $\mathcal{K}$ must be sequentially closed.

To show that $\mathcal{K}$ must be equicontinuous, let W_n be the set of all continuous maps f such that $\max_{x' \in B^d_{x,\leqslant 1/2^n}} d'(f(x), f(x')) < \epsilon$. Prove that every $f \in \mathcal{K}$ is in some W_n, and conclude using the Heine–Borel Property. You will need to justify that $\max_{x' \in B^d_{x,\leqslant 1/2^n}} d'(f(x), f(x'))$ makes sense, i.e., that it is a maximum, not just a least upper bound: what result have we seen that guarantees the existence of maximums?

Exercise 3.5.15 Let X, d be a compact metric space, and $(f_n)_{n \in \mathbb{N}}$ be a sequence of maps from X to $\mathbb{R}^m$, for some $m \in \mathbb{N}$. Assume that this sequence is pointwise bounded, i.e., that $\{f_n(x) \mid n \in \mathbb{N}\}$ is bounded in $\mathbb{R}^m$ or, equivalently, that, for every $x \in X$, there is a number $a_x \in \mathbb{R}^+$ such that $\{f_n(x) \mid n \in \mathbb{N}\} \subseteq [-a_x, a_x]^m$.

Show that if $(f_n)_{n \in \mathbb{N}}$ is equicontinuous, then it has a uniformly convergent subsequence.

From this, deduce that, for any fixed $c \in \mathbb{R}^+$, every pointwise bounded sequence of c-Lipschitz maps has a uniformly convergent subsequence.

4

Topology

4.1 Topology, topological spaces

Abstracting away from metrics, a topological space is a set, with a collection of so-called *open subsets* U, satisfying the following properties. We have already seen them, for sequentially open subsets of a metric space, in Proposition 3.2.7.

Definition 4.1.1 (Topology) Let X be a set. A *topology* on X is a collection of subsets of X, called the *opens* of the topology, such that:

- every union of opens is open (including the empty union, $\emptyset$);
- every finite intersection of opens is open (including the empty intersection, taken as X itself).

A *topological space* is a pair $(X, \mathcal{O})$, where $\mathcal{O}$ is a topology on X.

We often abuse the notation, and write X itself as the topological space, leaving $\mathcal{O}$ implicit. It is also customary to talk about the elements of a topological space X as *points*.

Example 4.1.2 The sequentially open subsets of a metric space form a topology. This is what Proposition 3.2.7 states exactly.

Given a metric space X, d, we shall call *metric topology* the topology whose opens are the sequentially open subsets.

Example 4.1.3 We shall often consider $\mathbb{R}$ with the metric topology of its L^1 metric. We shall either call this space $\mathbb{R}$, or $\mathbb{R}$ *with its metric topology* when there is a risk of confusion as to which topology is intended. There are indeed other natural candidates, as we shall see in Example 4.2.19, for instance.

Example 4.1.4 It is useful in understanding a definition to give examples of objects that do *not* satisfy the definition. In $\mathbb{R}$ with its metric topology, every open interval (a, b) (with $a < b$) is open, since it is just the open ball with center $\frac{a+b}{2}$ and radius $\frac{b-a}{2}$. But the closed interval $[a, b]$ is not open, nor are the half-open intervals $(a, b]$ and $[a, b)$. (Exercise: prove this.)

Figure 3.4 is meant to give an illustration of similar situations in the real plane.

Much as the sequentially closed subsets were the complements of the sequentially open subsets, we define:

Definition 4.1.5 (Closed) A *closed* subset F of a topological space X is the complement $X \smallsetminus U$ of an open subset of X.

"Closed" does *not* mean "not open." First, the half-open intervals in Example 4.1.4 are neither open nor closed. Second, a space X also contains some subsets that are *clopen*, that is, both open and closed. For example, in any space X, the empty subset $\emptyset$, and the whole space X, are clopen. There may be more: see Exercise 4.1.34, for example.

4.1.1 Bases and subbases

Proposition 3.2.4, Item 3, states that the metric topology of a metric space consists of a union of open balls: we say that open balls form a *base* of the metric topology. This is useful, since open balls are easier to understand than general opens.

Definition 4.1.6 (Base) Let X be a topological space, and $\mathcal{B}$ be a family of open subsets of X. $\mathcal{B}$ is a *base* of the topology if and only if every open is a union of elements of $\mathcal{B}$.

One can generalize the concept as follows. Order topologies on a set X by inclusion. A smaller topology (i.e., one that has less opens) is *coarser*, while a larger topology is *finer*. If we understand open subsets U as *tests* – whether, given a point x, x belongs to U – then a finer topology gives us more tests by which to classify points.

Lemma 4.1.7 *Given any collection $\mathcal{B}$ of subsets of a set X, there is a coarsest topology containing $\mathcal{B}$, i.e., such that every element of $\mathcal{B}$ is open. This is the collection of all unions of finite intersections of elements of $\mathcal{B}$. That is, the opens are of the form $\bigcup_{i \in I} \bigcap_{j \in J_i} U_{ij}$ where I is an arbitrary index set, J_i is finite for each $i \in I$, and U_{ij} is an element of $\mathcal{B}$.*

Proof Let us say that a subset is open if and only if it is the union of finite intersections of elements of $\mathcal{B}$. That every union of open subsets is open is trivial. To check that every finite intersection of opens is open, it is enough to show that X itself is open (empty intersection, obtained with I a one-element set $\{i\}$, $J_i = \emptyset$), and that the intersection of two opens is open. Now $\left(\bigcup_{i \in I} \bigcap_{j \in J_i} U_{ij}\right) \cap \left(\bigcup_{i' \in I'} \bigcap_{j' \in J'_{i'}} U'_{i'j'}\right) = \bigcup_{\substack{i \in I \\ i' \in I'}} \left(\bigcap_{j \in J_i} U_{ij} \cap \bigcap_{j' \in J'_{i'}} U'_{i'j'}\right)$. So the opens indeed form a topology. It is clearly the coarsest possible one. □

Definition 4.1.8 (Generation, subbase) Let X be a set, and $\mathcal{B}$ be a family of subsets of X. The coarsest topology on X that contains $\mathcal{B}$ is the topology *generated by* $\mathcal{B}$.

Then we say that $\mathcal{B}$ is a *subbase* of the topology. That is, a subbase of a topology $\mathcal{O}$ is a family $\mathcal{B}$ such that every open is a union of finite intersections of elements from $\mathcal{B}$.

The difference between base and subbase is: in a base, we don't need finite intersections to generate all opens.

Fact 4.1.9 *Any subbase $\mathcal{B}$ containing X itself, and in which the intersection of two elements is again in $\mathcal{B}$, is a base.*

Indeed, in this case, each finite intersection $\bigcap_{j \in J_i} U_{ij}$ with each U_{ij} in $\mathcal{B}$ can be rewritten as an element of $\mathcal{B}$. However, there are cases where the base is not closed under intersections: e.g., the base of open balls in the real plane $\mathbb{R}^2$ with the metric topology of its L^2 metric, where the intersection of two open balls in general has the shape of a lens.

Lemma 4.1.10 *A family $\mathcal{B}$ of subsets of a topological space X is a base of the topology iff, for every open U and every element x of U, there is an element V of $\mathcal{B}$ such that $x \in V$ and $V \subseteq U$.*

Proof If: pick $V_x \in \mathcal{B}$ such that $x \in V_x$ and $V_x \subseteq U$ for each $x \in U$, using the Axiom of Choice; then $U = \bigcup_{x \in U} V_x$. Only if: write U as $\bigcup_{i \in I} V_i$ for some family of elements V_i of $\mathcal{B}$. For every $x \in U$, by definition there is an $i \in I$ such that $x \in V_i$: let $V = V_i$. □

Trick 4.1.11 The if part of the above proof illustrates a useful trick to show that some subset V is open: just show that, for every $x \in X$, there is an open subset V_x of x included in V.

See Figure 4.1: in (i), we can find an open ball around $x \in U$; in (ii), a few (not all!) of the open balls obtained this way are displayed.

Since it is generated by open balls, the metric topology on a metric space is usually called the *open ball topology*.

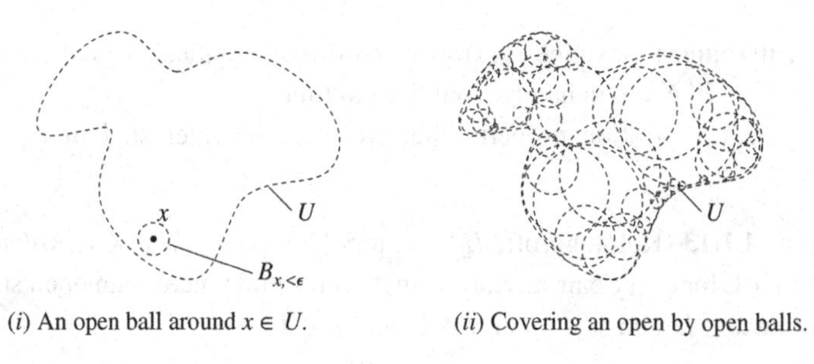

(*i*) An open ball around $x \in U$. (*ii*) Covering an open by open balls.

Figure 4.1 Opens and open balls.

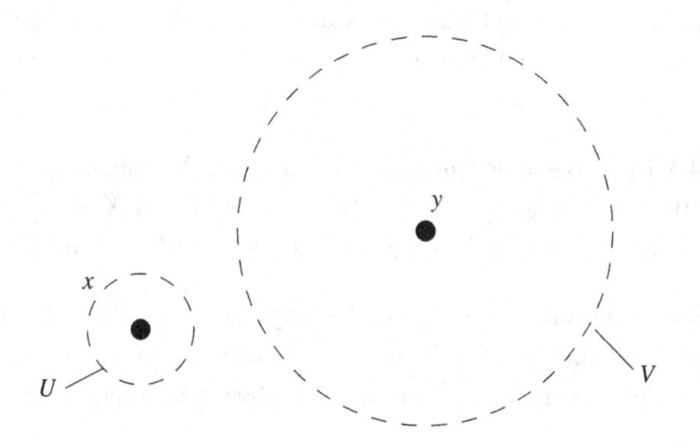

Figure 4.2 The Hausdorff Separation Property.

4.1.2 Separation properties

The metric topology has a rather remarkable property, which is illustrated in Figure 4.2:

Definition 4.1.12 (Hausdorff, T_2) A topology on a space X is *Hausdorff*, or T_2, iff, for every pair x, y of distinct points in X, there is a pair of open subsets U and V such that $x \in U$, $y \in V$, and $U \cap V = \emptyset$.

This is a *separation property*: it states that one can separate distinct points by two tests U and V that no point can satisfy simultaneously.

In most topology books, the Hausdorff Separation Property is assumed from the very start. Most topological spaces in mainstream mathematics are indeed T_2 (one rare exception is given by spectra of rings with their so-called Zariski topology, which we shall meet in Exercise 9.5.38), and T_2 spaces are a bit more comfortable to work with.

We shall usually *not* assume the Hausdorff Separation Property. Typically, one of the topologies that will interest us most, the so-called *Scott topology* of a poset, will (almost) never be T_2. However, the mathematical arsenal available on T_2 spaces will come in handy from time to time.

The typical separation property that we shall be interested in is much weaker:

Definition 4.1.13 (Kolmogoroff, T_0) A topology on a space X is *Kolmogoroff*, or T_0, iff, for every pair x, y of distinct points in X, there is an open subset U such that $x \in U$ and $y \notin U$, or $y \in U$ and $x \notin U$.

That is, we require any two distinct points to be separated by at least one test U, in the sense that one point must pass the test (be in U) and the other must fail it (be outside U). Note that every T_2 space is T_0: T_0 is a *weaker* requirement than T_2.

Example 4.1.14 (Coarse topology) On any set X, the *coarse topology*, also called the *indiscrete topology*, only contains $\emptyset$ and X as open subsets. (Exercise: for what sets X is the coarse topology T_0? When is it T_2?)

Example 4.1.15 (Discrete topology) On any set X, the *discrete topology* is $\mathbb{P}(X)$, the set of all subsets of X. That is, every subset of X is considered open. The discrete topology is always T_2: we can always separate x and y by the opens $\{x\}$ and $\{y\}$.

There is an intermediate property:

Definition 4.1.16 (T_1) A space is *accessible*, or T_1, iff any two distinct points x and y can be separated by two open sets, one U containing x but not y, the other V containing y but not x.

Every T_2 space is T_1, and every T_1 space is T_0.

Exercise 4.1.17 Show that, in a T_1 space X, the *singleton* $\{x\}$ is closed for every element x. More: show that a space X is T_1 if and only if every singleton is closed.

Exercise 4.1.18 (Sierpiński space, $\mathbb{S}$) Let $\mathbb{S}$ be the space $\{0, 1\}$, with opens $\emptyset$, $\mathbb{S}$, and $\{1\}$, but not $\{0\}$. This is *Sierpiński space*. Show that $\mathbb{S}$ is T_0 but not T_1.

There are also stronger properties. The following, for example, will be used when we come back to the study of metric spaces (Chapter 6).

Definition 4.1.19 (Regular, T_3) A space is *regular* iff one can separate points from *closed sets*, in a Hausdorff-like way, i.e., iff, for every point x and every closed set F not containing x, there are disjoint opens U and V, where U contains x and V contains F. A space is T_3 iff it is T_0 and regular.

Example 4.1.20 Every T_3 space is T_2. But there are regular spaces that are not T_2, and not even T_0, such as $\{0, 1\}$ with the coarse topology. (Exercise: prove this.)

Exercise 4.1.21 (Locally closed = regular) Show that the regular spaces are the locally closed spaces (in the same sense as the locally compact spaces; see Section 4.8), i.e., the spaces where, for every point x and every open subset U containing x, there is a closed subset F and a further open subset V such that $x \in V \subseteq F \subseteq U$.

Definition 4.1.22 (Normal, T_4) A space X is *normal* iff one can separate any two disjoint closed subsets, in a Hausdorff-like fashion, i.e., iff, for any two disjoint closed subsets F and F', there are two disjoint open subsets U and U' such that $F \subseteq U$ and $F' \subseteq U'$. A space is T_4 iff it is T_1 and normal.

Every T_1 normal space is regular, using Exercise 4.1.17. So every T_4 space is T_3. Remember (Example 4.1.20) that every T_3 space is also T_2.

Exercise 4.1.23 Show that every T_4 space is T_3, but that there are normal spaces that are not even T_0.

4.1.3 Interior, closure, and density

For every subset A of a space X, the union of all the opens contained in A is open, and is the *largest* open U contained in A; i.e., for every other open $V \subseteq A$, V is included in U. Conversely, the intersection of all the closed sets that contain A is closed, and is the *smallest* closed set F containing A: every other closed subset containing A must contain F.

Definition 4.1.24 (Interior, closure) Let X be any topological space. For each subset A of X, the *interior* $int(A)$ is the largest open included in A. The *closure* $cl(A)$ is the smallest closed subset containing A.

Note that this definition of closure agrees with our previous one, in metric spaces (Chapter 3).

The interior and the closure of A are defined relative to a given space X, which is left implicit in the notation. The ambient space X will usually be clear from context. When it is not, we shall make it precise by saying "the interior of A *in* X," or "the closure of A *in* X."

Exercise 4.1.25 In $\mathbb{R}$, with its metric topology, what is the interior of $(0, 1]$, of $[0, 1)$? What are their closures?

Exercise 4.1.26 Let X be a topological space. Show that $int(A \cap B)$ is equal to $int(A) \cap int(B)$, and that $int(A \cup B)$ contains $int(A) \cup int(B)$, but is in general strictly larger. You may consider the metric topology on $\mathbb{R}$, and the subset $\mathbb{Q}$ of all rational numbers. (Hint: between any two real numbers a and b with $a < b$, there is an irrational number, i.e., one outside $\mathbb{Q}$, in (a, b); there is also a rational number in (a, b).)

Here is a useful property of closures, which we shall use often to show that a point x is in some closure $cl(A)$.

Lemma 4.1.27 *Let X be a topological space, x be a point of X, and A be a subset of X. Then $x \in cl(A)$ iff every open subset U containing x meets A.*

Proof If $x \in cl(A)$, then every open subset U that contains x must meet A: otherwise A would be included in the complement F of U, and since F is closed, $cl(A)$ would be contained in F, so that x would be in F, a contradiction. Conversely, if every open subset U containing x meets A, then this is in particular the case when U is the complement of $cl(A)$. Since U does not meet A, it cannot contain x, whence $x \in cl(A)$. □

Corollary 4.1.28 *Let X be a topological space, and A be a subset of X. For every open subset U of X, U meets $cl(A)$ if and only if U meets A.*

Definition 4.1.29 (Dense) A subset A of a topological space X is *dense* in X iff $cl(A) = X$.

Density is a natural property. We shall see that A is dense in X iff every element of X is obtained as a limit of elements of A (Proposition 4.7.9). However, it is a weak property in non-T_2 spaces; see Exercise 4.2.10.

We shall usually show that a subset A is dense by using the following Lemma.

Lemma 4.1.30 *Let X be a topological space, and A be a subset of X. Then A is dense in X if and only if every non-empty open subset of X meets A, i.e., iff every non-empty open subset U of X is such that $U \cap A$ is non-empty.*

Proof If A is dense in X, then consider any non-empty open subset U of X. If $U \cap A$ were empty, then A would be included in the complement F of U in X. Since U is open, F is closed. However, $cl(A)$ is the smallest closed subset containing A, so $cl(A) \subseteq F$. By assumption $cl(A) = X$, so $F = X$, hence U is empty: contradiction.

Conversely, assume that every non-empty open subset of X meets A. If A were not dense in X, then the complement of $cl(A)$ would be non-empty, and open. So it would meet A. But this is impossible, since $A \subseteq cl(A)$. □

Example 4.1.31 The set $\mathbb{Q}$ of rational numbers is dense in $\mathbb{R}$ with its metric topology. (Exercise: prove this.)

Exercise 4.1.32 One can describe the opens U of $\mathbb{R}$ in a synthetic way: they are exactly the disjoint unions of countably many open intervals (a, b), $a < b$ (where $a \in \mathbb{R} \cup \{-\infty\}$, and $b \in \mathbb{R} \cup \{+\infty\}$). Prove this. You may start by considering the largest open interval included in U containing a given rational number $q \in \mathbb{Q}$.

An *open neighborhood* of x in X is any open subset U containing a given point x. The T_2 property then requires any two distinct points to have disjoint open neighborhoods. In general:

Definition 4.1.33 (Neighborhood) Let X be a topological space. A *neighborhood* of a point $x \in X$ is any subset A of X such that $x \in int(A)$.

Equivalently, a neighborhood of x is any subset A such that there is an open subset U containing x and included in A. For example, $(0, 1)$ and $(0, 1]$ are neighborhoods of $\frac{1}{2}$ in $\mathbb{R}$, but $(0, 1]$ is not a neighborhood of 1 in $\mathbb{R}$.

Exercise 4.1.34 (Sorgenfrey line) The *Sorgenfrey line* $\mathbb{R}_\ell$ (Sorgenfrey, 1947) is the set $\mathbb{R}$ of real numbers with the topology generated by the half-open intervals $[a, b)$, with $a < b$ in $\mathbb{R}$. Show that the topology of the Sorgenfrey line is finer than the metric topology of $\mathbb{R}$, and that $\mathbb{R}_\ell$ is T_2.

But $\mathbb{R}_\ell$ is also a space with rather strange properties. Show that $[b, +\infty)$ and $(-\infty, a)$ are open in $\mathbb{R}_\ell$ for all reals a, b, and deduce that $[a, b)$ is *clopen*, i.e., both open and closed. It follows that $\mathbb{R}_\ell$ is *zero-dimensional*, i.e., that its topology has a base consisting of clopen subsets.

Exercise 4.1.35 Show that every zero-dimensional space is regular. Hence every zero-dimensional T_2 space is T_3. Deduce that the Sorgenfrey line $\mathbb{R}_\ell$ is T_3. (We shall see in Exercise 6.3.32 that it is T_4, and even more.)

Exercise 4.1.36 Show that $\mathbb{Q}$ is dense in the Sorgenfrey line $\mathbb{R}_\ell$.

4.2 Order and topology

There is a deep connection between order and topology, of which we give a first taste.

4.2.1 The specialization quasi-ordering

One can always extract a quasi-ordering from any topology.

Definition 4.2.1 (Specialization quasi-ordering) Let X be any topological space. The *specialization quasi-ordering* $\leqslant$ on X is defined by $x \leqslant y$ iff every open subset containing x also contains y.

We shall almost always write $\leqslant$ for the specialization quasi-ordering of X, without saying that it is so. Understanding open subsets as tests, this says that $x \leqslant y$ iff y passes all the tests that x passes. In computer science, one would say that x is simulated by y.

Example 4.2.2 The specialization quasi-ordering of Sierpiński space $\mathbb{S}$ (Exercise 4.1.18) is given by $0 \leqslant 1$ (and $1 \not\leqslant 0$).

The notion of specialization quasi-ordering helps one elucidate the T_0 and T_1 properties:

Proposition 4.2.3 *Let X be any topological space, and $\leqslant$ its specialization quasi-ordering. Then X is T_0 iff $\leqslant$ is an ordering; and X is T_1 iff $\leqslant$ is equality.*

Proof Let X be T_0, and assume $x \leqslant y$ and $y \leqslant x$, but $x \neq y$. By T_0, there is a subset U containing x but not y, contradicting $x \leqslant y$, or a subset U containing y but not x, contradicting $y \leqslant x$. So $\leqslant$ is an ordering. Conversely, if $\leqslant$ is an ordering, then whenever $x \neq y$, either $x \not\leqslant y$ or $y \not\leqslant x$. In the first case, there is an open subset U containing x but not y. In the second case, there is an open subset U containing y but not x.

Let X be T_1, assume $x \leqslant y$ but $x \neq y$. By T_1, there is an open subset U containing x but not y, a contradiction. So $x \leqslant y$ implies $x = y$. Conversely, assume that $\leqslant$ is equality. For any two distinct points x, y in X, $x \neq y$ so $x \not\leqslant y$ and $y \not\leqslant x$. By definition, there is an open subset U containing x but not y, and there is an open subset U containing y but not x. □

To compute the specialization quasi-ordering of a topological space, it is enough to look at its subbasic opens:

Proposition 4.2.4 *Let X be a topological space, $\leqslant$ be its specialization quasi-ordering, and $\mathcal{B}$ be a subbase of its topology. Then $x \leqslant y$ iff, for every $U \in \mathcal{B}$, if $x \in U$ then $y \in U$.*

Proof If: for every open U containing x, write U as $\bigcup_{i \in I} \bigcap_{j \in J_i} U_{ij}$ where I is an arbitrary index set, J_i is finite for each $i \in I$, and U_{ij} is an element of $\mathcal{B}$. For some $i \in I$, x is in $\bigcap_{j \in J_i} U_{ij}$. Since x is in U_{ij} for every $j \in J_i$, y is too. So $y \in \bigcap_{j \in J_i} U_{ij} \subseteq U$. The only if direction is obvious. □

The definition of $\leqslant$ yields the following immediately.

Lemma 4.2.5 *Let X be a topological space. Every open subset U of X is upward closed in the specialization quasi-ordering $\leqslant$.*

The downward closed subsets are exactly the complements of upward closed subsets. So:

Lemma 4.2.6 *Let X be a topological space. Every closed subset U of X is downward closed in the specialization quasi-ordering $\leqslant$.*

In T_1 spaces, and in particular in T_2 spaces and therefore in metric spaces, every subset is both upward closed and downward closed: the specialization quasi-ordering is just equality in this case. In the more general spaces we shall be interested in, the above lemmas show a fundamental difference between open and closed subsets. The former are upward closed, while the latter are downward closed.

Taking negations in the definition of the specialization quasi-ordering, we have that $y \not\leqslant x$ iff there is an open subset U containing y but not x. Figure 4.3 shows two points x and y in this situation, plus an open subset U that witnesses this fact. Note that U is upward closed, as the figure suggests. On the other hand, the downward closure $\downarrow x$ of a point x is of course downward closed; topologically, $\downarrow x$ is exactly the smallest closed subset containing x:

Lemma 4.2.7 *Let X be a topological space, with specialization quasi-ordering $\leqslant$. For all $x, y \in X$, $x \leqslant y$ iff $x \in cl(\{y\})$. The closure $cl(\{y\})$ of a single point y is its downward closure $\downarrow y$.*

Proof The point x belongs to $cl(\{y\})$ if and only if every open neighborhood U of x meets $\{y\}$, by Lemma 4.1.27, i.e., iff every open neighborhood U of x contains y. The latter is the definition of $x \leqslant y$.

So $cl(\{y\})$ is the set of all points x such that $x \leqslant y$, namely $\downarrow y$. □

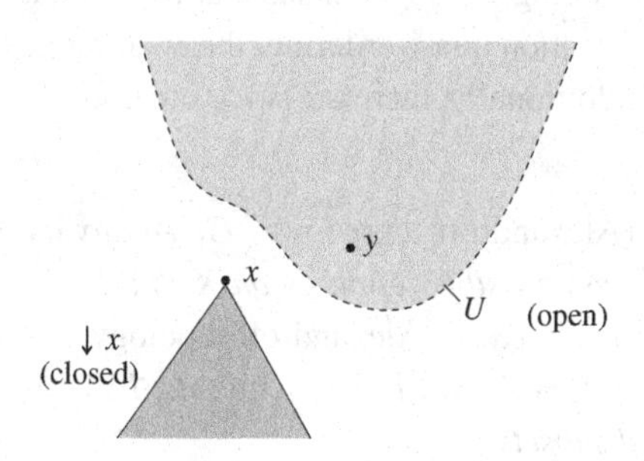

Figure 4.3 Open and closed subsets, and the specialization quasi-ordering.

Write $\downarrow E$ for the downward closure of the set E, i.e., the set of all elements below some element of E.

Exercise 4.2.8 (Finitary closed) Show that $cl(E) = \downarrow E$ for every finite set E, but not in general when E is infinite. The subsets of the form $\downarrow E$, E finite, are called the *finitary closed* subsets of X.

Proposition 4.2.9 (Saturation) *Let X be a topological space. The* saturation *of a subset A is the intersection of all opens U that contain A. We say that A is* saturated *iff it equals its saturation.*

For any subset A of X, the saturation of A coincides with $\uparrow A$.

Proof Clearly, every open subset U containing A, being upward closed by Lemma 4.2.5, contains $\uparrow A$. Conversely, for any point x not in $\uparrow A$, there is an open set U containing $\uparrow A$ but not x: take U to be the complement of $\downarrow x$ (Lemma 4.2.7). Then U must contain A, hence $\uparrow A$; otherwise some element below x would be in A, meaning that $x \in \uparrow A$, a contradiction. So the saturation of A cannot contain any of the points outside $\uparrow A$. □

Exercise 4.2.10 Show that density is a very weak property in non-T_2 spaces: assume X is any topological space with a largest element $\top$ in its specialization quasi-ordering $\leqslant$, then show that $\{\top\}$ is already dense in X.

4.2.2 The Alexandroff, Scott, and upper topologies

The specialization quasi-ordering yields a quasi-ordering from any topology. Can we get back the original topology from the quasi-ordering alone? The answer is no in general: there are many topologies on $\mathbb{R}$ with equality as specialization quasi-ordering: its metric topology, the discrete topology (Example 4.1.15), and the topology of the Sorgenfrey line $\mathbb{R}_\ell$ (Exercise 4.1.34), for example.

Given any quasi-ordering $\leqslant$, we shall see that there are at least two topologies with $\leqslant$ as specialization quasi-ordering, the coarsest possible one and the finest possible one. Additionally, there are some other interesting topologies in between.

Proposition 4.2.11 (Alexandroff topology) *Given any quasi-ordering $\leqslant$ on a set X, there is a* finest *possible topology on X that has $\leqslant$ as specialization quasi-ordering. This is called the* Alexandroff topology *of $\leqslant$.*

Its opens are exactly the upward closed subsets. Its closed sets are exactly the downward closed subsets.

We write X_a for X topologized with the Alexandroff topology of $\leqslant$.

Proof Consider the collection $\mathcal{O}$ of all upward closed subsets of X. This is a topology. Call $\preceq$, temporarily, its specialization quasi-ordering. If $x \preceq y$, then every upward closed subset containing x must contain y, in particular $\uparrow x$; so $x \leqslant y$. Conversely, if $x \leqslant y$, then clearly $x \preceq y$. So $\leqslant$ is indeed the specialization quasi-ordering of $\mathcal{O}$. It is the finest topology with $\leqslant$ as specialization quasi-ordering, by Lemma 4.2.5. That its closed sets are exactly the downward closed subsets is because the downward closed subsets are exactly the complements of upward closed subsets. □

There is also a coarsest one:

Proposition 4.2.12 (Upper topology) *Given any quasi-ordering $\leqslant$ on a set X, there is a* coarsest *possible topology on X that has $\leqslant$ as specialization quasi-ordering. This is called the* upper topology *of $\leqslant$.*

A subbase is given by the complements of sets of the form $\downarrow x$, $x \in X$. A base is given by the complements of sets of the form $\downarrow E$, for E a finite subset of X.

We write X_u for X topologized with the upper topology of $\leqslant$.

Proof Let $\mathcal{O}$ be the topology generated by the complements of sets of the form $\downarrow x$, $x \in X$. The finite intersections of such sets are those that are complements of sets of the form $\downarrow E$, E finite. Let $\preceq$ be the specialization quasi-ordering of $\mathcal{O}$. By Lemma 4.1.10, $x \preceq y$ iff, for every $z \in X$, if $x \notin \downarrow z$ then $y \notin \downarrow z$; equivalently, for every z, if $y \leqslant z$ then $x \leqslant z$. Taking $z = y$, $x \preceq y$ then implies $x \leqslant y$. Conversely, if $x \leqslant y$ then $y \leqslant z$ implies $x \leqslant z$, so $x \preceq y$. So $\leqslant$ is the specialization quasi-ordering of the upper topology.

Any topology with $\leqslant$ as specialization quasi-ordering must make $\downarrow x$ closed (Lemma 4.2.7). So any such topology must be finer than the upper topology $\mathcal{O}$: the upper topology is indeed the coarsest possible one. □

Exercise 4.2.13 Call a topological space X *Alexandroff-discrete* iff every intersection (even infinite) of opens is again open. Show that the Alexandroff-discrete spaces are exactly the spaces whose topology is the Alexandroff topology of some quasi-ordering.

Exercise 4.2.14 Let X be a set, with a quasi-ordering $\leqslant$. Show that there is a unique topology on X whose specialization quasi-ordering is $\leqslant$ if and only if $X_a = X_u$.

Exercise 4.2.15 (Finite quasi-orders) Let X be a finite set, with a quasi-ordering $\leqslant$. Show that there is a unique topology on X whose specialization quasi-ordering is $\leqslant$.

An important example of a topology that sits in between is the *Scott topology*. One can explain it by imitating the definition of sequentially closed subsets (Section 3.2, in the case of metric spaces), replacing limits by certain sups: a Scott-closed subset should be a downward closed subset F (it should be downward closed; see Lemma 4.2.6), such that the sup of every ascending sequence $x_0 \leqslant x_1 \leqslant \cdots \leqslant x_n \leqslant \cdots$ of elements of F is still in F. Dually, a subset U should be open in the Scott topology iff it is upward closed, and every ascending sequence $(x_n)_{n \in \mathbb{N}}$ that has a least upper bound in U should be such that $x_n \in U$ for n large enough. This is not quite the definition we shall take: we shall use directed families instead of ascending sequences, but the above is a step in the right direction.

This topology arose from studies in logic and computer science, where one may think of $x_0, x_1, \ldots, x_n, \ldots$ as snapshots of an on-going computation, where x_n is taken at time n, and representing successive approximations to the final result x. For example, think of the successive states of a crossword puzzle solved by a computer, with access to a dictionary. Initially, the crossword puzzle x_0 is empty. The computer then finds the solution to one definition, fills the corresponding row or column accordingly: the crossword puzzle is now in state x_1. After finding n words, one may imagine that the crossword puzzle is in state x_n. Once the puzzle has been completely filled out, say at time n, the computer remains at state x_n, so that $x_n = x_{n+1} = \cdots = x$.

The ordering $\leqslant$ is the so-called *information ordering*: $x_n \leqslant x_{n+1}$, as one knows more in state x_{n+1} than in state x_n.

In the crossword puzzle example, the least upper bound is reached after finitely many steps: the computer program terminates. In general, computer programs may fail to terminate. Let us take the example of an automated theorem proving program in first-order logic – an undecidable problem, i.e., one for which no program exists that accepts exactly the provable formulae and terminates on all inputs. (See Goubault-Larrecq and Mackie (1997) for more on this domain. Many such programs do exist. But none can terminate on all inputs.) Given a set of axioms Ax and a formula φ, such a program will attempt to derive φ from Ax by using the rules of first-order logic. At each time n, the automated theorem prover will have derived more and more consequences of Ax, or of the negation of φ, until it reaches a conclusion… or runs for infinitely long.

Our point here is that there is a deep connection between computability and the Scott topology. Notably, all computable tests must be Scott-open—this observation, once suitably formalized, is the essence of the Myhill–Shepherdson Theorem (Myhill and Shepherdson, 1955); see Streicher (2006, theorem 16.7) for a modern exposition, or Scott (1975) for a clear and deep account of the relationship between computability and topology. Let us call a test any subset U of configurations x. We call it computable if and only if there is a (not necessarily terminating) program π that takes a configuration x as input, and returns 1 if and only if $x \in U$. Since programs can only examine a finite portion of x in any finite amount of time, if it returns 1 then it must only have explored a finite part of x, produced as x_n at some finite time n already. Since we only add new information (new entries in the crossword puzzle, new logical consequences of Ax, or the negation of φ) at each computation step, π would also return 1 on seeing $x_{n+1}, x_{n+2}, \ldots$ So, if $x \in U$, then $x_n \in U$ for some n large enough: i.e., U is open in the Scott topology (assuming it is upward closed).

Although computers can only run for countably many steps, we shall replace ascending sequences in the definition of the Scott topology by directed, not necessarily countable, families; see below. As Abramsky and Jung (1994, section 2.2.4) note, clinging to ascending sequences would produce a mathematical theory that becomes rather bizarre, whence our move to directed families.

Definition 4.2.16 (Directed, Scott-open) Let X be a quasi-ordered set. A subset D of X is *directed* iff D is non-empty, and any two points in D have an upper bound in D. In other words, a family $(x_i)_{i \in I}$ is directed iff $I \neq \emptyset$ and for all $i, i' \in I$, there is an $i'' \in I$ such that $x_i \leqslant x_{i''}$ and $x_{i'} \leqslant x_{i''}$.

A *Scott-open* subset of X is any upward closed subset U of X such that, for every directed family $(x_i)_{i \in I}$ such that $\sup_{i \in I} x_i$ exists and is in U, there is an $i \in I$ such that $x_i \in U$.

Example 4.2.17 Every ascending sequence $(x_n)_{n \in \mathbb{N}}$ is directed: given x_i and $x_{i'}$, one takes $i'' = \max(i, i')$. More generally, every chain is directed. But the converse fails. For example, the collection $\mathbb{P}_{\text{fin}}(\mathbb{N})$ of all finite subsets of $\mathbb{N}$ is directed (given A and A' in $\mathbb{P}_{\text{fin}}(\mathbb{N})$, $A \cup A'$ is again in $\mathbb{P}_{\text{fin}}(\mathbb{N})$), but is very far from being a chain. Its least upper bound is its union, $\mathbb{N}$.

It will often be useful to realize that directedness is equivalent to the apparently stronger statement: D is directed iff, for every finite collection of elements $x_1, \ldots, x_n$ in D, there is an even larger element x in D. (Use induction on n when $n \geqslant 1$, and the fact that D is non-empty when $n = 0$.)

Proposition 4.2.18 (Scott topology) *Let X be a poset, with ordering $\leqslant$. The collection of all Scott-opens is a topology, called the* Scott topology *on X. We write X_σ for the set X with the Scott topology.*

Its closed subsets are the Scott-closed *subsets, defined as those downward closed subsets F such that, for every directed family $(x_i)_{i \in I}$ in F such that $\sup_{i \in I} x_i$ exists, $\sup_{i \in I} x_i$ is in F again.*

The specialization ordering of the Scott topology is the original ordering $\leqslant$. In particular, the Scott topology is finer than the upper topology and coarser than the Alexandroff topology.

Proof Let $U_1, \ldots, U_n$ be n Scott-opens, and consider any directed family D having a least upper bound in $\bigcap_{i=1}^n U_i$. Since each U_i is Scott-open, there is an $x_i \in D \cap U_i$. By directedness, there is an even larger element x in D. Since each U_i is upward closed, x is in U_i for each i. So D meets $\bigcap_{i=1}^n U_i$ at x, whence $\bigcap_{i=1}^n U_i$ is Scott-open. It is clear that any union of Scott-opens is again Scott-open. So the Scott-opens form a topology.

The characterization of closed subsets as the Scott-closed subsets is elementary.

Since each Scott-open is upward closed in $\leqslant$, if $x \leqslant y$ then $x \in U$ implies $y \in U$ for every Scott-open U, i.e., x is below y in the specialization quasi-ordering. Conversely, realize that $\downarrow y$ is Scott-closed. So, if x is below y in the specialization quasi-ordering, then if x were in the complement U of $\downarrow y$, then y would be in U. This is impossible, so $x \in \downarrow y$, i.e., $x \leqslant y$. So the specialization quasi-ordering of the Scott topology is indeed $\leqslant$. Since the upper topology is the coarsest one with this property (Proposition 4.2.12), and the Alexandroff topology is the finest one (Proposition 4.2.11), we conclude. □

Exercise 4.2.19 Let $\mathbb{R}_\sigma$ be the set $\mathbb{R}$ of all real numbers, with the Scott topology of its natural ordering $\leqslant$. Show that the Scott-opens are exactly the open intervals $(t, +\infty)$, $t \in \mathbb{R}$, plus $\mathbb{R}$ and $\emptyset$.

Show that the (Scott) topology of $\mathbb{R}_\sigma$ is strictly coarser than the Alexandroff topology, but coincides with the upper topology on $\mathbb{R}$.

Exercise 4.2.20 What are the opens in the Scott topology of $\mathbb{R}^+_\sigma$, of $[0, 1]_\sigma$?

Exercise 4.2.21 On any set X, the *cofinite topology* has as opens all complements of finite subsets of X, plus $\emptyset$. In other words, the closed sets are X itself and all finite subsets of X. Show that the cofinite topology is the upper topology of the (rather trivial) ordering $=$.

In the same situation, show that the Scott and Alexandroff topologies both coincide with the discrete topology (defined in Example 4.1.15).

In $\mathbb{R}$ with $=$ as ordering, show that the upper topology is distinct from the Scott (= Alexandroff = discrete) topology. Show also that none of these topologies coincides with the metric topology.

A particularly interesting class of posets in this respect is the following.

Definition 4.2.22 (dcpo) A poset X is *directed-complete* iff $\sup D$ exists for every directed family D of X. A *dcpo* is any directed-complete poset.

Exercise 4.2.23 Which of the following posets are dcpos? (i) $\mathbb{R}$, (ii) $[0, 1]$, (iii) $\mathbb{N}$, (iv) $\mathbb{P}(E)$ (in this case, for which sets E?), (v) $\mathbb{P}_{\text{fin}}(E)$ with the inclusion ordering (again, for which sets E?), (vi) the set $\mathbb{Z}^*$ of all non-zero integers, with the funny ordering $\preceq$ defined by: $m \preceq n$ iff m and n have the same sign and $m \leqslant n$, or $m > 0$ and $n < 0$; and (vii) the collection of all closed subsets of a topological space X.

Hint: for (vii), why is this a complete lattice? Observe that every complete lattice is a dcpo. Then describe what least upper bounds are in concrete terms. You should then conclude that, in a lattice of sets, least upper bounds may be different from mere unions.

One may wonder why we required directed subsets in the definition of the Scott topology, or of dcpos. The situation is subtle. First, Markowsky's Theorem (Markowsky, 1976, corollary 2) states that the dcpos are *exactly* the chain-complete posets. However, its proof is rather complex, and in particular the following proof strategy does not work. We say that a family D is *cofinal* in the directed family D' iff $D \subseteq D'$ and for every $x' \in D'$ there is an $x \in D$ above x'. Note that, if D is directed and cofinal in D', and $\sup D'$ exists, then $\sup D$ exists and equals $\sup D'$. One might think that one could certainly extract a cofinal chain C from every directed subset D. If so, then a subset U of X would be Scott-open iff it is upward closed and every chain having a least upper bound in U meets U. However, this claim is wrong; see Exercise 4.2.25.

Exercise 4.2.24 Show that, if D is cofinal in D' and $\sup D'$ exists, then $\sup D$ exists and equals $\sup D'$.

Exercise 4.2.25 Consider the poset $\mathbb{P}(\mathbb{R})$ of all subsets of $\mathbb{R}$, ordered by inclusion. Let D be the collection $\mathbb{P}_{\text{fin}}(\mathbb{R})$ of all finite subsets of $\mathbb{R}$. Show that D is directed, and that $\sup D = \mathbb{R}$.

On the other hand, show that no chain C included in $\mathbb{P}_{\text{fin}}(\mathbb{R})$ can be cofinal in D, and more generally no chain C in D can be such that $\sup C = \sup D$. Hint: show that, if C is infinite, then C must be a chain of the form $(A_i)_{i \in \mathbb{N}}$, where $A_0 \subsetneq A_1 \subsetneq \cdots \subsetneq \cdots \subsetneq A_n \subsetneq \cdots$ is a countable, increasing sequence of finite sets. (We write $A \subsetneq B$ for $A \subseteq B$ and $A \neq B$.) Then use countability arguments (Section 2.2.)

Although we won't give a formal proof of Markowsky's Theorem, here is the idea. Let X be a chain-complete poset. We wish to show that X has sups of all directed subsets. We use Iwamura's Lemma (Iwamura, 1944), which states that every infinite directed set D is the union of a chain (with respect to inclusion) of directed subsets $(D_i)_{i \in I}$ of strictly smaller cardinalities than D. Then we use well-founded induction on cardinalities (a distant consequence of the Axiom of Foundation). If D is directed and finite, then it has a largest element, which is then its sup. If D is infinite, find $(D_i)_{i \in I}$ as above, let $x_i = \sup D_i$, which is guaranteed by the induction hypothesis, and realize that $(x_i)_{i \in I}$ is a chain in X: then $\sup_{i \in I} x_i$ must be the sup of D.

Exercise 4.2.26 On a dcpo X, define the *chain-open topology* by declaring opens all upward closed subsets U such that every chain C such that $\sup C$ is in U meets U. Show that this is nothing else than the Scott topology. (Hint: use Markowsky's Theorem, applied to the complement of any given chain-open subset.)

Example 4.2.27 One example that relates ordinary T_2 topology with non-T_2 topologies (such as the Scott topology of an ordering) is Scott's dcpo $\mathbf{I}\mathbb{R}$; see Exercise 4.2.28. This is an example of a *model* (here, of $\mathbb{R}$), a notion we shall explore in Section 7.7.2.

Exercise 4.2.28 ($\mathbf{I}\mathbb{R}$) Let $\mathbf{I}\mathbb{R}$ be the poset of all non-empty closed intervals $[a, b]$ with $a, b \in \mathbb{R}$, $a \leqslant b$. Order it by reverse inclusion $\supseteq$. Show that $\mathbf{I}\mathbb{R}$ is a dcpo, and that the map $\eta_{\mathbf{I}\mathbb{R}} : x \mapsto [x, x]$, which equates the points of $\mathbb{R}$ with certain elements of $\mathbf{I}\mathbb{R}$, has the following properties:

1. for every open subset V of $\mathbf{I}\mathbb{R}$, $\eta_{\mathbf{I}\mathbb{R}}^{-1}(V)$ is open in $\mathbb{R}$, with its metric topology (we shall see in Section 4.3 that this means that $\eta_{\mathbf{I}\mathbb{R}}$ is continuous);
2. conversely, for every open subset U of $\mathbb{R}$, there is an open subset V of $\mathbf{I}\mathbb{R}$ such that $U = \eta_{\mathbf{I}\mathbb{R}}^{-1}(V)$.

In other words, and anticipating Section 4.9, this means that, up to the identification of points $x \in \mathbb{R}$ with elements of the form $[x, x]$ in $\mathbf{I}\mathbb{R}$, the metric topology on $\mathbb{R}$ is induced from the Scott topology on the larger space $\mathbf{I}\mathbb{R}$. See Figure 4.4 for an illustration.

4.3 Continuity

We have already defined what a continuous function is – between metric spaces. The key to extending this notion to more general topological spaces is Proposition 3.5.2, Item 3.

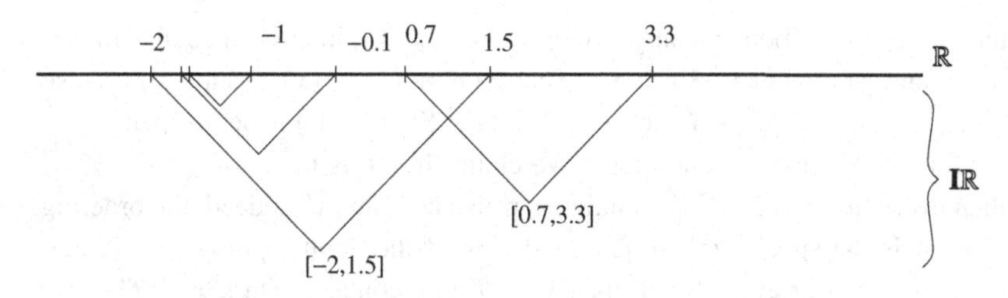

Figure 4.4 Scott's dcpo $\mathbb{IR}$.

Definition 4.3.1 (Continuity) Let X, Y be two topological spaces. A map $f: X \to Y$ is *continuous* iff the inverse image $f^{-1}(V)$ of every open subset V of Y is open in X.

Exercise 4.3.2 Show that $f: X \to Y$ is continuous iff the inverse image $f^{-1}(F)$ of every closed subset F of Y is closed in X. This is meant to remind you of Proposition 3.5.2, Item 5.

The following is useful in proving the continuity of functions. This should remind you of Proposition 3.5.2, Item 2.

Trick 4.3.3 Let $\mathcal{B}$ be a subbase of the topology of Y. To show that $f: X \to Y$ is continuous, it is enough to show that the inverse image $f^{-1}(V)$ of every *subbasic* open V (i.e., in $\mathcal{B}$) is open in X.

Exercise 4.3.4 Prove the above claim. More precisely, let $\mathcal{B}$ be a subbase of the topology of Y. Show that $f: X \to Y$ is continuous iff the inverse image $f^{-1}(V)$ of every *subbasic* open V (i.e., in $\mathcal{B}$) is open in X.

4.3.1 Continuity and monotonicity

We automatically get a notion of continuity for posets equipped with their Scott topologies. Much as the (sequentially) continuous maps were the ones that preserved limits (Definition 3.5.1), we define Scott-continuous maps as preserving directed sups.

Proposition 4.3.5 (Scott-continuity) *Let X, Y be two posets. A map $f: X_\sigma \to Y_\sigma$ is continuous iff $f: X \to Y$ is* Scott-continuous, *i.e., f is monotonic and, for every directed family $(x_i)_{i\in I}$ in X that has a least upper bound x in X, $f(x)$ is the least upper bound of the directed family $(f(x_i))_{i\in I}$.*

Proof If: let V be any Scott-open of Y. Since f is monotonic and V is upward closed, $f^{-1}(V)$ is upward closed. For every directed family $(x_i)_{i\in I}$ in X that has a least upper bound x in $f^{-1}(V)$, $(f(x_i))_{i\in I}$ is directed since f is monotonic, and $f(x)$ is in V. Since f is Scott-continuous, $f(x) = \sup_{i\in I} f(x_i)$, so $f(x_i) \in V$ for some $i \in I$, i.e., $x_i \in f^{-1}(V)$. So $f^{-1}(V)$ is Scott-open.

Only if: assume f continuous. We claim that f is monotonic. If $x \leqslant x'$ then every Scott-open of X containing x also contains x'. Indeed, the ordering $\leqslant$ of X is the specialization quasi-ordering of the Scott topology, by Proposition 4.2.18. For every Scott-open V of Y that contains $f(x)$, $x \in f^{-1}(V)$. Since $f^{-1}(V)$ is open, $x' \in f^{-1}(V)$, so $f(x') \in V$. So $f(x) \leqslant f(x')$, since the ordering $\leqslant$ of Y is the specialization quasi-ordering of the Scott topology.

Let us show that, for every directed family $(x_i)_{i\in I}$ in X that has a least upper bound x in X, $f(x)$ is the least upper bound of the directed family $(f(x_i))_{i\in I}$. The latter is indeed directed, since f is monotonic. Moreover, $f(x_i) \leqslant f(x)$ for every $i \in I$ by monotonicity, so $f(x)$ is an upper bound of $(f(x_i))_{i\in I}$. To show that it is the least one, consider any other upper bound y, and show that $f(x) \leqslant y$. It is enough to show that every Scott-open V containing $f(x)$ must contain y. And indeed, if V contains $f(x)$, then $x \in f^{-1}(V)$. Since $f^{-1}(V)$ is Scott-open, $x_i \in f^{-1}(V)$ for some $i \in I$. Then $f(x_i) \in V$. Since V is upward closed, $y \in V$, and we are done. □

Scott-continuous functions have fixed points, under mild conditions. The following is sometimes called the *Kleene Fixed Point Theorem*.

Proposition 4.3.6 (Least fixed point) *Let X be a* pointed *dcpo, i.e., one with a least element $\perp$. Every continuous map $f : X \to X$ has a least fixed point, obtained as* $\sup_{n\in\mathbb{N}} f^n(\perp)$.

Proof By Proposition 4.3.5, f is Scott-continuous. By induction on n, we show that $f^n(\perp) \leqslant f^{n+1}(\perp)$. This is obvious if $n = 0$, and otherwise this is because f is monotonic. In particular, the sequence $(x_n)_{n\in\mathbb{N}}$ is a chain, hence a directed subset. Since f is Scott-continuous, $f(\sup_{n\in\mathbb{N}} f^n(\perp)) = \sup_{n\in\mathbb{N}} f^{n+1}(\perp) = \sup(\sup_{n\geqslant 1} f^n(\perp), \perp) = \sup_{n\in\mathbb{N}} f^n(\perp)$, so $\sup_{n\in\mathbb{N}} f^n(\perp)$ is indeed a fixed point. Given any fixed point x of f, we must have $\perp \leqslant x$, trivially, hence $f(\perp) \leqslant f(x) = x$, and, by induction on n, $f^n(\perp) \leqslant x$. So $\sup_{n\in\mathbb{N}} f^n(\perp) \leqslant x$, i.e., $\sup_{n\in\mathbb{N}} f^n(\perp)$ is the least fixed point of f. □

The proof of Proposition 4.3.6 is simpler than that of Tarski's Theorem (Theorem 2.3.2, or Exercise 2.3.4), although the statement is similar. Note that Tarski's Theorem requires f to be monotonic only, not continuous.

The Kleene Fixed Point Theorem is essential in denotational semantics, where it is used to give meaning to recursive function definitions in programming languages (Winskel, 1993, chapter 9). For example, the definition of factorial as the function that maps $n \in \mathbb{N}$ to $f(n) = \texttt{if}\ n = 0\ \texttt{then}\ 1\ \texttt{else}\ n \times f(n-1)$ (note that f occurs both to the left and to the right of the equality sign) is obtained as the least fixed point of the functional F, mapping any function f to the function f' defined by $f'(n) = \texttt{if}\ n = 0\ \texttt{then}\ 1\ \texttt{else}\ n \times f(n-1)$, in suitable dcpos.

Example 4.3.7 Sum $+$ is Scott-continuous from $\mathbb{R} \times \mathbb{R}$ to $\mathbb{R}$, and from $\mathbb{R}^+ \times \mathbb{R}^+$ to $\mathbb{R}^+$. Product, written ., is also Scott-continuous from $\mathbb{R}^+ \times \mathbb{R}^+$ to $\mathbb{R}^+$, but not from $\mathbb{R} \times \mathbb{R}$ to $\mathbb{R}$. (Exercise: prove all this.)

Example 4.3.8 Equip $\mathbb{R}$ with its metric topology. Then $-: \mathbb{R} \to \mathbb{R}$, which maps x to $-x$, is continuous. (Exercise: prove this) But it is not Scott-continuous, as it is not even monotonic: see Proposition 4.3.9 below.

Proposition 4.3.9 *Let f be a continuous map from the topological space X to the topological space Y. Then f is monotonic with respect to the specialization quasi-orderings of X and Y.*

Proof Let $x \leqslant x'$ in X, and consider any open neighborhood V of $f(x)$ in Y. Then x is in $f^{-1}(V)$, which is open since f is continuous, hence upward closed (Lemma 4.2.5). So $x' \in f^{-1}(V)$, whence $f(x') \in V$. So $f(x) \leqslant f(x')$. □

This is vacuous in T_2 spaces, or even T_1 spaces, where the specialization quasi-ordering is equality—every function from a T_1 space to a T_1 space is monotonic.

Exercise 4.3.10 Let X be a poset, and Y be a topological space. Show that $f: X_a \to Y$ is continuous iff f is monotonic. In particular, continuity in Alexandroff-discrete spaces is nothing else than monotonicity.

Exercise 4.3.11 (Lower, upper semi-continuity) Let X be a topological space. $\mathbb{R} \cup \{-\infty, +\infty\}$ is a poset, with ordering $\leqslant$, hence it defines a topological space $(\mathbb{R} \cup \{-\infty, +\infty\})_\sigma$ (see Proposition 4.2.18). The continuous maps from X to $(\mathbb{R} \cup \{-\infty, +\infty\})_\sigma$ are usually called *lower semi-continuous*, or *lsc*. Conversely, a map from X to $\mathbb{R} \cup \{-\infty, +\infty\}$ is *upper semi-continuous*, or *usc*, iff $-f$, the function that maps x to $-f(x)$, is lsc. Show that a map $f: X \to \mathbb{R} \cup \{-\infty, +\infty\}$ is continuous (where $\mathbb{R} \cup \{-\infty, +\infty\}$ is equipped with the topology generated by open intervals (a, b), $(a, +\infty]$, and $[-\infty, b)$) if and only if it is both lsc and usc.

Instead of proving that a function is continuous, we shall often use the following result, without even citing it.

Proposition 4.3.12 *The identity map* $\mathrm{id}_X \colon X \to X$*, defined by* $\mathrm{id}_X(x) = x$ *for every* $x \in X$*, is always continuous.*

Whenever $f \colon X \to Y$ *and* $g \colon Y \to Z$ *are continuous, then so is their* composition $g \circ f$*, defined by* $(g \circ f)(x) = g(f(x))$ *for every* $x \in X$.

Proof For every open U of X, $\mathrm{id}_X^{-1}(U) = U$ is open. For every open W of Z, $(g \circ f)^{-1}(W) = f^{-1}(g^{-1}(W))$ is open. □

When we seek to prove that some complicated function is continuous, it will be simpler to rewrite this function as a composition of simpler functions, say $f_n \circ f_{n-1} \circ \cdots \circ f_1$, and to show that $f_1, \ldots, f_{n-1}, f_n$ are all continuous.

4.3.2 Continuity at a point

There is also a notion of being continuous *at a given point*. Recall that $f[U]$ is the image of U under f, i.e., the set of all elements of the form $f(x)$, $x \in U$.

Definition 4.3.13 (Continuity at x) Let X, Y be two topological spaces, and x be a point in X. The map $f \colon X \to Y$ is *continuous at* x iff, for every open neighborhood V of $f(x)$, there is an open neighborhood U of x such that $f[U] \subseteq V$.

Proposition 4.3.14 *Let* X, Y *be two topological spaces. A map* $f \colon X \to Y$ *is continuous iff it is continuous at every point* x *of* X.

Proof If: given any open subset V of Y, we must show that $f^{-1}(V)$ is open. For every element x of $f^{-1}(V)$, V is an open neighborhood of $f(x)$. By assumption, x has an open neighborhood U such that $f[U] \subseteq V$, i.e., such that $U \subseteq f^{-1}(V)$. By Trick 4.1.11, $f^{-1}(V)$ is open, hence f is continuous.

Only if: fix $x \in X$, and let V be an open neighborhood of $f(x)$. Note that $x \in f^{-1}(V)$. Since f is continuous, $f^{-1}(V)$ is open. Let $U = f^{-1}(V)$. Clearly $f[U] \subseteq V$, and we are done. □

We can then refine Proposition 4.3.12.

Proposition 4.3.15 *Let* X, Y, Z *be topological spaces, and* x *be a point of* X*. If* $f \colon X \to Y$ *is continuous at* x *and* $g \colon Y \to Z$ *is continuous at* $f(x)$*, then* $g \circ f$ *is continuous at* x.

Proof Let W be any open neighborhood of $(g \circ f)(x) = g(f(x))$ in Z. Since g is continuous at $f(x)$, there is an open neighborhood V of $f(x)$ such that $g[V] \subseteq W$. Since f is continuous at x, there is an open neighborhood U of x such that $f[U] \subseteq V$. Then $(g \circ f)[U] = g[f[U]] \subseteq g[V] \subseteq W$. □

Exercise 4.3.16 There is a deep relationship between continuity and openness. For every subset A of X, let $\chi_A : X \to \mathbb{S}$ (see Exercise 4.1.18) be the *characteristic map* of A, which maps every element of A to 1, and all others to 0. Show that χ_A is continuous iff A is open. More precisely, show that the continuous maps from X to $\mathbb{S}$ are exactly the maps χ_U, U open in X.

4.4 Compactness

Compactness also extends from metric to topological spaces. The key is the Heine–Borel (a.k.a. Borel–Lebesgue) Property already mentioned in Proposition 3.3.7, Item 2.

Definition 4.4.1 (Compact) A *cover* of a subset K of a topological space X is a family $(A_i)_{i \in I}$ of subsets of X such that $K \subseteq \bigcup_{i \in I} A_i$. A *subcover* of such a cover is just a subset, i.e., a family $(A_j)_{j \in J}$, where $J \subseteq I$. An *open cover* is one that consists entirely of open sets.

A subset K of a topological space X is *compact* iff every open cover of K has a finite subcover.

Thinking of opens as tests, a compact set K is one such that, if you want to check whether every element of K passes one among many tests, you know that you will need only finitely many.

Example 4.4.2 Every finite subset of a topological space is compact. Indeed, if $E = \{x_1, \ldots, x_n\}$ is finite, and $(U_i)_{i \in I}$ is any open cover of E, then pick $i_1 \in I$ such that $x_1 \in U_{i_1}$, pick $i_2 \in I$ such that $x_2 \in U_{i_2}$, ..., pick $i_n \in I$ such that $x_n \in U_{i_n}$, so that $(U_i)_{i \in \{i_1, i_2, \ldots, i_n\}}$ is a finite subcover.

Example 4.4.3 Since sequentially compact subsets and compact subsets are one and the same notion in metric spaces (this is Proposition 3.3.7), the Borel–Lebesgue Theorem states that the compact subsets of $\mathbb{R}^n$ with its metric topology (whatever L^p metric is taken, $p \geqslant 1$) are exactly the closed bounded subsets of $\mathbb{R}^n$.

Example 4.4.4 In $\mathbb{R}$ with its metric topology, any closed interval $[a, b]$ is compact. This is a special case of the Borel–Lebesgue Theorem.

Contrast this with $\mathbb{R}$ with its *Scott* topology:

Lemma 4.4.5 *The compact saturated subsets of* $\mathbb{R}_\sigma$ *are the subsets of the form* $[t, +\infty)$, $t \in \mathbb{R}$, *plus the empty set.*

Proof Let Q be any non-empty compact saturated subset of $\mathbb{R}_\sigma$. Since Q is upward closed, Q is of the form $(t, +\infty)$ or $[t, +\infty)$. However, $(t, +\infty)$ is not compact since the collection of all Scott-opens $(t+\frac{1}{n}, +\infty)$, $n \in \mathbb{N}$, $n \geqslant 1$, is an open cover of it, but any finite subcover, say given by those with n in some finite set E, would fail to contain $t + \frac{1}{2\max_{n\in E} n}$.

Conversely, $[t, +\infty)$ is saturated. To show that it is compact, assume that $(U_i)_{i\in I}$ is an open cover of $[t, +\infty)$. If some U_i is the whole of $\mathbb{R}$, then this single U_i forms a cover. Otherwise, and excluding those U_is that are empty, write U_i as $(t_i, +\infty)$, $i \in I$. Since $[t, +\infty) \subseteq \bigcup_{i\in I} U_i$, t is in $\bigcup_{i\in I}(t_i, +\infty) = (\inf_{i\in I} t_i, +\infty)$, so $t > \inf_{i\in I} t_i$. This is equivalent to requiring that $t > t_i$ for some $i \in I$, so that we can find a finite (indeed, a one-element) subcover, by just U_i. □

There is an alternative definition of compact subsets, which only requires directed unions instead of general unions. We use the following trick.

Trick 4.4.6 Every union $\bigcup_{i\in I} U_i$ can be written as a *directed* union, namely as $\bigcup_{J\in\mathbb{P}_{\text{fin}}(I)} U_J$, where each U_J is the finite union $\bigcup_{j\in J} U_j$.

Proposition 4.4.7 *A subset K of a topological space X is* compact *iff, for every directed family $(U_i)_{i\in I}$ of opens such that $K \subseteq \bigcup_{i\in I} U_i$, there is an $i \in I$ such that $K \subseteq U_i$ already.*

Here the assumption that $(U_i)_{i\in I}$ is *directed* is important.

Proof If K is compact, then from every directed family $(U_i)_{i\in I}$ that covers K, one can extract a finite subcover, say $U_{i_1}, \ldots, U_{i_n}$. Since the family is directed, there is a further index $i \in I$ such that $U_{i_1}, \ldots, U_{i_n} \subseteq U$. So $K \subseteq U_i$.

In the if direction, let $(U_i)_{i\in I}$ by any open cover of K. By Trick 4.4.6 and using the assumption, $K \subseteq U_J$ for some $J \in \mathbb{P}_{\text{fin}}(I)$. The family $(U_j)_{j\in J}$ is a finite subcover, showing that K is compact. □

Another characterization of compactness goes through filtered intersections of closed subsets.

Definition 4.4.8 (Filtered) Let X be a poset. A subset D of X is *filtered* iff D is directed in X^{op}, i.e., if D is non-empty, and any two points in D have a lower bound in D.

Proposition 4.4.9 *Let X be a topological space, and K be a subset of X. Then K is compact iff, for every filtered family $(F_i)_{i\in I}$ of closed subsets of X such that K meets every F_i, $i \in I$, K meets $\bigcap_{i\in I} F_i$.*

Proof Let U_i be the complement of F_i in X. Then use Proposition 4.4.7. □

Corollary 4.4.10 *Let X be a topological space, K be a compact subset of X, and F be a closed subset of X. Then $K \cap F$ is compact.*

Proof For every filtered family $(F_i)_{i \in I}$ of closed subsets of X such that $K \cap F$ meets each F_i, $(F_i \cap F)_{i \in I}$ is a filtered family of closed subsets of X that meet K. So $\bigcap_{i \in I} F_i \cap F$ meets K, whence $\bigcap_{i \in I} F_i$ meets $K \cap F$. □

The result in the following exercise is useful in proving the existence of points. One observes that there must be a point in $\bigcap_{i \in I} F_i$, assuming that there is at least one in each F_i—although not necessarily the same one in each. It is important that each F_i be closed and X be compact.

Exercise 4.4.11 A topological space X is *compact* iff X, seen as a subset of X itself, is compact. Show that X is compact iff any filtered intersection $\bigcap_{i \in I} F_i$ of non-empty closed subsets F_i is non-empty.

Proposition 4.4.12 (Finite unions of compacts are compact) *Let X be a topological space, and $Q_1, \ldots, Q_n$ be n compact (resp., compact saturated) subsets of X. Then $Q_1 \cup \cdots \cup Q_n$ is compact (resp., and saturated).*

Proof Saturation is obvious, using Proposition 4.2.9. Let $(U_i)_{i \in I}$ be any open cover of $Q_1 \cup \cdots \cup Q_n$. For every j, $1 \leqslant j \leqslant n$, this is an open cover of Q_j. Since Q_j is compact, one can find a finite subcover $(U_i)_{i \in I_j}$ of Q_j. Then $(U_i)_{i \in I_1 \cup \cdots \cup I_n}$ is a finite subcover of $Q_1 \cup \cdots \cup Q_n$. □

The single most important property that relates compactness and continuity is the following generalization of Proposition 3.5.5.

Proposition 4.4.13 (Image of a compact subset) *Recall that, for every map $f : X \to Y$ and every subset A of X, $f[A]$ is the* image $\{f(x) \mid x \in A\}$ *of A under f.*

Let X, Y be two topological spaces, f be a continuous map from X to Y, and K be a compact subset of X. Then $f[K]$ is compact in Y.

Proof Let $(V_i)_{i \in I}$ be any open cover of $f[K]$. From $f[K] \subseteq \bigcup_{i \in I} V_i$, we obtain $K \subseteq f^{-1}(\bigcup_{i \in I} V_i) = \bigcup_{i \in I} f^{-1}(V_i)$, so that $(f^{-1}(V_i))_{i \in I}$ is an open cover of the compact subset K. Extract a finite subcover $(f^{-1}(V_i))_{i \in J}$: then $(V_i)_{i \in J}$ is a finite subcover of $f[K]$. □

In non-T_1 spaces, not every compact subset will be equally interesting. The only ones really deserving some attention are those that are both compact and saturated, as we shall see. This does not make much difference, because saturation preserves compactness:

Proposition 4.4.14 *Let X be any topological space, and K be a compact subset of X. Then its saturation $\uparrow K$ is both saturated and compact. Conversely, if $\uparrow K$ is compact, then so is K.*

Proof Any open cover of $\uparrow K$ is one of K. Extract a finite subcover, and note that this covers $\uparrow K$, since every open subset is upward closed (Lemma 4.2.5). Conversely, any open cover of K is one of $\uparrow K$, since every open is upward closed. Any finite subcover is a subcover of K. □

We shall often reserve the letter K for compacts, while Q will usually denote *saturated* compacts.

In T_2 spaces, there is a strong relationship between compact and closed subsets, as witnessed by the following result. Recall from Exercise 4.4.11 that a space X is compact if and only if it is compact as a subset of itself.

Proposition 4.4.15 *Let X be a topological space. If X is compact, then every closed subset of X is compact. If X is T_2, then every compact subset of X is closed. In particular, if X is both compact and T_2, then the closed and the compact subsets of X coincide.*

Proof Let F be a closed subset of X, and assume X compact. Let U be the complement of F: U is open. Given any open cover $(U_i)_{i \in I}$ of F, adjoining it U yields an open cover of X. Since X is compact, extract a finite subcover. Without loss of generality, we may assume that U belongs to this subcover, i.e., it is of the form $(U_i)_{i \in J}$ plus U, for some finite subset J of I. Then $(U_i)_{i \in J}$ is a finite subcover of F, so F is compact.

Assume now that X is T_2, and let K be a compact subset of X. Fix an element $x \in X$ that is not in K. For every $y \in K$, there is an open subset U_{xy} containing x and an open subset V_{xy} containing y, such that $U_{xy} \cap V_{xy} = \emptyset$, because X is T_2. The family $(V_{xy})_{y \in K}$ is an open cover of K. Since K is compact, extract a finite subcover $(V_{xy})_{y \in E}$ of K, where E is a finite subset of K. Now $U_x = \bigcap_{y \in E} U_{xy}$ is open, contains x, and does not intersect $\bigcup_{y \in E} V_{xy}$, hence does not intersect the smaller set K. For every $x \in X$, we have therefore found an open neighborhood U_x of x in the complement of K. Using Trick 4.1.11, the complement of K is open, whence K is closed. □

Example 4.4.16 One should realize that compactness alone is a weak property. For example, any topological space that has a least element $\bot$ in its specialization quasi-ordering is compact: in any open cover, there must be an open containing $\bot$, and this open must then be the whole space.

This is in particular the case for all *pointed* posets, i.e., all posets with a least element $\bot$, in any of the three standard topologies of the ordering (Alexandroff, Scott, or upper).

Being T_2 is a rather weak property as well, as most geometric spaces from mainstream mathematics are T_2.

But spaces that are both compact and T_2 are special, and enjoy many interesting properties. One of these is normality.

Proposition 4.4.17 (Compact $T_2 \Rightarrow T_4$) *Every compact T_2 space is normal, hence T_4.*

Proof Let X be a compact T_2 space. We first show that X is T_3. Let F be a closed subset of X and x be an element of X outside of F. By Proposition 4.4.15, F is compact. As in the proof of this proposition, for every $y \in F$, there is an open subset U_{xy} containing x, an open subset V_{xy} containing y, such that $U_{xy} \cap V_{xy} = \emptyset$. Extract a finite subcover $(V_{xy})_{y \in E}$ of F, and observe that $U = \bigcap_{y \in E} U_{xy}$ is an open neighborhood of x, $V = \bigcup_{y \in E} V_{xy}$ is an open subset containing F, and $U \cap V = \emptyset$.

Now let F, F' be two disjoint closed subsets of X. Using a similar technique, for each $x \in F$, find an open neighborhood U_x of x, and an open subset V_x containing F' such that $U_x \cap V_x = \emptyset$ (using the fact that X is T_3). F' is compact, so extract a finite subcover $(V_x)_{x \in E}$ of F'. Then $U = \bigcup_{x \in E} U_x$ is an open subset containing F, $V = \bigcap_{x \in E} V_x$ is an open subset containing F', and U and V are disjoint. □

Proposition 4.4.15 also implies that, in compact T_2 spaces, the compact subsets and the closed subsets coincide. Here is an application.

Exercise 4.4.18 Let X be a T_2 space, $(K_i)_{i \in I}$ be a filtered family of compact subsets of X, and U be an open subset of X. Show that $\bigcap_{i \in I} K_i \subseteq U$ if and only if $K_i \subseteq U$ for some $i \in I$. (This states that every T_2 space is well-filtered, a notion we shall explore in Section 8.3.1.) In particular, every filtered family of non-empty compact subsets of a T_2 space X has non-empty intersection.

In non-T_2 spaces, compact subsets and closed sets can be very different. For one thing, as we said, the interesting compact subsets will be the saturated ones, that is, the upward closed ones, while closed sets are downward closed (Lemma 4.2.6).

Example 4.4.19 In $\mathbb{R}_\sigma$, Lemma 4.4.5 implies that $[0, +\infty)$ is compact (and saturated). It cannot be closed, since it is not downward closed.

Conversely, $(-\infty, 0]$ is closed but not compact in $\mathbb{R}_\sigma$. (Exercise: prove this.)

Exercise 4.4.20 Show that, if X is T_2, then the intersection of any two compact subsets of X is compact. We shall see in Exercise 4.4.23 that this is wrong when X is not T_2, even for compact saturated subsets.

We have seen that the finite subsets are always compact (Example 4.4.2). Using Proposition 4.4.14, we obtain:

Proposition 4.4.21 (Finitary compact) *Let X be any topological space. The subsets of X of the form $\uparrow E$, E a finite subset of X, are all compact saturated, and are called the* finitary compact *subsets of X.* □

Note how this looks like the finitary closed subsets introduced in Exercise 4.2.8, only with $\uparrow$ instead of $\downarrow$.

Exercise 4.4.22 Show that the finitary compacts are the *only* compact saturated subsets of X, whenever the topology of X is Alexandroff.

In particular, the finite subsets are the only compact subsets of a space with the discrete topology.

Exercise 4.4.23 (Followup to Exercise 4.4.20.) Show that, if X is not T_2, then the intersection of any two compact saturated subsets of X may fail to be compact. Exercise 4.4.22 should help.

Exercise 4.4.24 Show that *every* subset of any set X, equipped with the cofinite topology (Exercise 4.2.21), is compact. This is a paradoxical case of a space in which there are many more compact subsets than there are open, or closed, subsets.

Exercise 4.4.25 (Lifting) Consider an arbitrary topological space X. Let the *lifting* $X_\perp$ of X be X union a fresh element $\perp$, and equip $X_\perp$ with the topology whose opens are those of X, plus $X_\perp$ itself. Show that $X_\perp$ is *always* compact.

Certainly $X_\perp$ is not T_2, or even T_1. This is often taken as a proof that compactness, for non-T_2 spaces, is a very weak property.

Exercise 4.4.26 Show that the specialization quasi-ordering of $X_\perp$ is given by $x \leqslant y$ iff $x = \perp$, or x, y are in X and x is below y in X.

Show that the compact saturated subsets of $X_\perp$ are those of X plus $X_\perp$ itself.

Here is another testimony to the fact that compactness and the T_2 property interact nicely: any compact topology that is finer than some T_2 topology must coincide with it.

Theorem 4.4.27 (Compact T_2 topologies are minimal) *Let X be a space with two topologies, $\mathcal{O}_1$ and $\mathcal{O}_2$. Assume $\mathcal{O}_1$ is coarser than $\mathcal{O}_2$, that X is T_2 with the topology $\mathcal{O}_1$, and that X is compact with the topology $\mathcal{O}_2$. Then $\mathcal{O}_1$ and $\mathcal{O}_2$ are the same topology, and X is both compact and T_2.*

Proof We first notice that, since $\mathcal{O}_1$ has less opens than $\mathcal{O}_2$, it has more compacts. That is, every subset of X that is compact with respect to $\mathcal{O}_2$ is also compact with respect to $\mathcal{O}_1$: this is because there are less open covers in $\mathcal{O}_1$ to use in testing for compactness.

Since X is compact with respect to $\mathcal{O}_2$, it is then also compact with respect to $\mathcal{O}_1$, and T_2 by assumption. X is also T_2 with respect to $\mathcal{O}_2$, in addition to being compact, since the opens in $\mathcal{O}_1$ used to separate distinct points are also open in $\mathcal{O}_2$. So X is compact T_2 with any of the two topologies.

Proposition 4.4.15 then states that the compact subsets of X coincide with the closed subsets, in each topology. Together with the fact that $\mathcal{O}_1$ has more compacts than $\mathcal{O}_2$, it follows that $\mathcal{O}_1$ has more closed subsets than $\mathcal{O}_2$. Taking the complements of these closed subsets, $\mathcal{O}_1$ has more opens than $\mathcal{O}_2$. But we started from the assumption that $\mathcal{O}_1$ had less opens than $\mathcal{O}_2$. So $\mathcal{O}_1 = \mathcal{O}_2$. □

Corollary 4.4.28 *Any two topologies on X that make X compact and T_2 are either equal or incomparable.*

A useful trick to show that a subset is compact is the following result. The point is that we need only test for covers taken from a given *subbase*.

Theorem 4.4.29 (Alexander's Subbase Lemma) *Let X be a topological space, and $\mathcal{B}$ be a subbase for its topology. For any subset K of X, K is compact if and only if one can extract a finite subcover from any cover $(U_i)_{i \in I}$ of K such that each U_i is taken from $\mathcal{B}$.*

Proof The only if direction is obvious. Conversely, assume that one can extract a finite subcover from any cover of K made of subbasic opens, but that K is not compact. So there is an open cover of K from which one cannot extract any finite subcover. The family $\mathcal{C}$ of all open covers of K from which one cannot extract any finite subcover, ordered by inclusion, is inductive, as the union of any family of such covers is again such a cover. By Zorn's Lemma (Theorem 2.4.2), there is a maximal element of $\mathcal{C}$. This is a maximal open cover $(U_i)_{i \in I}$ of K that has no finite subcover. Note that maximality means that: $(*)$ for every open V not among $(U_i)_{i \in I}$, K has a finite cover made of V and finitely many opens from $(U_i)_{i \in I}$.

Let J be the subset of those indices $i \in I$ such that U_i is in $\mathcal{B}$. If $(U_i)_{i \in J}$ was a cover of K, then one could extract a finite subcover by assumption. So there must be a point $x \in K$ such that x is not in $\bigcup_{i \in J} U_i$. Since $(U_i)_{i \in I}$ is a cover of K, there is an index $i_0 \in I \setminus J$ such that $x \in U_{i_0}$. Writing U_{i_0} as a union of finite intersections of subbasic opens, there must be a finite intersection $V_1 \cap V_2 \cap \cdots \cap V_m$ of subbasic opens included in U_{i_0} such that $x \in V_1 \cap V_2 \cap \cdots \cap V_m$.

If V_1 were among $(U_i)_{i \in I}$, then it would be of the form U_i with $i \in J$, by definition of J. But this contradicts the fact that $x \in V_1$, but $x \notin \bigcup_{i \in J} U_i$. Similarly for $V_2, \ldots, V_m$. So $(*)$ applies: for each j, $1 \leqslant j \leqslant m$, K has a finite cover made of V_j and $(U_i)_{i \in J_j}$ for some finite subset J_j of I. It follows that

$V_1 \cap V_2 \cap \cdots \cap V_m$ plus $(U_i)_{\substack{1 \leqslant j \leqslant m \\ i \in J_j}}$ forms a cover of K, hence also U_{i_0} plus $(U_i)_{\substack{1 \leqslant j \leqslant m \\ i \in J_j}}$. That is, the finite family of all U_i, where $i \in \bigcup_{j=1}^{m} J_j \cup \{i_0\}$, is a finite subcover of K, a contradiction. ☐

This will be instrumental in proving Tychonoff's Theorem below.

4.5 Products

Given any family $(X_i)_{i \in I}$ of sets, the *product* $\prod_{i \in I} X_i$ is the collection of tuples of elements $(x_i)_{i \in I}$ where, for each $i \in I$, x_i is in X_i. By convention, we write $\vec{x}$ for such tuples, and x_i for its i component. We write $X \times Y$ for the product of two sets: this is the special case where I contains exactly two elements.

There is a standard way of putting a topology on any product.

Definition 4.5.1 (Product topology) Let X_i be topological spaces, $i \in I$. The *product topology* on $\prod_{i \in I} X_i$ has a base of open sets of the form $\prod_{i \in I} U_i$, where U_i is open in X_i for every $i \in I$, and all but finitely many are such that $U_i = X_i$.

One checks that this is indeed a base, not just a subbase, because any finite intersection of such basic sets is again of this form.

When I is finite, the opens are just products $\prod_{i \in I} U_i$ of open sets U_i from X_i, $i \in I$. Such products are called *open rectangles*. When I is infinite, we require the additional condition that only finitely many opens U_i are allowed to be different from X_i. This may seem strange. The main reason is that the above definition is reverse-engineered from the actual, abstract one; namely, the product topology is the coarsest one that makes all projections continuous. Let us put this as a theorem.

Proposition 4.5.2 *Let X_i be topological spaces, $i \in I$. Let the* projection map *$\pi_i : \prod_{j \in I} X_j \to X_i$ map each element $\vec{x} = (x_j)_{j \in I}$ to x_i. The product topology is the coarsest one on $\prod_{j \in I} X_j$ such that π_i is continuous for every $i \in I$.*

Proof Clearly, π_i is continuous, since, for every open subset U of X_i, $\pi_i^{-1}(U)$ is the product all open sets U_j, $j \in I$, where $U_j = X_j$ whenever $j \neq i$, and $U_i = U$. Conversely, if $\mathcal{O}$ is any topology on $\prod_{j \in I} X_j$ such that every π_i is continuous, then, for any family $(U_i)_{i \in I}$ where each U_i is open in X_i, and only finitely many are different from X_i, let J be the finite subset of all indices $i \in I$ such that $U_i \neq X_i$. Then $\prod_{i \in I} U_i$ is the finite intersection $\bigcap_{j\in I} \pi_j^{-1}(U_j)$, so must be open in $\mathcal{O}$. So $\mathcal{O}$ is finer than the product topology. □

So the subsets of the form $\pi_i^{-1}(U)$, $i \in I$, U open in X_i, form a subbase of the product topology.

Exercise 4.5.3 Let X_i be topological spaces, $i \in I$, and A_i be subsets of X_i, one for each $i \in I$. Show that $cl(\prod_{i\in I} A_i) = \prod_{i\in I} cl(A_i)$.

Trick 4.5.4 To show that a map $f : Y \to \prod_{i\in I} X_i$ is continuous, it is sufficient to show that $\pi_i \circ f : Y \to X_i$ is continuous for each $i \in I$.

(The condition is necessary, by Proposition 4.3.12.) Indeed, assume that $\pi_i \circ f$ is continuous for each $i \in I$. Then $f^{-1}(\pi_i^{-1}(V))$ is open for every open subset V of X_i, for every $i \in I$. That is, $f^{-1}(U)$ is open for every subbasic open $\pi_i^{-1}(V)$ of the product topology. It follows that f is continuous, by Trick 4.3.3.

Conversely, there is no way of simplifying the statement that f is continuous from a product $\prod_{i\in I} X_i$ to Y, in general. This is already the case for binary products; see Exercise 4.5.11.

Exercise 4.5.5 (Product of two maps) Let $f : X \to Y$ and $g : X' \to Y'$ be two maps. Show that $f \times g : X \times X' \to Y \times Y'$, defined by $(f \times g)(x, x') = (f(x), g(x'))$, is continuous whenever f and g are. The map $f \times g$ is called the *product* of the two maps f and g. Show that $f \times g$ is continuous at (x, y) whenever f is continuous at x and g is continuous at y.

Exercise 4.5.6 (Pairing of two maps) Let $f : X \to Y$ and $g : X \to Y'$ be two maps. Define $\langle f, g\rangle : X \to Y \times Y'$ by $\langle f, g\rangle(x) = (f(x), g(x))$. This is the *pairing* of f and g. Show that $\langle f, g\rangle$ is continuous (resp., at x) whenever f and g are. (Hint: a short argument goes through the study of the *diagonal map* $\Delta : X \to X \times X$, defined by $\Delta(x) = (x, x)$, and using composition with functions defined in the preceding exercises.)

Example 4.5.7 (Polynomial functions are continuous) Sum $+ : \mathbb{R} \times \mathbb{R} \to \mathbb{R}$ and product $. : \mathbb{R} \times \mathbb{R}$ are continuous. Using Example 4.3.8, it follows that every polynomial function from $\mathbb{R}^n$ to $\mathbb{R}$, i.e., every function expressible using just variables, constants, $+$, $-$, and $.$, is continuous. (Exercise: prove all this.)

Example 4.5.8 (Rational fractions are continuous) Division $/ : \mathbb{R} \times \mathbb{R} \to \mathbb{R}$ is continuous at every point where it is defined, i.e., at every point (x, y) such that $y \neq 0$. (Exercise: prove this.)

Using Example 4.3.8 and Example 4.5.7, we conclude that every function from $\mathbb{R}^n$ to $\mathbb{R}$ that is expressible using just variables, constants, $+$, $-$, $.$, and $/$, is continuous at every point where it is defined. In particular, *rational fractions*, i.e., functions expressible as the quotient p/q of two polynomials p and q, are continuous at every point where q does not take the value 0.

Exercise 4.5.9 We now have several candidate topologies on $\mathbb{R}^n$: (a) the product topology of n copies of $\mathbb{R}$, each with its metric topology, and (b) the metric topology of the L^p metric, for each $p \geqslant 1$. Show that they all coincide.

Exercise 4.5.10 Let X, Y be two topological spaces. For any open subset W of $X \times Y$, show that $W_{|x} = \{y \in Y \mid (x, y) \in W\}$ and $W^{|y} = \{x \in X \mid (x, y) \in W\}$ are open in Y and X, respectively, for every $x \in X$, $y \in Y$.

Exercise 4.5.11 (Separate continuity) A function $f : X \times Y \to Z$ is continuous *in its second argument* (resp., at x) iff the function $f(x, \_) : Y \to Z$, which maps each y to $f(x, y)$, is continuous for every $x \in X$ (resp., at x). Similarly, f is continuous *in its first argument* (resp., at y) iff $f(\_, y) : X \to Z$ is continuous for every $y \in Y$ (resp., at y). Finally, f is *separately continuous* (resp., at (x, y)) iff it is continuous in both its arguments (resp., in its first argument at y and in its second argument at x).

Show that, if $f : X \times Y \to Z$ is continuous (resp., at (x, y)), then it is separately continuous (resp., at (x, y)). Show that the converse may fail, even for T_2 spaces. You may want to consider the example of the *wavy cross* map $f : \mathbb{R} \times \mathbb{R} \to \mathbb{R}$ defined by:

$$f(x, y) = \frac{(y^2 - x^2)^2 - 4x^2y^2}{(x^2 + y^2)^2} \tag{4.1}$$

when $(x, y) \neq (0, 0)$, and $f(0, 0) = 1$. (See Figure 4.5; in polar coordinates, where $x = r\cos\theta$ and $y = r\sin\theta$, $f(x, y)$ is just $\cos(4\theta)$.) Show that f is continuous at every point except at $(0, 0)$, but that f is separately continuous at every point of $\mathbb{R}^2$, including $(0, 0)$.

To make the difference clearer, one sometimes speaks of *joint continuity* to denote continuity, as opposed to separate continuity.

4.5.1 Products of compact spaces, and Tychonoff's Theorem

We arrive at Tychonoff's Theorem, which is probably the most important result in topology. This generalizes Proposition 3.3.8 to topological spaces, and to arbitrary, not just finite, products.

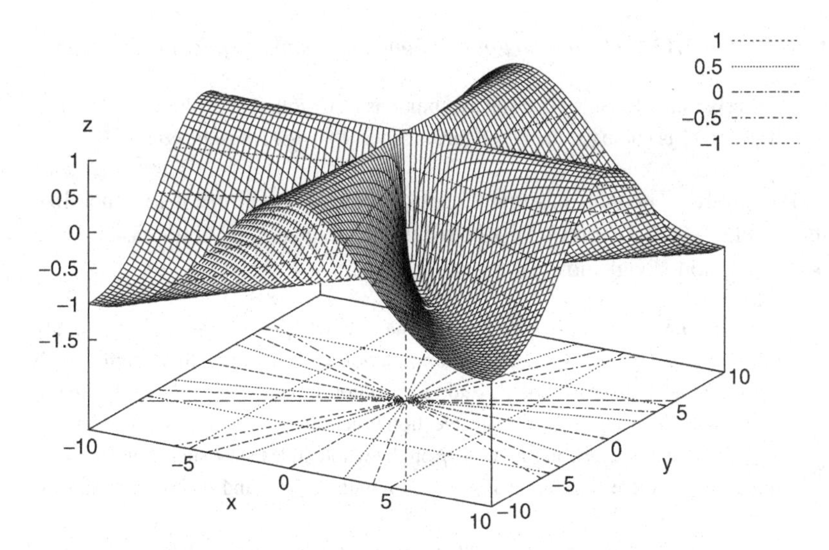

Figure 4.5 The wavy cross: a separately continuous function that is not jointly continuous.

Theorem 4.5.12 (Tychonoff) *Let $(X_i)_{i \in I}$ be any family of topological spaces, and K_i be a compact subset of X_i for each $i \in I$. Then $\prod_{i \in I} K_i$ is compact in $\prod_{i \in I} X_i$.*

Proof Assume that $\prod_{i \in I} K_i$ is not compact. By Alexander's Subbase Lemma (Theorem 4.4.29), there is an open cover $(U_j)_{j \in J}$ of $\prod_{i \in I} K_i$ consisting of subbasic opens that has no finite subcover. As subbasic open sets U_j, we take $U_j = \pi_{i_j}^{-1}(V_j)$ where $i_j \in I$ and V_j is open in X_{i_j}, using Proposition 4.5.2.

For each $i \in I$, consider the set J_i of all indices $j \in J$ such that $i_j = i$. There cannot be any finite subset J'_i of J_i such that $(V_j)_{j \in J'_i}$ would be a cover of K_i. Otherwise $(U_j)_{j \in J'_i}$ would be a finite cover of $\prod_{i \in I} K_i$. So $(V_j)_{j \in J_i}$ cannot be a cover of K_i either: otherwise, one could extract a finite subcover from it, since K_i is compact. It follows that there is a point $x_i \in K_i$ such that x_i is not in $\bigcup_{j \in J_i} V_j$. (Picking one x_i for each $i \in I$ requires the Axiom of Choice.)

Consider the point $\vec{x} = (x_i)_{i \in I}$. By construction, $\vec{x}$ is in $\prod_{i \in I} K_i$. Since $(U_j)_{j \in J}$ is a cover of $\prod_{i \in I} K_i$, $\vec{x}$ is in some U_j, i.e., in some $\pi_{i_j}^{-1}(V_j)$. This implies that x_{i_j} is in V_j. Since $i_j \in J_{i_j}$, x_{i_j} would then be in $\bigcup_{j \in J_{i_j}} V_j$, a contradiction. □

By taking $K_i = X_i$ for every $i \in I$, we get the usual form of Tychonoff's Theorem.

Corollary 4.5.13 *The product of any family of compact spaces is compact.*

Exercise 4.5.14 Show that the assumptions of Corollary 4.5.13 are needed; i.e., if $\prod_{i \in I} X_i$ is compact, then each factor space X_i, $i \in I$, must be compact.

Tychonoff's Theorem has many applications. We shall use it to generalize the Arzelà–Ascoli Theorem 3.5.12 beyond metric spaces in Section 5.4.2. Here is another, non-trivial and curious, application.

Exercise 4.5.15 (Rado's Selection Lemma) Let $(A_i)_{i \in I}$ be an arbitrary family of finite sets. Given $J \subseteq I$, a *selection function* s_J is a map with domain J such that $s_J(i) \in A_i$ for each $i \in J$. Rado's Selection Lemma (Rado, 1949, lemma 1) states that, given any collection of selection functions s_J, one for each finite subset $J \subseteq I$, there is a selection function s_I on the whole index set I such that, for every finite $J \subseteq I$, there is a finite subset $J' \subseteq I$ with $J \subseteq J'$ and such that s_I and $s_{J'}$ coincide on J.

To prove it, equip each A_i with the discrete topology, and realize that $\prod_{i \in I} A_i$ must be compact. (No, the topology on $\prod_{i \in I} A_i$ is not the discrete topology in general, unless I is finite.) Then show that, for every finite subset J of I, the set S_J of all tuples $\vec{x}$ such that, for some finite subset J' of I containing J, $x_i = s_{J'}(i)$ for each $i \in J$, is closed, and use Exercise 4.4.11.

The product $\prod_{i \in I} X_i$ of spaces X_i that are all equal to the same space X is often written X^I. This is just the space of all functions from I to X. Tychonoff's Theorem has the following important trivial corollary.

Proposition 4.5.16 (Powers of compact spaces) *Any power X^I of a compact space X is compact.*

4.5.2 Products and orderings

The specialization quasi-ordering of a product is the product of the specialization quasi-orderings.

Lemma 4.5.17 *Let $(X_i)_{i \in I}$ be any family of topological spaces, and $\leqslant_i$ be the specialization quasi-ordering of X_i, $i \in I$. The specialization quasi-ordering $\leqslant$ of $\prod_{i \in I} X_i$ is the* product quasi-ordering $\prod_{i \in I} \leqslant_i$, *namely $\vec{x} \leqslant \vec{y}$ iff $x_i \leqslant_i y_i$ for each $i \in I$.*

Proof Let $\leqslant$ be the specialization quasi-ordering of $\prod_{i \in I} X_i$. Since each projection π_i is continuous, it is monotonic, by Proposition 4.3.9. So, if $\vec{x} \leqslant \vec{y}$, then $x_i \leqslant_i y_i$ for each $i \in I$, that is, $\vec{x} \, (\prod_{i \in I} \leqslant_i) \, \vec{y}$. Conversely, if $\vec{x} \, (\prod_{i \in I} \leqslant_i) \, \vec{y}$, then $x_i \leqslant_i y_i$ for each $i \in I$. In particular, if $\vec{x}$ is in the subbasic open set $\pi_i^{-1}(U)$, where U is open in X_i, then $\vec{y}$ is, too. By Proposition 4.2.4, $\vec{x} \leqslant \vec{y}$. □

In the binary case, Lemma 4.5.17 states that (x, y) is below (x', y') in the specialization quasi-ordering of $X \times Y$ iff $x \leqslant x'$ and $y \leqslant y'$, where we use the same symbol $\leqslant$ for the specialization quasi-ordering of X and of Y.

Let $(X_i)_{i \in I}$ be a family of posets, and write $\leqslant_i$ for the ordering on X_i, $i \in I$. Their *poset product*, still written $\prod_{i \in I} X_i$, is also the collection of tuples of elements $(x_i)_{i \in I}$ where, for each $i \in I$, x_i is in X_i, but with the *pointwise ordering* $\leqslant = \prod_{i \in I} \leqslant_i$ defined by $\vec{x} \leqslant \vec{y}$ iff, for every $i \in I$, $x_i \leqslant y_i$. In particular, the poset product, still written $X_1 \times X_2$, of two posets X_1 (with ordering $\leqslant_1$) and X_2 (with ordering $\leqslant_2$) is ordered by $(x_1, x_2) \leqslant (y_1, y_2)$ iff $x_1 \leqslant_1 y_1$ and $x_2 \leqslant_2 y_2$.

Exercise 4.5.18 Show that any product of T_0 spaces is T_0, that any product of T_1 spaces is T_1, and that any product of T_2 spaces is T_2.

Exercise 4.5.19 Let X_1, X_2 be two posets. Show that $(X_1 \times X_2)_a = X_{1\,a} \times X_{2\,a}$. (The product on the left is the poset product, while the one on the right is the topological product.)

Show that, on the other hand, the topologies of $(X_1 \times X_2)_u$ and $X_{1\,u} \times X_{2\,u}$ are in general distinct; the product topology (of the latter) is always finer than the upper topology (of the former).

Finally, show that the (product) topology on $X_{1\,\sigma} \times X_{2\,\sigma}$ is coarser than the (Scott) topology on $(X_1 \times X_2)_\sigma$. (It is strictly coarser in general, but we are not asking for a proof of this right now.)

Lemma 4.5.17 states that, given topological spaces, it does not matter whether we compute the topological product, then take the specialization quasi-ordering, or take the specialization quasi-ordering, then the poset product. Exercise 4.5.19 shows that a similar coincidence is harder to obtain in the reverse direction, from posets to topological spaces, except in the Alexandroff case. It is terribly easy to write a product sign $\prod$ or $\times$ without thinking whether it is poset product or topological product, and this leads to deep and subtle errors. Let us stress this in the important case of the Scott topology—although a similar statement could be made for the upper topology.

Given two posets X_1, X_2, ordered respectively by $\leqslant_1$ and $\leqslant_2$, the (Scott) topology of the product poset $(X_1 \times X_2)_\sigma$ is finer, and *in general strictly finer*, than the product topology on $X_{1\,\sigma} \times X_{2\,\sigma}$.

We don't have the right tools yet to produce a counterexample: see Exercise 5.2.16. However, it is easy to show that this kind of commutation already fails for infinite products:

Exercise 4.5.20 Let X_i be a poset, ordered by $\leqslant_i$, $i \in I$. Show that the (Alexandroff) topology on $(\prod_{i \in I} X_i)_a$ is in general strictly finer than the (product) topology on $\prod_{i \in I} X_{i\,a}$. (You may use the poset $\{0, 1\}^{\mathbb{N}} = \prod_{i \in \mathbb{N}} X_i$ where $X_i = \{0, 1\}$ for each $i \in \mathbb{N}$, and $\{0, 1\}$ is ordered by equality $=$. In which topologies is this compact?) Show a similar statement for the Scott topology.

Exercise 4.5.21 The pathologies of Exercise 4.5.11 do not occur with Scott topologies. Show indeed that, given three posets X, Y, and Z, then a map $f : X \times Y \to Z$ is Scott-continuous iff f is *separately* Scott-continuous, i.e., iff $f(x, \_)$ and $f(\_, y)$ are Scott-continuous for all $x \in X$, $y \in Y$. (Warning: in one direction, if you think this follows from the fact that every Scott-continuous map is continuous, then you have probably fallen into the trap mentioned above, right after Exercise 4.5.19.)

Exercise 4.5.22 Show that the poset product of any family of dcpos is a dcpo.

4.6 Coproducts

Given any family $(X_i)_{i \in I}$ of sets, the *coproduct* $\coprod_{i \in I} X_i$ is the disjoint union of X_i, $i \in I$. Formally, $\coprod_{i \in I} X_i$ is the space of pairs (i, x) where $i \in I$ and $x \in X_i$.

Definition 4.6.1 (Coproduct topology) Let X_i be topological spaces, $i \in I$. The *coproduct topology* on $\coprod_{i \in I} X_i$ is defined by declaring open exactly those subsets U of $\coprod_{i \in I} X_i$ such that $\{x \mid (i, x) \in U\}$ is open in X_i for every $i \in I$.

It is customary to equate (i, x) with x, relying on the context to know whether we mean x as an element of X_i or as an element of $\coprod_{i \in I} X_i$. This is meaningful when the spaces X_i, $i \in I$, are pairwise disjoint. In this case, $\coprod_{i \in I} X_i$ can be equated with the union $\bigcup_{i \in I} X_i$, and a subset U of the latter is open if and only if $U \cap X_i$ is open for every $i \in I$. In other words, the open subsets of the coproduct are the disjoint unions of open subsets taken from each X_i.

When $I = \{1, 2\}$, we write $X_1 + X_2$ instead of $\coprod_{i \in \{1,2\}} X_i$.

Proposition 4.6.2 *Let X_i be topological spaces, $i \in I$. Let the* injection map *$\iota_i : X_i \to \coprod_{j \in I} X_j$ map each element $x \in X_i$ to $(i, x) \in \coprod_{j \in I} X_j$. The coproduct topology is the finest one on $\coprod_{j \in I} X_j$ such that ι_i is continuous for every $i \in I$.*

Proof The finest topology that makes ι_i continuous is by definition the one such that U is open in $\coprod_{j \in I} X_j$ iff $\iota_i^{-1}(U)$ is open in X_i for every $i \in I$. This is just the definition of the coproduct topology. □

Trick 4.6.3 To show that a map $f : \coprod_{i \in I} X_i \to Y$ is continuous, it is sufficient to show that $f \circ \iota_i : X_i \to Y$ is continuous for each $i \in I$.

Indeed, f is continuous iff $\iota_i^{-1}(f^{-1}(V))$ is open for every open subset V of Y and every $i \in I$, iff $(f \circ \iota_i)^{-1}(V)$ is open for every V and i.

Exercise 4.6.4 (Coproduct of two maps) Let $f : X \to Y$ and $g : X' \to Y'$ be two maps. Show that $f + g : X + X' \to Y + Y'$, defined by $(f+g)(1, x) = (1, f(x))$ and $(f + g)(2, x') = (2, g(x'))$, is continuous whenever f and g are. The map $f + g$ is called the *coproduct* of the two maps f and g. Show that $f + g$ is continuous at $(1, x)$ whenever f is continuous at x, and at $(2, x')$ whenever g is continuous at x'.

Exercise 4.6.5 (Copairing of two maps) Let $f : X \to Y$ and $g : X' \to Y$ be two maps. Define $[f, g] : X + X' \to Y$ by $[f, g](1, x) = f(x), [f, g](2, x') = g(x')$. This is the *copairing* of f and g. Show that $[f, g]$ is continuous (resp., at $(1, x)$, at $(2, x')$) whenever f and g are. (Hint: a short argument goes through the study of the *codiagonal map* $\nabla : Y + Y \to Y$, defined by $\nabla(1, y) = y, \nabla(2, y) = y$.)

Exercise 4.6.6 (Coproducts of compacts) Let $(X_i)_{i \in I}$ be a family of topological spaces. Show that the compact subsets of $\coprod_{i \in I} X_i$ are the finite disjoint unions of compact subsets of each X_i, i.e., the subsets of the form $\bigcup_{i \in J}(\{i\} \times K_i)$, where J is a finite subset of I and K_i is compact in X_i for each $i \in J$.

In particular, any finite coproduct of compact spaces is compact, and any compact coproduct $\coprod_{i \in I} X_i$ is such that X_i is compact for every $i \in I$, and empty for all but finitely many.

Lemma 4.6.7 *Let $(X_i)_{i \in I}$ be any family of topological spaces, and $\leqslant_i$ be the specialization quasi-ordering of X_i, $i \in I$. The specialization quasi-ordering $\leqslant$ of $\prod_{i \in I} X_i$ is the* coproduct quasi-ordering $\coprod_{i \in I} \leqslant_i$, *defined by (i, x) $(\coprod_{i \in I} \leqslant_i)$ (j, y) iff $i = j$ and $x \leqslant_i y$.*

Proof Let $\leqslant$ be the specialization quasi-ordering of $\prod_{i \in I} X_i$. Since each injection ι_i is continuous, it is monotonic, by Proposition 4.3.9. So, if $x \leqslant_i y$, then $(i, x) \leqslant (i, y)$. Conversely, if $(i, x) \leqslant (j, y)$, then we first claim that $i = j$. Otherwise indeed, $\{i\} \times X_i$ would be an open subset containing (i, x) but not (j, y). Next, every open subset of the form $\{i\} \times U$, where U is open in X_i and $(i, x) \in \{i\} \times U$, would also contain (i, y). So any open subset U of X_i containing x also contains y, i.e., $x \leqslant_i y$. □

In the binary case, Lemma 4.6.7 states that $(1, x) \leqslant (1, y)$ iff $x \leqslant y$, $(2, x) \leqslant (2, y)$ iff $x \leqslant y$, and $(1, x)$ and $(2, y)$ are incomparable. (We use the same symbol $\leqslant$ for all the specialization quasi-orderings.)

Exercise 4.6.8 Let $(X_i)_{i\in I}$ be any family of posets, where $\leqslant_i$ is the quasi-ordering of X_i. Define their poset coproduct as the set $\coprod_{i\in I} X_i$ of all pairs (i, x) with $(i, x) \leqslant (i', x')$ iff $i = i'$ and $x \leqslant_i x'$.

Show that $(\coprod_{i\in I} X_i)_a = \coprod_{i\in I} X_{i\,a}$, and that $(\coprod_{i\in I} X_i)_\sigma = \coprod_{i\in I} X_{i\,\sigma}$.

Show that $(\coprod_{i\in I} X_i)_u = \coprod_{i\in I} X_{i\,u}$ if and only if all but finitely many X_i, $i \in I$, are empty, and that in general the topology of $\coprod_{i\in I} X_{i\,u}$ is finer than that of $(\coprod_{i\in I} X_i)_u$.

4.7 Convergence and limits

Convergence also generalizes to topological spaces. Imitating the definition in metric spaces, or more precisely Proposition 3.2.4, Item 3, we say that the sequence $(x_n)_{n\in\mathbb{N}}$ converges to x if and only if every open neighborhood U of x contains x_n for n large enough. This is fine, but not enough to characterize the open subsets – equivalently, the closed subsets – as the following example shows.

Example 4.7.1 Consider the space $X = \{0, 1\}^{\mathbb{R}}$, with the product topology, and where $\{0, 1\}$ comes with the discrete topology. Let A be the subset of all $\vec{x} \in X$ such that $x_i = 1$ for countably many values of $i \in \mathbb{R}$, and B be its complement.

A satisfies the property that, for every sequence $(\vec{x}_n)_{n\in\mathbb{N}}$ of elements of A that converges to $\vec{x} \in X$, $\vec{x}$ is in A. (We might say that A is *sequentially closed*.) Indeed, writing $\vec{x}_n$ as $(x_{ni})_{i\in\mathbb{R}}$ and $\vec{x}$ as $(x_i)_{i\in\mathbb{R}}$, for every $i \in \mathbb{R}$, x_{ni} must equal x_i for n large enough. It follows that $\{i \in \mathbb{R} \mid x_i = 1\} \subseteq \bigcup_{n\in\mathbb{N}}\{i \in \mathbb{R} \mid x_{ni} = 1\}$ is countable, so $\vec{x}$ is in A.

Nevertheless, A is not closed, i.e., B is not open: otherwise, for every $\vec{x} \in B$, using Lemma 4.1.10, $\vec{x}$ would be in some basic open $\prod_{i\in\mathbb{R}} U_i$ where each U_i is open, and such that there is a finite set $J \subseteq \mathbb{R}$ such that $U_i = \{0, 1\}$ whenever $i \notin J$; but then the tuple $\vec{x}'$ obtained by letting $x'_i = x_i$ for every $i \in J$ and $x'_i = 0$ otherwise would be in $\prod_{i\in\mathbb{R}} U_i$, too: this contradicts the fact that $\vec{x}'$ is in A, hence not in B.

Moore and Smith (1922) corrected the problem by replacing sequences by *nets*, where points are indexed, not by natural numbers, but by elements of a general, possibly uncountable, directed quasi-order.

Definition 4.7.2 (Net) Let X be a set. A *net* on X is a family $(x_i)_{i \in I, \sqsubseteq}$, where $\sqsubseteq$ is a quasi-ordering on I that turns I into a directed set.

Given any subset U of X, the net is *eventually in U* iff there is an index $i \in I$ such that, for every $j \in I$ with $i \sqsubseteq j$, x_j is in U. Similarly, the net *eventually satisfies* a property P iff there is an index $i \in I$ such that, for every $j \in I$ with $i \sqsubseteq j$, P holds for x_j. We also say that x_i is in U, resp. x_i satisfies P or P holds for x_i, *for i large enough.*

Definition 4.7.3 (Convergence, limit) Let X be a topological space. The net $(x_i)_{i \in I, \sqsubseteq}$ *converges to* $x \in X$ if and only if, for every open neighborhood U of x, x_i is in U for i large enough. We then say that x is *a limit* of the net.

One usually thinks of *the* limit of a sequence, and indeed limits, when they exist, are unique, provided the ambient space X is T_2. The T_2 assumption is necessary (see Exercise 4.7.13).

Proposition 4.7.4 *Let X be a T_2 space. Every net in X has at most one limit.*

Proof Let $(x_i)_{i \in I, \sqsubseteq}$ be a net with two distinct limits x and y. Since X is T_2, there are two disjoint opens U and V containing respectively x and y. The net is then both eventually in U and V, hence in $U \cap V$, a contradiction. ☐

Example 4.7.5 (Limits in $\mathbb{R}$) Let $(x_i)_{i \in I, \sqsubseteq}$ be a net in $\mathbb{R}$, with its metric topology. This net converges to $x \in \mathbb{R}$ if and only if, for every $\epsilon > 0$, $|x - x_i| < \epsilon$ for i large enough. (Exercise: prove this.)

Exercise 4.7.6 (Limits in $\mathbb{R}_\ell$) Let $(x_i)_{i \in I, \sqsubseteq}$ be a net in the Sorgenfrey line $\mathbb{R}_\ell$ (see Exercise 4.1.34). Show that this net converges to $x \in \mathbb{R}_\ell$ if and only if, for every $\epsilon > 0$, $x \leqslant x_i < x + \epsilon$ for i large enough. That is, not only does x_i come arbitrarily close to x, but also, eventually x_i will be above x: we say that $(x_i)_{i \in I, \sqsubseteq}$ converges to x *from the right.*

However, in general, limits fail to be unique. For example:

Exercise 4.7.7 Let X be a quasi-ordered set. Every directed family $(x_i)_{i \in I}$ defines a canonical net $(x_i)_{i \in I, \sqsubseteq}$, where we define $\sqsubseteq$ by $i \sqsubseteq j$ iff $x_i \leqslant x_j$.

Consider now the case where X is a dcpo. Show that $\sup_{i \in I} x_i$ is a limit of the net in X_σ. This much shows how taking least upper bounds of directed sets indeed corresponds to taking a limit.

Show, next, that $\sup_{i \in I} x_i$ is the *largest* limit of the net, and more generally that a point y is a limit of the net if and only if it is below $\sup_{i \in I} x_i$. In particular, limits in dcpos in general are far from being unique.

The following exercise stresses the fact that limits are non-unique in non-T_2-spaces.

Exercise 4.7.8 Show that the set of all limits of any net is always closed, hence downward closed. Show that, given any *monotone net* $(x_i)_{i \in I, \sqsubseteq}$, i.e., one where $i \sqsubseteq j$ implies $x_i \leqslant x_j$, every element x_i, $i \in I$, is a limit of the net.

Proposition 4.7.9 (Closure = set of limits) *Let X be a topological space. For every subset A of X, $cl(A)$ is the set of all limits of nets $(x_i)_{i \in I, \sqsubseteq}$ whose elements x_i are in A, for every $i \in I$.*

Proof Assume that x is a limit of such a net. Every open subset U of X containing x must contain x_i for i large enough, hence U must intersect A. By Lemma 4.1.27, x must be in $cl(A)$. Conversely, let $x \in cl(A)$. Let $\mathcal{N}_x$ be the collection of all open neighborhoods of x, ordered by reverse inclusion. This is directed. For each $U \in \mathcal{N}_x$, since U intersects $cl(A)$ (at x), U also intersects A, by Corollary 4.1.28. Pick an element x_U of $U \cap A$, for each $U \in \mathcal{N}_x$, using the Axiom of Choice. By construction, $(x_U)_{U \in \mathcal{N}_x, \supseteq}$ converges to x, and is a net composed of elements of A. □

The closed sets are then exactly the subsets that are closed under taking limits—of nets; recall that sequences are not enough for this (Example 4.7.1).

Corollary 4.7.10 *Let X be a topological space. For every subset F of X, F is closed if and only if, for every net $(x_i)_{i \in I, \sqsubseteq}$ that converges to x in X, and such that $x_i \in F$ for every $i \in I$, x is in F.*

Corollary 4.7.11 (Opens in terms of nets) *Let X be a topological space. For every subset U of X, U is open if and only if every net $(x_i)_{i \in I, \sqsubseteq}$ that converges to an element x of U contains some x_i in U already.*

Corollary 4.7.12 (Comparing topologies through convergence) *Let $\mathcal{O}_1$, $\mathcal{O}_2$ be two topologies on a set X. Then $\mathcal{O}_2$ is finer than $\mathcal{O}_1$ if and only if every net that converges to a point x in X with respect to $\mathcal{O}_2$ also converges to x with respect to $\mathcal{O}_1$.*

In particular, two topologies that have the same notion of convergence are equal.

Proof If $\mathcal{O}_2$ is finer than $\mathcal{O}_1$, then, for every net $(x_i)_{i \in I, \sqsubseteq}$ that converges to x in $\mathcal{O}_2$, for every open neighborhood U of x in $\mathcal{O}_1$, U is also an open neighborhood of x in $\mathcal{O}_2$, so $x_i \in V$ for i large enough. Conversely, assume that convergence with respect to $\mathcal{O}_2$ implies convergence with respect to $\mathcal{O}_1$, and let U be an open in $\mathcal{O}_1$. Let $(x_i)_{i \in I, \sqsubseteq}$ be an arbitrary net that converges to some element x of U with respect to $\mathcal{O}_2$. By assumption, it converges to x with respect to $\mathcal{O}_1$ as well, so $x_i \in U$ for i large enough. By Corollary 4.7.11, U is also open in $\mathcal{O}_2$. □

Exercise 4.7.13 Show that the assumption that X is T_2 is necessary in Proposition 4.7.4. In other words, let X be a non-T_2 space, and let x, y be two distinct points such that one cannot find any pair of disjoint open neighborhoods U of x, and V of y. Exhibit a net that has two distinct limits—x and y, as you may have guessed. You may get some inspiration from the proof of Proposition 4.7.9.

Although sequences are not enough in general, they are in the case of first-countable spaces:

Exercise 4.7.14 (First-countable spaces) A space X is *first-countable* iff every point has a countable neighborhood base, i.e., for every point $x \in X$, there are countably many open neighborhoods $U_0, U_1, \ldots, U_n, \ldots$ of x such that every open neighborhood U of x contains some U_i. For example, every metric space is first-countable: take the open ball centered at x with radius $1/2^n$ for U_n.

Show that we obtain an equivalent notion if we additionally assume that $(U_i)_{i \in \mathbb{N}}$ is *descending*, i.e., that $U_0 \supseteq U_1 \supseteq \cdots \supseteq U_n \supseteq \cdots$.

Show that, when X is first-countable, for every subset A of X, A is the set of all limits of *sequences* $(x_i)_{i \in \mathbb{N}}$, and that a subset F is closed iff it contains all limits of *sequences* in F. That is, in a first-countable space, the sequentially closed subsets coincide with the closed subsets.

Example 4.7.15 $\mathbb{R}$, with its metric topology, is first-countable, as is every metric space. $\mathbb{R}_\sigma$, with its Scott topology, is also first-countable (show this).

Example 4.7.16 The space $\{0, 1\}^{\mathbb{R}}$ is not first-countable. Otherwise, by Exercise 4.7.14 all its sequentially closed subsets would be closed. But the subset A of Example 4.7.1 is a counterexample.

Exercise 4.7.17 Show that the Sorgenfrey line $\mathbb{R}_\ell$ (Exercise 4.1.34) is first-countable.

4.7.1 Continuity and limits

Continuous maps preserve all limits, and are the only maps that preserve all limits.

Proposition 4.7.18 (Continuous = preserves limits) *Let X, Y be two topological spaces, and f be a function from X to Y. Then f is continuous at $x \in X$ if and only if, for every net $(x_i)_{i \in I, \sqsubseteq}$ that converges to x, the net $(f(x_i))_{i \in I, \sqsubseteq}$ converges to $f(x)$.*

In particular, f is continuous if and only if it preserves all limits, *i.e., the image of any net that converges to any point x converges to $f(x)$.*

Proof Assume f is continuous at x. For every open subset V of Y such that $f(x) \in V$, there is an open neighborhood U of x such that $f[U] \subseteq V$. If $(x_i)_{i \in I, \sqsubseteq}$ converges to x, then x_i is in U for i large enough. So $f(x_i)$ is in $f[U]$, hence in V, for i large enough.

Conversely, assume that f preserves limits of nets that converge to x. For every closed subset F of Y, consider any net $(x_i)_{i \in I, \sqsubseteq}$ composed of elements of $f^{-1}(F)$, and which converges to some $x \in X$. Then $(f(x_i))_{i \in I, \sqsubseteq}$ is a net composed of elements of F, which converges to $f(x)$ by assumption. Since F is closed, $f(x)$ is in F by Corollary 4.7.10, so $x \in f^{-1}(F)$. So $f^{-1}(F)$ is closed, again by Corollary 4.7.10. By the result of Exercise 4.3.2, f is continuous. □

Exercise 4.7.19 (Convergence in coproducts) Show that a net $(x_i)_{i \in I, \sqsubseteq}$ converges in a coproduct $\coprod_{k \in K} X_k$ if and only if there is a $k \in K$ such that x_i is of the form (k, y_i) with $y_i \in X_k$ for i large enough, say $i_0 \sqsubseteq i$, and such that $(y_i)_{i \in \uparrow i_0, \sqsubseteq}$ converges in X_k. Then the limits of $(x_i)_{i \in I, \sqsubseteq}$ are the points (k, y) with y a limit of $(y_i)_{i \in \uparrow i_0, \sqsubseteq}$. That is, equating each summand $\{k\} \times X_k$ with X_k, a net converges in $\coprod_{k \in K} X_k$ iff it is eventually in some summand X_k, and converges there.

Exercise 4.7.20 (Convergence in products) Show that a net $(\vec{x}_i)_{i \in I, \sqsubseteq}$ converges in a product $\prod_{k \in K} X_k$ if and only if $(x_{ik})_{i \in I, \sqsubseteq}$ converges in X_k for every $k \in K$, where $x_{ik} = \pi_k(\vec{x}_i)$. Then the limits of $(\vec{x}_i)_{i \in I, \sqsubseteq}$ are exactly those tuples $\vec{x}$ whose kth component $x_k = \pi_k(x)$ is a limit of $(x_{ik})_{i \in I, \sqsubseteq}$ for each $k \in K$. In other words, tuples converge componentwise.

4.7.2 Compactness, limits, and cluster points

Limits can also be used to characterize compactness.

Definition 4.7.21 (Cluster point) A *cluster point* of a net $(x_i)_{i \in I, \sqsubseteq}$ in a topological space X is a point $x \in X$ such that, for every open subset U of X containing x, for every $i \in I$, there is an $i' \in I$ with $i \sqsubseteq i'$ and such that $x_{i'}$ is in U.

When $I = \mathbb{N}$, we retrieve our previous definition (for sequences; see Section 3.3) that x_i should be in U for infinitely many values of $i \in \mathbb{N}$.

Proposition 4.7.22 (Compact = every net has a cluster point) *Let X be a topological space. The subset K of X is compact if and only if every net $(x_i)_{i\in I,\sqsubseteq}$ of elements of K has a cluster point in K.*

Proof Assume that K is compact, but that there is a net $(x_i)_{i\in I,\sqsubseteq}$ of elements of K with no cluster point in K. By definition, for every point $x \in K$, x is not a cluster point of the net, so there is an open neighborhood U_x of x and an index $i_x \in I$ such that x_j is not in U_x, for any j such that $i_x \sqsubseteq j$. (Use the Axiom of Choice to pick U_x, i_x for each $x \in K$.) Clearly, $(U_x)_{x\in K}$ is an open cover of K. By compactness, it has a finite subcover $U_{x_1}, \ldots, U_{x_n}$. Since I is directed, find $i \in I$ so that $i_{x_1}, \ldots, i_{x_n} \sqsubseteq i$. By definition x_i is not in U_x for any $x \in K$, so x_i is not in K, a contradiction.

Conversely, assume that every net in K has a cluster point in K, and assume that K is not compact. By Proposition 4.4.7 there is a directed family $(U_i)_{i\in I}$ of opens whose union contains K, but such that no U_i contains K. Using the Axiom of Choice, pick an element x_i of K outside U_i for each $i \in I$. Quasi-order I by $i \sqsubseteq i'$ iff $U_i \subseteq U_{i'}$, so that I is directed. So $(x_i)_{i\in I,\sqsubseteq}$ has a cluster point x in K. Since $K \subseteq \bigcup_{i\in I} U_i$, x is in some U_i. Since it is a cluster point, there is a $j \in I$ with $i \sqsubseteq j$ such that $x_j \in U_i$. However, $i \sqsubseteq j$ means that $U_i \subseteq U_j$, whence $x_j \in U_j$, a contradiction. □

Exercise 4.7.23 A *subsequence* of a sequence $(x_i)_{i\in\mathbb{N}}$ is a sequence of the form $x_{i_0}, x_{i_1}, \ldots, x_{i_j}, \ldots$, where $i_0 < i_1 < \cdots < i_j < \cdots$, i.e., we sample infinitely many elements from the original sequence. It is convergent iff it converges to some point. Generalize Lemma 3.3.1, and show that, in a first-countable space, x is a cluster point of $(x_i)_{i\in\mathbb{N}}$ if and only if the latter has a subsequence that converges to x.

It only seems natural that, in the absence of first-countability, the result of Exercise 4.7.23 would have an equivalent formulation in terms of nets, and *subnets* instead of subsequences. However, a subnet of $(x_i)_{i\in I,\sqsubseteq}$ will not be a net of the form $(x_j)_{j\in J,\sqsubseteq}$ for some directed, cofinal subset J of I, as one would have guessed. We need J to be, possibly, an entirely different index set, with a mapping to I, which we call a sampler. This is important already in Proposition 4.7.26 below.

Definition 4.7.24 (Subnet) Let $(x_i)_{i\in I,\sqsubseteq}$ be a net on a space X. A *sampler* of the net is a monotonic and cofinal map α from a directed quasi-ordered set J, say with quasi-ordering $\preceq$, to I. We say that α is *cofinal* if and only if, for every $i \in I$, there is a $j \in J$ such that $i \sqsubseteq \alpha(j)$.

The net $(x_{\alpha(j)})_{j \in J, \preceq}$ is then called the *subnet* of $(x_i)_{i \in I, \sqsubseteq}$ obtained from the sampler α.

Exercise 4.7.25 Show that any subnet of a net that converges to some point x also converges to x.

Proposition 4.7.26 *Let X be a topological space, x be a point of X, and $(x_i)_{i \in I, \sqsubseteq}$ be a net in X. Then x is a cluster point of the net if and only if there is a subnet that converges to x.*

Proof Assume some subnet $(x_{\alpha(j)})_{j \in J, \preceq}$ converges to x, with sampler $\alpha \colon J \to I$. For every open neighborhood U of x, $x_{\alpha(j)}$ is in U for j large enough, in particular for some $j \in J$. For every $i \in I$, by cofinality there is a $j' \in J$ such that $i \sqsubseteq \alpha(j')$. By directedness, we may assume $j' = j$. Then $x_{\alpha(j)}$ is in U. So x is a cluster point of the net.

Conversely, assume that x is a cluster point of the net. Let $J = \mathbb{P}_{\text{fin}}(I \times \mathcal{N}_x)$, where $\mathcal{N}_x$ is the collection of all open neighborhoods of x, and order J by inclusion. Clearly, J is directed. (Note that J is certainly much larger than I itself, in general.) Define $\alpha(E)$ for every $E \in J$, by induction on the cardinality of E. This will have the property that if $i'' = \alpha(E)$, then, for every $(i, U) \in E$, $x_{i''}$ is in U. When E is empty, let $\alpha(E)$ be any element i of I. Otherwise, observe that the set $B(E)$ of all elements E' of J that are strictly included in E is finite. Since I is directed, there is an $i' \in I$ such that $\alpha(E') \sqsubseteq i'$ for every $E' \in B(E)$, and $j \sqsubseteq i'$ for every element (j, U) of E. Consider the open neighborhood $\bigcap_{(i,U) \in E} U$ of x. Since x is a cluster point of $(x_i)_{i \in I, \sqsubseteq}$, there is an index $i'' \in I$ with $i' \sqsubseteq i''$ such that $x_{i''}$ is in $\bigcap_{(i,U) \in E} U$, i.e., such that, for every $(i, U) \in E$, $x_{i''}$ is in U. Then let $\alpha(E) = i''$. (Doing so for every E involves the Axiom of Choice.)

Since we chose i' such that $\alpha(E') \sqsubseteq i'$ for every $E' \in B(E)$, α is monotonic. It is also cofinal, because for every $i \in I$, $i \sqsubseteq \alpha(\{(i, X)\})$—we picked i' such that $j \sqsubseteq i'$ for every element (j, U) of $E = \{(i, X)\}$, i.e., $i \sqsubseteq i'$, and we picked $i'' = \alpha(\{(i, X)\})$ so that $i' \sqsubseteq i''$.

We claim that the subnet obtained from α converges to x. Fix $i \in I$. For every open neighborhood U of x, we just check that $x_{\alpha(E)}$ is in U for every $E \in J$ that is above $\alpha(\{(i, U)\})$. Indeed, since $\{(i, U)\}$ is below E, (i, U) is in E. Now α was built so that $i'' = \alpha(E)$ was such that $x_{i''}$ is in U, and we conclude. □

Corollary 4.7.27 (Compact = every net has a convergent subnet) *Let K be a subset of some topological space X. K is compact if and only if* *every net $(x_i)_{i \in I, \sqsubseteq}$ of elements of K has a subnet that converges to a point in K.*

Exercise 4.7.28 Let X be a topological space. Show that, for every closed subset F of X, every cluster point of every net $(x_i)_{i \in I, \sqsubseteq}$ of elements of F is in F.

Exercise 4.7.29 (Convergence and subbases) Let X be a topological space, and $\mathcal{B}$ be a subbase of open sets. Show that the net $(x_i)_{i \in I}$ converges to x if and only if, for every $U \in \mathcal{B}$, if $x \in U$ then $x_i \in U$ for i large enough.

Exercise 4.7.30 (Pointwise convergence) Let X_i be a family of topological spaces, $i \in I$. Show that the net $(\vec{x}^j)_{j \in J, \sqsubseteq}$ converges to $\vec{x}$ if and only if, for every $i \in I$, $(x_i^j)_{j \in J, \sqsubseteq}$ converges to x_i. (Recall that $\vec{x} = (x_i)_{i \in I}$. You may use Exercise 4.7.29.) The latter notion of convergence is called *pointwise convergence*: so pointwise convergence is the notion of convergence arising from the product topology.

Exercise 4.7.31 Show that the same thing happens when J is taken to be possibly different for each $i \in I$. That is, let X_i be a family of topological spaces, $i \in I$. For each $i \in I$, assume a net $(x_{ij})_{j \in J_i, \sqsubseteq_i}$ in X_i. For each $\vec{j} \in \prod_{i \in I} J_i$, let $\vec{x}^{\vec{j}}$ be the tuple whose ith component $x_i^{\vec{j}}$ is x_{ij_i}, $i \in I$. Show that $(\vec{x}_{\vec{j}})_{\vec{j} \in \prod_{i \in I} J_i, \prod_{i \in I} \sqsubseteq_i}$ converges to $\vec{x}$ if and only if $(x_{ij})_{j \in J_i, \sqsubseteq_i}$ converges to x_i in X_i for every $i \in I$.

4.7.3 Kelley's Theorem

Definition 4.7.32 (Ultranet) An *ultranet* on a set X is a net $(x_i)_{i \in I, \sqsubseteq}$ such that, for every subset A of X, either the net is eventually in A, or it is eventually in the complement of A in X.

This is a strange notion, admittedly. The only ultranets that one can reasonably imagine are the eventually constant nets, i.e., those nets $(x_i)_{i \in I, \sqsubseteq}$ such that, for some $i_0 \in I$, $x_i = x_{i_0}$ for all $i \in I$ with $i_0 \sqsubseteq i$. To prove that there are others, we must appeal to the Axiom of Choice, and Kelley's Theorem then shows that there are many. The proof relies on the following important concept, which we shall see again in more generality in Chapter 8.

Definition 4.7.33 (Filter of subsets) A *filter* of subsets of a set X is any filtered, upward closed family of subsets, or, equivalently, any non-empty family $\mathcal{F}$ of subsets of X such that, whenever $A \subseteq B$ and $A \in \mathcal{F}$, then $B \in \mathcal{F}$, and such that $A, B \in \mathcal{F}$ entails $A \cap B \in \mathcal{F}$.

A filter $\mathcal{F}$ is *non-trivial* if $\mathcal{F} \neq \mathbb{P}(X)$, or, equivalently, if $\mathcal{F}$ does not contain the empty set.

Filters provide an equivalent foundation for the notion of convergence: a filter $\mathcal{F}$ of opens converges to some point x iff every open neighborhood of x belongs to $\mathcal{F}$. We will not pursue this here, although we shall need the analogous notion of an ultranet in terms of filters, ultrafilters:

Definition 4.7.34 (Ultrafilter of subsets) An *ultrafilter* of subsets of a set X is a non-trivial filter of subsets $\mathcal{F}$ such that, for every subset A of X, either A or its complement is in $\mathcal{F}$.

For a natural generalization, see Exercise 9.5.22.

We can now state Kelley's Theorem (Kelley, 1950, theorem 20).

Theorem 4.7.35 (Kelley's Theorem) *Every net has a subnet that is an ultranet.*

Proof Given a net $(x_i)_{i \in I, \sqsubseteq}$, define the section s_i of $i \in I$ by $s_i = \{x_{i'} \mid i \sqsubseteq i'\}$. Then any maximal non-trivial filter (with respect to inclusion of filters) $\mathcal{F}_0$ containing all sections is called a *section filter*.

Section filters do exist, because of Zorn's Lemma (Theorem 2.4.2). We first need to build a filter $\mathcal{F}$ containing all sections. We take the smallest one. This is just the collection of all supersets of finite intersections $\bigcap_{i \in A} s_i$ of sections (A finite included in I). The only tricky point is to show that $\emptyset$ is not in $\mathcal{F}$. Otherwise, there would be a finite subset of I such that $\bigcap_{i \in A} s_i$ is empty. However, this is impossible: since I is directed, there is an $i' \in I$ such that $i \sqsubseteq i'$ for every $i \in A$, and then $x_{i'} \in \bigcap_{i \in A} s_i$, by the definition of sections.

We now need to show that the set of all filters containing all sections is inductive. Given any chain $\mathcal{C}$ of filters containing all sections, let $\mathcal{F}_1$ be the union of all the filters $\mathcal{F}$ in $\mathcal{C}$. This does not contain $\emptyset$; otherwise some element $\mathcal{F}$ would already contain it. Also, $\mathcal{F}_1$ is a filter containing all sections. The only non-trivial part is to show that the intersection of any two elements A, B of $\mathcal{F}_1$ is again in $\mathcal{F}_1$. A is in some $\mathcal{F}$ in $\mathcal{C}$, and B is in some $\mathcal{F}'$ in $\mathcal{C}$. Without loss of generality, $\mathcal{F}' \subseteq \mathcal{F}$ (or the converse inclusion), so A and B are both in $\mathcal{F}$, hence also $A \cap B$; so $A \cap B$ is in $\mathcal{F}_1$. So Zorn's Lemma applies, showing the existence of a section filter $\mathcal{F}_0$.

We now claim that any section filter $\mathcal{F}_0$ is an ultrafilter. Assume that $\mathcal{F}_0$ was not an ultrafilter. So there is a subset A that is not in $\mathcal{F}_0$, and whose complement is not in $\mathcal{F}_0$ either. Since $\mathcal{F}_0$ is maximal, the smallest filter containing all the elements of $\mathcal{F}_0$ and A would be trivial. Since the elements of this smallest filter are the supersets of finite intersections of elements among those of $\mathcal{F}_0$ and A, this would imply that $E \cap A = \emptyset$ for some $E \in \mathcal{F}_0$. Similarly, there would be an element $E' \in \mathcal{F}_0$ that does not meet the complement of A. In particular, $E \cap E'$ is also an element of $\mathcal{F}_0$, and does not meet either A or its complement. So $E \cap E'$ must be empty, a contradiction.

Given that $\mathcal{F}_0$ is an ultrafilter, we extract a subnet as follows. Let J be the set of all pairs (i, E) where $i \in I$, $E \in \mathcal{F}_0$, and $x_i \in E$, ordered by $(i, E) \preceq (i', E')$ iff $i \sqsubseteq i'$ and $E \supseteq E'$. Let $\alpha \colon J \to I$ be defined by $\alpha(i, E) = i$. We claim that α is a sampler, and that the subnet obtained from α is an ultranet.

We check that J is directed: given (i, E) and (i', E') in J, pick $i_1 \in I$ such that $i, i' \sqsubseteq i_1$. This is possible since I is directed. Let $E'' = E \cap E'$. Then $E'' \in \mathcal{F}_0$, since $\mathcal{F}_0$ is a filter. Since $\mathcal{F}_0$ contains all sections, it contains s_{i_1}, and therefore also $s_{i_1} \cap E''$. Now $\mathcal{F}_0$ does not contain the empty set, so there must be an element in $s_{i_1} \cap E''$. This element must be of the form $x_{i''}$ with $i_1 \sqsubseteq i''$, by definition of s_{i_1}, and in particular $i, i' \sqsubseteq i''$; also, $x_{i''} \in E''$, and $E, E' \supseteq E''$. So $(i, E), (i', E') \preceq (i'', E'')$.

The map α is obviously monotonic, and cofinal (since it is surjective). So α is a sampler.

We show that the subnet obtained from α is an ultranet. Let A be any subset of X. If A is in $\mathcal{F}_0$, then fix $i \in I$ such that $x_i \in A$. This must exist because, given any fixed $i_0 \in I$, $s_{i_0} \cap A$ must be in $\mathcal{F}_0$, so cannot be empty: any element in $s_{i_0} \cap A$ must be of the form x_i for some i, $i_0 \sqsubseteq i$. We now observe that, for every $(i', E') \in J$ such that $(i, A) \preceq (i', E')$, $x_{\alpha(i',E')} = x_{i'}$ is such that $x_{i'} \in E'$. Since $E' \subseteq A$, $x_{\alpha(i',E')}$ is in A. That is, the subnet $(x_{\alpha(i,E)})_{(i,E) \in J}$ is eventually in A.

If A is not in $\mathcal{F}_0$, since $\mathcal{F}_0$ is an ultrafilter, the complement of A is in $\mathcal{F}_0$, and by the same arguments, the subnet $(x_{\alpha(i,E)})_{(i,E) \in J}$ is eventually in the complement of A. □

Exercise 4.7.36 Let X be a topological space. Show that the subset K of X is compact if and only if every ultranet $(x_i)_{i \in I, \sqsubseteq}$ of elements of K converges to a point in K.

Exercise 4.7.37 From Exercise 4.7.36, deduce another proof of Tychonoff's Theorem, based on ultranets.

4.8 Local compactness

A space X is locally compact if and only if every point has sufficiently many small neighborhoods, where we again equate small with compact. This is an important property, which we shall encounter often.

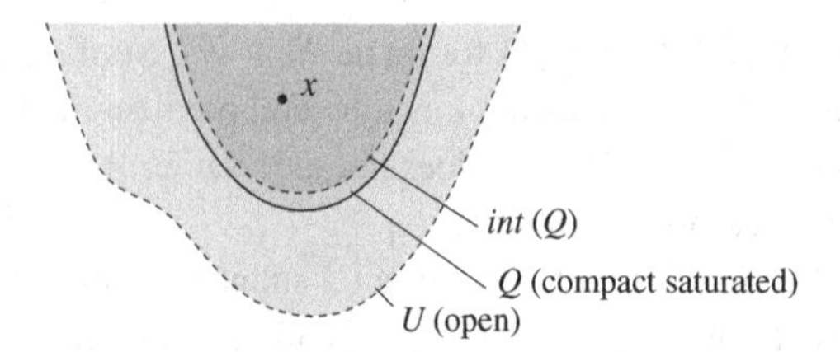

Figure 4.6 Local compactness.

Definition 4.8.1 (Locally compact) A topological space X is *locally compact* if and only if, for every $x \in X$ and every open neighborhood U of x, there is a compact saturated subset Q of X such that $x \in int(Q)$ and $Q \subseteq U$.

See Figure 4.6. So, for every open neighborhood U of x, there is a smaller compact saturated neighborhood of x. Using Lemma 4.1.10, a locally compact topological space therefore has a *base* consisting of the interiors of compact saturated (small) subsets.

Trick 4.8.2 To show that a space X is locally compact, it is enough to show that, for every $x \in X$, for every open neighborhood U of x in X, there is a compact set K and an open subset V such that $x \in V \subseteq K \subseteq U$.

Indeed, then $\uparrow K$ is compact saturated and still included in U (Proposition 4.4.14), and $V \subseteq int(\uparrow K)$.

Example 4.8.3 $\mathbb{R}$, $\mathbb{R}_\sigma$, $\mathbb{N}_\sigma$ are locally compact. Only the latter one is compact. (Exercise: prove this.)

Exercise 4.8.4 ($\mathbb{Q}$ is not locally compact) Consider the space $\mathbb{Q}$ of all rational numbers, with the topology having a base given by the intervals $(a, b) \cap \mathbb{Q}$, where $a < b$ in $\mathbb{R}$. (This is the subspace topology from $\mathbb{R}$; see Section 4.9.) Show that every compact subset Q of $\mathbb{Q}$ has empty interior. In other words, consider any compact subset Q of $\mathbb{Q}$, and show that it cannot contain any interval $[a, b] \cap \mathbb{Q}$ with $a < b$.

Conclude that $\mathbb{Q}$ is not locally compact.

Exercise 4.8.5 ($\mathbb{R}_\ell$ is not locally compact) Show that the Sorgenfrey line $\mathbb{R}_\ell$ (Exercise 4.1.34) is not locally compact either: as for $\mathbb{Q}$, every compact subset of $\mathbb{R}_\ell$ has empty interior.

Exercise 4.8.6 Show that the lifting $X_\perp$ (Exercise 4.4.25) of a space X is locally compact if and only if X is locally compact.

The standard definition of local compactness, used in the context of T_2 spaces, is that every point should have a compact neighborhood. Indeed:

Proposition 4.8.7 *A T_2 space X is locally compact if and only if every point $x \in X$ has a compact neighborhood.*

Proof The only if direction is trivial. Conversely, assume that every point x has a compact neighborhood K. Consider any open neighborhood U of x, and let $F = K \setminus U$. F is compact by Corollary 4.4.10. Since X is T_2, find disjoint opens U_y containing x and V_y containing y, for each $y \in F$ (using the Axiom of Choice to collect them as y ranges over F). The family $(V_y)_{y \in F}$ is an open cover of F, so there is a finite subset A of F such that $F \subseteq \bigcup_{y \in A} V_y$. Consider $U' = int(K) \cap \bigcap_{y \in A} U_y$. (Take this to denote $int(K)$ itself if A is empty.) Since A is finite, U' is open, and clearly contains x. Moreover, U' does not intersect $\bigcup_{y \in A} V_y$, so U' is included in $Q = K \setminus \bigcup_{y \in A} V_y$. Observe that Q is compact by Corollary 4.4.10. Q is saturated, because saturation is trivial in T_2 spaces. Finally, since $F \subseteq \bigcup_{y \in A} V_y$, Q is included in U. So $x \in U' \subseteq Q \subseteq U$, and we conclude. □

That every point $x \in X$ has a compact neighborhood is far from sufficient to prove that X is locally compact, when X is not T_2. For example, in $\mathbb{Q}_\perp$, every point has a compact neighborhood, namely the whole space itself, which is compact (Exercise 4.4.25). However, $\mathbb{Q}_\perp$ is *not* locally compact; otherwise $\mathbb{Q}$ would be, by Exercise 4.8.6, contradicting Exercise 4.8.4.

Proposition 4.8.8 *Every compact T_2 space is locally compact.*

Proof Every point has a compact neighborhood, namely the whole space itself. Then apply Proposition 4.8.7. □

That compact spaces are locally compact is certainly *wrong* in non-T_2 spaces. For example, $\mathbb{Q}_\perp$ is compact (Exercise 4.4.25), but not locally compact, as we have seen.

Exercise 4.8.9 (Alexandroff one-point compactification) By Exercise 4.4.25, one can embed any space X into a compact space $X_\perp$. But $X_\perp$ is never T_2. If one wishes to embed X into a compact T_2 space, X first needs to be T_2. We shall give a complete characterization of those spaces that embed in a compact T_2 space in Exercise 6.7.22. For the time being, let X be a locally compact T_2 space. Let αX be the space obtained from X by adding a new point ∞, called the *point at infinity*. The open subsets of αX are those of X, plus the sets of the form $\{\infty\} \cup (X \setminus K)$, where K is compact in X. Show that this is indeed a topology, and that αX is compact T_2: this is the Alexandroff *one-point compactification*.

Proposition 4.8.10 (Products of locally compact spaces) *Let $(X_i)_{i \in I}$ be a family of topological spaces. If every X_i is locally compact, and all but finitely many are compact, then $\prod_{i \in I} X_i$ is locally compact.*

Proof Assume that X_i is compact except when $i \in J$, where J is finite. Let $\vec{x} = (x_i)_{i \in I} \in \prod_{i \in I} X_i$, and U be an open neighborhood of $\vec{x}$. Then $\vec{x}$ belongs to some product $\prod_{i \in I} U_i \subseteq U$, where U_i is open for each $i \in I$, and all but finitely many are such that $U_i = X_i$. Let J' be any finite subset of I containing J and such that $U_i = X_i$ whenever $i \notin J'$. For each $i \in J'$, since X_i is locally compact, let Q_i be a compact saturated neighborhood of x_i included in U_i. For each $i \in I \setminus J'$, i is not in J, so X_i is compact, and also open: let $Q_i = X_i$. Then $\vec{x}$ is in the open $\prod_{i \in I} int(Q_i)$ (all but finitely many $int(Q_i)$ equal X_i), which is contained in the compact set $\prod_{i \in I} Q_i$ (Tychonoff's Theorem 4.5.12), which is contained in $\prod_{i \in I} U_i \subseteq U$. □

Corollary 4.8.11 *The product of two locally compact spaces is locally compact.*

It is in general false that the product of infinitely many locally compact spaces is locally compact:

Example 4.8.12 (Baire space $\mathcal{B}$ is not locally compact) The *Baire space* $\mathcal{B}$ is $\mathbb{N}^{\mathbb{N}}$, the product of countably many copies of $\mathbb{N}$ with the discrete topology. One can show that $\mathcal{B}$ is not locally compact, although $\mathbb{N}$ is. (Exercise: prove this.) As for $\mathbb{Q}$ (Exercise 4.8.4), the idea is to show that every compact subset of $\mathcal{B}$ has empty interior, and more precisely, that every compact subset of $\mathcal{B}$ is included in a product of finite subsets of $\mathbb{N}$.

Exercise 4.8.13 The continuous maps from Baire space $\mathcal{B}$ to $\mathbb{N}$ (the latter with the discrete topology) form an interesting class. Show that a map $f : \mathcal{B} \to \mathbb{N}$ is continuous iff, for every $\vec{x} \in \mathcal{B}$, $f(\vec{x})$ only depends on a finite prefix of $\vec{x}$; i.e., for every $\vec{x}$, there is an integer $k \in \mathbb{N}$ such that $f(\vec{y}) = f(\vec{x})$ for every $\vec{y}$ with the same first k components.

Local compactness, as pictured in Figure 4.6, allows one to interpolate between a point x and an open neighborhood U of x, by finding a compact saturated subset Q smaller than U, and another even smaller open subset $V = int(Q)$, still containing x. This kind of interpolation generalizes to the case where x is replaced by an arbitrary compact saturated subset included in U; see Figure 4.7.

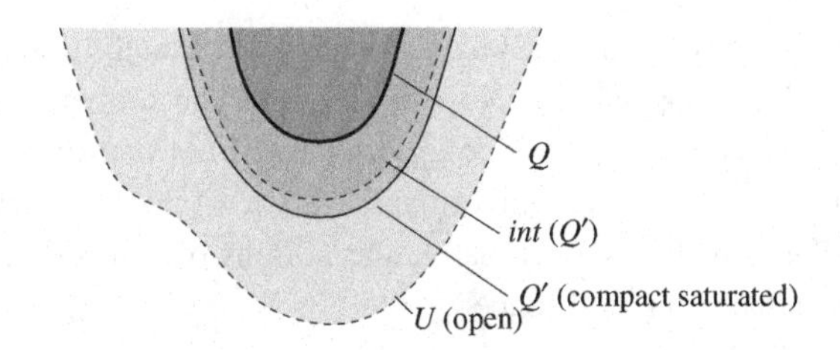

Figure 4.7 Interpolation in locally compact spaces.

Proposition 4.8.14 (Interpolation in locally compact spaces) *Let X be a locally compact space. For every compact subset Q of X, for every open subset U containing Q, there is an* interpolating *compact saturated subset Q', i.e., such that $Q \subseteq int(Q')$ and $Q' \subseteq U$.*

We also say that Q' is a compact saturated neighborhood *of Q when $Q \subseteq int(Q')$: so every open neighborhood of any compact saturated has a smaller compact saturated neighborhood.*

Proof For every $x \in Q$, pick a compact saturated neighborhood Q_x of x contained in U by local compactness (and using the Axiom of Choice). The family $(int(Q_x))_{x \in Q}$ is an open cover of Q. Since Q is compact, there is a finite subset A of Q such that $Q \subseteq \bigcup_{x \in A} int(Q_x)$. Let $Q' = \bigcup_{x \in A} Q_x$. By Proposition 4.4.12, Q' is compact saturated. Also, $Q \subseteq \bigcup_{x \in A} int(Q_x) \subseteq int(Q')$, since $\bigcup_{x \in A} int(Q_x)$ is an open subset of Q', and $int(Q)'$ is the largest one. Finally, $Q' \subseteq U$ since $Q_x \subseteq U$ for every $x \in Q$. □

4.9 Subspaces

A subspace Y of a topological space X is just a subset of X. However, one needs to impose a topology on Y to make it a sub*space*.

Definition 4.9.1 (Subspace topology) Let X be a topological space, and Y be a subset of X. The *subspace* topology on Y has as opens the subsets of Y of the form $U \cap Y$, U open in X.

One then says that Y, with the subspace topology, is a *subspace* of X.

There is a slightly more general definition.

Definition 4.9.2 (Induced topology) Given any map $f : Z \to X$, where Z is a set and X is a topological space, the topology on Z *induced* from that of X by f is the coarsest topology on Z that makes f continuous.

The induced topology is simple: take as opens in Z exactly those subsets of the form $f^{-1}(U)$, U open in X. Now observe that the subspace topology on a subset Y of X is exactly the topology induced from that of X through the inclusion map $\subseteq : Y \to X$, which maps every $x \in Y$ to x itself, as an element of X. We shall freely interchange the notions of subspace topology and of topology induced on a subspace.

Example 4.9.3 Exercise 4.2.28 shows that the metric topology on $\mathbb{R}$ is exactly the topology induced from the Scott topology of $\mathbf{I}\mathbb{R}$ by $\eta_{\mathbf{I}\mathbb{R}}$.

Proposition 4.9.4 *Let Y be a subspace of the topological space X. The closed subsets of Y are exactly those of the form $F \cap Y$, F closed in X.*

Proof The closed subsets F' of Y are the complements $Y \smallsetminus (U \cap Y)$ in Y of the subsets of the form $U \cap Y$, U open in Y. Letting $F = X \smallsetminus U$, one checks that $F' = Y \smallsetminus F$. □

A more interesting fact is that subspace topologies preserve the specialization quasi-ordering:

Proposition 4.9.5 *Let Y be a subspace of the topological space X. The specialization quasi-ordering $\preceq$ of Y coincides with the restriction of the specialization quasi-ordering $\leqslant$ of X to Y. That is, for every $y, y' \in Y$, $y \preceq y'$ iff $y \leqslant y'$.*

This is an easy consequence of the following more general result.

Proposition 4.9.6 *Let Z be a set, and $f : Z \to X$ be a map, where X is a topological space, with specialization quasi-ordering $\leqslant$. The specialization quasi-ordering $\preceq$ on Z (with the induced topology) is given by $z \preceq z'$ iff $f(z) \leqslant f(z')$.*

Proof We have $z \preceq z'$ iff, for every open $f^{-1}(U)$ of Z (U open in X), $z \in f^{-1}(U)$ implies $z' \in f^{-1}(U)$. Equivalently, for every open subset U of X, $f(z) \in U$ implies $f(z') \in U$, that is, $f(z) \leqslant f(z')$. □

Exercise 4.9.7 (Convergence in subspaces) Let X be a topological space, and Y be a subspace of X. Show that, for every net $(y_i)_{i \in I, \sqsubseteq}$ in Y, for every $y \in Y$, the following are equivalent: (1) $(y_i)_{i \in I, \sqsubseteq}$ converges to y in X; (2) $(y_i)_{i \in I, \sqsubseteq}$ converges to y in Y.

Exercise 4.9.8 Show that subspace topologies of Alexandroff-discrete spaces are again Alexandroff-discrete. In fact, if Y is a subset of the poset X, then the upward closed subsets of Y are exactly the intersections of upward closed subsets of X with Y.

Show that a similar statement fails in general for upper topologies and Scott topologies. More precisely, if Y is a subset of the poset X, then the topology of Y_u is coarser than the one induced from X_u, but not equal in general. Give examples of situations where the subspace topology on Y is strictly finer, resp. strictly coarser than, resp. incomparable with the Scott topology. (Hint: one can build elementary examples for most cases; for the next-to-last one, one may obtain some inspiration from Exercise 4.2.28.)

Exercise 4.9.9 There is still a situation where the topology induced by a Scott topology is Scott. Let X be a poset, with ordering $\leqslant$, and F be a closed subset of X_σ. Show that the subspace topology on F is the Scott topology of (the restriction to F of) $\leqslant$.

Exercise 4.9.10 Show that the (Scott) topologies of $\mathbb{R}^+_\sigma$ and of $[0, 1]_\sigma$ *are* induced from the Scott topology of $\mathbb{R}_\sigma$.

Whenever Y is a subset of a topological space X, consider the statement "Y is compact." There is risk of ambiguity here. This may mean that Y is a compact *subset* of X, or that, with the subspace topology, it is a compact *space*, i.e., a compact subset *of* Y. We shall say that Y is a compact *subspace* of X in the latter case. Exercise 4.9.11 shows that the two notions agree.

Exercise 4.9.11 (Compact subset = compact subspace) Let Y be a subset of the topological space X. Show that Y is a compact subset of X if and only if Y is a compact subspace of X.

Exercise 4.9.12 Let Y be a closed subset of the topological space X, and Q be a compact subset of Y with the subspace topology. Show that Q is also a compact subset of X.

4.10 Homeomorphisms, embeddings, quotients, retracts

There is no need to make fine distinctions between, e.g., the set $\mathbb{Q}$ of all rational numbers and the set Q of all pairs $(p, q) \in \mathbb{Z} \times (\mathbb{N} \smallsetminus \{0\})$ such that p and q have no common multiple. Indeed, any rational number can be written as p/q where $(p, q) \in Q$, and conversely, any pair $(p, q) \in Q$ gives rise to a rational number p/q. Concretely, there are two maps $f : \mathbb{Q} \to Q$ and $g : Q \to \mathbb{Q}$ that are inverses of each other, namely $f \circ g = \mathrm{id}_Q$ and $g \circ f = \mathrm{id}_{\mathbb{Q}}$. (Another example is given by the space $\mathbb{R}$ and the set of all elements of the form $[x, x]$ in $\mathbb{IR}$; see Exercise 4.2.28.)

In a more general setting, given two sets X, Y with some structure on it (of topological spaces, of posets, etc.), we define a *morphism* $f : X \to Y$ as a map from X to Y that preserves the structure. We shall make this more precise in Section 4.12. In the case of topological spaces, the structure that we want to preserve is that of convergence, so the morphisms will be the continuous maps (see Proposition 4.7.18). In the case of posets, the structure that we want to preserve is the ordering, so the morphisms will be the monotonic maps.

An *isomorphism* $X \cong Y$ between sets with structure X, Y is then a pair of morphisms $f: X \to Y$ and $g: X \to Y$ such that $f \circ g = \mathrm{id}_X$ and $g \circ f = \mathrm{id}_Y$. In the case of topological spaces, this is called a *homeomorphism*.

Definition 4.10.1 (Homeomorphism) A *homeomorphism* between two topological spaces X, Y is a pair of continuous maps $f: X \to Y$ and $g: Y \to X$ such that $f \circ g = \mathrm{id}_Y$ and $g \circ f = \mathrm{id}_X$.

If there is a homeomorphism between X and Y, then we say that X and Y are *homeomorphic*.

Similarly, an isomorphism between posets is called an *order isomorphism*. This is a pair of monotonic maps $f: X \to Y$ and $g: Y \to X$ such that $f \circ g = \mathrm{id}_Y$ and $g \circ f = \mathrm{id}_X$.

It is usual to define an isomorphism by only giving one of the morphisms f or g, leaving the other one implicit. We shall then sometimes say that f itself is the isomorphism, homeomorphism, or order isomorphism.

Example 4.10.2 It is not enough to check that f is bijective and monotonic to ensure that f is an order isomorphism. For example, take $X = \{0, 1\}$ with equality as ordering and $Y = \mathbb{S}$; that is, $\{0, 1\}$ with $0 \leqslant 1$, and $f = \mathrm{id}_{\{0,1\}}$. The inverse of f is not monotonic, since $0 \nleqslant 1$ in X.

Example 4.10.3 It is not enough to check that f is bijective and continuous to ensure that f is a homeomorphism. Indeed, use Example 4.10.2, equipping X and Y with their Alexandroff topologies: the inverse of f cannot be continuous, since it is not even monotonic (Proposition 4.3.9).

Proposition 4.10.4 (Homeomorphism = bijective + continuous + open) *Let X, Y be two topological spaces. Then $f: X \to Y$ is a homeomorphism (i.e., is one component of a homeomorphism in the sense of Definition 4.10.1) if and only if f is bijective, continuous, and* open.

A map $f: X \to Y$ is open *iff $f[U]$ is open for every open subset U of X.*

Proof Let $f: X \to Y$, $g: Y \to X$ be a homeomorphism. Since f and g are inverse to each other, they are both bijections. Also, f is continuous, and for every open subset U of X, $f[U] = g^{-1}(U)$ is open since g is continuous.

Conversely, let f be bijective, continuous, and open. Define $g(y)$ as the unique element $x \in X$ such that $f(x) = y$. Then g is continuous since $f[U] = g^{-1}(U)$ and f is open. □

Exercise 4.10.5 Let X_i, $i \in I$, be topological spaces, and $X = \prod_{i \in I} X_i$. Show that the projections $\pi_i : X \to X_i$ are examples of open, continuous maps that are not homeomorphisms in general.

It is customary to reason "up to isomorphism": if X and Y are isomorphic, then X and Y are the same, for all practical purposes. However, one has to take some care (see Exercise 4.10.6): two spaces may be the same when looked at from an order-theoretic angle, but differ when considered as topological spaces.

Exercise 4.10.6 Let X and Y be two topological spaces. Show that if X and Y are homeomorphic, then they are order isomorphic (considered as posets with their specialization quasi-orderings).

Let X and Y be two posets. Show that, if X and Y are order isomorphic, then X_a and Y_a are homeomorphic, X_u and Y_u are homeomorphic, and X_σ and Y_σ are homeomorphic.

However, give examples of two topological spaces X and Y that are order isomorphic in their specialization orderings but not homeomorphic.

Exercise 4.10.7 ($\mathbb{R} \cup \{-\infty, +\infty\} \cong [0, 1]$) Let $k : \mathbb{R} \cup \{-\infty, +\infty\} \to [-1, 1]$ be defined by $k(r) = \frac{r}{\sqrt{1+r^2}}$ if $r \neq -\infty, +\infty$, $k(-\infty) = -1$, $k(+\infty) = 1$ (see Figure 4.8). Show that k is a homeomorphism of $(\mathbb{R} \cup \{-\infty, +\infty\})_\sigma$ onto $[-1, 1]_\sigma$, and also of $\mathbb{R} \cup \{-\infty, +\infty\}$ onto $[-1, 1]$, where the latter is equipped

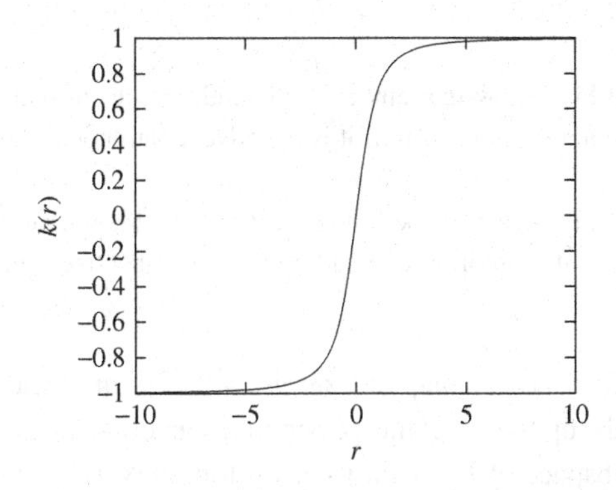

Figure 4.8 The k homeomorphism.

from the subspace topology of $\mathbb{R}$, and the basic open subsets of the former are (a, b) with $a < b$ in $\mathbb{R}$, $(a, +\infty]$ and $[-\infty, b)$, $a, b \in \mathbb{R}$ (see Exercise 4.3.11).

Deduce that $\mathbb{R} \cup \{-\infty, +\infty\}$ is homeomorphic to $[0, 1]$, and that $(\mathbb{R} \cup \{-\infty, +\infty\})_\sigma$ is homeomorphic to $[0, 1]_\sigma$, through the same homeomorphism.

Definition 4.10.8 (Embedding) Let X, Y be topological spaces, and f be a map from X to Y. Then f is an *embedding* of X *into* Y iff f is a homeomorphism of X onto $f[X]$ (with the subspace topology from Y).

Proposition 4.10.9 (Embedding = injective + continuous + almost open) *A map f of a topological space X into a topological space Y is an embedding if and only if f is injective, continuous, and* almost open, *meaning that, for every open subset U of X, there is an open subset V of Y such that $U = f^{-1}(V)$.*

Proof Consider the corestriction f' of f, defined as the map from X to $f[X]$ that sends x to $f(x)$, where $f[X]$ is equipped with the subspace topology from Y. Then f is an embedding iff f' is a homeomorphism, i.e., continuous, bijective, and open by Proposition 4.10.4. Now f' is continuous iff f is, f' is bijective iff f is injective, and f' is open iff, for every open subset U of X, there is an open subset V such that $f[U] = f[X] \cap V$. We show that the latter equality is equivalent to $U = f^{-1}(V)$, assuming f injective.

If: assume $f[U] = f[X] \cap V$. For every $x \in X$, if $x \in U$ then $f(x) \in f[U]$, so $f(x) \in V$, i.e., $x \in f^{-1}(V)$; and if $x \in f^{-1}(V)$, then $f(x)$ is both in $f[X]$ and in V, so $f(x) \in f[U]$. That is, there is an $x' \in U$ such that $f(x) = f(x')$ and $x' \in U$. Since f is injective, $x \in U$.

Only if: assume $U = f^{-1}(V)$. Then $f[U] = f[f^{-1}(V)]$, the set of elements of the form $f(x)$ that are in V, or, equivalently, $f[X] \cap V$. □

Exercise 4.10.10 Show that any continuous injective map f from a compact space X to a T_2 space Y is an embedding.

Exercise 4.10.11 Show that any surjective, almost open map is open. Deduce that a map is a homeomorphism iff it is bijective, continuous, and almost open.

Exercise 4.10.12 Let X, Y be topological spaces. Show that every embedding i of X into Y is also an order embedding for the respective specialization quasi-orderings.

Example 4.10.13 The $\eta_{\mathbf{I}\mathbb{R}}$ map of Exercise 4.2.28 is an embedding of $\mathbb{R}$ into $\mathbf{I}\mathbb{R}$. In other words, up to homeomorphism, one can consider $\mathbb{R}$, with its metric topology, as a subspace of $\mathbf{I}\mathbb{R}$, with its Scott topology. (Exercise: check this.) This was implicit in Figure 4.4.

Exercise 4.10.14 (T_0 = embeds into a power of $\mathbb{S}$) Let X be a topological space. Show that the map $\eta_{\mathbb{S}} : X \to \mathbb{S}^{\mathcal{O}(X)}$ that sends x to the family $(\chi_U(x))_{U \in \mathcal{O}(X)}$ is continuous and almost open. Show that it is injective if and only if X is T_0.

Conclude that X is T_0 if and only if there is an embedding of X into some power $\mathbb{S}^I$ of Sierpiński space $\mathbb{S}$.

Exercise 4.10.15 ($\mathcal{B} \cong [0,1] \smallsetminus \mathbb{Q} \cong [0,+\infty) \smallsetminus \mathbb{Q} \cong \mathbb{R} \smallsetminus \mathbb{Q}$) The purpose of this exercise is to show that Baire space $\mathcal{B}$ is homeomorphic to the space $\mathbb{R} \smallsetminus \mathbb{Q}$ of all irrational numbers with the subspace topology, induced from $\mathbb{R}$, as well as to $[0,1] \smallsetminus \mathbb{Q}$. We first define a homeomorphism between $(\mathbb{N} \smallsetminus \{0\})^{\mathbb{N}}$ and $[0,1] \smallsetminus \mathbb{Q}$, as follows.

Let $\lfloor z \rfloor$ be the *integer part* of real z, i.e., the largest integer below z. Given any irrational number x in $[0,1]$, define $\vec{n} = cf(x) \in (\mathbb{N} \smallsetminus \{0\})^{\mathbb{N}}$ and auxiliary irrational numbers x_k in $[0,1]$, $k \in \mathbb{N}$, by $x_0 = x$, $n_0 = \lfloor \frac{1}{x_0} \rfloor$, $x_1 = \frac{1}{x_0} - n_0$, $n_1 = \lfloor \frac{1}{x_1} \rfloor$, …, $x_{k+1} = \frac{1}{x_k} - n_k$, $n_{k+1} = \lfloor \frac{1}{x_{k+1}} \rfloor$, … You may find it useful to show that, if x were rational, then this process would be undefined, in the sense that x_k would equal 0 at some rank k.

Conversely, given $\vec{n} \in (\mathbb{N} \smallsetminus \{0\})^{\mathbb{N}}$, let $/\vec{n}/ \in \mathbb{R}^+$ be defined as follows. For every k-tuple of natural numbers $(n_0, n_1, \ldots, n_k)$, let $/n_0, n_1, \ldots, n_k/$ be defined as $\frac{1}{n_0}$ if $k = 0$, as $\frac{1}{n_0 + /n_1, \ldots, n_k/}$ otherwise; i.e., $/n_0, n_1, \ldots, n_k/ = \frac{1}{n_0 + \frac{1}{n_1 + \frac{1}{\ldots + \frac{1}{n_k}}}}$; such an expression is called a finite *continued fraction*. Show that, given $\vec{n} \in (\mathbb{N} \smallsetminus \{0\})^{\mathbb{N}}$, the sequence of intervals $[/n_0, n_1, \ldots, n_{2k}, n_{2k+1}/, /n_0, n_1, \ldots, n_{2k}/]$, $k \in \mathbb{N}$, is a chain in the dcpo $\mathbf{I}\mathbb{R}$ (see Exercise 4.2.28), and that its least upper bound is of the form $[x, x]$ for some $x \in [0,1] \smallsetminus \mathbb{Q}$: we let $/\vec{n}/$ be this x. One thinks of $\vec{n}$ as the infinite continued fraction $\frac{1}{n_0 + \frac{1}{n_1 + \frac{1}{\ldots + \frac{1}{n_k + \frac{1}{\ldots}}}}}$.

Show that the maps $cf : [0,1] \smallsetminus \mathbb{Q} \to (\mathbb{N} \smallsetminus \{0\})^{\mathbb{N}}$ and $/\_/ : (\mathbb{N} \smallsetminus \{0\})^{\mathbb{N}} \to [0,1] \smallsetminus \mathbb{Q}$ are continuous and inverse to each other. To show that cf is continuous, it is easier to show that $/\_/$ is open.

From this, deduce that $\mathcal{B}$ is homeomorphic to $[0,1] \smallsetminus \mathbb{Q}$.

Using the map $x \mapsto \frac{1}{x} - 1$, deduce that $\mathcal{B}$ is homeomorphic to $[0,+\infty) \smallsetminus \mathbb{Q}$ as well.

Finally, let Z be any bijection from $\mathbb{N}$ onto $\mathbb{Z}$ (see Section 2.2). Show that the map $ZR : x \mapsto r + Z(q)$, where $q = \lfloor x \rfloor$ and $r = x - q$, is a homeomorphism from $[0,+\infty) \smallsetminus \mathbb{Q}$ onto $\mathbb{R} \smallsetminus \mathbb{Q}$. Conclude that $\mathcal{B}$ is homeomorphic to the space $\mathbb{R} \smallsetminus \mathbb{Q}$ of irrational numbers.

Given any equivalence relation $\equiv$ on a set X, one may form the quotient $X/\equiv$ (see Section 2.3.1). When X is a topological space, a natural choice of a topology of $X/\equiv$ is one that makes the quotient map $q_\equiv : X \to X/\equiv$ continuous. Thus:

Definition 4.10.16 (Quotient topology) Let X be a topological space, and $\equiv$ be an equivalence relation on X. The *quotient topology* is the finest topology on $X/\equiv$ that makes $q_\equiv : X \to X/\equiv$ continuous.

Proposition 4.10.17 *Let X be a topological space, and $\equiv$ be an equivalence relation on X. The open subsets of $X/\equiv$ are exactly the subsets V such that $q_\equiv^{-1}(V)$ is open in X.*

Proof If V is open in $X/\equiv$, then $q_\equiv^{-1}(V)$ is open in X by definition of the quotient topology. Conversely, let V be any subset of $X/\equiv$ such that $q_\equiv^{-1}(V)$ is open in X. Consider the topology $\mathcal{O}$ on $X/\equiv$ generated by V and the quotient topology. Since $q_\equiv^{-1}$ commutes with unions and (finite) intersections, it is easy to check that $q_\equiv$ is again continuous from X to $X/\equiv$, where the latter comes with the topology $\mathcal{O}$. But the quotient topology is the finest such. So $\mathcal{O}$ is the quotient topology, hence V is open in the quotient topology. □

Another characterization of the open subsets in the quotient topology goes through so-called $\equiv$-saturated sets.

Definition 4.10.18 ($\equiv$-saturated set) Let X be a set, and $\equiv$ be an equivalence relation on X. A subset A of X is *$\equiv$-saturated* iff, for every $a \in A$, for every $x \in X$ such that $a \equiv x$, x is in A. Equivalently, A is $\equiv$-saturated iff it is a union of equivalence classes for $\equiv$.

Lemma 4.10.19 ($\equiv$-saturation) *Let X be a set, and $\equiv$ be an equivalence relation on X. For every subset A of X, there is a smallest $\equiv$-saturated subset of X containing A; call it the* $\equiv$-saturation *of A. This is the union of all equivalence classes $q_\equiv(a)$ of elements a of A, or equivalently $q_\equiv^{-1}(q_\equiv[A])$.*

Proof That this exists and is the union of all $q_\equiv(a)$, $a \in A$ is clear. Now realize that $q_\equiv^{-1}(q_\equiv[A])$ is the collection of elements x such that $q_\equiv(x) = q_\equiv(a)$ for some $a \in A$, i.e., such that $x \equiv a$ for some $a \in A$. This is the same set. □

Proposition 4.10.20 *Let X be a topological space, and $\equiv$ be an equivalence relation on X. The open subsets of $X/\equiv$ are exactly the images $q_\equiv[U]$ of $\equiv$-saturated open subsets U of X.*

Proof If U is open and $\equiv$-saturated, then $U = q_{\equiv}^{-1}(q_{\equiv}[U])$ by Lemma 4.10.19. By Proposition 4.10.17 applied to $V = q_{\equiv}[U]$, $q_{\equiv}[U]$ is open in $X/\equiv$. Conversely, if V is open in $X/\equiv$, then $U = q_{\equiv}^{-1}(V)$ is open in X, and $V = q_{\equiv}[U]$ since $q_{\equiv}$ is surjective. □

Proposition 4.10.17 also leads to the following definition.

Definition 4.10.21 (Quotient map) Let X, Y be two topological spaces, and f be a map from X to Y. Then f is *quotient* if and only if f is surjective, and for every subset V of Y, V is open in Y iff $f^{-1}(V)$ is open in X.

Then a map is quotient if and only if it is the quotient map $q_{\equiv}$ of some equivalence relation $\equiv$ on X, up to some canonical identification of Y with $X/\equiv$ obtained through a homeomorphism:

Proposition 4.10.22 *A map $f : X \to Y$ between topological spaces is quotient if and only if there is an equivalence relation $\equiv$ on X that makes Y homeomorphic to $X/\equiv$ (say through $\cong : X/\equiv \to Y$) in such a way that $f = \cong \circ q_{\equiv}$. One can define $x \equiv x'$ if and only if $f(x) = f(x')$.*

Proof If: $q_{\equiv}$ is quotient by Proposition 4.10.20; any homeomorphism is quotient, and the composition of two quotient maps is quotient, so $\cong \circ q_{\equiv}$ is quotient.

Only if: define $\equiv$ by $x \equiv x'$ iff $f(x) = f(x')$, and then let $\cong : X/\equiv \to Y$ map $q_{\equiv}(x)$ to $f(x)$. This is well defined, i.e., $f(x)$ only depends on $q_{\equiv}(x)$, not on the particular x chosen in the equivalence class $q_{\equiv}(x)$, by the definition of $\equiv$. This is also bijective, and the inverse maps every $y \in Y$ to $q_{\equiv}(x)$, where x is any element in X such that $f(x) = y$: such an x exists since f is surjective, and if x and x' are two, then $x \equiv x'$, hence $q_{\equiv}(x) = q_{\equiv}(x')$ by definition.

We claim that $\cong$ is continuous: for every open subset V of Y, $\cong^{-1}(V) = \{q_{\equiv}(x) \mid x \in f^{-1}(V)\} = q_{\equiv}[f^{-1}(V)]$. Now $f^{-1}(V)$ is open since f is continuous, and $\equiv$-saturated by definition of $\equiv$. By Proposition 4.10.20, $\cong^{-1}(V)$ is therefore open.

Finally, we claim that $\cong$ is open: for every open subset U of $X/\equiv$, $q_{\equiv}^{-1}(U)$ is open in X, and $\cong[U] = \{f(x) \mid q_{\equiv}(x) \in U\} = f[q_{\equiv}^{-1}(U)]$. Let V be the latter set. Since f is quotient, to show that V is open, it is enough to show that $f^{-1}(V)$ is open in X. But $f^{-1}(V) = f^{-1}(f[q_{\equiv}^{-1}(U)])$ is the set of elements x such that there is an x' in $q_{\equiv}^{-1}(U)$ with $f(x) = f(x')$; equivalently, $x \equiv x'$. So $f^{-1}(V) = q_{\equiv}^{-1}(U)$ is indeed open.

By Proposition 4.10.4, $\cong$ is then a homeomorphism. □

It is easy to show that a map is continuous from a quotient space $X/\equiv$ to some space Y:

Proposition 4.10.23 *Let X and Y be two topological spaces, $\equiv$ be an equivalence relation on X, and f be a map from $X/\equiv$ to Y. Then f is continuous if and only if $f \circ q_\equiv$ is continuous from X to Y.*

Proof If f is continuous, then so is $f \circ q_\equiv$, by Proposition 4.3.12. Conversely, if $f \circ q_\equiv$ is continuous, then, for every open subset V of Y, $f^{-1}(V)$ is a subset U of $X/\equiv$ such that $q_\equiv^{-1}(U)$ is open in X. So U is open, by Proposition 4.10.20. □

Exercise 4.10.24 Show that the quotient topology on $X/\equiv$ is the finest one such that, for every topological space Y, for every continuous map $f: X/\equiv \to Y$, $f \circ q_\equiv$ is continuous from X to Y. (Hint: what about $Y = \mathbb{S}$?)

Exercise 4.10.25 (T_0 quotient) Let X be a topological space, with specialization quasi-ordering $\leqslant$. Define an equivalence relation $=_0$ on X by $x =_0 y$ iff $x \leqslant y$ and $y \leqslant x$. Equivalently, $x =_0 y$ iff the set of open neighborhoods of x coincides with the set of open neighborhoods of y. The T_0 *quotient* of X is $X/{=_0}$. Show that the specialization quasi-ordering $\leqslant/{=_0}$ on $X/{=_0}$ is defined by $q_{=_0}(x)\ (\leqslant/{=_0})\ q_{=_0}(y)$ iff $x \leqslant y$. Deduce from this that $X/{=_0}$ is T_0.

Show that the T_0 quotient has the following universal property: for every T_0 space Y, and every continuous map $f: X \to Y$, there is a unique continuous map $f': X/{=_0} \to Y$ such that $f = f' \circ q_{=_0}$.

Exercise 4.10.26 ($\mathbb{R}/\mathbb{Q}$) Consider the equivalence relation $\equiv$ on $\mathbb{R}$ defined by $x \equiv y$ iff $x - y \in \mathbb{Q}$. The space $\mathbb{R}/\equiv$ will be written $\mathbb{R}/\mathbb{Q}$.

Equip $\mathbb{R}$ with its usual topology, generated by the open intervals (a, b), $a < b$. Show that the quotient topology on $\mathbb{R}/\mathbb{Q}$ is the coarse topology, i.e., its only opens are $\emptyset$ and the whole space. In particular, the quotient of a T_0 space (even T_2) may fail to be T_0 (in the worst possible way). You will need the fact that, given $a < b$, there is a rational number between a and b.

A retract X of a space Y is a space that both embeds into Y and arises as a quotient of Y, where the embedding and the quotient map interact nicely:

Definition 4.10.27 (Retract) A *retract* of a topological space Y is a topological space X such that there are two continuous maps $s: X \to Y$ (the *section*) and $r: Y \to X$ (the *retraction*) such that $r \circ s = \mathrm{id}_X$.

Note that we do not require $s \circ r = \mathrm{id}_Y$, which would imply that r, s form a homeomorphism. Naturally, every homeomorphism defines a section and a retraction. The converse fails. For example, $\{1\}$ is a retract of $\{1, 2\}$ with the discrete topology, through $s\colon 1 \mapsto 1$ and $r\colon 1, 2 \mapsto 1$.

Proposition 4.10.28 (Sections are embeddings) *Every section $s\colon X \to Y$ is an embedding.*

Proof Let r be some associated retraction. Clearly s is continuous. It is injective since, whenever $s(x) = s(x')$, then $x = r(s(x)) = r(s(x')) = x'$. It is almost open since, for every open subset U of X, $V = r^{-1}(U)$ is open in Y, and $s^{-1}(V) = (r \circ s)^{-1}(U) = U$. So s is an embedding by Proposition 4.10.9. □

Example 4.10.29 There are embeddings that are not sections. Take $X = \{a_1, a_2, b_1, b_2\}$ with $a_i < b_j$ for every i, j, a_1 and a_2 incomparable, b_1 and b_2 incomparable. Let Y be X plus one middle element c, above a_1, a_2 and below b_1, b_2 (see Figure 4.9). Equip X, Y with their Alexandroff topologies. Then the inclusion map from X into Y is an embedding, but cannot be a section: the retraction would be forced to map c to some element of X above a_1 and a_2, and below b_1 and b_2, but there is no such element in X.

Proposition 4.10.30 (Retractions are quotient) *Every retraction $r\colon Y \to X$ is quotient.*

Proof Let s be some associated section. First, r is surjective, since $s(x)$ is an element y such that $r(y) = x$. Second, let U be any subset of X such that $r^{-1}(U)$ is open in Y. Then $s^{-1}(r^{-1}(U))$ is open in X. But this is just U. So r is quotient. □

Example 4.10.31 There are also quotient maps that are not retractions. Take $X = \{a_1, a_2, b_1, b_2\}$ as in Example 4.10.29, and define Y to be the quotient of X by the equivalence relation $\equiv$ such that $b_1 \equiv a_2$, and which makes no other pair of distinct elements equivalent. Let b be the equivalence class $\{b_1, a_2\}$. There is no associated section s, since otherwise $s(a_1) = a_1$, $s(b_2) = b_2$, and $s(b)$ would need to be above a_1 and below b_2.

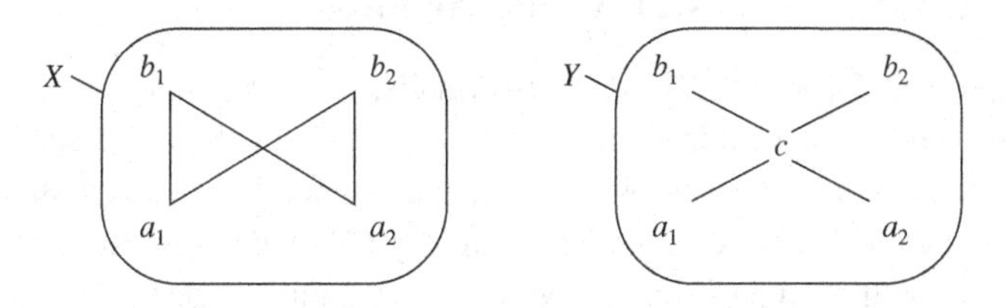

Figure 4.9 Embeddings, not sections.

Retracts X inherit many topological properties from the larger space Y. Here are a few examples.

Lemma 4.10.32 *Any retract of a T_0 space is T_0.*

Proof Because the section is an embedding (Proposition 4.10.28), the specialization quasi-ordering of the retract appears as the restriction of that of the larger space. □

Proposition 4.10.33 (Retracts of locally compact and of compact spaces) *Any retract of a compact space is compact. Any retract of a locally compact space is locally compact.*

Proof Let $r: Y \to X$ be the retraction, $s: X \to Y$ be the associated section. Since r is surjective (Proposition 4.10.30), $X = r[Y]$ is compact whenever Y is, by Proposition 4.4.13.

Assume now that Y is locally compact. Let $x \in X$, and U an open neighborhood of x in X. Then $r^{-1}(U)$ is an open neighborhood of $s(x)$ in Y. So there is a compact saturated subset Q of $r^{-1}(U)$ and an open subset W of Q such that $s(x) \in W$. It follows that $s^{-1}(W)$ is an open neighborhood of x; $s^{-1}(W)$ is included in $r[Q]$, since every element x of $s^{-1}(W)$ is such that $s(x) \in W \subseteq Q$, hence $x = r(s(x)) \in r[Q]$; $r[Q]$ is compact in X by Proposition 4.4.13; and $r[Q] \subseteq U$ since $Q \subseteq r^{-1}(U)$. By Trick 4.8.2 with $K = r[Q]$ and $V = s^{-1}(W)$, X is locally compact. □

Proposition 4.10.34 (Scott-retracts of dcpos) *For any two posets X and Y, we say that X is a* Scott-retract *of Y iff X_σ is a retract of Y_σ.*

Any poset X which is a Scott-retract of a dcpo Y is a dcpo.

Proof Let $r: Y \to X$ be the retraction, $s: X \to Y$ be the associated section. Let $(x_i)_{i \in I}$ be any directed family in X. Then $(s(x_i))_{i \in I}$ is a directed family in Y. Let y be its least upper bound. Since r is continuous, hence Scott-continuous (Proposition 4.3.5), $r(y)$ is the least upper bound of the family $(r(s(x_i)))_{i \in I}$. We conclude since $r(s(x_i)) = x_i$ for every $i \in I$. □

4.11 Connectedness

In a coproduct $X_1 + X_2$, X_1 and X_2 (formally, $\{1\} \times X_1$ and $\{2\} \times X_2$) are disconnected from each other, and just sit side by side, with no interaction. We wish to define *connected* spaces as spaces that cannot be equated with such a coproduct, with X_1 and X_2 non-empty, even up to homeomorphism. Since X_1 and X_2 are both open and non-empty in $X_1 + X_2$, this suggests the following definition.

Definition 4.11.1 (Connected) A topological space X is *connected* if and only if it cannot be written as the union $U \cup V$ of two disjoint, non-empty open subsets U, V.

Recall that a subset of X is *clopen* iff it is both open and closed. Equivalently, a space X is connected iff its only clopen subsets are $\emptyset$ and X itself.

Exercise 4.11.2 Let us check our intuition: show a space X is connected if and only if it is not homeomorphic to any non-trivial coproduct $X_1 + X_2$, where non-trivial means that both X_1 and X_2 are non-empty.

Definition 4.11.3 A subset A of a topological space X is *connected* if and only if it is connected as a subspace of X.

Lemma 4.11.4 (Intervals) *The connected subsets of* $\mathbb{R}$ *with its metric topology are its* intervals, *i.e., its subsets of the form* $[a, b]$, $[a, b)$, $(a, b]$, (a, b), $[a, +\infty)$, $(a, +\infty)$, $(-\infty, b]$, $(-\infty, b)$, *or* $(-\infty, +\infty) = \mathbb{R}$ *itself.*

More concisely, the intervals of $\mathbb{R}$ are its order-convex subsets, i.e., its subsets A such that, for all $x, z \in A$, every point y such that $x \leqslant y \leqslant z$ is also in A.

Proof If $A \subseteq \mathbb{R}$ is not an interval, then there are two points x, z in A and a third one y with $x \leqslant y \leqslant z$ but $y \notin A$. So $A \cap (-\infty, y)$ and $A \cap (y, +\infty)$ are disjoint non-empty subsets of A whose union is A. Therefore A is not connected.

Conversely, we claim that every interval is connected. By contradiction, let I be order-convex but not connected. So there are two open subsets U and V of $\mathbb{R}$ such that $I \cap U$ and $I \cap V$ are non-empty, and have empty intersection, but their union is I. Let a_0 be a point in $I \cap U$, and b_0 be one in $I \cap V$, and assume without loss of generality that $a_0 \leqslant b_0$. Since $I \cap U$ and $I \cap V$ do not meet, $a_0 < b_0$. We build a sequence of subintervals $[a_n, b_n]$ of $[a_0, b_0]$ by *dichotomy*. Given $[a_n, b_n]$ with $a_n \in I \cap U$, $b_n \in I \cap V$ and $a_n < b_n$, we consider $c_n = \frac{a_n+b_n}{2}$. Since I is order-convex, c_n is in I, hence either in $I \cap U$ or in $I \cap V$. In the first case, let $a_{n+1} = c_n$, $b_{n+1} = b_n$. In the second case, let $a_{n+1} = a_n$, $b_{n+1} = c_n$.

Let $c^- = \sup_{n \in \mathbb{N}} a_n$, $c^+ = \inf_{n \in \mathbb{N}} b_n$. Since $a_n < b_n$ for every $n \in \mathbb{N}$, $c^- \leqslant c^+$. On the other hand, we notice that $b_n - a_n = \frac{1}{2^n}(b_0 - a_0)$, so $c^+ - c^- = \inf_{m,n \in \mathbb{N}}(b_m - a_n) \leqslant \inf_{n \in \mathbb{N}} \frac{1}{2^n}(b_0 - a_0) = 0$. So $c^- = c^+$. Call this value simply c. For every $\epsilon > 0$, $c - a_n \leqslant b_n - a_n < \epsilon$ for n large enough, while $a_n \leqslant c$, so $|a_n - c| < \epsilon$ for n large enough. So c is the limit of $(a_n)_{n \in \mathbb{N}}$. Similarly, c is the limit of $(b_n)_{n \in \mathbb{N}}$.

Since $a_0 \leqslant c \leqslant b_0$ and I is order-convex, c is in I, hence in $I \cap U$ or in $I \cap V$. If $c \in I \cap U$, since U is open and is the limit of $(b_n)_{n \in \mathbb{N}}$, b_n is in U for n large enough. This is impossible since b_n is in V for every n. We reach a similar contradiction if c is in $I \cap V$. So I is connected. □

Using the Borel–Lebesgue Theorem (Proposition 3.3.4), Lemma 4.11.4 implies:

Corollary 4.11.5 *The compact connected subsets of $\mathbb{R}$, with its metric topology, are its closed intervals $[a, b]$, $a, b \in \mathbb{R}$, $a \leqslant b$.*

Example 4.11.6 Just like compactness or density, connectedness is a weak requirement for non-T_2-spaces: any space with a largest element $\top$ in its specialization quasi-ordering is connected. Indeed, any two non-empty open subsets must meet (at $\top$).

Just like compact subsets, connected subsets are preserved by continuous maps. We retrieve our initial intuition that continuous maps do not create jumps (Section 3.5).

Proposition 4.11.7 *Let X, Y be topological spaces. The image $f[A]$ of any connected subset A of X by a continuous map $f : X \to Y$ is a connected subset of Y.*

Proof Assume we could write $f[A]$ as the disjoint union of two non-empty open subsets W_1, W_2 in $f[A]$. Write W_1 as $f[A] \cap V_1$, W_2 as $f[A] \cap V_2$, where V_1 and V_2 are open in Y. Then $f^{-1}(V_1) \cap A \subseteq f^{-1}(W_1)$ and $f^{-1}(V_2) \cap A \subseteq f^{-1}(W_2)$ are open in A. They are non-empty and have empty intersection, and their union is A. But this contradicts the fact that A is connected. □

Theorem 4.11.8 (Intermediate Value Theorem) *Let f be any continuous map from a closed interval $[a, b]$ to $\mathbb{R}$, both with their metric topologies, and c be any value between $f(a)$ and $f(b)$, i.e., in the interval $[f(a), f(b)]$ if $f(a) \leqslant f(b)$, or in the interval $[f(b), f(a)]$ if $f(b) \leqslant f(a)$—an* intermediate value.

Then there is a point $x \in [a, b]$ such that $f(x) = c$.

Proof The image of $[a, b]$ by f is connected, by Proposition 4.11.7. So $f[[a, b]]$ is an interval, by Lemma 4.11.4. It contains $f(a)$ and $f(b)$, hence also c, from which the claim follows. □

Exercise 4.11.9 Show that Theorem 4.11.8 would fail if f is only assumed lsc (i.e., continuous from $[a, b]$ to $\mathbb{R}_\sigma$; see Exercise 4.3.11), or usc.

Exercise 4.11.10 (Path-connected) A *path* from x to y in a space X is a continuous map $\gamma : [0, 1] \to X$. A space is *path-connected* if and only if, for every two points $x, y \in X$, there is a path from x to y. Show that every path-connected space is connected.

The converse is false: there are connected spaces that are not path-connected, even among T_2 spaces. One example is given by the subspace of the real plane $\{(x, y) \in \mathbb{R} \mid (x \neq 0 \text{ and } y = \sin(1/x)) \text{ or } (x = 0 \text{ and } y \in [-1, 1])\}$ (see Figure 3.9 (*d*)). Here is a similar counterexample, which does not appeal to the sine function. Let the *comb* be the subspace X of $\mathbb{R}^2$ obtained as the union of a horizontal bar $(0, 1] \times \{1\}$, of countably many vertical bars $\{\frac{1}{2^n}\} \times [0, 1]$, and of one point, $\{(0, 0)\}$. Show that X is connected, but not path-connected.

To show that X is connected, it is useful to notice that any two points of X other than $(0, 0)$ are connected by a path (make a drawing of X). To show that X is not path-connected, show that any path γ in X with $\gamma(0) = (0, 0)$ must be constant. Hint: for small $\eta > 0$, realize that $\gamma[[0, \eta)]$ must be a connected subset of X that does not intersect the horizontal bar, and deduce that $\gamma[[0, \eta)]$ cannot contain points from two distinct vertical bars, hence is included in the union of $\{(0, 0)\}$ with some subset of a single vertical bar. Then show that the latter must be empty.

Exercise 4.11.11 Let X be a topological space. Show that the union of two connected subsets of X that meet at some point x is connected, and similarly for path-connected subsets. Show that the union of two disjoint connected (even path-connected) subsets may fail to be connected.

Exercise 4.11.12 Show that the closure of a connected subset of a topological space X is connected again.

Exercise 4.11.13 Let X be a topological space. Show that any directed union of connected subsets of X is again connected. So the family of all connected subsets of X is a dcpo, and is in particular inductive.

A *connected component* of a topological space X is a maximal connected subset of X. Show that every point of X is contained in some connected component, and that any two distinct connected components are disjoint, so that the connected components form a partition of X.

Show also that every connected component is closed. If X only has finitely many connected components, show that every connected component is also open, hence clopen. Show that this does not hold if X has infinitely many connected components. You may show that $\mathbb{Q}$, with its metric topology, has all one-point subsets $\{x\}$, $x \in \mathbb{Q}$, as connected components.

Products of path-connected spaces X_i, $i \in I$, are path-connected: for any two points $\vec{x}$ and $\vec{y}$ in $\prod_{i \in I} X_i$, there is a path γ_i from x_i to y_i, for every $i \in I$; so $\gamma(t) = (\gamma_i(t))_{i \in I}$ defines a path from $\vec{x}$ to $\vec{y}$. The same thing happens with connectedness.

Proposition 4.11.14 (Products of connected subsets) *Let X_i be topological spaces, $i \in I$. If each X_i is connected, $i \in I$, then the product $\prod_{i \in I} X_i$ is connected as well.*

Proof Let $X = \prod_{i \in I} X_i$, and assume X is not connected. So there are two open subsets U and V of X that are non-empty, disjoint, and whose union is X. We first claim that there is a point $\vec{x}$ in U and a point $\vec{y}$ in V such that $x_i = y_i$ for every $i \in I$ except exactly one.

Since $\vec{x}$ is in U, and by the definition of the product topology, $\vec{x}$ is in some basic open set $\prod_{i \in I} U_i$, where each U_i is open and there is a finite subset J of I such that $U_i = X_i$ for every $i \in I \smallsetminus J$. Replacing x_i by y_i for every $i \in I \smallsetminus J$ if necessary, we may assume that $\vec{x}$ is in U but $\vec{x}$ and $\vec{y}$ differ at only finitely many positions, namely those i in J.

We may then assume that $\vec{x} \in U$, $\vec{y} \in V$, and J were chosen so as to make the number of elements of J minimal. J cannot be empty, since $\vec{x} \neq \vec{y}$. If J contains just one index $i \in I$, then the claim is proved. Otherwise, pick $i \in J$, and let $\vec{x}'$ be obtained from $\vec{x}$ by changing x_i into y_i. If $\vec{x}'$ is in V, then the pair of points $\vec{x}, \vec{x}'$ proves the claim. Otherwise, $\vec{x}'$ is in U, $\vec{y}$ is in V, and they differ in one less position than $\vec{x}$ and $\vec{y}$, contradicting minimality. So the claim is proved.

Let $i \in I$ be the unique position where $x_i \neq y_i$. Let f be the map from X_i to X such that $f(z_i)$ is the tuple whose element at position i is z_i and whose elements at positions $j \neq i$ are x_j $(= y_j)$. By Trick 4.5.4, f is continuous, so $f^{-1}(U)$ and $f^{-1}(V)$ are open in X_i. Since U and V are disjoint, so are $f^{-1}(V)$ and $f^{-1}(V)$. Since $U \cup V = X$, $f^{-1}(U) \cup f^{-1}(V) = X_i$. Finally, $f^{-1}(U)$ is non-empty since it contains $\vec{x}$, and $f^{-1}(V)$ is non-empty since it contains $\vec{y}$. So X_i is not connected: contradiction. ☐

Exercise 4.11.15 Show that connectedness is not preserved under intersections. First, show that $F_1 = ([0, 1] \times [0, 3]) \cup ([0, 3] \times [0, 1])$ and $F_2 = ([2, 3] \times [0, 3]) \cup ([0, 3] \times [2, 3])$ are (path-)connected in $\mathbb{R}^2$, but their intersection is not. (Draw a picture first.) Second, show that filtered intersections of connected subsets also fail to be connected in general: e.g., in $\mathbb{R}^2$, let F_n be the complement of $(-n-1, n+1) \times (-1, 1)$, and show that each F_n is (path-)connected, but that $\bigcap_{n \in \mathbb{N}} F_n$ is not.

Exercise 4.11.16 In contrast to Exercise 4.11.15, show that intersections $(K_i)_{i \in I}$ of compact connected subsets of a T_2 space X are (compact and) connected. You may use Exercise 4.4.18. To solve the technical difficulties you will certainly encounter, show that you can reason inside some fixed K_{i_0} instead of the whole of X; observe that, as a compact T_2 subspace of X, K_{i_0} is normal (Proposition 4.4.17); and finally notice that if K is not connected, then it is the union of two disjoint non-empty closed subsets of K_{i_0}.

4.12 A bit of category theory I

It is time we introduced category theory. See the classic book by Mac Lane (1971), or the nicely written book by Adámek *et al.* (2009). The language of category theory, despite being sometimes overly abstruse, is a marvelous organizing tool, which we would be foolish to ignore forever.

The purpose of this section is to introduce some of the basic concepts, and to explain how the constructions of previous sections arise naturally from a categorical perspective. Most of these arise as *limits*, and *colimits*. Beware that limits are not limits of nets of points in a space here, but limits of so-called diagrams of spaces.

A *graph* **G** (see Figure 4.10) consists of a collection of *vertices* (e.g., A, B, C, etc.) and a collection of *edges* (e.g., f, g, h, etc.; not all edges have been given a name in Figure 4.10). Each edge f has a *source* vertex and a *target* vertex; e.g., in Figure 4.10, the source of f is A, its target is B, the source of g is B, its target is C. We say that f is an edge *from A to B*, and write $f : A \to B$ or $A \xrightarrow{f} B$.

We do not require graphs to be finite. In fact, we do not even require the vertices to form a set. For example, there are graphs whose vertices are *all* sets. But we shall require collections of sets between two vertices to be sets. A *small graph* is a graph whose collection of vertices is small, i.e., a set.

Formally, then, a graph **G** is a pair $(V, (E_{AB})_{A,B \in V})$ where V is a collection of so-called *vertices*, and E_{AB} is a set of *edges* from A to B, for each pair of vertices A, B.

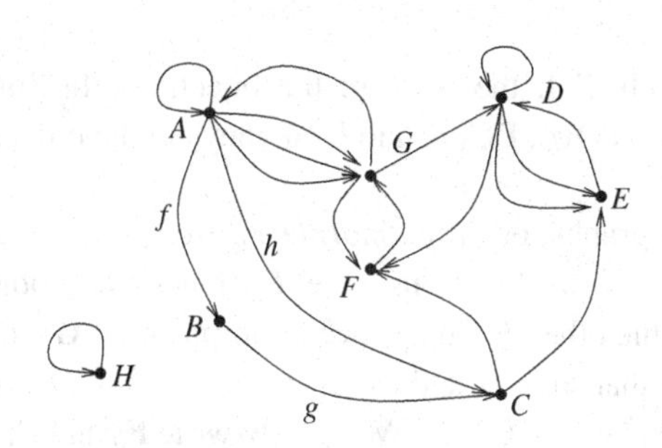

Figure 4.10 A finite graph.

A *category* **C** is a graph with a partial operation, called *composition* $\circ$, such that $g \circ f$ is defined for every $f: A \to B$ and $g: B \to C$ and is an edge from A to C, and which obeys some obvious laws. In the case of categories, it is more traditional to call the vertices *objects*, and the edges *morphisms*.

The obvious laws are as follows. First, for each object A, there is an *identity* morphism $\mathrm{id}_A : A \to A$, such that $f \circ \mathrm{id}_A = \mathrm{id}_B \circ f = f$ for every $f: A \to B$. Second, composition is *associative* when defined, i.e., if $f: A \to B$, $g: B \to C$, $h: C \to D$, then $h \circ (g \circ f) = (h \circ g) \circ f$.

Example 4.12.1 (Free category) Given a graph **G**, the graph whose morphisms are all finite paths $A_0 \xrightarrow{f_1} A_1 \xrightarrow{f_2} \cdots \xrightarrow{f_n} A_n$ ($n \geqslant 0$) obtained by concatenating edges $f_1: A_0 \to A_1$, $f_1: A_1 \to A_2, \ldots, f_n: A_{n-1} \to A_n$, where id_A is the length 0 path from A to A, and composition is concatenation, forms a category. This is the *free category* over the graph **G**.

Example 4.12.2 The category **Set** has all sets as objects, and all maps from A to B as morphisms from A to B. The identity morphism id_A is the identity map from A to A, and composition is function composition.

The category **Top** of topological spaces has topological spaces as objects, and continuous maps as morphisms. For short, we say that **Top** is the category of topological spaces and continuous maps.

Similarly, we define the categories **Cpo** of dcpos and continuous maps, **PCpo** of pointed dcpos (i.e., with a least element $\bot$) and continuous maps, and **Ord** of posets and monotonic maps.

An *isomorphism* between objects A and B in a category **C**, or *iso* for short, is a pair of morphisms $f: A \to B$, $g: B \to A$ such that $g \circ f = \mathrm{id}_A$ and $f \circ g = \mathrm{id}_B$. If such an iso exists between A and B, then A and B are *isomorphic*. We often mention only one of the morphisms, leaving the other one implicit.

Example 4.12.3 In **Set**, the isos are the bijections. In **Top**, they are the homeomorphisms. In **Ord**, **PCpo**, and **Cpo**, they are the order isomorphisms.

A *morphism* of graphs, or *graph morphism*, from some graph $\mathbf{G}_1$ to some graph $\mathbf{G}_2$, is a pair of two functions. One, $\mathbf{F}_0$, maps every object of $\mathbf{G}_1$ to an object of $\mathbf{G}_2$, and the other, $\mathbf{F}_1$, maps every morphism of $\mathbf{G}_1$ to a morphism of $\mathbf{G}_2$, in such a way that sources and targets are preserved, i.e., if $f: A \to B$ in $\mathbf{G}_1$, then $\mathbf{F}_1(f): \mathbf{F}_0(A) \to \mathbf{F}_0(B)$. We usually write $\mathbf{F}_0$ and $\mathbf{F}_1$ using the same symbol, e.g., F.

4.12.1 Functors

A *functor* from a category $\mathbf{C}$ to a category $\mathbb{C}$ is a graph morphism F that also preserves identities and composition, namely $F(\mathrm{id}_A) = \mathrm{id}_{F(A)}$, and if $f: A \to B$ and $g: B \to C$ in $\mathbf{C}$, then $F(g \circ f) = F(g) \circ F(f)$.

Example 4.12.4 There are many functors around. E.g., the *forgetful functor* U from **Top** to **Set** maps each topological space X (in **Top**) to the underlying set (in **Set**), and each continuous map $f: X \to Y$ to the underlying function $f: X \to Y$.

There are similar forgetful functors from **Ord** to **Set**, from **Cpo** to **Set**, and also from **Cpo** to **Ord**, or from $\mathbf{Top}_0$ (the category of T_0 topological spaces and continuous maps) to **Ord**. The latter maps T_0 spaces X to X seen as a poset, equipped with its specialization ordering, and continuous maps $f: X \to Y$ to the underlying poset map (which is monotonic by Proposition 4.3.9).

Here are some others. There is a functor from **Ord** to **Top**, which maps each poset X to X_a, and each monotonic map $f: X \to Y$ to the map $f: X_a \to Y_a$. Note that the latter is continuous by definition of the Alexandroff topologies.

Some functors have the same source and target categories: we call them *endofunctors*. E.g., there is an endofunctor $\mathbb{P}$ on **Set** that maps each set X to $\mathbb{P}(X)$, and each function $f: X \to Y$ to the image functor $\mathbb{P}(f): \mathbb{P}(X) \to \mathbb{P}(Y)$, defined by $\mathbb{P}(f)(A) = f[A]$.

4.12.2 Limits

Most categories have extra structure. A *terminal object* 1 in a category $\mathbf{C}$ is an object such that there is a unique morphism $!_A: A \to 1$ for every object A of $\mathbf{C}$ to 1. In **Top**, the terminal objects are the singletons $\{*\}$, with the unique possible topology.

In general, terminal objects are defined *up to iso*, namely, every object that is isomorphic to a terminal object is terminal, and any two terminal objects are isomorphic. Moreover, the isomorphism is itself unique.

A *(binary) product* of two objects A and B is an object $A \times B$ with two morphisms $\pi_1: A \times B \to A$ and $\pi_2: A \times B \to B$ (first and second *projection*) such that, for all morphisms $f_1: C \to A$ and $f_2: C \to B$, there is a unique morphism $\langle f_1, f_2 \rangle: C \to A \times B$ such that $\pi_1 \circ \langle f_1, f_2 \rangle = f_1$ and $\pi_2 \circ \langle f_1, f_2 \rangle = f_2$. The requirement for uniqueness implies, and is in fact equivalent to, the equations $\langle \pi_1, \pi_2 \rangle = \mathrm{id}_{A \times B}$ and $\langle f_1, f_2 \rangle \circ g = \langle f_1 \circ g, f_2 \circ g \rangle$. Again, products of A and B are unique up to iso.

In case $\mathbf{C}$ has binary products, for every pair of morphisms $f_1 : A_1 \to B_1$ and $f_2 : A_2 \to B_2$, there is also a morphism $f_1 \times f_2$ from $A_1 \times A_2$ to $B_1 \times B_2$, defined by $f_1 \times f_2 = \langle f_1 \circ \pi_1, f_2 \circ \pi_2 \rangle$.

Example 4.12.5 In **Top**, the usual product $X \times Y$ of two spaces X and Y, with the product topology (Section 4.5), is a product. Then $\langle f_1, f_2 \rangle(z) = (f_1(z), f_2(z))$, and $(f_1 \times f_2)(x_1, x_2) = (f_1(x_1), f_2(x_2))$. (See Exercises 4.5.5, 4.5.6.) It is futile to look for another notion of product in **Top**, since products are unique up to iso.

Similarly, the notion of product of **Ord** is, up to iso, poset product.

The properties of the binary product in a category $\mathbf{C}$ are summed up in the following three-part diagram:

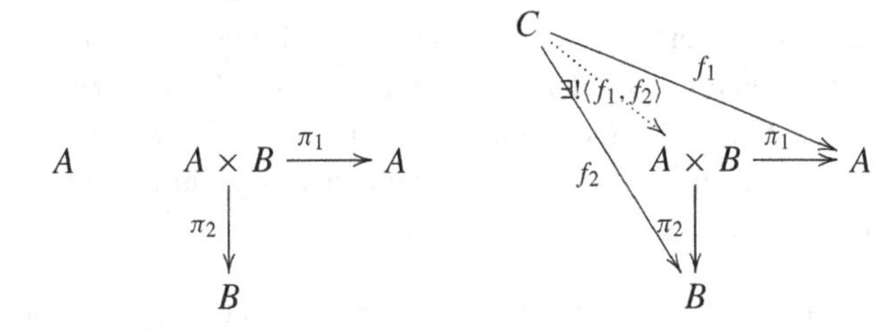

This reads as follows. Given any two objects A and B (on the left), the object $A \times B$ and the two morphisms π_1 and π_2 exist (middle), such that in every situation depicted on the right (looking only at the solid arrows), there is a unique dotted morphism that makes all created triangles commutative. A diagram is *commutative* if any two paths with the same source object and the same target object denote the same morphism. Here, this means that $\pi_1 \circ \langle f_1, f_2 \rangle = f_1$ and $\pi_2 \circ \langle f_1, f_2 \rangle = f_2$, which we have already stated above.

These constructions are special cases of so-called *limits* in a category $\mathbf{C}$. Let $\mathbf{J}$ be an arbitrary category, e.g., the free category over some small graph $\mathbf{G}$. One can think of $\mathbf{J}$ as the shape of a diagram. A functor F from $\mathbf{J}$ to $\mathbf{C}$ is a model of this diagram, as realized inside $\mathbf{C}$. In the case of products, for example, the shape $\mathbf{J}$ of the diagram on the left is the (free category over) the graph with just two vertices $*_1, *_2$, and no edge, and it is realized inside $\mathbf{C}$ by picking one object for each vertex (here, $F(*_1) = A$ and $F(*_2) = B$), and one morphism for each edge (here, none).

A *cone* for F is an object X of $\mathbf{C}$, together with morphisms $\pi_J : X \to F(J)$ for each object J of $\mathbf{J}$, such that, for every morphism $j : J \to J'$ in $\mathbf{J}$, $F(j) \circ \pi_J = \pi_{J'}$. We call X the *apex* of the cone. So, for example, $A \times B$ together with π_1 and π_2 form a cone in the case of products. C, together with f_1 and f_2, also form a cone.

A cone X, π_J (J object of **J**), is *universal* if and only if, for any other cone X', π'_J (J object of **J**), for the same functor F, there is a unique morphism $h\colon X' \to X$ in **C** such that $\pi'_J = \pi_J \circ h$ for every object J of **J**. So the cone $A \times B, \pi_1, \pi_2$, in the example of products, is a universal cone.

A *limit* $\lim F$ is the apex of a universal cone for F. This may fail to exist. When it exists, the universality requirement says that, in some sense, $\lim F$ is the "largest" object that can occur as apex of a cone for F. A product $A \times B$ is then a limit of the functor $F(*_1) = A$, $F(*_2) = B$.

Similarly, a terminal object in **C** is a limit of the unique functor from the empty category to **C**.

By the uniqueness requirement for h in the definition of limits, it is easy to see that limits are unique up to iso. We shall therefore talk about *the* limit of F, although this is only defined up to iso, and may also fail to exist.

Here are some more involved cases of limits. The (general) *product* $A = \prod_{i \in I} A_i$ of a family $(A_i)_{i \in I}$ in **C** is the limit of the functor $F\colon I \to$ **C**, where the set I is seen as a trivial category whose objects are the elements of I and the only morphisms are identities. This means that there are projection maps $\pi_i\colon A \to A_i$ for each $i \in I$, such that, for every family of morphisms $f_i\colon C \to A_i$, $i \in I$, there is a morphism $\langle f_i \rangle_{i \in I}\colon C \to A$ such that $\pi_j \circ \langle f_i \rangle_{i \in I} = f_j$ for every $j \in I$, $\langle \pi_i \rangle_{i \in I} = \mathrm{id}_A$, and $\langle f_i \rangle_{i \in I} \circ g = \langle f_i \circ g \rangle_{i \in I}$.

Example 4.12.6 In **Top**, this notion of product yields exactly Definition 4.5.1. In **Ord**, we obtain the poset product, with the pointwise ordering.

Here is a more involved notion of limit. Consider the free category over the graph $*_1 \to *_2 \leftarrow *_3$, and the functor mapping this to the situation depicted on the left, below:

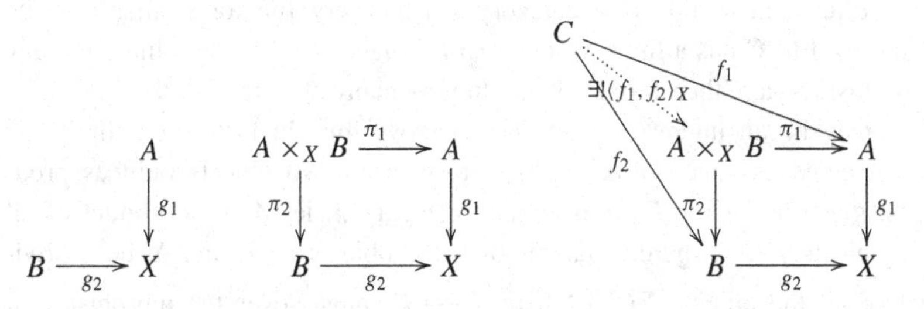

Then a limit of this functor is a *pullback* of A and B over X, and will be denoted by $A \times_X B$, leaving the morphisms g_1, g_2 implicit, or $\ker(g_1, g_2)$ if one wants to make g_1 and g_2 explicit. In **Top**, the canonical pullback is the space $A \times_X B = \{(x, y) \in A \times B \mid g_1(x) = g_2(y)\}$, with the subspace topology from $A \times B$. If f_1 and f_2 are such that the diagram made from the solid edges on the right is commutative, i.e., if $g_1(f_1(z)) = g_2(f_2(z))$ for every $z \in C$, then the dotted morphism $\langle f_1, f_2 \rangle_X$ is the function that maps each $z \in C$ to $(f_1(z), f_2(z)) \in A \times_X B$.

One says that π_1 is obtained by *pulling back* g_2 along g_1, or also that π_2 is obtained by pulling back g_1 along g_2.

An *equalizer* is similar, except now $A = B$, $\pi_1 = \pi_2$, and f_1 is required to equal f_2 in the right-hand diagram. More precisely, an equalizer of $g_1, g_2 \colon A \to X$ is an object $[g_1 = g_2]$ together with a morphism $[g_1 = g_2] \xrightarrow{\pi} A$ such that $g_1 \circ \pi = g_2 \circ \pi$, and such that, for any morphisms $f \colon C \to A$ such that $g_1 \circ f = g_2 \circ f$, there is a unique morphism $h \colon C \to [g_1 = g_2]$ such that $\pi \circ h = f$.

Alternatively, equalizers are limits of functors from the free category over the graph $*_1 \rightrightarrows *_2$, with two morphisms from $*_1$ to $*_2$.

Example 4.12.7 In **Top**, the canonical equalizer of $g_1, g_2 \colon A \to X$ is the subspace $[g_1 = g_2]$ of A consisting of all elements $z \in A$ such that $g_1(z) = g_2(z)$, and π is the canonical inclusion. We shall see later that the equalizers in **Top** are exactly the subspaces, up to iso.

A category **C** such that every functor from any small category **J** to **C** has a limit is called *complete*, or sometimes small-complete, to stress the fact that **J** must be small. It is a standard result of category theory that a category is complete if and only if it has all products (indexed by arbitrary sets) and equalizers (Adámek *et al.*, 2009, theorem 12.3).

Every limit is then expressible as an equalizer of two maps whose target space is a product. In **Top**, the limits are exactly the subspaces of products of spaces.

Similarly, a category with all *finite* products and equalizers is *finitely complete*. Equivalently, this is a category where every functor from any finite category **J** to **C** has a limit, where a *finite category* is one with finitely many morphisms—and therefore also only finitely many objects.

It may be an interesting exercise to prove this. In fact, every limit can be obtained as an equalizer of two maps from two objects built as products: given a functor F from a small category **J**, let A be a product of all the objects $F(J)$ where J ranges over the objects of **J**, and X be a product of all the objects $F(J')$ where $J \xrightarrow{j} J'$ ranges over the morphisms of **J**; define $g_1^{J \xrightarrow{j} J'} \colon A \to F(J')$ as projection $\pi_{J'}$ on the J'th component, $g_2^{J \xrightarrow{j} J'} \colon A \to F(J')$ as $F(j) \circ \pi_J$, then define g_1 as the unique morphism from A to X such that $\pi_{J'} \circ g_1 = g_1^{J \xrightarrow{j} J'}$ for every object J' of **J**, and similarly for g_2; then the limits of F are exactly the equalizers of g_1, g_2.

Example 4.12.8 **Top**, **Cpo**, **Ord**, and **Set** are complete categories. One can check, either directly or by the equalizer of product construction above, that a limit of a functor $F: \mathbf{J} \to \mathbf{Top}$ is the subspace (resp., subposet, subset) of $\prod_{J \text{ object of } \mathbf{J}} F(J)$ consisting of those vectors $\vec{x} = (x_J)_{J \text{ object of } \mathbf{J}}$ such that, for every morphism $j: J \to J'$ in $\mathbf{J}$, $F(j)(x_J) = x_{J'}$.

4.12.3 Colimits

Dual to limits are *colimits*. A quick definition is as follows. Every category $\mathbf{C}$ has an *opposite category* $\mathbf{C}^{op}$, whose objects are the same as that of $\mathbf{C}$ and whose morphisms from A to B are the morphisms from B to A in $\mathbf{C}$. Intuitively, $\mathbf{C}^{op}$ is obtained by reversing the direction of arrows. For every functor F from $\mathbf{J}$ to $\mathbf{C}$, or equivalently from $\mathbf{J}^{op}$ to $\mathbf{C}^{op}$, a *colimit* of F in $\mathbf{C}$ is by definition a limit of $F: \mathbf{J}^{op} \to \mathbf{C}^{op}$ in $\mathbf{C}^{op}$.

A slightly more concrete definition is obtained by expanding the definition. First, a cone in $\mathbf{C}^{op}$ is called a *cocone* in $\mathbf{C}$. This is an object A, the *apex*, with a family of morphisms $\iota_J: F(J) \to A$, one for each object J of $\mathbf{J}$, such that, for every morphism $j: J \to J'$ in $\mathbf{J}$, $\iota_J = \iota_{J'} \circ F(j)$. A colimit is a universal cocone $(A, (\iota_J)_{J \text{ object of } \mathbf{J}})$, i.e., a cocone such that, for any other cocone $(A', (\iota'_J)_{J \text{ object of } \mathbf{J}})$, there is a unique morphism $h: A \to A'$ in $\mathbf{C}$ such that $\iota'_J = h \circ \iota_J$ for every object J of $\mathbf{J}$. Roughly speaking, A is the "smallest" object that is the apex of a cocone. We write colim F for the colimit of F. Again, when they exist, colimits are unique up to iso.

Dually to terminal objects, *initial objects* are colimits of the unique functor from the empty category to $\mathbf{C}$. By definition, these are objects 0 such that there is a unique morphism $!_A: 0 \to A$ for every object A. In **Top**, **Ord**, **Set**, and **Cpo**, the initial object is the empty set. **PCpo** has no initial object.

The *(binary) coproduct* $A + B$ of two objects of $\mathbf{C}$, if it exists, is the colimit of the functor F defined by $F(*_1) = A$, $F(*_2) = B$, from the category with two objects $*_1$, $*_2$ and only identity morphisms to $\mathbf{C}$. This is the dual notion to products. More generally, the coproduct $\coprod_{i \in I} A_i$ of a family $(A_i)_{i \in I}$ of objects is the colimit of the functor $F: I \to \mathbf{C}$ defined by $F(i) = A_i$, $i \in I$.

Example 4.12.9 In **Top**, this is exactly the notion of coproducts that we introduced in Section 4.6. In **Ord**, this yields the poset coproducts of Exercise 4.6.8. For less immediately obvious notions of coproducts, see Exercise 8.4.28.

The notion dual to pullbacks is *pushouts*. Consider the free category $\mathbf{J}$ over the graph $*_1 \leftarrow *_2 \to *_3$, and the functor mapping it to the situation depicted below, left:

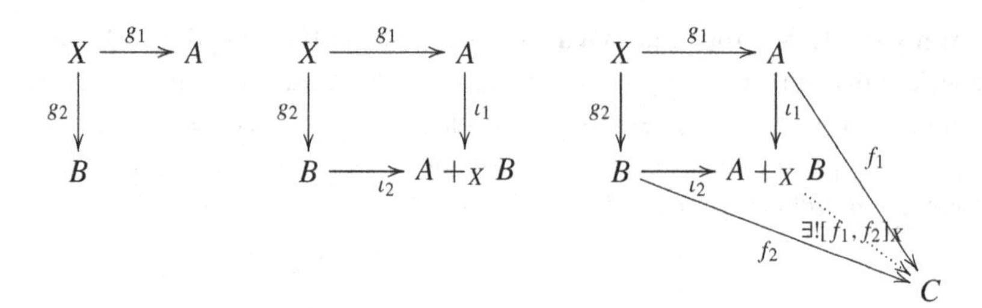

(We also write $\mathrm{coker}(g_1, g_2)$ instead of $A +_X B$, to make g_1 and g_2 explicit.)

Example 4.12.10 In **Top**, the canonical pushout $A +_X B$ is $(A + B)/\equiv$, where $\equiv$ is the smallest equivalence relation such that $(1, g_1(x)) \equiv (2, g_2(x))$ for every $x \in X$. That is, $z \equiv z'$ if and only if one can find elements $z_0 = z, z_1, \ldots, z_{n-1}, z_n = z'$ of $A + B$ ($n \in \mathbb{N}$) such that, for every i, $1 \leqslant i \leqslant n$, either $z_{i-1} = (1, g_1(x))$ and $z_i = (2, g_2(x))$ for some $x \in X$, or $z_i = (1, g_1(x))$ and $z_{i-1} = (2, g_2(x))$ for some $x \in X$. If f_1 and f_2 are given so as to make the solid lines on the right commute, i.e., if $f_1(g_1(x)) = f_2(g_2(x))$ for every $x \in X$, then $[f_1, f_2]_X$ maps the equivalence class of $(1, x)$ to $f_1(x)$, and the equivalence class of $(2, y)$ to $f_2(y)$.

Finally, dual to equalizers are *coequalizers*.

Example 4.12.11 In **Top**, the canonical coequalizer of $g_1, g_2 \colon X \to A$ is the quotient $A/\equiv$, where $\equiv$ is the smallest equivalence relation on A such that $g_1(x) = g_2(x)$ for every $x \in X$.

It turns out that, conversely, any quotient can be realized as a coequalizer. Let $f \colon X \to Y$ be any continuous map. Pull back f along itself, i.e., form the pullback $\ker(f, f)$, and consider the *kernel pair* (π_1, π_2) of f. That is, $\ker(f, f) = \{(x, x') \in X \times X \mid f(x) = f(x')\}, \pi_1(x, x') = x, \pi_2(x, x') = x'$. Then form the coequalizer of the kernel pair. This is $X/\equiv$, where $x \equiv x'$ iff $f(x) = f(x')$. Now, if f is quotient, then Y is homeomorphic to $X/\equiv$ by Proposition 4.10.22, say through $\cong \colon X/\equiv \to Y$, in such a way that $f = \cong \circ\, q_{\equiv}$. So the quotients are exactly the coequalizers in **Top**.

Similarly, we can dualize the construction to show that every subspace arises as an equalizer up to iso in **Top**. That is, whenever $f \colon X \to Y$ is an embedding (i.e., X is a subspace up to iso), then there is an equalizer $[g_1 = g_2]$, with map $\pi \colon [g_1 = g_2] \to Y$, and an isomorphism $\cong \colon X \to [g_1 = g_2]$ such that $f = \pi \circ \cong$. Indeed, build the pushout $\mathrm{coker}(f, f)$, i.e., $(Y + Y)/\equiv$, where $\equiv$ is the smallest equivalence relation such that $(1, f(x)) \equiv (2, f(x))$ for every $x \in X$. The canonical injections ι_1, ι_2 form the so-called *cokernel pair* of f. Note that $\iota_1 \colon Y \to (Y + Y)/\equiv$ maps y to the equivalence class of $(1, y)$, while

ι_2 maps y to the equivalence class of $(2, y)$. Form the equalizer $[\iota_1 = \iota_2]$ of the cokernel pair. This is the subspace of those elements y of Y such that $(1, y) \equiv (2, y)$. It is easy to see that $[\iota_1 = \iota_2]$ is the image $f[X]$ of f, seen as a subspace of Y. By definition, f itself defines a homeomorphism $\cong$ of X onto $[\iota_1 = \iota_2]$, and we are done: the equalizers in **Top** are the subspaces, up to iso, that is, the embeddings.

A category **C** such that every functor from any small category **J** to **C** has a colimit is called *cocomplete*. Equivalently, the cocomplete categories are those whose opposite is complete. In particular, a category is cocomplete if and only if it has all coproducts and coequalizers. And colimits are expressed as coequalizers of two maps whose source space is a coproduct.

Example 4.12.12 **Top** is cocomplete. All colimits are expressible as quotients of coproducts.

Similarly, a category with all *finite* coproducts and coequalizers is *finitely cocomplete*. Equivalently, this is a category where every functor from any finite category **J** to **C** has a colimit, or the opposite of a finitely complete category.

5

Approximation and function spaces

In Section 4.2.2, we discussed how computer programs could be thought of as computing values x obtained as $\sup_{n \in \mathbb{N}} x_n$, where x_n are values in a given dcpo, e.g., the dcpo of sets of formulae in the example of the automated theorem-proving computer. A distinguishing feature of the approximants x_n to x is that they are finite, and this particular relation between x_n and x can be described in arbitrary posets by saying that x_n is *way-below* x. The way-below relation $\ll$ is a fundamental notion, leading to so-called continuous and algebraic dcpos, and we define it and study it in Section 5.1.

Beyond dcpos, the way-below relation will be instrumental in studying the lattice of open subsets of a topological space (Section 5.2). This will lead us to investigate the spaces whose lattice of open subsets is continuous, the so-called core-compact spaces. We shall see that the core-compact spaces are exactly the spaces X that are exponentiable, that is, such that we can define a topology on the space $[X \to Y]$ of continuous maps from X to an arbitrary space Y so that application and currification are themselves continuous (Section 5.3). These are basic requirements in giving semantics to higher-order programming languages, and desirable features in algebraic topology.

5.1 The way-below relation

The *approximation*, or *way-below* relation $\ll$ on a poset X is of fundamental importance in the study of dcpos.

Definition 5.1.1 (Way-below, $\ll$) Let X be a poset, and x and y be two elements of X. Then x is *way-below* y, in notation $x \ll y$, if and only if, for every directed family z_i that has a a least upper bound z above y, there is an i in I such that $y \leqslant z_i$ already.

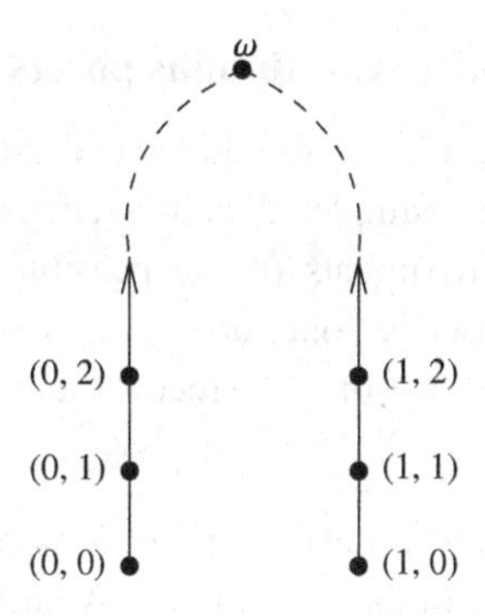

Figure 5.1 The non-continuous dcpo $\mathcal{N}_2$.

Note that $x \ll y$ implies $x \leqslant y$: take the directed family $\{y\}$. The converse fails; see Example 5.1.2.

To explain the notion through its complement, x is not way-below y if there is a way of taking a limit of elements all properly below x, but where the limit would jump above not just x but also y.

Example 5.1.2 Here are a few instances of the way-below relation $\ll$: (i) if $X = \mathbb{N}$, then $x \ll y$ iff $x \leqslant y$; (ii) if $X = \mathbb{N}_\omega$, where $\mathbb{N}_\omega$ is $\mathbb{N}$ plus a fresh element ω above all natural numbers, then $x \ll y$ iff $x \leqslant y$ and $x \neq \omega$; (iii) if $X = \mathbb{P}(Y)$, where Y is some set, possibly infinite, $A \ll B$ iff A is a *finite* subset of B; (iv) if $X = \mathbb{R}$, then $x \ll y$ iff $x < y$; (v) if $X = \mathbb{R}^+$ or $X = [0, 1]$, then $x \ll y$ iff $x = 0$ or $x < y$. (As an exercise, prove these claims.)

Exercise 5.1.3 Here is a case where the way-below relation is slightly less intuitive. Consider the dcpo $\mathcal{N}_2$ in Figure 5.1. Show that, in $\mathcal{N}_2$, $x \ll y$ iff $x \leqslant y$ and $x, y \neq \omega$. In particular, *no* element is way-below ω.

Let us state a few easy properties of $\ll$. Note that $\ll$ is neither reflexive nor irreflexive in general. E.g., some elements are way-below themselves (the elements of $\mathbb{N}$ in $X = \mathbb{N}_\omega$), and some others are not (ω in the same example).

Proposition 5.1.4 *If $x \ll y$, then $x \leqslant y$.*

If $x \ll y$ and $y \leqslant z$, then $x \ll z$.

If $x \leqslant y$ and $y \ll z$, then $x \ll z$. In particular, $\ll$ is transitive.

Proof We have already mentioned the first claim. For the second part, we show more generally that if $x_1 \leqslant x_2 \ll y_2 \leqslant y_1$, then $x_1 \ll y_1$; this follows easily from the definition. □

5.1.1 Continuous posets

The elements y such that $y \ll x$ should be understood as approximants of x, which in good cases are simpler than x itself. E.g., in the case of $\mathbb{P}(Y)$ (Example 5.1.2), the approximants of the possibly infinite sets B are the *finite* subsets of B. A poset is continuous if and only if there are enough approximants, in the sense that one can recover any element as a limit of its approximants.

Definition 5.1.5 (Continuous poset) A poset X is *continuous* if and only if, for every $x \in X$, the set $\twoheaddownarrow x$ of all elements $y \in X$ such that $y \ll x$ is directed, and has x as least upper bound.

It is a common mistake to forget to check that $\twoheaddownarrow x$ is *directed*, and to only verify that $x = \sup \twoheaddownarrow x$.

Example 5.1.6 The dcpo $\mathcal{N}_2$ of Figure 5.1 is not continuous, because $\twoheaddownarrow \omega$ is empty, hence not directed. The posets of Example 5.1.2, namely $\mathbb{N}$, $\mathbb{N}_\omega$, $\mathbb{P}(Y)$, $\mathbb{R}$, $\mathbb{R}^+$, $[0, 1]$, are all continuous.

Notice in particular that, in the case of $X = \mathbb{P}(Y)$, every subset B is the union (least upper bound) of its finite subsets, since it contains the union of its one-element subsets; and that the family of all finite subsets of B is directed. This is an important subcase of continuous posets:

Definition 5.1.7 (Algebraic poset) Let X be a poset. An element $y \in X$ is *finite* if and only if $y \ll y$. X is *algebraic* iff, for every element $x \in X$, the set of finite elements y below x is directed and has x as least upper bound.

Example 5.1.8 In $X = \mathbb{P}(Y)$, the finite elements are exactly the finite subsets of Y, where the second "finite" is taken with its usual meaning. Since every set is the directed sup of its finite subsets, $X = \mathbb{P}(Y)$ is algebraic.

We shall see later (Corollary 5.1.24) that every algebraic poset is continuous. For now, note that if y is finite and below x, then $y \ll x$, by Proposition 5.1.4.

Example 5.1.9 Conversely, there are continuous posets that are not algebraic, e.g., $\mathbb{R}$, $\mathbb{R}^+$, or $[0, 1]$: the first one has no finite element at all, and 0 is the only one in the other two.

Exercise 5.1.10 Consider the posets of Example 5.1.2. Show that $\mathbb{N}$ is an algebraic poset, $\mathbb{N}_\omega$ is an algebraic dcpo, $\mathbb{P}(Y)$ is an algebraic dcpo, $\mathbb{R}$, $\mathbb{R}^+$, are continuous posets that are not algebraic, and $[0, 1]$ is a continuous dcpo that is not algebraic.

Exercise 5.1.11 Let X be a poset, and $x \in X$. Show that the following are equivalent: (a) x is finite in X; (b) $\uparrow x$ is open in the Scott topology of X; (c) the map $\chi_{\uparrow x}$ from X to $\mathbb{S}$, defined by $\chi_{\uparrow x}(y) = 1$ iff $x \leqslant y$, is Scott-continuous. Recall that $\mathbb{S}$ is Sierpiński space (Exercise 4.1.18).

Exercise 5.1.12 Let X be a poset. If X has a least element $\bot$, show that $\bot$ is finite. If x and y are finite elements with a least upper bound, show that $\sup(x, y)$ is again finite.

Exercise 5.1.13 Let X be a poset. Let $(x_i)_{i \in I}$ be a directed family in X, and assume that, for each $i \in I$, x_i is the least upper bound of a directed family $(x_{ij})_{j \in J_i}$. One would be tempted to conclude that $(x_{ij})_{\substack{i \in I \\ j \in J_i}}$ is again a directed family, with the same least upper bound as $(x_i)_{i \in I}$, if any. This is wrong, however: find an explicit counterexample. The dcpo $[0, 1]^2$ with the ordering $(x, y) \preceq (x', y')$ iff $x = x'$ and $y \leqslant y'$, or $y' = 1$ and $x \leqslant x'$ is a good candidate for X.

On the other hand, show that if $x_{ij} \ll x_i$ for all $i \in I$, $j \in J_i$, then the family $(x_{ij})_{\substack{i \in I \\ j \in J_i}}$ is again directed; also, that if $(x_i)_{i \in I}$ has a least upper bound x, then x is also the least upper bound of $(x_{ij})_{\substack{i \in I \\ j \in J_i}}$.

Exercise 5.1.14 (Closures in continuous posets) Using Exercise 5.1.13, show that, in a continuous poset X, the closure $cl(E)$ of any subset E is exactly the set of least upper bounds of directed families $(x_i)_{i \in I}$ in $\downarrow E$. (That is, we obtain the closure by taking the downward closure first, then adding all directed sups, without having to iterate the process.)

Show that this would fail if X were not continuous: take $X = [0, 1]^2$ with the ordering $\preceq$ of Exercise 5.1.13, and consider $E = [0, 1)^2$.

In a continuous poset, $\ll$ has a fundamental property called *interpolation* (Mislove, 1998, lemma 4.16). In this form, it is due to Zhang (1993).

Proposition 5.1.15 (Interpolation) *Let X be a continuous poset. If $x \ll y$, then $x \ll z \ll y$ for some $z \in X$.*

Proof Consider $D = \{w \in X \mid \exists z \in X \cdot w \ll z \ll y\}$. D is non-empty, since $\twoheaddownarrow y$ is non-empty, and for every $z \in \twoheaddownarrow y$, $\twoheaddownarrow z$ is non-empty.

We claim that D is directed: if $w_1 \ll z_1 \ll y$ and $w_2 \ll z_2 \ll y$, since $\twoheaddownarrow y$ is directed, there is a $z \ll y$ such that $z_1, z_2 \leqslant z$. So $w_1 \ll z \ll y$ and $w_2 \ll z \ll y$. Since $\twoheaddownarrow z$ is directed, there is a w such that $w_1, w_2 \leqslant w \ll z$.

We now check that y is the least upper bound of D. It is clearly an upper bound, and we must show that any upper bound y' of D is above y. Since y' is an upper bound of D, for all w, z such that $w \ll z \ll y$, $w \leqslant y'$. Fix $z \ll y$: since $z = \sup_{w \ll z} w$, $z \leqslant y'$. Since every $z \ll y$ is below y', $y = \sup_{z \ll y} z$ is below y' as well.

So $y = \sup D$. Since $x \ll y$, by definition there is an element w of D such that $x \leqslant w$. That is, $x \leqslant w \ll z \ll y$, and we conclude. □

The definition of $\ll$ should remind us of the definition of Scott-opens. And indeed, we have:

Proposition 5.1.16 *Let X be a continuous poset, $x \in X$. Write $\twoheaduparrow x$ for the set of all elements $y \in X$ with $x \ll y$: $\twoheaduparrow x$ is always open in X_σ. More generally, for every subset E of X, the subset $\twoheaduparrow E = \bigcup_{x \in E} \twoheaduparrow x$ is open in X_σ.*

Proof First, $\twoheaduparrow x$ is upward closed since $x \ll y \leqslant z$ implies $x \ll z$. Second, given any directed family $(z_i)_{i \in I}$ with least upper bound z in $\twoheaduparrow x$, we first apply interpolation (Proposition 5.1.15) to obtain y such that $x \ll y \ll z$, the use the definition of $\ll$ to obtain an $i \in I$ such that $y \leqslant z_i$. Then $x \ll z_i$, i.e., $z_i \in \twoheaduparrow x$. So $\twoheaduparrow x$ is open in X_σ. □

We have more: on continuous posets, the Scott topology can be described completely through sets $\twoheaduparrow x$.

Theorem 5.1.17 (Base of the Scott topology, continuous posets) *Let X be a continuous poset. The Scott topology on X has a base of sets of the form $\twoheaduparrow x$, $x \in X$. More precisely, every open subset U of X_σ is the union of all the basic open subsets $\twoheaduparrow x$, $x \in U$.*

Proof Clearly, $\twoheaduparrow x \subseteq U$ for every $x \in U$. Conversely, let y be any element of U. Since X is continuous, y is the least upper bound of the directed family $\twoheaddownarrow y$. Since U is open in X_σ, some element x of $\twoheaddownarrow y$ is in U. That is, y is in $\twoheaduparrow x$ for some $x \in U$. □

> **Exercise 5.1.18** Show the following algebraic analogue of Theorem 5.1.17. Let X be an algebraic poset. The Scott topology on X has a base of sets of the form $\uparrow x$, x finite in X. More precisely, every open subset U of X_σ is the union of all the basic open subsets $\uparrow x$, x finite in U.

Another base of the Scott topology is given by Scott-open filtered sets. This is less useful, but is sometimes handy (see Proposition 8.3.19).

Proposition 5.1.19 *Let X be a continuous poset. The Scott topology on X has a base of Scott-open* filtered *subsets.*

Proof Let $x \in X$, and U be a Scott-open neighborhood of x in X. We must show that there is a Scott-open filtered neighborhood of x included in U, and use Lemma 4.1.10. Since X is continuous, x is the least upper bound of the directed family of all $y \ll x$, so some $y \ll x$ is already in U. By interpolation (Proposition 5.1.15), find x_0 such that $y \ll x_0 \ll x$. Repeat the process, and find a countable sequence of points x_n, $n \in \mathbb{N}$, such that $y \ll \cdots \ll x_n$

$\ll \cdots \ll x_1 \ll x_0 \ll x$. Let $V = \bigcup_{n \in \mathbb{N}} \uparrow x_n$. V is clearly filtered, contains x, and is contained in U. It only remains to show that V is Scott-open, and this will follow from the fact that V is also equal to $\bigcup_{n \in \mathbb{N}} \twoheaduparrow x_n$. Indeed, $\twoheaduparrow x_n \subseteq \uparrow x_n$ for every $n \in \mathbb{N}$, so $\bigcup_{n \in \mathbb{N}} \twoheaduparrow x_n \subseteq V$, and the converse inclusion follows from the fact that $\uparrow x_n \subseteq \twoheaduparrow x_{n+1}$. □

One can often describe every element x of a continuous poset, not as the least upper bound of all elements $y \ll x$, but of all elements $y \ll x$ taken from a specific set of approximants. For example, in $\mathbb{R}$, every real number x is the least upper bound of all *rational* numbers $y < x$. This is interesting in the sense that all rational numbers are representable on a machine, for example.

Trick 5.1.20 To show that a poset X is continuous, it is enough to find, for each $x \in X$, a directed family of elements $y \ll x$ whose least upper bound is x.

That is, we do not need to identify precisely *all* elements that are way-below x.

Proof Assume $(x_i)_{i \in I}$ is a directed family of elements way-below x whose least upper bound is x. We must show that $\twoheaddownarrow x$ is directed; that $\twoheaddownarrow x$ has x as least upper bound is then clear. Given $y, z \in \twoheaddownarrow x$, since $y \ll x$ and $x = \sup_{i \in I} x_i$, $y \leqslant x_i$ for some $i \in I$. Similarly, $z \leqslant x_i$ for some $i \in I$. Since I is directed, we may choose the same i in both cases. Then x_i is above y and z, and is in $\twoheaddownarrow x$. □

Exercise 5.1.21 Similarly, assume that every element x of the poset X is the least upper bound of some directed family $(x_i)_{i \in I}$ of finite elements below x. Show that X is algebraic.

Definition 5.1.22 Let X be a poset. A *basis* B of X is a subset of X such that, for every $x \in X$, the collection $B \cap \twoheaddownarrow x$ of all elements of the basis way-below x is directed, and has x as least upper bound.

Lemma 5.1.23 *A poset X has a basis if and only if it is continuous.*

Proof If X is continuous, then X itself is a basis (the largest one). Conversely, let B be a basis of X. Then x is the least upper bound of $B \cap \twoheaddownarrow x$, which is a directed family of elements way-below x: apply Trick 5.1.20. □

Corollary 5.1.24 *Every algebraic poset is continuous. The set of finite elements is a basis.*

Proof By Proposition 5.1.4, every finite element y below $x \in X$ is way-below x. So the set of finite elements is a basis. Then apply Lemma 5.1.23. □

Exercise 5.1.25 In an algebraic poset, the set of finite elements is a basis. Show that this is the smallest basis.

Conversely, assume that X is a continuous poset with a minimal basis $\mathcal{B}$. Show that if $b \in \mathcal{B}$ is not finite, then $\mathcal{B} \setminus \{b\}$ is another basis, contradicting minimality. Conclude that X must be algebraic, and $\mathcal{B}$ is the set of finite elements of X.

One can then refine the results obtained above. Their proofs are left as an exercise to the reader.

Proposition 5.1.26 (Interpolation, basis case) *Let X be a continuous poset with basis B. If $x \ll y$, then $x \ll z \ll y$ for some $z \in B$.*

Theorem 5.1.27 (Base of the Scott topology, basis case) *Let X be a continuous poset with basis B. The Scott topology on X has a base of sets of the form $\Uparrow x$, $x \in B$. More precisely, every open subset U of X_σ is the union of all the basic open subsets $\Uparrow x$, $x \in B \cap U$.*

Trick 5.1.28 To show that a subset B of a poset X is a basis, it is enough to find, for each $x \in X$, a directed family of elements of B way-below x whose least upper bound is x.

Exercise 5.1.29 Show that any basis B of a continuous poset X is dense in X_σ. However, explain why dense subsets of X_σ may fail to be bases.

Exercise 5.1.30 Show a form of converse of Proposition 5.1.26: if X is a continuous poset, and B is a subset of X such that, for any two elements x, y in X such that $x \ll y$, there is a $b \in B$ such that $x \ll b \ll y$, then B is a basis of X.

Exercise 5.1.31 (Interpolation, algebraic posets) Show that, in an algebraic poset X, interpolation takes the following simple form: $x \ll y$ if and only if $x \leqslant b \leqslant y$ for some finite element b of X.

One can also express axioms that any basis of a continuous poset should satisfy, and reconstruct a continuous poset, even a continuous dcpo, from it.

Lemma 5.1.32 (Abstract basis) *A binary relation $\prec$ on a set B is* interpolative *if and only if, for every finite subset E of B and every $b \in B$, if $E \prec b$—which we take to abbreviate the fact that $x \prec b$ for every $x \in E$—then there is a $b' \in B$ such that $E \prec b' \prec b$.*

An abstract basis *is a set B with a binary relation $\prec$ that is transitive and interpolative.*

Given any continuous poset X and any basis B of X, B together with the restriction $\prec$ of $\ll$ to B is an abstract basis.

Proof First, $\prec$ is transitive, because $\ll$ is. To show that $\prec$ is interpolative, let $E \prec b$, i.e., $x_i \ll b$ for every i, $1 \leqslant i \leqslant n$, where $E = \{x_1, \ldots, x_n\}$. By Proposition 5.1.26, there is a $b_i \in B$ such that $x_i \ll b_i \ll b$. Since X is a continuous poset, $b = \sup(B \cap \downarrow b)$, and $B \cap \downarrow b$ is directed. So there is a $b' \in B$ with $b' \ll b$ and $b_i \leqslant b'$ for every i, $1 \leqslant i \leqslant n$. So $b' \prec b$ and $E \prec b'$. □

Proposition 5.1.33 (Rounded ideal completion) *Let $B, \prec$ be an abstract basis. A* rounded ideal *of B is any subset D of B that is* downward closed *(if $b' \prec b$ and $b \in D$, then $b' \in D$) and* directed *(for every finite subset E of D, there is a $b \in D$ such that $E \prec b$). The* rounded ideal completion $\mathbf{RI}(B, \prec)$ *of $(B, \prec)$ is the set of all rounded ideals of $(B, \prec)$ ordered by inclusion.*

$\mathbf{RI}(B, \prec)$ is always a continuous dcpo, where least upper bounds of directed families are computed as unions. It has a basis B' consisting of the elements $\Downarrow b = \{b' \in B \mid b' \prec b\}$. In $\mathbf{RI}(B, \prec)$, $D \ll D'$ iff $D \subseteq \Downarrow b$ for some $b \in D'$.

Every continuous dcpo arises this way, in the sense that, for any continuous dcpo X, with basis B, X is order isomorphic to $\mathbf{RI}(B, \prec)$, where $\prec$ is the restriction of $\ll$ to B. Precisely, the order isomorphism maps $x \in X$ to $B \cap \downarrow x \in \mathbf{RI}(B, \prec)$, and, conversely, $D \in \mathbf{RI}(B, \prec)$ to $\sup D$.

The notion of a directed set with respect to $\prec$ is deeper than one may expect at first. Taking E empty yields the fact that any directed set is non-empty. Taking E of cardinality 2 gives back the usual definition for quasi-orderings (Definition 4.2.16). But $\prec$ is not necessarily reflexive, and the new feature is obtained by taking $E = \{b\}$: for each element b of the directed set D, there must be an element b' in D such that $b \prec b'$.

Proof First, $\mathbf{RI}(B, \prec)$ is a dcpo. It is enough to show that the union D of any directed family of rounded ideals $(D_i)_{i \in I}$ is again a rounded ideal. If $b' \prec b$ and $b \in D$, indeed $b \in D_i$ for some $i \in I$, so $b' \in D_i$, hence $b' \in D$. For every finite subset E of D, E is contained in some D_i, $i \in I$, because $(D_i)_{i \in I}$ is directed. Since D_i itself is directed, there is a $b \in D_i$ such that $E \prec b$. Since $b \in D$, D is directed.

Given any rounded ideal D, D is the least upper bound of the family of all $\Downarrow b$, $b \in D$. Indeed, first, $\Downarrow b \subseteq D$ for every $b \in D$ since D is downward closed; next, we must show that every $b' \in D$ is in some $\Downarrow b$ with $b \in D$. This is obtained since D is directed, applied to the finite subset $E = \{b'\}$. Finally, the family of all $\Downarrow b$, $b \in D$, is directed: it is non-empty because D is (D is directed, and consider $E = \emptyset$), and whenever $b', b'' \in D$, there is a $b \in D$ such that $b', b'' \prec b$ (since D is directed, taking $E = \{b', b''\}$); therefore $\Downarrow b', \Downarrow b'' \subseteq \Downarrow b$, using the fact that $\prec$ is transitive.

It follows that, if $D \ll D'$, where $\ll$ is the way-below relation on $\mathbf{RI}(B, \prec)$, then $D \subseteq \Downarrow b$ for some $b \in D'$. Conversely, if $D \subseteq \Downarrow b$ for some $b \in D'$, we claim that $D \ll D'$. For every directed family $(D_i)_{i \in I}$ of rounded ideals whose union contains D', $\bigcup_{i \in I} D_i$ contains b, so $b \in D_i$ for some $i \in I$. Since D_i is downward closed, $D \subseteq D_i$. So $D \ll D'$.

In particular, $\Downarrow b \ll D'$ whenever $b \in D'$. Since D' is the least upper bound of the directed family of all elements $\Downarrow b$, $b \in D'$, B' is a basis, as required.

Finally, let X be any continuous dcpo, with basis B. We have seen that $(B, \prec)$ is an abstract basis, where $\prec$ is the restriction of $\ll$ to B. Define $f : X \to \mathbf{RI}(B, \prec)$ as mapping x to $B \cap \twoheaddownarrow x$. Define $g : \mathbf{RI}(B, \prec) \to X$ as mapping D to $\sup D$, where the latter least upper bound is taken in X. Note that g is well defined, since D is directed with respect to $\prec$ in B, hence with respect to $\leqslant$ in X. It is clear that both f and g are monotonic, and that $g \circ f = \mathrm{id}_X$, since X is a continuous poset. It remains to show that $f \circ g = \mathrm{id}_{\mathbf{RI}(B, \prec)}$, i.e., that, for every rounded ideal D, $D = B \cap \twoheaddownarrow \sup D$. For every $b \in B \cap \twoheaddownarrow \sup D$, $b \ll \sup D$. By interpolation, $b \ll b' \ll \sup D$ for some $b' \in B$, so $b' \leqslant b''$ for some $b'' \in D$; so $b \ll b''$, i.e., $b \prec b''$; since $b'' \in D$ and D is downward closed (with respect to $\prec$), b is in D. Conversely, for every $b \in D$, b is in B by construction of the rounded ideals, and we claim that $b \ll \sup D$. Indeed, since D is directed, with $E = \{b\}$, $b \ll b'$ for some $b' \in D$. Clearly, $b' \leqslant \sup D$, so $b \ll \sup D$. □

We shall see later (Exercise 8.2.48) that the rounded ideal completion is a special case of a more general construction called sobrification.

Exercise 5.1.34 (Quasi-continuous posets) The quasi-continuous dcpos enjoy many of the properties of a continuous dcpo, while generalizing them. We start by examining quasi-continuous posets.

Let X be a poset. For any subset E of X, and any element x of X, we say that E is *way-below* x, and write $E \ll x$, if and only if, for every directed family $(z_i)_{i \in I}$ that has a least upper bound above x, $y \leqslant z_i$ for some $y \in E$ and some $i \in I$. This generalizes the usual way-below relation: $y \ll x$ in the usual sense iff $\{y\} \ll x$ in the new sense. However, it may be that $E \ll x$, i.e., the elements of E are (collectively) way-below x, but no y in E is (individually) way-below x, i.e., $y \ll x$ for no $y \in E$.

A *quasi-continuous poset* is a poset X such that, for every $x \in X$, $\uparrow x$ is the intersection of the filtered family of all sets of the form $\uparrow E$ with E finite, $E \ll x$. Show that every continuous poset is quasi-continuous, but that there are quasi-continuous posets that are not continuous (consider $\mathcal{N}_2$; see Figure 5.1). To show the former, first show that in a continuous poset, $E \ll x$ iff $y \ll x$ for some $y \in E$ – something that will not hold in every quasi-continuous poset.

Show the following generalized interpolation property: in a quasi-continuous poset X, for every $x \in X$ and every finite subset E such that $E \ll x$, there is a finite subset E' such that $E \ll E' \ll x$. We write $E \ll E'$ to mean $E \ll y$ for every $y \in E'$. Deduce that, when X is a quasi-continuous poset, for every finite subset E of X, the set of all elements $x \in X$ such that $E \ll x$ is open in X_σ.

The mysterious notation $\twoheaduparrow x$ has a clear topological meaning:

Proposition 5.1.35 *Let X be a continuous poset. For every $x \in X$, $\twoheaduparrow x$ is the interior of $\uparrow x$. More generally, for every finite subset E of X, $\twoheaduparrow E = int(\uparrow E)$.*

Proof First, $\twoheaduparrow E$ is open by Proposition 5.1.16, and included in $\uparrow E$. Conversely, for every open U included in $\uparrow E$, we must show that every $y \in U$ is such that $x \ll y$ for some $x \in E$. Since X is continuous, y is the least upper bound of the directed family of all $z \ll y$. So one of these elements z is in U. Since $U \subseteq \uparrow E$, there is an $x \in E$ such that $x \leqslant z$, whence $x \ll y$. □

Corollary 5.1.36 (Continuous posets are locally compact) *Every continuous poset X is locally compact in its Scott topology: for every $x \in X$, and every open neighborhood U of x in X_σ, there is a compact saturated subset Q of the form $\uparrow y$ with $x \in int(Q)$ and $Q \subseteq U$.*

Proof Let U be an open neighborhood of $x \in X$. By Theorem 5.1.17, there is a $y \in U$ such that $x \in \twoheaduparrow y$. By Proposition 5.1.35, we can take $Q = \uparrow y$ (a finitary compact), so that $int(Q) = \twoheaduparrow y$, and then $x \in int(Q)$ and $Q \subseteq U$. □

5.1.2 c-spaces

Corollary 5.1.36 shows that continuous posets enjoy a stronger form of local compactness, where Q can be taken of the form $\uparrow y$. These spaces are the c-spaces of Erné (1991). The T_0 c-spaces are Erné's *C-spaces* (notice the different capitalization), also called the locally supercompact spaces, or Ershov's (1973) A-spaces (equivalently, α-spaces; see Ershov (1997)).

Proposition 5.1.37 (c-spaces) *A* c-space *is any topological space X such that, for every $x \in X$, for every open neighborhood U of x, there is a point $y \in U$ such that x is in the interior of $\uparrow y$.*

Let X be any poset. Then X is continuous if and only if X_σ is a c-space.

Proof If X is continuous, then X_σ is a c-space, by Corollary 5.1.36. Conversely, fix some element x of X. By assumption: $(*)$ for every open neighborhood U of x in X_σ, there is a point $y \in U$ such that x is in the interior of $\uparrow y$.

Let D be the set of all elements $y \in X$ such that $x \in int(\uparrow y)$. D is nonempty by $(*)$, taking $U = X$. We claim that D is directed. If $y, y' \in D$, then $x \in int(\uparrow y) \cap int(\uparrow y') = int(\uparrow y \cap \uparrow y')$ (see Exercise 4.1.26). Using $(*)$ with $U = int(\uparrow y \cap \uparrow y')$, there is a point $y'' \in U$, in particular $y, y' \leqslant y''$, such that $x \in int(\uparrow y'')$, i.e., $y'' \in D$.

We now claim that every element y of D is way-below x. Every directed family $(y_i)_{i \in I}$ that has a least upper bound above x must be such that $y_i \in int(\uparrow y)$ for some $i \in I$, in particular $y \leqslant y_i$.

Finally, we claim that $x = \sup D$. First, if $y \in D$, then $x \in \uparrow y$, so x is an upper bound of D. Let z be any other upper bound of D. We show that $x \leqslant z$ by showing that every open neighborhood U of x contains z, and using the fact that $\leqslant$ is the specialization quasi-ordering of X_σ (Proposition 4.2.18). By $(*)$, since $x \in U$, there is an element $y \in D$ such that $y \in U$. Since z is an upper bound of D, $y \leqslant z$, and since U is upward closed, $z \in U$. □

We shall see later (Proposition 8.3.36) that the connection between c-spaces and continuous posets is even tighter: the sober c-spaces are exactly the continuous dcpos with their Scott topology.

Exercise 5.1.38 Show that X is a c-space iff, for every family $(A_i)_{i \in I}$ of upward closed subsets of X, $int(\bigcup_{i \in I} A_i) = \bigcup_{i \in I} int(A_i)$.

Exercise 5.1.39 (b-spaces) A *b-space* is any topological space X such that, for every $x \in X$, for every open neighborhood U of x, there is a point $y \in U$ such that $\uparrow y$ is open and contains x. It is clear that every b-space is a c-space.

Let X be any poset. Show that X_σ is a b-space if and only if X is algebraic.

Proposition 5.1.40 (Retracts of c-spaces) *Any retract of a c-space is a c-space.*

Proof Let $r: Y \to X$ be the retraction, $s: X \to Y$ be the associated section, where Y is a c-space. Let $x \in X$, U be an open neighborhood of x in X. Then $r^{-1}(U)$ is an open neighborhood of $s(x)$ in Y. Since Y is a c-space, there is an element $y \in Y$ and an open subset W of Y such that $s(x) \in W \subseteq \uparrow y$ and $y \in r^{-1}(U)$. Then $x \in s^{-1}(W)$, $s^{-1}(W) \subseteq \uparrow r(y)$ (every $x' \in s^{-1}(W)$ is such that $s(x') \in W$, so $y \leqslant s(x')$, so $r(y) \leqslant r(s(x')) = x'$, using the fact that r is continuous, hence monotonic by Proposition 4.3.9), and $r(y) \in U$. So X is a c-space. □

We shall also see that c-spaces and abstract bases are essentially the same thing (Exercise 8.3.46). Here is a first view into this.

Lemma 5.1.41 (C-space $\Rightarrow$ abstract basis) *Let X be a c-space. Define $x \lll y$ iff $y \in int(\uparrow x)$. Then $X, \lll$ is an abstract basis.*

Proof If $x \lll y$ and $y \lll z$, i.e., $y \in int(\uparrow x)$ and $z \in int(\uparrow y)$, then $z \in int(\uparrow int(\uparrow x)) = int(int(\uparrow x)) = int(\uparrow x)$, so $x \lll z$. So $\lll$ is transitive.

Let E be a finite subset of X, $y \in X$, and assume that every element z of E is such that $z \lll y$, i.e., y is in some open subset U_z included in $\uparrow z$ for each $z \in E$. So y is in the open subset $U = \bigcap_{z \in E} U_z$ (i.e., X itself if E is empty). Since X is a c-space, there is a point $x \in U$ such that $y \in int(\uparrow x)$, i.e., $x \lll y$. Also, $z \lll x$ for every $z \in E$ since $x \in U_z$ for every $z \in E$. So $\lll$ is interpolative. □

Exercise 5.1.42 (Locally finitary compact spaces) A topological space X is *locally finitary compact* if and only if, for every $x \in X$, for every open neighborhood U of x, there is a finitary compact $\uparrow E$ (i.e., with E finite) included in U such that x is in the interior of $\uparrow E$. Every c-space is clearly locally finitary compact.

Check that, for example, the dcpo $\mathcal{N}_2$ of Figure 5.1 is locally finitary compact in its Scott topology, but not a c-space. Give an example of a locally compact space that is not locally finitary compact. So local finitary compactness is a property in between being a c-space and being locally compact.

Show that every retract of a locally finitary compact space is locally finitary compact.

5.1.3 Retracts of continuous posets

Recall that, for any two posets X and Y, X is a Scott-retract of Y iff X_σ is a retract of Y_σ.

Lemma 5.1.43 (Scott-retracts of continuous posets) *Any poset X which is a Scott-retract of a continuous poset Y is a continuous poset.*

Proof By Proposition 5.1.37, Y_σ is a c-space. So X_σ is a c-space by Proposition 5.1.40. By Proposition 5.1.37 again, X is a continuous poset. □

We shall improve on this in Corollary 8.3.37.

Exercise 5.1.44 Let X be a poset which is a Scott-retract of a continuous poset Y. Let r be the retraction, and s be the section. Given a basis B of Y, show that $r[B]$ is a basis of X: show that if $b \ll s(y)$ in Y, then $r(b) \ll y$ in X, and use Trick 5.1.28. This yields another proof of Lemma 5.1.43.

To show the converse, we need to restrict ourselves to the case of dcpos X. The right construction is the ideal completion $\mathbf{I}(X)$. This is the special case of the rounded ideal completion (Proposition 5.1.33) $\mathbf{RI}(B, \prec)$, where B is a poset X, and $\prec$ is defined as the ordering $\leqslant$ of X.

Definition 5.1.45 (Ideal completion) Let X be a poset. An *ideal* of X is a downward closed directed subset of X. The *ideal completion* $\mathbf{I}(X)$ is the poset of all ideals of X, ordered by inclusion.

Proposition 5.1.46 (The ideal completion is algebraic) *Let X be a poset. Then $\mathbf{I}(X)$ is an algebraic dcpo, where least upper bounds of directed subsets are computed as unions. Its finite elements are those ideals of the form $\downarrow x$, $x \in X$.*

Proof $\mathbf{I}(X) = \mathbf{RI}(X, \leqslant)$ is a continuous dcpo by Proposition 5.1.33, and $D \ll D$ iff $D \subseteq \Downarrow x$ for some $x \in D$, i.e., iff $D = \downarrow x$ for some $x \in D$, since $\Downarrow x = \downarrow x$. Since the sets $\Downarrow x$ form a basis, we conclude. □

Exercise 5.1.47 (Algebraic dcpo = ideal completion of its finite elements) Let X be an algebraic dcpo, and $\mathcal{K}(X)$ be its poset of finite elements. Show that X arises as the ideal completion of $\mathcal{K}(X)$, up to order isomorphism. The order isomorphism maps $x \in X$ to $\downarrow x \cap \mathcal{K}(X) \in \mathbf{I}(\mathcal{K}(X))$, and conversely every $D \in \mathbf{I}(\mathcal{K}(X))$ to $\sup D$.

Theorem 5.1.48 (Continuous = retract of algebraic) *The continuous dcpos are exactly those posets that arise as Scott-retracts of algebraic dcpos.*

More precisely, every continuous dcpo X arises as a retract of the algebraic dcpo $\mathbf{I}(X)$, with section s defined by $s(x) = \twoheaddownarrow x$ and retraction r defined by $r(D) = \sup D$.

Proof By Corollary 5.1.24 and Lemma 5.1.43, every poset X that arises as a Scott-retract of an algebraic poset Y is continuous. Also, if Y is a dcpo, then so is X by Proposition 4.10.34.

Conversely, let X be a continuous dcpo, and define Y as $\mathbf{I}(X)$. $\mathbf{I}(X)$ is an algebraic dcpo by Proposition 5.1.46. Define $s(x) = \twoheaddownarrow x$, $r(D) = \sup D$, the latter being defined because X is a dcpo. (The theorem does not apply on mere continuous *posets*.) Clearly, $r \circ s = \mathrm{id}_X$, because X is continuous, i.e., $\sup_{y \in \twoheaddownarrow x} y = x$. Next, r is trivially Scott-continuous, and it remains to show that s is Scott-continuous to prove that X is a Scott-retract of Y. First, s is monotonic, since, whenever $x \leqslant x'$, every element of $\twoheaddownarrow x$ is way-below x' as well, by Proposition 5.1.4. Consider any directed family $(x_i)_{i \in I}$, and $x = \sup_{i \in I} x_i$. We have $\bigcup_{i \in I} s(x_i) \subseteq s(\sup_{i \in I} x_i)$ since s is monotonic. For the converse inclusion, every element y of $s(\sup_{i \in I} x_i)$ is way-below $\sup_{i \in I} x_i$, so $y \ll z \ll \sup_{i \in I} x_i$ for some $z \in X$ by interpolation (Proposition 5.1.15). So $y \ll z \leqslant x_i$ for some $i \in I$, hence $y \in s(x_i)$. □

Exercise 5.1.49 Let L be a pointed lattice. Show that the ideal completion $\mathbf{I}(L)$ is a complete lattice, in which greatest lower bounds are intersections. (Do not try to understand least upper bounds. We shall come back to them in Exercise 9.5.11.)

Exercise 5.1.50 Up to isomorphism, retracts are both subspaces (equalizers) and quotients (coequalizers). Show that quotients alone do not preserve continuous dcpos: show that $(\mathcal{N}_2)_\sigma$ (Exercise 5.1.3) is the topological quotient of a continuous dcpo, although $\mathcal{N}_2$ is not itself a continuous dcpo.

Exercise 5.1.51 We now demonstrate that subspaces of continuous dcpos also fail to define continuous dcpos in their Scott topology in general. By Exercise 5.1.10, $[0, 1]$ is a continuous dcpo. By Proposition 5.1.54 below, $[0, 1]^2$ with the pointwise ordering $(x, y) \leqslant (x', y')$ iff $x \leqslant x'$ and $y \leqslant y'$ is a continuous dcpo. (Admit this. Also, admit that $(a, 1] \times (b, 1]$ is open in $([0, 1]^2)_\sigma$ for all $a, b \in [0, 1]$.) Show that the subspace N_2 of $([0, 1]^2)_\sigma$ consisting of the points $(1, 1)$, $\left(0, 1 - \frac{1}{2^{n+1}}\right)$ and $\left(1 - \frac{1}{2^{n+1}}, 0\right)$, $n \in \mathbb{N}$, is not a continuous dcpo in the ordering $\leqslant$, restricted to N_2, and that its topology is not the Scott topology either.

The next exercise shows a case of well-behaved subspaces of continuous dcpos.

Exercise 5.1.52 (Scott-closed subsets of dcpos, of continuous posets) Let F be a Scott-closed subset of X_σ, for some poset X. Show that, if $(x_i)_{i \in I}$ is any directed family in F, and has a least upper bound x in X, then x is the least upper bound of $(x_i)_{i \in I}$ in F. Conclude that if X is a dcpo, then so is F.

Show that, if X is a continuous dcpo, then so is F, that if B is any basis of X, then $B \cap F$ is a basis of F, and that, for all $x, y \in F$, x is way-below y in F iff x is way-below y in X. Deduce that the subspace topology on F coincides with its Scott topology.

5.1.4 Products of continuous posets

We said in Exercise 4.5.19 that the Scott topology on the product of two posets was finer, and in general strictly finer, than the product topology. In continuous posets, the two topologies agree.

Lemma 5.1.53 *Let X_1, X_2 be two posets, with respective orderings $\leqslant_1$, $\leqslant_2$, and respective way-below relations $\ll_1$, $\ll_2$. The way-below relation $\ll$ on $X_1 \times X_2$ is $\ll_1 \times \ll_2$, i.e., $(x, x') \ll (y, y')$ iff $x \ll_1 y$ and $x' \ll_2 y'$.*

Proof If: let $(z_i, z'_i)_{i \in I}$ be any directed family of elements of $X_1 \times X_2$ with least upper bound (z, z') above (y, y'). Then $(z_i)_{i \in I}$ is a directed family of elements of X_1, with least upper bound z above y, so $x \leqslant z_i$ for some $i \in I$. Similarly, $x' \leqslant z'_{i'}$ for some $i' \in I$. By directedness, we may choose i' and i equal. So $(x, x') \ll (y, y')$.

Only if: let $(z_i)_{i \in I}$ be any directed family of elements of X_1 with a least upper z bound above y. Then $(z_i, y')_{i \in I}$ is directed, and its least upper bound (z, y') is above (y, y'), so $(x, x') \leqslant (z_i, y')$ for some $i \in I$, that is, $x \leqslant z_i$. So $x \ll_1 y$. Similarly, $x' \ll_2 y'$. □

Lemma 5.1.53 extends easily to the product of any finite number of posets.

Proposition 5.1.54 (Finite products of continuous posets) *The product of finitely many continuous posets $X_1, \ldots, X_n$ is a continuous poset. Whenever B_1 is a basis of $X_1, \ldots, B_n$ is a basis of X_n, then $B_1 \times \cdots \times B_n$ is a basis of $X_1 \times \cdots \times X_n$.*

Moreover, the Scott topology agrees with the product topology, i.e., $(X_1 \times \cdots \times X_n)_\sigma = X_{1\,\sigma} \times \cdots \times X_{n\,\sigma}$.

The product sign $\times$ on the left-hand side of the latter inequality denotes poset product, while $\times$ on the right denotes the topological product.

Proof Consider any element $\vec{x} = (x_1, \ldots, x_n)$ in $X = X_1 \times \cdots \times X_n$. For each i, $1 \leqslant i \leqslant n$, x_i is the least upper bound of the directed family $B_i \cap \Downarrow_i x_i$, where $\Downarrow_i$ is for $\Downarrow$ inside X_i. (We write similarly $\ll_i$ for the way-below relation in X_i, and $\Uparrow_i$ for $\Uparrow$ in X_i.) So $\vec{x}$ is the least upper bound of a family of elements $\vec{b} = (b_1, \ldots, b_n)$ in $(B_1 \cap \Downarrow_1 x_1) \times \cdots \times (B_n \cap \Downarrow_n x_n)$, i.e., in $B = B_1 \times \cdots \times B_n$, and where $b_1 \ll_1 x_1, \ldots, b_n \ll_n x_n$. By Lemma 5.1.53, $\vec{b} \ll \vec{x}$, where $\ll$ is the way-below relation in X. It remains to observe that $(B_1 \cap \Downarrow_1 x_1) \times \cdots \times (B_n \cap \Downarrow_n x_n)$, as a product of finitely many directed sets, is directed.

For the second part, every Scott-open of X is a union of sets of the form $\Uparrow \vec{b}$, $\vec{b} \in B$, by Theorem 5.1.27. Writing $\vec{b}$ as $(b_1, \ldots, b_n)$, Lemma 5.1.53 states that $\Uparrow \vec{b} = \Uparrow_1 b_1 \times \cdots \Uparrow_n b_n$. So $\Uparrow \vec{b}$ is an open rectangle. It follows that every Scott-open is open in the product topology. The converse is the last part of Exercise 4.5.19. □

This also works in the case of infinite products, modulo a few extra assumptions.

Lemma 5.1.55 *Let $(X_i)_{i \in I}$ be a family of posets, where X_i is ordered by $\leqslant_i$, and has $\ll_i$ as way-below relation.*

Assume that every X_i, except possibly finitely many, is pointed, *that is, has a least element $\bot_i$. To fix ideas, let J be the set of indices $i \in I$ such that X_i is not pointed.*

The way-below relation $\ll$ on the poset product $\prod_{i \in I} X_i$ is given by $\vec{x} \ll \vec{y}$ iff there is a finite subset K of I containing J such that $x_k \ll_k y_k$ for every $k \in K$, and $x_i = \bot_i$ for every $i \in I \smallsetminus K$.

Proof If: Let $(\vec{z}_\ell)_{\ell \in L}$ be a directed family of elements of $X = \prod_{i \in I} X_i$ having a least upper bound $\vec{z}$ above $\vec{y}$. For each $k \in K$, there is an index $\ell_k \in L$ such that the kth component $z_{\ell k}$ of $\vec{z}_\ell$ is above x_k. By directedness, and since K is finite, we may take the same ℓ_k, say ℓ, for each $k \in K$. Then $\vec{x} \leqslant \vec{z}_\ell$. Observe that $x_k \leqslant z_{\ell k}$ whenever $k \in K$, and that $x_i = \bot_i \leqslant z_{\ell i}$ for every $i \in I \smallsetminus K$.

Only if: For each finite subset K of I containing J, let $\vec{y}_{\downarrow K}$ be the tuple whose kth component is y_k for every $k \in K$, and whose ith component when $i \in I \smallsetminus K$ is $\bot$. The family of all such $\vec{y}_{\downarrow K}$ is directed, and admits $\vec{y}$ as least upper bound. Since $\vec{x} \ll \vec{y}$, there must be a finite subset K of I containing J such that $\vec{x} \leqslant \vec{y}_{\downarrow K}$. This entails that $x_i = \bot_i$ for every $i \in I \smallsetminus K$. Moreover, writing $\vec{y}_{|K}$ for the vector $(y_k)_{k \in K}$, $\vec{x}_{|K}$ is way-below $\vec{y}_{|K}$ in $\prod_{k \in K} X_k$, so $x_k \ll_k y_k$ for every $k \in K$, using Lemma 5.1.53. □

An equivalent statement is: $\vec{x} \ll \vec{y}$ iff $x_i \ll_i y_i$ for every $i \in I$, and $x_i = \bot_i$ for every $i \in I \smallsetminus J$ except possibly finitely many. Indeed, if $x_i = \bot_i$, then certainly $x_i \ll_i y_i$.

This is mostly used when every X_i is pointed, i.e., J is empty, in which case $\vec{x} \ll \vec{y}$ iff $x_i \ll_i y_i$ for every $i \in I$, and every x_i except finitely many is equal to the bottom element.

Proposition 5.1.56 (Products of continuous pointed posets) *The product of a family $(X_i)_{i \in I}$ of continuous posets, where all but finitely many are pointed, is a continuous poset.*

To fix ideas, let J be the set of indices $i \in I$ such that X_i is not pointed, and B_i be a basis of X_i, $i \in I$. If X_i is pointed, let $\bot_i$ be its least element. Then B is a basis of $X = \prod_{i \in I} X_i$, where B consists of tuples $\vec{b}$ such that, for some finite subset K of I containing J, $b_k \in B_k$ for every $k \in K$, and $b_i = \bot_i$ for every $i \in I \smallsetminus K$.

Moreover, the Scott topology agrees with the product topology: $(\prod_{i \in I} X_i)_\sigma = \prod_{i \in I} X_{i\,\sigma}$.

Note that $\prod$ on the left is the poset product, while $\prod$ on the right is the topological product.

Proof As for Proposition 5.1.54, except we use Lemma 5.1.55 instead of Lemma 5.1.53. To show that every Scott-open is open in the product topology, notice that $\twoheaduparrow \vec{b}$ is a product $\prod_{i \in I} \twoheaduparrow_i b_i$. For every i except finitely many, $b_i = \bot_i$, so $\twoheaduparrow_i b_i$ is the whole of X_i. This shows that $\twoheaduparrow \vec{b}$ is indeed open in the product topology. □

Exercise 5.1.57 (Products of algebraic posets) Show that the product of a family $(X_i)_{i \in I}$ of algebraic posets, where all but finitely many are pointed, is an algebraic poset. Let J be the set of indices $i \in I$ such that X_i is not pointed. If X_i is pointed, let $\bot_i$ be its least element. Show that the finite elements of $\prod_{i \in I} X_i$ are the tuples $\vec{b}$ of finite elements of each X_i, where there is a finite subset K of I containing J such that b_k is finite for every $k \in K$ and $b_i = \bot_i$ for every $i \in I \setminus K$.

Example 5.1.58 Let $X_n = \{0, 1\}$ with equality as ordering, for each $n \in \mathbb{N}$. X_n is trivially an algebraic, hence a continuous, dcpo (Corollary 5.1.24), since every element is finite. However, the Scott topology on the poset product $\prod_{n \in \mathbb{N}} X_n$ is the discrete topology, hence is strictly finer than the product topology. (Check this: think about compactness.) So the assumption that all but finitely many posets are pointed is necessary in Proposition 5.1.56.

The situation with coproducts is much simpler.

Proposition 5.1.59 (Coproducts of continuous posets) *The product of any family of continuous posets X_i, $i \in I$, is a continuous poset. If B_i is a basis of X_i for each $i \in I$, then $\coprod_{i \in I} B_i$ is a basis of $\coprod_{i \in I} X_i$. The way-below relation $\ll$ on $\coprod_{i \in I} X_i$ is given by $(i, x) \ll (j, y)$ iff $i = j$ and x is way-below y in X_i. Finally, the Scott topology on $\coprod_{i \in I} X_i$ agrees with the coproduct topology, i.e., $(\coprod_{i \in I} X_i)_\sigma = \coprod_{i \in I} X_{i\,\sigma}$.*

Proof That $(i, x) \ll (j, y)$ iff $i = j$ and x is way-below y in X_i is obvious, as is the fact that $\coprod_{i \in I} B_i$ is a basis. Then the Scott topology has a base of sets of the form $\Uparrow(i, x)$ where $x \in B_i$, by Theorem 5.1.27. This is also a base of the coproduct topology. □

5.1.5 Scott's Formula and other extension results

One of the nice things about continuous posets is that, given any monotonic map f on a basis B, we can produce a best possible continuous approximation $\mathfrak{r}(f)$ to f on the whole space.

Proposition 5.1.60 (Scott's Formula) *Let Y be a* bdcpo (bounded-directed-complete poset)*, i.e., a poset in which every directed family that has an upper bound has a least upper bound.*

Let X be a continuous poset, with basis B, and f be a monotonic map from B to Y. There is a largest Scott-continuous map $\mathfrak{r}(f)$ below f, i.e., such that $\mathfrak{r}(f)(b) \leqslant f(b)$ for every $b \in B$. This is given by Scott's Formula*:*

$$\mathfrak{r}(f)(x) = \sup_{b \in B, b \ll x} f(b).$$

The typical case of a bdcpo is $\mathbb{R}$, or $\mathbb{R}^+$.

Proof First, $\mathfrak{r}(f)$ is well defined, since $\{f(b) \mid b \in B, b \ll x\}$ is directed, using the fact that f is monotonic; and it has an upper bound, namely $f(x)$. So it has a least upper bound, since Y is a bdcpo. It is easy to see that $\mathfrak{r}(f)$ is monotonic, since whenever $x \leqslant x'$, $b \ll x$ implies $b \ll x'$. Let us show that $\mathfrak{r}(f)$ is Scott-continuous. Let $(x_i)_{i \in I}$ be any directed family in X that has a least upper bound, say x. Then $\mathfrak{r}(f)(x) = \sup_{b \in B, b \ll x} f(b)$. Using interpolation (Proposition 5.1.15), if $b \ll x$ then $b \ll z \ll x$ for some $z \in X$, so $b \ll z \leqslant x_i$ for some $i \in I$; conversely, if $b \ll x_i$, then clearly $b \ll x$. So $\mathfrak{r}(f)(x) = \sup_{b \in B, \exists i \in I \cdot b \ll x_i} f(b) = \sup_{i \in I} \sup_{b \in B, b \ll x_i} f(b) = \sup_{i \in I} \mathfrak{r}(f)(x_i)$.

Let f' be any other Scott-continuous map such that $f'(b) \leqslant f(b)$ for every $b \in B$. Since B is a basis, for every $x \in X$, $x = \sup_{b \in B, b \ll x} b$, and, since f' is continuous, $f'(x) = \sup_{b \in B, b \ll x} f'(b) \leqslant \sup_{b \in B, b \ll x} f(b) = \mathfrak{r}(f)(x)$.

It only remains to show that $\mathfrak{r}(f)(x) \leqslant f(x)$ for every $x \in B$. Since f is monotonic, $f(b) \leqslant f(x)$ for every $b \in B$ such that $b \ll x$. So $\mathfrak{r}(f)(x) = \sup_{b \in B, b \ll x} f(b) \leqslant f(x)$. □

Corollary 5.1.61 (Extension) *Let Y be a bdcpo, and X be a continuous poset with basis B. Every Scott-continuous map f from B to Y has a unique Scott-continuous extension to the whole of X, and this is $\mathfrak{r}(f)$. An* extension *g of f to X is any map $g \colon X \to Y$ such that $g(b) = f(b)$ for every $b \in B$.*

Proof Since f is Scott-continuous from B to Y, for every $b' \in B$, $\mathfrak{r}(f)(b') = \sup_{b \in B, b \ll b'} f(b) = f(\sup_{b \in B, b \ll b'} b) = f(b')$. So $\mathfrak{r}(f)$ is indeed an extension of f. Any Scott-continuous extension g of f to X must satisfy $g(x) = \sup_{b \in B, b \ll x} g(b) = \sup_{b \in B, b \ll x} f(b) = \mathfrak{r}(f)(x)$, so the extension is unique. □

Exercise 5.1.62 (Extension, algebraic case) When X is algebraic, show that Corollary 5.1.61 has a simpler formulation: let Y be a dcpo, X be an algebraic poset, and B be the (minimal) basis of finite elements of X; then every monotonic map f from B to Y has a unique Scott-continuous extension to the whole of X, and this is $\mathfrak{r}(f)$. Indeed, show that every monotonic map f from B to Y is already Scott-continuous.

A variant of Scott's Formula is obtained by relaxing the constraints on X, but imposing more constraints on Y.

Proposition 5.1.63 *Let X be a topological space, and Y be a continuous poset in which every family that has an upper bound has a least upper bound—for example, a continuous complete lattice.*

For every map $f: X \to Y$, there is a largest continuous map $\mathfrak{r}'(f)$ below f. For every $x \in X$, $\mathfrak{r}'(f)(x)$ is the least upper bound of all elements $y \in Y$ such that $x \in int(f^{-1}(\uparrow y))$.

When X is a continuous poset, and f is monotonic, $\mathfrak{r}'(f)$ coincides with $\mathfrak{r}(f)$, as defined in Proposition 5.1.60, taking $B = X$.

Proof The family of all elements $y \in Y$ such that $x \in int(f^{-1}(\uparrow y))$ has an upper bound, namely $f(x)$. Indeed, if $x \in int(f^{-1}(\uparrow y))$ then $x \in f^{-1}(\uparrow y)$, so $y \leqslant f(x)$. So the least upper bound of these elements y exists by assumption.

For every open subset V of Y, $\mathfrak{r}'(f)^{-1}(V) = \{x \in X \mid \exists y \in V \cdot x \in int(f^{-1}(\uparrow y))\} = \bigcup_{y \in V} int(f^{-1}(\uparrow y))$ is open. So $\mathfrak{r}'(f)$ is continuous.

We check that $\mathfrak{r}'(f)$ is below f: as we have seen, every $y \in Y$ such that $x \in int(f^{-1}(\uparrow y))$ must be such that $y \leqslant f(x)$. Now let f' be any other continuous map below f. To show that $f'(x) \leqslant \mathfrak{r}'(f)(x)$, consider any $y \ll f'(x)$. Then x is in $f'^{-1}(\twoheaduparrow y)$, and the latter is open since f' is continuous. Since $f' \leqslant f$, one checks easily that $f'^{-1}(\twoheaduparrow y) \subseteq f^{-1}(\twoheaduparrow y) \subseteq f^{-1}(\uparrow y)$. Let V be the open subset $f'^{-1}(\twoheaduparrow y)$. We have $x \in V \subseteq f^{-1}(\uparrow y)$, so x is in $int(f^{-1}(\uparrow y))$. By definition, then, $y \leqslant \mathfrak{r}'(f)(x)$. Since Y is continuous, $f'(x)$ is the least upper bound of all such y that are way-below $f'(x)$. So $f'(x) \leqslant \mathfrak{r}'(f)(x)$.

Finally, when X is a continuous poset and f is monotonic, letting $B = X$, $\mathfrak{r}'(f)$ and $\mathfrak{r}(f)$ both are the largest continuous map below f, using Proposition 5.1.60. So they coincide. □

Example 5.1.64 Consider $f: [0, 1]^{\mathbb{N}} \to [0, 1]$ that maps $\vec{x}$ to $\inf_{n \in \mathbb{N}} x_n$. Clearly f is monotonic. However, not only is $\mathfrak{r}(f) = \mathfrak{r}'(f) = 0$, but 0 is the only continuous map below f. (Exercise: prove this.)

To illustrate the need for the constructions to come, imagine X comes with an addition operation $+$, that $Y = \mathbb{R}$, and that we know f is additive, in the sense that $f(x + y) = f(x) + f(y)$. We would like to conclude that $\mathfrak{r}(f)$ is also additive. This will hold if $+$ preserves and reflects $\ll$, in the following sense.

Definition 5.1.65 (Preserving, reflecting $\ll$) For every map $g: X^k \to X$ on a poset X, we say that g *preserves* $\ll$ iff whenever $v_1 \ll u_1, \ldots, v_k \ll u_k$ then $g(v_1, \ldots, v_k) \ll g(u_1, \ldots, u_k)$.

We say that g *reflects* $\ll$ iff whenever $v \ll g(u_1, \ldots, u_k)$, there are elements $v_1, \ldots, v_k$ in X such that $v \leqslant g(v_1, \ldots, v_k)$, $v_1 \ll u_1, \ldots,$ and $v_k \ll u_k$.

Proposition 5.1.66 (Continuous maps reflect $\ll$) *Let X be a continuous poset. Every Scott-continuous map $g\colon X^k \to X$ reflects $\ll$.*

Proof Assume that $v \ll g(u_1, \ldots, u_k)$. Since X is continuous, X^k is too, by Proposition 5.1.54. Using Lemma 5.1.53, $(u_1, \ldots, u_k)$ is the least upper bound of the directed family of all elements $(v_1, \ldots, v_k)$ with $v_1 \ll u_1, \ldots, v_k \ll u_k$. Since g is Scott-continuous, $g(u_1, \ldots, u_k)$ is the least upper bound of the directed family of elements $g(v_1, \ldots, v_k)$ with $v_1 \ll u_1, \ldots, v_k \ll u_k$. We conclude by the definition of $v \ll g(u_1, \ldots, u_k)$. □

Proposition 5.1.67 *Let X be a continuous poset with basis B, Y be a bdcpo, and f be a monotonic map from B to Y. Also let g be any monotonic map from X^k to X that preserves and reflects $\ll$. Then, for every $x_1, \ldots, x_k \in X$,*

$$\mathfrak{r}(f)(g(x_1, \ldots, x_k)) = \sup_{b_1 \ll x_1, \ldots, b_k \ll x_k} f(g(b_1, \ldots, b_k)),$$

where $b_1, \ldots, b_k \in B$.

Notice also that the least upper bound on the right is directed, since X^k is a continuous poset, and using Lemma 5.1.53.

Proof By definition, $\mathfrak{r}(f)(g(x_1, \ldots, x_k)) = \sup_{b \in B, b \ll g(x_1, \ldots, x_k)} f(b)$. Since g reflects $\ll$, whenever $b \ll g(x_1, \ldots, x_k)$, there are elements $v_1 \ll x_1, \ldots, v_k \ll x_k$ such that $b \leqslant g(v_1, \ldots, v_k)$. Using interpolation (Proposition 5.1.26), we can find $b_1, \ldots, b_k \in B$ such that $v_1 \ll b_1 \ll x_1, \ldots, v_k \ll b_k \ll x_k$. So $b \leqslant g(b_1, \ldots, b_k)$ since g is monotonic, whence $f(b) \leqslant f(g(b_1, \ldots, b_k))$. Taking least upper bounds over b yields $\mathfrak{r}(f)(g(x_1, \ldots, x_k)) \leqslant \sup_{b_1 \ll x_1, \ldots, b_k \ll x_k} f(g(b_1, \ldots, b_k))$.

Conversely, if $b_1 \ll x_1, \ldots, b_k \ll x_k$, then $g(b_1, \ldots, b_k) \ll g(x_1, \ldots, x_k)$ since g preserves $\ll$. That is, $g(b_1, \ldots, b_k)$ is one possible $b \in B$ such that $b \ll g(x_1, \ldots, x_k)$. So $\sup_{b_1 \ll x_1, \ldots, b_k \ll x_k} f(g(b_1, \ldots, b_k)) \leqslant \sup_{b \in B, b \ll g(x_1, \ldots, x_k)} f(b) = \mathfrak{r}(f)(g(b_1, \ldots, b_k))$. □

Example 5.1.68 It follows that if f is additive from the continuous poset X to $\mathbb{R}$, assuming X is equipped with some addition $+$, then $\mathfrak{r}(f)$ is also additive whenever $+$ preserves and reflects $\ll$ on X. Indeed, $\mathfrak{r}(f)(x + y) = \sup_{b_1 \ll x, b_2 \ll y} f(b_1 + b_2)$ (Proposition 5.1.67) $= \sup_{b_1 \ll x, b_2 \ll y}(f(b_1) + f(b_2))$ (since f is additive) $= \sup_{b_1 \ll x} f(b_1) + \sup_{b_2 \ll y} f(b_2)$ (since addition on $\mathbb{R}$ is Scott-continuous; see Example 4.3.7) $= \mathfrak{r}(f)(x) + \mathfrak{r}(f)(y)$.

Proposition 5.1.69 ($\mathfrak{r}$ is a Scott-retraction) *Let X be a continuous poset with basis B, and Y be a bdcpo. Let Z be any set of monotonic maps from B to Y, and Z' be the set of all continuous maps f from X_σ to Y_σ whose restriction to B is in Z. (The restriction $f_{|B}$ of f to B is the map from B to Y that sends b to $f(b)$.) Equip each of these spaces with the topology of* pointwise convergence, *i.e., the subspace topologies from Y_σ^B, resp., Y_σ^X.*

Assume that $\mathfrak{r}(f)$ is in Z for every map f in Z. Then $\mathfrak{r}$ is a Scott-retraction of Z onto Z', and the associated section $\mathfrak{s}$ maps each $f \in Z'$ to $f_{|B}$.

Proof A subbase of the topology of Z is given by the subsets $Z \cap \pi_b^{-1}(V)$, $b \in B$, V open in Y_σ (where π_b is defined in Definition 4.5.2, i.e., $\pi_b(f) = f(b)$), and $\mathfrak{s}^{-1}(Z \cap \pi_b^{-1}(V)) = Z' \cap \pi_b^{-1}(V)$, which is open in Z'. Then apply Trick 4.3.3.

We claim that $\mathfrak{r}$ is also continuous. For every $x \in X$, for every open subset V of Y_σ, $\mathfrak{r}^{-1}(Z' \cap \pi_x^{-1}(V))$ is the set of all maps $f \in Z$ such that $\mathfrak{r}(f)(x) \in V$, i.e., such that $\sup_{b \in B, b \ll x} f(b) \in V$, or, equivalently, such that $f(b) \in V$ for some $b \in B$, $b \ll x$. So $\mathfrak{r}^{-1}(Z' \cap \pi_x^{-1}(V)) = \bigcup_{b \in B, b \ll x} \pi_b^{-1}(V)$ is open in Z. We conclude by Trick 4.3.3 again.

Finally, we claim that $\mathfrak{r} \circ \mathfrak{s} = \mathrm{id}_{Z'}$: for every $f \in Z'$, since f is continuous from X_σ to Y_σ, f is Scott-continuous by Proposition 4.3.5, so $\mathfrak{r}(\mathfrak{s}(f))(x) = \sup_{b \in B, b \ll x} f_{|B}(b) = f(x)$. □

Example 5.1.70 In particular, if X is a continuous poset and Y is a bdcpo, then the poset $[X \to Y]$ of all continuous maps from X to Y, with the pointwise ordering ($f \leqslant g$ iff $f(x) \leqslant g(x)$ for every $x \in X$), is a Scott-retract of the poset of all monotonic maps from X to Y.

Example 5.1.71 For a more involved example, if X is a continuous poset with some addition $+$, and $+$ preserves and reflects $\ll$, then the poset of all continuous additive maps from X to $\mathbb{R}_\sigma$ is a Scott-retract of the poset of all monotonic additive maps from X to $\mathbb{R}_\sigma$.

Retractions preserve many properties: compactness, local compactness, being a dcpo, and so on. It follows that, if one can prove that the space of all monotonic additive maps is, say, locally compact, then so will be the space of all continuous additive maps.

5.2 The lattice of open subsets of a space

We have given a first taste of the deep connection between order and topology in Section 4.2. Another one goes through the study of the complete lattice of open sets of a topological space.

Definition 5.2.1 ($\mathcal{O}(X)$) Given any topological space X, we let $\mathcal{O}(X)$ be the set of all open subsets of X, ordered by inclusion $\subseteq$.

Exercise 5.2.2 $\mathcal{O}(X)$ is a complete lattice (see Section 2.3.2). The least upper bound of a family of open subsets $(U_i)_{i \in I}$ is clearly its union $\bigcup_{i \in I} U_i$. What is its greatest lower bound?

As a poset, one may wonder when $\mathcal{O}(X)$ is continuous. We shall call X *core-compact* when this is the case. Core-compact spaces arise in many different contexts. We shall explore a few of them.

5.2.1 Core-compact spaces

Definition 5.2.3 (Core-compact) A topological space X is *core-compact* if and only if $\mathcal{O}(X)$ is a continuous poset.

Definition 5.2.4 ($\Subset$) Given any topological space, let $\Subset$ be the way-below relation on $\mathcal{O}(X)$. In other words, $U \Subset V$ iff, from any open covering $(U_i)_{i \in I}$ of V, one can extract a finite subcover of U.

The equivalence of the two sentences follows from Trick 4.4.6. Notice how the latter sentence characterizes $\Subset$ as a variant of the compactness property. We read $U \Subset V$ as: U is *relatively compact* in V.

Lemma 5.2.5 *Let X be a topological space, $U \in \mathcal{O}(X)$, and $(V_i)_{i=1}^n$ be a finite family of open subsets of X such that $V_i \Subset U$ for every i, $1 \leqslant i \leqslant n$. Then $\bigcup_{i=1}^n V_i \Subset U$.*

Proof Let $(U_j)_{j \in J}$ be an open covering of U. For each i, $1 \leqslant i \leqslant n$, extract a finite subcover $(U_j)_{j \in J_i}$ of V_i. Then $(U_j)_{j \in \bigcup_{i=1}^n J_i}$ is a finite subcover of $\bigcup_{i=1}^n V_i$, so $\bigcup_{i=1}^n V_i \Subset U$. ☐

So core-compactness can be characterized in a way that sounds like local compactness:

Proposition 5.2.6 *A space X is core-compact if and only if, for every $x \in X$ and for every open neighborhood V of x in X, there is an open subset U such that $x \in U \Subset V$.*

Proof If: Every open subset V is the union of all opens $U \Subset V$; indeed, for every $x \in V$, x is in one such U, by assumption. Then the family of all opens $U \Subset V$ is directed by Lemma 5.2.5. So $\mathcal{O}(X)$ is a continuous poset.

Only if: Let V be an open neighborhood of x. Since $\mathcal{O}(X)$ is a continuous poset, V is the union of the directed family of all opens $U \Subset V$. Then x must be in one of them. ☐

Here is another characterization of core-compact spaces.

Exercise 5.2.7 (Core-compact $\Leftrightarrow$ ($\Subset$) open) Let X be any topological space. Show that the subset ($\Subset$) of $X \times \mathcal{O}(X)_\sigma$ defined by $(\Subset) = \{(x, U) \in X \times \mathcal{O}(X) \mid x \in U\}$ is open in the product topology if and only if X is core-compact.

The connection with local compactness is deeper. We start with:

Lemma 5.2.8 *Let X be a topological space, $U, V \in \mathcal{O}(X)$. If $U \subseteq K \subseteq V$ for some compact subset K of X, then $U \Subset V$.*

Proof Any open cover containing V contains K: extract a finite subcover of K, and this will be a finite subcover of U. □

Theorem 5.2.9 (Locally compact $\Rightarrow$ core-compact) *Every locally compact space X is core-compact.*

In this case, for any two opens U, V of X, $U \Subset V$ if and only if $U \subseteq Q \subseteq V$ for some compact saturated subset Q of X.

Proof Let X be locally compact. Let $x \in X$, V be an open neighborhood of x. Then $x \in int(Q)$ for some compact saturated subset Q of V, so $int(Q) \Subset V$ by Lemma 5.2.8. So X is core-compact.

Next, assume $U \Subset V$. For each $x \in V$, use the Axiom of Choice and pick a compact saturated subset Q_x of V such that $x \in int(Q_x)$. Clearly, $(int(Q_x))_{x \in V}$ is an open cover of V. So there is a finite subset A of V such that $U \subseteq \bigcup_{x \in A} int(Q_x)$. So U is included in $Q = \bigcup_{x \in A} Q_x$, which is compact by Proposition 4.4.12, and $Q \subseteq V$. □

We shall see later that, conversely, every sober, core-compact space is locally compact; i.e., core-compactness and local compactness are the same notion for sober spaces (Theorem 8.3.10). In particular, since every T_2 space happens to be sober (Proposition 8.2.12 (*a*)), core-compactness and local compactness will be the same notion on T_2 spaces.

Exercise 5.2.10 (Coproducts of core-compact spaces) Show that any coproduct of core-compact spaces is core-compact.

Exercise 5.2.11 (Alexander's Subbase Lemma for $\Subset$) Here is another connection between compactness and the $\Subset$ relation. Show the following generalization of Alexander's Subbase Lemma (Theorem 4.4.29). Let X be a topological space, and $\mathcal{B}$ be a subbase for its topology. Let U, V be two open subsets of X. Show that $U \Subset V$ if and only if, for every family $(U_i)_{i \in I}$ of elements of $\mathcal{B}$ whose union contains V, there is is a finite subfamily whose union already contains U. As an indication, study closely the proof of Theorem 4.4.29.

Exercise 5.2.12 (Products of core-compact spaces) Using Exercise 5.2.11, show that if $U' \Subset U$ and $V' \Subset V$, then $U' \times V' \Subset U \times V$. Conclude that the product $X \times Y$ of two core-compact spaces X and Y is core-compact. We shall generalize this to infinite products in Exercise 8.4.9.

Exercise 5.2.13 (Relatively compact = every net has a cluster point) There are many connections between relative compactness and compactness. Show the analogue of Proposition 4.7.22: in a topological space X, given two open subsets U and V, $U \Subset V$ if and only if every net $(x_i)_{i \in I, \sqsubseteq}$ of elements of U has a cluster point in V. Note that we are not assuming core-compactness.

Deduce an analogue of Corollary 4.7.27: $U \Subset V$ iff every net of elements of U has a subnet that converges to a point in V.

Example 5.2.14 It is easy to find spaces that are not core-compact. One can, for example, check that $\mathbb{Q}$ (Exercise 4.8.4), the Sorgenfrey line $\mathbb{R}_\ell$ (Exercise 4.8.5), and Baire space $\mathcal{B}$ (Example 4.8.12) all fail to be core-compact. (Check this, by verifying that $U \Subset V$ if and only if U is empty, in all three cases.) Another example is given in Exercise 5.2.15. This was originally invented as an example of a dcpo that is not sober (see Exercise 8.2.14), but serves as a counterexample to several other claims.

Exercise 5.2.15 (Johnstone space) The *Johnstone space* $\mathbb{J}$ is $\mathbb{N} \times (\mathbb{N} \cup \{\omega\})$, with ordering defined by $(j, k) \leqslant (m, n)$ iff $j = m$ and $k \leqslant n$, or $n = \omega$ and $k \leqslant m$. (See Figure 5.2.) Intuitively, to go up in $\mathbb{J}$, starting from (j, k), either we

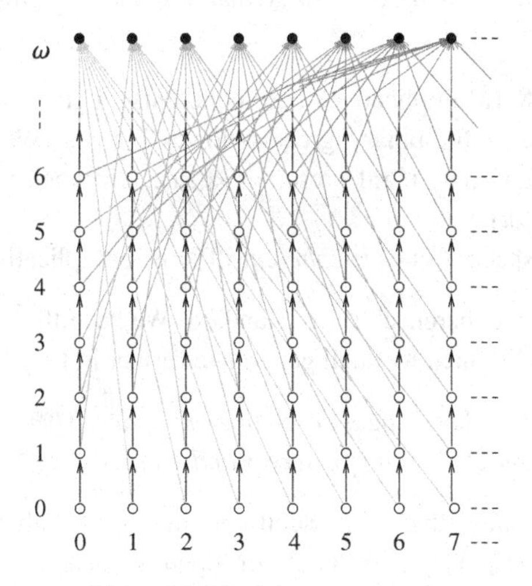

Figure 5.2 The Johnstone space $\mathbb{J}$.

increase the k component or we go directly to some point (m, ω) on the top row, with $m \geqslant k$.

Show that the Johnstone space $\mathbb{J}$ is a dcpo, and that the least upper bound of a directed family $(m_i, n_i)_{i \in I}$ is obtained as follows. Either $n_i = \omega$ for some $i \in I$, and the least upper bound is then (m_i, ω); or all m_is are equal, say to $m \in \mathbb{N}$, and the least upper bound is $(m, \sup_{i \in I} n_i)$.

Show that $U \Subset \mathbb{J}$ iff U is empty. Deduce that $\mathbb{J}$ is a dcpo that is not core-compact in its Scott topology.

This allows us to show the long-promised claim that the Scott topology of a product poset is in general strictly finer than the product of the Scott topologies (see Exercise 4.5.19).

Exercise 5.2.16 Using Exercise 5.2.15 and Exercise 5.2.7, show that given two dcpos X_1, X_2, the (Scott) topology of the product poset $(X_1 \times X_2)_\sigma$ is finer, and *in general strictly finer*, than the product topology on $X_{1\,\sigma} \times X_{2\,\sigma}$.

Proposition 5.2.17 *Let X be a core-compact space. Then $\cup$ preserves and reflects $\Subset$ on $\mathcal{O}(X)$.*

Proof Notice that $\cup$ is trivially Scott-continuous, so, by Proposition 5.1.66, $\cup$ reflects $\Subset$. To show that it preserves $\Subset$, let $V_1 \Subset U_1$ and $V_2 \Subset U_2$. In particular, $V_1, V_2 \Subset U_1 \cup U_2$. By Lemma 5.2.5, $V_1 \cup V_2 \Subset U_1 \cup U_2$. □

5.2.2 Coherence, core-coherence

It is tempting to find a condition on X so that $\cap$ would also preserve and reflect $\Subset$ on $\mathcal{O}(X)$.

Definition 5.2.18 (Multiplicative, core-coherent) Let X be an inf-semi-lattice, and write $\wedge$ for binary greatest lower bounds. We say that $\ll$ is *multiplicative* if and only if inf preserves $\ll$, i.e., iff whenever $v \ll u_1$ and $v \ll u_2$ then $v \ll u_1 \wedge u_2$.

A topological space X is *core-coherent* iff $\Subset$ is multiplicative on $\mathcal{O}(X)$.

The name "core-coherence" is not standard. We have the promised dual to Proposition 5.2.17 (which we shall generalize further in Exercise 8.1.3):

Proposition 5.2.19 *Let X be a core-coherent space. Then intersection $\cap$ is Scott-continuous on $\mathcal{O}(X)$, and preserves and reflects $\Subset$ on $\mathcal{O}(X)$.*

Proof We first show that $\cap$ is Scott-continuous, namely that, for every directed family $(U_i, V_i)_{i \in I}$ of pairs of open subsets of X, $\bigcup_{j \in I} U_j \cap \bigcup_{k \in I} V_k = \bigcup_{i \in I}(U_i \cap V_i)$. The right-to-left inclusion is obvious (take

$j = k = i$), while the converse is by directedness (pick i such that $U_j \subseteq U_i$, $V_k \subseteq V_i$). By Proposition 5.1.66, $\cap$ reflects $\Subset$.

On the other hand, if $V_1 \Subset U_1$ and $V_2 \Subset U_2$, then $V_1 \cap V_2 \Subset U_1, U_2$, so $V_1 \cap V_2 \Subset U_1 \cap U_2$ by core-coherence. So $\cap$ preserves $\Subset$. □

Exercise 5.2.20 We say that a map ν from $\mathcal{O}(X)$ to $\mathbb{R}^+$ is *modular* iff $\nu(U \cup V) + \nu(U \cap V) = \nu(U) + \nu(V)$ for all opens U, V of X. It is monotonic iff $U \subseteq V$ entails $\nu(U) \leqslant \nu(V)$. Show that, whenever X is core-compact and core-coherent, the map $\tau(\nu)$ (i.e., $\tau(\nu)(U) = \sup_{V \Subset U} \nu(V)$) is not only the largest continuous map below ν, but is also modular.

An important notion, which implies core-coherence in the presence of local compactness, is *coherence*, defined below. Some other authors use coherence for a much stronger notion, typically stable compactness (see Chapter 9) or the conjunction of coherence, compactness, and well-filteredness (see Proposition 8.3.5).

Definition 5.2.21 (Coherence) A topological space X is *coherent* iff, for any two compact saturated subsets Q_1 and Q_2 of X, $Q_1 \cap Q_2$ is compact (and then automatically saturated).

Remember that this may fail in some spaces, but that every T_2 space is coherent; see Exercise 4.4.23.

Fact 5.2.22 ($T_2 \Rightarrow$ coherent) *Every T_2 space is coherent.*

Exercise 5.2.23 Remember the lifting $X_\perp$ of a space X (Exercise 4.4.25). Show that X is coherent if and only if $X_\perp$ is coherent.

Lemma 5.2.24 *Every locally compact, coherent space is core-compact and core-coherent.*

Proof Core-compactness is by Theorem 5.2.9. Let $V \Subset U_1$, $V \Subset U_2$. By the same theorem, there are compact saturated subsets Q_1, Q_2 such that $V \subseteq Q_1 \subseteq U_1$, $V \subseteq Q_2 \subseteq U_2$. Then $V \subseteq Q_1 \cap Q_2 \subseteq U_1 \cap U_2$. Since X is coherent, $Q_1 \cap Q_2$ is compact. So, by Lemma 5.2.8, $V \Subset U_1 \cap U_2$. □

Proposition 5.2.34 below is a useful criterion that ensures that a poset is coherent in its Scott topology. At the same time, this gives us a good opportunity to introduce a useful tool:

Proposition 5.2.25 (Rudin's Lemma) *Let X be a poset, and let $\mathcal{Q}_{fin}(X)$ be the set of all finite non-empty subsets of X with the* Smyth quasi-ordering $\leqslant^\sharp$, *defined by $A \leqslant^\sharp B$ iff, for every $b \in B$, there is an $a \in A$ such that $a \leqslant b$.*

For any directed family $(E_i)_{i \in I}$ in $\mathcal{Q}_{\mathrm{fin}}(X)$, there is a directed subset D of X such that $D \subseteq \bigcup_{i \in I} E_i$ and D meets E_i for each $i \in I$.

Proof Quasi-order I by $i \preceq j$ iff $E_i \leqslant^\sharp E_j$. Consider the family $\mathcal{D}$ of all subsets D of $\bigcup_{i \in I} E_i$ such that (i) D meets every E_i, and such that (ii) $D \cap E_i \leqslant^\sharp D \cap E_{i'}$ whenever $i \preceq i'$. $\mathcal{D}$ is non-empty, since $\bigcup_{i \in I} E_i$ is in $\mathcal{D}$. Order $\mathcal{D}$ by reverse inclusion $\supseteq$.

We claim that $\mathcal{D}$ is inductive. Indeed, take any chain $(D_j)_{j \in J}$ in $\mathcal{D}$. We claim that $D = \bigcap_{j \in J} D_j$ is again in $\mathcal{D}$. (Note that inclusion is reversed, so this will give us an upper bound in $\supseteq$.) Clearly, D is included in $\bigcup_{i \in I} E_i$. Let us show (i). If D did not meet some E_i, $i \in I$, then, for every $x \in E_i$, there would be an index $j_x \in J$ such that $x \notin D_{j_x}$. Among the finitely many sets D_{j_x}, $x \in E_i$, one, say D_j, is included in the others, since $\mathcal{D}$ is a chain: so for every $x \in E_i$, $x \notin D_j$; but this contradicts the fact that D_j meets every E_i. Finally, let us show (ii). Assume $i \preceq i'$. Observe that $D \cap E_i$ is a finite set, and the intersection of a chain of finite sets $D_j \cap E_i$. So there is a smallest one. Fix $j \in J$ such that $D_j \cap E_i$ is minimal. In particular, $D \cap E_i = D_j \cap E_i$. For every element b of $D \cap E_{i'}$, b is in $D_j \cap E_{i'}$, so there is an element a of $D_j \cap E_i = D \cap E_i$ such that $a \leqslant b$. That is, $D \cap E_i \leqslant^\sharp D \cap E_{i'}$.

Now use Zorn's Lemma (Theorem 2.4.2), and pick a maximal element D of $\mathcal{D}$, i.e., a smallest one in the inclusion ordering $\subseteq$. Before we show that D is directed, we claim that, for every $x \in D$ there is an $i \in I$ such that $E_i \cap D \subseteq \uparrow x$. Otherwise, there would be an $x \in D$ such that, for every $i \in I$, $(E_i \cap D) \setminus \uparrow x$ would be non-empty. We claim that $D \setminus \uparrow x$ would then be in $\mathcal{D}$. Indeed, (i) is satisfied since, for every $i \in I$, $(D \setminus \uparrow x) \cap E_i = (E_i \cap D) \setminus \uparrow x$ is non-empty; and (ii) is satisfied, too: if $i \preceq i'$, then, for every $b \in (D \setminus \uparrow x) \cap E_{i'} = (E_{i'} \cap D) \setminus \uparrow x$, there is an $a \in E_i \cap D$ such that $a \leqslant b$, and clearly a cannot be in $\uparrow x$; otherwise b would also be in $\uparrow x$. Since D is minimal in $\mathcal{D}$ (for inclusion), $D \setminus \uparrow x$ would be equal to D, contradicting the fact that x is in D.

To show that D is directed, take $x, y \in D$. There is an $i \in I$ such that $E_i \cap D \subseteq \uparrow x$, and an $i' \in I$ such that $E_{i'} \cap D \subseteq \uparrow y$. By directedness, there is an $i'' \in I$ with $i \preceq i''$ and $i' \preceq i''$. From $E_i \cap D \subseteq \uparrow x$ and $E_i \cap D \leqslant^\sharp E_{i''} \cap D$ (property (ii)), we obtain that $E_{i''} \cap D \subseteq \uparrow x$, and similarly that $E_{i''} \cap D \subseteq \uparrow y$. Pick any element z from $E_{i''} \cap D$, using property (i): then z is both in $\uparrow x$ and in $\uparrow y$, i.e., $x \leqslant z$ and $y \leqslant z$, and also $z \in D$. □

Exercise 5.2.26 Show that $A \leqslant^\sharp B$ iff $\uparrow B \subseteq \uparrow A$. That is, there is a natural correspondence between $\mathcal{Q}_{\mathrm{fin}}(X)$ and the poset of all finitary compacts in X, ordered by reverse inclusion $\supseteq$, which you shall make explicit. This is related to the Smyth powerdomain (Proposition 8.3.25).

Exercise 5.2.27 One might want to conclude from Rudin's Lemma that D picks exactly one element x_i from each E_i, in such a way that $i \preceq i'$ would imply $x_i \preceq x_{i'}$. (Recall that $i \preceq i'$ iff $E_i \leqslant^\sharp E_{i'}$.) However, in the proof we have only showed that, for every $x \in D$ (so that $x \in E_{i'}$ for some $i' \in I$), there is an $i \in I$ such that $E_i \cap D \subseteq \uparrow x$ i.e., that there is possibly *another* index i such that the only elements picked by D in E_i are all above the sole element x. Using a counterexample, show that Rudin's Lemma cannot be improved in the desired way.

Proposition 5.2.28 *Let X be a dcpo. For every directed family $(E_i)_{i \in I}$ in $\mathcal{Q}_{fin}(X)$, for every Scott-open subset U, $\bigcap_{i \in I} \uparrow E_i \subseteq U$ iff $E_i \subseteq U$ for some $i \in I$.*

Proof The if direction is obvious. Conversely, assume that E_i is not included in U for any $i \in I$. So $E_i \setminus U$ is non-empty, and is clearly finite for each $i \in I$. Whenever $E_i \leqslant^\sharp E_{i'}$, we also have $E_i \setminus U \leqslant^\sharp E_{i'} \setminus U$: for every $b \in E_{i'} \setminus U$, there is an $a \in E_i$ such that $a \leqslant b$, in particular $a \notin U$, since U is upward closed. It follows that the family $(E_i \setminus U)_{i \in I}$ is directed. Now use Rudin's Lemma (Proposition 5.2.25) to find a directed subset D in $\bigcup_{i \in I}(E_i \setminus U)$ such that D meets each $E_i \setminus U$. Since X is a dcpo, $\sup D$ exists. Also, $\sup D$ is in $\uparrow E_i$ for every $i \in I$, since there is an element of D in E_i. So $\sup D$ is in $\bigcap_{i \in I} \uparrow E_i$, hence in U. By the definition of Scott-opens, there must be an element x of D in U. But this contradicts the fact that $x \in D \subseteq \bigcup_{i \in I}(E_i \setminus U) = (\bigcup_{i \in I} E_i) \setminus U$. □

Corollary 5.2.29 *Let X be a dcpo. For every directed family $(E_i)_{i \in I}$ in $\mathcal{Q}_{fin}(X)$, i.e., for every filtered family $(\uparrow E_i)_{i \in I}$ of finitary compacts, $\bigcap_{i \in I} \uparrow E_i$ is compact saturated in X_σ.*

Proof This is clearly saturated. Now let $(U_j)_{j \in J}$ be any open cover of $\bigcap_{i \in I} \uparrow E_i$. By Proposition 5.2.28, there is an $i \in I$ such that $\uparrow E_i \subseteq \bigcup_{j \in J} U_j$. Let $E_i = \{x_1, \dots, x_n\}$. For each k, $1 \leqslant k \leqslant n$, pick $j_k \in J$ such that $x_k \in U_{j_k}$. Then $(U_{j_k})_{k=1}^n$ is a finite subcover of $\uparrow E_i$, hence of $\bigcap_{i \in I} \uparrow E_i$. It follows that the latter is compact. □

The finitary compact subsets play a central role in continuous dcpos. Corollary 5.2.29 shows that the limit, $\bigcap_{i \in I} \uparrow E_i$, of finitary compacts is compact saturated. Conversely, we show that *every* compact saturated subset arises as a limit of finitary compacts.

Theorem 5.2.30 *Let X be a continuous poset. Every compact saturated subset Q of X_σ is a filtered intersection of finitary compacts $\uparrow E$ such that*

$Q \subseteq \uparrow E$. More precisely, let B be any basis of X. Then every compact saturated subset Q of X_σ is the intersection of the directed family of all finitary compacts $\uparrow E$ with $E \subseteq B$ and such that $Q \subseteq int(\uparrow E)$.

Proof Since Q is saturated, Q is the intersection of all open subsets U containing it (Proposition 4.2.9). Each such U is the union of all $\twoheaduparrow x$ with $x \in B \cap U$, by Theorem 5.1.27. Since Q is compact, Q is contained in a union of only finitely many of such sets, namely in $\bigcup_{x \in E} \twoheaduparrow x$, where E is a finite subset of $B \cap U$. By Proposition 5.1.35, $\bigcup_{x \in E} \twoheaduparrow x$ is then equal to $int(\uparrow E)$. To sum up, Q is the intersection of opens U, and for each we can find a finite set E such that $Q \subseteq int(\uparrow E)$, $E \subseteq B \cap U$. In particular, Q is the intersection of all sets of the form $\uparrow E$, and also the intersection of their interiors, where E ranges over the finite subsets of B such that $Q \subseteq int(\uparrow E)$.

It remains to show that the family of all such finitary compacts $\uparrow E$ is filtered. Given E_1 and E_2 as above, in particular such that $Q \subseteq int(\uparrow E_1)$ and $Q \subseteq int(\uparrow E_2)$, we apply the above construction to $U = int(\uparrow E_1) \cap int(\uparrow E_2)$ and get $\uparrow E$ such that $Q \subseteq int(\uparrow E)$, $\uparrow E \subseteq int(\uparrow E_1) \cap int(\uparrow E_2)$, and $E \subseteq B$. □

Exercise 5.2.31 Recall from Proposition 5.1.35 that, in a continuous poset, $\twoheaduparrow x$ is the interior of $\uparrow x$. Adapt this to quasi-continuous dcpos as follows: when X is a quasi-continuous dcpo, for every finite subset E of X, the set of elements $x \in X$ such that $E \ll x$ (see Exercise 5.1.34) is the interior of $\uparrow E$. Deduce that the sets $\{x \in X \mid E \ll x\}$, E finite, form a base of the Scott topology of the quasi-continuous dcpo X.

Show that the quasi-continuous dcpos are exactly the dcpos X such that X_σ is locally finitary compact (see Exercise 5.1.42).

Exercise 5.2.32 Show the following analogue of Lemma 5.1.43: any poset X which is a Scott-retract of a quasi-continuous dcpo Y, is a quasi-continuous dcpo.

Exercise 5.2.33 Show the analogue of Theorem 5.2.30 for quasi-continuous dcpos. That is, show that, when X is a quasi-continuous dcpo, then the compact saturated subsets of X_σ are exactly the filtered intersections of finitary compacts. More generally, when X is any locally finitary compact space, show that every compact saturated subset Q of X is the intersection of the filtered family of all finitary compacts $\uparrow E$ with $Q \subseteq int(\uparrow E)$.

Proposition 5.2.34 *Let X be a continuous dcpo. If $\uparrow x \cap \uparrow y$ is finitary compact in X_σ for all $x, y \in X$, then X_σ is coherent.*

Proof The assumption implies that the intersection of any two finitary compacts $\uparrow E_1 \cap \uparrow E_2$ is finitary compact: letting $E_1 = \{x_1, \ldots, x_m\}$, $E_2 = \{y_1, \ldots, y_n\}$, $\uparrow E_1 \cap \uparrow E_2 = \bigcup_{\substack{1 \leqslant i \leqslant m \\ 1 \leqslant j \leqslant n}} \uparrow x_i \cap \uparrow y_j$. Now consider two compact saturated subsets Q_1 and Q_2 of X_σ. By Theorem 5.2.30, Q_1 is the intersection of a filtered family of finitary compacts $\uparrow E_{1i}$, $i \in I$, and Q_2 is the intersection of a filtered family of finitary compacts $\uparrow E_{2j}$, $j \in J$. So $Q_1 \cap Q_2$ is the intersection of the finitary compacts $\uparrow E_{1i} \cap \uparrow E_{2j}$, $(i, j) \in I \times J$; and they form a filtered family. We conclude that $Q_1 \cap Q_2$ is compact by Corollary 5.2.29. □

Proposition 5.2.34 is only a sufficient criterion. We shall give a complete order-theoretic characterization of coherence in continuous dcpos only later; see Exercise 8.3.33. Furthermore, the proof makes it clear that the conclusion of Proposition 5.2.34 still holds if X is only a quasi-continuous dcpo: instead of invoking Theorem 5.2.30, use Exercise 5.2.33.

5.3 Spaces of continuous maps

We wish to endow the space $[X \to Y]$ of all continuous maps with a topology in such a way that application $\mathsf{App}\colon [X \to Y] \times X \to Y$ becomes continuous, at least, for every space Y.

Definition 5.3.1 (Application map) For each pair of topological spaces X, Y, the *application map* App maps pairs (f, x) of a continuous map $f\colon X \to Y$ and of an element $x \in X$ to $f(x)$.

We would also like the currification $\Lambda(f)$ of any continuous map f to be continuous.

Definition 5.3.2 (Currification) Let X, Y, Z be three topological spaces. For every continuous map $f\colon Z \times X \to Y$, the *currification* $\Lambda_X(f)$ is the map from Z to the set of continuous maps from X to Y, defined by $\Lambda_X(f)(z)(x) = f(z, x)$.

This is well defined: $\Lambda_X(f)(z)$ is a continuous map for every $z \in Z$, since f is continuous in its second argument. See Exercise 4.5.11, where we wrote $f(z, \_)$ for $\Lambda_X(f)(z)$: $\Lambda_X(f)$ is the function that maps z to $f(z, \_)$.

Requiring App and $\Lambda_X(f)$ to be continuous for all continuous f is natural. This is a basic requirement in giving semantics to higher-order programming languages (Winskel, 1993, chapter 11), which include explicit syntactic constructs for application (interpreted as App) and so-called λ-abstraction (function definition, interpreted through $\Lambda_X(f)$), and where we wish the values of all programs to depend continuously on its free parameters. This is also a simplifying requirement in algebraic topology, where so-called Kelley spaces, a.k.a. compactly generated spaces (Kelley, 1955; Steenrod, 1967), are used to ensure this property.

We notice that X must be core-compact for App to be continuous and currification to turn continuous maps into continuous maps, because this is already what we need when $Y = \mathbb{S}$.

Proposition 5.3.3 *Let X be a space such that $\mathsf{App}\colon [X \to \mathbb{S}] \times X \to \mathbb{S}$ is continuous, for some topology on the space $[X \to \mathbb{S}]$ of continuous maps from X to $\mathbb{S}$. Assume also that $\Lambda_X(f)$ is continuous from Z to $[X \to \mathbb{S}]$ for every continuous map $f\colon Z \times X \to \mathbb{S}$, for every topological space Z.*

Then X is core-compact.

Proof By Exercise 4.3.16, the elements of $[X \to \mathbb{S}]$ are the characteristic maps χ_U of open subsets U of X. Moreover, let $\leqslant$ be the pointwise ordering on $[X \to \mathbb{S}]$, i.e., $f \leqslant g$ iff, for every $x \in X$, if $f(x) = 1$ then $g(x) = 1$. Then $\chi_U \leqslant \chi_V$ iff $U \subseteq V$.

Let $\preceq$ be the specialization quasi-ordering of $[X \to \mathbb{S}]$, with the topology assumed in the statement of the proposition.

We first show that $f \leqslant g$ implies $f \preceq g$. Take $Z = \mathbb{S}$, and consider the map $h\colon Z \times X \to \mathbb{S}$ that sends $(0, x)$ to $f(x)$, $(1, x)$ to $g(x)$; h is continuous because for every open subset V of $\mathbb{S}$, $h^{-1}(V) = (\{0\} \times f^{-1}(V)) \cup (\{1\} \times g^{-1}(V)) = (\mathbb{S} \times f^{-1}(V)) \cup (\{1\} \times g^{-1}(V))$, using the fact that $f^{-1}(V) \subseteq g^{-1}(V)$. By assumption, $\Lambda_X(h)$ is continuous, hence monotonic by Proposition 4.3.9. This implies that $\Lambda_X(h)(0) \preceq \Lambda_X(h)(1)$, i.e., $f \preceq g$.

It follows that every open subset $\mathcal{U}$ of $[X \to \mathbb{S}]$ is upward closed in $\leqslant$. We claim that the Scott topology of $\leqslant$, on $[X \to \mathbb{S}]_\sigma$, is finer than the given topology on $[X \to \mathbb{S}]$. Otherwise, there is an open subset $\mathcal{U}$ in $[X \to \mathbb{S}]$ that is not Scott-open, so there is also a directed family $(f_i)_{i \in I}$ of elements of $[X \to \mathbb{S}]$ (directed with respect to $\leqslant$) whose least upper bound f is in $\mathcal{U}$, but such that no f_i is in $\mathcal{U}$. For $i, j \in I$, let $i \sqsubseteq j$ iff $f_i \leqslant f_j$, let Z be I plus a fresh element ω, and extend $\sqsubseteq$ so that $i \sqsubseteq \omega$ for every $i \in I$. Equip Z with the topology whose opens are all the upward closed subsets, with respect to $\sqsubseteq$, other than $\{\omega\}$. That is, the non-empty opens of Z must contain some element of I. One checks that this is a topology: in particular, the intersection of finitely many non-empty opens is non-empty and open, because I is directed.

Now define $h\colon Z \times X \to \mathbb{S}$ as mapping (i, x) to $f_i(x)$ for every $i \in I$, and (ω, x) to $f(x)$. We claim that h is continuous. It is enough to show that $h^{-1}\{1\}$ is open, since $\{1\}$ is a basis of $\mathbb{S}$ (Trick 4.3.3). We note that $h^{-1}\{1\} \subseteq \bigcup_{i \in I}(\uparrow i \times f_i^{-1}\{1\})$: every element of $h^{-1}\{1\}$ is either of the form (i, x) with $f_i(x) \in \{1\}$, which is trivially in the right-hand side, or of the form (ω, x) with

$f(x) = \sup_{i \in I} f_i(x) \in \{1\}$; in the latter case, $f_i(x) = 1$ for some $i \in I$, so $(\omega, x) \in \uparrow i \times f_i^{-1}\{1\}$. The converse inclusion $\bigcup_{i \in I}(\uparrow i \times f_i^{-1}\{1\}) \subseteq h^{-1}\{1\}$ is easy. So equality holds. This entails that $h^{-1}\{1\}$ is open, hence that h is continuous.

By assumption, $\Lambda_X(h)$ is continuous. So $\Lambda_X(h)^{-1}(\mathcal{U})$ is open in Z. It is non-empty since $f \in \mathcal{U}$, i.e., $\omega \in \Lambda_X(h)^{-1}(\mathcal{U})$. By definition of the topology on Z, $\Lambda_X(h)^{-1}(\mathcal{U})$ must contain some $i \in I$. So $f_i = \Lambda_X(h)(i)$ is in $\mathcal{U}$, a contradiction. So $\mathcal{U}$ is Scott-open, and the topology of $[X \to \mathbb{S}]_\sigma$ is indeed finer than that of $[X \to \mathbb{S}]$.

The argument until now has used Λ_X exclusively. Let us consider App. The argument is close to Exercise 5.2.7; in fact we could apply the result of this exercise directly to conclude that X is core-compact.

Let $x \in X$, and U be an open neighborhood of x. Let $(\ni)$ be $\mathsf{App}^{-1}(\{1\})$. This is open since App is continuous. Write $(\ni)$ as the union $\bigcup_{i \in I} \mathcal{W}_i \times U_i$ of open rectangles. Since $x \in U$, (χ_U, x) is in $(\ni)$, so there is an $i \in I$ such that $\chi_U \in \mathcal{W}_i$ and $x \in U_i$. Take $V = U_i$. It remains to show that $V \Subset U$, i.e., $U_i \Subset U$. Consider any directed family $(V_j)_{j \in J}$ of open subsets of X whose union contains U. The least upper bound of the directed family $(\chi_{V_j})_{j \in J}$ is $\chi_{\bigcup_{j \in J} V_j}$, which is above χ_U in $\leqslant$. Since $\chi_U \in \mathcal{W}_i$, since $\mathcal{W}_i$ is open in $[X \to \mathbb{S}]$, and since every open subset of the latter is Scott-open, χ_{V_j} must be in $\mathcal{W}_i$ for some $j \in J$. It remains to show that $\chi_{U_i} \leqslant \chi_{V_j}$, i.e., that $U_i \subseteq V_j$. For every $z \in U_i$, (χ_{V_j}, z) is in $\mathcal{W}_i \times U_i$, hence in $(\ni)$. That is, $z \in V_j$. Since z is arbitrary in U_i, $U_i \subseteq V_j$.

We have just shown that, given any open neighborhood U of x, there is an open neighborhood $V = U_i$ such that $x \in V \Subset U$. So X is core-compact, by Proposition 5.2.6. □

Exercise 5.3.4 Show that, under the assumptions of Proposition 5.3.3, $[X \to \mathbb{S}]$ has the Scott topology of its pointwise ordering, i.e., $[X \to \mathbb{S}] = [X \to \mathbb{S}]_\sigma$.

5.4 The exponential topology

When $Y \neq \mathbb{S}$, the right topology that one has to put on $[X \to Y]$ may be different from the Scott topology. A good candidate is the following.

Definition 5.4.1 (Core-open topology) The *core-open topology* on the space of continuous maps from X to Y is the one generated by subbasic open sets of the form $[U \Subset^{-1} V]$, U open in X, V open in Y, and where $[U \Subset^{-1} V]$ denotes the set of all continuous maps $f : X \to Y$ such that $U \Subset f^{-1}(V)$.

We let $[X \to Y]^{\circledcirc}$ be the space of all continuous maps from X to Y with the core-open topology.

In other words, we agree that the subbasic tests that one can use on continuous maps f are given by pairs of tests U and V, and consist in testing whether U is relatively compact in, i.e., way-below, $f^{-1}(V)$.

Proposition 5.4.2 (Application map) *If X is core-compact, then App is continuous from $[X \to Y]^{\circledcirc} \times X$ to Y.*

Proof Assume X is core-compact. For every open subset V of Y, $\mathsf{App}^{-1}(V)$ is the collection of pairs (f, x) such that $f(x) \in V$, i.e., such that $x \in U$ for some open subset U with $U \Subset f^{-1}(V)$, using Proposition 5.2.6. So $\mathsf{App}^{-1}(V) = \bigcup_{U \in \mathcal{O}(X)} [U \Subset^{-1} V] \times U$. This is open, so App is continuous. □

Proposition 5.4.3 (Currification) *If X is core-compact, then, for every continuous map $f : Z \times X \to Y$, $\Lambda_X(f)$ is continuous from Z to $[X \to Y]^{\circledcirc}$.*

Proof Fix an open subset U of X, and an open subset V of Y: we must show that $\Lambda_X(f)^{-1}[U \Subset^{-1} V]$ is open in Z. Using Trick 4.3.3, this will be enough to prove the claim.

Fix z in $\Lambda_X(f)^{-1}[U \Subset^{-1} V]$. We show that z has an open neighborhood W included in $\Lambda_X(f)^{-1}[U \Subset^{-1} V]$, and will conclude by Trick 4.1.11. Since $z \in \Lambda_X(f)^{-1}[U \Subset^{-1} V]$, $U \Subset \{x \in X \mid f(z, x) \in V\}$. Since f is continuous, $f^{-1}(V)$ is open in $Z \times X$, hence is a union of open rectangles $\bigcup_{i \in I} W_i \times U_i$. So $U \Subset \bigcup_{\substack{i \in I \\ z \in W_i}} U_i$. Since X is core-compact, use interpolation (Proposition 5.1.15) in $\mathcal{O}(X)$: there is an open subset U' such that $U \Subset U' \Subset \bigcup_{\substack{i \in I \\ z \in W_i}} U_i$. By the definition of $\Subset$, there is a finite set J of indices $i \in I$ with $z \in W_i$ such that $U' \subseteq \bigcup_{i \in J} U_i$. In particular, $U \Subset \bigcup_{i \in J} U_i$. Let $W = \bigcap_{i \in J} W_i$. This is open (because J is finite), and contains z. For every $z' \in W$, $\{x \in X \mid f(z', x) \in V\} = \bigcup_{\substack{i \in I \\ z' \in W_i}} U_i$ contains $\bigcup_{i \in J} U_i$, so $U \Subset \{x \in X \mid f(z', x) \in V\}$, i.e., $z' \in \Lambda_X(f)^{-1}[U \Subset^{-1} V]$. Since z' is arbitrary, $W \subseteq \Lambda_X(f)^{-1}[U \Subset^{-1} V]$. □

The choice of the core-open topology in Definition 5.4.1 may seem arbitrary, except for the fact that it fits the bill, i.e., that App is continuous (Proposition 5.4.2) and that currification preserves continuity (Proposition 5.4.3), whenever X is core-compact (which is forced by Proposition 5.3.3). One can show that the Isbell topology, which we shall examine in Exercise 5.4.9, would also fit the bill. But we shall check that it coincides with the core-open topology when X is core-compact. This is no accident:

Theorem 5.4.4 (Existence and uniqueness of the exponential topology) *Let X be a topological space. The following are equivalent:*

1. *For every topological space Y, there is an* exponential topology *on $[X \to Y]$, i.e., one that makes $\mathsf{App}\colon [X \to Y] \times X \to Y$ continuous, and such that $\Lambda_X(f)\colon Z \to [X \to Y]$ is continuous for every topological space Z and every continuous map $f\colon Z \times X \to Y$;*
2. *X is core-compact.*

If this is so, then there is only one topology on $[X \to Y]$ satisfying the constraints of Item 1, and this is the core-open topology.

Before we prove this, observe the following:

Lemma 5.4.5 (Splitting, conjoining topologies) *Call a topology on $[X \to Y]$* splitting *iff $\Lambda_X(f)\colon Z \to [X \to Y]$ is continuous for every topological space Z and every continuous map $f\colon Z \times X \to Y$. Call it* conjoining *iff $\mathsf{App}\colon [X \to Y] \times X \to Y$ is continuous. In particular, a topology is exponential if and only if it is both splitting and conjoining.*

Every conjoining topology is finer than every splitting topology.

Proof Let $\mathcal{O}_1$ be a splitting topology, and $\mathcal{O}_2$ be a conjoining topology. Write $[X \to Y]_1$ for $[X \to Y]$ with the topology $\mathcal{O}_1$, and similarly for $[X \to Y]_2$. By definition, App is continuous from $[X \to Y]_2 \times X$ to Y. So $\Lambda_X(\mathsf{App})$ is continuous from $[X \to Y]_2$ to $[X \to Y]_1$. However, $\Lambda_X(\mathsf{App}) = \mathrm{id}_{[X \to Y]}$. So, for every open subset $\mathcal{U}$ of $\mathcal{O}_1$, $\mathcal{U} = \Lambda_X(\mathsf{App})^{-1}(\mathcal{U})$ is open in $\mathcal{O}_2$. ☐

Proof (of Theorem 5.4.4) Only the last part remains to be proved. There is at most one topology that is both splitting and conjoining, by Lemma 5.4.5. So the core-open topology is the unique splitting and conjoining topology. ☐

Theorem 5.4.4 is another illustration of the versatility of the concept of core-compactness. One can reformulate it by saying that every conjoining topology is finer than every splitting topology, that there is one that is both if and only if X is core-compact, and that this topology is then the core-open topology. In particular:

Fact 5.4.6 (Core-open = exponential) *If X is core-compact, then the core-open topology on $[X \to Y]$ is both the finest splitting topology and the coarsest conjoining topology.*

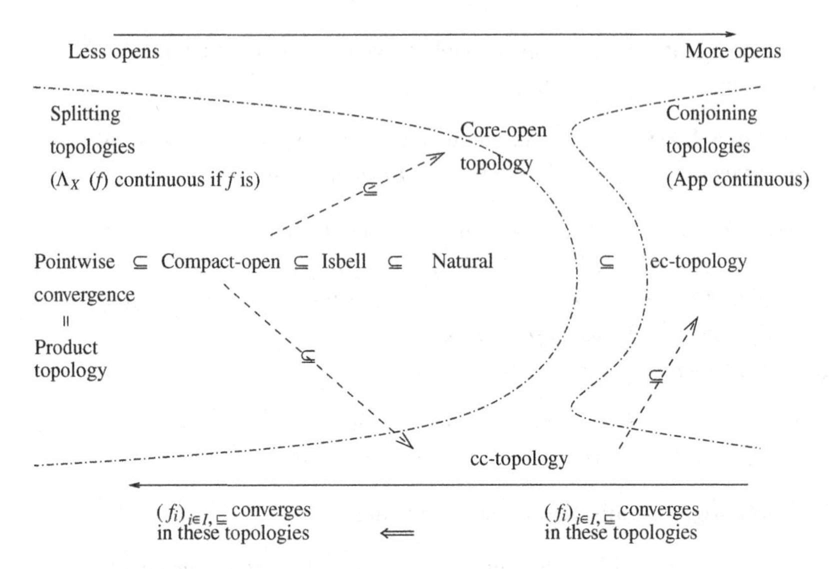

Figure 5.3 Topologies on the space $[X \to Y]$ of continuous functions.

The following is also immediate.

Lemma 5.4.7 *Every topology finer than a conjoining topology is conjoining. Every topology coarser than a splitting topology is splitting.*

There are many other candidates for an exponential topology, and the following exercises explore some of them. The compact-open topology is one of the most well-studied. Don't be frightened by how many topologies there are (Figure 5.3): as we shall see, all these topologies must coincide under mild assumptions (Figure 5.5; see also Figure 5.4). If this still does not reassure you, please skip to Section 5.4.1.

Exercise 5.4.8 (Compact-open topology) The *compact-open* topology on $[X \to Y]$ is generated by subsets of the form $[Q \subseteq V]$, where Q is compact saturated in X and V is open in Y; we define $[Q \subseteq V]$ as the set of all continuous maps $f : X \to Y$ such that $f[Q] \subseteq V$. That is, the tests $[Q \subseteq V]$ consist in checking that f maps the whole of Q inside V.

Show that the compact-open topology is splitting. Show that it coincides with the core-open topology whenever X is locally compact, hence is the exponential topology on $[X \to Y]$ in this case.

Exercise 5.4.9 (Isbell topology) The *Isbell topology* on $[X \to Y]$ is generated by subsets of the form $[{}^{-1}(V) \in \mathcal{U}]$, where V is open in Y and $\mathcal{U}$

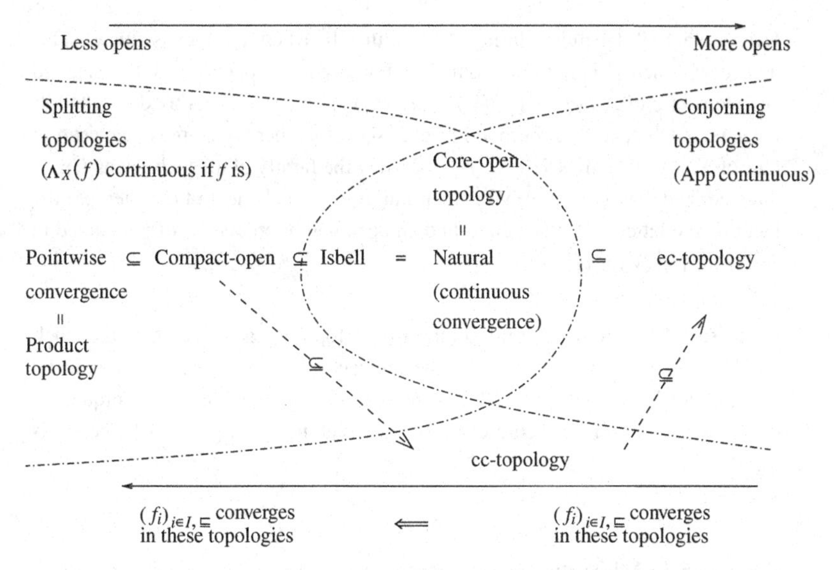

Figure 5.4 Topologies on $[X \to Y]$, if X is core-compact.

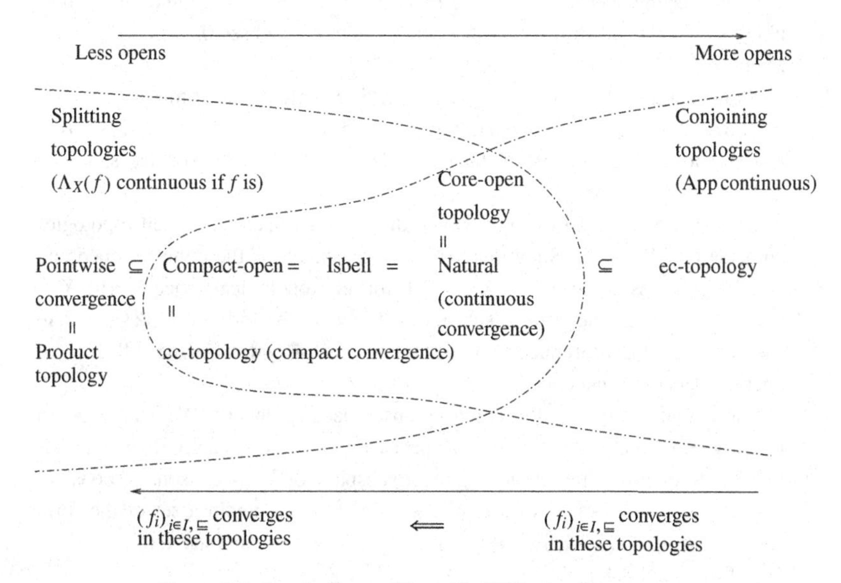

Figure 5.5 Topologies on $[X \to Y]$, if X is locally compact.

is Scott-open in $\mathcal{O}(X)$; we define $[\_^{-1}(V) \in \mathcal{U}] = \{f \in [X \to Y] \mid f^{-1}(V) \in \mathcal{U}\}$. (We also write $[X \to Y]^{\mathrm{I}}$ for $[X \to Y]$ with the Isbell topology.) Show that the Isbell topology coincides with the core-open topology when X is core-compact.

Exercise 5.4.10 (Isbell is splitting) Show that the Isbell topology is always splitting (even when X is not core-compact). For any Isbell open $[{}^{-1}(V) \in \mathcal{U}]$, and any element z of $\Lambda_X(f)^{-1}[{}^{-1}(V) \in \mathcal{U}]$, find an open neighborhood of z as follows. For each open neighborhood W of z, show that there is a largest open subset U_W of X such that $W \times U_W \subseteq f^{-1}(V)$, that the family of all such opens U_W is directed, and that $f(z, \_)^{-1}(V)$ is their union, and an element of $\mathcal{U}$. Then use the fact that the latter is Scott-open to find an open neighborhood W of z included in $\Lambda_X(f)^{-1}[{}^{-1}(V) \in \mathcal{U}]$.

Exercise 5.4.11 Show that the compact-open topology is coarser than the Isbell topology. So we retrieve the fact that the compact-open topology is splitting, using Exercise 5.4.8 and Lemma 5.4.7. Because exponential topologies are unique, the compact-open, Isbell, and core-open topologies coincide on $[X \to Y]$ when X is locally compact.

Exercise 5.4.12 (Locally compact $\Rightarrow$ consonant) Let X be a topological space. We consider the special case of the function space $[X \to \mathbb{S}]$, which is isomorphic to $\mathcal{O}(X)$, through the map that sends each open subset U of X to χ_U (see Exercise 4.3.16).

Show that the Isbell topology on $[X \to \mathbb{S}]$ is exactly the Scott topology.

Show that the compact-open topology on $[X \to \mathbb{S}]$ has a base (not just a subbase) of opens of the form $[Q \subseteq \{1\}]$, Q compact saturated in X.

X is called *consonant* if and only if the compact-open and Isbell topologies coincide on $[X \to \mathbb{S}]$. Show that: (i) X is consonant iff the compact-open and Isbell topologies coincide on $[X \to Y]$, for any topological space Y; (ii) X is consonant iff every Scott-open subset of $\mathcal{O}(X)$ is of the form $\bigcup_{i \in I} \blacksquare Q_i$, where each Q_i is compact saturated in X, and the notation $\blacksquare Q$ denotes the collection of open subsets of X that contain Q.

Show that every locally compact space is consonant. (We shall see in Exercise 8.3.4 that not all consonant spaces are locally compact.) Show that neither the Sorgenfrey line $\mathbb{R}_\ell$ nor $\mathbb{Q}$ (as a subspace of $\mathbb{R}$) is consonant: consider the collection $\mathcal{U}$ of open subsets of $\mathbb{R}_\ell$, resp. $\mathbb{Q}$ that contain a set of the form $(-a, 1+b)$ (resp., $(-a, 1+b) \cap \mathbb{Q}$) for some $a, b > 0$, and use Exercise 4.8.5, resp. Exercise 4.8.4.

Exercise 5.4.13 (Even convergence, compact convergence) Let X, Y be topological spaces. For any subset A of X, we say that the net $(f_i)_{i \in I, \sqsubseteq}$ of continuous maps from X to Y *converges evenly on* A to $f \colon X \to Y$ if and only if, for every open subset V of Y, there is an $i_0 \in I$ such that, for every $x \in A$, if $f(x) \in V$

then $f_i(x) \in V$ for every $i \in I$ with $i_0 \sqsubseteq i$. While this may seem complicated, observe that the net converges pointwise (Exercise 4.7.30) to f on A, i.e., $(f_i(x))_{i \in I, \sqsubseteq}$ converges to $f(x)$ for every $x \in A$, iff, for every open subset V, for every $x \in A$, there is an $i_0 \in I$ such that if $f(x) \in V$ then $f_i(x) \in V$ for every $i \in I$ with $i_0 \sqsubseteq i$. That is, the net converges evenly on A iff i_0 can be chosen the same for all $x \in A$. This is the same idea as with uniform convergence.

The net $(f_i)_{i \in I, \sqsubseteq}$ *converges compactly* to f if and only if $(f_i)_{i \in I, \sqsubseteq}$ converges evenly to f on every compact saturated subset Q of X.

The *cc-topology* (cc is for compact convergence) is defined by declaring open any set $\mathcal{U}$ of continuous functions such that, given any net $(f_i)_{i \in I, \sqsubseteq}$ that converges compactly to $f \in \mathcal{U}$, then $f_i \in \mathcal{U}$ for i large enough. (Beware that, although compact convergence implies convergence in the cc-topology, there is no reason yet to believe that the two notions of convergence coincide.)

Show that $[Q \subseteq V]$ is open in the cc-topology for every compact saturated subset Q of X and every open subset V of Y, and conclude that the cc-topology is finer than the compact-open topology.

Show that any net that converges compactly to f converges to f in the compact-open topology (see Exercise 5.4.8). Conversely, show that, if X is locally compact, then any net that converges to f in the compact-open topology also converges compactly to f. Conclude that the cc-topology coincides with the compact-open topology if X is locally compact.

Conclude also that, if X is locally compact, then convergence in the cc-topology (equivalently, the compact-open topology) is exactly compact convergence. For this reason, in the literature the cc-topology is sometimes called the *topology of compact convergence*.

Exercise 5.4.14 (Even convergence) Let X, Y be topological spaces. Define the *ec-topology* on $[X \to Y]$ by declaring open any set $\mathcal{U}$ of continuous functions such that, given any net $(f_i)_{i \in I, \sqsubseteq}$ that converges evenly to $f \in \mathcal{U}$, then $f_i \in \mathcal{U}$ for i large enough. (Beware again that, although even convergence implies convergence in the ec-topology, there is no reason yet to believe that the two notions of convergence coincide.)

Show that if the net $(f_i)_{i \in I, \sqsubseteq}$ of continuous maps from X to Y converges evenly to f, then it also converges to f evenly on every compact saturated subset of X. Conclude that the ec-topology is finer than the cc-topology.

Next, assume X compact. Show that compact convergence then coincides with even convergence. So, if X is both compact and locally compact, then the core-open topology, a.k.a. the compact-open topology, is the ec-topology. Conclude that, if X is compact and locally compact, then convergence in the compact-open topology is even convergence.

5.4.1 Continuous convergence and the natural topology

Let us now examine convergence in $[X \to Y]^{\circledcirc}$. Exercise 5.4.13 shows that this is compact convergence when X is locally compact. This also coincides with the following notion of *continuous convergence* (when X is core-compact).

Definition 5.4.15 (Continuous convergence) Let X, Y be two topological spaces. The net $(f_i)_{i \in I, \sqsubseteq}$ of continuous maps from X to Y is said to *converge continuously* to $f \in [X \to Y]$ if and only if, for every net $(x_j)_{j \in J, \preceq}$ in X that converges to some $x \in X$, the net $(f_i(x_j))_{(i,j) \in I \times J, \sqsubseteq \times \preceq}$ converges to $f(x)$.

Exercise 5.4.16 Let $(f_i)_{i \in I, \sqsubseteq}$ and $(x_i)_{i \in I, \sqsubseteq}$ be two nets *with the same index set* I, in $[X \to Y]$ and in X, respectively. Show that if $(f_i)_{i \in I, \sqsubseteq}$ converges continuously to $f \in [X \to Y]$, and $(f_i)_{i \in I, \sqsubseteq}$ converges to $x \in X$, then $(f_i(x_i))_{i \in I, \sqsubseteq}$ converges to $f(x)$ in Y.

Exercise 5.4.17 Show that a sequence $(f_n)_{n \in \mathbb{N}}$ of continuous functions converges continuously to f if and only if, for every convergent sequence $(x_n)_{n \in \mathbb{N}}$ of points of X, with limit x, $(f_n(x_n))_{n \in \mathbb{N}}$ converges to $f(x)$. It follows that the notion of continuous convergence defined in Definition 5.4.15 coincides with the notion of Exercise 3.5.8 for sequences.

Example 5.4.18 Let $X = Y = [0, 1]$, both with their usual T_2 topology, and consider the sequence of maps f_i, $i \in \mathbb{N}$, $i \geqslant 2$, from $[0, 1]$ to $[0, 1]$ defined by $f_i(x) = 1 - i.x$ if $x \leqslant \frac{1}{i}$, $f_i(x) = i.x - 1$ if $\frac{1}{i} \leqslant x \leqslant \frac{2}{i}$, and $f_i(x) = 1$ if $x \geqslant \frac{2}{i}$; see Figure 5.6. This sequence converges pointwise to the constant 1 map, in the sense that $(f_i(x))_{i \geqslant 2}$ converges to 1 for every $x \in X$ (exercise:

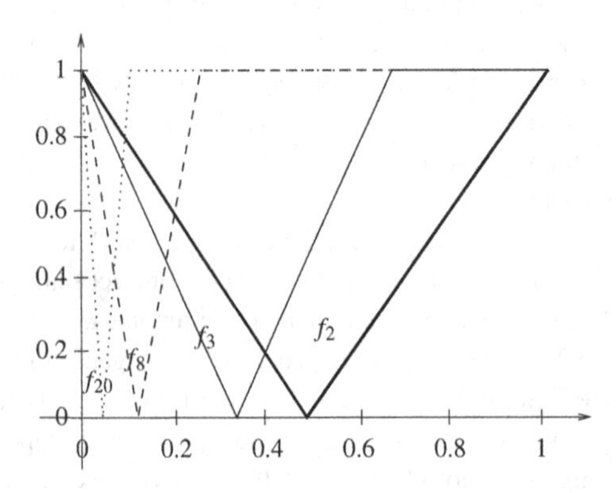

Figure 5.6 Functions that converge pointwise, not continuously.

check this), but it does not converge continuously to any function. If it did, it would converge continuously to the constant 1 map (check this, too). But letting $x_i = \frac{1}{i}$, one sees that $(f_i(x_i))_{i \geqslant 2}$, which is the constant 0 sequence, converges to 0 instead, contradicting Exercise 5.4.16.

Proposition 5.4.19 (Natural topology) *Let X, Y be two topological spaces. The* natural topology *on $[X \to Y]$ is the topology whose opens $\mathcal{U}$ are those subsets such that, for every net $(f_i)_{i \in I, \sqsubseteq}$ that converges continuously to $f \in \mathcal{U}$ in $[X \to Y]$, $f_i \in \mathcal{U}$ for i large enough.*

Write $[X \to Y]^\natural$ for $[X \to Y]$ with the natural topology.

If the net $(f_i)_{i \in I, \sqsubseteq}$ converges continuously to f in $[X \to Y]$, then it converges to f in the natural topology.

Proof It is easy to check that the natural topology is indeed a topology. Assume $(f_i)_{i \in I, \sqsubseteq}$ converges continuously to f in $[X \to Y]$. Let $\mathcal{U}$ be any open neighborhood of f in $[X \to Y]^\natural$. By definition, $f_i \in \mathcal{U}$ for i large enough. So $(f_i)_{i \in I, \sqsubseteq}$ converges to f in the natural topology. □

Despite what its definition may suggest, convergence in the natural topology is *not* continuous convergence—otherwise we would have called it simply the topology of continuous convergence. See Exercise 5.4.23. However, the two notions of convergence do coincide when X is core-compact; see Proposition 5.4.21.

Proposition 5.4.20 (Natural = exponential) *Let X, Y be two topological spaces.*

The natural topology is the finest splitting topology, and the intersection of all conjoining topologies.

When X is core-compact, the natural topology coincides with the core-open topology.

Proof We show that the natural topology is splitting. Let f be continuous from $Z \times X$ to Y. For every net $(z_i)_{i \in I, \sqsubseteq}$ that converges to z in Z, for every net $(x_j)_{j \in J, \preceq}$ that converges to x in X, the net $(z_i, x_j)_{(i,j) \in I \times J, \sqsubseteq \times \preceq}$ converges to (z, x) in $Z \times X$ by Exercise 4.7.31. Since continuous maps preserve limits (Proposition 4.7.18), $(f(z_i, x_j))_{(i,j) \in I \times J, \sqsubseteq \times \preceq}$ converges to $f(z, x)$. In other words, $(\Lambda_X f(z_i)(x_j))_{(i,j) \in I \times J, \sqsubseteq \times \preceq}$ converges to $\Lambda_X f(z)(x)$, for every net $(x_j)_{j \in J, \preceq}$ that converges to x in X, so $(\Lambda_X f(z_i))_{i \in I, \sqsubseteq}$ converges continuously to $\Lambda_X f(z)$. By Proposition 5.4.19, it converges to $\Lambda_X f(z)$ in $[X \to Y]^\natural$. So, using Proposition 4.7.18, $\Lambda_X f$ is continuous from Z to $[X \to Y]^\natural$.

We delay the task of showing that, among all splitting topologies, the natural topology is the finest. Instead, we proceed to show that the natural topology is the intersection of all conjoining topologies.

Given any net $(f_i)_{i \in I, \sqsubseteq}$ that converges continuously to f in $[X \to Y]$, build the topology $\mathcal{O}_{(f_i)_{i \in I, \sqsubseteq}, f}$ whose sole opens are those sets $\mathcal{U}$ of continuous maps such that $f \notin \mathcal{U}$, or $f_i \in \mathcal{U}$ for i large enough. Equivalently, if $f \in \mathcal{U}$ then $f_i \in \mathcal{U}$ for i large enough: this is the topology of a given net $(f_i)_{i \in I, \sqsubseteq}$ converging to a given limit f.

We claim that $\mathcal{O}_{(f_i)_{i \in I, \sqsubseteq}, f}$ is conjoining. Let V be an open subset of Y, and let $(g, x) \in \mathsf{App}^{-1}(V)$. We shall find an open neighborhood of (g, x) included in $\mathsf{App}^{-1}(V)$. This will show that $\mathsf{App}^{-1}(V)$ is open, by Trick 4.1.11, hence that App is continuous, which will prove the claim.

There are two cases. If $g \neq f$ (the simple case), then $\{g\}$ is open in $\mathcal{O}_{(f_i)_{i \in I, \sqsubseteq}, f}$, and $\{g\} \times g^{-1}(V)$ is an open neighborhood of (g, x) included in $\mathsf{App}^{-1}(V)$.

If $g = f$ (the hard case), then $f(x) \in V$. Let $\mathcal{N}_x$ be the collection of all open neighborhoods of x included in $f^{-1}(V)$. We say that $U \in \mathcal{N}_x$ is *good* iff, for i large enough, for every $x' \in U$, $f_i(x') \in V$; more precisely, iff, there is an $i \in I$ such that, for every $j \in I$ above i and every $x' \in U$, $f_j(x') \in V$. It is *bad* otherwise. Assume that every $U \in \mathcal{N}_x$ is bad. For every $i \in I$, there is a $j \in I$ above i and a point $x_{i,U}$ in U such that $f_j(x_{i,U}) \notin V$. (Pick them using the Axiom of Choice.) Consider the net $(x_{i,U})_{(i,U) \in I \times \mathcal{N}_x, \sqsubseteq \times \supseteq}$. This clearly converges to x. Since $(f_i)_{i \in I, \sqsubseteq}$ converges continuously to f, $(f_j(x_{i,U}))_{(j,i,U)) \in I \times I \times \mathcal{N}_x, \sqsubseteq \times \sqsubseteq \times \supseteq}$ converges to $f(x)$. So $f_j(x_{i,U})$ is in V for j, i, U large enough, contradicting the fact that U is bad. So there is a good $U \in \mathcal{N}_x$. By definition, there is an $i_0 \in I$ such that, for every i above i_0, for every $x' \in U$, $f_i(x') \in V$. Let $\mathcal{U}$ be the set consisting of f plus all f_i, with $i_0 \sqsubseteq i$. This is open in $\mathcal{O}_{(f_i)_{i \in I, \sqsubseteq}, f}$. $\mathcal{U} \times U$ is an open neighborhood of (f, x), and is included in $\mathsf{App}^{-1}(V)$, by construction. This finishes the proof that $\mathcal{O}_{(f_i)_{i \in I, \sqsubseteq}, f}$ is conjoining.

Every open subset in the natural topology is open in any conjoining topology, since the former is splitting, and by Lemma 5.4.5. Conversely, let $\mathcal{U}$ be an element of the intersection of all conjoining topologies. So $\mathcal{U}$ is in each of the topologies $\mathcal{O}_{(f_i)_{i \in I, \sqsubseteq}, f}$ as above. So, if $(f_i)_{i \in I, \sqsubseteq}$ is a net that converges continuously to $f \in \mathcal{U}$, $f_i \in \mathcal{U}$ for i large enough, so $\mathcal{U}$ is open in the natural topology. So this topology is the intersection of all conjoining topologies.

As such, it is the finest splitting topology: any splitting topology is coarser than every conjoining topology by Lemma 5.4.5, hence also coarser than their intersection.

When X is core-compact, the natural topology then coincides with the core-open topology by Fact 5.4.6. □

Proposition 5.4.21 (Convergence in the natural topology) *Let X be a core-compact space, and Y be a topological space. The net $(f_i)_{i \in I, \sqsubseteq}$ converges to f in $[X \to Y]^\natural$ (equivalently, in $[X \to Y]^{\circledcirc}$) if and only if $(f_i)_{i \in I, \sqsubseteq}$ converges continuously to f.*

Proof One direction is Proposition 5.4.19. Conversely, assume that $(f_i)_{i \in I, \sqsubseteq}$ converges to f in the natural topology. Let $(x_j)_{j \in J, \preceq}$ be a net that converges to x in X. Since X is core-compact, $[X \to Y]^\natural = [X \to Y]^{\circledcirc}$, by the last part of Proposition 5.4.20, and therefore App is continuous from $[X \to Y]^\natural \times X$ to Y (Proposition 5.4.2). Now $(f_i, x_j)_{(i,j) \in I \times J, \sqsubseteq \times \preceq}$ converges to (f, x) in $[X \to Y]^\natural \times X$ (Exercise 4.7.31), and continuous maps preserve limits (Proposition 4.7.18), so $(\mathsf{App}(f_i, x_j))_{(i,j) \in I \times J, \sqsubseteq \times \preceq}$ converges to $\mathsf{App}(f, x)$ in Y. In other words, $(f_i)_{i \in I, \sqsubseteq}$ converges continuously to f. □

Using Exercise 5.4.13, we obtain:

Corollary 5.4.22 (Continuous convergence and compact convergence) *Let X be a locally compact topological space, and Y be a topological space. For every net $(f_i)_{i \in I, \sqsubseteq}$ of continuous maps from X to Y, $(f_i)_{i \in I, \sqsubseteq}$ converges continuously to $f \in [X \to Y]$ if and only if $(f_i)_{i \in I, \sqsubseteq}$ converges compactly to f.*

Exercise 5.4.23 Show that the assumption that X is core-compact is necessary in Proposition 5.4.21. That is, assume X is a topological space such that, for every topological space Y, convergence in $[X \to Y]^\natural$ coincides with continuous convergence. Then show that X is core-compact. This can be shown by proving the more general claim that, for any fixed topological spaces X and Y, if convergence in $[X \to Y]^\natural$ coincides with continuous convergence, then the natural topology is conjoining.

Conclude that, if X is not core-compact, then there are spaces Y and nets in $[X \to Y]$ that converge in the natural topology to some map f, but do not converge to f continuously.

Exercise 5.4.24 (Even and continuous convergence) Let X, Y be topological spaces. Show that the ec-topology on $[X \to Y]$ is conjoining. From this, deduce that every net of continuous maps that converges evenly to $f \in [X \to Y]$ (or more generally, that converges to f in the ec-topology) also converges continuously to f.

Exercise 5.4.25 (Natural is strictly coarser than Isbell, compact-open) Since the Isbell topology is splitting, it is coarser than the natural topology. Let $X = \mathcal{B}$ be the Baire space $\mathbb{N}^{\mathbb{N}}$ (see Example 4.8.12), and $Y = \mathbb{N}$ with the discrete topology. Write $\overline{n}$ for the constant sequence in $\mathbb{N}^{\mathbb{N}}$ whose elements all equal $n \in \mathbb{N}$. Let $\mathcal{A}$ be the set of all continuous maps f from X to Y such that $f(f(\overline{0}), f(\overline{1}), \ldots, f(\overline{n}), \ldots) = 0$. Show that $\mathcal{A}$ is open in the natural topology.

Next, show that it is not open in the compact-open topology, as follows. For each continuous map $f\colon \mathcal{B} \to \mathbb{N}$, assume that f has an open neighborhood $\bigcap_{i=1}^{m} [Q_i \subseteq V_i]$ in the compact-open topology. Using Example 4.8.12, patch f so as to produce a map g that differs from f only on vectors $\vec{x} \in \mathcal{B}$ such that $\vec{x} \notin \bigcup_{i=1}^{m} Q_i$, so that g is still in $\bigcap_{i=1}^{m} [Q_i \subseteq V_i]$. Make sure that g is continuous, and use Exercise 4.8.13 to check it easily. The purpose of the patch is to ensure that $g(g(\overline{0}), g(\overline{1}), \ldots, g(\overline{n}), \ldots) = 1$ (the value 1 is arbitrary). Conclude that no point of $\mathcal{A}$ has any open neighborhood in the compact-open topology, hence that $\mathcal{A}$ is not open in this topology.

So the natural topology is in general strictly finer than the compact-open topology. We shall see in Exercise 8.3.4 that the compact-open topology coincides with the Isbell topology on $[\mathcal{B} \to \mathbb{N}]$, so that the natural topology is also in general strictly finer than the Isbell topology. This construction is due to Escardó *et al.* (2004, example 5.12).

Exercise 5.4.26 (Pointwise is coarser than compact-open) Show that the topology of *pointwise convergence* on $[X \to Y]$ (i.e., the subspace topology from Y^X) is always splitting, and coarser than the compact-open topology.

In case X is locally compact, deduce that if $(f_i)_{i \in I, \sqsubseteq}$ converges compactly to f in $[X \to Y]$, then $(f_i)_{i \in I, \sqsubseteq}$ converges pointwise to f, but that the converse fails in general, even when X is compact T_2. (One may consider the functions of Example 5.4.18; see Figure 5.6.) So the topology of pointwise convergence on $[X \to Y]$ is in general strictly finer than the compact-open topology, even when X is compact T_2.

Exercise 5.4.27 Let X, Y be two topological spaces. Show that $(f_i)_{i \in I, \sqsubseteq}$ converges to f pointwise in Y^X, or in $[X \to Y]$, if and only if $(f_i(x))_{i \in I, \sqsubseteq}$ converges to $f(x)$ for every $x \in X$. Deduce that if $(f_i)_{i \in I, \sqsubseteq}$ converges to f continuously, then it also converges to f pointwise.

One may look at Escardó and Heckmann (2001–2002) or Escardó *et al.* (2004), and the references therein, for further information on topologies on spaces of continuous functions.

Exercise 5.4.28 Let X be a topological space, and Y be a T_2 space. Show that Y^X is T_2, and that $[X \to Y]$ is T_2, whether it is equipped with the topology of pointwise convergence, the compact-open topology, the Isbell topology, the natural topology, the cc-topology, the ec-topology, or the core-open topology.

Here is yet another characterization of core-compactness.

Exercise 5.4.29 Let X be a topological space. Show that X is core-compact iff $q \times \mathrm{id}_X$ is a quotient map from $Y \times X$ to $Z \times X$ for every quotient map $q \colon Y \to Z$. The product of two maps was defined in Exercise 4.5.5.

Hint: do not try to use the definition of core-compactness. Use Theorem 5.4.4, and the fact that, when q is quotient, $f \circ q$ is continuous iff f is continuous. In the if direction, observe that, for every space Z and every continuous map $f \colon Z \times X \to Y$, there is a finest topology $\mathcal{O}_f$ on $[X \to Y]$ that makes $\Lambda_X(f)$ continuous. Let $[X \to Y]_f$ be $[X \to Y]$ with the topology $\mathcal{O}_f$. Let P_f be the image of f in $[X \to Y]_f$ with the subspace topology. Call the spaces P_f the *pieces*, and let $\mathcal{P}$ be the set of all pieces. Show that $[X \to Y]^\natural$ (i.e., $[X \to Y]$ with the natural topology) arises as a quotient of $\coprod_{P \in \mathcal{P}} P$, and deduce that App is continuous from $[X \to Y]^\natural \times X$ to Y.

5.4.2 Arzelà–Ascoli theorems and compact subsets of function spaces

It is natural to ask which subsets of $[X \to Y]$ are compact in the core-open topology, or in the natural topology. Assuming that Y is T_2, theorems to this effect are called *Arzelà–Ascoli theorems*, and are generalizations of the original Arzelà–Ascoli Theorem 3.5.12.

Our presentation takes its roots in Kelley (1955) and Noble (1969), and is a beautiful application of the minimality of compact T_2 topologies (see Theorem 4.4.27) and Tychonoff's Theorem 4.5.12, in addition to the above theory of core-compact spaces. We first notice that a key aspect of equicontinuous families of maps in metric spaces is that pointwise and uniform convergence coincide on equicontinuous families (Proposition 3.5.10). This suggests that one look at subsets of functions on which a similar coincidence of notions of convergence occurs:

Lemma 5.4.30 (Applicatively continuous) *Let X, Y be two topological spaces, and E be a subset of Y^X. We say that E is* applicatively continuous *if and only if the restriction of App to $E \times X$ is continuous, when we equip E with the subspace topology from Y^X, i.e., with the topology of pointwise convergence.*

The applicatively continuous subsets are exactly those subsets E of continuous maps (i.e., subsets of $[X \to Y]$) on which pointwise convergence and continuous convergence coincide: i.e., such that every net in E that converges pointwise to some $f \in E$ also converges continuously to f.

Proof Assume E is applicatively continuous.

Let $f \in E$. Since App is continuous from $E \times X$ to Y, it is separately continuous (Exercise 4.5.11), in particular $\mathsf{App}(f, \_)$ is continuous from X to Y. But $\mathsf{App}(f, \_)$ is just f. So f is continuous. We have shown that $E \subseteq [X \to Y]$.

Next, we show that pointwise convergence implies continuous convergence (so they will coincide, using Exercise 5.4.27). We must show that every net $(f_i)_{i \in I, \sqsubseteq}$ that converges pointwise to $f \in E$ also converges continuously to f. For any $(x_j)_{j \in J, \preceq}$ in X that converges to some $x \in X$, the net $(f_i, x_j)_{(i,j) \in I \times J, \sqsubseteq \times \preceq}$ converges to (f, x) in $E \times X$. Since App is continuous on $E \times X$ by assumption, $(f_i(x_j))_{(i,j) \in I \times J, \sqsubseteq \times \preceq}$ converges to $f(x)$. So $(f_i)_{i \in I, \sqsubseteq}$ converges to f continuously.

Conversely, assume that $E \subseteq [X \to Y]$ and that pointwise convergence and continuous convergence coincide. Let $(f_i, x_i)_{i \in I, \sqsubseteq}$ be a net that converges to (f, x) in $E \times X$. By Exercise 4.7.20, $(f_i)_{i \in I, \sqsubseteq}$ converges (pointwise) to f in E, and $(x_i)_{i \in I, \sqsubseteq}$ converges to x in X. By assumption, the former converges continuously to f. So $(f_i(x_j))_{(i,j) \in I \times I, \sqsubseteq}$ converges to $f(x)$. For every open neighborhood V of $f(x)$, one therefore obtains that $f_i(x_j) \in V$ for $(i, j) \in I \times I$ large enough, in particular $f_i(x_i)$ for $i \in I$ large enough. So $\mathsf{App}\colon E \times X \to Y$ preserves limits. By Proposition 4.7.18, App is continuous from $E \times X$ to Y. So E is applicatively continuous. □

Example 5.4.31 So the family E of all functions f_i, $i \geqslant 2$, considered in Example 5.4.18 (Figure 5.6), is not applicatively continuous. Indeed they converge pointwise to the constant 1 map, but they do not converge continuously, as we have seen.

Lemma 5.4.32 *Let X, Y be topological spaces, and assume X is core-compact. Then a subset E of $[X \to Y]$ is applicatively continuous if and only if the subspace topology of $[X \to Y]^\natural$ ($= [X \to Y]^{\circledcirc}$) on E coincides with the pointwise topology.*

Proof To fix ideas, let $E^\natural$ denote E with the subspace topology of $[X \to Y]^\natural$, and E^{pw} be E with the topology of pointwise convergence, i.e., the subspace topology of $[X \to Y]$ with its topology of pointwise convergence. By Proposition 5.4.21, convergence in $[X \to Y]^\natural$ is continuous convergence. It follows that convergence in $E^\natural$ is also continuous convergence.

If E is applicatively continuous, then, by Lemma 5.4.30, $E^\natural$ and E^{pw} have the same notions of convergence. So $E^\natural = E^{pw}$, using Corollary 4.7.12. Conversely, if $E^\natural = E^{pw}$, then convergence in $E^\natural$ is pointwise convergence. Since this is also continuous convergence, as we have seen above, Lemma 5.4.30 entails that E is applicatively continuous. □

Lemma 5.4.33 (Compact ⇒ applicatively continuous) *Let X be a core-compact space and Y be T_2. Every compact subset K of $[X \to Y]^{\circledcirc}$ is applicatively continuous.*

Proof Consider the topology $\mathcal{O}_2$ on K induced from $[X \to Y]^{\circledcirc}$. Consider also the topology $\mathcal{O}_1$ on K induced from the topology of pointwise convergence on $[X \to Y]$. By Exercise 5.4.26, $\mathcal{O}_1$ is coarser than $\mathcal{O}_2$. We claim that K is T_2 in the $\mathcal{O}_1$ topology. Indeed, if $f, g \in K$ are distinct, then there is an $x \in X$ such that $f(x) \neq g(x)$. Since Y is T_2, there are disjoint open neighborhoods U, V of, respectively, $f(x)$ and $g(x)$. Then $\pi_x^{-1}(U)$ and $\pi_x^{-1}(V)$ are disjoint open neighborhoods of f and g, respectively, where π_x is the projection $h \mapsto h(x)$, which is continuous by Proposition 4.5.2, and since $\mathcal{O}_1$ is induced from the product topology on Y^X.

By the minimality of compact T_2 topologies (Theorem 4.4.27), $\mathcal{O}_1 = \mathcal{O}_2$. By Proposition 5.4.20, $\mathcal{O}_2$ is also the topology induced from the natural topology. So the natural topology and the topology of pointwise convergence coincide on E. By Lemma 5.4.32, K is applicatively continuous. □

Example 5.4.34 So the set E of maps f_i, $i \geqslant 2$ of Figure 5.6, plus their pointwise limit, the constant 1 map, is not compact in $[[0,1] \to [0,1]]^{\circledcirc}$. And indeed one cannot extract any subnet of $(f_i)_{i \in \mathbb{N}}$ that converges continuously to any element of E, even of $[[0,1] \to [0,1]]^{\circledcirc}$. (Exercise: show this.) For the same reason, $[[0,1] \to [0,1]]^{\circledcirc}$ itself is not compact either.

Proposition 5.4.35 (General Arzelà–Ascoli Theorem) *Let X be a core-compact space, and Y be a T_2 space. Given any subset E of Y^X, and any point $x \in X$, let $E(x)$ denote the set $\{f(x) \mid f \in E\}$. We say that E is* pointwise closed *if and only if E is closed in Y^X, i.e., if and only if, for every net $(f_i)_{i \in I, \sqsubseteq}$ in E that converges pointwise to $f \in Y^X$, f is in E.*

The compact subsets K of $[X \to Y]^{\circledcirc}$ are exactly those that are applicatively continuous, pointwise closed, and such that $K(x)$ is compact for every $x \in X$.

Proof The characterization of pointwise closed subsets in terms of nets is by Exercise 4.7.30.

If K is compact, then it is applicatively continuous by Lemma 5.4.33. By Lemma 5.4.32, the topology of K, as a subspace from $[X \to Y]^{\circledcirc}$, is the topology of pointwise convergence, i.e., the one induced from the product topology on Y^X. So the projection $\pi_x : f \mapsto f(x)$ is continuous by Proposition 4.5.2. By Proposition 4.4.13, the image $\pi_x[K] = K(x)$ is compact. Finally, since K is a compact subset of $[X \to Y]^{\circledcirc}$, it is also a compact subspace of $[X \to Y]^{\circledcirc}$ (Exercise 4.9.11), hence a compact subspace of Y^X (since the subspace topologies coincide, by Lemma 5.4.32), hence a compact subset of Y^X. Using Exercise 4.5.18, the latter is T_2. So K, being compact, is closed, by Proposition 4.4.15. So K is pointwise closed.

Conversely, assume that $K(x)$ is compact for every x and K is applicatively continuous and pointwise closed. So K is closed in Y^X, and included in $\prod_{x \in X} K(x)$. The latter is compact by Tychonoff's Theorem 4.5.12, so K is compact in Y^X by Corollary 4.4.10. Since K is applicatively continuous, the subspace topologies from Y^X and from $[X \to Y]^{\circledcirc}$ coincide on K by Lemma 5.4.32, so K is compact in $[X \to Y]^{\circledcirc}$. □

Checking that E is applicatively continuous can be difficult. Kelley (1955, p.234) introduced the following notion of even continuity, which is a bit more usable. The definition is meant to resemble continuity at a point x, and can be described as: E is evenly continuous if and only if, for all $x \in X$, $y \in Y$, and $f \in E$, if $f(x)$ is near y then f maps points near x to points near y—and being "near x" is independent of $f \in E$ for this to work. (Compare with uniform continuity; or with even convergence, where continuity was uniform in x; now we require uniformness in f.)

Lemma 5.4.36 (Even continuity) *Let X, Y be topological spaces. We say that $E \subseteq [X \to Y]$ is* evenly continuous *if and only if, for every $x \in X$, for every $y \in Y$, and every open neighborhood V of y, there is an open neighborhood U of x and yet another open neighborhood W of y such that, for every $f \in E$ such that $f(x) \in W$, $f[U] \subseteq V$.*

If E is evenly continuous, then E is applicatively continuous. Conversely, if Y is T_2, if E is applicatively continuous, pointwise closed, and if $E(x)$ is compact for every $x \in X$, then E is evenly continuous.

Proof Assume E is evenly continuous. Consider any net $(f_i)_{i \in I, \sqsubseteq}$ in E that converges to $f \in E$, pointwise. Let $(x_j)_{j \in J, \preceq}$ be a net in X that converges to $x \in X$. Let $y = f(x)$, and V be any open neighborhood of y in Y. By definition, we can find an open neighborhood U of x and an open neighborhood W of y such that, for every $g \in E$ such that $g(x) \in W$, $g[U] \subseteq V$. Since $f(x) \in W$,

$f_i(x)$ is in W, too, for i large enough, i.e., we can take $g = f_i$: so $f_i[U] \subseteq V$ for i large enough. Also, $x_j \in U$ for j large enough. So $f_i(x_j) \in V$ for (i, j) large enough. Since $(x_j)_{j \in J, \preceq}$ is arbitrary, $(f_i)_{i \in I, \sqsubseteq}$ converges continuously to f. Hence E is applicatively continuous, by Lemma 5.4.30.

Conversely, assume that E is pointwise closed, and $E(x)$ is compact for every $x \in X$. Then E is closed in Y^X, and included in $\prod_{x \in X} E(x)$, which is compact by Tychonoff's Theorem 4.5.12. By Corollary 4.4.10, $E = E \cap \prod_{x \in X} E(x)$ is compact in Y^X.

Now assume by contradiction that Y is T_2, and E is applicatively continuous but not evenly continuous. So there is an $x \in X$, an $y \in Y$, and an open neighborhood V of y such that, for all open neighborhoods U of x and W of y, there is a map $f_{U,W} \in E$ such that $f_{U,W}(x) \in W$ and a point $x_{U,W} \in U$ such that $f_{U,W}(x_{U,W}) \notin V$. (Use the Axiom of Choice to collect the maps $f_{U,W}$ and the points $x_{U,W}$.) The collection I of all pairs of open neighborhoods U and W as above is directed when ordered by $\supseteq \times \supseteq$, so we can use I to index various nets. The net $(x_{U,W})_{(U,W) \in I, \supseteq \times \supseteq}$ converges to x, by definition. Because E is compact in Y^X, there is a sampler $\alpha : J \to I$, and a quasi-ordering $\preceq$ on J that makes J directed such that $(f_{\alpha(j)})_{j \in J, \preceq}$ converges (pointwise) to some element $f \in E$. This is by Corollary 4.7.27. Then $(x_{\alpha(j)})_{j \in J, \preceq}$ still converges to x.

However, we now claim that $(f_{\alpha(j)}(x_{\alpha(j)}))_{j \in J, \preceq}$ cannot converge to $f(x)$. To this end, we first show that $f(x) = y$: for every open neighborhood W of y, and fixing an open neighborhood U of x, $f_{\alpha(j)}(x)$ is in W for j large enough, namely for any $j \in J$ such that $\alpha(j)$ $(\subseteq \times \subseteq)$ (U, W); so y is a limit of $(f_{\alpha(j)}(x))_{j \in J, \preceq}$, and as $f(x)$ is another, and Y is T_2, we conclude that $f(x) = y$ by uniqueness of limits (Proposition 4.7.4). So, if $(f_{\alpha(j)}(x_{\alpha(j)}))_{j \in J, \preceq}$ converged to $f(x)$, it would converge to y. Consider the open neighborhood V of y. By construction $f_{U,W}(x_{U,W})$ is not in V for any $(U, W) \in I$, so $f_{\alpha(j)}(x_{\alpha(j)})$ is not in V for any $j \in J$. In particular, $(f_{\alpha(j)}(x_{\alpha(j)}))_{j \in J, \preceq}$ does not converge to $f(x)$. So $(f_{\alpha(j)})_{j \in J, \preceq}$ does not converge continuously, using Exercise 5.4.16, contradicting the fact that E is applicatively continuous (Lemma 5.4.30). $\square$

Example 5.4.37 The family E of all functions f_i, $i \geqslant 2$, considered in Example 5.4.18 (Figure 5.6), is therefore not evenly continuous, as it is not applicatively continuous. Intuitively, the problem lies near $x = 0$. Let us check this explicitly: let $x = 0$, $y = 1$, $V = (0, 1]$, then assume there is an open neighborhood U of x and an open neighborhood W of y such that, for every $i \in \mathbb{N}$ with $f_i(x) \in W$, $f_i[U] \subseteq V$. U must contain some interval $[0, a)$, and we can pick $i \geqslant 2$ such that $\frac{1}{i} < a$. Then $f_i[U] = [0, 1] \not\subseteq V$, so we must have $f_i(x) \notin W$. Since $f_i(x) = 1$, this implies that $y = 1$ is not in W, a contradiction: E is not evenly continuous.

Theorem 5.4.38 (First Arzelà–Ascoli–Kelley Theorem) *Let X be a core-compact space, and Y be a T_2 space.*

The compact subsets K of $[X \to Y]^{\circledcirc}$ are exactly those that are evenly continuous, pointwise closed, and such that $K(x)$ is compact for every $x \in X$.

Proof This is a direct consequence of Theorem 5.4.35 and Lemma 5.4.36. □

Some other classical forms of the Arzelà–Ascoli theorems are about relative compactness. By this one usually understands a different notion than $\Subset$ (Definition 5.2.4): a subset E of a space is *relatively compact* (in this sense) if and only if it is contained in some compact subset. In a T_2 space, this is equivalent to requiring that $cl(E)$ is compact, by Proposition 4.4.15. We won't use the term "relatively compact" in this sense, and will reserve it for the relation $\Subset$.

We now need Y to be regular for the following result.

Lemma 5.4.39 *Let X be a topological space, and Y be a regular space. If $E \subseteq Y^X$ is evenly continuous, then so is its closure $cl(E)$ in Y^X (i.e., for the topology of pointwise convergence).*

Proof Fix $x \in X$, $y \in Y$, and an open neighborhood V of y. Since Y is regular, it is locally closed (Exercise 4.1.21), so there is a closed neighborhood F of y included in V. Since E is evenly continuous, there is an open neighborhood U of x and an open neighborhood W of y such that, for every $f \in E$ such that $f(x) \in W$, $f[U] \subseteq int(F)$. Now let $f \in cl(E)$. By Proposition 4.7.9, f is the pointwise limit of some net $(f_i)_{i \in I, \sqsubseteq}$ of elements of E. If $f(x) \in W$, then $f_i(x) \in W$ for i large enough, say above i_0, so $f_i[U] \subseteq int(F) \subseteq F$. For every $x \in U$, $f(x)$ is the limit of $(f_i(x))_{i \in \uparrow i_0, \sqsubseteq}$, and is therefore again in F, hence in V. So $f[U] \subseteq V$. □

Theorem 5.4.40 (Second Arzelà–Ascoli–Kelley Theorem) *Let X be a core-compact space, and Y be a T_3 space.*

The subsets E of $[X \to Y]^{\circledcirc}$ that are included in some compact subset of $[X \to Y]^{\circledcirc}$ are exactly those that are evenly continuous and such that $E(x)$ is included in some compact subset of Y for every $x \in X$. In this case, the pointwise closure of E, i.e., its closure in the topology of pointwise convergence, is such a compact subset, and coincides with the closure of E in $[X \to Y]^{\circledcirc}$.

Proof Assume E is included in some compact subset K of $[X \to Y]^{\circledcirc}$. By Theorem 5.4.38, K is evenly continuous, hence its subset E is, too; $K(x)$

is compact for every $x \in X$, so $E(x)$ is trivially included in a compact subset of Y.

Conversely, assume that E is evenly continuous and $E(x)$ is included in some compact subset of Y for every $x \in X$; i.e., $cl(E(x))$ is compact. Let K be the pointwise closure of E. Clearly, K is pointwise closed. K is evenly continuous by Lemma 5.4.39, hence applicatively continuous by Lemma 5.4.36. As in the proof of Theorem 5.4.35, K is a closed subset of $\prod_{x \in X} cl(E(x))$, which is compact by Tychonoff's Theorem 4.5.12. So K is compact in the topology of pointwise convergence, hence also in the core-open topology, since the two topologies induce the same topology on K.

Finally, since K is compact and $[X \to Y]^{\circledcirc}$ is T_2 (by Exercise 5.4.28), K is closed in $[X \to Y]^{\circledcirc}$, by Proposition 4.4.15. So K contains the closure of E in $[X \to Y]^{\circledcirc}$. Since the topology of pointwise convergence is finer than the core-open topology (Exercise 5.4.26), the closure $cl(E)$ of E in Y^X is included in the closure of E in $[X \to Y]^{\circledcirc}$. Since $K = cl(E)$, the two closures coincide. ☐

We shall see special cases of these theorems in Section 6.6.1, where we shall eventually rederive the original Arzelà–Ascoli Theorem 3.5.12 from the latter.

5.5 A bit of category theory II

5.5.1 Exponential objects

Given any two objects X, Y in a category **C** with binary products, an *exponential* object, if it exists, is an object Y^X, together with a morphism $\mathsf{App}\colon Y^X \times X \to Y$, and a collection of morphisms $\Lambda_X(f)\colon Z \to Y^X$, one for each morphism $f\colon Z \times X \to Y$, where Z is an arbitrary object of **C**, satisfying the equations:

$$
\begin{array}{lrcl}
(\beta) & \mathsf{App} \circ (\Lambda_X(f) \times \mathrm{id}_X) & = & f \\
 & & & \text{for every } f\colon Z \times X \to Y \\
(\eta) & \Lambda_X(\mathsf{App}) & = & \mathrm{id}_{Y^X} \\
(\sigma) & \Lambda_X(f) \circ g & = & \Lambda_X(f \circ (g \times \mathrm{id}_X)) \\
 & & & \text{for all } f\colon Z \times X \to Y, g\colon Z' \to Z\,.
\end{array}
$$

An object X is *exponentiable* in **C** if and only if it has an exponential Y^X for every object Y of **C**.

The constructions of Section 5.3 show that the exponentiable objects of **Top** are exactly the core-compact spaces, and that the space $[X \to Y]$ is an exponential object, with $\mathsf{App}(h, x) = h(x)$ and $\Lambda_X(f) = f(\_, x)$, if and only if it is equipped with the core-open topology. Verifying (β), (η), and (σ) is straightforward.

More remarkably, the only exponential object Y^X in **Top**, up to iso, must be $[X \to Y]$, $\mathsf{App}(h, x)$ must be $h(x)$ (up to the iso), and $\Lambda_X(f)(z)$ must be $f(z, \_)$ (up to the iso). We prove this below (Theorem 5.5.1).

This is not only true in **Top**, but in many full subcategories of **Top**. A *subcategory* of a category **C** is a category $\mathbb{C}$ whose objects are objects of **C**, whose morphisms are morphisms of **C**, and whose identities and composition coincide with that of **C**. For example, **Cpo** can be seen as a subcategory of **Top**, once we agree to equate each dcpo X with the topological space X_σ. A subcategory $\mathbb{C}$ of **C** is *full* iff, for every two objects X, Y of $\mathbb{C}$, the morphisms from X to Y in $\mathbb{C}$ are exactly those from X to Y in **C**. For example, **Cpo** is a full subcategory of **Top**, while **Cpo** is a non-full subcategory of **Ord**.

Theorem 5.5.1 *Let* **C** *be any full subcategory of* **Top** *with finite products, and assume that* $1 = \{*\}$ *is an object of* **C**. *Let* X, Y *be two objects of* **C** *that have an exponential object* Y^X *in* **C**.

Then there is a unique homeomorphism $\theta \colon Y^X \to [X \to Y]$, *for some unique topology on* $[X \to Y]$, *such that* $\mathsf{App}(h, x) = \theta(h)(x)$ *for all* $h \in Y^X$, $x \in X$.

Moreover, $\Lambda_X(f)(z)$ *is the image by* θ^{-1} *of* $f(z, \_)$ *for all* $f \colon Z \times X \to Y$, $z \in Z$. *(Here* $Z \times X$ *denotes the product in* **C**, *which may or may not be the topological product.)*

Proof One should note that the product $\times$ in **C** may be different from the topological product. (E.g., if **C** = **Cpo**; see Exercise 5.2.16). Let $\langle f, g\rangle$ denote pairing, and π_1 and π_2 be the two projections relative to the product $\times$ in **C**. We first claim that, given any two objects A, B in **C**, up to homeomorphism $A \times B$ must be the set of all pairs (a, b) with $a \in A$, $b \in B$, i.e., the set-theoretic product of A and B, with some topology. Let us write this set-theoretic product $A.B$, for now. Indeed, there is a map $p \colon A \times B \to A.B$ that sends each $z \in A \times B$ to $(\pi_1(z), \pi_2(z))$. Equip $A.B$ with the coarsest topology that makes p continuous, i.e., W is open in $A.B$ iff $p^{-1}(W)$ is open in $A \times B$. Conversely, for any object X of **C**, and every element $x \in X$, there is a map $x^1 \colon 1 \to X$ that maps $*$ to x. (We use the fact that $1 = \{*\}$.) As a constant map, x^1 is continuous, and since **C** is a full subcategory of **Top**, x^1 is a morphism in **C**. In particular, there is a map $q \colon A.B \to A \times B$ defined by $q(a, b) = \langle a^1, b^1\rangle(*)$. One checks that p and q are inverse to each other: $p(q(a, b)) = (\pi_1 \circ \langle a^1, b^1\rangle(*), \pi_2 \circ \langle a^1, b^1\rangle(*)) = (a^1(*), b^1(*)) = (a, b)$, and $q(p(z)) = \langle \pi_1(z)^1, \pi_2(z)^1\rangle(*) = \langle \pi_1 \circ z^1, \pi_2 \circ z^1\rangle(*) = \langle \pi_1, \pi_2\rangle \circ z^1(*) = z^1(*) = z$. Finally, we check that q is continuous: the inverse image $q^{-1}(V)$ of an open subset V of $A \times B$ is open in $A.B$ iff $p^{-1}(q^{-1}(V))$ is open in $A \times B$, by definition; and this is trivial since $p^{-1}(q^{-1}(V)) = V$.

This allows us to identify $A \times B$ with the set-theoretic product of A and B, equipped with some topology. Also, we have the formulae $\langle f, g\rangle(z) = (f(z), g(z))$, $\pi_1(a, b) = a$, $\pi_2(a, b) = b$, $(f \times g)(a, b) = (f(a), g(b))$.

We claim that $1 = \{*\}$ is a terminal object in **C**: given any object A of **C**, there is indeed a unique continuous map $f: A \to 1$, and it is a morphism in **C** since **C** is full.

We now claim that there is an isomorphism $\ell_X : X \to 1 \times X$ in **C**. Define indeed $\ell_X = \langle !_X, \mathrm{id}_X\rangle$, where $!_X$ is the unique morphism from X to 1, and $\ell_X^{-1} = \pi_2$. They are indeed inverse to each other. First, $\ell_X^{-1} \circ \ell_X = \pi_2 \circ \langle !_X, \mathrm{id}_X\rangle = \mathrm{id}_X$, while $\ell_X \circ \ell_X^{-1} = \langle !_X, \mathrm{id}_X\rangle \circ \pi_2 = \langle !_X \circ \pi_2, \mathrm{id}_X \circ \pi_2\rangle$. Now $!_X \circ \pi_2$ is the unique morphism $!_{1\times X}$ from $1 \times X$ to 1; since π_1 is another, $!_X \circ \pi_2 = \pi_1$; so $\ell_X \circ \ell_X^{-1} = \langle \pi_1, \pi_2\rangle = \mathrm{id}_{1\times X}$. Since $1 \times X$ is the set-theoretic product of 1 and X, and $!_X$ maps every $x \in X$ to $*$, we have the explicit formula $\ell_X(x) = (*, x)$. The point in defining ℓ_X as $\langle !_X, \mathrm{id}_X\rangle$, rather than through the simpler formula $\ell_X(x) = (*, x)$, is that it makes it obvious that ℓ_X is a morphism, and in particular is continuous, although we make no assumption on the topology of $1 \times X$.

Define $\theta(h)$ as $\mathsf{App} \circ (h^1 \times \mathrm{id}_X) \circ \ell_X$. Concretely, $\theta(h)(x) = \mathsf{App}(h, x)$.

Conversely, for every $g \in [X \to Y]$, let $\theta'(g) = \Lambda_X(g \circ \ell_X^{-1})(*)$. We claim that θ' is inverse to θ. When $g = \theta(h)$, $\theta'(g) = \Lambda_X(\mathsf{App} \circ (h^1 \times \mathrm{id}_X))(*) = (\Lambda_X(\mathsf{App}) \circ h^1)(*)$ (by (σ)) $= h^1(*)$ (by (η)) $= h$. Conversely, when $h = \theta'(g)$, $h^1 = \Lambda_X(g \circ \ell_X^{-1})$, so $\theta(h) = \mathsf{App} \circ (\Lambda_X(g \circ \ell_X^{-1}) \times \mathrm{id}_X) \circ \ell_X = g$ by (β).

So θ and θ' are inverse to each other. Equip $[X \to Y]$ with the topology whose opens are the subsets $\theta'^{-1}(U)$, U open in Y^X. Then θ (and θ') define a homeomorphism of Y^X onto $[X \to Y]$ with this topology. Since **C** is full, this defines an iso in **C**.

Next, θ is determined uniquely from the equation $\mathsf{App}(h, x) = \theta(h)(x)$; hence $\theta^{-1} = \theta'$ is unique as well.

Finally, we check that $\theta(\Lambda_X(f)(z)) = f(z, \_)$ for all $f: Z \times X \to Y$, $z \in Z$. Let $g = f(z, \_)$. It is enough to show that $\Lambda_X(f)(z) = \theta'(g)$. Note that $g \circ \ell_X^{-1}$ maps $(*, x)$ to $f(z, x)$, so $g \circ \ell_X^{-1} = f \circ (z^1 \times \mathrm{id}_X)$. So $\theta'(g) = \Lambda_X(f \circ (z^1 \times \mathrm{id}_X))(*) = (\Lambda_X(f) \circ z^1)(*)$ (by (σ)) $= \Lambda_X(f)(z)$. □

Exponential objects can be characterized in the following equivalent way. In a category **C**, fix an object X such that the product $Z \times X$ exists for every object Z. For every object Y, an *exponential* Y^X, if it exists, is an object together with a morphism $\mathsf{App}: Y^X \times X \to Y$ such that, for every morphism $f\colon Z \times X \to Y$, there is a unique morphism $h\colon Z \to Y^X$ that makes the following diagram commute:

$$\begin{array}{ccc} Z \times X & \xrightarrow{h \times \mathrm{id}_X} & Y^X \times X \\ & \searrow^{f} & \downarrow \mathsf{App} \\ & & Y \end{array} \tag{5.1}$$

The two definitions are equivalent. If Y^X is an exponential in the former sense, then take $h = \Lambda_X(f)$ above: this makes Diagram (5.1) commute by (β). Moreover, h is unique: indeed, if $f = \mathsf{App} \circ (h \times \mathrm{id}_X)$ as required in Diagram (5.1), then $\Lambda_X(f) = \Lambda_X(\mathsf{App}) \circ h$ (by (σ)) $= h$ (by (η)), so one can recover h from f uniquely. Conversely, if Y^X is an exponential in the new sense, then define $\Lambda_X(f)$ as the morphism h given in Diagram (5.1), so that (β) is satisfied. Taking $Z = Y^X$ and $f = \mathsf{App}$ in the diagram, one sees that $h = \mathrm{id}_{Y^X}$ would make the diagram commute: since h is unique, $h = \Lambda_X(\mathsf{App})$, whence (η) must hold. Similarly, (σ) must hold, this time by appealing to the uniqueness of $h' = \Lambda_X(f \circ (g \times \mathrm{id}_X))$, in the outer triangle of the following diagram, where $h = \Lambda_X(f)$:

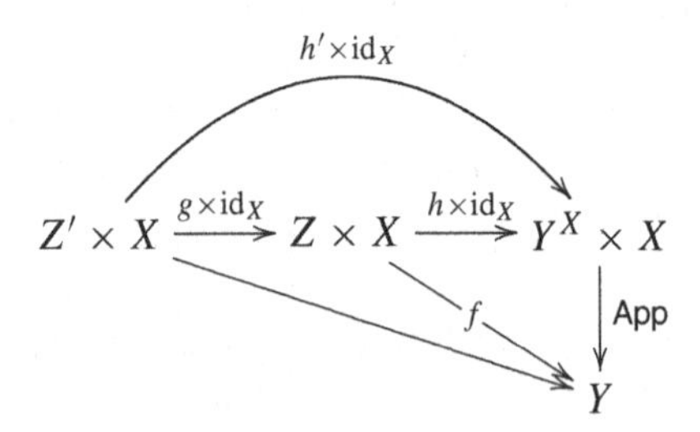

> **Exercise 5.5.2** (Retracts of exponential objects) Let **C** be a full subcategory of **Top**. Assume that finite products in **C** are computed as in **Top**, i.e. $1 = \{*\}$ is terminal in **C** and, for every two objects A, B in **C**, their product in **C** is the topological product $A \times B$. Let $\mathbf{C}'$ be the full subcategory of **Top** whose objects are retracts of objects of **C**.
>
> Given any object A' of $\mathbf{C}'$, let it be a retract of some object A of **C**, with retraction $r_A\colon A \to A'$ and section $s_A\colon A' \to A$; similarly for B', B, r_B, s_B. Assume that A and B have an exponential $[A \to B]$ in **C**. Let $[A' \to B']'$ be the space of continuous maps from A' to B', with the coarsest topology that makes s continuous, where $s\colon [A' \to B']' \to [A \to B]$ maps g to $s_B \circ g \circ r_A$, and $r\colon [A \to B] \to [A' \to B']'$ maps g to $r_B \circ g \circ s_A$. Show that $r \circ s$ is the identity map, that $s \circ r$ is continuous (use the fact that $[A \to B]$ is an exponential in **C**), so that $[A' \to B']'$ is an exponential of A' and B' in $\mathbf{C}'$.

5.5.2 Adjunctions

There is yet another way of describing what an exponential object is, as a particular instance of an *adjunction*. However, explaining adjunctions through the example of exponential objects is awkward. So let us introduce adjunctions with a general definition, and using different examples for some time.

Given two functors F and G from $\mathbf{C}$ to $\mathbb{C}$, a *transformation* from F to G is a collection η of morphisms $\eta_X \colon F(X) \to G(X)$, one for each object X of $\mathbf{C}$. It is *natural* iff the following diagram commutes for every morphism $f \colon X \to Y$ in $\mathbf{C}$:

$$\begin{array}{ccc} F(X) & \xrightarrow{\eta_X} & G(X) \\ {\scriptstyle F(f)}\downarrow & & \downarrow{\scriptstyle G(f)} \\ F(Y) & \xrightarrow[\eta_Y]{} & G(Y) \end{array}$$

We then usually write $\eta \colon F \dot{\to} G$ to state that η is a natural transformation from F to G.

Given two functors F, G, we usually write FG instead of $F \circ G$. We also let id be the identity functor. This structure builds a category **Cat** of all small categories (i.e., of all categories that have a set rather than a collection of objects), with functors as morphisms.

An *adjunction* between two categories $\mathbf{C}$ and $\mathbb{C}$ is a pair of two functors $F \colon \mathbf{C} \to \mathbb{C}$ and $U \colon \mathbb{C} \to \mathbf{C}$, written $F \dashv U$, with two connecting natural transformations $\eta \colon \mathrm{id} \dot{\to} UF$ in $\mathbf{C}$ (the *unit* of the adjunction) and $\epsilon \colon FU \dot{\to} \mathrm{id}$ in $\mathbb{C}$ (the *counit* of the adjunction), making the following two *coherence diagrams* commute for every object X of $\mathbf{C}$ and every object Y of $\mathbb{C}$:

$$\begin{array}{ccc} F(X) & \xrightarrow{F(\eta_X)} & FUF(X) \\ & {\scriptstyle \mathrm{id}_{F(X)}}\searrow & \downarrow{\scriptstyle \epsilon_{F(X)}} \\ & & F(X) \end{array} \qquad \begin{array}{ccc} U(Y) & \xrightarrow{\eta_{U(Y)}} & UFU(Y) \\ & {\scriptstyle \mathrm{id}_{U(Y)}}\searrow & \downarrow{\scriptstyle U(\epsilon_Y)} \\ & & U(Y) \end{array}$$

The functor F is called *left adjoint* to U, and U is *right adjoint* to F. It is useful to think of U as the "underlying" functor and F as the "free" functor, as the following example will illustrate. (The underlying functor U is often called the *forgetful* functor, but "forgetful" also starts with an "f".)

The formal definition is of course impenetrable. The typical example, which hopefully should give a better intuition of the concept, is given by $\mathbf{C} = \mathbf{Set}$, the category of sets, and $\mathbb{C} = \mathbf{Mon}$, the category of monoids and monoid homomorphisms. A *monoid* M is any set $|M|$ with a binary operation $\cdot$ (multiplication) and a unit element 1, such that $1 \cdot m = m \cdot 1 = m$ for every $m \in |M|$, and satisfying $(m \cdot m') \cdot m'' = m \cdot (m' \cdot m'')$ (associativity). E.g., $\mathbb{Z}$ is a monoid, either with $+$ or ordinary multiplication as multiplication, and Σ^*, the set of finite words over some alphabet Σ, with concatenation as multiplication and empty word ε as unit, is also a monoid. We shall usually drop parentheses, e.g., $m \cdot m' \cdot m''$ stands for $(m \cdot m') \cdot m''$, or $m \cdot (m' \cdot m'')$. The *monoid homomorphisms* from M to M' are those maps $f: |M| \to |M'|$ that respect unit and product, i.e., such that $f(1) = 1$ and $f(m_1 \cdot m_2) = f(m_1) \cdot f(m_2)$.

The typical underlying functor U : **Mon** $\to$ **Set** maps each monoid M to its underlying set $|M|$, and each monoid homomorphism $f: M \to M'$ to f seen as a mere function. This functor U has a left adjoint F, which is described as follows. Given any set Σ, $F(\Sigma)$ is just the monoid Σ^*: Σ^* is the *free monoid* over the set Σ. Given any map $f: \Sigma \to \Sigma'$, which one can see as a substitution of letters in Σ' for letters in Σ, $F(f)$ is defined as mapping every word $a_1 a_2 \ldots a_n$ to the word $f(a_1) f(a_2) \ldots f(a_n)$. One checks that $F \dashv U$ is indeed an adjunction. The unit η_Σ maps each letter $a \in \Sigma$ to the one-letter word a, and the counit ϵ_M maps each word $m_1 m_2 \ldots m_n$ in $FU(M) = |M|^*$ to the product $m_1 \cdot m_2 \cdot \cdots \cdot m_n$. We invite the reader to check that the coherence diagrams above indeed commute.

Exercise 5.5.3 (Ideal completion = free dcpo) Here is an example of an adjunction that we have already encountered, without saying so. Recall that **Ord** is the category of posets and monotonic maps, and **Cpo** is the category of dcpos and Scott-continuous maps. Given any poset X, show that its ideal completion $\mathbf{I}(X)$ (Definition 5.1.45) is the free dcpo over X.

In other words, let U be the obvious underlying functor from **Cpo** to **Ord**, mapping every dcpo X to X as poset, and every Scott-continuous map f to f as a monotonic map. Let $\mathbf{I}$ be the functor defined on objects as the ideal completion, and on monotonic maps $f: X \to Y$ by $\mathbf{I}(f)(D) = \downarrow f[D]$. (Hint: use Scott's Formula, and recall Proposition 5.1.46.) Check that $\mathbf{I} \dashv U$ is an adjunction, with unit $\eta_X : X \to U(\mathbf{I}(X))$ defined by $\eta_X(x) = \downarrow x$, and counit $\epsilon_Y : \mathbf{I}(U(Y)) \to Y$ defined by $\epsilon_Y(D) = \sup D$.

There are many alternative ways to define adjunctions. Given a category $\mathbf{C}$, and two objects X, X' in $\mathbf{C}$, let the *homset* $\mathrm{Hom}_{\mathbf{C}}(X, X')$ be the set of all morphisms from X to X'. Then $F \dashv U$ is an adjunction iff there is a bijection $\mathrm{lan}_{X,Y}$ from $\mathrm{Hom}_{\mathbb{C}}(F(X), Y)$ to $\mathrm{Hom}_{\mathbf{C}}(X, U(Y))$ for all objects X of $\mathbf{C}$ and Y of $\mathbb{C}$, and $\mathrm{lan}_{X,Y}$ is natural in X and Y, meaning that the following diagram commutes for all morphisms $f: X' \to X$ in $\mathbf{C}$ and $g: Y \to Y'$ in $\mathbb{C}$:

$$\begin{array}{ccc} \mathrm{Hom}_{\mathbb{C}}(F(X), Y) & \xrightarrow{\mathrm{lan}_{X,Y}} & \mathrm{Hom}_{\mathbf{C}}(X, U(Y)) \\ \Big\downarrow {\scriptstyle \mathrm{Hom}_{\mathbb{C}}(F(f),g)} & & \Big\downarrow {\scriptstyle \mathrm{Hom}_{\mathbf{C}}(f,U(g))} \\ \mathrm{Hom}_{\mathbb{C}}(F(X'), Y') & \xrightarrow{\mathrm{lan}_{X',Y'}} & \mathrm{Hom}_{\mathbf{C}}(X', U(Y')) \end{array}$$

Here $\mathrm{Hom}_{\mathbb{C}}(F(f), g)$ maps any morphism $h\colon F(X) \to Y$ to $g \circ h \circ F(f)$, and $\mathrm{Hom}_{\mathbf{C}}(f, U(g))$ maps $h\colon X \to U(Y)$ to $U(g) \circ h \circ f$. (This is the same notion of naturality as before, realizing that we have just defined $\mathrm{Hom}_{\mathbb{C}}(F(\_), \_)$ and $\mathrm{Hom}_{\mathbf{C}}(\_, U(\_))$ as functors from $\mathbf{C}^{op} \times \mathbb{C}$ to **Set**.)

We usually write $\mathrm{ran}_{X,Y}$ for the inverse of $\mathrm{lan}_{X,Y}$.

The connection with the previous definition of adjunctions is given by $\eta_X = \mathrm{lan}_{X,F(X)}(\mathrm{id}_{F(X)})$, $\epsilon_Y = \mathrm{ran}_{U(Y),Y}(\mathrm{id}_{U(Y)})$, and conversely $\mathrm{lan}_{X,Y}(h) = U(h) \circ \eta_X$, $\mathrm{ran}_{X,Y}(h) = \epsilon_Y \circ F(h)$.

In our example of sets and monoids, $\mathrm{lan}_{\Sigma,M}$ maps any monoid homomorphism $f\colon \Sigma^* \to M$ to the map $\mathrm{lan}_{\Sigma,M}(f)\colon \Sigma \to |M|$ that sends each letter $a \in \Sigma$ to $f(a) \in M$. The inverse $\mathrm{ran}_{\Sigma,M}$ of $\mathrm{lan}_{\Sigma,M}$ is more interesting: given any map f from letters in Σ to elements of the monoid M, $\mathrm{ran}_{\Sigma,M}(f)$ maps every word $a_1a_2 \ldots a_n$ to the product $f(a_1) \cdot f(a_2) \cdot \cdots \cdot f(a_n)$. That is, $\mathrm{ran}_{\Sigma,M}(f)$ is the unique monoid homomorphism from Σ^* to M that extends f, that is, which maps every one-letter word a to $f(a)$.

The latter characterizes the free monoids. In general, given any (underlying) functor $U\colon \mathbb{C} \to \mathbf{C}$, and any object X of $\mathbf{C}$, a *free object* of $\mathbb{C}$ over X (with respect to U) is any object $F(X)$ of $\mathbb{C}$ together with a morphism $\eta_X : X \to U(F(X))$ such that, for every object Y of $\mathbb{C}$ and every morphism $f\colon X \to U(Y)$, there is a unique morphism $h\colon F(X) \to Y$ such that the following diagram commutes:

$$\begin{array}{ccc} U(F(X)) & & \\ {\scriptstyle \eta_X}\Big\uparrow & \searrow^{U(h)} & \\ X & \xrightarrow[f]{} & U(Y) \end{array} \tag{5.2}$$

If F is left adjoint to U, then $F(X)$ is a free object over X with respect to U, and h is just $\mathrm{ran}_{X,Y}(f)$. Conversely, if there is a free object $F(X)$ over every object X of $\mathbf{C}$, then decide to call $\mathrm{ran}_{X,Y}(f)$ the unique morphism h given in the diagram above. We can turn F into a functor by letting $F(f)$ be $\mathrm{ran}_{X,F(X')}(\eta_{X'} \circ f)$, for every morphism $f\colon X \to X'$, and check that $\mathrm{ran}_{X,Y}$ is a natural bijection of homsets.

One can, for example, redo Exercise 5.5.3 in a much simpler way, by checking that the ideal completion $\mathbf{I}(X)$ is the free dcpo over the poset X directly.

One only has to check that, for every dcpo Y, every monotonic map $f: X \to Y$ has a unique Scott-continuous *extension* $h: \mathbf{I}(X) \to Y$, in the sense that $h \circ \eta_X = f$, where $\eta_X(x) = \downarrow x$. (We omit mention of the underlying functor U here.) Indeed, if such an h exists, then one must have $h(D) = \sup_{x \in D} f(x)$, since every ideal D is the least upper bound of the directed family $(\downarrow x)_{x \in D}$, by Scott-continuity. Since $x \in D$ is equivalent to $\downarrow x \subseteq D$, hence to $\downarrow x \ll D$ (because $\downarrow x$ is finite in $\mathbf{I}(X)$, by Proposition 5.1.46), we may write $h(D) = \sup_{\downarrow x \ll D} f(x)$, which we recognize as Scott's Formula (Proposition 5.1.60; the basis B is $\{\downarrow x \mid x \in X\}$, and the map we extend is the one mapping $\downarrow x$ to $f(x)$). So h is indeed Scott-continuous. We check that $h \circ \eta_X = f$: for every $x \in X$, $h(\eta_X(x)) = \sup_{x' \in \downarrow x} f(x') = f(x)$, because f is monotonic. If you have done Exercise 5.5.3, you should be convinced that this is a much shorter argument.

An important property of adjoints is that:

Right adjoints U preserve all existing limits.
Left adjoints F preserve all existing colimits.

By this we mean that, given a left adjoint F, if $(X, (\iota_J)_{J \text{ object of } \mathbf{J}})$ is a colimit of some functor $G: \mathbf{J} \to \mathbf{C}$ in $\mathbf{C}$, then $(F(X), (F(\iota_J))_{J \text{ object of } \mathbf{J}})$ is a colimit of $F \circ G: \mathbf{J} \to \mathbb{C}$ in $\mathbb{C}$, and similarly for limits with right adjoints U.

This property follows by diagrammatic reasoning. Namely, assume that $(X, (\iota_J)_{J \text{ object of } \mathbf{J}})$ is a colimit of some functor $G: \mathbf{J} \to \mathbf{C}$ in $\mathbf{C}$, and show that $(F(X), (F(\iota_J))_{J \text{ object of } \mathbf{J}})$ is a colimit of $F \circ G: \mathbf{J} \to \mathbb{C}$ in $\mathbb{C}$. Given any other cocone $(Y, (\iota'_J)_{J \text{ object of } \mathbf{J}})$ over the diagram $F \circ G$, $(U(Y), (\mathrm{lan}_{G(J),Y}(\iota'_J))_{J \text{ object of } \mathbf{J}})$ is a cocone over G. Since $(X, (\iota_J)_{J \text{ object of } \mathbf{J}})$ is a universal cocone, there is a unique morphism $h: X \to U(Y)$ such that $\mathrm{lan}_{G(J),Y}(\iota'_J) = h \circ \iota_J$ for every morphism $j: J \to J'$ in $\mathbf{J}$. Note that $\mathrm{lan}_{G(J),Y}(\iota'_J) = h \circ \iota_J$ is equivalent to $\iota'_J = \mathrm{ran}_{G(J),Y}(h \circ \iota_J)$, and by naturality, to $\iota'_J = \mathrm{ran}_{X,Y}(h) \circ F(\iota_J)$. So $\mathrm{ran}_{X,Y}(h)$ is the unique morphism h' from $F(X)$ to Y such that $\iota'_J = h' \circ F(\iota_J)$ for every morphism $j: J \to J'$ in $\mathbf{J}$, whence $(F(X), (F(\iota_J))_{J \text{ object of } \mathbf{J}})$ is indeed a colimit.

That right adjoints preserve all existing limits can be obtained from the latter by realizing that $F \dashv U$ is an adjunction if and only $U' \dashv F'$ is an adjunction, where $U': \mathbb{C}^{op} \to \mathbf{C}^{op}$ coincides with $U: \mathbb{C} \to \mathbf{C}$ and $F': \mathbf{C}^{op} \to \mathbb{C}^{op}$ coincides with F.

Using the fact that left adjoints preserve all existing colimits entails, for example, that the free monoid of any colimit of sets is the corresponding colimit of monoids, and we do not even have to know what these colimits may look like in **Mon** to be able to conclude this.

This is rather non-trivial, if one looks at it. Note that even binary coproducts in **Mon** are rather complex objects, as we shall illustrate. Despite this, the above result shows in particular that the coproduct of two free monoids Σ_1^* and Σ_2^* in **Mon** must be $(\Sigma_1 + \Sigma_2)^*$, up to natural iso. Concretely, binary coproducts in **Mon**, or *free products* as they are sometimes, somehow misleadingly, called, are built as follows. Given two monoids M_1, M_2, their coproduct $M_1 + M_2$ is $(|M_1| + |M_2|)^*/\equiv$, the set of finite words $m_1 m_2 \dots m_n$ in $(|M_1| + |M_2|)^*$, modulo the least equivalence relation $\equiv$ such that $m_1 \dots m_{i-1} 1 m_i \dots m_n \equiv m_1 \dots m_{i-1} m_i \dots m_n$ (where 1 is the unit element of M_1, or of M_2) and $m_1 \dots m_{i-1} m_i m_i' m_{i+1} \dots m_n \equiv m_1 \dots m_{i-1} (m_i \cdot m_i') m_{i+1} \dots m_n$ if m_i and m_i' are both elements of M_1, or both elements of M_2. The unit element is the equivalence class of the empty word, and multiplication is concatenation.

Exercise 4.10.25 provides another example of an adjunction. Let $\mathbf{Top}_0$ be the full subcategory of **Top** consisting of the T_0 spaces. As a subcategory, it has an *inclusion functor* $\subseteq: \mathbf{Top}_0 \to \mathbf{Top}$, which is the identity on objects and on morphisms. The universal property of the T_0 quotient can be recast by saying that, for every topological space X, $X/{=_0}$ is the free T_0 space over X. In other words, the map $X \mapsto X/{=_0}$ is the object part of a functor F, left adjoint to U.

Such particular adjunctions, where the right adjoint is an inclusion of categories, are called *reflections*. We also say that $\mathbf{Top}_0$ is a *reflective subcategory* of **Top**. Because right adjoints preserve all limits, we immediately see that any limit in $\mathbf{Top}_0$ must be defined exactly as in **Top**. And because left adjoints preserve all colimits, we deduce that $\mathbf{Top}_0$ has all colimits, and that they are computed as the T_0 quotients of the corresponding colimits computed in **Top**.

Symmetrically, if an inclusion of categories is left adjoint, we talk about *coreflections*, and *coreflective subcategories*. We shall see an example of a coreflective subcategory of **Top** in Lemma 5.6.13.

A degenerate case of adjunction is the notion of equivalence of categories. An *equivalence* of categories $\mathbf{C}$, $\mathbb{C}$ is a pair of functors $f: \mathbf{C} \to \mathbb{C}$ and $G: \mathbb{C} \to \mathbf{C}$, together with natural isomorphisms $\epsilon: FG \dot{\to} \mathrm{id}$ in $\mathbb{C}$ and $\eta: \mathrm{id} \dot{\to} GF$ in $\mathbf{C}$. (A *natural isomorphism* is a collection of isos that altogether form a natural transformation.) If such an equivalence exists, then we say that $\mathbf{C}$ and $\mathbb{C}$ are *equivalent categories*. In this case, $\mathbf{C}$ and $\mathbb{C}$ are the same categories for all practical purposes, except for details pertaining to the representations of objects. The definition above makes it clear that an equivalence is just an adjunction $F \dashv G$ whose unit and counit are iso. In this case, $G \dashv F$ is also an equivalence of categories.

5.5.3 Cartesian-closed categories

Let's return to exponential objects. They will arise from an adjunction, although one that tends to look strange compared to the typical picture obtained in the case of sets and monoids.

Observe that in any category **C** with binary products, for any fixed object X, one can define a functor $\_ \times X$ from **C** to **C**. (We take $\mathbb{C} = \mathbf{C}$ here.) This amounts to picking a fixed product $Z \times X$ for each Z (there are usually many, defined up to iso), say the topological product if $\mathbf{C} = \mathbf{Top}$, and defining the action of $\_ \times X$ on morphisms $g\colon Z' \to Z$ as $g \times \mathrm{id}_X$.

We claim that there are exponential objects Y^X for every object Y if and only if $\_ \times X$ has a right adjoint. This right adjoint is then a functor $\_^X$, mapping each object Y to Y^X, and with some action on morphisms that we shall make explicit.

Indeed, assume first that exponentials Y^X exist for every object Y. To stick with the definition, let $F(Z) = Z \times X$, and define U on objects by $U(Y) = Y^X$, and on morphisms by $U(h) = \Lambda_X(h \circ \mathsf{App})$ for every morphism $h\colon Y \to Y'$, where $h \circ \mathsf{App}\colon Y^X \times X \to Y'$. The unit $\eta_Z\colon Z \to (Z \times X)^X$ is $\Lambda_X(\mathrm{id}_{Z\times X})$, and the counit $\epsilon_Y\colon Y^X \times X \to Y$ is just application App. We invite the reader to check the naturality conditions and the coherence diagrams.

Conversely, assume $F = \_ \times X$ has a right adjoint U. Define Y^X as $U(Y)$, $\mathsf{App}\colon Y^X \times X \to Y$ as ϵ_Y, and, for every morphism $f\colon Z \times X \to Y$, i.e., $f\colon F(Z) \to Y$, let $\Lambda_X(f) = U(f) \circ \eta_Z = \mathrm{lan}_{Z,Y}(f)$.

If X is an exponentiable object in **C**, $\_ \times X$, as a left adjoint, must preserve all existing colimits. In particular, it must preserve all existing coproducts and all existing coequalizers. In **Top**, the first condition implies that $(\coprod_{i\in I} Y_i) \times X$ should be isomorphic to $\coprod_{i\in I}(Y_i \times X)$, and is clearly true. The second one implies that, whenever $q\colon Y \to Z$ is quotient, then $q \times \mathrm{id}_X$ should be quotient. We have seen in Exercise 5.4.29 that this condition was equivalent to requiring X to be core-compact. We now see that this is no accident: if X is exponentiable, then $\_ \times X$ must preserve coequalizers. The converse direction, in **Top**, was dealt with using non-categorical means in Exercise 5.4.29. This can also be done categorically, using Freyd's special adjoint functor theorem (Escardó *et al.*, 2004, remark 2.7).

The category **Cpo** has the remarkable property that *every* object of **Cpo** is exponentiable. Such categories have a name: a category **C** is *Cartesian-closed*, or we may say it is a *Cartesian-closed category (CCC)*, if and only if it has all finite products, and every object is exponentiable.

Proposition 5.5.4 (**Cpo** is Cartesian-closed) *Given any two dcpos X, Y, the dcpo $[X \to Y]$ of all continuous maps from X to Y, with the* pointwise ordering, *defined by $f \leqslant g$ iff $f(x) \leqslant g(x)$ for every $x \in X$, is an exponential* *in* **Cpo**. *The least upper bound f of any directed family $(f_i)_{i \in I}$ of continuous maps is defined pointwise: $f(x) = \sup_{i \in I} f_i(x)$ for every $x \in X$.*

Application App: $[X \to Y] \times X \to Y$ *is defined by* $\mathsf{App}(f, x) = f(x)$, *and currification* $\Lambda_X(f)$ *is defined for every continuous map* $f : Z \times X \to Y$ *by* $\Lambda_X(f)(z)(x) = f(z, x)$.

Proof We first show that the least upper bound f of any directed family $(f_i)_{i \in I}$ of continuous maps is defined pointwise: $f(x) = \sup_{i \in I} f_i(x)$ for every $x \in X$. For this, it is enough to show that f is (Scott-)continuous, since then it will necessarily be the least of all upper bounds. Given any directed family $(x_j)_{j \in J}$ of elements of X, $f(\sup_{j \in J} x_j) = \sup_{i \in I} f_i(\sup_{j \in J} x_j) = \sup_{i \in I} \sup_{j \in J} f_i(x_j)$ (since each f_i is Scott-continuous, being continuous; see Proposition 4.3.5) $= \sup_{j \in J} \sup_{i \in I} f_i(x_j) = \sup_{j \in J} f(x_j)$. So $[X \to Y]$ is a dcpo.

Next, we need to show that App is Scott-continuous. Given any directed family $(f_i, x_i)_{i \in I}$ in $[X \to Y] \times X$, $\mathsf{App}(\sup_{i \in I}(f_i, x_i)) = \mathsf{App}(\sup_{i \in I} f_i, \sup_{j \in I} x_j) = \sup_{i \in I} f_i(\sup_{j \in I} x_j) = \sup_{i,j \in I} f_i(x_j)$, since each f_i is Scott-continuous. This is certainly above $\sup_{i \in I} f_i(x_i)$, i.e., above $\sup_{i \in I} \mathsf{App}(f_i, x_i)$ (take $j = i$). Conversely, for all $i, j \in I$, there is a $k \in I$ such that (f_i, x_i) and (f_j, x_j) are both below (f_k, x_k), by directedness. In particular $f_i \leqslant f_k$ and $x_j \leqslant x_k$, so $f_i(x_j) \leqslant f_k(x_k)$. So $\sup_{i,j \in I} f_i(x_j) \leqslant \sup_{k \in I} f_k(x_k) = \sup_{k \in I} \mathsf{App}(f_k, x_k)$, whence the equality.

Finally, given a Scott-continuous map $f : Z \times X \to Y$, f is Scott-continuous in each argument separately, whence $\Lambda_X(f)$ is well defined and continuous. The equations (β), (η), and (σ) are obvious. □

Top is *not* Cartesian-closed, since any non-core-compact space, such as, e.g., the Johnstone space $\mathbb{J}$ (Exercise 5.2.15), fails to be exponentiable. Even the subcategory of T_2 spaces fails to be Cartesian-closed; see Exercise 6.7.25 and Exercise 6.7.26.

It may seem paradoxical that **Cpo**, which one may see as a full subcategory of **Top**, *is* Cartesian-closed. That is, even non-core-compact spaces such as $\mathbb{J}$ are exponentiable in **Cpo**. The source of the paradox lies in the fact that product in **Cpo** does not coincide with product in **Top** (see Exercise 5.2.16), so that the notions of exponentials necessarily differ somewhat between the two categories.

In particular, given two dcpos X and Y, there is a topological space $[X \to Y]_\sigma$. If X_σ is core-compact (e.g., when X is a continuous dcpo), one can also build an exponential in **Top**, namely $[X_\sigma \to Y_\sigma]^{\circledcirc}$ with the core-open topology. While the two spaces $[X \to Y]_\sigma$ and $[X_\sigma \to Y_\sigma]^{\circledcirc}$ contain the same elements, they usually do not have the same topology.

Exercise 5.5.5 (Cartesian-closed category of retracts) Let $\mathbf{C}$ be a Cartesian-closed full subcategory of $\mathbf{Top}$. Assume that finite products in $\mathbf{C}$ are computed as in $\mathbf{Top}$, and let $\mathbf{C}'$ be the full subcategory of $\mathbf{Top}$ whose objects are retracts of objects of $\mathbf{C}$. Show that $\mathbf{C}'$ is also Cartesian-closed.

5.6 $\mathcal{C}$-generated spaces

A famous, and pretty large, Cartesian-closed category of topological spaces is given by Kelley spaces. These are particularly convenient for algebraic topology (Kelley, 1955; Steenrod, 1967).

That Kelley spaces are Cartesian-closed is far from obvious, and Strickland (2009) gives one of the clearest introductions to the topic. Kelley spaces are only a special case of so-called $\mathcal{C}$-generated spaces, which also form a Cartesian-closed category of topological spaces. Our presentation is based on Escardó *et al.* (2004), a very comprehensive paper from which we have borrowed a lot.

We start by introducing a category $\mathbf{Map}_\mathcal{C}$, whose objects are the topological spaces, but with more morphisms than just continuous maps. Showing that it is Cartesian-closed is not too hard, provided one keeps in mind that products, exponentials, and various other notions are defined categorically, and may or may not coincide with their topological counterparts.

Definition 5.6.1 ($\mathbf{Map}_\mathcal{C}$) Let $\mathcal{C}$ be a class of topological spaces. Given any topological space X, a *$\mathcal{C}$-probe* on X is any continuous map $k: C \to X$, for some $C \in \mathcal{C}$.

The category $\mathbf{Map}_\mathcal{C}$ has all topological spaces as objects (just like $\mathbf{Top}$), but its morphisms from X to Y are the *$\mathcal{C}$-continuous maps*, i.e., the maps $f: X \to Y$ such that $f \circ k: C \to Y$ is continuous for every $\mathcal{C}$-probe $k: C \to X, C \in \mathcal{C}$.

The identity map from X to X is trivially $\mathcal{C}$-continuous. Given two $\mathcal{C}$-continuous maps $f: X \to Y$ and $g: Y \to Z$, we check that $g \circ f$ is $\mathcal{C}$-continuous: for every $\mathcal{C}$-probe $k: C \to X$, $f \circ k$ is also a $\mathcal{C}$-probe, so $g \circ (f \circ k)$ is continuous, i.e., $(g \circ f) \circ k$ is continuous. So $\mathbf{Map}_\mathcal{C}$ defines a category.

$\mathbf{Map}_\mathcal{C}$ has more morphisms than $\mathbf{Top}$, in the following sense.

Fact 5.6.2 *Every continuous map from X to Y is $\mathcal{C}$-continuous, hence a morphism in* $\mathbf{Map}_\mathcal{C}$.

Lemma 5.6.3 $\mathbf{Map}_\mathcal{C}$ *has all finite products, and they are defined exactly as in* $\mathbf{Top}$*—including projections and pairing maps.*

Proof We first check that $\{*\}$ is a terminal object in $\mathbf{Map}_{\mathcal{C}}$: for every space X, there is a unique map from X to $\{*\}$, and it is obviously $\mathcal{C}$-continuous, since it is continuous.

Then, we show that, given two topological spaces X, Y, their ordinary topological product $X \times Y$ is their product in $\mathbf{Map}_{\mathcal{C}}$. We first need to check that the projections $\pi_1 \colon X \times Y \to X$ and $\pi_2 \colon X \times Y \to Y$ are $\mathcal{C}$-continuous (since they are continuous). Given $\mathcal{C}$-continuous maps $f \colon Z \to X$ and $g \colon Z \to Y$, we claim that $\langle f, g \rangle$ is $\mathcal{C}$-continuous from Z to $X \times Y$: for every $\mathcal{C}$-probe $k \colon C \to Z$, $\langle f, g \rangle \circ k = \langle f \circ k, g \circ k \rangle$ is continuous. The equations $\langle \pi_1, \pi_2 \rangle = \mathrm{id}_{A \times B}$ and $\langle f_1, f_2 \rangle \circ g = \langle f_1 \circ g, f_2 \circ g \rangle$ hold, as in **Top**. □

The following is the *Escardó–Lawson–Simpson Lemma* (Escardó *et al.*, 2004, lemma 3.12 (iv)), except that they are assuming a weaker condition of productivity. The cited lemma will be an easy consequence of our theorem; see Exercise 5.6.19.

Theorem 5.6.4 ($\mathbf{Map}_{\mathcal{C}}$ is Cartesian-closed) *A class of topological space $\mathcal{C}$ is* strongly productive *iff every space in $\mathcal{C}$ is core-compact, and every binary topological product of spaces in $\mathcal{C}$ is again in $\mathcal{C}$.*

Whenever $\mathcal{C}$ is strongly productive, $\mathbf{Map}_{\mathcal{C}}$ is Cartesian-closed. The exponential object between X and Y is $[X \to Y]_{\mathcal{C}}$, defined as the set of all $\mathcal{C}$-continuous maps from X to Y, with the coarsest topology that makes $\_ \circ k$ continuous from $[X \to Y]_{\mathcal{C}}$ to $[C \to Y]^{\circledcirc}$ for every $\mathcal{C}$-probe $k \colon C \to X$.

App is the ordinary function application, and currification is defined as usual by $\Lambda_X(f)(z)(x) = f(z, x)$.

Proof We first check that $\_ \circ k$ makes sense: for every $\mathcal{C}$-continuous map f from X to Y, $f \circ k$ is a continuous map from C to Y. And since $\mathcal{C}$ is strongly productive, C is core-compact, so $[C \to Y]^{\circledcirc}$ has the unique exponential topology (Theorem 5.4.4).

We then check that $\mathsf{App} \colon [X \to Y]_{\mathcal{C}} \times X \to Y$ is $\mathcal{C}$-continuous. This means that, for every $\mathcal{C}$-probe $k \colon C \to [X \to Y]_{\mathcal{C}} \times X$, $\mathsf{App} \circ k$ must be continuous from C to Y. Necessarily, $k = \langle k_1, k_2 \rangle$ for some continuous maps $k_1 \colon C \to [X \to Y]_{\mathcal{C}}$ and $k_2 \colon C \to X$. By definition of $[X \to Y]_{\mathcal{C}}$, and since k_2 is a $\mathcal{C}$-probe, $\_ \circ k_2$ is continuous from $[X \to Y]_{\mathcal{C}}$ to $[C \to Y]^{\circledcirc}$, so $(\_ \circ k_2) \circ k_1$ is continuous from C to $[C \to Y]^{\circledcirc}$. By Theorem 5.4.4, and since C is core-compact, App is continuous from $[C \to Y]^{\circledcirc} \times C$ to Y. So $\mathsf{App} \circ \langle (\_ \circ k_2) \circ k_1, \mathrm{id}_C \rangle$ is also continuous. This maps every $c \in C$ to $\mathsf{App}((\_ \circ k_2)(k_1(c)), c) = \mathsf{App}(k_1(c) \circ k_2, c) = k_1(c)(k_2(c)) = \mathsf{App}(k(c))$, i.e., is equal to $\mathsf{App} \circ k$. So $\mathsf{App} \circ k$ is continuous. So App is $\mathcal{C}$-continuous.

Let $f: Z \times X \to Y$ be $\mathcal{C}$-continuous. We must show that $\Lambda_X(f): Z \to [X \to Y]_{\mathcal{C}}$ is well defined and $\mathcal{C}$-continuous.

Well-definedness: We must show that, for any fixed $z \in Z$, $\Lambda_X(f)(z) = f(z, \_)$ is $\mathcal{C}$-continuous. Let k be any $\mathcal{C}$-probe from $C \in \mathcal{C}$ to X. Then $k': C \to Z \times X$ defined by $k'(c) = (z, k(c))$ is continuous, hence a $\mathcal{C}$-probe as well. By assumption, $f \circ k'$ is continuous. But $f \circ k' = f(z, \_) \circ k$, so $f(z, \_)$ is $\mathcal{C}$-continuous.

$\mathcal{C}$-continuity: Let $k: C \to Z$ be a $\mathcal{C}$-probe. We must show that $\Lambda_X(f) \circ k$ is continuous from C to $[X \to Y]_{\mathcal{C}}$. By definition of the topology of $[X \to Y]_{\mathcal{C}}$ as a coarsest topology, a subbase of it is given by the subsets of the form $(\_ \circ k')^{-1}(V)$, $k': C' \to X$ a $\mathcal{C}$-probe, V open in $[C' \to Y]^{\circledcirc}$. Equivalently, a map φ from C to $[X \to Y]_{\mathcal{C}}$ is continuous iff, for every $\mathcal{C}$-probe $k': C' \to X$, $(\_ \circ k') \circ \varphi$ is continuous from C to $[C' \to Y]^{\circledcirc}$. So it remains to show that $(\_ \circ k') \circ \Lambda_X(f) \circ k$ is continuous from C to $[C' \to Y]^{\circledcirc}$. This is the map sending $c \in C$ to the map sending $c' \in C'$ to $(\_ \circ k')(\Lambda_X(f)(k(c)))(c') = \Lambda_X(f)(k(c))(k'(c')) = f(k(c), k'(c'))$. So this is exactly $\Lambda_{C'}(f \circ (k \times k'))$. Since $\mathcal{C}$ is strongly productive, $C \times C'$ is in $\mathcal{C}$, so $k \times k'$ is a $\mathcal{C}$-probe. Since f is $\mathcal{C}$-continuous, $f \circ (k \times k')$ is therefore continuous. Also, C' is in $\mathcal{C}$ and is therefore core-compact. By Theorem 5.4.4, $\Lambda_{C'}(f \circ (k \times k'))$ is then continuous, and we are done. ☐

Map$_{\mathcal{C}}$ is not a subcategory of **Top**. But the following category **Top**$_{\mathcal{C}}$ is, and will have strong ties with **Map**$_{\mathcal{C}}$.

Definition 5.6.5 ($\mathcal{C}$-generated spaces, **Top**$_{\mathcal{C}}$) The *$\mathcal{C}$-generated topology* on X is the finest topology on X that makes all $\mathcal{C}$-probes on X continuous. We let $\mathcal{C}X$ be X with its $\mathcal{C}$-generated topology. A space X is *$\mathcal{C}$-generated* if and only if $X = \mathcal{C}X$.

Top$_{\mathcal{C}}$ is the category of $\mathcal{C}$-generated spaces and continuous maps.

Clearly, the $\mathcal{C}$-generated topology on X is finer than the original topology of X. X itself is a *$\mathcal{C}$-generated space* iff $\mathcal{C}X = X$, i.e., its topology coincides with the $\mathcal{C}$-generated topology. The following may provide a more concrete feel for the $\mathcal{C}$-generated topology.

Fact 5.6.6 *A subset U of X is open in the $\mathcal{C}$-generated topology on X if and only if $k^{-1}(U)$ is open in C for every $\mathcal{C}$-probe $k: C \to X$.*

Indeed, the collection of subsets U such that $k^{-1}(U)$ is open for every $\mathcal{C}$-probe forms a topology, and it makes every $\mathcal{C}$-probe continuous by definition. It is easy to see that it is the finest one.

If $f: C \to X$ is continuous from C to $\mathcal{C}X$, where $C \in \mathcal{C}$, then it is continuous from C to X, since the topology of $\mathcal{C}X$ is finer than that of X.

Conversely, if $C \in \mathcal{C}$, and f is continuous from C to X, then f is a $\mathcal{C}$-probe by definition, hence is continuous from C to $\mathcal{C}X$, by definition of the $\mathcal{C}$-generated topology. So:

Fact 5.6.7 *For every map $k: C \to X$ with $C \in \mathcal{C}$, k is continuous from C to X if and only if it is continuous from C to $\mathcal{C}X$.*

Example 5.6.8 As a trivial example of $\mathcal{C}$-generated space, we find the spaces $C \in \mathcal{C}$ themselves. Indeed the topology of $\mathcal{C}C$ is finer than that of C, and, conversely, by considering the identity $\mathcal{C}$-probe on C, which must be continuous from C to $\mathcal{C}C$ by Fact 5.6.7, the topology of C is finer than that of $\mathcal{C}C$

Lemma 5.6.9 *Every space of the form $\mathcal{C}X$ is $\mathcal{C}$-generated.*

Proof The topology of $\mathcal{C}\mathcal{C}X$ is the finest that makes all the $\mathcal{C}$-probes $k: C \to \mathcal{C}X$ continuous from C to $\mathcal{C}\mathcal{C}X$. Equivalently, that makes all the $\mathcal{C}$-probes $k: C \to X$ continuous from C to $\mathcal{C}X$, using Fact 5.6.7. So $\mathcal{C}\mathcal{C}X$ has the same topology as $\mathcal{C}X$. □

Lemma 5.6.10 *The $\mathcal{C}$-continuous maps from X to Y are exactly the continuous maps from $\mathcal{C}X$ to Y.*

Proof Assume that $f: \mathcal{C}X \to Y$ is continuous. For every $\mathcal{C}$-probe $k: C \to X$, by definition $k: C \to \mathcal{C}X$ is continuous, so $f \circ k$ is continuous. So f is $\mathcal{C}$-continuous.

Conversely, assume that $f \circ k$ is continuous for every $\mathcal{C}$-probe $k: C \to X$. Let $\mathcal{O}$ be the topology on X whose open subsets are the subsets of X of the form $f^{-1}(V)$, V open in Y. This is a topology, since f^{-1} commutes with all unions and all intersections. With this topology on X, every $\mathcal{C}$-probe $k: C \to X$ is continuous, since $k^{-1}(f^{-1}(V)) = (f \circ k)^{-1}(V)$ is open in C by assumption. But the $\mathcal{C}$-generated topology is the finest such topology, so it is finer than $\mathcal{O}$. In other words, every subset of the form $f^{-1}(V)$, V open in Y, is open in $\mathcal{C}X$. So f is continuous from $\mathcal{C}X$ to Y. □

$\mathcal{C}$ extends to a functor from $\mathbf{Map}_{\mathcal{C}}$ to $\mathbf{Top}_{\mathcal{C}}$, by defining $\mathcal{C}(f)$, for each $\mathcal{C}$-continuous map $f: X \to Y$, as f itself, seen as a continuous map from $\mathcal{C}X$ to $\mathcal{C}Y$:

Proposition 5.6.11 ($\mathbf{Top}_{\mathcal{C}} \equiv \mathbf{Map}_{\mathcal{C}}$) *$\mathcal{C}$ defines a functor from $\mathbf{Map}_{\mathcal{C}}$ to $\mathbf{Top}_{\mathcal{C}}$. On morphisms $f: X \to Y$ in $\mathbf{Map}_{\mathcal{C}}$, $\mathcal{C}(f)$ is defined as f itself.*

Together with the inclusion functor I from $\mathbf{Top}_{\mathcal{C}}$ to $\mathbf{Map}_{\mathcal{C}}$, the $\mathcal{C}$ functor defines an equivalence of categories. The unit and the counit of the adjunction $I \dashv \mathcal{C}$ are identity maps.

Proof Given any $\mathcal{C}$-continuous map $f : X \to Y$, we check that f is continuous from $\mathcal{C}X$ to $\mathcal{C}Y$. It suffices to check that $f \circ k$ is continuous from C to $\mathcal{C}Y$ for every $\mathcal{C}$-probe $k : C \to X$, by Lemma 5.6.10. But $f \circ k$ is already continuous from C to Y, and this is equivalent to the desired statement, by Fact 5.6.7. So the functor $\mathcal{C}$ is well defined – and it is clearly a functor.

Let us show that the identity map ϵ_X from $\mathcal{C}X$ to X is an isomorphism in $\mathbf{Map}_{\mathcal{C}}$. This is a morphism in $\mathbf{Map}_{\mathcal{C}}$: since the topology of $\mathcal{C}X$ is finer than that of X, ϵ_X is continuous, hence $\mathcal{C}$-continuous by Fact 5.6.2. Conversely, $\epsilon_X^{-1} = \mathrm{id}_X$ is $\mathcal{C}$-continuous from X to $\mathcal{C}X$, since it is continuous from $\mathcal{C}X$ to $\mathcal{C}X$, and using Lemma 5.6.10.

There is an inclusion functor I from $\mathbf{Top}_{\mathcal{C}}$ to $\mathbf{Map}_{\mathcal{C}}$, which maps each $\mathcal{C}$-generated space X to itself, and every continuous map f to itself, seen as a $\mathcal{C}$-continuous map (Fact 5.6.2).

Then ϵ defines a natural isomorphism from $I\mathcal{C}$ to the identity functor on $\mathbf{Map}_{\mathcal{C}}$.

The identity map $\eta_X : X \to \mathcal{C}X$ is an isomorphism in $\mathbf{Top}_{\mathcal{C}}$: since X is generated, $\mathcal{C}X = X$. Then η defines a natural isomorphism from the identity functor on $\mathbf{Top}_{\mathcal{C}}$ to $\mathcal{C}I$. □

The following follows immediately – at least the first part. This is *Day's Theorem* (Day, 1972). We shall slightly relax the conditions on $\mathcal{C}$ in Exercise 5.6.19.

Theorem 5.6.12 (**$\mathbf{Top}_{\mathcal{C}}$** is Cartesian-closed) *For every strongly productive class $\mathcal{C}$ of topological spaces, $\mathbf{Top}_{\mathcal{C}}$ is a Cartesian-closed full subcategory of* **Top**.

More explicitly, the terminal objects are the singletons $\{*\}$; the product $X \times_{\mathcal{C}} Y$ of two $\mathcal{C}$-generated spaces X and Y is $\mathcal{C}(X \times Y)$, where $\times$ is product in **Top**, *projections and pairing are defined as in* **Top**; *and their exponential is $\mathcal{C}([X \to Y]_{\mathcal{C}})$, with application and currification defined as in* **Top**.

Proof The first part follows from Theorem 5.6.4, and the equivalence of categories given in Proposition 5.6.11: since Cartesian-closedness is defined by universal properties, it is preserved through equivalence of categories. Let us be more explicit, and prove the second part of the theorem. Let I be the inclusion functor from $\mathbf{Top}_{\mathcal{C}}$ to $\mathbf{Map}_{\mathcal{C}}$, and η (resp., ϵ) be the unit (resp., counit) of the adjunction $I \dashv \mathcal{C}$.

Since $\{*\}$ is terminal in $\mathbf{Map}_{\mathcal{C}}$, $\mathcal{C}\{*\}$ is terminal in $\mathbf{Top}_{\mathcal{C}}$: the morphisms from X to $\mathcal{C}\{*\}$ in $\mathbf{Top}_{\mathcal{C}}$ are in one-to-one correspondence with those from IX to $\{*\}$ in $\mathbf{Map}_{\mathcal{C}}$, and there is exactly one. On the other hand, $\mathcal{C}\{*\} = \{*\}$ since there is only one topology on $\{*\}$. So $\{*\}$ is terminal in $\mathbf{Top}_{\mathcal{C}}$.

Let us turn to binary products. Let X, Y be $\mathcal{C}$-generated spaces. The product of IX and IY in $\mathbf{Map}_{\mathcal{C}}$ is $IX \times IY$, by Lemma 5.6.3. In particular, there is a projection morphism $\pi_1 \colon IX \times IY \to IX$ in $\mathbf{Map}_{\mathcal{C}}$. So $\mathcal{C}\pi_1$ is a morphism from $\mathcal{C}(IX \times IY) = \mathcal{C}(X \times Y)$ to $\mathcal{C}X = X$ (because X is $\mathcal{C}$-generated). This morphism is just first projection, so π_1 is continuous from $\mathcal{C}(X \times Y)$ to X. Similarly for second projection π_2. Given any two continuous maps $f \colon Z \to X$ and $g \colon Z \to Y$ (in $\mathbf{Top}_{\mathcal{C}}$), where X, Y, Z are $\mathcal{C}$-generated, $f = If$ is a morphism from $IZ = Z$ to $IX = X$, and g is a morphism from Z to Y, in $\mathbf{Map}_{\mathcal{C}}$. Since products in $\mathbf{Map}_{\mathcal{C}}$ are defined exactly as in $\mathbf{Top}$, the map $\langle f, g\rangle$ that sends $z \in Z$ to $(f(z), g(z)) \in X \times Y$ is the pairing of f and g in $\mathbf{Map}_{\mathcal{C}}$. So $\mathcal{C}\langle f, g\rangle = \langle f, g\rangle$ is a morphism from $\mathcal{C}Z = Z$ (since Z is $\mathcal{C}$-generated) to $X \times Y$. The equations defining products, $\pi_1 \circ \langle f, g\rangle = f$, $\pi_2 \circ \langle f, g\rangle = g$ and $\langle f, g\rangle \circ h = \langle f \circ h, g \circ h\rangle$, are then obviously satisfied.

Finally, let us deal with exponentials. Let X, Y, Z be $\mathcal{C}$-generated spaces, and $f \colon \mathcal{C}(Z \times X) \to Y$ be continuous. So $If \circ \eta_{Z \times X}$, which is just f since η is the identity map, is a morphism from $Z \times X$ to Y in $\mathbf{Map}_{\mathcal{C}}$. By Theorem 5.6.4, $\Lambda_X(f)$ is a morphism from Z to $[X \to Y]_{\mathcal{C}}$ in $\mathbf{Map}_{\mathcal{C}}$, where $\Lambda_X(f)(z) = f(z, \_)$. Reapplying $\mathcal{C}$, we obtain that $\Lambda_X(f)$ is a morphism in $\mathbf{Top}_{\mathcal{C}}$ from $\mathcal{C}Z = Z$ to $\mathcal{C}[X \to Y]_{\mathcal{C}}$. And $\mathcal{C}(\Lambda_X(f))$ is simply defined as mapping z to $\Lambda_X(f)(z) = f(z, \_)$, so the usual notion of currification $\Lambda_X(f)$ in $\mathbf{Top}$ defines a morphism, i.e., a continuous map, from Z to $\mathcal{C}[X \to Y]_{\mathcal{C}}$.

Also, $\mathsf{App} \colon [X \to Y]_{\mathcal{C}} \times X \to Y$ is a morphism in $\mathbf{Map}_{\mathcal{C}}$. We can form $\mathsf{App} \circ (\epsilon^{-1}_{[X \to Y]_{\mathcal{C}}} \times \mathrm{id}_X)$, a morphism from $\mathcal{C}[X \to Y]_{\mathcal{C}} \times X$ to Y in $\mathbf{Map}_{\mathcal{C}}$. Applying $\mathcal{C}$, we obtain that $\mathcal{C}(\mathsf{App} \circ (\epsilon^{-1}_{[X \to Y]_{\mathcal{C}}} \times \mathrm{id}_X))$ is a continuous map from $\mathcal{C}(\mathcal{C}[X \to Y]_{\mathcal{C}} \times X) = \mathcal{C}[X \to Y]_{\mathcal{C}} \times_{\mathcal{C}} X$ to Y. Since ϵ is the identity, and $\mathcal{C}\mathsf{App} = \mathsf{App}$, this continuous map is just ordinary application App. The equations (β), (η), and (σ) are then obvious. □

It may be felt as a nuisance that binary products in $\mathbf{Top}_{\mathcal{C}}$ are not topological products $X \times Y$, but $\mathcal{C}(X \times Y)$. We shall address this in Proposition 5.6.18, and see that these products coincide in many cases. As a first step, we observe that the $\mathcal{C}$-generated spaces have a more concrete description (see Proposition 5.6.16).

Lemma 5.6.13 *Let $\mathcal{C}$ be any class of topological spaces. Then* $\mathbf{Top}_{\mathcal{C}}$ *is a coreflective subcategory of* $\mathbf{Top}$*. The right adjoint to the inclusion functor is the functor, again written $\mathcal{C}$, that maps X to $\mathcal{C}X$ and each continuous map $f \colon X \to Y$ to $f \colon \mathcal{C}X \to \mathcal{C}Y$.*

Proof To check that $\mathcal{C}$ is a functor, we must check that $\mathcal{C}X$ is an object of $\mathbf{Top}_{\mathcal{C}}$ (Lemma 5.6.9), and that f is continuous from $\mathcal{C}X$ to $\mathcal{C}Y$ whenever it is continuous from X to Y: for every $\mathcal{C}$-probe $k\colon C \to X$, $f \circ k$ is a $\mathcal{C}$-probe, hence is continuous from C to $\mathcal{C}Y$ by Fact 5.6.7; so f is $\mathcal{C}$-continuous from X to $\mathcal{C}Y$, hence continuous from $\mathcal{C}X$ to $\mathcal{C}Y$ by Lemma 5.6.10.

Write I for the inclusion functor of $\mathbf{Top}_{\mathcal{C}}$ into $\mathbf{Top}$. Define the unit of the claimed adjunction, $\eta_X\colon X \to \mathcal{C}IX$ (for each $\mathcal{C}$-generated space X), as the identity map. This is a morphism since $\mathcal{C}IX = X$. We must check that IX is the free topological space above the $\mathcal{C}$-generated space X, i.e., that, for every space Y, for every continuous map $f\colon X \to \mathcal{C}Y$, there is a unique continuous map $h\colon IX \to Y$ such that $\mathcal{C}(h) \circ \eta_X = f$ (see Diagram (5.2)). This map h must be f itself, and the claim boils down to showing that f is continuous from X to Y. This is clear, since the topology of $\mathcal{C}Y$ is finer than that of Y. □

Since $\mathcal{C}$ is a left adjoint, it preserves all colimits, whence:

Fact 5.6.14 *Given any class $\mathcal{C}$ of topological spaces,* $\mathbf{Top}_{\mathcal{C}}$ *has all colimits, and they are computed as in* **Top**. *In particular, any colimit (finite or infinite coproduct, quotient) of $\mathcal{C}$-generated spaces in* **Top** *is again $\mathcal{C}$-generated.*

Let us now make an additional assumption, namely that $\mathcal{C}$ contains a non-empty space C_0.

Remark 5.6.15 All reasonable strongly productive classes $\mathcal{C}$ will contain a non-empty space C_0. The only classes that do not are the empty class and the class that contains just $\emptyset$. In both cases, the $\mathcal{C}$-generated spaces are the discrete spaces. The category of discrete spaces is equivalent to **Set**.

Proposition 5.6.16 ($\mathcal{C}$-generated = colimit of spaces in $\mathcal{C}$) *Let $\mathcal{C}$ be any class of topological spaces that contains a non-empty space C_0. Every $\mathcal{C}$-generated space X is the quotient of a coproduct of spaces in $\mathcal{C}$.*

It follows that the $\mathcal{C}$-generated spaces are exactly the colimits in **Top** *of spaces in $\mathcal{C}$.*

Proof Let X be a $\mathcal{C}$-generated space, and let I be the family of all non-open subsets A of X. By definition, each such subset A fails to be open in $\mathcal{C}X$, so there is a $\mathcal{C}$-probe $k_A\colon C_A \to X$ such that $k_A^{-1}(A)$ is not open in C_A. (We use the Axiom of Choice to collect them.) Let k_x be the constant map from C_0 to X such that $k_x(c) = x$ for every $c \in C_0$. Let $q\colon \coprod_{A \in I} C_A + \coprod_{x \in X} C_0 \to X$ be defined as mapping (A, c) to $k_A(c)$ (for $A \in I$) and (x, c) to $x = k_x(c)$ (for $x \in X$). We claim that q is quotient.

First, q is surjective: for every $x \in X$, $x = q(x, c)$ for any $c \in C_0$—and c exists because C_0 is non-empty.

Second, q is continuous: by Trick 4.6.3, this boils down to checking that k_A is continuous for every $A \in I$ and that k_x is continuous for every $x \in X$. The former is because every $\mathcal{C}$-probe is continuous by definition, and the latter is because every constant map is continuous.

Finally, we must check that every subset U of X such that $q^{-1}(U)$ is open in $\coprod_{A \in I} C_A + \coprod_{x \in X} C_0$ is open in X. Assume that U is not open. Then, taking $A = U$ in I, $\iota_A^{-1}(q^{-1}(U))$ is open in C_A, but $\iota_A^{-1}(q^{-1}(U)) = \{c \in C_A \mid (A, c) \in q^{-1}(U)\} = k_A^{-1}(U) = k_A^{-1}(A)$ is not open in C_A by definition of k_A: contradiction. So U is open in X. Hence q is quotient.

By Proposition 4.10.22, X arises as a quotient of the coproduct $\coprod_{A \in I} C_A + \coprod_{x \in X} C_0 \to X$.

Conversely, we must show that any colimit of spaces from $\mathcal{C}$ spaces, taken in **Top**, is again $\mathcal{C}$-generated. But every space in $\mathcal{C}$ is $\mathcal{C}$-generated (Example 5.6.8), and any colimit of $\mathcal{C}$-generated spaces is $\mathcal{C}$-generated by Fact 5.6.14. □

Exercise 5.6.17 (The largest category $\mathbf{Top}_{\mathcal{C}}$) Let $\mathcal{C}$ be a class of core-compact spaces, and let $\overline{\mathcal{C}}$ be the class of all core-compact, $\mathcal{C}$-generated spaces. Show that the $\overline{\mathcal{C}}$-generated spaces are exactly the $\mathcal{C}$-generated spaces, and that $\overline{\mathcal{C}}$ is the largest class of core-compact spaces with this property.

A space is *core-compactly generated* if and only if it is $\mathcal{C}$-generated, where $\mathcal{C}$ is the class of core-compact spaces. (Note that this is a strongly productive class, by Exercise 5.2.12.) Conclude that the category of core-compactly generated spaces is the largest category one can obtain through the $\mathbf{Top}_{\mathcal{C}}$ construction with $\mathcal{C}$ strongly productive, and consists exactly of the quotients of core-compact spaces.

Proposition 5.6.18 *Let $\mathcal{C}$ be a strongly productive class of spaces that contains some non-empty space C_0. For any two $\mathcal{C}$-generated spaces X and Y, the topological product $X \times Y$ is $\mathcal{C}$-generated whenever X or Y is core-compact. In other words, the product $X \times_{\mathcal{C}} Y$ in $\mathbf{Top}_{\mathcal{C}}$ coincides with the ordinary topological product $X \times Y$ whenever one of X or Y is core-compact.*

Proof We first show that, for every $\mathcal{C}$-generated space X and every $C \in \mathcal{C}$, $X \times C$ is $\mathcal{C}$-generated. By Proposition 5.6.16, X is the quotient of some coproduct $\coprod_{i \in I} C_i$ of spaces in $\mathcal{C}$; let q be the quotient map. By Exercise 5.4.29, and since $C \in \mathcal{C}$ is core-compact, $q \times \mathrm{id}_C$ is also a quotient map, this time from $(\coprod_{i \in I} C_i) \times C$ to $X \times C$. Now there is a homeomorphism from $(\coprod_{i \in I} C_i) \times C$ to $\coprod_{i \in I}(C_i \times C)$, which maps $((i, x), y)$ to $(i, (x, y))$. So $X \times C$ also arises as a quotient of $\coprod_{i \in I}(C_i \times C)$. For each $i \in I$, $C_i \times C$ is in $\mathcal{C}$, since binary products of spaces in $\mathcal{C}$ are in $\mathcal{C}$. So $X \times C$ is a colimit of spaces in $\mathcal{C}$, and therefore must be $\mathcal{C}$-generated by Proposition 5.6.16.

Symmetrically, $C \times Y$ is $\mathcal{C}$-generated for every $C \in \mathcal{C}$ and every $\mathcal{C}$-generated space Y.

Now let X and Y be $\mathcal{C}$-generated, and assume Y is core-compact. By Proposition 5.6.16, X is the quotient of some coproduct $\coprod_{i \in I} C_i$ of spaces in $\mathcal{C}$; let q be the quotient map. By Exercise 5.4.29, $q \times \mathrm{id}_Y$ is also a quotient map, this time from $(\coprod_{i \in I} C_i) \times Y$ to $X \times Y$. As above, we obtain a quotient map from $\coprod_{i \in I}(C_i \times Y)$ onto $X \times Y$. But we have seen that $C_i \times Y$ was $\mathcal{C}$-generated, and quotients of coproducts of $\mathcal{C}$-generated spaces are $\mathcal{C}$-generated by Proposition 5.6.16. So $X \times Y$ is $\mathcal{C}$-generated.

The case where X instead is $\mathcal{C}$-generated is symmetrical. □

Exercise 5.6.19 (Escardó–Lawson–Simpson Lemma, strong form) A class $\mathcal{C}$ of topological spaces is *weakly productive*, or *productive*, if and only if every space in $\mathcal{C}$ is core-compact, and every binary topological product of spaces in $\mathcal{C}$, while not necessarily in $\mathcal{C}$, is $\mathcal{C}$-generated.

Show that Proposition 5.6.18 still holds if we only assume $\mathcal{C}$ *weakly* productive instead of strongly, and containing some non-empty space C_0.

Show that Theorem 5.6.4 and Day's Theorem 5.6.12 still hold, in particular that $\mathbf{Map}_{\mathcal{C}}$ and $\mathbf{Top}_{\mathcal{C}}$ are Cartesian-closed, if $\mathcal{C}$ is merely assumed to be weakly productive. (Hint: use the relaxed form of Proposition 5.6.18 to show that the class $\overline{\mathcal{C}}$ introduced in Exercise 5.6.17 is strongly productive, and note that $\mathbf{Top}_{\mathcal{C}} = \mathbf{Top}_{\overline{\mathcal{C}}}$.)

Exercise 5.6.20 Let $\mathcal{C}$ be a strongly productive class of spaces. We are looking for simpler characterizations of the exponential $\mathcal{C}[X \to Y]_{\mathcal{C}}$ in $\mathbf{Top}_{\mathcal{C}}$. Let X, Y, Z be three topological spaces.

Show that, if Y is $\mathcal{C}$-generated and f is $\mathcal{C}$-continuous from $Z \times X$ to Y, then $\Lambda_X(f)$ is $\mathcal{C}$-continuous from Z to $[X \to Y]^{\natural}$. (You will need to use the fact that the natural topology is splitting, and Proposition 5.6.18.) Deduce that, if Y and Z are $\mathcal{C}$-generated and f is continuous from $Z \times_{\mathcal{C}} X$ to Y, then $\Lambda_X(f)$ is continuous from Z to $[X \to Y]^{\natural}$. Conclude that, if X and Y are $\mathcal{C}$-generated, then the topology of $[X \to Y]^{\natural}$, hence also that of $\mathcal{C}[X \to Y]^{\natural}$, is coarser than the topology of $\mathcal{C}[X \to Y]_{\mathcal{C}}$.

Show that the topology of $[X \to Y]_{\mathcal{C}}$ is generated by the subsets $[U \Subset (\circ k)^{-1}(V)]$, defined as $\{f \in [X \to Y] \mid U \Subset k^{-1}(f^{-1}(V))\}$, where V ranges over the open subsets of Y, k ranges over the $\mathcal{C}$-probes from the spaces $C \in \mathcal{C}$ to X, and U ranges over the open subsets of C. Show that the set $\mathcal{U}_{U,k}$ of all open subsets U' of X such that $U \Subset k^{-1}(U')$ is Scott-open, hence that $[U \Subset (\circ k)^{-1}(V)]$ is always open in the Isbell topology. Conclude that the topology of $[X \to Y]_{\mathcal{C}}$ is coarser than the Isbell topology on $[X \to Y]$, and therefore that the topology of $\mathcal{C}[X \to Y]_{\mathcal{C}}$ is coarser than than of $\mathcal{C}[X \to Y]^{I}$. (Recall that $[X \to Y]^{\mathrm{I}}$ is $[X \to Y]$ with its Isbell topology.)

From all this, conclude that $\mathcal{C}[X \to Y]_{\mathcal{C}} = \mathcal{C}[X \to Y]^{\mathrm{I}} = \mathcal{C}[X \to Y]^{\natural}$.

Exercise 5.6.21 Using a similar line of reasoning as in Exercise 5.6.20, show that, if $\mathcal{C}$ is a strongly productive class of spaces, and if every space in $\mathcal{C}$ is consonant (e.g., locally compact; see Exercise 5.4.12), then $\mathcal{C}[X \to Y]_{\mathcal{C}} = \mathcal{C}[X \to Y]^{co}$, where $[X \to Y]^{co}$ is $[X \to Y]$ with its compact-open topology.

Definition 5.6.22 (Compactly generated space, Kelley space) A *compactly generated space* is a $\mathcal{C}$-generated space, where $\mathcal{C}$ is the class of all compact T_2 spaces. A *Kelley space* is a compactly generated T_2 space.

Every compact T_2 space is locally compact by Proposition 4.8.8, hence core-compact by Theorem 5.2.9. The binary product of two compact T_2 spaces is compact by Tychonoff's Theorem 4.5.12, and T_2 by Exercise 4.5.18. So compact T_2 spaces form a strongly productive class. By the above results:

Fact 5.6.23 *Compactly generated spaces form a Cartesian-closed category. Colimits are computed as in* **Top**, *and products coincide with ordinary products $X \times Y$ if X or Y is core-compact (e.g., locally compact). Moreover, the compactly generated spaces are exactly the quotients of coproducts of compact T_2 spaces.*

Exercise 5.6.24 Let X be a locally compact T_2 space. For each point $x \in X$, fix a compact neighborhood Q_x of x, and consider the map $q\colon \coprod_{x \in X} Q_x \to X$ that sends (x, y) with $y \in Q_x$ to y, as a point of X. Show that q is quotient. (To show that U is open whenever $q^{-1}(U)$ is open, let x be any point in U, and show that $int(Q_x) \cap U$ is an open neighborhood of x included in U.) Conclude that every locally compact T_2 space is compactly generated.

Deduce that the compactly generated spaces are exactly the quotients of locally compact T_2 spaces.

In the literature one often finds the following property as a definition of compactly generated spaces.

Proposition 5.6.25 *Given a topological space X, a k-*closed *subset of X is any subset F such that $F \cap K$ is closed in X for every compact subset K of X. A subset is k-*open *if and only if its complement is k-closed. The k-open subsets of X form a topology, called the k-*topology *on X, and we write kX for X with its k-topology.*

If X is T_2, then $kX = \mathcal{C}X$ where $\mathcal{C}$ is the class of compact T_2 spaces. In particular, a T_2 space X is Kelley if and only if $kX = X$, if and only if every k-closed subset is closed.

If X is T_2, note that every closed subset F of X is k-closed: for every compact subset K, $F \cap K$ is compact by Corollary 4.4.10, hence closed by Proposition 4.4.15.

Proof Let $\mathcal{C}$ be the class of compact T_2 spaces, and let X be a T_2 space. By Fact 5.6.6, and passing to complements, a subset F of X is closed in $\mathcal{C}X$ if and only if $k^{-1}(F)$ is closed in C for every compact T_2 space C and every continuous map $k \colon C \to X$.

If F is closed in $\mathcal{C}X$, then, for every compact subset K of X, K defines a compact subspace of X (Exercise 4.9.11), which is clearly T_2. Let k be the inclusion map from K to X. So $k^{-1}(F)$ is closed in K. But $k^{-1}(F) = F \cap K$. So F is k-closed.

Conversely, if F is k-closed in X, then let k be any continuous map from a compact T_2 space C to X. By Proposition 4.4.13, $k[C]$ is compact, so $F \cap k[C]$ is closed in X, using Corollary 4.4.10 and Proposition 4.4.15. Then $k^{-1}(F) = k^{-1}(F \cap k[C])$ is closed in C. So F is closed in $\mathcal{C}X$.

Since $\mathcal{C}X$ and kX have the same closed subsets, they coincide. Given a T_2 space X, the fact that X is Kelley if and only if $kX = X$ is then obvious. Since every closed subset is clearly k-closed, an equivalent condition is that every k-closed subset be closed. □

Theorem 5.6.26 *The category of Kelley spaces, or, equivalently, of T_2 quotients of locally compact T_2 spaces, is Cartesian-closed. The one-element space $\{*\}$ is terminal, the product of two Kelley spaces X and Y is $k(X \times Y)$, and their exponential is $k([X \to Y])$, where $[X \to Y]$ is topologized with the compact-open topology.*

Proof That Kelley spaces are the T_2 quotients of locally compact T_2 spaces is by Exercise 5.6.24. To show that the category of Kelley spaces is Cartesian-closed, apply Day's Theorem 5.6.12. It is then enough to check that the terminal objects, binary products $X \times_{\mathcal{C}} Y = \mathcal{C}(X \times Y)$, and exponentials $\mathcal{C}([X \to Y]_{\mathcal{C}})$ given there, where $\mathcal{C}$ is the class of compact T_2 spaces, are T_2 whenever X and Y are.

For products, recall that products of T_2 spaces are T_2 (Exercise 4.5.18). It is enough to check that $\mathcal{C}Z$ is T_2 whenever Z is: indeed the topology of $\mathcal{C}Z$ is finer than that of T_2, so any two disjoint open subsets of Z used to separate two distinct points must also serve to separate them in $\mathcal{C}Z$. Now apply this to the case $Z = X \times Y$ to obtain that $\mathcal{C}(X \times Y)$ is Kelley whenever X and Y are, and use Proposition 5.6.25 to conclude that $k(X \times Y) = \mathcal{C}(X \times Y)$ is the product in the category of Kelley spaces.

For terminal objects, $\{*\}$ is T_2, hence Kelley.

For exponentials, we should check that $\mathcal{C}[X \to Y]_{\mathcal{C}}$ is Kelley. It is easier to show directly that $\mathcal{C}[X \to Y]$ is an exponential, where $[X \to Y]$ is equipped with the compact-open topology.

We must first show that App is continuous from $\mathcal{C}(\mathcal{C}[X \to Y] \times X)$ to Y, i.e., $\mathcal{C}$-continuous from $\mathcal{C}[X \to Y] \times X$ to Y. Let $k = \langle k_1, k_2 \rangle$ be any $\mathcal{C}$-probe from C to $\mathcal{C}[X \to Y] \times X$. We claim that $\mathsf{App} \circ k$ is continuous from C to Y. Note that k_2 is continuous from C to X, and k_1 is continuous from C to $[X \to Y]$ by Fact 5.6.7. Let V be an open subset of Y, then $(\mathsf{App} \circ k)^{-1}(V) = \{c \in C \mid k_1(c)(k_2(c)) \in V\}$. Let c be any element of the latter set. Since $k_1(c)$ is continuous, $k_2(c)$ is in the open subset $U = k_1(c)^{-1}(V)$. Since k_2 is continuous, c is also in the open subset $k_2^{-1}(U)$. But C is compact T_2, hence locally compact by Proposition 4.8.8. So we can find a compact neighborhood Q of c included in $k_2^{-1}(U)$. By Proposition 4.4.13, $k_2[Q]$ is compact in X, hence also $\uparrow k_2[Q]$ (Proposition 4.4.14); also, $\uparrow k_2[Q]$ contains $k_2(c)$, and is included in U. Consider the open subset $W = k_1^{-1}([\uparrow k_2[Q] \subseteq V]) \cap int(Q)$ of C—this is open in particular because $[\uparrow k_2[Q] \subseteq V]$ is open in the compact-open topology, and k_1 is continuous. W contains c: indeed $c \in int(Q)$ by definition, and $c \in k_1^{-1}([\uparrow k_2[Q] \subseteq V])$ because $k_1(C)$ maps every element of $\uparrow k_2[Q] \subseteq U = k_1(c)^{-1}(V)$ to elements of V. Moreover, W is included in $(\mathsf{App} \circ k)^{-1}(V)$: for every $c' \in W$, $k_1(c')$ is in $[\uparrow k_2[Q] \subseteq V]$, and since $c' \in int(Q)$, $k_2(c')$ is in $\uparrow k_2[Q]$, so that $k_1(c')(k_2(c'))$ is in V. We have found an open neighborhood W of c included in $(\mathsf{App} \circ k)^{-1}(V)$. So, by Trick 4.1.11, $(\mathsf{App} \circ k)^{-1}(V)$ is open, whence $\mathsf{App} \circ k$ is continuous.

Next, let f be a continuous map from $\mathcal{C}(Z \times X)$ to Y, i.e., a $\mathcal{C}$-continuous map from $Z \times X$ to Y. We must check that $\Lambda_X(f)$ is continuous from Z to $\mathcal{C}[X \to Y]$. Since Z is $\mathcal{C}$-generated, it is equivalent to show that $\Lambda_X(f)$ is $\mathcal{C}$-continuous from Z to $\mathcal{C}[X \to Y]$. Let k be a $\mathcal{C}$-probe from C to Z. We must show that $\Lambda_X(f) \circ k$ is continuous from C to $\mathcal{C}[X \to Y]$. By Fact 5.6.7, it is equivalent to show that it is continuous from C to $[X \to Y]$. But $\Lambda_X(f) \circ k = \Lambda_X(f \circ (k \times \mathrm{id}_X))$, and $f \circ (k \times \mathrm{id}_X)$ is continuous, from which the claim follows by using the fact that the compact-open topology on $[X \to Y]$ is splitting (Exercise 5.4.8).

We conclude by noticing that the compact-open topology is T_2: see Exercise 5.4.28. So $\mathcal{C}[X \to Y]$ is T_2, hence Kelley, and also equal to $k([X \to Y])$, by Proposition 5.6.25. □

Remark 5.6.27 Another way of proving that the exponential in Kelley spaces is $k([X \to Y])$, where $[X \to Y]$ is topologized with the compact-open topology, is to use Exercises 5.6.20 and 5.6.21. This shows that the choice of the compact-open topology is somewhat arbitrary, as one could use the Isbell or the natural topology instead: $k([X \to Y]) = k([X \to Y]^{\mathrm{I}}) = k([X \to Y]^{\natural})$. The same argument shows that the exponential objects in the category of compactly generated (not necessarily T_2) spaces is $\mathcal{C}([X \to Y]) = \mathcal{C}([X \to Y]^I) = \mathcal{C}([X \to Y]^\natural)$, where $\mathcal{C}$ is the class of compact T_2 spaces. A similar three-way equality holds more generally in $\mathbf{Top}_\mathcal{C}$ whenever every object of $\mathcal{C}$ is consonant.

One may wonder why we chose compact T_2 spaces to define compactly generated spaces. Apart from historical reasons, the following exercise may provide an answer.

Exercise 5.6.28 Show that every compactly generated space is $\mathcal{C}$-generated when $\mathcal{C}$ is the class of locally compact T_2 spaces (use Exercise 5.6.24), hence also when $\mathcal{C}$ is the class of core-compact spaces. Using Exercise 5.6.17, conclude that Kelley spaces form the largest category of $\mathcal{C}$-generated T_2 spaces, for any strongly productive class $\mathcal{C}$.

Exercise 5.6.29 Using the relevant theorem, show that the ordinary topological product of a locally compact T_2 space with a compactly generated (resp., Kelley) space is compactly generated (resp., Kelley).

The following proposition confirms that compactly generated spaces form a large Cartesian-closed category of topological spaces. Another confirmation is given in Exercise 5.6.31.

Proposition 5.6.30 (dcpo $\Rightarrow$ compactly generated) *Every dcpo is compactly generated in its Scott topology.*

Proof Let X be a dcpo, and fix an open subset U of $\mathcal{C}X_\sigma$, where $\mathcal{C}$ is the class of compact T_2 spaces. We claim that U is Scott-open: this will prove that U is open in X_σ, hence that the topology of X_σ is finer than that of $\mathcal{C}_\sigma$. Since it is also coarser, $X_\sigma = \mathcal{C}X_\sigma$, which will prove the claim.

We first show that U is upward closed. Let $x \in U$, and y be a point above x in X. Let C be the compact T_2 space $[0, 1]$ (Proposition 3.3.4), and define $k: C \to X$ by $k(0) = x$, $k(t) = y$ for every $t \in (0, 1]$. This is a continuous map: the inverse image of open subsets of X, which are upward-closed, are $\emptyset$, $(0, 1]$, or $[0, 1]$. So k is a $\mathcal{C}$-probe. By Fact 5.6.6, $k^{-1}(U)$ is open in $[0, 1]$. This excludes the case where y would not be in U, since then $k^{-1}(U)$ would be $\{0\}$, which is not open. So y is in U.

Let D be any directed family of elements such that $x = \sup D$ is in U. Let C be $\mathbb{P}(D)$, ordered by inclusion. C is an algebraic complete lattice, with $\mathbb{P}_{\text{fin}}(D)$ as the basis of finite elements (see Example 5.1.8). Let C_λ be C with the so-called *Lawson topology* of $\subseteq$, whose subbasic open sets are the subbasic Scott-open subsets $\uparrow E$, $E \in \mathbb{P}_{\text{fin}}(D)$ (see Exercise 5.1.18), plus the complements $C \smallsetminus \uparrow E = \{E' \in C \mid E \not\subseteq E'\}$, $E \in C$. We observe the following, which will be special cases of the results of Chapter 9, and in particular of Theorem 9.1.32.

1. The topology of C_λ is finer than that of C_σ.
2. C_λ is T_2. Indeed, given distinct points E, E' of C, assume without loss of generality that $E \not\subseteq E'$. So there is an element E_0 of $\mathbb{P}_{\text{fin}}(D)$ such that $E_0 \subseteq E$ and $E_0 \not\subseteq E'$: otherwise every $E_0 \subseteq E$ would be below E', and since E is the least upper bound of such elements E_0, E' would be above E. It follows that $C \smallsetminus \uparrow E_0$ is an open neighborhood of E', that $\uparrow E_0$ is an open neighborhood of E, and they are disjoint.
3. C_λ is compact. To show this, we rely on Alexander's Subbase Lemma (Theorem 4.4.29). Assume $C \subseteq \bigcup_{i \in I} \uparrow E_i \cup \bigcup_{j \in J}(C \smallsetminus \uparrow E'_j)$, where $E_i \in \mathbb{P}_{\text{fin}}(D)$ and $E'_j \in \mathbb{P}(D)$. So $\bigcap_{j \in J} \uparrow E'_j \subseteq \bigcup_{i \in I} \uparrow E_i$, i.e., every subset of C that contains every E'_j contains some E_i. In particular, $\bigcup_{j \in J} E'_j$ contains some E_i. For each element of E_i, find some E'_j containing it. Since E_i is finite, there is a finite subset J_0 of J such that $E_i \subseteq \bigcup_{j \in J_0} E'_j$. It follows that $C \subseteq \uparrow E_i \cup \bigcup_{j \in J_0}(C \smallsetminus \uparrow E'_j)$: for every $E \in C$, if E is not in $\bigcup_{j \in J_0}(C \smallsetminus \uparrow E'_j)$, then it is in $\bigcap_{j \in J_0} \uparrow E'_j$ and therefore contains $\bigcup_{j \in J_0} E'_j$, hence E_i.
4. Every upward closed (in $\subseteq$) open subset V of C_λ is open in C_σ. Indeed, let E be any element of V. Then E must be in some basic open subset W included in V, where basic opens are obtained as finite intersections of subbasic opens. So we can write W as $\uparrow E_0 \cap \bigcap_{i=1}^n (C \smallsetminus \uparrow E_i)$, where E_0 is finite. Alternatively, $W = \uparrow E_0 \smallsetminus \bigcup_{i=1}^n \uparrow E_i$. Then E_0 is in W; it suffices to check that E_0 contains no E_i, $1 \leqslant i \leqslant n$: if it did, then E, which contains E_0, would contain E_i, and therefore fail to be in W. So E_0 is in V. Since E is arbitrary, V is included in the union of the sets $\uparrow E_0$, $E_0 \in V \cap \mathbb{P}_{\text{fin}}(D)$. The converse inclusion follows from the fact that V is upward closed. Since each set $\uparrow E_0$, E_0 finite, is Scott-open (Exercise 5.1.11), V is Scott-open.

We build a $\mathcal{C}$-probe $k\colon C_\lambda \to X$ as follows. By induction on the number of elements of $E \in \mathbb{P}_{\text{fin}}(D)$, define $k(E)$ as any element of D that is above every element of E and above every element $k(E')$ with E' strictly included in E: this is possible since D is directed, and we are asking for an element in D above finitely many. By construction, this map is monotonic, and therefore extends to a unique Scott-continuous map $k\colon C_\sigma \to X$, by Exercise 5.1.62. (Explicitly, $k(E)$ is defined for $E \in \mathbb{P}(D)$ as $\sup_{E' \text{ finite} \subseteq E} k(E')$.) Since the Lawson topology is finer than the Scott topology (Claim 1 above), k is also continuous from C_λ to X. So k is a $\mathcal{C}$-probe.

By Fact 5.6.6, $k^{-1}(U)$ is open in C_λ. Since k is Scott-continuous, k is monotonic, so $k^{-1}(U)$ is also upward closed in C. By Claim 4 above, $k^{-1}(U)$ must be open in C_σ. Now recall that $x = \sup(D)$, so x is the image by k of the largest element, D itself, of C. So $k^{-1}(U)$ is non-empty, and must contain a finite element of C, namely an element E of $\mathbb{P}_{\text{fin}}(D)$. So $k(E)$ is in U. By definition of k, $k(E)$ is an element of D. So D meets U. This concludes our argument that every directed family D whose sup is in U meets U. So U is Scott-open. □

Exercise 5.6.31 (Sequential space) A *sequential space* is a space in which every sequentially closed subset is closed, i.e., in which every subset F that contains all limits of *sequences* (not nets) of elements of F is closed. Exercise 4.7.14 showed that every first-countable space is sequential.

Let $\mathcal{C}_{\text{seq}}$ consist of just $\alpha\mathbb{N}$, the Alexandroff compactification of the discrete space $\mathbb{N}$ of natural numbers. (The open subsets are all the subsets of $\mathbb{N}$, plus all sets of the form $A \cup \{\infty\}$ where A is cofinite. See Exercise 4.8.9.) Show that the sequence $(n)_{n\in\mathbb{N}}$ converges to ∞ in $\alpha\mathbb{N}$. By realizing that, in a sequential space X, the closed subsets of $\mathcal{C}_{\text{seq}}X$ are the sequentially closed subsets, show that the sequential spaces are exactly the $\mathcal{C}_{\text{seq}}$-generated spaces.

From this, deduce that all colimits taken in **Top** of sequential spaces are sequential.

Show that $\alpha\mathbb{N} \times \alpha\mathbb{N}$ is first-countable, hence sequential. Deduce that $\mathcal{C}_{\text{seq}}$ is weakly productive, hence that the sequential spaces form a Cartesian-closed category $\mathbf{Top}_{\mathcal{C}_{\text{seq}}}$. (Beware, though: the products in the category of sequential spaces are rather different from topological products.)

Exercise 5.6.32 Using Proposition 5.6.16, show that every sequential space is compactly generated.

So every first-countable space (Exercise 4.7.14), in particular every metric space, is compactly generated, hence a Kelley space. We shall see in Lemma 6.3.6 that, more generally, every hemi-metric space is first-countable, hence sequential, hence compactly generated. Finally, we shall see that every stably compact space is also compactly generated (Exercise 9.1.41).

5.7 bc-domains

Certain further subcategories of **Top** are also Cartesian-closed. Particularly important are those subcategories consisting of continuous dcpos. We explore below the category of so-called bc-domains.

We first note that the category of continuous dcpos itself is *not* Cartesian-closed:

Example 5.7.1 Consider the dcpo $\mathbb{Z}^-$ of all non-positive integers, i.e., of all elements $-n$, $n \in \mathbb{N}$. This is a continuous dcpo, but $[\mathbb{Z}^- \to \mathbb{Z}^-]$ is not continuous, as no continuous self-map can be way-below the identity. This is the topic of the next exercise.

Exercise 5.7.2 Show that there is no element way-below $\mathrm{id}_{\mathbb{Z}^-}$ in $[\mathbb{Z}^- \to \mathbb{Z}^-]$. Conclude that the category of continuous dcpos and continuous maps is not Cartesian-closed; and that the category of algebraic dcpos and continuous maps is not Cartesian-closed either.

Definition 5.7.3 (Bounded-complete) A poset is *bounded-complete* if and only if it is pointed, and every pair of elements x, y that have an upper bound have a least upper bound $\sup(x, y)$.

Example 5.7.4 In particular, all sup-semi-lattices are bounded-complete. The difference is that, in a bounded-complete poset, two elements may fail to have an upper bound; but if they do have one, then they have a least one. An example of what can go wrong is $\mathbb{Z}^-$, the set of non-positive integers, with two incomparable elements a, b added so that $a, b \leqslant -n$ for every $n \in \mathbb{N}$: a and b have (plenty of) upper bounds, but no least one.

Exercise 5.7.5 Let X be a bdcpo (in particular, this will apply to dcpos). Show that the following are equivalent:

1. every pair of elements that have an upper bound have a least upper bound;
2. every finite non-empty subset E of X that has an upper bound has a least upper bound;
3. every non-empty subset A of X that has an upper bound has a least upper bound;

Show that the following are equivalent, too:

1. X is bounded-complete;
2. every finite subset E of X that has an upper bound has a least upper bound;
3. X is *consistently complete*, i.e., every subset A of X that has an upper bound has a least upper bound;
4. every non-empty subset A of X has a greatest lower bound.

For the last part, remember why every set that has least upper bounds of all families also has greatest lower bounds of all families (Section 2.3.2).

Exercise 5.7.6 (Topless bounded lattices) A *bounded* poset is one that has a least element $\bot$ and a largest element $\top$.

Show that the bounded-complete posets are exactly the "topless bounded lattices": given a bounded lattice L (with largest element $\top$), show that $L \smallsetminus \{\top\}$ is bounded-complete.

Conversely, show that, given any bounded-complete poset X, the poset $X^\top$ obtained by formally adding a new largest element $\top$ to X is a bounded lattice. ($X^\top$ is the disjoint union of X and $\{\top\}$, with ordering defined by $x \leqslant y$ iff $x \leqslant y$ in X, or $y = \top$.)

Definition 5.7.7 (bc-domain) A *bc-domain* is a continuous, bounded-complete dcpo.

Example 5.7.8 Among the posets of Example 5.1.2 (which are all continuous posets), $\mathbb{N}$, $\mathbb{R}$, and $\mathbb{R}^+$ are not bc-domains, since they are not even dcpos. The others, $\mathbb{N}_\omega$, $\mathbb{P}(Y)$, and $[0, 1]$, are in fact continuous complete lattices, so they are bc-domains, in particular.

The continuous lattices (but not the continuous dcpos; see Exercise 5.7.2) will also form a Cartesian-closed category, as we shall see. The difficult point in showing this lies in finding a basis for $[X \to Y]$, where X and Y are two bc-domains. This is given by so-called *step functions*. We first describe a simpler notion.

Definition 5.7.9 (Elementary step function) let X be a topological space, and Y be a pointed poset (i.e., with a least element $\bot$). For every open subset U of X, for every element y of Y, let $U \searrow y$ be the function that maps every $x \in U$ to y and all other elements to $\bot$. Such a function is called an *elementary step function.*

It is not too hard to see that every elementary step function is continuous. The general case is given as follows.

Lemma 5.7.10 (Step function) *Let X be a topological space, and Y be a poset. A finite collection of pairs $(U_i, y_i) \in \mathcal{O}(X) \times Y$, $1 \leqslant i \leqslant n$, is* admissible *if and only if, for every subset $I \subseteq \{1, \dots, n\}$ such that $\bigcap_{i \in I} U_i \smallsetminus \bigcup_{i \notin I} U_i$ is non-empty, the family $(y_i)_{i \in I}$ has a least upper bound. (When I is empty, we take $\bigcap_{i \in I} U_i$ to denote X.)*

In this case, the function $\sup_{i=1}^n U_i \searrow y_i$ that maps each $x \in X$ to $\sup_{i \in I_x} y_i$, where $I_x = \{i \in I \mid x \in U_i\}$, is defined and continuous. Such a function is called a step function.

Note that we do not require Y to be pointed. However, an admissible collection of elementary step functions $U_i \searrow y_i$, $1 \leqslant i \leqslant n$, can only exist provided either $(U_i)_{1\leqslant i\leqslant n}$ is a cover of X, or Y is pointed: if $(U_i)_{1\leqslant i\leqslant n}$ is not a cover of X, then $\bigcap_{i\in I} U_i \smallsetminus \bigcup_{i\notin I} U_i$ is not empty when $I = \emptyset$ (where we take $\bigcap_{i\in\emptyset} U_i$ to denote X), and we require the empty family $(y_i)_{i\in\emptyset}$ to have a least upper bound: this must be the least element of Y.

Example 5.7.11 Consider three open subsets U_1, U_2, U_3 as in Figure 5.7, where $U_1 \cap U_2 \smallsetminus U_3$ is non-empty, so we require $\sup(y_1, y_2)$ to exist, and $U_2 \cap U_3 \smallsetminus U_1$ is non-empty, so we require $\sup(y_2, y_3)$ to exist, but we do not require either $\sup(y_1, y_3)$ or $\sup(y_1, y_2, y_3)$ to exist.

Proof (of Lemma 5.7.10.) To show that $\sup_{i=1}^n U_i \searrow y_i$ is defined, we have to show that $\sup_{i\in I_x} y_i$ is defined for every x. We use admissibility, noticing that $\bigcap_{i\in I_x} U_i \smallsetminus \bigcup_{i\notin I_x} U_i$ is non-empty, since it contains x.

To show that the step function is continuous, let V be any open subset of Y. Let $\mathcal{I}$ be the (finite) collection of subsets of the form I_x, $x \in X$, such that $\sup_{i\in I_x} y_i$, which is defined by admissibility, is in V. We claim that the inverse image of V is $\bigcup_{I\in\mathcal{I}} \bigcap_{i\in I} U_i$. Since this is open, we shall conclude that the step function is continuous.

Given any $x \in \bigcup_{I\in\mathcal{I}} \bigcap_{i\in I} U_i$, there is an I in $\mathcal{I}$ such that x is in every U_i with $i \in I$. In particular, $I \subseteq I_x$. So $\sup_{i\in I_x} y_i$ is above $\sup_{i\in I} y_i$, which is in V since $I \in \mathcal{I}$. Conversely, if $\sup_{i\in I_x} y_i$ is in V, then I_x is in $\mathcal{I}$; since $x \in \bigcap_{i\in I_x} U_i$, $x \in \bigcup_{I\in\mathcal{I}} \bigcap_{i\in I} U_i$. $\square$

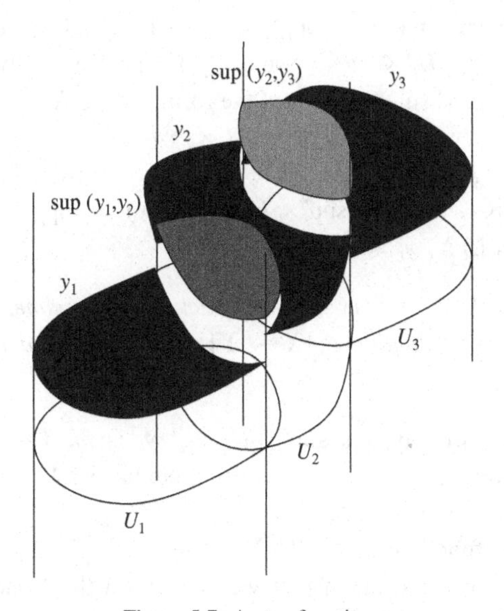

Figure 5.7 A step function.

Recall that we order the space $[X \to Y]$ of continuous maps from X to Y pointwise, i.e., $f \leqslant g$ iff $f(x) \leqslant g(x)$ for every $x \in X$. The following gives a sufficient condition for being way-below an element of $[X \to Y]$.

Lemma 5.7.12 *Let X be a core-coherent topological space, and Y be a continuous poset. For each continuous map f from X to Y_σ, for each admissible collection of pairs $(U_i, y_i) \in \mathcal{O}(X) \times Y$, $1 \leqslant i \leqslant n$, such that $U_i \Subset f^{-1}(\twoheaduparrow y_i)$ for every i, $\sup_{i=1}^n U_i \searrow y_i$ is way-below f in $[X \to Y]$.*

Proof Let $(g_j)_{j \in J}$ be any directed family of elements of $[X \to Y]$ such that $f \leqslant \sup_{j \in J} g_j$. For every open subset V of Y, $f^{-1}(V) \subseteq \bigcup_{j \in J} g_j^{-1}(V)$. Indeed, any element x of $f^{-1}(V)$ is such that $f(x) \in V$, so $\sup_{j \in J} g_j(x) \in V$, whence $g_j(x) \in V$ for some $j \in J$, by definition of Scott-opens.

By assumption, for every subset I of $\{1, \ldots, n\}$, $\bigcap_{i \in I} U_i \Subset f^{-1}(\bigcap_{i \in I} \twoheaduparrow y_i)$. Indeed, $\cap$ preserves $\Subset$, since X is core-coherent (Proposition 5.2.19). In particular, this is the case if $I = I_x$ for some $x \in X$. Then $\sup_{i \in I} y_i$ exists by admissibility, and $\bigcap_{i \in I} \twoheaduparrow y_i = \twoheaduparrow \sup_{i \in I} y_i$. Indeed, any y such that $\sup_{i \in I} y_i \ll y$ is such that $y_i \ll y$ for every $i \in I$, and conversely, if $y_i \ll y$ for every $i \in I$, then, for every directed family $(z_k)_{k \in K}$ that has a least upper bound above y, z_k is above each y_i for k large enough, whence z_k is above all of them, hence above $\sup_{i \in I} y_i$ for k large enough. So, when $I = I_x$ for some $x \in X$, $\bigcap_{i \in I} U_i \Subset f^{-1}(\twoheaduparrow \sup_{i \in I} y_i)$.

Together with the fact that $f^{-1}(V) \subseteq \bigcup_{j \in J} g_j^{-1}(V)$ for every open subset V of Y, in particular for $V = \twoheaduparrow \sup_{i \in I} y_i$, there is an index $j \in J$, depending on I, such that $\bigcap_{i \in I} U_i \subseteq g_j^{-1}(\twoheaduparrow \sup_{i \in I} y_i)$. There are only finitely many subsets I of $\{1, \ldots, n\}$ (in particular of the form I_x, $x \in X$), and J is directed, so we may choose j independent of $I = I_x$.

For every $x \in X$, letting $I = I_x$, x is in $\bigcap_{i \in I} U_i$ by definition of I_x, so x is in $g_j^{-1}(\twoheaduparrow \sup_{i \in I} y_i)$, i.e., $(\sup_{i=1}^n U_i \searrow y_i)(x) = \sup_{i \in I} y_i \ll g_j(x)$. In particular, $\sup_{i=1}^n U_i \searrow y_i \leqslant g_j$. □

Proposition 5.7.13 *For every core-compact and core-coherent topological space X, and every bc-domain Y, $[X \to Y]$ is a continuous dcpo, with a basis of step functions.*

Proof For short, we say that $\sup_{i=1}^n U_i \searrow y_i$ is *far below* f iff $U_i \Subset f^{-1}(\twoheaduparrow y_i)$ for every i, $1 \leqslant i \leqslant n$. By Lemma 5.7.12, this entails that $\sup_{i=1}^n U_i \searrow y_i$ is way-below f.

Given two step functions $\sup_{i=1}^m U_i \searrow y_i$ and $\sup_{i=m+1}^{m+n} U_i \searrow y_i$ far below $f \in [X \to Y]$, we first claim that we can find a third one far below f, and above the given two. To show this, we observe that the collection of all pairs (U_i, y_i), $1 \leqslant i \leqslant m+n$, is admissible. Indeed, for every subset I of $\{1, \ldots, m+n\}$ such that $\bigcap_{i \in I} U_i \setminus \bigcup_{i \in \{1,\ldots,m+n\} \setminus I} U_i$ is non-empty, write I as the union of $J = I \cap \{1, \ldots, m\}$ and $K = I \cap \{m+1, \ldots, m+n\}$. Fix $x \in \bigcap_{i \in I} U_i \setminus \bigcup_{i \in \{1,\ldots,m+n\} \setminus I} U_i$. By assumption, the value $\sup_{j \in J} y_j$ of the first step function $\sup_{i=1}^m U_i \searrow y_i$ at x is below $f(x)$, and similarly $\sup_{k \in K} y_k \leqslant f(x)$. Since Y is a bc-domain, $\sup_{j \in J} y_j$ and $\sup_{k \in K} y_k$ have a least upper bound, which must then be $\sup_{i \in I} y_i$. So the collection $\sup_{i=1}^m U_i \searrow y_i$ is indeed admissible. Clearly, the step function $\sup_{i=1}^{m+n} U_i \searrow y_i$ is far below f.

Next, there is a step function far below f, say the one obtained from the empty admissible collection of pairs, i.e., the constant $\perp$ map. So the family of all step functions far below f is directed.

For every $x \in X$, and every $y \ll f(x)$, the elementary step function $U \searrow y$, where U is any open subset of X such that $x \in U \Subset f^{-1}(\Uparrow y)$, is far below f. Note that U exists because X is core-compact, and using Proposition 5.2.6. Also, $(U \searrow y)(x) = y$. Since Y is continuous, $f(x)$ is the least upper bound of all $y \ll f(x)$, hence also the least upper bound of all $(U \searrow y)(x)$, where $U \searrow y$ ranges over the elementary step functions far below f. It follows that f is the least upper bound of all step functions (not just the elementary ones, which would not form a directed family) far below f. Since every step function far below f is way-below f, we conclude that $[X \to Y]$ is continuous, using Trick 5.1.20. □

Proposition 5.7.14 *For every poset X that is core-compact and core-coherent in its Scott topology, and for every bc-domain Y, $[X \to Y]$ is a bc-domain, with a basis of step functions.*

Proof $[X \to Y]$ is a continuous dcpo, with a basis of step functions, by Proposition 5.7.13. It is pointed, since the constant $\perp$ map is least, where $\perp$ is the least element of Y. It remains to show that every pair of continuous maps f, g that have an upper bound h have a least upper bound.

For each $x \in X$, $f(x)$ and $g(x)$ are below $h(x)$, so $\sup(f(x), g(x))$ exists for each $x \in X$, using the fact that Y is a bc-domain. We claim that $\sup(f, g)$ is the map k sending each $x \in X$ to $\sup(f(x), g(x))$, and to show this, it is enough to show that k is continuous, or, equivalently, that k is Scott-continuous, by Proposition 4.3.5. Now, for every directed family $(x_i)_{i \in I}$ that has a least upper bound x in X, $k(x) = \sup(\sup_{i \in I} f(x_i), \sup_{i \in I} g(x_i))$ (since f and g are Scott-continuous) $= \sup_{i \in I} k(x_i)$. □

Lemma 5.7.15 *Every bc-domain is compact, locally compact, and coherent.*

Proof Let X be a bc-domain. As a pointed poset, it is compact. As a continuous dcpo, X is locally compact (Corollary 5.1.36). Given any two elements x and y of X, $\uparrow x \cap \uparrow y$ is either empty or of the form $\uparrow \sup(x, y)$, and is certainly finitary compact. By Proposition 5.2.34, X is therefore coherent. □

In particular, by Lemma 5.2.24, every bc-domain is core-compact and core-coherent.

Theorem 5.7.16 (**BCDom** is Cartesian-closed) *The category* **BCDom** *of bc-domains and continuous maps is Cartesian-closed. Finite products are defined as in* **Cpo**, *and the exponential object is* $[X \to Y]$ *with the pointwise ordering.*

Proof The statement about finite products is easy. $[X \to Y]$ is a bc-domain by Proposition 5.7.14, which applies by Lemma 5.7.15. Application and currification are defined as expected. □

Exercise 5.7.17 (**CCLat** is Cartesian-closed) Show that the category **CCLat** of continuous complete lattices and continuous maps is Cartesian-closed.

Exercise 5.7.18 (Scott domain) A *Scott domain* is an algebraic bounded-complete dcpo. In particular, every Scott domain is a bc-domain. Show that, in a Scott domain X, the subsets of the form $\uparrow x$, $x \in X$, form a base of the Scott topology, and are *compact-open*, i.e., both compact and open.

Show the analogue of Proposition 5.7.14: for all Scott domains X and Y, $[X \to Y]$ is a Scott domain. You will need to consider least upper bounds of finite admissible families of elementary step functions of the form $(\uparrow x) \searrow y$, and show that they form a basis of finite elements in $[X \to Y]$.

Deduce that the category of Scott domains and Scott-continuous maps is Cartesian-closed, the exponential objects being $[X \to Y]$. Similarly, show that the category of algebraic complete lattices and Scott-continuous maps is Cartesian-closed.

While we are dealing with bc-domains, we should mention a few additional properties they enjoy. First, while being a bc-domain seems to be the juxtaposition of two independent properties, continuity and bounded-completeness, we show that bc-domains can be characterized using another, nicer set of interwoven properties. Recall (Exercise 5.7.5) that a bc-domain has greatest lower bounds of all non-empty families. Moreover, it is a dcpo, i.e., it has least upper bounds of all directed families.

Proposition 5.7.19 (bc-domain = directed sups + non-empty infs + infinite distributivity) *Let X be a bounded-complete dcpo, or, equivalently, a dcpo with greatest lower bounds of all non-empty families. Then X is continuous,* *i.e., a bc-domain, if and only if the following doubly infinite distributivity law is satisfied: for all directed families* $(x^i_j)_{j \in J_i}$*, one for each* $i \in I$*, where* I *is a non-empty index set,* $\inf_{i \in I} \sup_{j \in J_i} x^i_j = \sup_{f \in \prod_{i \in I} J_i} \inf_{i \in I} x^i_{f(i)}$.

Proof We first realize that the family $(\inf_{i \in I} x^i_{f(i)})_{f \in \prod_{i \in I} J_i}$ is directed, so that the right-hand side makes sense: if $\inf_{i \in I} x^i_{f(i)}$ and $\inf_{i \in I} x^i_{g(i)}$ are two elements of the family, then they are below $\inf_{i \in I} x^i_{h(i)}$, where h is obtained (using the Axiom of Choice) by letting $h(i) \in J_i$ be any element such that $x^i_{f(i)}, x^i_{g(i)} \leqslant x^i_{h(i)}$, for every $i \in I$.

Let X be continuous. The inequality $\sup_{f \in \prod_{i \in I} J_i} \inf_{i \in I} x^i_{f(i)} \leqslant \inf_{i \in I} \sup_{j \in J_i} x^i_j$ is clear, and follows from the fact that, for every $f \in \prod_{i \in I} J_i$, for every $i \in I$, $\inf_{i \in I} x^i_{f(i)} \leqslant x^i_{f(i)} \leqslant \sup_{j \in J_i} x^i_j$. To show the converse direction, we use the fact that X is continuous: $\inf_{i \in I} \sup_{j \in J_i} x^i_j$ is the least upper bound of the directed family $(y_k)_{k \in K}$ of all elements $y_k \ll \inf_{i \in I} \sup_{j \in J_i} x^i_j$. For each $k \in K$, $y_k \ll \sup_{j \in J_i} x^i_j$ for every $i \in I$. So $y_k \leqslant x^i_j$ for some $j \in J_i$ depending on k and i: for fixed k, using the Axiom of Choice, define $f \in \prod_{i \in I} J_i$ by letting $f(i)$ be such a j. So $y_k \leqslant \inf_{i \in I} x^i_{f(i)}$, where f depends on k. So $\sup_{k \in K} y_k \leqslant \sup_{f \in \prod_{i \in I} J_i} \inf_{i \in I} x^i_{f(i)}$. But the left-hand side is $\inf_{i \in I} \sup_{j \in J_i} x^i_j$.

Conversely, assume the doubly infinite distributivity law. Consider any $x \in L$. Consider the family $(D_i)_{i \in I}$ of all directed subsets D_i of X whose least upper bound is above x. Write $D_i = (x^i_j)_{j \in J_i}$. Among the families D_i, one can find the singleton $\{x\}$; i.e., there is an $i \in I$ such that $x^i_j = x$ for every $j \in J_i$. It follows that $x = \inf_{i \in I} \sup_{j \in J_i} x^i_j$. It also follows that I is non-empty. By the doubly infinite distributivity law, $x = \sup_{f \in \prod_{i \in I} J_i} \inf_{i \in I} x^i_{f(i)}$. We claim that the family $(\inf_{i \in I} x^i_{f(i)})_{f \in \prod_{i \in I} J_i}$ is a directed family of elements way-below x, which will show that X is indeed continuous, by Lemma 5.1.23.

Consider any directed family D whose least upper bound is above x. We must show that one of its elements is already above some element of the family $(\inf_{i \in I} x^i_{f(i)})_{f \in \prod_{i \in I} J_i}$. This is easy: by definition $D = D_i$ for some $i \in I$, so the least upper bound of D is $\sup_{j \in J_i} x^i_j$; this is certainly above $x^i_{f(i)}$, where f is any element of $\prod_{i' \in I} J_{i'}$ such that $f(i) = j$, hence above $\inf_{i \in I} x^i_{f(i)}$. ☐

We also note that bc-domains are robust in the following sense.

Lemma 5.7.20 *Every poset* Z *that arises as a Scott-retract of a bc-domain (resp., a continuous complete lattice)* Y *is itself a bc-domain (resp., a continuous complete lattice).*

Proof Let $s\colon Z \to Y$ be the section and $r\colon Y \to Z$ be the retraction, where Z is a bc-domain.

Since Y is a continuous dcpo, by Proposition 4.10.34 and Lemma 5.1.43 Z is also a continuous dcpo. It only remains to show that Z is bounded-complete. Since Y has a least element $\bot$, $r(\bot)$ is least in Z: for every $z \in Z$, $\bot \leqslant s(z)$, so $r(\bot) \leqslant r(s(z)) = z$. Let z_1, z_2 be any two elements of Z with an upper bound. So $s(z_1)$ and $s(z_2)$ also have an upper bound in Y. Since Y is bounded-complete, $\sup(s(z_1), s(z_2))$ exists. Let $z = r(\sup(s(z_1), s(z_2)))$. Then $z_1 = r(s(z_1)) \leqslant z$, and similarly $z_2 \leqslant z$, so z is an upper bound of z_1 and z_2. Let us show that it is the least one. Let z' be another one. Then $s(z_1), s(z_2) \leqslant s(z')$, so $z \leqslant r(s(z')) = z'$.

When Y is a continuous complete lattice, let $\top$ be its top element. Then $r(\top)$ is the largest element of Y: for every $y \in Y$, $y = r(s(y)) \leqslant r(\top)$. So Z is a continuous complete lattice. □

Exercise 5.7.21 (bc-domains = retracts of Scott domains) Show that the bc-domains are exactly the Scott-retracts of Scott domains. In the hard direction, show that the ideal completion of a bc-domain is a Scott domain.

Exercise 5.7.22 Extending Exercise 5.1.52, show that every non-empty Scott-closed subset F (in either the Scott or the subspace topology) of a bc-domain X is again a bc-domain.

There is still more magic with continuous complete lattices and bc-domains. We shall see in Exercise 9.3.9 and in Exercise 9.3.12 that they are respectively the injective and the densely injective topological spaces, for example.

6

Metrics, quasi-metrics, hemi-metrics

6.1 Metrics, hemi-metrics, and open balls

A natural instance of topological space is given by *metric spaces*, as already studied in Chapter 3.

Note that metrics are symmetric: the distance between x and y is the same as the distance between y and x. There is no need to assume so in all generality, and Lawvere (2002) has already rejected the symmetry axiom as being artificial in various situations – although, as we shall see, symmetry certainly makes our lives easier in several important cases, notably in the study of completeness (see Section 7.1). We are led to the following definition.

Definition 6.1.1 (Hemi-metric, quasi-metric) Let X be a set. A *hemi-metric* d on X is a map from $X \times X$ to $\overline{\mathbb{R}^+}$ such that:

- $d(x, x) = 0$ for every $x \in X$;
- (Triangular inequality) $d(x, y) \leqslant d(x, z) + d(z, y)$ for all $x, y, z \in X$.

A set X with a hemi-metric is called a *hemi-metric space*.

A hemi-metric d, or a hemi-metric space, is said to be T_0 if and only if $d(x, y) = d(y, x) = 0$ implies $x = y$. It is T_1 if and only if the sole equality $d(x, y) = 0$ already implies $x = y$.

A *quasi-metric* is a T_0 hemi-metric. A T_0 hemi-metric space is called a *quasi-metric space*.

The quantity $d(x, y)$ is now called the distance *from x to y*. Dispensing with the axiom of symmetry $d(x, y) = d(y, x)$ allows us to distinguish between the distance from x to y and that from y to x. This is intuitively useful if $d(x, y)$ is to measure the amount of effort to go from point x to point y in a region of mountains. It indeed usually takes less effort to go down than up. A number of other constructions of metrics in the mathematical literature are also naturally described as symmetrized version of otherwise non-symmetric hemi-metrics: see Lawvere (2002) for a list.

Example 6.1.2 The simplest example of a hemi-metric that is not symmetric is $\mathsf{d}_{\mathbb{R}}(x, y) = \max(x - y, 0)$ on $\mathbb{R}$. This is a quasi-metric. Note that $\mathsf{d}_{\mathbb{R}}(3, 1) = 2$, while $\mathsf{d}_{\mathbb{R}}(1, 3) = 0$.

Every hemi-metric defines a topology, the *open ball* topology. This is defined exactly as in the metric case.

Definition 6.1.3 (Open ball) Let d be a hemi-metric on a set X. The *open ball* $B^d_{x,<\epsilon}$ with *center* x and *radius* $\epsilon \in \mathbb{R}^+$, $\epsilon > 0$, is the set of all points $y \in X$ such that $d(x, y) < \epsilon$.

The *topology of* d, also called the *open ball topology* $\mathcal{O}^d$, is the topology generated by the open balls.

For short, whenever we say "for some $\epsilon > 0$" or "for every $\epsilon > 0$," this will always mean "for some $\epsilon \in \mathbb{R}^+$ with $\epsilon > 0$," resp., "for every $\epsilon \in \mathbb{R}^+$ with $\epsilon > 0$." That is, implicitly, we shall always assume ϵ to be a real number, and certainly not $+\infty$; similarly for η.

Example 6.1.4 In hemi-metric spaces, the open balls are allowed to be asymmetric. E.g., in $\mathbb{R}$ with the hemi-metric $\mathsf{d}_{\mathbb{R}}$ of Example 6.1.2, the open ball with center x and radius r is $(x - r, +\infty)$.

Lemma 6.1.5 (Open balls form a base) *Let X, d be a hemi-metric space. For every open U in the open ball topology, and every point x of U, there is an $\epsilon > 0$ such that the open ball $B^d_{x,<\epsilon}$ is included in U.*

In particular, the open balls form a base of the open ball topology: every open is a union of open balls.

Proof It is enough to show that, given any finite family $B^d_{x_i,<\epsilon_i}$, $1 \leqslant i \leqslant n$, of open balls, and any point x in their intersection, there is an open ball $B^d_{x,<\epsilon}$ included in this intersection. Let $\epsilon = \min_{1 \leqslant i \leqslant n}(\epsilon_i - d(x_i, x))$. By definition, for every i, $1 \leqslant i \leqslant n$, $d(x_i, x) \leqslant \epsilon_i - \epsilon$. For every $y \in B^d_{x,<\epsilon}$, $d(x, y) < \epsilon$, so, by the triangular inequality, $d(x_i, y) < \epsilon_i$. So $B^d_{x,<\epsilon} \subseteq B^d_{x_i,<\epsilon_i}$. We conclude by Lemma 4.1.10. □

Trick 6.1.6 Similarly to Trick 4.1.11, it follows that to show that a subset U of a hemi-metric space X, d is open, it is equivalent to show that, for every $x \in U$, there is an open ball $B^d_{x,<\epsilon}$ centered at x and included in U.

For every $\epsilon > 0$, there is a natural number $n \in \mathbb{N}$ such that $1/2^n \leqslant \epsilon$. Then $B^d_{x,<1/2^n}$ still contains x and is included in $B^d_{x,<\epsilon}$. It follows that one can refine Lemma 6.1.5 to:

Fact 6.1.7 *Let X, d be a hemi-metric space. For every open U in the open ball topology, and every point x of U, there is an $n \in \mathbb{N}$ such that the open ball $B^d_{x,<1/2^n}$ is included in U.*

The next proposition characterizes the specialization quasi-ordering of a hemi-metric.

Proposition 6.1.8 ($\leqslant^d$ means distance zero) *Let X, d be a hemi-metric space, and $\leqslant^d$ be the specialization quasi-ordering of X with its open ball topology. Then $x \leqslant^d y$ if and only if $d(x, y) = 0$.*

Proof By Proposition 4.2.4, $x \leqslant^d y$ iff, for every $z \in X$, for every $\epsilon > 0$, $x \in B^d_{z,<\epsilon}$ implies $y \in B^d_{z,<\epsilon}$, or equivalently $d(z, x) < \epsilon$ implies $d(z, y) < \epsilon$ for all $z \in X$, $\epsilon > 0$. This certainly holds if $d(x, y) = 0$, using the triangular inequality, as then $d(z, y) \leqslant d(z, x) + d(x, y) < \epsilon + 0$. Conversely, if $x \leqslant^d y$, then taking $z = x$, we obtain that $d(x, y) < \epsilon$ for every $\epsilon > 0$, whence $d(x, y) = 0$. □

This justifies our definitions of T_0 and T_1 hemi-metrics above.

Lemma 6.1.9 *Let X, d be a hemi-metric space. With its open ball topology, X is T_0 if and only if d is T_0 (i.e., d is a quasi-metric). X is T_1 if and only if d is T_1.*

Proof By Proposition 4.2.3, X is T_0 if and only if $\leqslant^d$ is an ordering, i.e., if and only if $x \leqslant^d y$ and $y \leqslant^d x$ imply $x = y$. By Proposition 6.1.8, this is equivalent to: $d(x, y) = 0$ and $d(y, x) = 0$ imply $x = y$, which is our definition of a T_0 hemi-metric. Similarly, X is T_1 iff $\leqslant^d$ is just equality, i.e., iff $d(x, y) = 0$ is equivalent to $x = y$, which is again our definition of a T_1 hemi-metric. □

If d is T_1 and also symmetric, then X is in fact T_2 in its open ball topology.

Lemma 6.1.10 (Metric $\Rightarrow T_2$) *Every metric space is T_2 in its open ball topology.*

Proof Assume $x \neq y$. In particular, $d(x, y) \neq 0$. Let $\epsilon = d(x, y)/2$. Then $B^d_{x,<\epsilon}$ and $B^d_{y,<\epsilon}$ are open neighborhoods respectively of x and y, and they are disjoint: if z were in their intersection, then we would have $d(x, z) < \epsilon$ and $d(z, y) = d(y, z) < \epsilon$, whence $d(x, y) \leqslant d(x, z) + d(z, y) < 2\epsilon = d(x, y)$ by the triangle inequality, leading to a contradiction. □

One can say more, and claim that every metric space is normal, hence T_4. For now, we shall only show the weaker property that every metric space is regular, hence T_3. One needs the following notion of distance from a point to a *subset*, which allows us to generalize Proposition 6.1.8.

Lemma 6.1.11 (Closure = distance zero) *Let X, d be a hemi-metric space. For every subset A of X, for every $x \in X$, let $d(x, A) = \inf_{y \in A} d(x, y)$. We take this to be $+\infty$ if A is empty.*

The set of points x such that $d(x, A) = 0$ is exactly the closure of A in the open ball topology.

Proof Assume $d(x, A) = 0$, and consider any open neighborhood U of x. By Lemma 6.1.5, $B^d_{x,<\epsilon}$ is included in U for some $\epsilon > 0$. Since $d(x, A) = 0$, $d(x, y) < \epsilon$ for some $y \in A$. So U meets A. By Lemma 4.1.27, x is in $cl(A)$. Conversely, if $x \in cl(A)$, then, for every $\epsilon > 0$, $B^d_{x,<\epsilon}$ must meet A, so $d(x, A) < \epsilon$. As $\epsilon > 0$ is arbitrary, $d(x, A) = 0$. □

Definition 6.1.12 (Fattening) Let X, d be a hemi-metric space. For every subset A of X, for every $\epsilon > 0$, the *ϵ-fattening* $A^{d,+(\epsilon)}$ of A is the union of all the open balls $B^d_{x,<\epsilon}$, $x \in A$.

Note that any fattening is automatically open.

Proposition 6.1.13 (Metric $\Rightarrow T_3$) *Every metric space is regular, hence T_3 in its open ball topology.*

Proof Let X, d be metric. Let $x \in X$, F be a closed set not containing x. By Lemma 6.1.11, $\epsilon = d(x, F)$ is non-zero. Then $B^d_{x,<\epsilon/2}$ is an open neighborhood of x, and the fattening $F^{d,+(\epsilon/2)}$ is an open subset containing F. We claim that they do not intersect. Otherwise there would be a point y with $d(x, y) < \epsilon/2$ and a point $z \in F$ such that $y \in B^d_{z,<\epsilon/2}$; so $d(y, z) = d(z, y) < \epsilon/2$, hence $d(x, z) \leqslant d(x, y) + d(y, z) < \epsilon$. This would imply $d(x, F) < \epsilon$, a contradiction. □

Exercise 6.1.14 Let A be a subset of a metric space X, d. Given any family $(\epsilon_i)_{i \in I}$ of positive real numbers with $\inf_{i \in I} \epsilon_i = 0$, show that $\bigcap_{i \in I} A^{d,+(\epsilon_i)}$ is exactly $cl(A)$. Why does this not hold in arbitrary quasi-metric spaces?

Conclude that every closed subset of a metric space is the intersection of countably many opens.

Exercise 6.1.15 shows that the open ball topology is similar to the Alexandroff topology.

Exercise 6.1.15 ($d_\leqslant$, open ball and Alexandroff topologies) Let X be any set, with a quasi-ordering $\leqslant$. The latter defines a canonical hemi-metric $d_\leqslant$, defined by $d_{\leqslant}(x, y) = 0$ iff $x \leqslant y$, $d_{\leqslant}(x, y) = +\infty$ otherwise. Check that this is indeed a hemi-metric. Show that the open ball topology of $d_{\leqslant}$ is just the Alexandroff topology of $\leqslant$.

There is no real gap between quasi-metrics and metrics, as one can always symmetrize a quasi-metric and get a metric.

Lemma 6.1.16 (d^{op}, d^{sym}) *Given any hemi-metric d on a set X, its* opposite *d^{op} is defined by $d^{op}(x, y) = d(y, x)$, and its* symmetrization *d^{sym} by $d^{sym}(x, y) = \max(d(x, y), d(y, x))$.*

If d is T_0 (resp., T_1), then so is d^{op}. If d is T_0, then d^{sym} is a metric.

One can extend the quasi-metric $\mathsf{d}_{\mathbb{R}}$ (see Example 6.1.2) to any subset of $\mathbb{R} \cup \{-\infty, +\infty\}$:

Definition 6.1.17 ($\mathsf{d}_{\mathbb{R}}$) The quasi-metric $\mathsf{d}_{\mathbb{R}}$ on $\mathbb{R} \cup \{-\infty, +\infty\}$ is defined by: (*a*) $\mathsf{d}_{\mathbb{R}}(-\infty, t) = 0$; (*b*) $\mathsf{d}_{\mathbb{R}}(s, +\infty) = 0$; (*c*) $\mathsf{d}_{\mathbb{R}}(s, t) = \max(s - t, 0)$ otherwise, i.e., when $s \neq -\infty$ and $t \neq +\infty$.

A more synthetic definition is to let $\mathsf{d}_{\mathbb{R}}(s, t)$ be $\max(s - t, 0)$ in any case, taking by convention $(-\infty) - (-\infty) = 0$ and $(+\infty) - (+\infty) = 0$.

The symmetrization of $\mathsf{d}_{\mathbb{R}}$ on $\mathbb{R}$ is the usual L^1 metric (Example 3.1.4). On $\mathbb{R}$, $\mathbb{R} \cup \{-\infty, +\infty\}$, or on $\overline{\mathbb{R}^+}$, $\mathsf{d}_{\mathbb{R}}^{sym}(s, t) = |s - t|$, agreeing again that $(-\infty) - (-\infty) = 0$ and $(+\infty) - (+\infty) = 0$.

Exercise 6.1.18 (Open ball topology $=$ Scott on $\mathbb{R}$) Show that the open ball topology of $\mathsf{d}_{\mathbb{R}}$ on $\mathbb{R}$ is its Scott topology, and similarly for $[a, +\infty)$, $(-\infty, b]$, and $[a, b]$ (with $a \leqslant b$ in $\mathbb{R}$).

Show, however, that the open ball topology is strictly finer than the Scott topology on $\overline{\mathbb{R}^+}$ and on $\mathbb{R} \cup \{-\infty, +\infty\}$.

Proposition 6.1.19 *Let X, d be a hemi-metric space. The open ball topology of d^{sym} is generated by the open ball topologies of d and of d^{op}. That is, it is the coarsest topology containing all opens of $\mathcal{O}^d$ and of $\mathcal{O}^{d^{op}}$.*

Proof We first show that the topology of d is coarser than that of d^{sym}—the case of d^{op} is symmetric. It is enough to show that every open ball $B^d_{x,<\epsilon}$ is open in $\mathcal{O}^{d^{sym}}$. Let y be any point of $B^d_{x,<\epsilon}$, and let $a = d(x, y) < \epsilon$. For every $z \in B^{d^{sym}}_{y,<\epsilon-a}$, $d(y, z) \leqslant d^{sym}(y, z) < \epsilon - a$, so $d(x, z) \leqslant d(x, y) + d(y, z) < \epsilon$. So $B^{d^{sym}}_{y,<\epsilon-a}$ is an open ball for d^{sym} containing y and included in $B^d_{x,<\epsilon}$. By Trick 4.1.11, $B^d_{x,<\epsilon}$ is open in $\mathcal{O}^{d^{sym}}$.

Conversely, every open ball $B^{d^{sym}}_{x,<\epsilon}$ is the set of points y such that $d(x, y) < \epsilon$ and $d^{op}(x, y) < \epsilon$, hence equals the intersection $B^d_{x,<\epsilon} \cap B^{d^{op}}_{x,\epsilon}$, so every open of $\mathcal{O}^{d^{sym}}$ is in any topology containing all opens of $\mathcal{O}^d$ and of $\mathcal{O}^{d^{op}}$. □

Exercise 6.1.20 An *ultra-hemi-metric* on X is any map d from $X \times X$ to $\overline{\mathbb{R}^+}$ such that:

- $d(x, x) = 0$ for every $x \in X$;
- $d(x, y) \leqslant \max(d(x, z), d(z, y))$.

An *ultra-quasi-metric* is a T_0 ultra-hemi-metric. An *ultra-metric* is a symmetric ultra-quasi-metric, namely $d(x, y) = d(y, x)$ for all x, y. Show that every ultra-hemi-metric is a hemi-metric, every ultra-quasi-metric is a quasi-metric, and every ultra-metric is a metric.

Show that, for every quasi-ordering (resp., ordering) $\leqslant$ on X, $d_{\leqslant}$ is an ultra-hemi-metric (resp., an ultra-quasi-metric).

Exercise 6.1.21 (The ultra-quasi-metric space of words) Let Σ be a non-empty set of so-called *letters*. Σ itself is called an *alphabet*, and Σ^* denotes the set of finite words $a_1a_2 \dots a_n$ over Σ, i.e., of finite sequences of letters $a_1, a_2, \dots, a_n \in \Sigma$. The number n is the *length* of the latter word. We write ε for the empty word (i.e., the unique word with length zero), and ww' for the concatenation of two words w and w'.

Given two words w, w' over Σ, w is a *prefix* of w' if and only if w' can be written ww'' for some word w''. Let $\leqslant^{\mathrm{pref}}$ be the relation "is a prefix of"; this is an ordering. Note that if w is not a prefix of w', then w can be written uniquely as w_1aw_2 where w_1 is the longest prefix of w that is also a prefix of w', a is a letter – and w_1a is not a prefix of w'.

Define $d_{\Sigma^*}(w, w')$ as zero if w is a prefix of w', and as 2^{-n} otherwise, where n is the length of the longest prefix of w that is also a prefix of w'. Show that d_{Σ^*} is a T_0 ultra-hemi-metric, whose specialization ordering is $\leqslant^{\mathrm{pref}}$.

Exercise 6.1.22 Show that the opposite of any ultra-hemi-metric is an ultra-hemi-metric, and that the symmetrization of any ultra-quasi-metric (i.e., of a T_0 ultra-hemi-metric) is an ultra-metric. What is the symmetrization of d_{Σ^*}?

Exercise 6.1.23 (Ultra-metric spaces are zero-dimensional) Ultra-metric spaces have strange properties. Let X be a set with an ultra-metric d. Show that, given any point y outside an open ball $B^d_{x,<\epsilon}$, the whole open ball $B^d_{y,<\epsilon}$ around y is disjoint from $B^d_{x,<\epsilon}$. Deduce from this that every open ball is closed.

Conclude that the open ball topology of any ultra-metric space is zero-dimensional (Exercise 4.1.34).

6.2 Continuous and Lipschitz maps

In hemi-metric spaces, as in metric spaces (Proposition 3.5.2), the notion of continuous map specializes to the following ϵ-η formula.

Lemma 6.2.1 *Let f be any map from the hemi-metric space X, d to the hemi-metric space Y, d'. Then f is continuous at $x \in X$, where both X and Y are equipped with their open ball topologies, if and only if:*

for every $\epsilon > 0$, there is an $\eta > 0$ such that, for every $x' \in X$ with
$$d(x, x') < \eta,\ d'(f(x), f(x')) < \epsilon.$$

As usual, the map f is continuous from X to Y iff the above ϵ-η formula holds for every $x \in X$: see Proposition 4.3.14.

Proof The ϵ-η formula is equivalent to: for every $\epsilon > 0$, there is an $\eta > 0$ such that $B^d_{x,<\eta} \subseteq f^{-1}(B^d_{f(x),<\epsilon})$. If so, Proposition 6.1.5 tells us that $f^{-1}(B^d_{f(x),<\epsilon})$ is open, whence f is continuous by Trick 4.3.3. Conversely, if f is continuous, $f^{-1}(B^d_{f(x),<\epsilon})$ is open, so $B^d_{x,<\eta} \subseteq f^{-1}(B^d_{f(x),<\epsilon})$ for some $\eta > 0$, using Proposition 6.1.5. □

6.2.1 Lipschitz and uniformly continuous maps

An important class of continuous maps, which has no equivalent in general topological spaces, is the Lipschitz maps. The following is a natural generalization of the similarly named notion in Section 3.4.

Definition 6.2.2 (Lipschitz) Let f be a map from the hemi-metric space X, d to the hemi-metric space Y, d'. Given $c \in \mathbb{R}^+$, f is *c-Lipschitz* at $x \in X$ if and only if $d(f(x), f(x')) \leqslant c.d(x, x')$ for every $x' \in X$.

We say that f is *c-Lipschitz* if and only if it is c-Lipschitz at every point $x \in X$, and that it is *Lipschitz* (resp., at x) if and only if it is c-Lipschitz (resp., at x) for some $c \in \mathbb{R}^+$.

Fact 6.2.3 *Every Lipschitz map (resp., at x) is continuous (resp., at x) with respect to the open ball topologies.*

Indeed, use Lemma 6.2.1, defining η as ϵ/c if $c \neq 0$, or as an arbitrary number if $c = 0$.

An intermediate class of maps is the following.

Definition 6.2.4 (Uniformly continuous) Let X, d and Y, d' be hemi-metric spaces. A map $f : X \to Y$ is *uniformly continuous* if and only if:

$$\text{for every } \epsilon > 0, \text{ there is an } \eta > 0 \text{ such that, for all } x, x' \in X \text{ with } d(x, x') < \eta, d'(f(x), f(x')) < \epsilon.$$

Note the similarity with Lemma 6.2.1. Here η must be chosen *independently* of x.

Lemma 6.2.5 *Every Lipschitz map is uniformly continuous. Every uniformly continuous map is continuous.*

Proof If f is c-Lipschitz, then define η as ϵ/c if $c \neq 0$, or as an arbitrary number if $c = 0$, as for Fact 6.2.3. The second claim is obvious. □

Example 6.2.6 In general, not all continuous maps are uniformly continuous. Consider the map $f(x) = x^2$ on $\mathbb{R}^+$. This is continuous (Example 4.5.7), but not uniformly continuous. For any fixed $\epsilon > 0$, there is no $\eta > 0$ such that $|x^2 - x'^2| < \epsilon$ whenever $|x - x'| < \eta$: take $x' = \epsilon/\eta$, and $x = x' + \eta/2$, then $|x^2 - x'^2| = \eta x' + \eta^2/4 > \epsilon$.

There is an important case where continuity and uniform continuity coincide. This is sometimes called *Heine's Theorem.*

Lemma 6.2.7 (Heine's Theorem) *Let X, d and Y, d' be hemi-metric spaces, and assume X compact and d' metric. Every continuous map $f : X \to Y$ is uniformly continuous.*

Proof Fix $\epsilon > 0$. By Lemma 6.2.1, for every $x \in X$, there is an $\eta_x > 0$ such that, for every $x' \in X$ with $d(x, x') < \eta_x$, $d'(f(x), f(x')) < \epsilon/2$. (We use the Axiom of Choice to build the map $x \mapsto \eta_x$.) Since X is compact, there is a finite subset E of X such that $(B^d_{y,<\eta_y/2})_{y \in E}$ covers X. Let $\eta = \min_{y \in E} \eta_y/2$. For all $x, x' \in X$ with $d(x, x') < \eta$, find $y \in E$ such that $x \in B^d_{y,<\eta_y/2}$. Since $d(y, x) < \eta_y/2$, $d(y, x') \leqslant d(y, x) + d(x, x') < \eta_y/2 + \eta \leqslant \eta_y$. Then $d'(f(x), f(x')) \leqslant d'(f(x), f(y)) + d'(f(y), f(x')) = d'(f(y), f(x)) + d'(f(y), f(x'))$ (since d' is a metric) $< \epsilon/2 + \epsilon/2 = \epsilon$. □

Example 6.2.8 There are also uniformly continuous maps that are not c-Lipschitz, for any $c \in \mathbb{R}^+$. Consider the following oscillating function from $\mathbb{R}$ to $\mathbb{R}$: $o(x) = x - \lfloor x \rfloor$ if $\lfloor x \rfloor$ is even, $o(x) = \lfloor x \rfloor + 1 - x$ otherwise. This is a continuous map such that $o(2i) = 0$ and $o(2i + 1) = 1$ for every $i \in \mathbb{N}$. Now the function f that maps each $x \in (0, 1]$ to $x.o(\frac{1}{x^2})$ and 0 to 0 is continuous (at every point other than 0, but also at 0, since, for every $\epsilon > 0$, one can let $\eta = \epsilon$ so that, whenever $|0 - x| < \eta$, then $|0 - f(x)| < \epsilon$, since indeed $0 \leqslant f(x) \leqslant x$

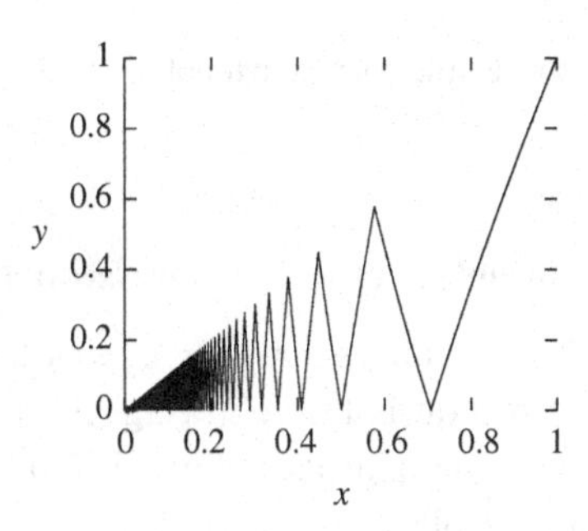

Figure 6.1 A uniformly continuous, not Lipschitz map.

for every x). So f is uniformly continuous on $[0, 1]$, since $[0, 1]$ is compact (Lemma 6.2.7). See Figure 6.1. But f is not c-Lipschitz for any $c \in \mathbb{R}^+$. Indeed, take any $i \in \mathbb{N}$ strictly above $\frac{c^2}{2(1+2c)}$, $x = \frac{1}{\sqrt{2i}}$, $x' = \frac{1}{\sqrt{2i+1}}$, so that $f(x) = 0$, $f(x') = x'$; then one can check that $|f(x) - f(x')| = x'$ is no smaller than $c.|x - x'|$. (Exercise: prove this.)

Exercise 6.2.9 Consider the function $f : \mathbb{R}^+ \to \mathbb{R}^+$ defined by $f(x) = \sqrt{x}$. When $\mathbb{R}^+$ is equipped with the hemi-metric $\mathsf{d}_{\mathbb{R}}$, show that f is Lipschitz at every $x \in \mathbb{R}^+$, but not Lipschitz, i.e., not c-Lipschitz for any $c \in \mathbb{R}^+$. So there are non-Lipschitz maps that are Lipschitz at every point.

When $\mathbb{R}^+$ is equipped with the metric $\mathsf{d}_{\mathbb{R}}^{sym}$, show that f is continuous at every $x \in \mathbb{R}^+$, and in fact Lipschitz at every non-zero point x, but that it is not Lipschitz at 0.

Exercise 6.2.10 Let X, d be a hemi-metric space. Show that $d(\_, y)$ is 1-Lipschitz from X to $\overline{\mathbb{R}^+}, \mathsf{d}_{\mathbb{R}}$ for every $y \in X$, and that $d(x, \_)$ is 1-Lipschitz from X, d^{op} (not X, d! Show that $d(x, \_)$ is not even monotonic from X to $\overline{\mathbb{R}^+}$ in general) to $\overline{\mathbb{R}^+}$ for every $x \in X$.

Exercise 6.2.11 (Open hole and upper topologies) Let d be a hemi-metric on a set X. The *open hole* $T^d_{x,>\epsilon}$ with *center* x and *radius* $\epsilon \in \mathbb{R}^+$, $\epsilon > 0$, is the set of points $y \in X$ such that $d(y, x) > \epsilon$. (Note: $d(y, x)$, not $d(x, y)$; and $> \epsilon$, not $< \epsilon$. Note also that the centers of open holes are never in the holes themselves.) The *open hole topology* is the one generated by open holes.

Show that the open hole topology is always coarser than the open ball topology, and has the same specialization quasi-ordering. (Hint for the first part: $d(\_, x)$ is Lipschitz.) Given any quasi-ordering $\leqslant$ on a set X, show that the open hole topology of $d_{\leqslant}$ (see Exercise 6.1.15) is the upper topology of $\leqslant$. In particular, the open hole topology is in general strictly coarser than the open ball topology.

Exercise 6.2.12 Show that the open hole topology of $\mathsf{d}_{\mathbb{R}}$ on $\mathbb{R}$ coincides with its open ball topology.

6.2.2 Isometric embeddings, isometries

There are several possible candidates for categories of hemi-metric or metric spaces. One of the most natural is the following. Another one would be to allow all continuous maps as morphisms, but this would be best understood as subcategories of hemi-metrizable, resp., metrizable, *topological* spaces. Hemi-metrizability and metrizability questions will be examined in Section 6.3.

Definition 6.2.13 (HMet, QMet, Met) Let **HMet** be the category whose objects are hemi-metric spaces, and whose morphisms are 1-Lipschitz maps, and similarly for **QMet** (quasi-metric spaces), and **Met** (metric spaces).

The isos in each of these categories are the *isometries*.

Definition 6.2.14 (Isometric embedding, isometry) A map f from a hemi-metric space X, d to a hemi-metric space X', d' is an *isometric embedding* if and only if $d'(f(x), f(x')) = d(x, x')$ for all $x, x' \in X$.

An *isometry* is a bijective isometric embedding.

Isometric embeddings may fail to be injective, but will be whenever d is T_0. Note that the isometries are exactly those bijections that are isometric embeddings and whose inverse is also an isometric embedding, or equivalently those bijections that are 1-Lipschitz and whose inverse is also 1-Lipschitz.

Lemma 6.2.15 *Let X, d be a quasi-metric space and X', d' be a hemi-metric space.*

Every isometric embedding $f: X \to X'$ is an embedding (in the topological sense) with respect to the underlying open ball topologies.

Every isometry is a homeomorphism with respect to the underlying open ball topologies.

Proof Assume that f is an isometric embedding of X into X'. The map f defines a bijection from X to $f[X]$. It is injective because, whenever $f(x) = f(y)$, $d(x, y) = d'(f(x), f(y)) = 0$ and similarly $d(y, x) = 0$, so $x = y$ since d is T_0.

Next, f is continuous since it is Lipschitz (Fact 6.2.3). In fact, any open ball $B^d_{x,<\epsilon}$ is the inverse image of the open ball $B^{d'}_{f(x),<\epsilon}$. As every open subset U of X is a union of open balls, by Lemma 6.1.5 U is the inverse image of the corresponding union V of open balls in X', showing that f is almost open. So f is a (topological) embedding, by Proposition 4.10.9.

When f is also bijective, then f is open, hence defines a homeomorphism by Proposition 4.10.4. □

When there is an isometric embedding f from some quasi-metric space X, d to some hemi-metric space X', d', we also say that X, d *isometrically embeds* into X', d' *through* f.

6.2.3 Separation properties of metric spaces

One application of the fact that Lipschitz maps are continuous is that every metric space is T_4.

Lemma 6.2.16 *Let X, d be a hemi-metric space. For every subset A of X, for all $x, x' \in X$, $d(x, A) \leqslant d(x, x') + d(x', A)$. In particular, the map $d(\_, A)$ that sends x to $d(x, A)$ is 1-Lipschitz from X, d to $\overline{\mathbb{R}^+}, \mathsf{d}_{\mathbb{R}}$.*

If d is a metric, then $d(\_, A)$ is 1-Lipschitz from X, d to $\overline{\mathbb{R}^+}, \mathsf{d}_{\mathbb{R}}^{sym}$.

Proof Clearly $d(x, x') + d(x', A) = d(x, x') + \inf_{y \in A} d(x', y) = \inf_{y \in A}(d(x, x') + d(x', y)) \geqslant \inf_{y \in A} d(x, y)$. So $\mathsf{d}_{\mathbb{R}}(d(x, A), d(x', A)) \leqslant d(x, x')$, whence $d(\_, A)$ is 1-Lipschitz from X, d to $\overline{\mathbb{R}^+}, \mathsf{d}_{\mathbb{R}}$. The final claim is obtained by symmetrization. □

In particular, by Fact 6.2.3, $d(\_, A)$ is continuous from X to $\overline{\mathbb{R}^+}$, equipped with $\mathsf{d}_{\mathbb{R}}$ or $\mathsf{d}_{\mathbb{R}}^{sym}$ respectively.

Proposition 6.2.17 (Metric $\Rightarrow T_4$) *Every metric space is T_4.*

Proof Let X, d be a metric space, and F, F' be two disjoint closed subsets. Let $g(x) = \min(d(x, F), 1) + \min(d(x, F'), 1)$. If $g(x) = 0$, then $d(x, F) = d(x, F') = 0$, so $x \in cl(F) = F$ and similarly $x \in F'$ by Lemma 6.1.11. This is impossible since $F \cap F'$ is empty. So g never takes the value zero. By Lemma 6.2.16, $d(\_, F)$ and $d(\_, F')$ are continuous, so the formula $f(x) = \min(d(x, F), 1)/g(x)$ defines a continuous map from X to $[0, 1]$ (use Example 4.5.8). Let $U = f^{-1}[0, 1/3)$, $U' = f^{-1}(2/3, 1]$. U and U' are open, since f is continuous, and are disjoint; U contains F since $f(x) = 0$ for every $x \in F$; and U' contains F' since $f(x) = 1$ for every $x \in F'$. □

Exercise 6.2.18 (Completely regular, $T_{3\frac{1}{2}}$) A space X is *completely regular* iff one can separate points from closed sets by continuous maps, i.e., iff, for every point x and every closed set F not containing x, there is a continuous map $f : X \to [0, 1]$ such that $f(x) = 1$ and $f(y) = 0$ for every $y \in F$. (Note that f has to be continuous from X to $[0, 1]$, not to $[0, 1]_\sigma$. Requiring f to be continuous from X

to $[0,1]_\sigma$ instead would merely yield an alternate definition of regular spaces.) A $T_{3\frac{1}{2}}$ space is a completely regular T_0 space.

Show that every metric space is completely regular in its open ball topology.

We shall see later that every T_4 space is $T_{3\frac{1}{2}}$, whence the latter exercise arises as a trivial consequence of Proposition 6.2.17. Exercise 6.2.19 completes the explanation of the funny name "$T_{3\frac{1}{2}}$".

Exercise 6.2.19 Show that every completely regular space is regular. So every $T_{3\frac{1}{2}}$ space is T_3.

6.3 Topological equivalence, hemi-metrizability, metrizability

Every hemi-metric defines an open ball topology, but several hemi-metrics can define the same topology. Lemma 6.3.2 below gives a fairly general way of modifying a hemi-metric without changing the topology.

Definition 6.3.1 (Topological equivalence) Two hemi-metrics d and d' on a set X are *topologically equivalent* if and only if they define the same open ball topology.

The lemma below states that topologies only care about close points. We get the same open ball topologies from hemi-metrics d and d' that may be very different, except that $d(x,y)$ should be close to 0 if and only if $d'(x,y)$ is close to 0. Note that turning to specialization orderings loses even more information, as $\leqslant^d$ only cares about which pairs of points are at distance exactly 0 (Proposition 6.1.8).

Lemma 6.3.2 *Let d be any hemi-metric (resp., a quasi-metric, a metric) on a set X, and $a>0$ be any positive real number. Define $d'(x,y)$ as $\min(d(x,y),a)$. Then d' is a hemi-metric (resp., a quasi-metric, a metric), and is topologically equivalent to d.*

Proof We check the triangular inequality: $d'(x,y)=\min(d(x,y),a)$ is less than or equal to $\min(d(x,z)+d(z,y),a)$, which is clearly below $d'(x,z)+d'(z,y)=\min(d(x,z),a)+\min(d(z,y),a)$. Next, $d'(x,x)=\min(0,a)=0$. If d is a quasi-metric, then $d'(x,y)=0$ implies $d(x,y)=0$, hence $x=y$, so d' is a quasi-metric. And clearly, if d is a metric, then so is d'.

To show topological equivalence, we first refine Lemma 6.1.5 and observe that, whichever hemi-metric we are considering, the open balls of radius $\epsilon\leqslant a$ still form a base of the open ball topology. Indeed, for every open neighborhood

U of a point x, there is an open ball, say of radius r, centered at x and included in U. Then the open ball of center x and radius $\min(r, a)$ is also included in U. However, the open balls of radius at most a are the same for d and d'. So they have identical open ball topologies. □

Exercise 6.3.3 A map φ from $\overline{\mathbb{R}^+}$ to $\overline{\mathbb{R}^+}$ is *strict* if and only if $\varphi(0) = 0$, and *subadditive* if and only if $\varphi(s + t) \leqslant \varphi(s) + \varphi(t)$ for all $s, t \in \overline{\mathbb{R}^+}$.

Let d be any hemi-metric (resp., a quasi-metric, a metric) on a set X, and let φ be a strict, monotonic, and subadditive map from $\overline{\mathbb{R}^+}$ to $\overline{\mathbb{R}^+}$, and assume that there is an interval $[0, a]$ with $a \in \mathbb{R}^+, a > 0$, such that φ restricts to a homeomorphism from $[0, a]_\sigma$ onto $[0, \varphi(a)]_\sigma$.

Show that $\varphi \circ d$ is a hemi-metric (resp., a quasi-metric, a metric), and is topologically equivalent to d. This generalizes Lemma 6.3.2, which is the particular case $\varphi = \min(\_, a)$.

It follows that every hemi-metric is topologically equivalent to a bounded one. In particular, requiring hemi-metrics to take values in $\overline{\mathbb{R}^+}$, including $+\infty$, is not really required if all that interests us is the underlying open ball topology. Lemma 6.3.2 indeed immediately implies:

Corollary 6.3.4 (Bounded hemi-metrics) *A hemi-metric d' on a set X is* a-bounded *if and only if $d(x, y) \leqslant a$ for all pairs of points $x, y \in X$. It is* bounded *if and only if it is a-bounded for some real number $a \in \mathbb{R}^+$.*

For every $a > 0$, every hemi-metric d on X is topologically equivalent to some a-bounded hemi-metric on X, e.g., $\min(d(\_, \_), a)$.

6.3.1 Hemi-metrizable spaces and Wilson's Theorem

A question related to topological equivalence is the existence of a hemi-metric whose open ball topology is a given topology.

Definition 6.3.5 (Hemi-metrizability, metrizability) A topological space X is *hemi-metrizable* (resp., *metrizable*) if and only if there is a hemi-metric d (resp., a metric) such that the topology of X is exactly the open ball topology of d.

We note that a necessary condition for a space to be hemi-metrizable is first countability (see Exercise 4.7.14):

Lemma 6.3.6 *Every hemi-metric space is first-countable.*

Proof Direct application of Fact 6.1.7. □

Wilson's Theorem (Wilson, 1931; and see Theorem 6.3.13 below) will state that every *second*-countable space is hemi-metrizable. Exercise 6.3.9 below shows that this is a stronger property than first countability.

Definition 6.3.7 (Countably based, second-countable) A topological space is *second-countable*, a.k.a. *countably based*, if and only if it has a countable subbase of opens.

Exercise 6.3.8 Show that a topological space is countably-based if and only if it has a countable *base* of opens.

Exercise 6.3.9 Show that every countably-based space is first-countable.

Exercise 6.3.10 Show that the Sorgenfrey line $\mathbb{R}_\ell$ (Exercise 4.1.34) is not countably-based. Show also that $\mathbb{R}_\ell$ is hemi-metrizable, through the T_1 quasi-metric d_ℓ defined by $\mathsf{d}_\ell(r,s) = r - s$ if $r \geqslant s$, $\mathsf{d}_\ell(r,s) = +\infty$ if $r < s$. Conclude that there are first-countable spaces, and even quasi-metric spaces, that are not second-countable.

The main idea in the proof of Wilson's Theorem is to define topologies not through one hemi-metric, but by several simultaneously, and to realize that the case of countably many hemi-metrics reduces to just one:

Lemma 6.3.11 *Let X be a set, and $(d_i)_{i\in I}$ be a family of hemi-metrics on X. The topology* defined *by this family is the coarsest one containing the open balls $B^{d_i}_{x,<\epsilon}$, $x \in X$, $\epsilon \in \mathbb{R}^+ \smallsetminus \{0\}$, $i \in I$.*

The topology defined by a countable family of hemi-metrics (resp., quasi-metrics, resp., metrics) $(d_i)_{i\in I}$, where $I \subseteq \mathbb{N}$ without loss of generality, is hemi-metrizable (resp., quasi-metrizable, resp., metrizable), e.g., through $\bigvee_{i\in I} d_i$ defined by:

$$\left(\bigvee_{i\in I} d_i\right)(x,y) = \sup_{i\in I} \frac{1}{2^i}\min(d_i(x,y),1).$$

Proof Let $d = \bigvee_{i\in I} d_i$. It is easy to see that d is a hemi-metric (resp., quasi-metric, metric) whenever all the d_is are. To simplify things, we shall assume that each d_i is 1-bounded (Corollary 6.3.4), replacing d_i by $\min(d_i, 1)$ if necessary. So $d(x,y) = \sup_{i\in I} \frac{1}{2^i} d_i(x,y)$.

Given any $i \in I$, for every $x \in B^{d_i}_{x,<\epsilon}$, the open ball $B^d_{x,<\epsilon/2^i}$ is an open neighborhood of x included in $B^{d_i}_{x,<\epsilon}$, so $B^{d_i}_{x,<\epsilon}$ is open in $\mathcal{O}^d$ by Trick 4.1.11. So the topology defined by $(d_i)_{i\in I}$ is coarser than $\mathcal{O}^d$.

Conversely, we claim that $B^d_{x,<\epsilon}$ is open in the topology defined by $(d_i)_{i\in I}$. For every $y \in B^d_{x,<\epsilon}$, let $\eta < \epsilon - d(x,y)$. For $i \in I$ large enough, say $i \geqslant i_0$, we have $1/2^i < \eta$. Let $U = \bigcap_{i \in I, i < i_0} B^{d_i}_{y,<\eta}$. For every $z \in U$, for every $i < i_0$, $d_i(y, z) < \eta$, so $\frac{1}{2^i} d_i(x, z) < \frac{1}{2^i} d_i(x, y) + \eta \leqslant d(x, y) + \eta$; for every $i \geqslant i_0$, $\frac{1}{2^i} d_i(x, z) \leqslant \frac{1}{2^i} < \eta$. In any case, $\frac{1}{2^i} d_i(x, z) < d(x, y) + \eta$ for every $i \in I$, so $d(x, z) < \epsilon$. In other words, $U \subseteq B^d_{x,<\epsilon}$. By Trick 4.1.11, and since U is open in the topology defined by $(d_i)_{i \in I}$, $B^d_{x,<\epsilon}$ is open in this same topology. So this topology is finer than $\mathcal{O}^d$. □

Example 6.3.12 The quasi-metric d_ℓ on $\mathbb{R}_\ell$ is defined by the family consisting of the two quasi-metrics $\mathsf{d}_\mathbb{R}$ and $d_\geqslant$ on $\mathbb{R}$ (see Exercise 6.3.10).

Above, we restricted ourselves to countably many hemi-metrics. The topologies defined by uncountable families of metrics may fail to be metrizable. These are exactly the topologies arising from so-called *uniformities*, a topic that we won't touch: see (Bourbaki, 1971, chapter 2).

Theorem 6.3.13 (Wilson's Theorem) *Every countably-based topological space is hemi-metrizable.*

Proof Let $(U_n)_{n \in \mathbb{N}}$ be a countable subbase of the topology of the countably-based space X. (We may assume the indexes to range over the whole of $\mathbb{N}$, not just a subset, adding entries equal to $\emptyset$ if necessary. This is only meant to make the rest of the proof more readable.)

Let d_n be the hemi-metric defined by $d_n(x, y) = 1$ iff $x \in U_n$ and $y \notin U_n$, and $d_n(x, y) = 0$ otherwise. Alternatively, $d_n(x, y) = \max(\chi_{U_n}(x) - \chi_{U_n}(y), 0)$. This is a hemi-metric on X. Notably, the triangular inequality is obtained by: $d_n(x, y) + d_n(y, z) = \max(\chi_{U_n}(x) - \chi_{U_n}(y), 0) + \max(\chi_{U_n}(y) - \chi_{U_n}(z), 0) = \max(\chi_{U_n}(x) - \chi_{U_n}(z), \chi_{U_n}(x) - \chi_{U_n}(y), \chi_{U_n}(y) - \chi_{U_n}(z), 0) \geqslant d_n(x, z)$.

The open balls $B^{d_n}_{x,<\epsilon}$ are U_n if $\epsilon \leqslant 1$ and $x \in U_n$, or X itself otherwise. The topology defined by $(d_n)_{n \in \mathbb{N}}$ is then the topology generated by $(U_n)_{n \in \mathbb{N}}$. By Lemma 6.3.11, this is $\mathcal{O}^{\bigvee_{n \in \mathbb{N}} d_n}$. So, explicitly, the desired hemi-metric $d = \bigvee_{n \in \mathbb{N}} d_n$ is defined by $d(x, y) = \sup_{\substack{n \in \mathbb{N} \\ x \in U_n, y \notin U_n}} 1/2^n$. □

We shall refine this in Theorem 6.3.50. Note that Wilson's Theorem does not characterize the hemi-metrizable spaces: the Sorgenfrey line $\mathbb{R}_\ell$ is a quasi-metric space that is not second-countable (Exercise 6.3.10).

Exercise 6.3.14 (Ponomarev's Theorem) A kind of converse to Lemma 6.3.6 is provided by Ponomarev's Theorem (Ponomarev, 1960), which states that a T_0 space X is first-countable if and only if it is the continuous open image of a metric space, i.e., iff there is a metric space Y, d and a continuous, open, and surjective map $f: Y \to X$. Additionally, one may assume d to be a ultra-metric, hence Y to be zero-dimensional (see Exercise 6.1.23).

Prove this. In one direction, show that every continuous open image of a first-countable space is first-countable. In the other direction, assume X first-countable. Each point $x \in X$ has a descending countable neighborhood base $\vec{U} = (U_0 \supseteq U_1 \supseteq \cdots \supseteq U_n \supseteq \cdots)$, by Exercise 4.7.14. Let Y be the set of all descending countable neighborhood bases of points of X. Equip Y with the ultra-metric d defined by: $d(\vec{U}, \vec{V}) = 0$ if $\vec{U} = \vec{V}$, $1/2^n$ otherwise, where n is the largest number such that the first n components of $\vec{U}$ and $\vec{V}$ coincide. (Notice the similarity with $d^{sym}_{\Sigma^*}$; see Exercises 6.1.21 and 6.1.22.)

Show that, for every $\vec{U} \in Y$ as above, $\vec{U}$ is a neighborhood base of a unique point $x \in X$. You may show that the intersection of all elements of $\vec{U}$ is $\uparrow x$, where x is any point such that $\vec{U}$ is a neighborhood base of x. Let $f: Y \to X$ map $\vec{U}$ to this point x. Show that f is continuous, open, and surjective.

6.3.2 The Urysohn Property

A similar theorem for metrizability is the Urysohn-Tychonoff Theorem below. The construction is about the same, except that we must replace the $1/2^n$ summand obtained if $x \in U_n$ and $y \notin U_n$ by some quantity $1/2^n|f_n(x) - f_n(y)|$ where f_n is a continuous approximation of the characteristic function χ_{U_n} of U_n. (Note that χ_{U_n} is continuous from X to $[0, 1]_\sigma$, not to $[0, 1]$.) We need quite a bit of mathematics here, exploring relationships between regular, normal, and so-called paracompact spaces.

The first result we require is the celebrated Urysohn Lemma.

Theorem 6.3.15 (Urysohn Lemma) *Let X be a topological space. Then X is normal if and only if it satisfies the* Urysohn Property*: one can separate disjoint closed subsets F, F' by a continuous function, in the sense that if $F \cap F' = \emptyset$ then there is a continuous map $f: X \to [0, 1]$ such that $f(x) = 0$ for every $x \in F$, and $f(y) = 1$ for every $y \in F'$.*

Remember how we proved that every metric space was normal (Proposition 6.2.17): we actually exhibited such a continuous map f, and therefore showed that every metric space had the Urysohn Property.

Proof If X has the Urysohn Property, then we can separate any two disjoint closed subsets F, F' by the opens $U = f^{-1}[0, 1/3)$, $U' = f^{-1}(2/3, 1]$, where f is as above.

Conversely, assume X is normal. We shall use the following equivalent characterization of normality, called *interpolation*: for every closed subset F and

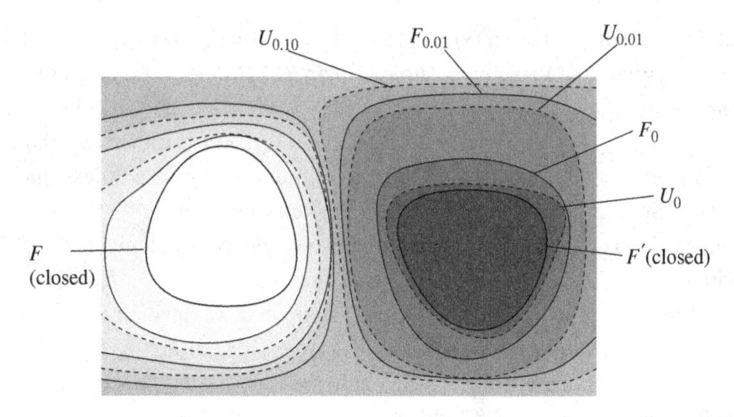

Figure 6.2 The Urysohn construction.

every open subset U such that $F \subseteq U$, there is a closed subset F_1 and an open subset U_1 such that $F \subseteq U_1 \subseteq F_1 \subseteq U$. This is close to the interpolation property obtained for locally compact spaces (Proposition 4.8.14), replacing compact saturated subsets by closed subsets.

Call *dyadic* any number α of the form $k/2^n$, with $k \in \mathbb{Z}$, $n \in \mathbb{N}$. Let I_2 be the set of all dyadic numbers in $[0, 1]$. Every element α of I_2 other than 1 can be written in binary notation as $0.b_1b_2 \dots b_n$, where each b_i, $1 \leqslant i \leqslant n$, is 0 or 1; i.e., $\alpha = \sum_{i=1}^{n} b_i 2^{-i}$. If n is least such that α can be written as above, we call n the *length* of α. In other words, the length of 0 is 0, and the length of $0.b_1b_2 \dots b_n$, where $b_n = 1$, is n. We also agree that the length of the number 1 is 0.

We build a family of open subsets U_α and a family of closed subsets F_α, $\alpha \in I_2$, such that $U_\alpha \subseteq F_\alpha$ for every $\alpha \in I_2$, such that $F_\alpha \subseteq U_\beta$ whenever $\alpha < \beta$, such that $F' \subseteq U_\alpha$ and $F_\alpha \subseteq X \smallsetminus F$ for every $\alpha \in I_2$. E.g., $F' \subseteq U_0 \subseteq F_0 \subseteq U_{0.01} \subseteq F_{0.01} \subseteq U_{0.10} \subseteq F_{0.10} \subseteq U_{0.11} \subseteq F_{0.11} \subseteq U_1 \subseteq F_1 \subseteq X \smallsetminus F$, looking only at U_α and F_α for α of length at most 2 (these are again written in binary). See Figure 6.2.

We build U_α and F_α by induction on the length of α. When α is of length 0, then $\alpha = 0$ or $\alpha = 1$. Let $U = X \smallsetminus F$. Since $F \cap F' = \emptyset$, $F' \subseteq U$. Since X is normal, use interpolation twice to obtain two closed subsets F_0 and F_1 and two open subsets U_0 and U_1 such that $F' \subseteq U_0 \subseteq F_0 \subseteq U_1 \subseteq F_1 \subseteq U$. When α is of length n, $n \geqslant 1$, we can write α in binary as $0.b_1b_2 \dots b_{n-1}b_n$, with $b_n = 1$. Let $\beta = 0.b_1b_2 \dots b_{n-1}0 = \alpha - 2^{-n}$, $\gamma = \alpha + 2^{-n}$. Both β and γ have lengths at most $n - 1$, so we have already built $U_\beta \subseteq F_\beta \subseteq U_\gamma \subseteq F_\gamma$. Use interpolation to find a closed subset F_α and an open subset U_α such that $F_\beta \subseteq U_\alpha \subseteq F_\alpha \subseteq U_\gamma$.

Let $f(x) = \sup_{\alpha \in I_2} (1-\alpha)\chi_{U_\alpha}$, and $g(x) = 1 - \sup_{\alpha \in I_2} \alpha \chi_{X \smallsetminus F_\alpha}$. We claim that $f = g$. First, $f(x) - g(x) = \sup_{\beta, \alpha \in I_2} [\beta \chi_{X \smallsetminus F_\beta}(x) + (1-\alpha)\chi_{U_\alpha}(x)] - 1$. When $\beta \leqslant \alpha$, $\beta \chi_{X \smallsetminus F_\beta}(x) + (1-\alpha)\chi_{U_\alpha}(x) \leqslant \beta + 1 - \alpha \leqslant 1$. Otherwise, $\alpha < \beta$, so $F_\alpha \subseteq U_\beta$, so $\beta \chi_{X \smallsetminus F_\beta}(x) + (1-\alpha)\chi_{U_\alpha}(x)$ is either less than or equal to β, if $x \notin F_\alpha$, or to $1 - \alpha$, if $x \in F_\alpha$, and in any case is less than or equal to 1. So $f(x) - g(x) \leqslant 1 - 1 = 0$. Therefore $f \leqslant g$.

Conversely, fix $x \in U_1 \smallsetminus F_0$, and define two sequences of elements $\alpha_n, \beta_n \in I_2$, with $\alpha_n \leqslant \beta_n$, $\beta_n \leqslant \alpha_n + (3/4)^n$, and such that $x \in U_{\beta_n} \smallsetminus F_{\alpha_n}$ for every $n \in \mathbb{N}$. First, let $\alpha_0 = 0$, $\beta_0 = 1$. Next, assume that we have already defined α_n and β_n. Let $\gamma = (3\alpha_n + \beta_n)/4$, $\delta = (\alpha_n + 3\beta_n)/4$, so $F_{\alpha_n} \subseteq U_\gamma \subseteq F_\gamma \subseteq U_\delta \subseteq F_\delta \subseteq U_{\beta_n}$. Since $x \in U_{\beta_n} \smallsetminus F_{\alpha_n}$, either $x \in U_{\beta_n} \smallsetminus F_\gamma$ or $x \in U_\delta \smallsetminus F_{\alpha_n}$. In the first case, let $\beta_{n+1} = \beta_n$ and $\alpha_{n+1} = \gamma$, else $\beta_{n+1} = \delta$ and $\alpha_{n+1} = \alpha_n$. In any case, $x \in U_{\beta_{n+1}} \smallsetminus F_{\alpha_{n+1}}$, $\alpha_{n+1} \leqslant \beta_{n+1}$, $\beta_{n+1} \leqslant \alpha_{n+1} + 3/4(\beta_n - \alpha_n) \leqslant \alpha_{n+1} + (3/4)^{n+1}$. So $f(x) - g(x) \geqslant \alpha_n + (1 - \beta_n) - 1 \geqslant -(3/4)^n$. Since this holds for every $n \in \mathbb{N}$, $f(x) - g(x) \geqslant 0$. So $f = g$ on $U_1 \smallsetminus F_0$.

We observe that, since $F_0 \subseteq U_\alpha$ for every $\alpha \in I_2$, $\alpha > 0$, for every $x \in F_0$, $f(x) = \sup_{\alpha \in I_2, \alpha > 0}(1 - \alpha) = 1$. Since $F_\alpha \subseteq U_1$ for every $\alpha \in I_2, \alpha < 1$, for every $x \in X \smallsetminus U_1$, $g(x) = 1 - \sup_{\alpha \in I_2} \alpha = 0$. So $f = g$ on the whole of X, and $f(x) = 1$ for $x \in F'$, $g(x) = 0$ for $x \in F$.

We now claim that f is continuous. The inverse image of $(t, 1]$, $0 \leqslant t \leqslant 1$, by f is $\{x \in X \mid \exists \alpha \in I_2 \cdot (1 - \alpha)\chi_{U_\alpha}(x) > t\} = \bigcup_{\alpha \in I_2, 1-\alpha > t} U_\alpha$, which is open. The inverse image of $[u, 1]$, $0 \leqslant u \leqslant 1$, by f (i.e., by g) is $\{x \in X \mid \sup_{\alpha \in I_2} \alpha \chi_{X \smallsetminus F_\alpha}(x) \leqslant 1 - u\} = \bigcap_{\alpha \in I_2, \alpha > 1-u} F_\alpha$, which is closed. (When the intersection is empty as well: in this case, the inverse image is the whole of X.) In particular, the inverse image of the basic open subset (t, u) of $[0, 1]$, which is $f^{-1}(t, 1] \smallsetminus f^{-1}[u, 1]$, is open. □

Exercise 6.3.16 ($T_4 \Rightarrow T_{3\frac{1}{2}} \Rightarrow T_3$) Show that every completely regular space is regular, and that every T_1 normal space is completely regular. In particular, every T_4 space is $T_{3\frac{1}{2}}$, and every $T_{3\frac{1}{2}}$ space is T_3.

The form of Urysohn's repeated interpolation argument deserves to be given a more abstract treatment – one that we will reuse later on. The proof of the following is as above, replacing "open" by "element of $\mathcal{U}$" and "closed" by "element of $\mathcal{F}$."

Proposition 6.3.17 (Dual scales and the Urysohn Property) *Let X be a set. A family $\mathcal{U}$ of subsets of X is an* upward scale *if and only if it contains the empty set, and, for every chain $(U_i)_{i \in I}$ of elements of $\mathcal{U}$, $\bigcup_{i \in I} U_i$ is in $\mathcal{U}$. A family $\mathcal{F}$ of subsets of X is a* downward scale *if and only if the family of complements of elements of $\mathcal{F}$ is an upward scale, i.e., iff it contains X itself as well as the intersection of any chain of elements of $\mathcal{F}$.*

A dual scale *is a pair* $(\mathcal{U}, \mathcal{F})$ *of an upward and a downward scale satisfying the following* interpolation *property: if* $F \in \mathcal{F}$ *and* $U \in \mathcal{U}$ *are such that* $F \subseteq U$*, then there is an* $F_1 \in \mathcal{F}$ *and an* $U_1 \in \mathcal{U}$ *such that* $F \subseteq U_1 \subseteq F_1 \subseteq U$.

Given any dual scale $(\mathcal{U}, \mathcal{F})$ *on a set* X*, if* $F \in \mathcal{F}$ *and* $U \in \mathcal{U}$ *are such that* $F \subseteq U$*, then there is map* $f : X \to [0, 1]$ *such that* $f(x) = 1$ *for every* $x \in F$*,* $f(x) = 0$ *for every* $x \notin U$*,* $f^{-1}(t, 1] \in \mathcal{U}$ *for every* $t \in [0, 1]$*, and* $f^{-1}[u, 1] \in \mathcal{F}$ *for every* $u \in [0, 1]$.

Exercise 6.3.18 (Urysohn's original metrization theorem) The original form of the Urysohn Metrization Theorem (Urysohn, 1925) states: every countably-based T_4 space is metrizable.

Although this will be a corollary of Proposition 6.3.39 below, prove it directly, imitating the proof of Wilson's Theorem 6.3.13. Given a countable base $(U_n)_{n\in\mathbb{N}}$, define $d_{mn}(x, y) = |f_{mn}(x) - f_{mn}(y)|$ for each pair of numbers m, n such that $cl(U_m) \subseteq U_n$, and where f_{mn} is a continuous map from X to $[0, 1]$ such that $f_{mn}(x) = 0$ for every $x \in cl(U_m)$ and $f_{mn}(x) = 1$ for every $x \in X \setminus U_n$. Then show that the topology defined by $(d_{mn})_{\substack{m,n\in\mathbb{N}\\ cl(U_m)\subseteq U_n}}$ coincides with the original topology of X.

Exercise 6.3.19 Let X be a set. For every map $f : X \to \mathbb{R}$, show that $\mathcal{U}_f = (f^{-1}(t, +\infty))_{t\in\mathbb{R}} \cup \{\emptyset, X\}$ is an upward scale, that $\mathcal{F}_f = (f^{-1}[t, +\infty))_{t\in\mathbb{R}} \cup \{\emptyset, X\}$ is a downward scale, and that $(\mathcal{U}_f, \mathcal{F}_f)$ is a dual scale.

Exercise 6.3.20 (Separation of F_σ and G_δ subsets) A G_δ *subset* of a topological space X is the intersection of countably many open subsets. An F_σ *subset* is the union of countably many closed subsets, i.e., the complement of a G_δ subset.

Show that, if X is normal, H is an F_σ subset of X, G is a G_δ subset of X, and $H \subseteq G$, then there is a continuous map $f : X \to [0, 1]$ such that $f(x) = 0$ for every $x \in H$ and $f(y) = 1$ for every y outside G. As a hint, given countably many maps $f_n : X \to [0, 1]$, show that $f(x) = \sum_{n\in\mathbb{N}} \frac{1}{2^{n+1}} f_n(x)$ (i.e., the least upper bound of $\left(\sum_{n=0}^{m} \frac{1}{2^{n+1}} f_n(x)\right)_{m\in\mathbb{N}}$) defines a continuous map; also observe that $\sum_{n\in\mathbb{N}} \frac{1}{2^{n+1}} = 1$.

As a corollary, show that in a normal space X, if H is an F_σ subset of a G_δ subset G, then there is an open subset U and a closed subset F of X such that $H \subseteq U \subseteq F \subseteq G$.

Exercise 6.3.21 (Katětov–Tong Insertion Theorem) Let X be a normal space. Given an lsc map $h : X \to [0, 1]$ (see Exercise 4.3.11), let H_α be the closed set $h^{-1}(-\infty, \alpha]$, for each dyadic number $\alpha \in I_2$. Similarly, given a usc map $g : X \to [0, 1]$, let G_α be the open set $g^{-1}(-\infty, \alpha)$, for each $\alpha \in I_2$.

Assume that $g \leqslant h$. Build open subsets U_α and closed subsets F_α such that $U_\alpha \subseteq F_\alpha$ and, whenever $\alpha < \beta$ are in I_2, $H_\alpha \subseteq U_\beta$, $F_\alpha \subseteq U_\beta$, and $F_\alpha \subseteq G_\beta$. Exercise 6.3.20, last part, should help: e.g., $H'_\alpha = \bigcup_{\alpha' \in I_2, \alpha' < \alpha} H_{\alpha'}$ is F_σ and $G'_\beta = \bigcap_{\beta' \in I_2, \beta' > \beta} G_{\beta'}$ is G_δ. The construction is otherwise similar to the Urysohn Lemma (Theorem 6.3.15).

Deduce the *Katětov–Tong Insertion Theorem*: X is a normal space if and only if, for any two maps $g, h\colon X \to \mathbb{R} \cup \{-\infty, +\infty\}$ with g usc, h lsc, and $g \leqslant h$, there is a continuous map f from X to $\mathbb{R} \cup \{-\infty, +\infty\}$ (with the topology of Exercise 4.10.7) such that $g \leqslant f \leqslant h$.

Exercise 6.3.22 (Scale insertion) Let X be a set. Given an upward scale $\mathcal{U}$ on X, we say that a map $h\colon X \to \mathbb{R} \cup \{-\infty, +\infty\}$ is *$\mathcal{U}$-lsc* iff $h^{-1}(t, +\infty) \in \mathcal{U}$ for every $t \in \mathbb{R}$. Given a downward scale $\mathcal{F}$ on X, a map $g\colon X \to \mathbb{R} \cup \{-\infty, +\infty\}$ is *$\mathcal{F}$-usc* iff $g^{-1}[t, +\infty) \in \mathcal{F}$ for every $t \in \mathbb{R}$. A map that is both $\mathcal{U}$-lsc and $\mathcal{F}$-usc is *$(\mathcal{U}, \mathcal{F})$-continuous.*

An upward scale $\mathcal{U}$ is *stable* if and only the intersection of finitely many elements of $\mathcal{U}$ is again in $\mathcal{U}$. A downward scale $\mathcal{F}$ is stable if and only if it is the set of complements of elements of a stable upward scale, i.e., iff the union of finitely many elements of $\mathcal{F}$ is again in $\mathcal{F}$. A dual scale $(\mathcal{U}, \mathcal{F})$ is stable if and only if both $\mathcal{U}$ and $\mathcal{F}$ are stable.

Show that if $(\mathcal{U}, \mathcal{F})$ is a stable dual scale on X, g is a $\mathcal{F}$-usc map and h is a $\mathcal{U}$-lsc map such that $g \leqslant h$, then there is a $(\mathcal{U}, \mathcal{F})$-continuous map f such that $g \leqslant f \leqslant h$.

Exercise 6.3.23 (Tietze–Urysohn Extension Theorem) Let X be a normal space, F be a closed subset of X, and f be a continuous map from F (with the subspace topology) to some closed interval $[a, b]$. Show that f has a continuous extension to the whole of X, i.e., there is a continuous map f' from X to $[a, b]$ that coincides with f on F. To show this, find a usc extension g of f to the whole of X, and an lsc extension h of f to the whole of X such that $g \leqslant h$, and apply the Katětov–Tong Insertion Theorem.

Using Exercise 4.10.7, show the *Tietze–Urysohn Extension Theorem*: let X be a normal space, F be a closed subset of X, then every continuous map f from F to $\mathbb{R} \cup \{-\infty, +\infty\}$ has a continuous extension to the whole of X.

6.3.3 Paracompact spaces

We shall see that metric spaces are not just normal, but also paracompact. This new notion is instrumental in solving the metrization question, and is also interesting in itself. Here is the definition, which should remind you of compactness: replace subcovers by refinements, and finiteness by local finiteness.

Definition 6.3.24 (Paracompact) Let X be a space. A family of subsets $(V_j)_{j \in J}$ is a *refinement* of, or *refines*, a family of subsets $(U_i)_{i \in I}$ if and only if, for every $j \in J$, there is an $i \in I$ such that $V_j \subseteq U_i$.

A family of subsets $(U_i)_{i \in I}$ is *locally finite* in X if and only if every point $x \in X$ has an open neighborhood U that intersects only finitely many U_is, i.e., such that $\{i \in I \mid U \cap U_i \neq \emptyset\}$ is finite.

A space X is *paracompact* if and only if every open cover $(U_i)_{i \in I}$ of X is refined by some locally finite open cover.

Example 6.3.25 Clearly, every compact space is paracompact. The converse fails: as we shall see (Theorem 6.3.31), every metric space, e.g., $\mathbb{R}$, even non-compact, is paracompact.

Paracompactness is a stronger property than normality, up to a few details (Dieudonné, 1944):

Proposition 6.3.26 (Dieudonné's Theorem) *Every paracompact T_2 space is T_4.*

Proof The proof is similar to Proposition 4.4.17. Let X be paracompact T_2. We first show that X is regular. Let F be closed in X, $x \in X \smallsetminus F$. For every $y \in F$, there is an open neighborhood U_{xy} of x and an open neighborhood V_{xy} of y such that $U_{xy} \cap V_{xy} = \emptyset$, since X is T_2. (Pick them using the Axiom of Choice.) Consider the family of all V_{xy}, $y \in F$, plus $X \smallsetminus F$: this is an open cover of X. Since X is paracompact, it is refined by a locally finite open cover $(W_j)_{j \in J}$. Let J' be the subset of all $j \in J$ such that $W_j \not\subseteq X \smallsetminus F$. Then $(W_j)_{j \in J'}$ is an open cover of F: every $x \in F$ is in some W_j, $j \in J$, and clearly W_j cannot be included in $X \smallsetminus F$. Let $V = \bigcup_{j \in J'} W_j$—an open subset containing F. Then, for every $j \in J'$, W_j is included in V_{xy} for some $y \in F$. Using the Axiom of Choice, pick $y_j \in F$ such that $W_j \subseteq V_{xy_j}$ for each $j \in J$. Since $(W_j)_{j \in J}$ is locally finite, x has an open neighborhood U_0 such that $J'' = \{j \in J \mid U_0 \cap W_j \neq \emptyset\}$ is finite. Let $U = U_0 \cap \bigcap_{j \in J''} U_{xy_j}$. We claim that $U \cap V$ is empty: for any point z of $U \cap V$, since $z \in V$, z is in W_j for some $j \in J'$, hence in V_{xy_j}, hence outside U_{xy_j}; since $z \in U$, j cannot be in J'' (otherwise z would be in U_{xy_j}), so $U_0 \cap W_j$ is empty; since $z \in W_j$, z cannot be in U_0, hence in U, a contradiction. To sum up, we have found an open neighborhood U of x, and an open subset V containing F, and $U \cap V$ is empty. So X is regular.

Let F, F' be two disjoint closed subsets of X. For each $x \in F$, there is an open neighborhood U_x of x and an open subset V_x containing F' such that $U_x \cap V_x = \emptyset$, by regularity. (Again, use the Axiom of Choice.) Find a locally finite open cover $(W_j)_{j \in J}$ that refines the open cover consisting of all

U_x, $x \in F$, plus $X \smallsetminus F$. Let J' be the set of all indices $j \in J$ such that F intersects W_j. For each $j \in J'$, there must be a point $x_j \in F$ such that $W_j \subseteq U_{x_j}$; otherwise W_j would be included in $X \smallsetminus F$, a contradiction. Let $U = \bigcup_{j \in J'} W_j$. This is an open subset containing F. Indeed, otherwise there would be a point $x \in F$ such that $x \in W_j$ for some $j \in J$, but j cannot be taken in J'; then $W_j \cap F = \emptyset$, a contradiction.

We now build an open subset V containing F' whose intersection with U is empty. For every $y \in F'$, there is an open neighborhood V'_y of y that meets only finitely many W_j, $j \in J$. Let J'_y be the set of indices $j \in J$ such that W_j meets both V'_y and F. Note that J'_y is finite, so $V_y = V'_y \cap \bigcap_{j \in J'_y} V_{x_j}$ is open.

We claim that $V_y \cap U$ is empty. Otherwise, let $z \in V_y \cap U$. By definition of U, z is in W_j for some $j \in J'$, hence in U_{x_j}, hence outside V_{x_j}. Since $z \in V_y \subseteq \bigcap_{j \in J'_y} V_{x_j}$, j cannot be in J'_y. So $W_j \cap V'_y$ or $W_j \cap F$ is empty. Since $z \in W_j$ and $z \in V_y \subseteq V'_y$, the first case is impossible. However, in the second case j is not in J', another contradiction.

So, for each $y \in F'$, we have found an open neighborhood V_y of y that does not meet U. Now take $V = \bigcup_{y \in F'} V_y$. □

The real reason why paracompact spaces matter is the existence of so-called partitions of unity subordinate to any open cover. We shall make this precise in Proposition 6.3.35 below.

Definition 6.3.27 (Partition of unity) A *partition of unity* on a given topological space X is a family $(f_i)_{i \in I}$ of continuous maps from X to $[0, 1]$ such that $\sum_{i \in I} f_i(x) = 1$ for every $x \in X$.

Such a family is *weakly subordinate* to a cover $(U_i)_{i \in I}$ of X if and only if $f_i^{-1}(0, 1] \subseteq U_i$ for every $i \in I$, and *subordinate* to $(U_i)_{i \in I}$ iff $cl(f_i^{-1}(0, 1]) \subseteq U_i$ for every $i \in I$.

Again we insist that f_i be continuous from X to $[0, 1]$, not $[0, 1]_\sigma$. The sum $\sum_{i \in I} f_i(x)$ is defined as the least upper bound of all the finite sums $\sum_{i \in J} f_i(x)$, where J ranges over all finite subsets of I. This expression is much simpler when the partition of unity is (weakly) subordinate to a locally finite open cover: then only finitely many summands are non-zero, i.e., the sum $\sum_{i \in I} f_i(x)$ really is a finite sum.

An illustration is given in Figure 6.3, where we have drawn a (finite) partition of unity $f_1, \ldots, f_5$ subordinate to an open cover $U_1, \ldots, U_5$.

If a partition of unity is subordinate to an open cover, it is also weakly subordinate to the latter. The converse fails, but we can always massage a weakly

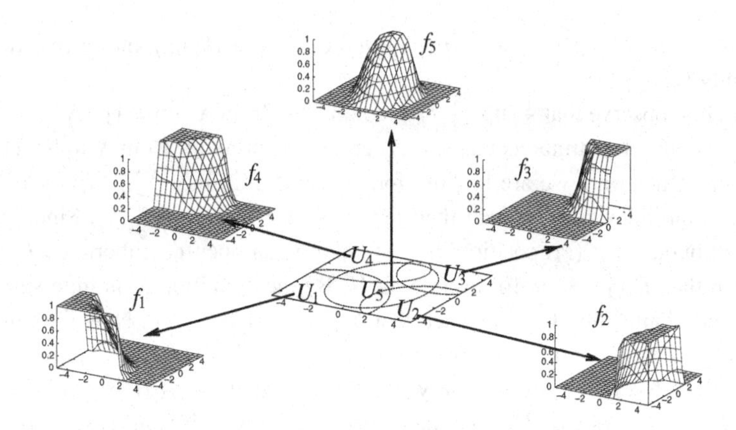

Figure 6.3 A partition of unity on five open subsets of a square.

subordinate one into a subordinate one, as shown in Lemma 6.3.30 below. We start with an important observation.

Fact 6.3.28 *Let X, Y be two topological spaces, and $f : X \to Y$ be any map. Assume that, for every $x \in X$, there is a continuous map $f_x : X \to Y$, and an open neighborhood U_x of x such that f and f_x coincide on U_x. Then f is continuous.*

Indeed, the assumption implies that f is continuous at every point $x \in X$, and we apply Proposition 4.3.14.

A typical application is as follows.

Exercise 6.3.29 (Gluing along a partition of unity) Let $(U_i)_{i \in I}$ be a locally finite open cover of X, and assume a family of continuous maps $g_i : U_i \to \mathbb{R}$, $i \in I$. Note that each g_i is only defined on the "small" set U_i, not on the whole of X. If $(f_i)_{i \in I}$ is a partition of unity that is subordinate to $(U_i)_{i \in I}$, then we can glue together the f_is, and define the map $f = \sum_{i \in I} f_i g_i$ from X to $\mathbb{R}$, where g_i is first extended to the whole of X by letting it take an arbitrary (and irrelevant) real value outside U_i. Show that $f_i g_i$ is then continuous from the whole of X to $\mathbb{R}$. Using Fact 6.3.28, deduce that f must be continuous.

Lemma 6.3.30 *Let X be a topological space, and $(f_i)_{i \in I}$ be a partition of unity that is weakly subordinate to some open cover $(U_i)_{i \in I}$. Then there is a partition of unity $(g_i)_{i \in I}$ that is subordinate to $(U_i)_{i \in I}$, and which is locally finite in the sense that $(cl(g_i^{-1}(0, 1]))_{i \in I}$ is locally finite.*

Proof The idea is to define $h_i(x) = \max(f_i(x) - \frac{1}{2} \sup_{j \in I} f_j(x), 0)$. By continuity arguments, we shall see that $(cl(h_i^{-1}(0, 1]))_{i \in I}$ is locally finite.

We then define $g_i(x) = h_i(x)/\sum_{j\in I} h_j(x)$ to turn $(h_i)_{i\in I}$ into a partition of unity $(g_i)_{i\in I}$.

We first observe that $\frac{1}{2}\sup_{j\in I} f_j$ is continuous from X to $[0, 1]$. (As a least upper bound of continuous maps, it is certainly continuous from X to $[0, 1]_\sigma$, but this is not what we are looking for.) Fix $x \in X$. Since $\sum_{i\in I} f_i(x) = 1$, there is an index $i_0 \in I$ such that $f_{i_0}(x) > 0$. Let $\epsilon = \frac{1}{4}f_{i_0}(x)$. Since f_{i_0} is continuous at x (Proposition 4.3.14), there is an open neighborhood U of x such that $f_{i_0}(y) > 2\epsilon$ for every $y \in U$. By the definition of infinite sums, there is a finite subset J of I such that $\sum_{i\in J} f_i(x) > 1 - \epsilon$. Now the finite sum $\sum_{i\in J} f_i$ is continuous, so there is an open neighborhood V of x such that $\sum_{i\in J} f_i(y) > 1 - \epsilon$ for every $y \in V$; in particular, $f_i(y) < \epsilon$ for every $i \in I \setminus J$. So, if $y \in U \cap V$, then $\sup_{j\in I} f_j(y)$, which is at least 2ϵ (take $j = i_0$), equals $\max_{j\in J} f_j(y)$. The function that maps y to $\max_{j\in J} f_j(y)$ is continuous, and coincides with $\sup_{j\in I} f_j$ on the open neighborhood $U \cap V$ of x. By Fact 6.3.28, $\sup_{j\in I} f_j$ is continuous.

The function h_i defined above is then continuous, too, from X to $[0, 1]$. We claim that $(cl(h_i^{-1}(0, 1]))_{i\in I}$ is locally finite. For every $x \in X$, taking U, V, ϵ, and J as above, for every $y \in U \cap V$, for every $i \in I \setminus J$, $f_i(y) < \epsilon < \frac{1}{2}\sup_{j\in I} f_j(y)$, so h_i is zero on $U \cap V$; i.e., $h_i^{-1}(0, 1] \cap (U \cap V)$ is empty, from which $cl(h_i^{-1}(0, 1]) \cap (U \cap V) = \emptyset$, using Corollary 4.1.28. In other words, the open neighborhood $U \cap V$ of x meets $cl(h_i^{-1}(0, 1])$ only when i is in the finite set J.

We now claim that $(h_i)_{i\in I}$ is subordinate to $(U_i)_{i\in I}$. Assume otherwise, i.e., that, for some $i \in I$, there is a point $x \in X$ outside U_i but in $cl(h_i^{-1}(0, 1])$. Since $f_i^{-1}(0, 1] \subseteq U_i$, $f_i(x)$ must equal 0. So $f_i(x) - \frac{1}{2}\sup_{j\in I} f_j(x) < 0$, whence $f_i - \frac{1}{2}\sup_{j\in I} f_j$ is below 0 on some open neighborhood of x by continuity; in other words, there is an open neighborhood W of x on which h_i is identically 0. Since $x \in cl(h_i^{-1}(0, 1])$, W must intersect $h_i^{-1}(0, 1]$ by Corollary 4.1.28, a contradiction.

Since $(cl(h_i^{-1}(0, 1]))_{i\in I}$ is locally finite, for every $x \in X$, $\sum_{j\in I} h_j(x)$ is a finite sum, and is therefore well defined. Moreover, the map that sends x to $\sum_{j\in I} h_j(x)$ is then continuous, as it coincides with a continuous finite sum of continuous maps on each small enough open subset. By Example 4.5.8, division is continuous at every point where it is defined, so g_i is continuous. Clearly, $\sum_{i\in I} g_i = 1$, and $(g_i)_{i\in I}$ satisfies the announced claims. □

The following is due to the British mathematician Arthur Harold Stone (Stone, 1948), not to be confused with the American mathematician Marshall Harvey Stone – the Stone of the so-called Stone duality, which we shall see in Chapter 8.

Theorem 6.3.31 (A.H. Stone) *Every metric space is paracompact in its open ball topology.*

Proof Let $(U_i)_{i \in I}$ be an open cover of the metric space X, d. We may assume d bounded by 1, by Corollary 6.3.4. The map $d(\_, X \smallsetminus U_i)$ is then 1-Lipschitz from X to $[0, 1]$ (with metric $\mathrm{d}_{\mathbb{R}}^{sym}$) by Lemma 6.2.16.

We shall show that: $(*)$ there is a family of continuous maps $(g_i)_{i \in I}$ from X to $[0, 1]$ such that $g_i^{-1}(0, 1] \subseteq U_i$ and $\sum_{i \in I} g_i(x) = \sup_{i \in I} d(x, X \smallsetminus U_i)$ for every $x \in X$. To do so, let $\mathcal{G}$ be the family of all pairs $(J, (g_j)_{j \in J})$ where $J \subseteq I$, each g_j, $j \in J$, is continuous from X to $[0, 1]$, $g_j^{-1}(0, 1] \subseteq U_j$ for every $j \in J$, and $\sum_{j \in J} g_j(x) = \sup_{j \in J} d(x, X \smallsetminus U_j)$ for every $x \in X$. Order $\mathcal{G}$ by extension, i.e., let $(J, (g_j)_{j \in J}) \sqsubseteq (J', (g'_j)_{j \in J'})$ iff $J \subseteq J'$ and $g_j = g'_j$ for every $j \in J$.

We claim that $\mathcal{G}$ is inductive. Given any chain $(J_k, (g_j^k)_{j \in J_k})_{k \in K}$, let $J = \bigcup_{k \in K} J_k$, and for every $j \in J$, let $g_j = g_j^k$ for any given $k \in K$ such that $j \in J_k$ (then g_j^k is independent of the chosen k: if j is both in J_k and in $J_{k'}$, then $J_k \subseteq J_{k'}$ or $J_{k'} \subseteq J_k$ and $g_j^k = g_j^{k'}$). Clearly each g_j is continuous, and $g_j^{-1}(0, 1] \subseteq U_j$. Next, for every $x \in X$, $\sum_{j \in J} g_j(x) = \sup_{J' \text{ finite } \subseteq J} \sum_{j \in J'} g_j(x)$. Since any finite subset J' of J must be contained in some J_k, this is equal to $\sup_{k \in K} \sup_{J' \text{ finite } \subseteq J_k} \sum_{j \in J'} g_j(x) = \sup_{k \in K} \sum_{j \in J_k} g_j(x) = \sup_{k \in K} \sup_{j \in J_k} d(x, X \smallsetminus U_j) = \sup_{j \in J} d(x, X \smallsetminus U_j)$.

$\mathcal{G}$ is trivially non-empty. By Zorn's Lemma (Theorem 2.4.2), there is a maximal element $(J, (g_j)_{j \in J})$ in $\mathcal{G}$. Assume $J \neq I$, and pick i from $I \smallsetminus J$. Then let $J' = J \cup \{i\}$, $g'_j = g_j$ for every $j \in J$ and $g'_i(x) = \sup_{j \in J'} d(x, X \smallsetminus U_j) - \sup_{j \in J} d(x, X \smallsetminus U_j)$. (This is the central construction.) The map $\sup_{j \in J} d(\_, X \smallsetminus U_j)$ is 1-Lipschitz: since $d(\_, X \smallsetminus U_j)$ is 1-Lipschitz for every $j \in J$, $d(x, X \smallsetminus U_j) \leqslant d(y, X \smallsetminus U_j) + |x - y|$ for every $x, y \in X$, so $\sup_{j \in J} d(x, X \smallsetminus U_j) \leqslant \sup_{j \in J} d(x, X \smallsetminus U_j) + |x - y|$. Similarly with J' instead of J. So g'_i is 2-Lipschitz, hence continuous (Fact 6.2.3). Next, $\sum_{j \in J'} g'_j(x) = \sup_{J'' \text{ finite} \subseteq J} \sum_{j \in J''} g'_j(x) = \sup_{J'' \text{ finite} \subseteq J} \sum_{j \in J'' \cup \{i\}} g'_j(x)$ (since every finite sum indexed by some $J'' \subseteq J$ is less than or equal to the same sum indexed by $J'' \cup \{i\}$) $= g'_i(x) + \sup_{J'' \text{ finite} \subseteq J} \sum_{j \in J''} g'_j(x) = g'_i(x) + \sum_{j \in J} g'_j(x) = g'_i(x) + \sup_{j \in J} d(x, X \smallsetminus U_j) = \sup_{j \in J' \cup \{i\}} d(x, X \smallsetminus U_j)$. So $(J', (g'_j)_{j \in J'})$ is a strictly larger element of $\mathcal{G}$, leading to a contradiction. So $(*)$ is proved.

The map g defined by $g(x) = \sum_{i \in I} g_i(x) = \sup_{i \in I} d(x, X \smallsetminus U_i)$ is 1-Lipschitz, hence continuous. Also, $g(x) = 0$ iff, for every $i \in I$, $d(x, X \smallsetminus U_i) = 0$, iff $x \in \bigcap_{i \in I} (X \smallsetminus U_i)$ (using Lemma 6.1.11) $= X \smallsetminus \bigcup_{i \in I} U_i = \emptyset$. So $g(x)$ does not take the value zero. Then $(g_i / g)_{i \in I}$ is a partition of unity,

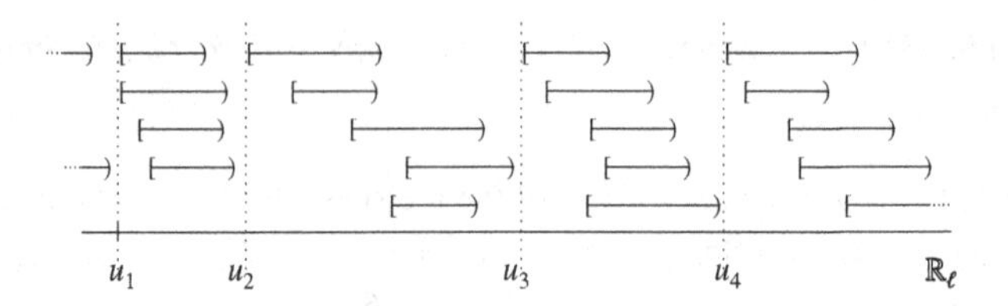

Figure 6.4 Open covers in the Sorgenfrey line $\mathbb{R}_\ell$.

and since $g^{-1}(0, 1] \subseteq U_i$ for each $i \in I$, it is weakly subordinate to $(U_i)_{i \in I}$. By Lemma 6.3.30, there is a partition of unity $(g_i)_{i \in I}$ that is subordinate to $(U_i)_{i \in I}$, and which is locally finite. Then $(g_i^{-1}(0, 1])_{i \in I}$ is a locally finite open cover that refines $(U_i)_{i \in I}$. □

Exercise 6.3.32 ($\mathbb{R}_\ell$ is paracompact) Show that the Sorgenfrey line $\mathbb{R}_\ell$ (Exercise 4.1.34) is paracompact, hence T_4. To show that it is paracompact, consider any open cover of $\mathbb{R}_\ell$ by basic open sets $([a_i, b_i))_{i \in I}$.

One should first observe that the general form of such a cover looks like Figure 6.4: there are countably many reals u_n, $n \in N$ (with $N \subseteq \mathbb{N}$) which separate the real line in open intervals (u_n, v_n) (where possibly $u_n = -\infty$, and either $v_n = +\infty$ or $v_n = u_m$ for some other $m \in N$) plus the endpoints u_n, $n \in N$, and each $[a_i, b_i)$ falls into one of these intervals. To show this, consider $U = \bigcup_{i \in I} (a_i, b_i)$, and use Exercise 4.1.32.

Let J_n be the set of indices $i \in I$ such that $(a_i, b_i) \subseteq (u_n, v_n)$. If $u_n = -\infty$, use the fact that $(-\infty, v_n)$ is paracompact in the subspace topology of $\mathbb{R}$ (why is this so?) to obtain a locally finite open refinement of $(a_i, b_i)_{i \in J_n}$ in $(-\infty, v_n)$ with the subspace topology of $\mathbb{R}$ and that is a cover of $(-\infty, v_n)$; then show that it is also a locally finite open cover of $(-\infty, v_n)$ with the subspace topology of $\mathbb{R}_\ell$. If $u_n \neq -\infty$, show that there is an index $j \in J_n$ such that $a_j = u_n$ and $b_j \leqslant v_n$, and use the fact that $[b_j, v_n)$ is paracompact in the subspace topology of $\mathbb{R}$ (again, why is this so?) and adapt the previous argument.

Exercise 6.3.33 (Sorgenfrey plane, products of paracompact spaces, of normal spaces) The *Sorgenfrey plane* is the space $\mathbb{R}_\ell \times \mathbb{R}_\ell$. Consider the subset $L = \{(x, y) \in \mathbb{R}_\ell \times \mathbb{R}_\ell \mid x + y = 0\}$. Show that L is closed in $\mathbb{R}_\ell \times \mathbb{R}_\ell$. As a subspace of $\mathbb{R}_\ell \times \mathbb{R}_\ell$, show that the topology of L is discrete. Deduce that every subset of L is closed in $\mathbb{R}_\ell \times \mathbb{R}_\ell$. By looking at the subset of the points of L with rational coordinates, and its complement, conclude that the Sorgenfrey plane is not normal.

Deduce that the product of two paracompact T_2 spaces may fail to be paracompact, and that the product of two normal (even T_4) spaces may fail to be normal.

Lemma 6.3.34 *Let X be a normal space. Every locally finite open cover $(U_i)_{i\in I}$ of X has a subordinate partition of unity.*

Proof Let $\mathcal{G}$ be the family of all pairs $(J, (g_j)_{j\in J})$ where $J \subseteq I$, each g_j, $j \in J$, is continuous from X to $[0, 1]$, $g_j^{-1}(0, 1] \subseteq U_j$ for each $j \in J$, and $(g_j^{-1}(0, 1])_{j\in J}$ is a cover of $X \smallsetminus \bigcup_{i\in I\smallsetminus J} U_i$. Order $\mathcal{G}$ by extension, i.e., let $(J, (g_j)_{j\in J}) \sqsubseteq (J', (g'_j)_{j\in J'})$ iff $J \subseteq J'$ and $g_j = g'_j$ for every $j \in J$.

We claim that $\mathcal{G}$ is inductive. Given any chain $(J_k, (g_j^k)_{j\in J_k})_{k\in K}$, let $J = \bigcup_{k\in K} J_k$, and, for every $j \in J$, let $g_j = g_j^k$ for any given $k \in K$ such that $j \in J_k$ (then g_j^k is independent of the chosen k). We must show that $(g_j^{-1}(0, 1])_{j\in J}$ is a cover of $X \smallsetminus \bigcup_{i\in I\smallsetminus J} U_i$. Let $x \in X \smallsetminus \bigcup_{i\in I\smallsetminus J} U_i$. Since $(U_i)_{i\in I}$ is locally finite, there is an open neighborhood U of x that meets only finitely many U_js, $j \in I$. In particular x itself lies in finitely many U_js, say $U_{j_1}, \ldots, U_{j_\ell}$, where $j_1, \ldots, j_\ell \in J$, and no other. Since $(J_k, (g_j^k)_{j\in J_k})_{k\in K}$ is a chain, $j_1, \ldots, j_\ell$ are all in some J_k, $k \in K$. In particular, x cannot be in U_i for $i \notin J_k$, i.e., x is in $X \smallsetminus \bigcup_{i\in I\smallsetminus J_k} U_i$. But $(g_j^{-1}(0, 1])_{j\in J_k}$ is a cover of the latter, so $x \in g_j^{-1}(0, 1]$ for some $j \in J_k$, in particular $j \in J$.

By Zorn's Lemma (Theorem 2.4.2), there is a maximal element $(J, (g_j)_{j\in J})$ in $\mathcal{G}$. Assume $J \neq I$, and pick i from $I \smallsetminus J$. Let $J' = J \cup \{i\}$. Let F be the complement of U_i, F' be the complement of $\bigcup_{j\in J} g_j^{-1}(0, 1] \cup \bigcup_{j\in I\smallsetminus J'} U_j$. Both F and F' are closed, and are disjoint: this means that $U_i \cup (\bigcup_{j\in J} g_j^{-1}(0, 1] \cup \bigcup_{j\in I\smallsetminus J'} U_j) = \bigcup_{j\in J} g_j^{-1}(0, 1] \cup \bigcup_{j\in I\smallsetminus J} U_j$ is the whole of X, which follows from the fact that $\bigcup_{j\in J} g_j^{-1}(0, 1]$ is a cover of $X \smallsetminus \bigcup_{j\in I\smallsetminus J} U_j$.

By Urysohn's Theorem 6.3.15, there is a continuous map $f: X \to [0, 1]$ such that $f(x) = 0$ for every $x \in F$ and $f(x) = 1$ for every $x \in F'$. Define $g'_j = g_j$ for every $j \in J$, and $g'_i = f$. Then $(g_j'^{-1}(0, 1])_{j\in J'}$ is a cover of $X \smallsetminus \bigcup_{j\in I\smallsetminus J'} U_j$: for every $x \in X \smallsetminus \bigcup_{j\in I\smallsetminus J'} U_j$, either x is in some $g_j^{-1}(0, 1] = g_j'^{-1}(0, 1]$ for some $j \in J$, or $x \in (X \smallsetminus \bigcup_{j\in I\smallsetminus J'} U_j) \cap (X \smallsetminus \bigcup_{j\in J} g_j^{-1}(0, 1]) = F'$. In the latter case, $g'_i(x) = f(x) = 1$, so $x \in g_i'^{-1}(0, 1]$. Moreover, by construction $(J, (g_j)_{j\in J}) \sqsubseteq (J', (g'_j)_{j\in J'})$, which contradicts the maximality of $(J, (g_j)_{j\in J})$.

So $J = I$. That is, there is a family $(g_i)_{i\in I}$ of continuous maps from X to $[0, 1]$ such that $g_i^{-1}(0, 1] \subseteq U_i$ and $(g_i^{-1}(0, 1])_{i\in I}$ is a cover of X. In particular, $g(x) = \sum_{i\in I} g_i(x)$ is never 0. Moreover, since $(U_i)_{i\in I}$ is locally finite, so is $(g_i^{-1}(0, 1])_{i\in I}$, so g is defined on some neighborhood of each point x as a finite sum. By Fact 6.3.28, g is continuous. The family $(g_i/g)_{i\in I}$ is then a partition of unity that is weakly subordinate to $(U_i)_{i\in I}$. We conclude by Lemma 6.3.30. ☐

Paracompactness allows us to state a result similar to Lemma 6.3.34, without the assumption that the cover $(U_i)_{i \in I}$ is locally finite.

Proposition 6.3.35 (Paracompact T_2 = open covers have subordinate partitions of unity)
Let X be a T_2 topological space. X is paracompact if and only if every open cover $(U_i)_{i \in I}$ of X has a subordinate partition of unity.

Proof Let $(U_i)_{i \in I}$ be an open cover of X. If X is paracompact, there is a locally finite cover $(V_j)_{j \in J}$ that refines $(U_i)_{i \in I}$. For each $j \in J$, use the Axiom of Choice to pick an index $i_j \in I$ such that $V_j \subseteq U_{i_j}$. Since X is paracompact and T_2, X is normal (Proposition 6.3.26), so $(V_j)_{j \in J}$ has a subordinate partition of unity $(g_j)_{j \in J}$ by Lemma 6.3.34. For each $i \in I$, let $f_i(x) = \sum_{\substack{j \in J \\ i_j = i}} g_j(x)$. Every $x \in X$ has an open neighborhood U_x that meets $cl(g_j^{-1}(0, 1])$ for only finitely many values of $j \in J$, say $j \in J_x$, J_x finite: so f_i coincides with $\sum_{\substack{j \in J_x \\ i_j = i}} g_j$ on U_x. By Fact 6.3.28, f_i is continuous from X to $[0, 1]$. Moreover, $\sum_{i \in I} f_i(x) = \sum_{j \in J} g_j(x) = 1$, so $(f_i)_{i \in I}$ is a partition of unity. Finally, $f_i^{-1}(0, 1] \subseteq \bigcup_{\substack{j \in J \\ i_j = i}} g_j^{-1}(0, 1] \subseteq \bigcup_{\substack{j \in J \\ i_j = i}} cl(g_j^{-1}(0, 1]) \subseteq \bigcup_{\substack{j \in J \\ i_j = i}} V_j \subseteq U_i$. So $(f_i)_{i \in I}$ is weakly subordinate to $(U_i)_{i \in I}$. We obtain another partition of unity, subordinate to $(U_i)_{i \in I}$, by Lemma 6.3.30.

Conversely, if $(U_i)_{i \in I}$ has a subordinate partition of unity $(f_i)_{i \in I}$, then we may assume that it is locally finite by Lemma 6.3.30. Then $(f_i^{-1}(0, 1])_{i \in I}$ is a locally finite open cover that refines $(U_i)_{i \in I}$. $\square$

We need a final observation and a technical lemma.

Fact 6.3.36 *Let $(F_k)_{k \in K}$ be a locally finite family of closed subsets of a topological space X. Then $\bigcup_{k \in K} F_k$ is closed.*

This is similar to Fact 6.3.28. Every point x in $X \smallsetminus \bigcup_{k \in K} F_k$ has an open neighborhood V that meets only finitely many F_ks, say $F_{k_1}, \ldots, F_{k_n}$. Then $V \smallsetminus \bigcup_{i=1}^{n} F_{k_i}$ is another open neighborhood of x that is contained in $X \smallsetminus \bigcup_{k \in K} F_k$. So the latter is open by Trick 4.1.11, hence $\bigcup_{k \in K} F_k$ is closed.

Lemma 6.3.37 *Let X be a topological space. Then the following are equivalent if X is regular:*

1. *Every open cover $(U_i)_{i \in I}$ of X is refined by some locally finite closed cover – where a closed cover is a cover all of whose elements are closed subsets.*
2. *X is paracompact.*

3. *Every open cover $(U_i)_{i \in I}$ of X is refined by some locally finite (not necessarily open) cover.*

In the general case, i.e., without assuming X regular, $1 \Rightarrow 2 \Rightarrow 3$.

Proof Let us first show $3 \Rightarrow 1$, assuming X regular. Let $(U_i)_{i \in I}$ be any open cover of X. For every $x \in X$, x is in some U_i. Since X is regular, i.e., locally closed (Exercise 4.1.21), we can find an open neighborhood U_x of x and a closed subset F_x such that $x \in U_x \subseteq F_x \subseteq U_i$. (We use the Axiom of Choice here to build the families U_x, F_x.) Let $(V_j)_{j \in J}$ be a locally finite cover that refines $(U_x)_{x \in X}$. For every $j \in J$, V_j is included in some U_x, $x \in X$, so $cl(V_j)$ is included in some F_x, $x \in X$, hence in some U_i, $i \in I$. So $(cl(V_j))_{j \in J}$ is a closed cover that refines $(U_i)_{i \in I}$. It is also locally finite: for every $x \in X$, there is an open neighborhood of x that meets only finitely many V_js, hence also only finite many closures of V_j, using Corollary 4.1.28.

$1 \Rightarrow 2$. Let $(U_i)_{i \in I}$ be any open cover of X, and consider a locally finite cover $(A_j)_{j \in J}$ that refines it. (That each A_j can be taken closed is irrelevant for now.) Using the Axiom of Choice, pick $i_j \in I$ be such that $A_j \subseteq U_{i_j}$. For every $x \in X$, there is an open neighborhood V_x that meets only finitely many A_js (using the Axiom of Choice again to pick some V_x for each x). Consider a locally finite closed cover $(F_k)_{k \in K}$ that refines $(V_x)_{x \in X}$.

For each $j \in J$, let $W_j = U_{i_j} \setminus \bigcup_{\substack{k \in K \\ A_j \cap F_k = \emptyset}} F_k$. By Fact 6.3.36, W_j is open. Next, $(W_j)_{j \in J}$ is a cover of X: for every $x \in X$, there is a $j \in J$ such that $x \in A_j$. Since $A_j \subseteq U_{i_j}$, and since x cannot be in any F_k such that $A_j \cap F_k = \emptyset$, x is in W_j. Also, $W_j \subseteq U_{i_j}$ entails that $(W_j)_{j \in J}$ refines $(U_i)_{i \in I}$. Finally, by construction, if V_x meets W_j, say at y, then for some $k \in K$, $y \in F_k \subseteq V_x$, and since $y \in W_j$, $A_j \cap F_k \neq \emptyset$: so $V_x \supseteq F_k$ also meets A_j. Since this can happen for only finitely many $j \in J$, it follows that $(W_j)_{j \in J}$ is locally finite.

The implication $2 \Rightarrow 3$ is clear. □

6.3.4 Metrizability and the Nagata–Smirnov Theorem

All this heavy machinery allows us to obtain the following famous characterization of metrizable spaces (Nagata, 1950; Smirnov, 1951). A similar characterization is due to Bing (1951).

Theorem 6.3.38 (Nagata–Smirnov) *Let X be a topological space. A family $\mathcal{A}$ of subsets of X is* σ-locally finite *if and only if it is a countable union of locally finite families.*

A topological space is metrizable if and only if it is T_3 and its topology has a σ-locally finite base.

Proof If X is metrizable, say with metric d, then it is T_3 by Proposition 6.1.13. For each $n \in \mathbb{N}$, the open cover $(B^d_{x,<1/2^n})_{x \in X}$ is refined by some locally finite open cover $\mathcal{A}_n$, since X is paracompact by Theorem 6.3.31. So $\mathcal{A} = \bigcup_{n \in \mathbb{N}} \mathcal{A}_n$ is σ-locally finite. We check that $\mathcal{A}$ is a base, namely that every open subset U is a union of elements of $\mathcal{A}$. For every $x \in U$, there is an $n \in \mathbb{N}$ such that $B^d_{x,<1/2^n} \subseteq U$ by Fact 6.1.7. There is also an element V_x of $\mathcal{A}_{n+1}$ such that $x \in V_x$, since $\mathcal{A}_{n+1}$ is a cover of X. Also, $V_x \subseteq B^d_{y,<1/2^{n+1}}$ for some $y \in X$. Then, for every $z \in V_x$, $d(x, z) \leqslant d(x, y) + d(y, z) = d(y, x) + d(y, z) < 1/2^n$, so $z \in B^d_{x,<1/2^n} \subseteq U$. So V_x is an open neighborhood of x contained in U, and is in $\mathcal{A}$. Since $U = \bigcup_{x \in U} V_x$, $\mathcal{A}$ is a base of the topology.

Conversely, assume that X is T_3 and has a σ-locally finite base $\mathcal{A} = \bigcup_{n \in \mathbb{N}} \mathcal{A}_n$, where each $\mathcal{A}_n$ is a locally finite family of open subsets. Without loss of generality, we may assume $\mathcal{A}_0 \subseteq \mathcal{A}_1 \subseteq \cdots \subseteq \mathcal{A}_n \subseteq \cdots$; otherwise replace $\mathcal{A}_n$ by $\bigcup_{m=0}^n \mathcal{A}_m$.

Let $(U_i)_{i \in I}$ be an open cover of X. Define $\mathcal{A}'_n$ as the family of all opens $U \in \mathcal{A}_n$ such that $U \subseteq U_i$ for some $i \in I$, and $\mathcal{A}' = \bigcup_{n \in \mathbb{N}} \mathcal{A}'_n$. Since $\mathcal{A}_n$ is locally finite, so is $\mathcal{A}'_n$, and since $\mathcal{A}$ is a base and $(U_i)_{i \in I}$ a cover, $\mathcal{A}'$ is also an open cover of X, which clearly refines $(U_i)_{i \in I}$. For each $n \in \mathbb{N}$, let U'_n be the union of all the open subsets in $\mathcal{A}'_n$. In particular, $U'_0 \subseteq U'_1 \subseteq \cdots \subseteq U'_n \subseteq \cdots$. Let $A_n = U'_n \setminus U'_{n-1}$. Consider the family $\mathcal{B}$ of all subsets of the form $V \cap A_n$ with $V \in \mathcal{A}'_n$, $n \in \mathbb{N}$. This is a (not necessarily open) cover of X: for every $x \in X$, let n be the smallest number such that $x \in U'_n$; then $x \in A_n$ and x is in some subset V of $\mathcal{A}'_n$, by definition. Also, we claim that $\mathcal{B}$ is locally finite. Indeed, for every $x \in X$, there is an open neighborhood V of x that meets only finitely many elements of $\mathcal{A}'$, say $V_1, \ldots, V_p$. These must all be in the same $\mathcal{A}'_q$, for some $q \in \mathbb{N}$, and in particular must all be included in U'_q. It follows that V can only meet those subsets $V \cap A_n$ in $\mathcal{B}$ where V is among $V_1, \ldots, V_p$ and $n \leqslant q$. So $\mathcal{B}$ is a locally finite cover that refines $\mathcal{A}'$, hence also $(U_i)_{i \in I}$.

By the $3 \Rightarrow 2$ implication of Lemma 6.3.37, X is paracompact. Since X is also T_2, by Proposition 6.3.26, X is normal.

Fix $m, n \in \mathbb{N}$. For every open subset U in $\mathcal{A}_n$, $n \in \mathbb{N}$, let F^U_m be the union of all subsets of U of the form $cl(V)$ where $V \in \mathcal{A}_m$. Since $\mathcal{A}_m$ is locally finite, the family of all $cl(V)$, $V \in \mathcal{A}_m$ is also locally finite: for each $x \in X$, there is an open neighborhood W of x that meets only finitely many such Vs, and W meets V iff W meets $cl(V)$ by Corollary 4.1.28. By Fact 6.3.36, F^U_m is closed. Also, $F^U_m \subseteq U$, so F^U_m and $X \setminus U$ are disjoint closed sets. Since X is normal, by Urysohn's Theorem 6.3.15 there is a continuous map $f^U_{mn}\colon X \to [0, 1]$ such that $f^U_{mn}(x) = 0$ for every $x \in F^U_m$ and $f^U_{mn}(x) = 1$ for every $x \in X \setminus U$.

Let $d_{mn}(x, y) = \sum_{U \in \mathcal{A}_n} |f^U_{mn}(x) - f^U_{mn}(y)|$. For every $x \in X$, there is an open neighborhood V of x that meets only finitely many elements of $\mathcal{A}_n$. So, for all $U \in \mathcal{A}_n$ except finitely many, say $U_1, \ldots, U_k$, f^U_{mn} coincides with 1 on V. Similarly, for every $y \in X$, there is an open neighborhood W that meets only finitely many elements $U_{k+1}, \ldots, U_{k+\ell}$ of $\mathcal{A}_n$; on all other $U \in \mathcal{A}_n$, f^U_{mn} coincides with 1 on W. So, on $V \times W$, d_{mn} coincides with the finite sum $\sum_{i=1}^{k+\ell} |f^{U_i}_{mn}(x) - f^{U_i}_{mn}(y)|$—all other summands equal $|1-1| = 0$. In particular, d_{mn} is well defined, and continuous from X to $\mathbb{R}^+$, using Fact 6.3.28.

Since d_{mn} is continuous from X to $\mathbb{R}^+$, $B^{d_{mn}}_{x,<\epsilon} = (d_{mn}(x, \_))^{-1}[0, \epsilon)$ is open in the topology of X. So the topology defined by $(d_{mn})_{m,n \in \mathbb{N}}$ is coarser than the topology of X.

Conversely, let U_0 be any open subset of X. For every $x \in U_0$, x has an open neighborhood $U \subseteq U_0$ taken from some $\mathcal{A}_n$, since $\mathcal{A}$ is a base. Since X is normal and T_2, X is regular; so there is an open neighborhood V of x such that $cl(V) \subseteq U$ (use, e.g., Exercise 4.1.21). Since $\mathcal{A}$ is a base, we may as well assume that V is in some $\mathcal{A}_m$. Then $x \in F^U_m$, so $f^U_{mn}(x) = 0$. For every element y of $B^{d_{mn}}_{x,<1}$, $d_{mn}(x, y) < 1$, in particular $|f^U_{mn}(x) - f^U_{mn}(y)| < 1$, i.e., $f^U_{mn}(y) < 1$. This is only possible if y is in U. So $B^{d_{mn}}_{x,<1} \subseteq U$. By Trick 4.1.11, U is open in the topology defined by $(d_{mn})_{m,n \in \mathbb{N}}$. So the latter is finer than the topology of X.

It follows that the topology of X is exactly defined by $(d_{mn})_{m,n \in \mathbb{N}}$, hence by the single (symmetric) hemi-metric $d = \bigvee_{m,n \in \mathbb{N}} d_{mn}$ (Lemma 6.3.11). Since X is T_0, d must be T_0, hence a metric, by Lemma 6.1.9. □

The following is a direct corollary.

Proposition 6.3.39 *Every countably-based T_3 space is metrizable.*

Proof If X has a countable base $(U_n)_{n \in \mathbb{N}}$, then $\mathcal{A}_n = \{U_n\}$ is locally finite. Then apply Theorem 6.3.38 with $\mathcal{A} = \bigcup_{n \in \mathbb{N}} \mathcal{A}_n$. □

Urysohn's original metrization theorem (Exercise 6.3.18) follows immediately, since every T_4 space is T_3.

6.3.5 Separable metrizable spaces

Proposition 6.3.39 will turn out to *characterize* metrizability in the presence of separability.

Definition 6.3.40 (Separable) A hemi-metric space X, d is *separable* if and only if its symmetrization X, d^{sym} has a countable dense subset.

Example 6.3.41 Since $\mathbb{Q}$ is dense in $\mathbb{R}$ (Example 4.1.31), $\mathbb{R}, \mathsf{d}_{\mathbb{R}}$ is separable. For the same reason, $\mathbb{R}, \mathsf{d}_{\mathbb{R}}^{sym}$ is separable.

The standard definition of separability only requires X, d to have a countable dense subset, not its symmetrization. This is equivalent for metric spaces. However, density is an extremely weak property for non-T_2 spaces, as we have already seen, and we shall need the refined form above later, e.g., in Theorem 6.7.13.

Lemma 6.3.42 *Every countably-based topological space has a countable dense subset.*

Proof Let X be countably-based. So X has a countable base $(U_n)_{n \in \mathbb{N}}$, using Exercise 6.3.8. Using the Axiom of Choice, pick an element x_n from those U_ns that are non-empty, $n \in \mathbb{N}$. Given any non-empty open subset U of X, let $x \in U$, and find U_n so that $x \in U_n \subseteq U$. Then x_n is in U. So $(x_n)_{\substack{n \in \mathbb{N} \\ U_n \neq \emptyset}}$ is dense, by Lemma 4.1.30. ☐

> **Exercise 6.3.43** Show that the converse of Lemma 6.3.42 fails, by considering a poset X with a largest element $\top$ in its Alexandroff topology, say $X = [0, 1]$, and showing, by elementary arguments, that $[0, 1]_a$ is not countably-based. We shall improve upon this in Exercise 6.3.49.

We say that a topological space is *separable metrizable* if and only if it can be equipped with a metric that makes it separable.

Theorem 6.3.44 (Urysohn–Tychonoff) *A topological space is separable metrizable if and only if it is countably-based and T_3.*

Proof Assume X, d metric and separable. X is T_3 by Proposition 6.1.13. Let $(x_n)_{n \in \mathbb{N}}$ be a countable dense subset of X. We claim that $(B^d_{x_n, <1/2^m})_{n,m \in \mathbb{N}}$ is a base of the topology of X. Since it is countable, X will be countably-based. Let U be any open subset of X, and $x \in U$. By Fact 6.1.7, $B_{x,<1/2^m} \subseteq U$ for $m \in \mathbb{N}$ large enough. Since $(x_n)_{n \in \mathbb{N}}$ is dense, and using Lemma 4.1.30, some x_n is in $B^d_{x,<1/2^{m+1}}$, so $d(x, x_n) < 1/2^{m+1}$. For every $y \in B^d_{x_n,<1/2^{m+1}}$, $d(x, y) \leqslant d(x, x_n) + d(x_n, y) < 1/2^m$. So $B^d_{x_n,<1/2^{m+1}} \subseteq U$. Moreover, x is in $B^d_{x_n,<1/2^{m+1}}$, because $d(x_n, x) = d(x, x_n) < 1/2^{m+1}$. By Lemma 4.1.10, $(B^d_{x_n,<1/2^{m+1}})_{n,m \in \mathbb{N}}$ is indeed a base of the topology of X.

Conversely, assume X countably-based and T_3. There is a metric d on X whose open ball topology is that of X, by Proposition 6.3.39. It is separable by Lemma 6.3.42. ☐

In passing, Theorem 6.3.44 establishes that every separable metric space is countably-based. The converse is Lemma 6.3.42:

Fact 6.3.45 *For a metric space, separability is equivalent to having a countable base.*

Exercise 6.3.46 ($\mathbb{R}_\ell$ is not metrizable) Using the Urysohn–Tychonoff Theorem, show that the Sorgenfrey line (Exercise 4.1.34) is *not* metrizable, despite the fact that it is T_4, and even paracompact (see Exercise 6.3.32). You may use Exercise 4.1.36 and Exercise 6.3.10.

6.3.6 Separable hemi-metrizable spaces

For hemi-metric spaces, there is also a relationship between separability and second-countability, but it is slightly more complex. We first need the following refinement of Corollary 6.3.4.

Lemma 6.3.47 *Let X, d be a hemi-metric space, and I be a dense subset of X, d^{sym}. Then the set of all open balls $B^d_{y,<\epsilon}$, $y \in I$, $\epsilon > 0$, is a base of the open ball topology of X, d.*

Proof For every $x \in X$, for every open neighborhood U of x, by Corollary 6.3.4 there is an $\eta > 0$ such that $B^d_{x,<\eta} \subseteq U$. Let $\epsilon = \eta/2$. Since I is dense in X, d^{sym}, there is a point $y \in I$ that is also in $B^{d^{sym}}_{x,<\epsilon}$. Since $d^{sym}(x, y) < \epsilon$, x is in $B^d_{y,<\epsilon}$. Moreover, for every $z \in B^d_{y,<\epsilon}$, $d(x, z) \leqslant d(x, y) + d(y, z) \leqslant d^{sym}(x, y) + d(y, z) < 2\epsilon = \eta$, so $z \in B^d_{x,<\eta} \subseteq U$. We conclude by Lemma 4.1.10. ☐

Note that Lemma 6.3.47 is certainly no longer true if we only assume I dense in X, d instead of d^{sym}. For example, letting d be $d_\leqslant$ for some ordering $\leqslant$ on a set X with a largest element $\top$, $\{\top\}$ alone is dense in X, d, but the only open ball with center $\top$ is $\{\top\}$ itself.

The following refines Fact 6.3.45. The first part is due to Künzi (2009).

Proposition 6.3.48 *For a hemi-metric space X, d, separability is equivalent to having a countable base, and having a countable base implies having a countable dense subset.*

Proof The second part is true from Lemma 6.3.42. For the first part, if X, d^{sym} has a countable dense subset I, Lemma 6.3.47 implies that the set of open balls $B^d_{y,<1/2^n}$, $y \in I$, $n \in \mathbb{N}$, is a countable base of the open ball topology. Conversely, assume that X, d has a countable base $(U_i)_{i \in \mathbb{N}}$. For all $i, n \in \mathbb{N}$, let $S_{i,n}$ be the set of points $x \in X$ such that $x \in U_i \subseteq B^d_{x,<1/2^n}$. Let us say that a pair i, n is *good* if and only if $S_{i,n}$ is non-empty. Using the Axiom of Choice, fix a point $x_{i,n}$ in $S_{i,n}$ for every good pair i, n. The set of all points $x_{i,n}$ thus obtained is countable, and we claim that it is dense in X, d^{sym}. Let y be any point of X, and $\epsilon > 0$. Find $n \in \mathbb{N}$ such that $\frac{1}{2^n} \leqslant \epsilon$. Since $(U_i)_{i \in \mathbb{N}}$ is a base, there is an $i \in \mathbb{N}$ such that $y \in U_i$ and $U_i \subseteq B^d_{y,<1/2^n}$. So y is in $S_{i,n}$, and i, n is a good pair. Since both $x_{i,n}$ and y are in $S_{i,n}$, U_i is included in $B^d_{x_{i,n},<1/2^n}$ and in $B^d_{y,<1/2^n}$. Also, $x_{i,n}$ and y are both in U_i. So $x_{i,n}$ is in $B^d_{y,<1/2^n}$ and y is in $B^d_{x_{i,n},<1/2^n}$, whence $d(y, x_{i,n}) < \frac{1}{2^n}$ and $d(x_{i,n}, y) < \frac{1}{2^n}$. It follows that $d^{sym}(x_{i,n}, y) < \frac{1}{2^n} \leqslant \epsilon$. ☐

Exercise 6.3.49 Using Exercise 6.3.43, show that there are quasi-metric spaces that have a countable dense subset but are not countably-based.

However, we have the following analogue of the Urysohn–Tychonoff Theorem for hemi-metric spaces. This is simpler than in the metric case—we just omit the T_3 requirement – and improves on Wilson's Theorem.

Theorem 6.3.50 (Separable hemi-metrizable = countably-based) *A topological space is separable hemi-metrizable if and only if it is countably-based.*

Proof Assume X is countably-based, with countable subbase $(U_n)_{n\in\mathbb{N}}$. As in the proof of Wilson's Theorem 6.3.13, the topology of X is defined by the hemi-metric $d = \bigvee_{n\in\mathbb{N}} d_n$, with $d_n(x, y) = \max(\chi_{U_n}(x) - \chi_{U_n}(y), 0)$. Then $d^{sym}(x, y) = \sup_{n\in\mathbb{N}} \frac{1}{2^n}|\chi_{U_n}(x) - \chi_{U_n}(y)|$. Given any point $x \in X$, $B^{d^{sym}}_{x,<1/2^n}$ is the set of points y such that $\chi_{U_i}(x) = \chi_{U_i}(y)$ for every $i \in \mathbb{N}$ with $0 \leqslant i \leqslant n$. Writing U_i^x for U_i if $x \in U_i$, and for its complement $X \smallsetminus U_i$ otherwise, $B^{d^{sym}}_{x,<1/2^n}$ therefore equals $U_0^x \cap U_1^x \cap \cdots \cap U_n^x$. By Fact 6.1.7, these balls form a base of the open ball topology of X, d^{sym}. But there are only countably many sets that are finite intersections of sets that are equal to U_i or to its complement. So X, d^{sym} is countably-based. By Fact 6.3.45, X, d^{sym}, hence also X, d, is separable.

The converse direction is Proposition 6.3.48. ☐

6.4 Coproducts, quotients

We shall see that the categories of hemi-metric, quasi-metric, and metric spaces have all coproducts and products, and more generally all colimits and all limits. We start with coproducts, which are particularly nice: the topology of a coproduct in **HMet** is the coproduct topology.

Lemma 6.4.1 (Coproducts of hemi-metric spaces) *Let X_i, d_i be hemi-metric spaces, $i \in I$. The* coproduct hemi-metric *$\coprod_{i\in I} d_i$ is the hemi-metric d defined on the set $\coprod_{i\in I} X_i$ by: $d((i, x), (j, y)) = d_i(x, y)$ if $i = j$, $+\infty$ otherwise.*

Then the open ball topology of $\coprod_{i\in I} X_i, \coprod_{i\in I} d_i$ is exactly the coproduct topology of the open ball topologies of X_i, d_i, $i \in I$.

So there is no ambiguity in writing $\coprod_{i \in I} X_i$, which may mean either the topological or the hemi-metric coproduct: they have the same topology.

Proof The open ball $B^d_{(i,x),<\epsilon}$ equals $\{i\} \times B^{d_i}_{x,<\epsilon}$, hence is open in the coproduct topology. Conversely, every open in the coproduct is a union of basic opens $\{i\} \times U_i$, where each open U_i is a union of open balls $B^{d_i}_{x,<\epsilon}$, $i \in I$, $x \in X_i$, $\epsilon > 0$. So every open in the coproduct is a union of sets of the form $\{i\} \times B^{d_i}_{x,<\epsilon}$, which are open balls. So each open in the coproduct topology is in the open ball topology. □

Exercise 6.4.2 Show that **HMet** (resp., **QMet**, **Met**) has all coproducts, and that they are defined as in Lemma 6.4.1.

Exercise 6.4.3 (Quotient hemi-metrics, coequalizers, cocompleteness) Let X, d be a hemi-metric space, and $\equiv$ be an equivalence relation on X. The *quotient hemi-metric* $d/\equiv$ is defined as follows. A *trip* from y to y' in $X/\equiv$ is a finite sequence $(x_1, x'_1), (x_2, x'_2), \ldots, (x_n, x'_n)$ of pairs of points of X, $n \geqslant 1$, where $y = q_\equiv(x_1)$, $y' = q_\equiv(x'_n)$, and $x'_1 \equiv x_2, x'_2 \equiv x_3, \ldots, x'_{n-1} \equiv x_n$. The *duration* of the trip is $d(x_1, x'_1) + d(x_2, x'_2) + \cdots + d(x_n, x'_n)$. Then $d/\equiv(y, y')$ is defined as the greatest lower bound of all durations of trips from y to y'.

Show that $d/\equiv$ is indeed a hemi-metric, and is symmetric whenever d is (i.e., $d(x, x') = d(x', x)$). Show that $q_\equiv$ is 1-Lipschitz from X, d to $X/\equiv, d/\equiv$. Show that, given any c-Lipschitz map $h: X \to Y$ such that $x \equiv x'$ implies $h(x) = h(x')$, where Y, d' is some hemi-metric space, the map $h': X/\equiv \to Y$ defined by $h'(q_\equiv(x)) = h(x)$ is c-Lipschitz.

Deduce from this that **HMet** has all coequalizers. The coequalizer of two 1-Lipschitz maps $g_1, g_2: Z \to X$ is $X/\equiv$, where $\equiv$ is the smallest equivalence relation on X such that $g_1(z) = g_2(z)$ for every $z \in Z$. Conclude that **HMet** is cocomplete.

Exercise 6.4.4 (T_0 quotient of hemi-metric spaces) Let X, d be a hemi-metric space. Define an equivalence relation $=_0$ on X by $x =_0 y$ iff $d(x, y) = d(y, x) = 0$. Check that this is exactly the equivalence relation defined in Exercise 4.10.25, namely $x =_0 y$ iff $x \leqslant y$ and $y \leqslant x$, where $\leqslant$ is the specialization quasi-ordering of X. Show that $d/=_0$ (Exercise 6.4.3) can be equivalently defined by $d/=_0(q_{=_0}(x), q_{=_0}(x')) = d(x, x')$. Show that $d/=_0$ is a quasi-metric, and a metric if d is symmetric.

Show that the T_0 quotient of the topological space underlying X, d coincides with the topological space underlying the T_0 quotient of X, d. That is, show that the open ball topology of $X/=_0, d/=_0$ coincides with the topology of the T_0 quotient (in the sense of Exercise 4.10.25) of X, the latter with its open ball topology.

Finally, show that the T_0 quotient of hemi-metric spaces we just defined has the following universal property: for every $c \in \mathbb{R}^+$, for every quasi-metric space Y, d', and every c-Lipschitz map $f: X \to Y$, there is a unique c-Lipschitz map $f': X/{=_0} \to Y$ such that $f = f' \circ q_{=_0}$. In particular, for $c = 1$, the T_0 quotient $X/{=_0}, d/{=_0}$ is the free quasi-metric space over the hemi-metric space X, d, resp., the free metric space over the symmetric hemi-metric space X, d. It follows that the category of quasi-metric spaces and 1-Lipschitz maps is reflective in **HMet**, and the category of metric spaces and 1-Lipschitz maps is reflective in the category of symmetric hemi-metric spaces and 1-Lipschitz maps.

Exercise 6.4.5 (Double quotient, coequalizers, cocompleteness) Let X, d be a hemi-metric space, and $\equiv$ be an equivalence relation on X. The *double quotient* of X, d by $\equiv$ is the T_0 quotient (in the sense of Exercise 6.4.3) of the quotient $X/{\equiv}, d/{\equiv}$.

When X, d is quasi-metric (resp., metric), and $\equiv$ is the smallest equivalence relation on X such that $f(y) \equiv g(y)$ for every $y \in Y$, where f and g are 1-Lipschitz maps from some quasi-metric (resp., metric) space Y, d' to X, d, show that the double quotient of X, d by $\equiv$ is exactly the coequalizer of f, g in **QMet** (resp., **Met**). Conclude that **QMet** and **Met** are cocomplete. There is a very short categorical answer.

Exercise 6.4.6 (Coequalizers = quotients, double quotients) Every coequalizer in **HMet** is a quotient, and every coequalizer in **QMet**, **Met** is a double quotient. Conversely, show that every quotient in **HMet** is a coequalizer, and that every double quotient in **QMet**, **Met** is a coequalizer. As in **Top**, form the coequalizer of the respective kernel pairs (see Section 4.12).

The fact that coequalizers in **QMet** and **Met** have to be defined as double quotients is not specific to quasi-metric and metric spaces. As we have seen in Section 5.5.2, colimits in $\mathbf{Top}_0$ are also computed as T_0 quotients of the corresponding colimits in **Top**. As quotients are coequalizers, the coequalizer of $f, g: Y \to X$ in $\mathbf{Top}_0$ is, similarly, the T_0 quotient of the quotient $X/{\equiv}$, where $\equiv$ is the smallest equivalence relation such that $f(y) \equiv g(y)$ for every $y \in Y$.

Exercise 6.4.7 ($\mathbb{R}/\mathbb{Q}$) Consider the set $\mathbb{R}/\mathbb{Q}$, already studied from a topological point of view in Exercise 4.10.26. Let $\equiv$ be the equivalence relation on $\mathbb{R}$ defined by $x \equiv y$ iff $x - y \in \mathbb{Q}$. Show that $\mathsf{d}_{\mathbb{R}}^{sym}/{\equiv}$ is the trivial hemi-metric on $\mathbb{R}/\mathbb{Q}$, i.e., $\mathsf{d}_{\mathbb{R}}^{sym}/{\equiv}(q_{\equiv}(x), q_{\equiv}(y)) = 0$ for all $x, y \in \mathbb{R}$. Deduce that the double quotient of $\mathbb{R}, \mathsf{d}_{\mathbb{R}}^{sym}$ by $\equiv$ contains only one element.

6.5 Products, subspaces

The open ball topology of coproducts is the topological coproduct, but a similar result for products will only hold for finite products. Hence we start with this case. We shall also define various variants of products of hemi-metric, quasi-metric, and metric spaces whose open ball topology will be the product topology. This will work for countable products, but not uncountable products.

There are several ways we can define a product hemi-metric so that the open ball topology is the product topology. The following defines the category-theoretic product in **HMet**.

Lemma 6.5.1 (Products of hemi-metric spaces) *Let X_i, d_i be finitely many hemi-metric spaces, $1 \leqslant i \leqslant n$. The* product hemi-metric $\prod_{i=1}^n d_i$ *is the hemi-metric d on the set $\prod_{i=1}^n X_i = X_1 \times \cdots \times X_n$ defined by $d((x_1, \ldots, x_n), (y_1, \ldots, y_n)) = \max(d_1(x_1, y_1), \ldots, d_n(x_n, y_n))$.*

Then the open ball topology of $\prod_{i=1}^n X_i, \prod_{i=1}^n d_i$ is exactly the product topology of the open ball topologies of X_i, d_i, $1 \leqslant i \leqslant n$.

If $n = 0$, we take the max above to be 0.

Proof The open ball $B^d_{(x_1,\ldots,x_n),<\epsilon}$ is equal to the product $B^{d_1}_{x_1,<\epsilon} \times \cdots \times B^{d_n}_{x_n,<\epsilon}$, and is therefore open in the product topology. Conversely, every open U in the product topology is a union of opens $U_1 \times \cdots \times U_n$, where each U_i is a union of open balls in X_i. So U is a union of products of open balls $B^{d_1}_{x_1,<\epsilon} \times \cdots \times B^{d_n}_{x_n,<\epsilon}$, which are themselves open balls in X, d, hence U is in the open ball topology. □

Exercise 6.5.2 Show that the finite products of Lemma 6.5.1 are the finite products in the categories **HMet**, **QMet**, and **Met**.

Exercise 6.5.3 (Tensor product of hemi-metric spaces) Show that the hemi-metric $\prod_{i=1}^n d_i$ is not the only possible hemi-metric d such that the conclusion of Lemma 6.5.1 holds. (This is the only one that defines the categorical product, by Exercise 6.5.2, since categorical products are unique up to iso.) In particular, show that the hemi-metric $d' = d_1 \otimes \cdots \otimes d_n$ defined by $d'((x_1, \ldots, x_n), (y_1, \ldots, y_n)) = d_1(x_1, y_1) + \cdots + d_n(x_n, y_n)$ defines an open ball topology that coincides with the product topology on $X_1 \times \cdots \times X_n$.

$X_1 \times \cdots \times X_n, d'$ is the *tensor product* of the hemi-metric spaces $X_1, d_1, \ldots, X_n, d_n$.

Now that we have defined products and tensor products, we can refine Exercise 6.2.10.

Lemma 6.5.4 (Distances are Lipschitz) *Let X, d be a hemi-metric space. Then the distance function d is 1-Lipschitz from $X \times X, d \otimes d^{op}$ to $\overline{\mathbb{R}^+}, \mathsf{d}_{\mathbb{R}}$, and 2-Lipschitz from $X \times X, d \times d^{op}$ to $\overline{\mathbb{R}^+}, \mathsf{d}_{\mathbb{R}}$.*

In any case, d is continuous from $X \times X^{op}$ to $\overline{\mathbb{R}^+}_\sigma$, where X^{op} denotes X with the open ball topology of d^{op}.

Proof Use the triangular inequality: $d(x, y) \leqslant d(x, x') + d(x', y') + d(y', y) = (d \otimes d^{op})((x, y), (x', y')) + d(x', y')$, so $\mathsf{d}_{\mathbb{R}}(d(x, y), d(x', y')) \leqslant (d \otimes d^{op})((x, y), (x', y'))$. Observe now that $(d \otimes d^{op}) \leqslant 2(d \times d^{op})$, since $a + b \leqslant 2\max(a, b)$ for any $a, b \in \overline{\mathbb{R}^+}$, so d is also 2-Lipschitz from $X \times X, d \times d^{op}$ to $\overline{\mathbb{R}^+}, \mathsf{d}_{\mathbb{R}}$. Continuity follows by Fact 6.2.3. □

Exercise 6.5.5 Show that **HMet**, **QMet**, and **Met** have infinite products. Given hemi-metric (resp., quasi-metric, metric) spaces $X_i, d_i, i \in I$, the product hemi-metric is the *sup hemi-metric* defined by $d(\vec{x}, \vec{y}) = \sup_{i \in I} d_i(x_i, y_i)$. However, show that when I is infinite, the open ball topology of $\prod_{i \in I} X_i, d$ is not the product topology. You may consider, e.g., $X_i = \{0, 1\}$ with the metric $d_i = d_=$ for each $i \in \mathbb{N}$. (Remember from Exercise 6.1.15 that $d_=(x, y) = 0$ if $x = y$, $+\infty$ otherwise.)

Exercise 6.5.6 (Subspaces, equalizers, completeness) Let X, d be a hemi-metric (resp., quasi-metric, metric) space, and Y be a subset of X. By abuse of language, we again write d for the restriction of d to Y, and call Y, d a *hemi-metric subspace* (resp., *quasi-metric subspace, metric subspace*) of X, d. Show that the open ball topology of Y, d coincides with the subspace topology of X, d.

Given two 1-Lipschitz maps g_1, g_2 from X, d to Z, d', show that $[g_1 = g_2]$, defined as $\{x \in X \mid g_1(x) = g_2(x)\}$, with hemi-metric d, is their equalizer in **HMet** (resp., **QMet**, **Met**).

Conclude that **HMet**, **QMet**, and **Met** are complete categories.

By Exercise 6.5.5, infinite products of hemi-metric spaces in general do not coincide with topological products. One can define another notion of product of hemi-metric spaces (which cannot be categorical products), and whose open ball topology will coincide with the product topology. There are several possibilities; however, this can only really be done for countable families of spaces.

Lemma 6.5.7 (Topological product of hemi-metric spaces) *Let $(X_i, d_i)_{i \in I}$ be a countable family of hemi-metric spaces. Without loss of generality, I is a subset of $\mathbb{N}$. Define $\prod_{i \in I} d_i$ by:*

$$\widetilde{\prod}_{i \in I} d_i(\vec{x}, \vec{y}) = \sup_{i \in I} \frac{1}{2^i} \min(d_i(x_i, y_i), 1).$$

Then $\prod_{i \in I} X_i, \widetilde{\prod}_{i \in I} d_i$ *is a hemi-metric space, whose open ball topology is the product topology. Accordingly, this is called the* topological product *of the countable family of hemi-metric spaces* X_i, d_i.

Proof Let $\widetilde{d_i}$ be the hemi-metric on $X = \prod_{j \in I} X_j$ defined by $\widetilde{d_i}(\vec{x}, \vec{y}) = d_i(x_i, y_i)$. Then $d = \widetilde{\prod}_{i \in I} d_i$ is just $\bigvee_{i \in I} \widetilde{d_i}$. By Lemma 6.3.11, $\mathcal{O}^d$ is the coarsest topology containing all the open balls $B^{\widetilde{d_i}}_{\vec{x}, < \epsilon}$. Since the latter are just $\pi_i^{-1}(B^{d_i}_{x_i, < \epsilon})$, where $\pi_i : X \to X_i$ is the ith projection, $\mathcal{O}^d$ is also the coarsest topology containing all subsets $\pi_i^{-1}(U_i)$, U_i open in X_i. This is just the product topology. □

Exercise 6.5.8 As in Exercise 6.5.3, show that one could also have taken the hemi-metric

$$d(\vec{x}, \vec{y}) = \sum_{i \in I} \frac{1}{2^i} \min(d_i(x_i, y_i), 1)$$

in Lemma 6.5.7, and that $\prod_{i \in I} X_i, d$ would still have the product topology, for $I \subseteq \mathbb{N}$ countable.

Exercise 6.5.9 Show that the assumption that I is countable in Lemma 6.5.7 is necessary. Assume I uncountable, say $\mathbb{P}(\mathbb{N})$, and show that there is no hemi-metric on $\{0, 1\}^I$ whose open ball topology is the product topology, where we equip $\{0, 1\}$ with the discrete topology. Hint: $\{0, 1\}^I$ is not even first-countable.

6.6 Function spaces

6.6.1 The sup hemi-metric and uniform convergence

Let us generalize the sup metric (Proposition 3.5.7) to the hemi-metric case:

Definition 6.6.1 (Sup hemi-metric) Let X be any set, and Y, d be any hemi-metric space. The *sup hemi-metric* $\sup_X d$ is defined on any space of maps from X to Y by $\sup_X d(f, g) = \sup_{x \in X} d(f(x), g(x))$.

Clearly, $\sup_X d$ is a quasi-metric, resp. a metric, whenever d is.

We also generalize uniform convergence from the metric to the hemi-metric case.

Lemma 6.6.2 (Uniform convergence) *Let X be a set, and Y, d be a hemi-metric space. On any set of maps from X to Y, the topology of* uniform convergence *is the open ball topology of the sup hemi-metric* $\sup_X d$.

A net $(f_i)_{i \in X}$ of maps from X to Y converges uniformly *to f, i.e., converges to f in the topology of uniform convergence, if and only if, for every $\epsilon > 0$, there is an $i_0 \in I$ such that, for every $i \in I$ with $i_0 \sqsubseteq i$, for every $x \in X$, $d(f(x), f_i(x)) < \epsilon$.*

Proof If $(f_i)_{i \in X}$ converges uniformly, then, for every $\epsilon > 0$, f_i is eventually in $B^{\sup_X d}_{f,<\epsilon}$. That is, for i large enough, $\sup_X d(f, f_i) < \epsilon$, hence $d(f(x), f_i(x)) < \epsilon$ for every $x \in X$. In the if direction, every open neighborhood W of f contains some open ball $B^{\sup_X d}_{f,<2\epsilon}$ by Lemma 6.1.5, $\epsilon > 0$. By assumption, for i large enough, for every $x \in X$, $d(f(x), f_i(x)) < \epsilon$, so $\sup_X d(f, f_i) \leqslant \epsilon < 2\epsilon$, i.e., $f_i \in W$. □

Lemma 6.6.3 *Let X be a set, and Y, d be a hemi-metric space. On the space Y^X of all functions from X to Y, the topology defined by the sup hemi-metric is in general finer than the product topology, i.e., the topology of pointwise convergence. It coincides with it when X is finite, and is in general strictly finer otherwise.*

In particular, every net of maps that converges uniformly to f also converges pointwise to f.

Proof The projections $\pi_x : Y^X \to Y$, which map f to $f(x)$, are 1-Lipschitz, because $d(f(x), g(x)) \leqslant \sup_X d(f, g)$, trivially. So π_x is continuous by Fact 6.2.3. Since the product topology is the coarsest one that makes each π_x continuous (Proposition 4.5.2), it is coarser than the one defined by the sup hemi-metric.

When X is finite, $\sup_X d$ is just the product hemi-metric, and therefore defines the product topology by Lemma 6.5.1. Otherwise, it does not; see Exercise 6.5.5.

The last claim is obvious. □

On spaces of continuous maps $[X \to Y]$, we defined several topologies in Section 5.3. They all coincided when X was locally compact (and compact in the case of the ec-topology), and were all finer than the topology of pointwise convergence. We show that the topology of uniform convergence (Proposition 5.4.20) is even finer.

Remember that $\mathsf{App}\colon [X \to Y] \times X \to Y$ maps (f, x) to $f(x)$. Remember also that a topology on the space $[X \to Y]$ of all continuous maps from X to Y is conjoining iff $\mathsf{App}\colon [X \to Y] \times X \to Y$ is continuous (see Theorem 5.4.4).

Proposition 6.6.4 (Uniform convergence $\Rightarrow$ continuous convergence) *Let X be a topological space and Y, d' be a hemi-metric space.*

Every net $(f_i)_{i \in I, \sqsubseteq}$ of continuous maps from X to Y that converges uniformly to $f \in [X \to Y]$ also converges continuously to f.

The topology of uniform convergence is conjoining. That is, writing $[X \to Y]^{\text{sup}}$ for the space $[X \to Y]$ of all continuous maps from X to Y with the topology of uniform convergence, the map $\mathsf{App}\colon [X \to Y]^{\text{sup}} \times X \to Y$ is continuous.

Proof Let $(f_i)_{i \in I, \sqsubseteq}$ be a net that converges uniformly to f in $[X \to Y]$, and $(x_j)_{j \in J, \preceq}$ be a net that converges to x in X. Fix an open neighborhood V of $f(x)$. By Lemma 6.1.5, V contains an open ball $B^{d'}_{f(x), <\epsilon}$, for some $\epsilon > 0$. Since f is continuous, $f^{-1}(B^{d'}_{f(x), <\epsilon/2})$ is an open neighborhood of x. So x_j is in it for j large enough, i.e., $d'(f(x), f(x_j)) < \epsilon/2$ for j large enough, say j above j_0. By Lemma 6.6.2, since $(f_i)_{i \in I, \sqsubseteq}$ is a net that converges uniformly to f, there is an $i_0 \in I$ such that, for every i above i_0, $d'(f(x_j), f_i(x_j)) < \epsilon/2$—with i_0 independent of j. So, for (i, j) above (i_0, j_0), $d'(f(x), f_i(x_j)) \leqslant d'(f(x), f(x_j)) + d'(f(x_j), f_i(x_j)) < \epsilon$. In other words, for (i, j) large enough, $f_i(x_j)$ is in $B^{d'}_{f(x), <\epsilon} \subseteq V$. So $(f_i)_{i \in I, \sqsubseteq}$ converges continuously to f.

Now let $(f_i, x_i)_{i \in I, \sqsubseteq}$ be a net that converges to (f, x) in $[X \to Y]^{\text{sup}} \times X$. So $(f_i)_{i \in I, \sqsubseteq}$ converges uniformly to f (Exercise 4.7.30), hence continuously, and similarly $(x_i)_{i \in I, \sqsubseteq}$ converges to x. Using Exercise 5.4.16, $(f_i(x_i))_{i \in I, \sqsubseteq}$ converges to $f(x)$ in Y. So App preserves limits, hence is continuous (Proposition 4.7.18). So the topology of uniform convergence is conjoining. □

Exercise 6.6.5 (Even, compact, and uniform convergence) Let X be a topological space, and Y, d be a metric space. Show that every net $(f_i)_{i \in I, \sqsubseteq}$ of continuous maps from X to Y that converges uniformly to $f \in [X \to Y]$ also converges compactly to f. (See Exercise 5.4.13. We are *not* assuming X locally compact here.)

In particular, if X is compact, then uniform convergence in $[X \to Y]$ implies even convergence. Show that the converse also holds if d is a metric: if $(f_i)_{i \in I, \sqsubseteq}$ converges evenly to $f \in [X \to Y]$, X is compact and d is a metric, then $(f_i)_{i \in I, \sqsubseteq}$ converges uniformly to f.

6.6.2 Compactwise uniform convergence

Recall that, if X is locally compact, then the net $(f_i)_{i \in I}$ of continuous maps from X to Y converges compactly to f if and only if it converges to f in the compact-open topology (Exercise 5.4.13); and that the compact-open topology is the unique topology that is both splitting and conjoining in this case, hence coincides with the core-open topology, the Isbell topology, and the natural topology. When Y, d is, additionally, a metric space, this topology can be characterized more simply.

Definition 6.6.6 (Compactwise uniform convergence) Let X be a topological space, and Y, d be a hemi-metric space. A net $(f_i)_{i \in I, \sqsubseteq}$ of continuous maps from X to Y is said to *converge uniformly on every compact saturated subset*, or to converge *compactwise uniformly* to f, if and only if $(f_i)_{i \in I, \sqsubseteq}$ converges uniformly to f on every compact saturated subset Q of X, i.e., if and only if, for every $\epsilon > 0$, $\sup_{x \in Q} d(f(x), f_i(x)) < \epsilon$ for i large enough.

Proposition 6.6.7 (Continuous convergence $\Rightarrow$ compactwise uniform convergence) *Let X be a topological space, and Y, d be a metric space. Every net $(f_i)_{i \in I, \sqsubseteq}$ of continuous maps from X to Y that converges continuously to f in $[X \to Y]$ also converges compactwise uniformly.*

Proof Assume the contrary. There is a compact saturated subset Q of X and an $\epsilon > 0$ such that, for every $i_1 \in I$, there is an $i \in I$ above i_1 such that $\sup_{x \in Q} d(f(x), f_i(x)) \geqslant \epsilon$. So, for every $i_1 \in I$, there is an element x_{i_1} in Q such that $d(f(x_{i_1}), f_i(x_{i_1})) > \epsilon/2$ for some i above i_1. (Use the Axiom of Choice to collect them.) By Corollary 4.7.27, the net $(x_i)_{i \in I, \sqsubseteq}$ has a subnet $(x_{\alpha(j)})_{j \in J, \preceq}$ that converges to some element x of Q. Since $(f_i)_{i \in I, \sqsubseteq}$ converges continuously to f, $(f_i(x_{\alpha(j)}))_{(i,j) \in I \times J, \sqsubseteq \times \preceq}$ converges to $f(x)$. So, for (i, j) large enough, $d(f(x), f_i(x_{\alpha(j)})) < \epsilon/4$. Also, $(f(x_{\alpha(j)}))_{j \in J, \preceq}$ converges to $f(x)$, since f is continuous (Proposition 4.7.18). So, for j large enough, $d(f(x_{\alpha(j)}), f(x)) = d(f(x), f(x_{\alpha(j)})) < \epsilon/4$. Hence, for (i, j) large enough, say above $(i_0, j_0) \in I \times J$, $d(f(x_{\alpha(j)}), f_i(x_{\alpha(j)})) < \epsilon/4 + \epsilon/4 = \epsilon/2$. Fix j above j_0, large enough that $\alpha(j)$ is above i_0: this is possible since α is cofinal. Let $i_1 = \alpha(j)$. There is an i above i_1 such that $d(f(x_{\alpha(j)}), f_i(x_{\alpha(j)})) > \epsilon/2$. Since $i_1 = \alpha(j)$ is above i_0, i is also above i_0; this contradicts the fact that, whenever i is above i_0 and j is above j_0, $d(f(x_{\alpha(j)}), f_i(x_{\alpha(j)})) < \epsilon/2$. □

Recall that the compact-open topology on $[X \to Y]$ is generated by subsets of the form $[Q \subseteq V]$, where Q is compact saturated in X and V is open in Y. When X is locally compact, this coincides with the core-open topology (Exercise 5.4.8), hence with the natural topology, by Proposition 5.4.20. Recall also that, in this case, convergence in the latter topology is exactly continuous convergence (Proposition 5.4.21).

Proposition 6.6.8 *Let X be a topological space, and Y, d be a hemi-metric space. Every net $(f_i)_{i \in I, \sqsubseteq}$ of continuous maps from X to Y that converges compactwise uniformly to $f \in [X \to Y]$ also converges to f in the compact-open topology.*

Proof Assume $(f_i)_{i \in I, \sqsubseteq}$ converges to f compactwise uniformly. Using Exercise 4.7.29, it is enough to show that if $f \in [Q \subseteq V]$, where Q is compact saturated in X and V is open in Y, then $f_i \in [Q \subseteq V]$ for i large enough. For every $y \in V$, there is an $\epsilon_y > 0$ such that $B^d_{y, <\epsilon_y} \subseteq V$, by Lemma 6.1.5. (Use the Axiom of Choice to collect these ϵ_y as y varies.) The open balls $B^d_{y, <\epsilon_y/2}$, $y \in V$, form an open cover of V, hence of $f[Q]$. Since f is continuous, $f[Q]$ is compact (Proposition 4.4.13), so there is a finite subset E of Q such that the open balls $B^d_{f(x'), <\epsilon_{f(x')}/2}$, $x' \in E$, cover $f[Q]$. Let $\epsilon = \min_{x' \in E} \epsilon_{f(x')} > 0$. For i large enough, $\sup_{x \in Q} d(f(x), f_i(x)) < \epsilon/2$. For every $x \in Q$, find $x' \in E$ so that $f(x) \in B^d_{f(x'), <\epsilon_{f(x')}/2}$. Then $d(f(x'), f_i(x)) \leqslant d(f(x'), f(x)) + d(f(x), f_i(x)) < \epsilon_{f(x')}/2 + \epsilon/2 \leqslant \epsilon_{f(x')}$. Since $B^d_{f(x'), <\epsilon_{f(x')}} \subseteq V$, $f_i(x)$ is in V. So $f_i \in [Q \subseteq V]$. □

Putting it all together, we obtain:

Theorem 6.6.9 (Compactwise = exponential) *Let X be a locally compact topological space, and Y, d be a metric space. The notions of convergence in the core-open topology, the natural topology, and the compact-open topology all coincide both with continuous convergence and with compactwise uniform convergence on $[X \to Y]$.*

In other words, $(f_i)_{i \in I, \sqsubseteq}$ converges to f uniformly on every compact saturated subset of X if and only if $(f_i)_{i \in I, \sqsubseteq}$ converges to f in the compact-open topology, if and only if, for every net $(x_j)_{j \in J, \preceq}$ that converges to x in X, $(f_i(x_j))_{(i,j) \in I \times J, \sqsubseteq \times \preceq}$ converges to $f(x)$ in Y.

Let us revisit the second Arzelà–Ascoli–Kelley Theorem 5.4.40 in case the target space Y is not only T_2, but metric. The following notion of equicontinuity is the natural generalization of the eponymous notion we used in the original Arzelà–Ascoli Theorem 3.5.12. This is just continuity of every $f \in E$, uniformly in f: the open neighborhood U of x below is independent of $f \in E$.

Lemma 6.6.10 (Equicontinuity) *Let X be a topological space, and Y, d be a hemi-metric space. A subset E of $[X \to Y]$ is* equicontinuous *at $x \in X$ if and only if, for every $\epsilon > 0$, there is an open neighborhood U of x such that, for every $f \in E$, for every $x' \in U$, $d(f(x), f(x')) < \epsilon$.*

E is equicontinuous *if and only if it is equicontinuous at every point of X. Every equicontinuous subset of $[X \to Y]$ is evenly continuous. Conversely,*

if E is evenly continuous, Y, d is metric, and $E(x)$ is compact, then E is equicontinuous at x.

Proof Assume E equicontinuous. Fix $x \in X$, $y \in Y$, and an open neighborhood V of y. By Lemma 6.1.5, $B^d_{y,<\epsilon} \subseteq V$ for some $\epsilon > 0$. Since E is equicontinuous, there is an open neighborhood U of x such that, for every $f \in E$, for every $x' \in U$, $d(f(x), f(x')) < \epsilon/2$. Let $W = B^d_{y,<\epsilon/2}$. Then, for every $f \in E$ such that $f(x) \in W$, for every $x' \in U$, $d(y, f(x')) \leqslant d(y, f(x)) + d(f(x), f(x')) < \epsilon$, so $f(x') \in B^d_{y,<\epsilon} \subseteq U$.

Conversely, assume that E is evenly continuous, and that $E(x)$ is compact. Fix $\epsilon > 0$. For every $y \in Y$, letting $V = B^d_{y,<\epsilon/2}$, there is an open neighborhood U_y of x and an open neighborhood W_y of y such that, for every $f \in E$ such that $f(x) \in W_y$, $f[U_y] \subseteq B^d_{y,<\epsilon/2}$. By Lemma 6.1.5, we can take W_y of the form $B^d_{y,<\epsilon_y}$ for some $\epsilon_y > 0$. We can also require that $\epsilon_y < \epsilon/2$. The collection of all opens W_y, $y \in Y$, is an open cover of Y, hence of $E(x)$. Find a finite set E' of points $y \in Y$ such that every element $f(x)$, $f \in E$, is in some W_y, $y \in E'$. Let $U = \bigcap_{y \in E'} U_y$. This is an open neighborhood of x, since E' is finite. For every $f \in E$, and every $x' \in U$, $f(x)$ is in some $W_y = B^d_{y,<\epsilon_y}$ for some $y \in E'$, and since $f[U_y] \subseteq B^d_{y,<\epsilon/2}$, $f(x')$ is in $B^d_{y,<\epsilon/2}$. So $d(f(x), f(x')) \leqslant d(f(x), y) + d(y, f(x')) = d(y, f(x)) + d(y, f(x')) < \epsilon_y + \epsilon/2 < \epsilon$. ☐

We obtain the following by using Theorem 5.4.38 for the first part, and Theorem 5.4.40 for the second part. The latter applies because every metric space is T_3 (Proposition 6.1.13).

Theorem 6.6.11 (Arzelà–Ascoli Theorem) *Let X be a core-compact space, and Y, d be a metric space.*

The compact subsets K of $[X \to Y]^{©}$ are exactly those that are equicontinuous, pointwise closed, and such that $K(x)$ is compact for every $x \in X$.

The subsets E of $[X \to Y]^{©}$ that are included in a compact subset are exactly those that are equicontinuous and such that $cl(E(x))$ is compact for every $x \in X$. In this case, the pointwise closure of E, i.e., its closure in the topology of pointwise convergence, is such a compact subset, and coincides with the closure of E in $[X \to Y]^{©}$.

If X is compact, then, if $(f_i)_{i \in I, \sqsubseteq}$ converges compactwise uniformly to f, it converges uniformly to f. The converse is obvious. So, and using Theorem 6.6.9 and Corollary 4.7.12 for the second part:

Fact 6.6.12 *Let Y, d be a metric space. If X is a compact topological space, then uniform convergence is equivalent to compactwise uniform convergence. If X is compact and locally compact (e.g., compact and T_2), then the topology of uniform convergence coincides with the compact-open, the core-open, and the natural topology.*

So, if X is a compact, locally compact space, and Y, d is a metric space, then Theorem 6.6.11 still holds, when one replaces $[X \to Y]^{\circledcirc}$ by the space $[X \to Y]$ with the topology of uniform convergence. In particular, using Corollary 4.7.27, and letting $E = \{f_i \mid i \in I\}$, we obtain the following (mild generalization) of Arzelà and Ascoli's original Theorem 3.5.12:

Corollary 6.6.13 (Original Arzelà–Ascoli Theorem) *Let X be a compact, locally compact space, and Y, d be a metric space. Every net $(f_i)_{i \in I, \sqsubseteq}$ of maps that is equicontinuous and such that $cl\{f_i(x) \mid i \in I\}$ is compact for $x \in X$ has a uniformly convergent subnet.*

Exercise 6.6.14 Let X be a core-compact space. Show that the subsets E of $[X \to \mathbb{R}]^{\circledcirc}$ that are included in a compact subset are exactly the equicontinuous, pointwise bounded subsets. A set E is *pointwise bounded* if and only if $\sup_{y \in E(x)} |y| < +\infty$.

Deduce that, if X is compact and locally compact, then every equicontinuous net of maps $(f_i)_{i \in I, \sqsubseteq}$ such that $\sup_{i \in I} |f_i(x)| < +\infty$ for every $x \in X$ has a uniformly convergent subnet.

6.6.3 A bit of category theory III: SMCCs

Remember that, for every map $f : Z \times X \to Y$, $\Lambda_X(f)$ maps $z \in Z$ to $f(z, \_) : X \to Y$. The tensor product of two hemi-metrics was defined in Exercise 6.5.3.

Proposition 6.6.15 *Let X, d and Y, d' be two hemi-metric spaces. For each $c \in \mathbb{R}^+$, let $[X \to Y]_c$ be the set of all c-Lipschitz maps from X, d to Y, d', with the sup hemi-metric $\sup_X d'$.*

The map $\mathsf{App}: [X \to Y]_1 \times X \to Y$ is 1-Lipschitz, where $[X \to Y]_1 \times X$ is equipped with the tensor product hemi-metric $\sup_X d' \otimes d$.

For every hemi-metric space Z, d'', and every 1-Lipschitz map f from $Z \times X$ to Y, where $Z \times X$ is equipped with the tensor product hemi-metric $d'' \otimes d$, the map $\Lambda_X(f)$ is 1-Lipschitz from Z to $[X \to Y]_1$.

Proof The first claim amounts to $d'(\mathsf{App}(f, x), \mathsf{App}(g, x')) \leqslant \sup_X d'(f, g) + d(x, x')$. Indeed, $d'(\mathsf{App}(f, x), \mathsf{App}(g, x')) = d'(f(x), g(x')) \leqslant$

$d'(f(x), f(x')) + d'(f(x'), g(x'))$ (by the triangular inequality) $\leqslant d(x, x') + \sup_X d'(f, g)$ (because f is 1-Lipschitz, and by definition of $\sup_X d'$). For the second claim, $\sup_X d'(\Lambda_X(f)(z), \Lambda_X(f)(z')) = \sup_{x \in X} d'(f(z, x), f(z', x)) \leqslant \sup_{x \in X} (d'' \otimes d)((z, x), (z', x)) = d''(z, z')$. □

Proposition 6.6.15 does *not* state that the category **HMet** is Cartesian-closed, because the tensor product of two hemi-metric spaces is not the product in **HMet**. One can repair this by turning to ultra-hemi-metric spaces (see Exercise 6.6.20), or by weakening the requirements of Cartesian-closed categories. To accomplish the latter, we must first weaken the notion of products to that of a monoidal structure on a category.

A *monoidal structure* on a category **C** is a 4-tuple $(\otimes, \mathbf{I}, \ell, \mathbf{r}, \alpha)$ where the following conditions are satisfied. The reader is invited to check that any notion of finite product on a category yields a monoidal structure $(\times, 1, \ell, \mathbf{r}, \alpha)$ where 1 is a terminal object (in **Set**, **Top**, or **HMet**, $1 = \{*\}$), $\ell_A : 1 \times A \to A$ equals π_2 (in **Set**, **Top**, or **HMet**, $\ell_A(*, x) = x$), $\mathbf{r}_A : A \times 1 \to A$ equals π_1 (in **Set**, **Top**, or **HMet**, $\mathbf{r}_A(x, *) = x$), and $\alpha_{A,B,C} : (A \times B) \times C \to A \times (B \times C)$ equals $\langle \pi_1 \circ \pi_1, \langle \pi_2 \circ \pi_1, \pi_2 \rangle \rangle$ (in **Set**, **Top**, or **HMet**, $\alpha_{A,B,C}((x, y), z) = (x, (y, z))$).

- $\otimes$ is a functor from $\mathbf{C} \times \mathbf{C}$ to $\mathbf{C}$. In general, the category $\mathbf{C} \times \mathbf{C}$ has as objects all pairs (A, B) of an object A of **C** and an object B of $\mathbb{C}$, and as morphisms from (A, B) to (C, D) all pairs of morphisms $(A \xrightarrow{f} B, C \xrightarrow{g} D)$, with the obvious identities and composition. Concretely, that $\otimes$ is a functor means that given any two morphisms $f : A \to B$ and $g : C \to D$, there is a morphism $f \otimes g : (A \otimes C) \to (B \otimes D)$ satisfying $\mathrm{id}_A \otimes \mathrm{id}_C = \mathrm{id}_{A \otimes C}$ and $(f' \otimes g') \circ (f \otimes g) = (f' \circ f) \otimes (g' \circ g)$.

  Given two objects A, B of **C**, $A \otimes B$ is called the *tensor product* of A and B.
- **I** is an object of **C**, called the *tensor unit*.
- The *left unit* ℓ is a natural isomorphism from $\mathbf{I} \otimes \_$ to id. That is, for every morphism $f : A \to B$, $f = \ell_B \circ (\mathrm{id}_{\mathbf{I}} \otimes f) \circ {\ell_A}^{-1}$.
- The *right unit* $\mathbf{r}$ is a natural isomorphism from $\_ \otimes \mathbf{I}$ to id: for every morphism $f : A \to B$, $f = \mathbf{r}_B \circ (f \otimes \mathrm{id}_{\mathbf{I}}) \circ {\mathbf{r}_A}^{-1}$.
- The *associator* α is a natural isomorphism from $(\_ \otimes \_) \otimes \_$ to $\_ \otimes (\_ \otimes \_)$. That is, $\alpha_{A,B,C} : (A \otimes B) \otimes C \to A \otimes (B \otimes C)$ is such that, for any three morphisms $f : A \to A'$, $g : B \to B'$, and $h : C \to C'$, $(f \otimes (g \otimes h)) \circ \alpha_{A,B,C} = \alpha_{A',B',C'} \circ ((f \otimes g) \otimes h)$.
- The following *coherence conditions* are satisfied:

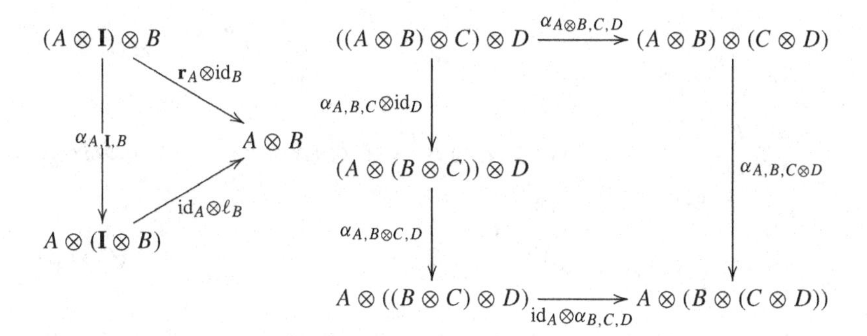

Contrarily to products, tensor products are not determined uniquely, even up to iso.

In **HMet** (and this was the whole point here), there is a monoidal structure where $\otimes$ is the (aptly named) tensor product of Exercise 6.5.3, $\mathbf{I}$ is any one-element hemi-metric space $\{*\}$, $\ell_A(*, x) = x$, $\mathbf{r}_A(x, *) = x$, and $\alpha_{A,B,C}((x, y), z) = (x, (y, z))$.

A category with a monoidal structure is called a *monoidal category*. An *exponential* object in a monoidal category, generalizing the notion in categories with binary products, is an object Y^X, together with a morphism $\mathsf{App}\colon Y^X \times X \to Y$, and a a collection of morphisms $\Lambda_X(f)\colon Z \to Y^X$, one for each morphism $f\colon Z \otimes X \to Y$, where Z is an arbitrary object of $\mathbf{C}$, satisfying the equations

$$
\begin{array}{llcl}
(\beta) & \mathsf{App} \circ (\Lambda_X(f) \otimes \mathrm{id}_X) & = & f \\
 & & & \text{for every } f\colon Z \otimes X \to Y \\
(\eta) & \Lambda_X(\mathsf{App}) & = & \mathrm{id}_{Y^X} \\
(\sigma) & \Lambda_X(f) \circ g & = & \Lambda_X(f \circ (g \otimes \mathrm{id}_X)) \\
 & & & \text{for all } f\colon Z \otimes X \to Y, g\colon Z' \to Z\,.
\end{array}
$$

These are the exact same equations (β), (η), and (σ) as in Section 5.5.1, with $\times$ replaced by $\otimes$.

A *monoidal closed category* is defined as a Cartesian-closed category, only replacing binary products by tensor products, i.e., as a monoidal category where every pair of objects has an exponential; equivalently, where the functor $\_ \otimes X$ is a left adjoint, for every object X, the right adjoint being $\_^X$, the counit being App, and the unit being $\Lambda_X(\mathrm{id}_X)$.

A *symmetric monoidal* structure is a monoidal structure as above, together with a *commutator* natural isomorphism $\mathbf{c}_{A,B} : A \otimes B \to B \otimes A$ (in **Set**, **Top**, or **HMet**, we will have $\mathbf{c}_{A,B}(x, y) = (y, x)$), satisfying the following additional coherence conditions:

$$A \otimes B \xrightarrow{\mathbf{c}_{A,B}} B \otimes A \xrightarrow{\mathbf{c}_{B,A}} A \otimes B$$

$$A \otimes \mathbf{I} \xrightarrow{\mathbf{r}_A} A \qquad A \otimes \mathbf{I} \xrightarrow{\mathbf{c}_{A,\mathbf{I}}} \mathbf{I} \otimes A \xrightarrow{\ell_A} A$$

$$\begin{array}{ccc}
 & (A \otimes B) \otimes C \xrightarrow{\alpha_{A,B,C}} A \otimes (B \otimes C) & \\
 & \downarrow \mathbf{c}_{A \otimes B, C} & \downarrow \mathrm{id}_A \otimes \mathbf{c}_{B,C} \\
(C \otimes A) \otimes B \xrightarrow{\alpha_{C,A,B}} C \otimes (A \otimes B) & & \\
\downarrow \mathbf{c}_{C,A} \otimes \mathrm{id}_C & & \\
(A \otimes C) \otimes B & \xrightarrow{\alpha_{A,C,B}} & A \otimes (C \otimes B)
\end{array}$$

A *symmetric monoidal category* is a category with a symmetric monoidal structure, and a *symmetric monoidal closed category*, a.k.a. an *SMCC*, is one which is also monoidal closed. To sum up, we have shown:

Fact 6.6.16 *The categories* **HMet**, **QMet**, *and* **Met** *are SMCC.*

Also, every CCC is an SMCC. Fact 6.6.16 gives an example of an SMCC that has finite products, but which is not a CCC, as Exercise 6.6.18 shows.

Exercise 6.6.17 Let X, d and Y, d' be two hemi-metric spaces. Show that the map $\mathsf{App}\colon [X \to Y]_c \times X \to Y$ is $(c+1)$-Lipschitz, where $[X \to Y]_c \times X$ is equipped with the *topological* product hemi-metric $\sup_X d' \times d$.

Exercise 6.6.18 (**HMet**, **QMet**, **Met** not CCC, exponentiable $\Leftrightarrow$ quasi-ordering) Let X, d be a hemi-metric space. As in Section 5.5.1, X is *exponentiable* in **HMet** (resp., **QMet**, **Met** – each with its Cartesian product as symmetric monoidal structure) if and only if the exponential object Y^X exists for each object Y, or, equivalently, iff the functor $\_ \times X$ is a left adjoint. Let X be such an exponentiable object. Show that, for every hemi-metric (resp., quasi-metric, metric) space Z, d', for every equivalence relation $\equiv$ on Z, the map i that sends $(q_\equiv(z), x)$ to $q_{\equiv'}(z, x)$ must be 1-Lipschitz from $(Z/\equiv) \times X$ to $(Z \times X)/\equiv'$, where $\equiv'$ is the equivalence relation on $Z \times X$ such that $(z, x) \equiv' (z', x')$ iff $z \equiv z'$ and $x = x'$.

Now assume there are two points x, x' in X such that $d(x, x') > 0$ and $d(x, x') \neq +\infty$, and let Z, d' be the sub-metric space $\{-a, -a+\epsilon, a-\epsilon, a\}$ of $\mathbb{R}, \mathsf{d}_{\mathbb{R}}^{sym}$, where $0 < \epsilon \leqslant d(x, x')/2 < a < d(x, x')$. Let $\equiv$ be the equivalence relation on Z defined as the smallest such that $-a+\epsilon \equiv a-\epsilon$. Looking at the points $(-a, x)$ and (a, x'), deduce a contradiction.

Conclude that the exponentiable objects in **HMet** (resp., **QMet**, **Met**) are exactly the quasi-orders (resp., the orders, the discrete metric spaces), namely the spaces $X, d_\leqslant$ where $\leqslant$ is a quasi-ordering (resp., an ordering, resp. the equality relation) – see Exercise 6.1.15. The exponential Y^X is then exactly $[X \to Y]_1$, with the sup hemi-metric.

As a consequence, show that none of **HMet**, **QMet**, and **Met** is Cartesian-closed.

Exercise 6.6.19 Prove the following analogue of Theorem 5.5.1. Let **C** be any full subcategory of **HMet** with monoidal structure $(\otimes, \mathbf{I}, \ell, \mathbf{r}, \alpha)$. Assume that: (a) **I** is some one-element set $\{*\}$; (b) for all objects Z, X in **C**, the elements of $Z \otimes X$ are the pairs (z, x) with $z \in Z, x \in X$; (c) $h \otimes \mathrm{id}_X$ satisfies $(h \otimes \mathrm{id}_X)(z, x) = (h(z), x)$ for all $h: Z \to Z'$, $z \in Z$, $x \in X$; and (d) ℓ_X maps $(*, x) \in \mathbf{I} \otimes X$ to $x \in X$.

Let X, Y be two objects of **C**, and assume that there is an exponential object Y^X in **C**. Show that there is a unique isometry $\theta: Y^X \to [X \to Y]_1$, where $[X \to Y]_1$ is equipped with a uniquely determined hemi-metric.

Show also that application and currification are as expected; more precisely, $\mathsf{App}(h, x) = \theta(h)(x)$, and $\theta(\Lambda_X(f)(z)) = f(z, \_)$ for every $f: Z \otimes X \to Y$, $z \in Z$.

Let X be any object in **HMet** (resp., **QMet**, **Met**), with symmetric monoidal structure arising from its tensor product, as defined in Exercise 6.5.3. By Proposition 6.6.15, this category is an SMCC, so there is an exponential Y^X for every object Y. We found Y^X as $[X \to Y]_1$, with the sup hemi-metric. There is essentially no other choice: since exponentials are obtained as right adjoints, they are unique up to iso.

Exercise 6.6.20 (**UHMet**, **UQMet**, **UMet** are Cartesian-closed) Let **UHMet** (resp., **UQMet**, **UMet**) be the category of ultra-hemi-metric (resp., ultra-quasi-metric, resp., ultra-metric) spaces, with 1-Lipschitz maps as morphisms. Show that **UHMet**, **UQMet**, and **UMet** are Cartesian-closed. Start by checking that the finite products in these categories are defined as in **HMet**, then that $[X \to Y]_1$ (with the sup hemi-metric) is the exponential of X and Y.

6.7 Compactness and symcompactness

Consider any compact subset K of a hemi-metric space X, d. For every $\epsilon > 0$, the open balls $B^d_{x,<\epsilon}$, $x \in K$, form an open cover of K. This must have a finite subcover. We have just proved:

Lemma 6.7.1 (Precompact) *Every compact subset of a hemi-metric space X, d is precompact.*

Here, as in the metric case, we call a subset K of X *precompact* if and only if, for every $\epsilon > 0$, K is included in a union of finitely many open balls $B^d_{x_i,<\epsilon}$ of radius ϵ, $1 \leqslant i \leqslant n$, where $x_1, \ldots, x_n \in K$.

Compactness is a very weak property, and precompactness is even weaker.

Example 6.7.2 Every hemi-metric space X with a least element $\perp$ (i.e., such that $d(\perp, x) = 0$ for every $x \in X$) is compact, hence precompact. Actually, for every $\epsilon > 0$, this space as a whole is covered by the single open ball $B^d_{\perp,<\epsilon}$.

On the other hand, in *metric* spaces, precompactness plus completeness is equivalent to compactness (Theorem 3.4.8).

One can also characterize compactness exactly, using a similar idea. Instead of covering K with open balls with the same radius ϵ, we allow the radius to vary. Namely, we consider open balls of radius $\delta(x)$ around each point x:

Exercise 6.7.3 (Gauge) A *gauge* on a set K is an arbitrary map $\delta: K \to \mathbb{R}^+ \setminus \{0\}$.

Let X, d be a hemi-metric space. Show that a subset K of X is compact if and only if, for every gauge δ on K, there is a finite set of elements $x_1, \ldots, x_n$ of K such that every element y of K is at distance less than $\delta(x_i)$ to some x_i, $1 \leqslant i \leqslant n$; i.e., $d(x_i, y) < \delta(x_i)$.

6.7.1 Symcompact spaces

Definition 6.7.4 (Totally bounded, symcompact) Let d be a hemi-metric on a set X. The hemi-metric space X, d is *totally bounded* if and only if X, d^{sym} is precompact. It is *symcompact* if and only if X, d^{sym} is compact.

Example 6.7.5 Every closed interval $[a, b]$, with quasi-metric $\mathsf{d}_{\mathbb{R}}$, is symcompact, by the Borel–Lebesgue Theorem 3.3.4.

Example 6.7.6 Expanding on Example 6.7.5, any countable topological product of symcompact spaces, hence any countable product of closed intervals $\prod_{i\in I}[a_i, b_i], \widetilde{\prod_{i\in I}}\mathsf{d}_{\mathbb{R}}$, $I \subseteq \mathbb{N}$, is symcompact, by Tychonoff's Theorem 4.5.12. We shall study some of these spaces in more detail in Section 6.7.2.

Precompactness is in general useless, except on metric spaces, where it coincides with total boundedness. Clearly, if X is symcompact, then X is totally bounded.

Totally bounded hemi-metrics may fail to be bounded. E.g., take $X = \{1, 2\}$, with 1 and 2 at distance $+\infty$ from each other.

Total boundedness is a rather demanding property, as the following exercise shows – less demanding than requiring the symmetrization X, d^{sym} to be compact in general, though.

Exercise 6.7.7 Let X be a poset, seen as a hemi-metric space with hemi-metric $d_{\leqslant}$ (see Exercise 6.1.15). Show that the following are equivalent: (1) X is totally bounded; (2) X is symcompact; (3) X is finite.

Exercise 6.7.8 Let $X = \{1/(n+1) \mid n \in \mathbb{N}\}$ with metric $d = \mathsf{d}_{\mathbb{R}}^{sym}$. Show that X, d is totally bounded but not symcompact, and not even compact.

We shall see in Theorem 9.1.39 another characterization of symcompact spaces, where the duality between d and d^{op} will agree with another important duality, de Groot duality.

Lemma 6.7.9 (Totally bounded $\Rightarrow$ separable) *Let X, d be a quasi-metric space. If X is totally bounded, then X is separable.*

Proof For every $n \in \mathbb{N}$, let E_n be a finite subset of points of X such that, for every $x \in X$, there is a $y \in E_n$ such that $d^{sym}(x, y) < 1/2^n$. Let $A = \bigcup_{n=0}^{+\infty} E_n$, a countable set. We claim that A is dense in X, d^{sym}. For every non-empty open subset U of X, d^{sym}, let $x \in U$, and $n \in \mathbb{N}$ be such that $B^{d^{sym}}_{x,<1/2^n}$ is included in U (using Fact 6.1.7). Since E_n meets $B^{d^{sym}}_{x,<1/2^n}$, A also meets $B^{d^{sym}}_{x,<1/2^n}$, hence U. We conclude using Lemma 4.1.30. ☐

Theorem 6.7.10 *A compact space is metrizable if and only if it is countably-based and T_2. Then it is also separable.*

Proof Let X be compact. If X is metrizable, say with metric d, then it is T_2, and precompact by Lemma 6.7.1, hence totally bounded, hence separable by Lemma 6.7.9. Since d is a metric, X is countably-based by Fact 6.3.45.

Conversely, if X is countably-based and T_2, remember that X is T_4 since compact T_2 (Proposition 4.4.17), hence T_3. Then apply the Urysohn–Tychonoff Theorem 6.3.44. ☐

6.7.2 The directed and undirected Hilbert cubes

An important example of symcompact space is the *directed Hilbert cube* $[0, 1]^{\mathbb{N}}$ with the topological product quasi-metric $\widetilde{\prod}_{n\in\mathbb{N}}\mathsf{d}_{\mathbb{R}}$ (see Lemma 6.5.7), namely:

$$\widetilde{\prod}_{n\in\mathbb{N}}\mathsf{d}_{\mathbb{R}}(\vec{x}, \vec{y}) = \sup_{n\in\mathbb{N}} \frac{1}{2^n} \max(x_n - y_n, 0).$$

This follows from the following proposition, which also shows that the symmetrization of the directed Hilbert cube is the *Hilbert cube*, i.e., $[0, 1]^{\mathbb{N}}$ with the metric $\widetilde{\prod}_{n\in\mathbb{N}}\mathsf{d}_{\mathbb{R}}^{sym}$, explicitly:

$$\widetilde{\prod}_{n\in\mathbb{N}} \mathsf{d}_{\mathbb{R}}^{sym}(\vec{x}, \vec{y}) = \sup_{n\in\mathbb{N}} \frac{1}{2^n}|x_n - y_n|.$$

These notions are slightly generalized by replacing $[0, 1]^{\mathbb{N}}$ by $[0, 1]^I$, I countable.

Proposition 6.7.11 (Directed Hilbert cube, Hilbert cube) *For every countable set I, $[0, 1]^I$, equipped with the topological product quasi-metric $\widetilde{\prod}_{i\in I} \mathsf{d}_{\mathbb{R}}$, is a symcompact quasi-metric space, called a* directed Hilbert cube.

Its symmetrization is $[0, 1]^I$ with the metric $\widetilde{\prod}_{i\in I} \mathsf{d}_{\mathbb{R}}^{sym}$; this is called a Hilbert cube.

Proof We again assume that I is a subset of $\mathbb{N}$.

We first check that

$$\begin{aligned}(\widetilde{\prod}_{i\in I} \mathsf{d}_{\mathbb{R}})^{sym}(\vec{x}, \vec{y}) &= \max(\sup_{i\in I} 1/2^i \mathsf{d}_{\mathbb{R}}(x_i, y_i), \sup_{i\in I} 1/2^i \mathsf{d}_{\mathbb{R}}(y_i, x_i)) \\ &= \sup_{i\in I} \max(1/2^i \mathsf{d}_{\mathbb{R}}(x_i, y_i), 1/2^i \mathsf{d}_{\mathbb{R}}(y_i, x_i)) \\ &= \sup_{i\in I} 1/2^i \mathsf{d}_{\mathbb{R}}^{sym}(x_i, y_i) = (\widetilde{\prod}_{i\in I} \mathsf{d}_{\mathbb{R}}^{sym})(\vec{x}, \vec{y}).\end{aligned}$$

By Lemma 6.5.7, the topology of $[0, 1]^I$, $(\widetilde{\prod}_{i\in I} \mathsf{d}_{\mathbb{R}})^{sym}$ is then the product topology on $[0, 1]^I$, where $[0, 1]$ comes with its usual metric topology. Then $[0, 1]$ is compact by the Borel–Lebesgue Theorem (Proposition 3.3.4), and we conclude by Tychonoff's Theorem 4.5.12. □

Exercise 6.7.12 Let I be any countable set. Show that the topological space underlying the directed Hilbert cube $[0, 1]^I$, $\widetilde{\prod}_{i\in I} \mathsf{d}_{\mathbb{R}}$, i.e., the one obtained as $[0, 1]^I$ with the open ball topology of $\widetilde{\prod}_{i\in I} \mathsf{d}_{\mathbb{R}}$, is exactly $([0, 1]_\sigma)^I$.

Show that it is also the dcpo $([0, 1]^I)_\sigma$ obtained as the poset product of I copies of $[0, 1]$, with the Scott topology.

Exercise 6.7.12 allows us to equate the directed Hilbert cube with the continuous pointed dcpo $[0, 1]^I_\sigma$, qua topological spaces.

The directed Hilbert cube has an important role, if only because every hemi-metric space embeds (topologically, not isometrically) into it, provided it is separable.

Theorem 6.7.13 (Separable quasi-metric ⇒ embeds into a directed Hilbert cube) *Let X, d be a hemi-metric space, and I be any subset of X. The map $\eta^I_d : X \to [0, 1]^I_\sigma$ defined by $\eta^I_d(x) = (\max(1 - d(y, x), 0))_{y\in I}$ is continuous* *from X, d to $[0,1]^I_\sigma$, and also from X, d^{op} to $([0,1]^{op})^I_\sigma$, and from X, d^{sym} to $[0,1]^I$ with the product topology of the metric topology on $[0,1]$.*

If X is T_0 and I is dense in X, d^{sym}, then η^I_d is an embedding (in the topological sense) of X, d into $[0,1]^I_\sigma$.

Proof By Corollary 6.3.4, we can assume that d is 1-bounded. Then $\eta^I_d(x)$ is the family $(1-d(y,x))_{y\in I}$.

To show that η^I_d is continuous from X, d to $[0,1]^I_\sigma$, using Trick 4.5.4, it is enough to show that $x \mapsto 1-d(y,x)$ is continuous. By the triangular inequality, $(1-d(y,x)) - (1-d(y,x')) \leqslant d(x,x')$, so $x \mapsto 1-d(y,x)$ is 1-Lipschitz. So it is continuous by Fact 6.2.3.

We show that η^I_d is also continuous from X, d^{op} to $([0,1]^{op})^I_\sigma$ in a similar way, realizing that the map $x \mapsto d(y,x)$ is 1-Lipschitz from X, d^{op} to $[0,1], \mathrm{d}^{op}_{\mathbb{R}}$ for every $y \in I$, and that the open ball topology of $[0,1], \mathrm{d}^{op}_{\mathbb{R}}$ is the same as that of $([0,1]^{op})_\sigma$. It follows that η^I_d is also continuous from X, d^{sym} to $[0,1]^I$, using Proposition 6.7.11.

Assume now that I is dense in X, d^{sym}.

We claim that any open ball $B^d_{x,<\epsilon}$ of X, d, with x in I, is the inverse image of $U_{x,>1-\epsilon}$ by η^I_d, where $U_{x,>1-\epsilon}$ is the set of all elements $\vec{s} = (s_y)_{y\in I}$ of $[0,1]^I_\sigma$ such that $s_x > 1-\epsilon$. (This only makes sense if $x \in I$, which is why we assumed so.) Indeed, $\eta^I_d(x') \in U_{x,>1-\epsilon}$ iff $1-d(x,x') > 1-\epsilon$, iff $d(x,x') < \epsilon$. It is easy to see that $U_{x,>1-\epsilon}$ is Scott-open. For every open subset U of X, d, U is a union of open balls $B^d_{x,<\epsilon}$ with $x \in I$ by Lemma 6.3.47, hence the inverse image of the corresponding union of opens $U_{x,>1-\epsilon}$, which is again Scott-open. So η^I_d is almost open from X, d to $[0,1]^I_\sigma$.

We claim that $\eta^I_d(x) \leqslant \eta^I_d(x')$ iff $x \leqslant^d x'$. The if direction is because every continuous map is monotonic (Proposition 4.3.9). Conversely, let $x, x' \in X$ be such that $\eta^I_d(x) \leqslant \eta^I_d(x')$. For every $\epsilon > 0$, there is an element y of I in $B^{d^{sym}}_{x,<\epsilon/2}$. Since $\eta^I_d(x) \leqslant \eta^I_d(x')$, $1-d(y,x) \leqslant 1-d(y,x')$, so $d(y,x') \leqslant d(y,x)$. Since $d(y,x) \leqslant d^{sym}(x,y) < \epsilon/2$, $d(y,x') < \epsilon/2$ as well. Since $d(x,y) \leqslant d^{sym}(x,y) < \epsilon/2$, $d(x,x') \leqslant d(x,y) + d(y,x') < \epsilon$. Since $\epsilon > 0$ is arbitrary, $d(x,x') = 0$, so $x \leqslant^d x'$ by Proposition 6.1.8.

In particular, if X is also T_0, then η^I_d is an order embedding, and in particular injective. Since η^I_d is continuous and almost open, by Proposition 4.10.9 η^I_d is a (topological) embedding. □

Exercise 6.7.14 For every countable set I, show that the directed Hilbert cube $[0,1]^I, \prod_{i\in I} \mathrm{d}_{\mathbb{R}}$ is separable.

Corollary 6.7.15 *Every separable quasi-metric space embeds (topologically) into $[0,1]^{\mathbb{N}}_\sigma$.*

Proof Let X, d be a separable quasi-metric space, and I be a countable dense subset of X, d^{sym}. If I is infinite, then there is a bijection $f : I \to \mathbb{N}$, and $[0, 1]^I_\sigma$ is homeomorphic to $[0, 1]^{\mathbb{N}}_\sigma$ through the map that sends $(t_i)_{i \in I}$ to $(t_{f^{-1}(n)})_{n \in \mathbb{N}}$. Then apply Theorem 6.7.13.

If I is finite, then, up to another bijection, we may assume that $I = \{0, 1, \ldots, n-1\}$, for some $n \in \mathbb{N}$. Let $\eta(t_0, t_1, \ldots, t_{n-1}) = (t_0, t_1, \ldots, t_{n-1}, 0, 0, \ldots)$. We claim that η is both a topological and an order embedding of $[0, 1]^I_\sigma$ into $[0, 1]^{\mathbb{N}}_\sigma$. This is continuous, using Trick 4.5.4, since the projection maps and the constant null maps are constant. This is also clearly injective, and in fact $\eta(t_0, t_1, \ldots, t_{n-1}) \leqslant \eta(t'_0, t'_1, \ldots, t'_{n-1})$ iff $(t_0, t_1, \ldots, t_{n-1}) \leqslant (t'_0, t'_1, \ldots, t'_{n-1})$, showing that η is an order embedding. Any basic open $\Uparrow(t_0, t_1, \ldots, t_{n-1}) = \Uparrow t_0 \times \Uparrow t_1 \times \cdots \times \Uparrow t_{n-1}$ (the equality being by Lemma 5.1.53) in the continuous dcpo $[0, 1]^I_\sigma$ is the inverse image of $\Uparrow t_0 \times \Uparrow t_1 \times \cdots \Uparrow t_{n-1} \times [0, 1] \times [0, 1] \times \cdots$, which is open. So η is almost open, hence a topological embedding by Proposition 4.10.9.

Then $\eta \circ \eta^I_d$ is an embedding of X into $[0, 1]^{\mathbb{N}}_\sigma$. □

We have a similar statement for metric spaces.

Theorem 6.7.16 (Separable metric ⇒ embeds into a Hilbert cube) *Let X, d be a metric space, and I be any subset of X. The map η^I_d defined by $\eta^I_d(x) = (\max(1 - d(y, x), 0))_{y \in I}$ is continuous from X, d to $[0, 1]^I$.*

If I is dense in X, d, then η^I_d is a (topological) embedding of X, d into the Hilbert cube $[0, 1]^I$.

Proof Using Theorem 6.7.13, η^I_d is continuous from X, d^{sym} to $[0, 1]^I$, i.e., from X, d to $[0, 1]$. Now assume I dense in X, d. Since X, d is metric, X is T_2 (Lemma 6.1.10) hence T_0, so η^I_d is injective, by Theorem 6.7.13. Also, η^I_d is almost open from X, d to $[0, 1]^I_\sigma$. Since every open of the latter is open in $[0, 1]^I$, η^I_d is almost open from X, d to $[0, 1]^I$, so η^I_d is an embedding by Proposition 4.10.9. □

Corollary 6.7.17 *Every separable metric space embeds (topologically) into the Hilbert cube* $[0, 1]^{\mathbb{N}}$.

Exercise 6.7.18 Using Exercise 6.5.9, show that $[0, 1]^I$ is metrizable if and only if I is countable, thus justifying the countability restrictions in the definition of the Hilbert cube.

Similarly, show that $\mathbb{S}^I$ is not first-countable, hence not hemi-metrizable when I is uncountable (see Exercise 4.1.18 for Sierpiński space $\mathbb{S}$). Deduce that $[0, 1]^I_\sigma$

is hemi-metrizable (hence quasi-metrizable) iff I is countable. This justifies the countability restriction in the definition of the directed Hilbert cube.

Exercise 6.7.19 (Urysohn–Carruth Theorem) A hemi-metric d on a set X is called *radially convex* iff, for all $x, y, z \in X$ such that $z \leqslant^d y \leqslant^d x$, the equality $d(x, z) = d(x, y) + d(y, z)$ holds. Show that every separable quasi-metric space X, d has a topologically equivalent radially convex quasi-metric d'. Observe that $\widetilde{\prod}_{i \in \mathbb{N}} \mathrm{d}_{\mathbb{R}}$ is a radially convex quasi-metric on the directed Hilbert cube.

Exercise 6.7.20 (Wilson's Theorem again) Modifying Exercise 4.10.14 slightly, show that, for every countably-based space X, there is a continuous and almost open map η of X to $\mathbb{S}^{\mathbb{N}}$. Exhibit an embedding of $\mathbb{S}^{\mathbb{N}}$ into the directed Hilbert cube. Show that Wilson's Theorem is then a simple consequence of the fact that the directed Hilbert cube is hemi-metrizable. Is this really a different proof than the one we gave for Theorem 6.3.13?

Exercise 6.7.21 Let X be a countably-based T_3 space. Show that there is an embedding of X into the Hilbert cube $[0, 1]^{\mathbb{N}}$.

If we do not assume I to be countable any longer, we can also characterize the subspaces of the cubes $[0, 1]^I$. They won't be metrizable in general, but are exactly the $T_{3\frac{1}{2}}$ spaces.

Exercise 6.7.22 ($T_{3\frac{1}{2}} \Leftrightarrow$ embeds into a compact T_2 space) Let X be a topological space. Equip $[0, 1]$ with its usual, metric topology, and let $Z = [0, 1]^{[X \to [0,1]]}$ be the space of all maps (even non-continuous) from $[X \to [0, 1]]$ to $[0, 1]$. Let $\eta_X^{\text{SČ}} : X \to Z$ be defined by $\eta_X^{\text{SČ}}(x)(f) = f(x)$. Show that Z is compact T_2, that $\eta_X^{\text{SČ}}$ is continuous, and that $\eta_X^{\text{SČ}}$ is an embedding whenever X is $T_{3\frac{1}{2}}$.

Conclude that the following properties are equivalent: (a) X is $T_{3\frac{1}{2}}$; (b) X embeds into a *cube*, i.e., some power $[0, 1]^I$ of $[0, 1]$, for some set I; (c) X embeds into a compact T_2 space.

Exercise 6.7.23 (Stone–Čech compactification) Given any $T_{3\frac{1}{2}}$ space X, its *Stone–Čech compactification* βX is the closure of $\eta_X^{\text{SČ}}[X]$ in $Z = [0, 1]^{[X \to [0,1]]}$. Show that $\eta_X^{\text{SČ}}$ is a homeomorphism from X onto βX when X is compact. Show that β defines a functor from the category of $T_{3\frac{1}{2}}$ spaces and continuous maps to itself, where, for every morphism $g : X \to Y$, $\beta g : \beta X \to \beta Y$ is defined by

$\beta g(F)(f) = F(f \circ g)$ for every $F \in \beta X$ and every $f \in [Y \to [0,1]]$. Show that $\eta^{S\check{C}}$ defines a natural transformation from id to β.

Deduce that βX is the free compact T_2 space over the category of $T_{3\frac{1}{2}}$ spaces. In other words, βX satisfies the following universal property: for every compact T_2 space Y, every continuous map $g : X \to Y$ extends to a unique continuous map $g' : \beta X \to Y$, i.e., $g = g' \circ \eta_X^{S\check{C}}$.

Exercises 6.7.22 and 6.7.23 allow us to consider any $T_{3\frac{1}{2}}$ space X as a subspace of the compact T_2 space βX. Then, by construction, X is a dense subset of βX. The last part of Exercise 6.7.23 allows us to state that any continuous map g from X to a compact T_2 space Y extends to a unique continuous map g' from βX to Y, in the ordinary sense that $g'_{|X} = g$.

Exercise 6.7.24 Using Alexandroff compactifications (Exercise 4.8.9), show that every locally compact T_2 space must be $T_{3\frac{1}{2}}$.

Exercise 6.7.25 (Arens' Theorem) The aim of this exercise is to show that the $T_{3\frac{1}{2}}$ spaces X that are exponentiable objects in the category of T_2 spaces and continuous maps are exactly the locally compact T_2 spaces (Arens, 1946).

In one direction, show that every locally compact T_2 space is $T_{3\frac{1}{2}}$, and exponentiable in the category of T_2 spaces.

Conversely, fix a $T_{3\frac{1}{2}}$ space X, and let $Y = [0,1]$ with its metric topology. Fix an exponential topology $\mathcal{O}$ on $[X \to Y]$. Let f_0 be the constant 0 map from X to Y, and let V_0 be an open neighborhood of 0 in Y that does not contain 1. Fixing $x_0 \in X$, show that there is an open neighborhood $W \in \mathcal{O}$ of f_0 and an open neighborhood U of x_0 such that $W \times U \subseteq \mathsf{App}^{-1}(V_0)$. The main point of the exercise is to show that $cl(U)$ is compact.

Fix an open cover $(U_i)_{i \in I_0}$ of $cl(U)$. Build an open cover $(U_i)_{i \in I}$ of X by adding the complement of $cl(U)$ to it. That is, let I be I_0 plus a fresh index $*$, and let U_* be the complement of $cl(U)$. Call a closed subset F of X *located* iff $F \subseteq U_i$ for some $i \in I$.

Define another topology $\mathcal{O}'$ on $[X \to Y]$, whose subbasic open sets are of the form $[F \subseteq V] = \{f \in [X \to Y] \mid f[F] \subseteq V\}$, where F is a located, closed subset of X, and V is open in Y. Show that $\mathcal{O}'$ is a conjoining topology. (If $(f, x) \in \mathsf{App}^{-1}(V)$, show that one can find a closed set F with $F \subseteq U_i$ for some $i \in I$ such that $(f, x) \in [F \subseteq V] \times int(F) \subseteq \mathsf{App}^{-1}(V)$. You will need to use the fact that X is regular for that.)

From this, deduce that $\mathcal{O}$ is finer than $\mathcal{O}'$, and that W must contain a finite intersection $\bigcap_{k=1}^{n} [F_k \subseteq V_k]$ containing (f_0, x_0), where each F_k is located and closed, and V_k is open in Y. Using complete regularity, show that $\bigcup_{k=1}^{n} F_k$ contains U.

Deduce that $cl(U)$ is contained in finitely many opens U_i, $i \in I_0$ (in particular, $i \neq *$). So $cl(U)$ is a compact neighborhood of x_0.

Argue that X must therefore be locally compact, and conclude.

Exercise 6.7.26 Using Arens' Theorem (Exercise 6.7.25), show that the category of T_2 spaces and continuous maps is not Cartesian-closed.

7

Completeness

Completeness is a handy property in metric spaces. In hemi-metric spaces, several definitions of completeness have been proposed, which all coincide with this notion on metric spaces, but are different in general. We shall explore two. The stronger one, Smyth-completeness, was introduced by Smyth (1988), the weaker notion of Yoneda-completeness by Bonsangue *et al.* (1998). We shall unite these notions of completeness through an extension of the formal ball construction of Weihrauch and Schneider (1981), whose potential was revealed by Edalat and Heckmann (1998). Finally, we shall explore Choquet-completeness, an incomparable notion, except (again) on metric spaces. This will lead us to Martin's theorems on spaces that have models, and to the study of so-called Polish spaces.

7.1 Limits, d-limits, and Cauchy nets

On metric and on general hemi-metric spaces, the notion of limit specializes to the following:

Lemma 7.1.1 *Let X, d be a hemi-metric space. The net $(x_i)_{i \in I, \sqsubseteq}$ converges to $x \in X$ if and only if, for every $\epsilon > 0$, $d(x, x_i) < \epsilon$ for i large enough, i.e., if and only if $(d(x, x_i))_{i \in I, \sqsubseteq}$ converges to 0 (in $\overline{\mathbb{R}^+}$, $\mathsf{d}_{\mathbb{R}}^{sym}$).*

Proof If $(x_i)_{i \in I, \sqsubseteq}$ converges to $x \in X$, then, by definition, for every open subset U containing x, x_i is in U for i large enough. Taking $U = B^d_{x, <\epsilon}$ yields $d(x, x_i) < \epsilon$ for i large enough. Conversely, assume that, for every $\epsilon > 0$, $d(x, x_i) < \epsilon$ for i large enough. Consider any open neighborhood U of x. By Lemma 6.1.5, there is an open ball $B^d_{x, <\epsilon}$ with center x included in U. By assumption, x_i is in $B^d_{x, <\epsilon}$, hence in U, for i large enough. □

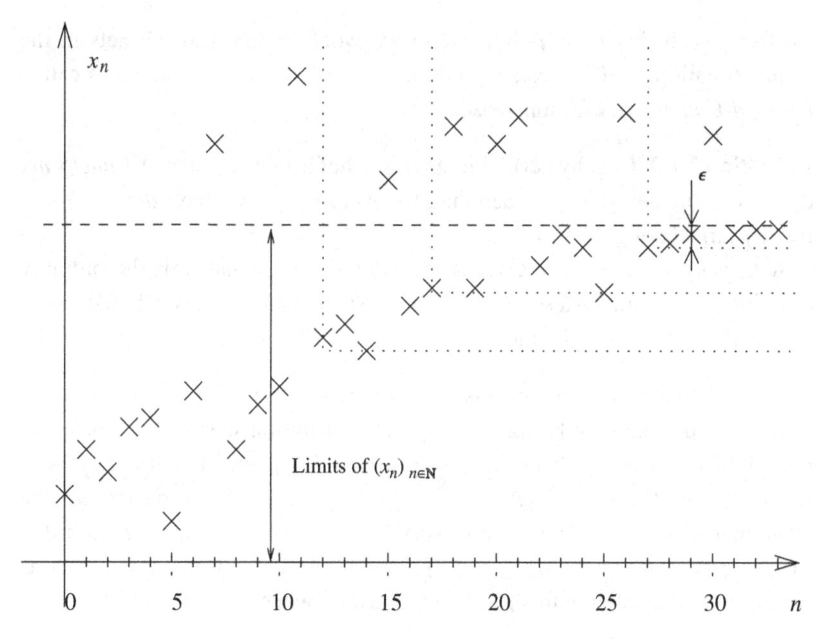

Figure 7.1 Limits in hemi-metric spaces.

We have noticed that notions of limits could be simplified on first-countable spaces (Exercise 4.7.14). This applies here, by Lemma 6.3.6. So, for example, using Exercise 4.7.14:

Fact 7.1.2 *A subset F of a hemi-metric space is closed if and only if it contains all limits of sequences of elements of F.*

So closed and sequentially closed subsets coincide in hemi-metric spaces.

One may refine one's intuition about limits in hemi-metric spaces by looking at Figure 7.1, representing a typical case of convergence in $\mathbb{R}$, $\mathsf{d}_{\mathbb{R}}$. In metric spaces, $(x_n)_{n\in\mathbb{N}}$ converges to x iff x_n becomes as close as we want when n becomes large. In a hemi-metric space that is not necessarily metric, we only require x_n to become larger than x, up to an error ϵ that we can make as small as we wish.

7.1.1 Cauchy nets

Completeness is defined in terms of Cauchy sequences, in metric spaces. We need to replace these by Cauchy *nets*, in general. See Definition 7.1.3 below. Moreover, we need to require $d(x_i, x_j) < \epsilon$ for i, j large enough, but only for those such that $i \sqsubseteq j$—otherwise we would define Cauchy nets in the symmetrization X, d^{sym} instead of X, d. The following are sometimes called *forward-Cauchy nets* for this reason.

Definition 7.1.3 (Cauchy net) Let X, d be a hemi-metric space. A *Cauchy net* on X, d is any net $(x_i)_{i \in I, \sqsubseteq}$ such that, for every $\epsilon > 0$, we have $d(x_i, x_j) < \epsilon$ for i, j large enough, $i \sqsubseteq j$.

A *Cauchy sequence* is a Cauchy net where $I = \mathbb{N}$ and $\sqsubseteq$ is the ordinary ordering on $\mathbb{N}$: namely, a sequence $(x_n)_{n \in \mathbb{N}}$ such that, for every $\epsilon > 0$, there is a natural number n_0 such that $d(x_i, x_j) < \epsilon$ for all $i, j \in \mathbb{N}$ with $n_0 \leqslant i \leqslant j$.

In the definition of Cauchy nets, the pairs i, j with $i \sqsubseteq j$ are quasi-ordered in the product quasi-ordering $\sqsubseteq \times \sqsubseteq$. The definition means that, for every $\epsilon > 0$, there are two indices i_0, j_0 in I with $i_0 \sqsubseteq j_0$ such that $d(x_i, x_j) < \epsilon$ for all $i, j \in I$ with $i_0 \sqsubseteq i \sqsubseteq j$ and $j_0 \sqsubseteq j$. Since I is directed, we can even choose $j_0 = i_0$. That is, for every $\epsilon > 0$, there is an $i_0 \in I$ such that $d(x_i, x_j) < \epsilon$ whenever $i_0 \sqsubseteq i \sqsubseteq j$. The definition of Cauchy sequence is obtained by specializing this formula to the case where $I = \mathbb{N}$.

Exercise 7.1.4 So Definition 7.1.3 states that a Cauchy sequence $(x_n)_{n \in \mathbb{N}}$ is one such that, for every $\epsilon > 0$, there is an $n_0 \in \mathbb{N}$ such that, for all $m, n \in \mathbb{N}$ with $n_0 \leqslant m \leqslant n$, $d(x_m, x_n) < \epsilon$. In Section 3.4, we defined Cauchy sequences in a metric space X, d as those such that, for every $\epsilon > 0$, there is an $n_0 \in \mathbb{N}$ such that, for all $m, n \in \mathbb{N}$ with $n_0 \leqslant m$ and $n_0 \leqslant n$, $d(x_m, x_n) < \epsilon$. Show that, in the case that d is a metric, i.e., satisfies the symmetry axiom, the two definitions are equivalent.

In a metric space, all convergent nets are Cauchy. This is not the case in general hemi-metric spaces; however, we have the following.

Lemma 7.1.5 *Let X, d be a hemi-metric space. Every convergent net in X, d^{sym} is Cauchy in X, d, and in X, d^{sym}.*

Proof Assume that $(x_i)_{i \in I}$ converges to $x \in X$. By Lemma 7.1.1, for every $\epsilon > 0$, there is an $i_0 \in I$ such that $d^{sym}(x, x_i) < \epsilon/2$ for every $i \in I$ with $i_0 \sqsubseteq i$. Note that $d^{sym}(x, x_i) = d^{sym}(x_i, x)$. For all $i, j \in I$ with $i_0 \sqsubseteq i, j$, $d^{sym}(x_i, x_j) \leqslant d^{sym}(x_i, x) + d^{sym}(x, x_j) < \epsilon$. So $(x_i)_{i \in I}$ is Cauchy in X, d^{sym}. Since $d(x_i, x_j) \leqslant d^{sym}(x_i, x_j)$, it is also Cauchy in X, d. □

Example 7.1.6 In $\mathbb{R}$, $\mathsf{d}_{\mathbb{R}}$, the sequence of numbers $r_n = n$, $n \in \mathbb{N}$, is Cauchy: whenever $m \leqslant n$, $\mathsf{d}_{\mathbb{R}}(r_m, r_n) = 0$. In general, every monotone net is Cauchy in any hemi-metric space.

Example 7.1.7 Every quasi-ordered set can be seen as a hemi-metric space, by letting $d_{\leqslant}(x, y) = 0$ if $x \leqslant y$, $+\infty$ otherwise. Then every monotone net is Cauchy. We study this example more deeply in Exercise 7.1.8.

> **Exercise 7.1.8** Let X be a set with a quasi-ordering $\leqslant$. Show that the Cauchy sequences $(x_n)_{n \in \mathbb{N}}$ in $X, d_{\leqslant}$ are exactly those that are *eventually ascending*, i.e., such that there is an $n_0 \in \mathbb{N}$ such that, for all $i, j \in \mathbb{N}$ with $n_0 \leqslant i \leqslant j$, $x_i \leqslant x_j$. (This can help one in obtaining some intuition about Cauchy sequences in hemi-metric spaces: these are the ones that are eventually ascending, up to some allowable, arbitrarily small error ϵ. See Figure 7.1.)
>
> Let us now prove some counter-intuitive facts. Show that $(x_n)_{n \in \mathbb{N}}$ in $X, d_{\leqslant}$ converges to x if and only if it is eventually above x, i.e., there is an $n_0 \in \mathbb{N}$ such that $x \leqslant x_n$ for every $n \geqslant n_0$.
>
> Conclude that we cannot replace d^{sym} by d in Lemma 7.1.5: in a hemi-metric space X, d, there may be convergent sequences (in X, d) that are not Cauchy in X, d.

Not all Cauchy nets converge in general, even in metric spaces. In Section 3.4, we saw that, for every irrational number r, such as $\sqrt{2}$, the sequence $(\frac{1}{2^n}\lfloor 2^n r \rfloor)_{n \in \mathbb{N}}$ is a Cauchy sequence in $\mathbb{Q}$, $\mathrm{d}_{\mathbb{R}}^{sym}$ (i.e., with the L^1 metric) which has no limit in $\mathbb{Q}$.

In hemi-metric spaces, we shall define complete spaces as those where every Cauchy net converges. But the difficulty is in finding the right notion of limit. There are several natural candidates, which we explore now.

Figure 7.1 gives the impression that, among all limits, there is a largest one. This is not always the case; however, limits taken not in X, d but in X, d^{sym}, when they exist, are indeed largest limits in X, d:

Lemma 7.1.9 *Let X, d be a hemi-metric space, and $(x_i)_{i \in I, \sqsubseteq}$ be a net in X. If $(x_i)_{i \in I, \sqsubseteq}$ has a limit x in X, d^{sym}, then x is a largest limit of $(x_i)_{i \in I, \sqsubseteq}$ in X, d, with respect to $\leqslant^d$.*

Proof Use Lemma 7.1.1 to characterize limits. For every $\epsilon > 0$, $d^{sym}(x, x_i) < \epsilon$ for i large enough. In particular, $d(x, x_i) < \epsilon$ for i large enough, so x is a limit of $(x_i)_{i \in I, \sqsubseteq}$ in X, d. For any other limit y of $(x_i)_{i \in I, \sqsubseteq}$ in X, d, for every $\epsilon > 0$, $d(y, x_i) < \epsilon/2$ for i large enough since y is a limit in X, d, and $d(x_i, x) < \epsilon/2$ for i large enough since x is a limit in X, d^{sym}. So $d(y, x) \leqslant d(y, x_i) + d(x_i, x) < \epsilon$ for i large enough. It follows that $d(y, x) = 0$, so $y \leqslant^d x$ by Proposition 6.1.8. □

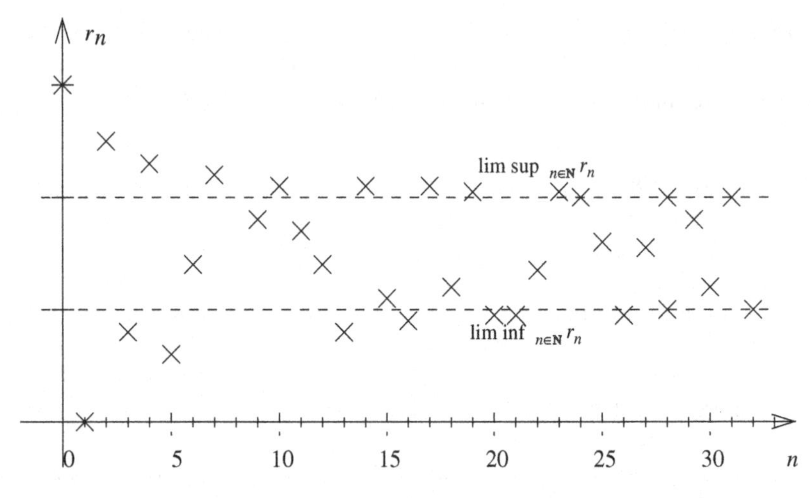

Figure 7.2 Limsups and liminfs of real numbers.

7.1.2 d-limits

In this chapter, we shall need other notions of limits. Let us start with various notions of limits in $\overline{\mathbb{R}^+}$, and in general in $\mathbb{R} \cup \{-\infty, +\infty\}$.

Definition 7.1.10 (Limsup, liminf) Consider any net $(r_i)_{i\in I,\sqsubseteq}$ of elements of $\mathbb{R} \cup \{-\infty, +\infty\}$. Its *limit superior*, a.k.a. *limsup*, is

$$\limsup_{i\in I,\sqsubseteq} r_i = \inf_{i\in I} \sup_{j\in I, i\sqsubseteq j} r_j.$$

Its *limit inferior*, a.k.a. *liminf*, is

$$\liminf_{i\in I,\sqsubseteq} r_i = \sup_{i\in I} \inf_{j\in I, i\sqsubseteq j} r_j.$$

See Figure 7.2 for an illustration. Note that limsups and liminfs always exist, even for non-convergent nets.

Exercise 7.1.11 Show that $\inf_{i\in I} r_i \leqslant \liminf_{i\in I,\sqsubseteq} r_i \leqslant \limsup_{i\in I,\sqsubseteq} r_i \leqslant \sup_{i\in I} r_i$, and that the inequalities are strict in general.

Show that $\liminf_{i\in I,\sqsubseteq} r_i = \limsup_{i\in I,\sqsubseteq} r_i$ if and only if $(r_i)_{i\in I,\sqsubseteq}$ converges in $\mathbb{R} \cup \{-\infty, +\infty\}$, and that, then, the limit $\lim_{i\in I,\sqsubseteq} r_i$ satisfies $\lim_{i\in I,\sqsubseteq} r_i = \liminf_{i\in I,\sqsubseteq} r_i = \limsup_{i\in I,\sqsubseteq} r_i$. The topology we take on $\mathbb{R} \cup \{-\infty, +\infty\}$ is the smallest one containing the intervals (a, b), $(a, +\infty]$, $[-\infty, b)$ with $a, b \in \mathbb{R}$, $a < b$. (Beware that this is not the open ball topology of $\mathsf{d}_{\mathbb{R}}^{sym}$, in which

$\{-\infty\} = B^{\mathsf{d}_{\mathbb{R}}^{sym}}_{-\infty,<1}$ and $\{+\infty\} = B^{\mathsf{d}_{\mathbb{R}}^{sym}}_{+\infty,<1}$ would be open as well. On $\mathbb{R}$, without $-\infty$ or $+\infty$, this topology does coincide with the familiar open ball topology, though.)

Exercise 7.1.12 Show that max is lim sup-continuous. That is, let a_{ik} be elements of $\mathbb{R} \cup \{-\infty, +\infty\}$, $1 \leqslant k \leqslant m$, $i \in I$, where I is directed in some quasi-ordering $\sqsubseteq$; then show that $\max_{k=1}^{m} \limsup_{i \in I, \sqsubseteq} a_{ik} = \limsup_{i \in I, \sqsubseteq} \max_{k=1}^{m} a_{ik}$.
Extend this to show that least upper bounds of countable families taken from $\prod_{k \in \mathbb{N}} [0, \frac{1}{2^k}]$, or more generally from any countable product $\prod_{k \in \mathbb{N}} [0, r_k]$ where $(r_k)_{k \in \mathbb{N}}$ converges to 0, are also lim sup-continuous. That is, letting $a_{ik} \in [0, r_k]$, $i \in I$, $k \in \mathbb{N}$, where I is directed in $\sqsubseteq$, show that $\sup_{k \in \mathbb{N}} \limsup_{i \in I, \sqsubseteq} a_{ik} = \limsup_{i \in I, \sqsubseteq} \sup_{k \in \mathbb{N}} a_{ik}$.

Exercise 7.1.13 Show that min is *not* lim sup-continuous. Consider, for example, $I = \mathbb{N}, a_{i1} = (-1)^i, a_{i2} = (-1)^{i+1} = -a_{i1}$ for every $i \in \mathbb{N}$. However, show that, for any net $(a_{i2})_{i \in I, \sqsubseteq}$ that converges in $\mathbb{R} \cup \{-\infty, +\infty\}$ with the topology of Exercise 7.1.11, say to a_2, the expected equality holds: $\min(\limsup_{i \in I, \sqsubseteq} a_{i1}, a_2) = \limsup_{i \in I, \sqsubseteq} \min(a_{i1}, a_{i2})$.

Exercise 7.1.14 Let $(r_i)_{i \in I, \sqsubseteq}$ be a net in $\mathbb{R} \cup \{-\infty, +\infty\}$. Show that, for any $i_0 \in I$, $\limsup_{i \in I, \sqsubseteq} r_i = \limsup_{i \in \uparrow i_0, \sqsubseteq} r_i$, and similarly for liminfs.

The important notion of limit used by Bonsangue *et al.* (1998) is the *d-limit* – which they call the forward-limit.

Definition 7.1.15 (d-limit) Let X, d be a hemi-metric space, and $(x_i)_{i \in I, \sqsubseteq}$ be a net in X, d. The point $x \in X$ is a *d-limit* of $(x_i)_{i \in I, \sqsubseteq}$ in X if and only if, for every $y \in X$, $d(x, y)$ is the limit superior $\limsup_{i \in I, \sqsubseteq} d(x_i, y)$. We also say that $(x_i)_{i \in I, \sqsubseteq}$ *d-converges* to x.

The definition of d-limits may look arbitrary. The following exercise may clarify the notion.

Exercise 7.1.16 Consider $X = \mathbb{R} \cup \{+\infty\}$ (resp., $X = \overline{\mathbb{R}^+}$), with quasi-metric $d = \mathsf{d}_{\mathbb{R}}$. Show that every net $(x_i)_{i \in I, \sqsubseteq}$ has a $\mathsf{d}_{\mathbb{R}}$-limit, namely $\limsup_{i \in I, \sqsubseteq} x_i$. (Extra question: show that a similar claim fails for $X = \mathbb{R} \cup \{-\infty, +\infty\}$, in the sense that there are sequences that do not have a $\mathsf{d}_{\mathbb{R}}$-limit in the latter space. But the limsup of any *Cauchy* net in $X, \mathsf{d}_{\mathbb{R}}$ is a $\mathsf{d}_{\mathbb{R}}$-limit.)

Limits in X, d and d-limits are different notions. A trivial difference is that d-limits are unique in every quasi-metric space X, d, while limits are far from being so.

Lemma 7.1.17 *Let X, d be a quasi-metric space. Every net in X has at most one d-limit.*

Proof Assume $(x_i)_{i \in I, \sqsubseteq}$ has two d-limits, x and y. Since both x and y are d-limits, $d(x, z) = d(y, z) = \limsup_{i \in I, \sqsubseteq} d(x_i, z)$, for every $z \in X$. For $z = x$, we obtain $d(y, x) = 0$; and for $z = y$, $d(x, y) = 0$. Since d is T_0, $x = y$. □

The notion of d-limit is even further removed from the notion of limit: some d-limits are indeed not limits at all, at least with respect to the open ball topology. Consider, for example, $X = \overline{\mathbb{R}^+}$, $d = \mathsf{d}_{\mathbb{R}}$, and let $(x_n)_{n \in \mathbb{N}}$ be an increasing sequence of positive reals, say $x_n = n$. Then the d-limit of $(x_n)_{n \in \mathbb{N}}$ is its limsup (Exercise 7.1.16), namely $+\infty$, but $+\infty$ is not a limit of $(x_n)_{n \in \mathbb{N}}$ in $\overline{\mathbb{R}^+}, \mathsf{d}_{\mathbb{R}}$ with its open ball topology: using Lemma 7.1.1, $\mathsf{d}_{\mathbb{R}}(+\infty, x_n)$ should converge to 0, but $\mathsf{d}_{\mathbb{R}}(+\infty, x_n) = +\infty$ for every $n \in \mathbb{N}$.

One way around this problematic state of affairs is to replace the open ball topology by another one, the d-Scott topology, which we shall define and study in Section 7.4.5.

7.1.3 d^{op}-limits

Another one is to turn to d^{op}-limits. Note that, using Exercise 7.1.16, the $\mathsf{d}_{\mathbb{R}}^{op}$-limits on $\mathbb{R} \cup \{-\infty\}$ are the liminfs. In general, every d^{op}-limit is a limit, and in fact the largest one.

Proposition 7.1.18 *Let X, d be a hemi-metric space, and $(x_i)_{i \in I, \sqsubseteq}$ be a net in X. If $(x_i)_{i \in I, \sqsubseteq}$ has a d^{op}-limit x, then x is a largest (with respect to $\leqslant^d$) limit of $(x_i)_{i \in I, \sqsubseteq}$ in X, d.*

Proof Assume $(x_i)_{i \in I, \sqsubseteq}$ has a d^{op}-limit x. In particular, note that $d^{op}(x, x) = \limsup_{i \in I, \sqsubseteq} d^{op}(x_i, x) = \limsup_{i \in I, \sqsubseteq} d(x, x_i)$. Since $d(x, x) = 0$, for every $\epsilon > 0$ $\limsup_{i \in I, \sqsubseteq} d(x, x_i) < \epsilon$, meaning that $d(x, x_i) < \epsilon$ for i large enough. So x is a limit of $(x_i)_{i \in I, \sqsubseteq}$, by Lemma 7.1.1.

Assume it has another limit y. By Lemma 7.1.1, $(d(y, x_i))_{i \in I, \sqsubseteq}$ converges to 0, in particular $\limsup_{i \in I, \sqsubseteq} d(y, x_i) = 0$ (see Exercise 7.1.11). Since x is a d^{op}-limit, $d(y, x) = d^{op}(x, y) = \limsup_{i \in I, \sqsubseteq} d^{op}(x_i, y) = \limsup_{i \in I, \sqsubseteq} d(y, x_i) = 0$, so $y \leqslant^d x$ by Proposition 6.1.8. □

And all notions coincide for limits in X, d^{sym}:

Proposition 7.1.19 *Let X, d be a hemi-metric space, and $(x_i)_{i \in I, \sqsubseteq}$ a net that converges to x in X, d^{sym}. Then x is also a d-limit, a d^{op}-limit, and a d^{sym}-limit of $(x_i)_{i \in I, \sqsubseteq}$.*

Proof Let x be a limit of $(x_i)_{i \in I, \sqsubseteq}$ in X, d^{sym}. For every $\epsilon > 0$, $x_i \in B^{d^{sym}}_{x, < \epsilon}$ for i large enough, i.e., $d(x_i, x) < \epsilon$ and $d(x, x_i) < \epsilon$ for i large enough. Using the triangular inequality for every $y \in X$, $d(x, y) - \epsilon < d(x_i, y) < d(x, y) + \epsilon$ for i large enough. So $d(x, y) - \epsilon \leqslant \limsup_{i \in I, \sqsubseteq} d(x_i, y) \leqslant d(x, y) + \epsilon$. As ϵ is arbitrary, $d(x, y) = \limsup_{i \in I, \sqsubseteq} d(x_i, y)$, so x is a d-limit of $(x_i)_{i \in I, \sqsubseteq}$. It is also a d^{op}-limit and a d^{sym}-limit, since the same argument works with d^{op} and d^{sym} instead of d. □

One can say more for Cauchy nets: d^{op}-convergence and convergence in X, d^{sym} are the *same* notion for Cauchy nets.

Proposition 7.1.20 *Let X, d be a hemi-metric space. Every limit of any net $(x_i)_{i \in I, \sqsubseteq}$ is a d^{op}-limit, and, conversely, if $(x_i)_{i \in I, \sqsubseteq}$ is a Cauchy net, then every d^{op}-limit of $(x_i)_{i \in I, \sqsubseteq}$ is a limit in X, d^{sym}.*

Proof The first part is Proposition 7.1.19. Conversely, let $(x_i)_{i \in I, \sqsubseteq}$ be a Cauchy net, and let x be a d^{op}-limit of $(x_i)_{i \in I, \sqsubseteq}$. Taking $y = x$ in the definition of d-limits, $0 = \limsup_{i \in I, \sqsubseteq} d(x, x_i)$. So, for every $\epsilon > 0$, (a): $d(x, x_i) < \epsilon$ for i large enough. On the other hand, taking $y = x_i$ for some i large enough that $d(x_i, x_j) < \epsilon$ for every $j \in I$ such that $i \sqsubseteq j$, which is possible since $(x_i)_{i \in I, \sqsubseteq}$ is Cauchy, we obtain $d(x_i, x) = \limsup_{j \in I, \sqsubseteq} d(x_i, x_j) < \epsilon$, for i large enough. Combined with (a), we deduce that $d^{sym}(x, x_i) < \epsilon$ for i large enough, i.e., x is a limit of $(x_i)_{i \in I, \sqsubseteq}$ in X, d^{sym}, using Lemma 7.1.1. □

Corollary 7.1.21 (Metric $\Rightarrow$ limit = d-limit) *In a metric space X, d, the notions of limit and d^{op}-limit of Cauchy nets coincide. That is, the following are equivalent: (a) x is the limit of $(x_i)_{i \in I, \sqsubseteq}$; (b) $(x_i)_{i \in I, \sqsubseteq}$ is Cauchy and x is the d-limit of $(x_i)_{i \in I, \sqsubseteq}$; (c) $(x_i)_{i \in I, \sqsubseteq}$ is Cauchy and x is the d^{op}-limit of $(x_i)_{i \in I, \sqsubseteq}$.*

Proof The equivalence of (b) and (c) is obvious, since $d = d^{op}$. Assume (a). Then $(x_i)_{i \in I, \sqsubseteq}$ is Cauchy by Lemma 7.1.5, and admits x as d^{op}-limit by Proposition 7.1.20. So (c) holds. Conversely, if (c) holds, then x is a limit of $(x_i)_{i \in I, \sqsubseteq}$ in X, d^{sym}, i.e., in X, d. □

7.2 A strong form of completeness: Smyth-completeness

We now explore our first notion of completeness for hemi-metric spaces. This was introduced by Smyth (1988), and is accordingly called Smyth-completeness.

Definition 7.2.1 (Smyth-complete) A hemi-metric space X, d is called *Smyth-complete* if and only if d is T_0 and every Cauchy net in X, d converges in X, d^{sym}.

In particular, we require Smyth-complete hemi-metric spaces to be quasi-metric, so that limits are unique in X, d^{sym}. Note that it is equivalent to require that every Cauchy net in X, d has a d^{op}-limit, by Proposition 7.1.20. Limits in X, d^{sym} are easier to work with.

Since every Cauchy sequence (or even net) in X, d^{sym} is also Cauchy in X, d, we obtain:

Fact 7.2.2 *For every Smyth-complete quasi-metric space X, d, the metric space X, d^{sym} is complete.*

The converse fails: we shall see that $\mathbb{R}, \mathsf{d}_{\mathbb{R}}^{sym}$ is complete (Fact 7.2.17, but see Proposition 3.4.2 already), but $\mathbb{R}, \mathsf{d}_{\mathbb{R}}$ is not Smyth-complete.

Exercise 7.2.3 ($\mathbb{R}, \mathsf{d}_{\mathbb{R}}$, $\overline{\mathbb{R}^+}, \mathsf{d}_{\mathbb{R}}$ are not Smyth-complete) Considering the sequence $(x_n)_{n \in \mathbb{N}}$ where $x_n = n$ for each $n \in \mathbb{N}$, show that none of $\mathbb{R}$, $\mathbb{R}^+$, $\mathbb{R} \cup \{+\infty\}$, $\mathbb{R} \cup \{-\infty, +\infty\}$, or $\overline{\mathbb{R}^+}$ is Smyth-complete when equipped with the quasi-metric $\mathsf{d}_{\mathbb{R}}$.

Exercise 7.2.4 ($\mathbb{R}_\ell$ is not Smyth-complete) Similarly, show that the Sorgenfrey line $\mathbb{R}_\ell$, with the quasi-metric d_ℓ (see Exercise 6.3.10: $\mathsf{d}_\ell(r, s) = r - s$ if $r \geqslant s$, $+\infty$ otherwise), is not Smyth-complete. To do so, show that the topology of $\mathbb{R}_\ell, \mathsf{d}_\ell^{sym}$ is discrete, so that the only convergent nets are the eventually constant nets.

Lemma 7.2.5 (Metric $\Rightarrow$ Smyth-complete = complete) *A metric space X, d is Smyth-complete if and only if it is complete, i.e., iff every Cauchy sequence in X, d converges.*

Proof Let $(x_i)_{i \in I, \sqsubseteq}$ be a Cauchy net in X, d, and assume every Cauchy *sequence* converges. Define $i_k \in I$ by induction on $k \in \mathbb{N}$ as follows. For $i \sqsubseteq j$ large enough, $d(x_i, x_j) < \frac{1}{2^k}$. Using the fact that I is directed, we can therefore find $i_k \in I$ above $i_0, i_1, \ldots, i_{k-1}$ such that this holds whenever $i_k \sqsubseteq i \sqsubseteq j$. The sequence $(x_{i_k})_{k \in \mathbb{N}}$ (which may fail to be a subsequence: there is no reason that $(i_k)_{k \in \mathbb{N}}$ be cofinal in I, i.e., that every element of I is below some i_k, $k \in \mathbb{N}$) is Cauchy: for every $\epsilon > 0$, find $k_0 \in \mathbb{N}$ so that $\frac{1}{2^{k_0}} \leqslant \epsilon$, then $d(x_{i_k}, x_{i_\ell}) < \epsilon$ whenever $k_0 \leqslant k \leqslant \ell$. So $(x_{i_k})_{k \in \mathbb{N}}$ has a limit x (in X, d, or, equivalently, in X, d^{sym}). For every $\epsilon > 0$, $d(x, x_{i_k}) < \epsilon/2$ for k large enough. Fix k so large that this holds, and also $\frac{1}{2^k} \leqslant \epsilon/2$. Then, for every $i \in I$ such that $i_k \sqsubseteq i$, $d(x, x_i) \leqslant d(x, x_{i_k}) + d(x_{i_k}, x_i) < \epsilon$. So x is also the limit of the larger net $(x_i)_{i \in I, \sqsubseteq}$, hence X, d is complete. □

7.2.1 Cauchy-weightable nets

The following notion will be instrumental in understanding completeness.

Definition 7.2.6 (Cauchy-weightable net) Let X, d be a hemi-metric space. Call a *Cauchy-weighted net* any net $(x_i, r_i)_{i \in I, \sqsubseteq}$ in $X \times \mathbb{R}^+$ such that:

- $\inf_{i \in I} r_i = 0$;
- if $i \sqsubseteq j$, then $d(x_i, x_j) \leqslant r_i - r_j$.

A net $(x_i)_{i \in I, \sqsubseteq}$ is *Cauchy-weightable* if and only if it arises from some Cauchy-weighted net $(x_i, r_i)_{i \in I, \sqsubseteq}$.

Note that, since $i \sqsubseteq j$ implies $d(x_i, x_j) \leqslant r_i - r_j$, in particular $r_j \leqslant r_i$ whenever $i \sqsubseteq j$. The link with Cauchy nets is given by the following two lemmas.

Lemma 7.2.7 *Let X, d be a hemi-metric space. Every Cauchy-weightable net in X, d is Cauchy.*

Proof Let $(x_i, r_i)_{i \in I, \sqsubseteq}$ be a Cauchy-weighted net. For every $\epsilon > 0$, since $\inf_{i \in I} r_i = 0$, we must have $r_i < \epsilon$ for i large enough, say $i_0 \sqsubseteq i$. Whenever $i_0 \sqsubseteq i \sqsubseteq j$, $d(x_i, x_j) \leqslant r_i - r_j \leqslant r_i < \epsilon$, so $(x_i)_{i \in I, \sqsubseteq}$ is Cauchy. □

Lemma 7.2.8 *Let X, d be a hemi-metric space. Every Cauchy net in X, d has a Cauchy-weightable subnet.*

Proof Let $(x_i)_{i \in I, \sqsubseteq}$ be a Cauchy net in X, d. For each finite subset E of I, build an element $\alpha(E)$ of I by induction on the cardinality $|E|$ of E, as follows. First, since $(x_i)_{i \in I, \sqsubseteq}$ is Cauchy, $d(x_i, x_j) < 1/2^{|E|+1}$ for $i \sqsubseteq j$ large enough, say above i_0. Since I is directed, it contains an element above i_0 and above $\alpha(E')$ for every proper subset E' of E: the latter are indeed finitely many, and defined by induction hypothesis. Let $\alpha(E)$ be this element. We collect all elements $\alpha(E)$, $E \in \mathbb{P}_{\text{fin}}(I)$, in a function α by using the Axiom of Choice. Clearly, α is a sampler, where $\mathbb{P}_{\text{fin}}(E)$ is ordered by inclusion $\subseteq$.

We check that $(x_{\alpha(E)})_{E \in \mathbb{P}_{\text{fin}}(I), \subseteq}$ is Cauchy-weightable. We do so by verifying that $(x_{\alpha(E)}, \frac{1}{2^{|E|}})_{E \in \mathbb{P}_{\text{fin}}(I), \subseteq}$ is Cauchy-weighted. Assume $E, E' \in \mathbb{P}_{\text{fin}}(I)$ are such that $E \subseteq E'$. If $E = E'$, then $d(x_{\alpha(E)}, x_{\alpha(E')}) = 0 \leqslant \frac{1}{2^{|E|}} - \frac{1}{2^{|E'|}}$. Otherwise, $d(x_{\alpha(E)}, x_{\alpha(E')}) < \frac{1}{2^{|E|+1}}$ by construction, and we must check that this is at most $\frac{1}{2^{|E|}} - \frac{1}{2^{|E'|}}$. Since $|E'| \geqslant |E| + 1$, $\frac{1}{2^{|E|}} - \frac{1}{2^{|E'|}} \geqslant \frac{1}{2^{|E|}} - \frac{1}{2^{|E|+1}} = \frac{1}{2^{|E|+1}}$. □

Lemma 7.2.9 *Let X, d be a hemi-metric space, and $(x_i)_{i \in I, \sqsubseteq}$ be a Cauchy net in X, d. If some subnet $(x_{\alpha(j)})_{j \in J, \preceq}$ has a limit x in X, d^{sym}, then x is also a limit of $(x_i)_{i \in I, \sqsubseteq}$ in X, d^{sym}.*

Proof For every $\epsilon > 0$, $d^{sym}(x, x_{\alpha(j)}) < \epsilon/2$ for j large enough, say $j_0 \preceq j$. Find $i_0 \in I$ such that $d(x_i, x_{i'}) < \epsilon/2$ whenever $i_0 \sqsubseteq i \sqsubseteq i'$, using the fact that $(x_i)_{i \in I, \sqsubseteq}$ is Cauchy. Let $j \in J$ be such that $j_0 \preceq j$ and $i_0 \sqsubseteq \alpha(j)$. Then, for every $i \in I$ with $\alpha(j) \sqsubseteq i$, (a): $d(x, x_i) \leqslant d(x, x_{\alpha(j)}) + d(x_{\alpha(j)}, x_i) < \epsilon$. Similarly, find $j_1 \in J$ such that $d(x_{\alpha(j)}, x) < \epsilon/2$ for every $j \in J$ with $j_1 \preceq j$. For every $i \in I$ with $i_0 \sqsubseteq i$, there is a $j \in J$ such that $i \sqsubseteq \alpha(j)$ and $j_1 \preceq j$, so (b): $d(x_i, x) \leqslant d(x_i, x_{\alpha(j)}) + d(x_{\alpha(j)}, x) < \epsilon$. Combining (a) and (b), we obtain $d^{sym}(x, x_i) < \epsilon$ for i large enough. So x is a limit of $(x_i)_{i \in I, \sqsubseteq}$ in X, d^{sym}. □

So we obtain the following refinement of Definition 7.2.1.

Proposition 7.2.10 *Let X, d be a quasi-metric space. Then X, d is Smyth-complete if and only if every Cauchy-weightable net in X, d converges in X, d^{sym}.*

Proof If X, d is Smyth-complete, then every Cauchy-weightable net in X, d is Cauchy by Lemma 7.2.7, hence converges in X, d^{sym}. Conversely, assume that every Cauchy-weightable net in X, d converges in X, d^{sym}, and let $(x_i)_{i \in I, \sqsubseteq}$ be a Cauchy net in X, d. By Lemma 7.2.8, it has a Cauchy-weightable subnet $(x_{\alpha(j)})_{j \in J, \preceq}$. By assumption, the latter has a limit x in X, d^{sym}, which must also be a limit of $(x_i)_{i \in I, \sqsubseteq}$ in X, d^{sym}, by Lemma 7.2.9. □

Exercise 7.2.11 Show that the definition of Cauchy-weightable net can be simplified to the following in the case of sequences: a sequence $(x_n)_{n \in \mathbb{N}}$ is Cauchy-weightable if and only if $\sum_{n \in \mathbb{N}} d(x_n, x_{n+1}) < +\infty$.

Exercise 7.2.12 Show that Lemma 7.2.8 cannot be improved upon, in the sense that one cannot avoid taking subnets: even in complete metric spaces, there are Cauchy sequences $(x_n)_{n \in \mathbb{N}}$ (even convergent) that are not Cauchy-weightable. Consider, for example, $\mathbb{R}$ with its metric $\mathrm{d}_{\mathbb{R}}^{sym}$, and $x_n = \frac{(-1)^n}{n+1}$. As a small help, you will have to study the nth *harmonic number* $H_n = \sum_{k=1}^{n} \frac{1}{k}$, and to replay Nicolas Oresme's (fourteenth-century) argument that $H_{2^n} \geqslant n/2$ for every $n \in \mathbb{N}$.

Exercise 7.2.13 Let X, d be a quasi-metric space. Let $(x_i, r_i)_{i \in I, \sqsubseteq}$ be a Cauchy-weighted net in X, d, and assume that $(x_i)_{i \in I, \sqsubseteq}$ has a limit x in X, d^{sym}. Show that this gives us some information about the rate of convergence in X, d: $d(x_i, x) \leqslant r_i$ for every $i \in I$.

7.2.2 The Banach Fixed Point Theorem

The Banach Fixed Point Theorem 3.4.5 generalizes smoothly.

Theorem 7.2.14 (Generalized Banach Fixed Point Theorem) *Let X, d be a Smyth-complete quasi-metric space, and $f : X \to X$ be any c-Lipschitz map, where $c \in [0, 1)$. Assume that there is a point $x_0 \in X$ such that $d(x_0, f(x_0)) < +\infty$.*

Then f has a fixed point x, obtained as the limit of the sequence of iterates $(f^n(x_0))_{n\in\mathbb{N}}$ in X, d^{sym}. Any two fixed points x and y of f are at infinite distance from one another: if $d(y, x) < +\infty$ then $y \leqslant^d x$, and if $y \neq x$ then $d^{sym}(x, y) = +\infty$.

Proof Fix $x_0 \in X$ with $d(x_0, f(x_0)) < +\infty$, and let $r_i = \frac{c^i}{1-c} d(x_0, f(x_0))$. By assumption, r_i is in $\mathbb{R}^+$. Since f is c-Lipschitz, $d(f^i(x_0), f^{i+1}(x_0)) \leqslant c^i d(x_0, f(x_0)) = r_i - r_{i+1}$. So, if $i \leqslant j$ then $d(f^i(x_0), f^j(x_0)) \leqslant d(f^i(x_0), f^{i+1}(x_0)) + d(f^{i+1}(x_0), f^{i+2}(x_0)) + \cdots + d(f^{j-1}(x_0), f^j(x_0)) \leqslant r_i - r_j$. So $(f^n(x_0), r_n)_{n\in\mathbb{N}}$ is a Cauchy-weighted sequence. Since X, d is Smyth-complete, by Proposition 7.2.10 $(f^n(x_0))_{n\in\mathbb{N}}$ has a limit x in X, d^{sym}.

Since f is c-Lipschitz from X, d to X, d, it is easy to see that it is also c-Lipschitz from X, d^{sym} to X, d^{sym}. By Fact 6.2.3, f is continuous from X, d^{sym} to X, d^{sym}, so $f(x)$ is also the limit of $(f(f^n(x_0)))_{n\in\mathbb{N}}$ in X, d^{sym} (Proposition 4.7.18), i.e., of $(f^n(x_0))_{n\geqslant 1}$ in X, d^{sym}. Using Lemma 7.2.9, for example, $f(x)$ is also the limit of $(f^n(x_0))_{n\in\mathbb{N}}$. Hence $f(x) = x$, as limits are unique in T_2 spaces (Proposition 4.7.4), and metric spaces are T_2 (Lemma 6.1.10).

Let y be any other fixed point of f, i.e., $y = f(y)$ and $y \neq x$. If $d(y, x) < +\infty$, then $d(y, x) = d(f(y), f(x)) \leqslant c.d(y, x)$. So $d(y, x) = 0$, i.e., $y \leqslant^d x$. Similarly, if $d(x, y) < +\infty$, then $x \leqslant^d y$. In particular, if $d^{sym}(x, y) < +\infty$, then both hold, so $x = y$. ☐

Corollary 7.2.15 *Let X, d be a Smyth-complete quasi-metric space such that $d(x, y) < +\infty$ for all $x, y \in X$. Let $f : X \to X$ be any c-Lipschitz map, where $c \in [0, 1)$. Then f has a unique fixed point x. Given any point $x_0 \in X$, x is the limit of the sequence of iterates $(f^n(x_0))_{n\in\mathbb{N}}$ in X, d^{sym}.*

Exercise 7.2.16 Show that the assumption that there is a point $x_0 \in X$ such that $d(x_0, f(x_0)) < +\infty$ is needed in Theorem 7.2.14. For example, let X, d be $\mathbb{N}$, $d_=$ and $f(n) = n + 1$. (Remember from Exercise 6.1.15 that $d_=(x, y) = 0$ if $x = y$, $+\infty$ otherwise.)

7.2.3 Building Smyth-complete spaces

An important example of a complete metric space is $\mathbb{R}$: this is Proposition 3.4.2.

Fact 7.2.17 ($\mathbb{R}$ is complete) *$\mathbb{R}$ with its L^1 metric $\mathsf{d}_{\mathbb{R}}^{sym}(x, y) = |x - y|$ is (Smyth) complete.*

However, recall that $\mathbb{R}$ with the quasi-metric $\mathsf{d}_{\mathbb{R}}$ instead of $\mathsf{d}_{\mathbb{R}}^{sym}$ is *not* Smyth-complete (Exercise 7.2.3).

Lemma 7.2.18 ($[0, 1]_\sigma$ is Smyth-complete) *The following subsets of $\mathbb{R} \cup \{-\infty\}$ are Smyth-complete quasi-metric spaces with the $\mathsf{d}_{\mathbb{R}}$ quasi-metric:* (1) *$[-\infty, b]$ for any $b \in \mathbb{R}$, meaning $(-\infty, b] \cup \{-\infty\}$;* (2) *$(-\infty, b]$ for any $b \in \mathbb{R}$;* (3) *$[a, b]$ for all $a \leqslant b$ in $\mathbb{R}$.*

Proof Let X be any of the spaces given. Let $(x_i)_{i \in I}$ be a Cauchy-weightable net in $X, \mathsf{d}_{\mathbb{R}}$, and let $x = \liminf_{i \in I, \sqsubseteq} x_i$. It is easy to see that x is in X, whichever X was chosen. For every $\epsilon > 0$, $x_i > x - \epsilon$ for i large enough, so $\mathsf{d}_{\mathbb{R}}(x, x_i) < \epsilon$ for i large enough. On the other hand, $(x_i)_{i \in I}$ is Cauchy by Lemma 7.2.7, so $\sup_{j \in I, i \sqsubseteq j} \mathsf{d}_{\mathbb{R}}(x_i, x_j) < \epsilon$ for i large enough. So $x_i < \inf_{j \in I, i \sqsubseteq j} x_j + \epsilon$ for i large enough. In particular, $x_i < x + \epsilon$, hence $\mathsf{d}_{\mathbb{R}}(x_i, x) < \epsilon$ for i large enough. So $d^{sym}(x, x_i) < \epsilon$ for i large enough, whence x is the limit of $(x_i)_{i \in I}$ in $X, \mathsf{d}_{\mathbb{R}}^{sym}$ by Lemma 7.1.1. □

Exercise 7.2.19 Let X, d be a hemi-metric space. Show that every subsequence of a Cauchy sequence is again Cauchy. (See Exercise 4.7.23 for the notion of a subsequence.) This is complementary to the fact that if a sequence converges to some point x, then any subsequence also converges to x; see Exercise 4.7.25.

Exercise 7.2.20 Let X, d be a hemi-metric space, and $(x_n)_{n \in \mathbb{N}}$ be a Cauchy sequence in X, d. Assume that some subsequence $(x_{n_k})_{k \in \mathbb{N}}$ converges to some point x in X, d. Show that $(x_n)_{n \in \mathbb{N}}$ also converges to x.

In general, let $(x_i)_{i \in I, \sqsubseteq}$ be a Cauchy net that has a subnet $(x_{\alpha(j)})_{j \in J, \preceq}$ that converges to x. Then show that $(x_i)_{i \in I, \sqsubseteq}$ converges to x.

Lemma 7.2.21 (Symcompact $\Rightarrow$ Smyth-complete) *Every symcompact quasi-metric space is Smyth-complete.*

Proof Let X, d be symcompact quasi-metric, and let $(x_i)_{i \in I, \sqsubseteq}$ be a Cauchy net in X, d. By Corollary 4.7.27, there is a subnet $(x_{\alpha(j)})_{j \in J, \preceq}$, where $\alpha : J \to I$ is a sampler, that converges to a point x in X, d^{sym}. Then x is a limit of $(x_i)_{i \in I, \sqsubseteq}$ in X, d^{sym} by Lemma 7.2.9. □

We already know that the metric spaces that are compact are exactly those that are precompact and complete (Theorem 3.4.8). Similarly:

Proposition 7.2.22 (Symcompact = totally bounded + Smyth-complete) *Let X, d be a quasi-metric space. Then X, d is symcompact if and only if it is Smyth-complete and totally bounded.*

Proof The only if direction is by Lemma 7.2.21 and the fact that every symcompact space is totally bounded. Conversely, we first show that: $(*)$ for every finite cover $(U_k)_{k \in K}$ of X, for every net $(x_i)_{i \in I, \sqsubseteq}$ there is a cofinal subset J of I such that the subset $(x_j)_{j \in J}$ is contained in some U_k, $k \in K$. (Recall that J is cofinal iff every element of I is below some element of J.) If this were not so, for every $k \in K$ the set $\{i \in I \mid x_i \in U_k\}$ would not be cofinal. So there would be an $i_k \in I$ such that x_i is outside U_k for every $i \in I$ with $i_k \sqsubseteq i$. Since I is directed, and K is finite, there is a $j \in I$ above each i_k, $k \in K$. Then x_j would be in no U_k, a contradiction.

So let X, d be totally bounded and Smyth-complete quasi-metric. Let $(x_i)_{i \in I, \sqsubseteq}$ be any net. For every $n \in \mathbb{N}$, by total boundedness X is covered by finitely many open balls of radius $1/2^n$ (with respect to d^{sym}). By induction on n, use $(*)$ to pick one of these balls, $B^{d^{sym}}_{y_n, <1/2^n}$, and a cofinal subset J_n, either of elements from I if $n = 0$ or from J_{n-1} otherwise, such that $x_j \in B^{d^{sym}}_{y_n, <1/2^n}$ for every $j \in J_n$. Define a subnet as follows. Let $J = \coprod_{n \in \mathbb{N}} J_n$, quasi-ordered by the product ordering $(n, j) \preceq (n', j')$ iff $n \leqslant n'$ and $j \sqsubseteq j'$. J is directed, because, given $(n, j), (n', j') \in J$, there is a j'' above $j, j' \in J_{\max(n,n')}$, since I is directed and $J_{\max(n,n')}$ is cofinal in I. Let the sampler $\alpha : J \to I$ be defined by $\alpha(n, j) = j$. Since every J_n is cofinal, α is cofinal; the fact that *some* J_n is cofinal is even sufficient.

We claim that $(x_{\alpha(j)})_{j \in J, \preceq}$ is Cauchy. For every $\epsilon > 0$, find n such that $2/2^n \leqslant \epsilon$. Fix some $i_0 \in I$. For every $j \in J$ such that $(n, i_0) \preceq j$, j is of the form (n', i') with $n \leqslant n'$ and $i' \in J_{n'}$. In particular, $i' \in J_n$, so $x_{\alpha(j)} = x_{i'}$ is in $B^{d^{sym}}_{y_n, <1/2^n}$. In particular, for all $j, k \in J$ with $(n, i_0) \preceq j \preceq k$, $d(x_{\alpha(j)}, x_{\alpha(k)}) \leqslant d(x_{\alpha(j)}, y_n) + d(y_n, x_{\alpha(k)}) < 2/2^n \leqslant \epsilon$. So $(x_{\alpha(j)})_{j \in J, \preceq}$ is Cauchy. By Smyth-completeness, it converges in X, d^{sym}.

So every net $(x_i)_{i \in I, \sqsubseteq}$ has a convergent subnet (even subsequence) $(x_{\alpha(j_k)})_{k \in \mathbb{N}}$. So X, d^{sym} is compact, by Corollary 4.7.27. ☐

Smyth-completeness is preserved by several constructions. First, one can replace the quasi-metric by a bounded one.

Lemma 7.2.23 *Let X, d be a quasi-metric space, and define d' as the quasi-metric $d'(x, y) = \min(d(x, y), 1)$. Then X, d is Smyth-complete if and only* *if X, d' is. Also, the notions of convergence are the same in X, d^{sym} and in X, d'^{sym}, i.e., X, d^{sym} and X, d'^{sym} have the same open ball topology.*

Proof X, d and X, d' have the same open ball topology, by Corollary 6.3.4. Similarly, X, d^{op} and X, d'^{op} define the same topology. So X, d^{sym} and X, d'^{sym} also define the same topology, which is the one generated by the previous two, by Proposition 6.1.19. So convergence in one coincides with convergence in the other.

Assume X, d' is Smyth-complete. Let $(x_i)_{i \in I, \sqsubseteq}$ be a Cauchy net in X, d. For every $\epsilon > 0$, $d(x_i, x_j) < \epsilon$ for $i \sqsubseteq j$ large enough. Since $d'(x_i, x_j) \leqslant d(x_i, x_j)$, $(x_i)_{i \in I, \sqsubseteq}$ is also Cauchy in X, d'. So it has a limit in X, d'^{sym}. This limit is also one in X, d^{sym}, since the two spaces have the same open ball topology.

Conversely, assume X, d is Smyth-complete. Let $(x_i)_{i \in I, \sqsubseteq}$ be a Cauchy net in X, d'. For every $\epsilon > 0$, $d'(x_i, x_j) < \epsilon$ for $i \sqsubseteq j$ large enough. This is in particular the case if we replace ϵ by $\min(\epsilon, 1)$: $d'(x_i, x_j) < \min(\epsilon, 1)$ for $i \sqsubseteq j$ large enough. But if $d'(x_i, x_j) < \min(\epsilon, 1)$ then $d'(x_i, x_j) < 1$ so $d'(x_i, x_j) = d(x_i, x_j)$. Then $d(x_i, x_j) < \min(\epsilon, 1)$, hence $d(x_i, x_j) < \epsilon$, for $i \sqsubseteq j$ large enough. So $(x_i)_{i \in I, \sqsubseteq}$ is also Cauchy in X, d, hence has a limit in X, d^{sym}. This limit is also one in X, d^{sym}, as above. □

Lemma 7.2.24 *Every coproduct of Smyth-complete quasi-metric spaces is Smyth-complete.*

Proof Use Proposition 7.2.10. Let X_k, d_k be Smyth-complete quasi-metric spaces, $k \in K$. Any Cauchy-weighted net $(z_i, r_i)_{i \in I, \sqsubseteq}$ in $\coprod_{k \in K} X_k, \coprod_{k \in K} d_k$ must lie entirely in the same X_k. In other words, writing z_i as (k_i, x_i), every k_i must be equal to some k. Indeed, given k_i and k_j in K, if $i \sqsubseteq j$ then $\coprod_{k \in K} d_k((k_i, x_i), (k_j, x_j)) \leqslant r_i - r_j < +\infty$, so $k_i = k_j$; and given arbitrary k_i and $k_{i'}$, by directedness there is a $j \in I$ such that $i, i' \sqsubseteq j$, so $k_i = k_j = k_{i'}$. Then $(x_i, r_i)_{i \in I, \sqsubseteq}$ is Cauchy-weighted in X_k, d_k, hence has a limit x in X_k, d_k^{sym}. It follows that (k, x) is a limit of the initial net in $\coprod_{k \in K} X_k, \coprod_{k \in K} d_k^{sym}$, and we conclude since $\coprod_{k \in K} d_k^{sym} = (\coprod_{k \in K} d_k)^{sym}$. □

Lemma 7.2.25 *Let X_k, d_k be finitely many Smyth-complete quasi-metric spaces, $1 \leqslant k \leqslant n$. Then $X = \prod_{k=1}^n X_k$ with the quasi-metric $d = \prod_{k=1}^n d_k$ is Smyth-complete.*

The limit $\vec{z}$ of any Cauchy net $(\vec{x}_i)_{i \in I, \sqsubseteq}$ in X, d^{sym} is computed componentwise, i.e., for each k, $1 \leqslant k \leqslant n$, z_k is the limit of $(x_{ik})_{i \in I, \sqsubseteq}$ in X_k, d_k^{sym}.

Proof By Lemma 6.5.1, the topology of X, d is the product topology, and similarly the topology of X, d^{sym} is the product topology of X_k, d_k^{sym}, $1 \leqslant k \leqslant n$ – note that $d^{sym} = \prod_{k=1}^n d_k^{sym}$. This entails the second part of the Lemma, using Exercise 4.7.20. Given any Cauchy net $(\vec{x}_i)_{i \in I, \sqsubseteq}$ in X, d^{sym}, with $\vec{x}_i = (x_{i1}, \ldots, x_{in})$, for every $\epsilon > 0$ we have $d_k(x_{ik}, x_{jk}) < \epsilon$ for every k, $1 \leqslant k \leqslant n$, and for all $i, j \in I$ with $i \sqsubseteq j$ large enough. So each $(x_{ik})_{i \in I, \sqsubseteq}$ is Cauchy in X_k, d_k, hence converges in X, d^{sym}, say to x_k. Letting $\vec{x} = (x_1, \ldots, x_n)$, the Cauchy net $(\vec{x}_i)_{i \in I, \sqsubseteq}$ then converges to $\vec{x}$ in X, d^{sym}. □

Example 7.2.26 So not only $(-\infty, 0]$ (see Lemma 7.2.18) but also $(-\infty, 0]^n$, for example (for any $n \in \mathbb{N}$), is Smyth-complete. Lemma 7.2.25 allows us to conclude that $[0, 1]^n$ is Smyth-complete as well, but not that $[0, 1]^{\mathbb{N}}$ is Smyth-complete, although the latter is true: indeed, the directed Hilbert cube $[0, 1]^{\mathbb{N}}$ is symcompact (Proposition 6.7.11), hence Smyth-complete by Lemma 7.2.21.

Infinite products of Smyth-complete spaces are not Smyth-complete in general, even countable ones. A counterexample is $(-\infty, 0]^{\mathbb{N}}$. Consider $\vec{x}_n = (\min(n - m, 0))_{m \in \mathbb{N}}$. Then $(\vec{x}_n)_{n \in \mathbb{N}}$ is ascending, i.e., $\vec{x}_n \leqslant \vec{x}_{n'}$ whenever $n \leqslant n'$. In particular, $\sup_{\mathbb{N}} \mathrm{d}_{\mathbb{R}}(\vec{x}_n, \vec{x}_{n'}) = 0$, so $(\vec{x}_n)_{n \in \mathbb{N}}$ is Cauchy. But $(\vec{x}_n)_{n \in \mathbb{N}}$ does not converge in $(-\infty, 0]^{\mathbb{N}}, \sup_{\mathbb{N}} \mathrm{d}_{\mathbb{R}}^{sym}$. If it did, it would be Cauchy in $(-\infty, 0]^{\mathbb{N}}, \sup_{\mathbb{N}} \mathrm{d}_{\mathbb{R}}^{op}$; however, $\sup_{\mathbb{N}} \mathrm{d}_{\mathbb{R}}^{op}(\vec{x}_n, \vec{x}_{n'}) = n' - n$ when $n \leqslant n'$, and this cannot be made smaller than an arbitrary $\epsilon > 0$.

This argument also shows that function spaces to a Smyth-complete quasi-metric space, with the sup quasi-metric, are in general not Smyth-complete either.

Exercise 7.2.27 Let X_k, d_k be finitely many Smyth-complete quasi-metric spaces, $1 \leqslant k \leqslant n$. Show that the tensor product $X = \prod_{k=1}^n X_k$, with the quasi-metric $d = d_1 \otimes \cdots \otimes d_n$ (Exercise 6.5.3), is Smyth-complete. (You may first show that d^{sym} and $d_1^{sym} \otimes \cdots \otimes d_n^{sym}$ define the same open ball topology, for example by showing that the identity map is 1-Lipschitz in one direction, and $\frac{1}{n}$-Lipschitz in the other direction.) Show also that, as for finite products, the limit $\vec{z}$ of any Cauchy net $(\vec{x}_i)_{i \in I, \sqsubseteq}$ in X, d^{sym} is computed componentwise, i.e., for each k, $1 \leqslant k \leqslant n$, z_k is the limit of $(x_{ik})_{i \in I, \sqsubseteq}$ in X_k, d_k^{sym}.

Still, infinite products of Smyth-complete spaces are Smyth-complete in two cases.

Lemma 7.2.28 *Let X_k, d_k be Smyth-complete quasi-metric spaces, $k \in K$, and $X = \prod_{k \in K} X_k$.*

1. *If K is countable and X_k, d_k is symcompact for every $k \in K$, then the topological product X with quasi-metric $d = \widetilde{\prod}_{k \in K} d_k$ is Smyth-complete (indeed, symcompact).*
2. *If X_k, d_k is metric for every $k \in K$, then the categorical product X with the sup quasi-metric $d(\vec{x}, \vec{y}) = \sup_{k \in K} d_k(x_k, y_k)$ is Smyth-complete (and metric).*

In both cases, limits in X, d^{sym} are computed componentwise.

Proof 1. By Lemma 7.2.23, we may as well replace d_k by $\min(d_k, 1)$. Let $d = \widetilde{\prod}_{k \in K} d_k$, so $d(\vec{x}, \vec{y}) = \sup_{k \in K} \frac{1}{2^k} d_k(x_k, y_k)$. Clearly, $d^{sym} = \widetilde{\prod}_{k \in K} d_k^{sym}$. The symmetrization X, d^{sym} of the topological product is then equal to the topological product of the symmetrizations X_k, d_k^{sym}, and has the product topology (Lemma 6.5.7). The latter is compact by Tychonoff's Theorem 4.5.12, so X, d^{sym} is compact, i.e., X, d is symcompact. In particular, X, d is Smyth-complete by Lemma 7.2.21. Limits in X, d^{sym} are topological limits, hence are computed componentwise (Exercise 4.7.20).

2. Let d be the sup quasi-metric. Let $(\vec{x}_i)_{i \in I, \sqsubseteq}$ be a Cauchy net in X, d (equivalently, in X, d^{sym}). In particular, $(x_{ik})_{i \in I, \sqsubseteq}$ is Cauchy in X_k, d_k, for every $k \in K$. Let x_k be its limit in X_k, d_k^{sym} (equivalently, in X_k, d_k). We claim that $\vec{x} = (x_k)_{k \in K}$ is the limit of $(\vec{x}_i)_{i \in I, \sqsubseteq}$ in X, d, i.e., that limits are componentwise. Fix $\epsilon > 0$. Since $(\vec{x}_i)_{i \in I, \sqsubseteq}$ is Cauchy, find $i_0 \in I$ such that $d(\vec{x}_i, \vec{x}_j) < \epsilon/3$ whenever $i_0 \sqsubseteq i \sqsubseteq j$. Since d is a metric, $d(\vec{x}_j, \vec{x}_i) < \epsilon/3$ for $i_0 \sqsubseteq i \sqsubseteq j$. In particular, $d_k(x_{jk}, x_{ik}) < \epsilon/3$ for every $k \in K$. For i fixed above i_0, by Lemma 7.1.1 there is a j above i such that $d_k(x_k, x_{jk}) < \epsilon/3$. So $d_k(x_k, x_{ik}) \leqslant d(x_k, x_{jk}) + d(x_{jk}, x_{ik}) < 2\epsilon/3$. As k is arbitrary, $d(\vec{x}, \vec{x}_i) \leqslant 2\epsilon/3 < \epsilon$. This holds for every i above i_0. By Lemma 7.1.1, $\vec{x}$ is the limit of $(\vec{x}_i)_{i \in I, \sqsubseteq}$. □

By Lemma 7.2.28 2, Y^X is Smyth-complete with the sup quasi-metric, whenever Y is metric. The following can be used to show that the spaces of continuous maps, of uniformly continuous maps, and of c-Lipschitz maps from X to Y (under natural assumptions on X) are again (Smyth) complete in this case.

Lemma 7.2.29 *Let X be a topological space, and Y, d' be a hemi-metric space. Let $(f_i)_{i \in I, \sqsubseteq}$ be a net of maps from X to Y that converges to a map f in $Y^X, \sup_X d'^{sym}$. Then: (a) If each f_i is continuous at $x \in X$, then so is f; (b) if each f_i is continuous, then so is f; (c) if X, d is a hemi-metric space, and if each f_i is uniformly continuous, then so is f; (d) if X, d is a hemi-metric space, and if each f_i is c-Lipschitz for some $c \in \mathbb{R}^+$, then so is f.*

Proof (a) Assume that each f_i is continuous at $x \in X$. For every neighborhood V of $f(x)$ in Y, there is an $\epsilon > 0$ such that $B^{d'}_{f(x), < \epsilon} \subseteq V$ by

Lemma 6.1.5. Let $i_0 \in I$ be such that $\sup_X d'^{sym}(f, f_i) < \epsilon/3$ for every $i \in I$ with $i_0 \sqsubseteq i$. Since f_{i_0} is continuous at x, $U = f_{i_0}^{-1}(B^{d'}_{f_{i_0}(x),<\epsilon/3})$ is open. For every $x' \in U$, $d'(f_{i_0}(x), f_{i_0}(x')) < \epsilon/3$. Since $\sup_X d'^{sym}(f, f_{i_0}) < \epsilon/3$ and d'^{sym} is symmetric, $d'(f(x), f_{i_0}(x)) < \epsilon/3$ and $d'(f_{i_0}(x'), f(x')) < \epsilon/3$. So $d'(f(x), f(x')) < \epsilon$ for every $x' \in U$. It follows that $f[U] \subseteq B^{d'}_{f(x),<\epsilon} \subseteq V$. So f is continuous at x.

Item (b) follows from (a), since a continuous map is one that is continuous at every point $x \in X$, by Proposition 4.3.14.

(c) Assume X is equipped with some hemi-metric d. For every $\epsilon > 0$, as above find $i_0 \in I$ such that $\sup_X d'^{sym}(f, f_{i_0}) < \epsilon/3$. Since f_{i_0} is uniformly continuous, there is an $\eta > 0$ such that, whenever $d(x, x') < \eta$, $d'(f_{i_0}(x), f_{i_0}(x')) < \epsilon/3$. Since $\sup_X d'^{sym}(f, f_{i_0}) < \epsilon/3$, we have $d'(f(x), f_{i_0}(x)) < \epsilon/3$, and also $d'(f_{i_0}(x'), f(x')) < \epsilon/3$, so $d'(f(x), f(x')) < \epsilon$. So f is uniformly continuous.

(d) Assume X is equipped with some hemi-metric d, and each f_i is c-Lipschitz. For every $\epsilon > 0$, find $i_0 \in I$ such that $\sup_X d'^{sym}(f, f_{i_0}) < \epsilon/2$. For all $x, x' \in X$, $d'(f(x), f(x')) \leqslant d'(f(x), f_{i_0}(x)) + d'(f_{i_0}(x), f_{i_0}(x')) + d'(f_{i_0}(x'), f(x')) < \epsilon/2 + c.d(x, x') + \epsilon/2 = c.d(x, x') + \epsilon$. Since ϵ is arbitrary, $d'(f(x), f(x')) \leqslant c.d(x, x')$. □

Corollary 7.2.30 (Uniform convergence preserves continuity, in metric spaces) *Let X be a topological space, and Y, d be a hemi-metric space. Let $(f_i)_{i \in I, \sqsubseteq}$ be a net of maps from X to Y that converges uniformly to a map f. If d is a metric, then: (a) if each f_i is continuous at $x \in X$, then so is f; (b) if each f_i is continuous, then so is f; (c) assuming X is equipped with some hemi-metric, if each f_i is uniformly continuous, then so is f; (d) again assuming X hemi-metric, and letting $c \in \mathbb{R}^+$, if each f_i is c-Lipschitz, then so is f.*

We sum up the results of this section in the following theorem. The last item refers to a notion we have not seen yet, formal balls. They are the subject of the upcoming section.

Theorem 7.2.31 (Constructing Smyth-complete spaces)

1. *Every sym-closed subset of a Smyth-complete quasi-metric space is Smyth-complete – a subset F of X, d is* sym-closed *if and only if it is closed in X, d^{sym}.*

   Since every closed subset is sym-closed, in particular, every closed subset of a Smyth-complete quasi-metric space is Smyth-complete.
2. *Every coproduct of Smyth-complete quasi-metric spaces is Smyth-complete.*
3. *Every finite product and every finite tensor product of Smyth-complete quasi-metric spaces is Smyth-complete.*

4. *Every symcompact space (and in particular every countable topological product of symcompact spaces) is Smyth-complete.*
5. *Categorical products of (Smyth) complete metric spaces are (Smyth) complete.*
6. *For every quasi-metric space Y, d', if Y^X, $\sup_X d'$ is Smyth-complete (which happens notably if d' is a metric), then the following subspaces are again Smyth-complete: (a) the space $[X \to Y]$ of all continuous maps from X to Y, where X is a topological space; (b) the space $[X \to Y]_u$ of all uniformly continuous maps from X to Y, if X, d is a hemi-metric space; (c) the space $[X \to Y]_c$ of all c-Lipschitz maps from X to Y, for any $c \in \mathbb{R}^+$, where X, d is a hemi-metric space.*
7. *The space of formal balls of a Smyth-complete quasi-metric space is Smyth-complete.*

Proof 1. If F is sym-closed, then any Cauchy net $(x_i)_{i \in I, \sqsubseteq}$ in F converges in X, d^{sym}, since X, d is complete. Call x the limit. Then x is in F, by Corollary 4.7.10. This is the case in particular if F is closed. Indeed, every closed subset is sym-closed, since the open ball topology of X, d^{sym} is finer than that of X, d, using Proposition 6.1.19, for example.

2 is true by Lemma 7.2.24. 3 is true by Lemma 7.2.25 and Exercise 7.2.27 for tensor products. 4 and 5 are true by Lemma 7.2.21 and Lemma 7.2.28. 6 is true because all the considered subspaces are sym-closed in Y^X, $\sup_X d'$ by Lemma 7.2.29, and using Item 1. (To be precise, the only thing to check is that $\sup_X d'^{sym} = (\sup_X d')^{sym}$, which is clear.) 7 is true by Lemma 7.3.8, which we will see shortly. □

Exercise 7.2.32 Show the converse to Theorem 7.2.31, Item 1: in a Smyth-complete quasi-metric space X, d, every Smyth-complete quasi-metric subspace must be sym-closed.

Theorem 7.2.31 (1) states that sym-closed subsets of Smyth-complete spaces are again Smyth-complete. If we agree to change the quasi-metric for a topologically equivalent one, this is also the case for open subsets.

Lemma 7.2.33 *Let X, d be a Smyth-complete quasi-metric space, and U be an open subset of X, d. Then U is Smyth-completely quasi-metrizable: there is a quasi-metric d' on U such that the open ball topology of U, d' is the subspace topology, and such that U, d' is Smyth-complete.*

Proof By Lemma 7.4.11, we may assume that d is 1-bounded. Let F be the closed set $X \smallsetminus U$, and remember that $x \in U$ iff $d(x, F) \neq 0$ by Lemma 6.1.11. So the map $f : y \mapsto (y, 1 - \frac{1}{d(y,F)})$ is well defined from U to $X \times (-\infty, 0]$. The map $y \mapsto d(y, F)$ is 1-Lipschitz from U, d to $(0, 1], \mathsf{d}_{\mathbb{R}}$ by Lemma 6.2.16, hence continuous (Fact 6.2.3). On the other hand, the map $r \mapsto 1 - \frac{1}{r}$ is continuous from $(0, 1], \mathsf{d}_{\mathbb{R}}$ to $(-\infty, 0], \mathsf{d}_{\mathbb{R}}$, using the ϵ-η formula (Lemma 6.2.1): given any $r \in (0, 1]$, one has $\mathsf{d}_{\mathbb{R}}(1 - \frac{1}{r}, 1 - \frac{1}{s}) < \epsilon$ whenever $\mathsf{d}_{\mathbb{R}}(r, s) < \eta$, where $\eta = \frac{r^2\epsilon}{1+r\epsilon}$. So $\pi_2 \circ f : y \mapsto 1 - \frac{1}{d(y,F)}$ is continuous from U to $(-\infty, 0]_\sigma$. (Remember that the latter is $(-\infty, 0]$ with the open ball topology of $\mathsf{d}_{\mathbb{R}}$, by Exercise 6.1.18.) By Trick 4.5.4, f is continuous.

Let $Z = f[U]$. The first projection map π_1 is continuous from $X \times (-\infty, 0]_\sigma$ to X, hence restricts to a continuous map from Z to X. Since $\pi_1(z)$ is in U whenever $z \in Z$, π_1 is also a continuous map from Z to U. It is easy to see that π_1 and f are mutual inverses, so U is homeomorphic to Z.

We claim that Z is sym-closed in $X \times (-\infty, 0]_\sigma$—precisely, closed in $X \times (-\infty, 0], d^{sym} \times \mathsf{d}_{\mathbb{R}}^{sym}$: notice that $d^{sym} \times \mathsf{d}_{\mathbb{R}}^{sym}$ coincides with $(d \times \mathsf{d}_{\mathbb{R}})^{sym}$. To this end, consider any point $(x, -r)$ not in Z. So $1 - \frac{1}{d(x,F)} \neq -r$. Let $\epsilon = \left|1 - \frac{1}{d(x,F)} + r\right|$, $a = d(x, F)$. Note that $\epsilon > 0$, $a > 0$. Find $\eta > 0$ small enough so that $\eta < a$, $\frac{1}{a-\eta} - \frac{1}{a} < \epsilon/2$, and $\frac{1}{a+\eta} - \frac{1}{a} > -\epsilon/2$ (e.g., $\eta < \frac{a^2\epsilon}{2+a\epsilon}$). If $y \in B^{d^{sym}}_{x,<\eta}$ and $1 - \frac{1}{d(y,F)} = -s$ (i.e., $(y, -s) \in Z$), then, by the triangular inequality, $a - \eta < d(y, F) < a + \eta$, so $1 - \frac{1}{a-\eta} < -s < 1 - \frac{1}{a+\eta}$. So $\frac{1}{a+\eta} - \frac{1}{a} < 1 - \frac{1}{a} + s < \frac{1}{a-\eta} - \frac{1}{a}$, hence $-\epsilon/2 < 1 - \frac{1}{a} + s < \epsilon/2$. If additionally $-s \in B^{\mathsf{d}_{\mathbb{R}}^{sym}}_{-r,<\epsilon/2}$, then $-\epsilon/2 < r - s < \epsilon/2$, so $-\epsilon < 1 - \frac{1}{a} + r < \epsilon$. This contradicts $\epsilon = \left|1 - \frac{1}{a} + r\right|$. So we cannot have $(y, -s) \in Z$ and $(y, -s) \in B^{d^{sym}}_{x,<\eta} \times B^{\mathsf{d}_{\mathbb{R}}^{sym}}_{-r,<\epsilon/2}$ at the same time. In other words, the open neighborhood $B^{d^{sym}}_{x,<\eta} \times B^{\mathsf{d}_{\mathbb{R}}^{sym}}_{-r,<\epsilon/2}$ of $(x, -r)$ in $X \times (-\infty, 0], d^{sym} \times \mathsf{d}_{\mathbb{R}}^{sym}$ does not meet Z. By Trick 4.1.11, Z is closed in $X \times (-\infty, 0], d^{sym} \times \mathsf{d}_{\mathbb{R}}^{sym}$.

Since X, d is Smyth-complete, and $(-\infty, 0], \mathsf{d}_{\mathbb{R}}$ is Smyth-complete (this is Lemma 7.2.18), their product is too (Lemma 7.2.25). So $Z, d \times \mathsf{d}_{\mathbb{R}}$ is Smyth-complete by Theorem 7.2.31. Since U is homeomorphic to the latter, it is Smyth-completely quasi-metrizable. Concretely, U is Smyth-complete with the quasi-metric d' defined by $d'(x, y) = (d \times \mathsf{d}_{\mathbb{R}})(f(x), f(y))$. □

7.3 Formal balls

There is an elegant way of characterizing Smyth-completeness, using Weihrauch and Schneider (1981)'s notion of formal ball, later refined by Edalat and Heckmann (1998).

Definition 7.3.1 (Formal balls, $\mathbf{B}(X, d)$) Let X, d be a hemi-metric space. A *formal ball* on X, d is a pair (x, r) with $x \in X$ and $r \in \mathbb{R}^+$: x is the *center* and r is the *radius*.

The set $\mathbf{B}(X, d)$ of formal balls is equipped with a hemi-metric d^+, defined by $d^+((x, r), (y, s)) = \mathrm{d}_{\mathbb{R}}(d(x, y), r - s) = \max(d(x, y) - r + s, 0)$.

We check that d^+ is indeed a hemi-metric, and in particular the triangular inequality: $d^+((x, r), (z, t)) = \max(d(x, z) - r + t, 0) \leqslant \max(d(x, y) + d(y, z) - r + t, 0) = \max((d(x, y) - r + s) + (d(y, z) - s + t), 0)$, while $d^+((x, r), (y, s)) + d^+((y, s), (z, t)) = \max((d(x, y) - r + s) + (d(y, z) - s + t), d(x, y) - r + s, d(y, z) - s + t, 0)$ is greater than or equal to the former.

Using Proposition 6.1.8, we obtain:

Fact 7.3.2 *Let X, d be a hemi-metric space. The specialization quasi-ordering $\leqslant^{d^+}$ on $\mathbf{B}(X, d)$ is defined by $(x, r) \leqslant^{d^+} (y, s)$ iff $d(x, y) \leqslant r - s$.*

Note in particular that this implies $r \geqslant s$.

Figure 7.3 illustrates how formal balls compare in $\leqslant^{d^+}$. X isometrically embeds into $\mathbf{B}(X, d)$ by equating the formal balls $(x, 0)$ with $x \in X$. Each formal ball (x, r) is a point, and the space of all formal balls above (x, r) is drawn as a pair of rays extending upward. The intersection of the cone delimited by these rays with X is the ordinary *closed ball* $B^d_{x, \leqslant r} = \{y \in X \mid d(x, y) \leqslant r\}$. (Beware that closed balls are usually not closed in X, d—they are closed in X, d^{op}, though.) Figure 7.3, left, indicates the general situation of a formal ball (x, r) below another one, (y, s). It is easy to check that, in this case where $(x, r) \leqslant^{d^+} (y, s)$, $B^d_{x, \leqslant r} \supseteq B^d_{y, \leqslant s}$. Figure 7.3, right, describes the general case, where $d^+((x, r), (y, s))$ is not necessarily 0.

Lemma 7.3.3 (Embedding into formal balls) *Let X, d be a hemi-metric space. The map $\eta^{\mathrm{EH}}_X \colon X \to \mathbf{B}(X, d)$ defined by $\eta^{\mathrm{EH}}_X(x) = (x, 0)$ is an isometric embedding.*

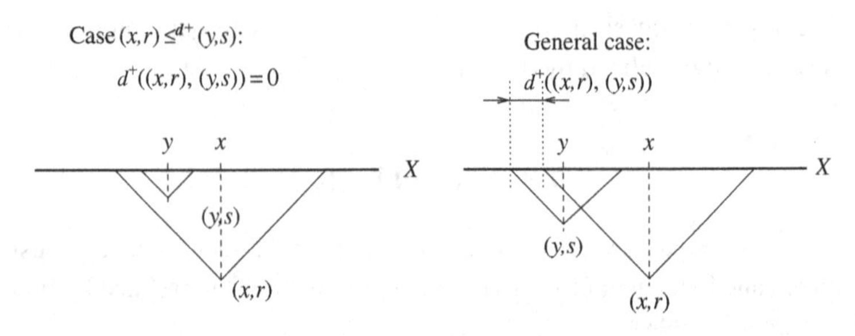

Figure 7.3 The space of formal balls.

Proof Since $d^+((x,0),(y,0)) = \max(d(x,y) - 0 + 0, 0) = d(x,y)$. □

Figure 7.3 is typical of the case where d is a metric, not a general hemi-metric, and more specifically where $X = \mathbb{R}$. This may give the false impression that $(x,r) \leqslant^{d^+} (y,s)$ if and only if $B^d_{x,\leqslant r} \supseteq B^d_{y,\leqslant s}$. Exercise 7.3.4 should help in grasping the extent to which this is or is not the case.

Exercise 7.3.4 Equip $\mathbb{R}^n$ with the Euclidean (L^2) metric $d(\vec{x},\vec{y}) = ||\vec{x} - \vec{y}||_2$, where subtraction $\vec{x} - \vec{y}$ is taken componentwise, and the L_2 *norm* is defined by $||(x_1,\ldots,x_n)||_2 = \sqrt{\sum_{i=1}^n x_i^2}$. Show that $B^d_{\vec{x},\leqslant r} \supseteq B^d_{\vec{y},\leqslant s}$ iff $(\vec{x},r) \leqslant^{d^+} (\vec{y},s)$ in the formal ball ordering. This in fact holds in every normed vector space, with coefficients in $\mathbb{R}$.

On the other hand, consider $\mathbb{N}$ with the metric d, defined by $d(m,n) = 1 + \frac{1}{2^{m+n}}$ if $m \neq n$, $d(m,n) = 0$ if $m = n$. Show that the equivalence $B^d_{x,\leqslant r} \supseteq B^d_{y,\leqslant s}$ iff $(x,r) \leqslant^{d^+} (y,s)$ fails in this space.

Exercise 7.3.5 Equip $\mathbb{N}$ with the metric d of Exercise 7.3.4, second part. Show that every Cauchy sequence $(m_n)_{n\in\mathbb{N}}$ in $\mathbb{N}, d$ is eventually constant, i.e., that there is an $n_0 \in \mathbb{N}$ such that $m_n = m_{n_0}$ for every $n \geqslant n_0$. Deduce that $\mathbb{N}, d$ is complete. However, $B^d_{n,\leqslant 1+1/2^{2n}}$, $n \in \mathbb{N}$, is a descending family of closed balls with empty intersection.

Exercise 7.3.6 (Cantor) Let X, d be a complete metric space, and $(F_i)_{i\in I}$ be a filtered family of non-empty closed subsets of X. Assume that $\inf_{i\in I} \delta(F_i) = 0$, where $\delta(F) = \sup_{x,x'\in F} d(x,x')$ is the *diameter* of F. Then show that $\bigcap_{i\in I} F_i$ contains a single point. This is *Cantor's Theorem* – not to be confused with Theorem 2.2.1, also due to G. Cantor.

Now explain the following apparent paradox: the closed balls $B^d_{n,\leqslant 1+1/2^{2n}}$ of Exercise 7.3.5, $n \in \mathbb{N}$, form a filtered family of non-empty and closed subsets in a complete metric space, but their intersection *is* empty.

Exercise 7.3.7 Let X, d be a quasi-metric space. Show that d^+ is T_0, i.e., $\mathbf{B}(X,d), d^+$ is also a quasi-metric space.

Lemma 7.3.8 *Let X, d be a Smyth-complete quasi-metric space.*

Then $\mathbf{B}(X,d), d^+$ is also a Smyth-complete quasi-metric space. The limit of any Cauchy-weightable net $(x_i,r_i)_{i\in I,\sqsubseteq}$ of formal balls in $\mathbf{B}(X,d), d^{+sym}$ is (x,r), where x is the limit of $(x_i)_{i\in I,\sqsubseteq}$ in X, d^{sym} and $r = \inf_{i\in I} r_i$.

Proof First, $\mathbf{B}(X,d), d^+$ is quasi-metric by Exercise 7.3.7. Let $((x_i,r_i),s_i)_{i\in I,\sqsubseteq}$ be a Cauchy-weighted net of formal balls. For all $i, j \in I$ such that $i \sqsubseteq j$, $d^+((x_i,r_i),(x_j,r_j)) \leqslant s_i - s_j$, i.e., $d(x_i,x_j) \leqslant r_i + s_i - r_j - s_j$ and $s_i \geqslant s_j$. Also, $\inf_{i\in I} s_i = 0$. Let $r = \inf_{i\in I} r_i$. Then $(x_i, r_i - r + s_i)_{i\in I,\sqsubseteq}$

is a Cauchy-weighted net in X, d. By assumption, $(x_i)_{i \in I, \sqsubseteq}$ has a limit x in X, d^{sym}. We claim that (x, r) is the limit of $(x_i, r_i)_{i \in I, \sqsubseteq}$ in $\mathbf{B}(X, d), d^{+sym}$.

For every $\epsilon > 0$, $d^{sym}(x_i, x) < \epsilon/2$ for i large enough. Also, $0 \leqslant r_i - r < \epsilon/2$ for i large enough. So $d^+((x_i, r_i), (x, r)) = \max(d(x_i, x) - r_i + r, 0) < \epsilon/2 < \epsilon$ and $d^+((x, r), (x_i, r_i)) = \max(d(x, x_i) - r + r_i, 0) < \epsilon$, i.e., $d^{+sym}((x_i, r_i), (x, r)) < \epsilon$ for i large enough. □

7.3.1 The c-space of formal balls, and the Romaguera–Valero Theorem

The closed ball $B^{d^+}_{(x,r), \leqslant \epsilon}$ is the set of formal balls (y, s) such that $d(x, y) - r + s \leqslant \epsilon$, or, equivalently, $d(x, y) \leqslant (r + \epsilon) - s$. So $B^{d^+}_{(x,r), \leqslant \epsilon}$ is just the upward closure of $(x, r + \epsilon)$ in $\mathbf{B}(X, d)$. In particular, all closed balls in $\mathbf{B}(X, d)$ are compact. But more holds:

Proposition 7.3.9 (The c-space of formal balls) *Let X, d be a hemi-metric space.*

Then $\mathbf{B}(X, d)$, with the open ball topology of d^+, is a c-space.

Proof For every $(x, r) \in \mathbf{B}(X, d)$, for every open neighborhood U of (x, r) in $\mathbf{B}(X, d)$, U contains an open ball $B^{d^+}_{(x,r), <\epsilon}$ (Lemma 6.1.5). The latter contains the closed ball $B^{d^+}_{(x,r), \leqslant 2\epsilon/3}$, which is the upward closure of $(x, r + 2\epsilon/3)$ in $\mathbf{B}(X, d)$. In turn, $\uparrow(x, r + 2\epsilon/3) = B^{d^+}_{(x,r), \leqslant 2\epsilon/3}$ contains $B^{d^+}_{(x,r), <\epsilon/3}$, so (x, r) is in the interior of $\uparrow(x, r + 2\epsilon/3)$. □

We shall see later (Exercise 8.3.46) that c-spaces and abstract bases $B, \prec$ (see Lemma 5.1.32) are essentially the same thing, where $b \prec b'$ is defined as $b' \in int(\uparrow b)$. This relation has a very simple description in $\mathbf{B}(X, d)$, which one should compare to the definition of $\leqslant^{d^+}$ (Fact 7.3.2).

Lemma 7.3.10 (The abstract basis of formal balls) *Let X, d be a hemi-metric space. Define the relation $\prec^{d^+}$ on $\mathbf{B}(X, d)$ by*

$$(x, r) \prec^{d^+} (y, s) \text{ iff } d(x, y) < r - s.$$

Then $(x, r) \prec^{d^+} (y, s)$ if and only if (y, s) is in the interior of $\uparrow(x, r)$ in $\mathbf{B}(X, d)$.

$\mathbf{B}(X, d), \prec^{d^+}$ is an abstract basis.

Proof If $(x, r) \prec^{d^+} (y, s)$, then let $\epsilon = r - s - d(x, y)$. Then $\epsilon > 0$, $B^{d^+}_{(y,s), <\epsilon}$ is an open neighborhood of (y, s), and is included in $\uparrow(x, r)$. Indeed, for every $(z, t) \in B^{d^+}_{(y,s), <\epsilon}$, $d(y, z) < s - t + \epsilon = r - t - d(x, y)$, so $d(x, z) \leqslant$

$d(x, y) + d(y, z) < r - t$, meaning that $(x, r) \leqslant^{d^+} (z, t)$. So (y, s) is in $B^{d^+}_{(y,s),<\epsilon}$, which is itself included in the interior of $\uparrow(x, r)$.

Conversely, assume (y, s) is in the interior of $\uparrow(x, r)$. By Lemma 6.1.5, $B^{d^+}_{(y,s),<\epsilon} \subseteq \uparrow(x, r)$ for some $\epsilon > 0$. In particular, $(y, s + \epsilon/2)$ is in $\uparrow(x, r)$, i.e., $d(x, y) \leqslant r - (s + \epsilon/2) < r - s$.

The last claim follows from Proposition 7.3.9 and Lemma 5.1.41. □

Although we cannot state it right now, we draw the attention of the reader to the following elegant characterization:

A quasi-metric space X, d is Smyth-complete iff $\mathbf{B}(X, d)$ is *sober* in its open ball topology.

This will come out of the study of sobriety (Chapter 8), and will be stated as Theorem 8.3.40. For now, we shall use the *Romaguera–Valero Theorem* (Romaguera and Valero, 2010) below as a substitute characterization—which will turn out to be equivalent to the above.

The Cauchy-weighted nets are exactly those nets $(x_i, r_i)_{i \in I, \sqsubseteq}$ in $\mathbf{B}(X, d)$ such that $\inf_{i \in I} r_i = 0$ and $i \sqsubseteq j$ implies $(x_i, r_i) \leqslant^{d^+} (x_j, r_j)$. The latter condition means that $(x_i, r_i)_{i \in I, \sqsubseteq}$ is a *monotone* net (see Exercise 4.7.8).

Theorem 7.3.11 (Romaguera–Valero) *Let X, d be a quasi-metric space. Then X, d is Smyth-complete if and only if $\mathbf{B}(X, d)$ is a continuous dcpo in its specialization quasi-ordering $\leqslant^{d^+}$, and $\prec^{d^+}$ is its way-below relation.*

If so, the least upper bound of a directed family of formal balls $(x_i, r_i)_{i \in I}$ is (x, r), where x is the limit of $(x_i)_{i \in I, \sqsubseteq}$ in X, d^{sym} (where $\sqsubseteq$ is such that $i \sqsubseteq j$ implies $(x_i, r_i) \leqslant^{d} (x_j, r_j)$), and $r = \inf_{i \in I} r_i$.

Proof Assume $\mathbf{B}(X, d)$ is a continuous dcpo with $\prec^{d^+}$ as way-below relation. Given any monotone net $(x_i, r_i)_{i \in I, \sqsubseteq}$ of formal balls, the family $(x_i, r_i)_{i \in I}$ is directed, since I is directed. Let (x, r) be its least upper bound in $\mathbf{B}(X, d)$.

We first show that $r = \inf_{i \in I} r_i$. Since $(x_i, r_i) \leqslant^{d^+} (x, r)$, $r_i \geqslant r$ for every $i \in I$, so $r \leqslant \inf_{i \in I} r_i$. Let $r_\infty = \inf_{i \in I} r_i$. Then $(x_i, r_i - r_\infty)_{i \in I}$ also has a least upper bound (x', r'). By the same argument, $r' \leqslant \inf_{i \in I}(r_i - r_\infty) = 0$. Now (x', r_∞) is also an upper bound of $(x_i, r_i)_{i \in I, \sqsubseteq}$: for every $i \in I$, $(x_i, r_i - r_\infty) \leqslant^{d^+} (x', 0)$, so $d(x_i, x') \leqslant r_i - r_\infty$, i.e., $(x_i, r_i) \leqslant^{d^+} (x', r_\infty)$. Since (x, r) is the least upper bound, $(x, r) \leqslant^{d^+} (x', r_\infty)$, in particular $r \geqslant r_\infty$. So $r = r_\infty = \inf_{i \in I} r_i$.

We claim that x is the limit of $(x_i)_{i \in I, \sqsubseteq}$ in X, d^{sym}. For each $i \in I$, $(x_i, r_i) \leqslant^{d^+} (x, r)$, so $d(x_i, x) \leqslant r_i - r$. For every $\epsilon > 0$, $r_i < \epsilon$ for i large enough, so (a): $d(x_i, x) < \epsilon$ for i large enough. Also, $(x, r + \epsilon) \prec^{d^+} (x, r)$.

Since $\prec^{d^+}$ is the way-below relation on $\mathbf{B}(X, d)$, there is an index $i_0 \in I$ such that $(x, r + \epsilon) \prec^{d^+} (x_{i_0}, r_{i_0})$, i.e., $d(x, x_{i_0}) < \epsilon + r - r_{i_0}$. For every $i \in I$ with $i_0 \sqsubseteq i$, $d(x, x_i) \leqslant d(x, x_{i_0}) + d(x_{i_0}, x_i) \leqslant d(x, x_{i_0}) + r_{i_0} - r_i < \epsilon + r - r_i \leqslant \epsilon$. So (b): $d(x, x_i) < \epsilon$ for i large enough. Together with (a), we obtain that $d^{sym}(x, x_i) < \epsilon$ for i large enough. Using Lemma 7.1.1, x is the limit of $(x_i)_{i \in I, \sqsubseteq}$ in X, d^{sym}.

So X, d is Smyth-complete. Moreover, we have shown the desired characterization of least upper bounds in $\mathbf{B}(X, d)$.

Conversely, assume that X, d is Smyth-complete, and let $(x_i, r_i)_{i \in I, \sqsubseteq}$ be any monotone net of formal balls. (In particular, for any directed family $(x_i, r_i)_{i \in I}$, defining $i \sqsubseteq j$ iff $(x_i, r_i) \leqslant^{d^+} (x_j, r_j)$ yields such a monotone net.) Let $r = \inf_{i \in I} r_i$, so that $(x_i, r_i - r)_{i \in I, \sqsubseteq}$ is a Cauchy-weighted net. Since X, d is Smyth-complete, $(x_i)_{i \in I, \sqsubseteq}$ has a limit x in X, d^{sym}, by Proposition 7.2.10. We claim that (x, r) is the least upper bound of $(x_i, r_i)_{i \in I}$.

Given any $i \in I$, and every $j \in I$ with $i \sqsubseteq j$, $d(x_i, x) \leqslant d(x_i, x_j) + d(x_j, x) \leqslant r_i - r_j + d(x_j, x)$. For every $\epsilon > 0$, there is a $j \in I$ such that $d^{sym}(x, x_j) < \epsilon$, in particular $d(x_j, x) < \epsilon$. Since I is directed, we can moreover assume that $i \sqsubseteq j$. Then $d(x_i, x) < r_i - r_j + \epsilon \leqslant r_i - r + \epsilon$. Since ϵ is arbitrary, $d(x_i, x) \leqslant r_i - r$. That is, $(x_i, r_i) \leqslant^{d^+} (x, r)$.

Let (y, s) be any other upper bound of $(x_i, r_i)_{i \in I}$. So $d(x_i, y) \leqslant r_i - s$ for every $i \in I$. Since x is a limit of $(x_i)_{i \in I}$ in X, d^{sym}, it is also a limit in X, d by Lemma 7.1.9, so for every $\epsilon > 0$, $d(x, x_i) < \epsilon$ for i large enough (Lemma 7.1.1). So $d(x, y) \leqslant d(x, x_i) + d(x_i, y) < r_i - s + \epsilon$ for i large enough. Taking greatest lower bounds, $d(x, y) \leqslant r - s + \epsilon$. Since ϵ is arbitrary, $d(x, y) \leqslant r - s$, i.e., $(x, r) \leqslant^{d^+} (y, s)$.

So (x, r) is the least upper bound of $(x_i, r_i)_{i \in I}$. In particular, $\mathbf{B}(X, d)$ is a dcpo.

We claim that (y, s) is way-below (y', s') if and only if $(y, s) \prec^{d^+} (y', s')$. If (y, s) is way-below (y', s'), then, considering the directed family $(y', s' + \epsilon)_{\epsilon > 0}$, there is an $\epsilon > 0$ such that $(y, s) \leqslant^{d^+} (y', s' + \epsilon)$, i.e., $d(y, y') \leqslant s - s' - \epsilon$. In particular, $d(y, y') < s - s'$, so $(y, s) \prec^{d^+} (y', s')$. Conversely, if $(y, s) \prec^{d^+} (y', s')$, then consider any directed family $(x_i, r_i)_{i \in I, \sqsubseteq}$ whose least upper bound (x, r) is above (y', s'). Recall that $r = \inf_{i \in I} r_i$ and x is the limit of $(x_i)_{i \in I, \sqsubseteq}$ in X, d^{sym}, and also that $d(y', x) \leqslant s' - r$. Since $(y, s) \prec^{d^+} (y', s')$, $d(y, x) \leqslant d(y, y') + d(y', x) < (s - s') + (s' - r) = s - r$. Let $\epsilon = s - r - d(y, x)$. For each $i \in I$, $d(y, x_i) \leqslant d(y, x) + d(x, x_i) = s - r - \epsilon + d(x, x_i)$. Since x is the limit of $(x_i)_{i \in I, \sqsubseteq}$ in X, d^{sym}, $d(x, x_i) < \epsilon/2$ for i large enough. And since $r = \inf_{i \in I} r_i$, $r > r_i - \epsilon/2$ for i large enough. So $d(y, x_i) < s - r_i$ for i large enough. That is, $(y, s) \leqslant^{d^+} (x_i, r_i)$ for i large enough. In particular, (y, s) is way-below (y', s').

Given any formal ball (x, r), (x, r) is now the least upper bound of the directed family (a chain, even) $(x, r + \epsilon)_{\epsilon > 0}$, consisting of elements way-below (x, r). So $\mathbf{B}(X, d)$ is a continuous dcpo, by Trick 5.1.20. □

The two conditions, that the space of formal balls is a continuous dcpo and that $\prec^{d^+}$ is the way-below relation, are needed. The following exercise shows that we cannot dispense with the latter, and Exercise 7.3.14 implies that we cannot dispense with the requirement that the space of formal balls is a dcpo.

Exercise 7.3.12 Show that $\mathbf{B}(\mathbb{R}_\ell, \mathsf{d}_\ell), \leqslant^{\mathsf{d}_\ell^+}$ is a continuous dcpo. It is easiest to show that it is order isomorphic to a Scott-closed subset of $(\mathbb{R} \times \mathbb{R})_\sigma$, through the map that sends the formal ball (x, r) to the pair $(-x, x - r)$. Concretely, show that the least upper bound of any monotone net $(x_i, r_i)_{i \in I, \sqsubseteq}$ is $(\inf_{i \in I} x_i, \inf_{i \in I} r_i)$, and that (x, r) is way-below (y, s) in $\mathbf{B}(\mathbb{R}_\ell, \mathsf{d}_\ell)$ iff $x > y$ and $x - y < r - s$.

Conclude that there are quasi-metric spaces X, d whose space of formal balls is a continuous dcpo, but whose way-below relation is not $\prec^{d^+}$.

In Theorem 7.3.11, we do not require the Scott topology on $\mathbf{B}(X, d)$ to coincide with its open ball topology. However, they do coincide in case X, d is Smyth-complete.

Lemma 7.3.13 *Let X, d be a Smyth-complete quasi-metric space. On $\mathbf{B}(X, d)$, the open ball topology of d^+ coincides with the Scott topology of $\leqslant^{d^+}$.*

Proof By the Romaguera–Valero Theorem 7.3.11, $\mathbf{B}(X, d)$ is a continuous dcpo with $\prec^{d^+}$ as way-below relation. By Theorem 5.1.17, the Scott topology has a base of open subsets of the form $\Uparrow(x, r) = \{(y, s) \mid d(x, y) < r - s\}$. If $r = 0$, this is empty. Otherwise, this is the open ball $B^{d^+}_{(x,0), <r}$. Conversely, each open ball $B^{d^+}_{(x,r), <\epsilon}$ is the Scott-open $\Uparrow(x, r + \epsilon)$. □

Exercise 7.3.14 (Formal balls, metric case) Let X, d be a metric space. Given a monotone net of formal balls $(x_i, r_i)_{i \in I, \sqsubseteq}$ on X, show that $(x_i)_{i \in I, \sqsubseteq}$ is Cauchy. Using this, show that, if it has a least upper bound (x, r) in $\mathbf{B}(X, d), \leqslant^{d^+}$, then $r = \inf_{i \in I} r_i$, and then that x is the limit of $(x_i)_{i \in I, \sqsubseteq}$ in X, d. Conversely, show that, if $(x_i)_{i \in I, \sqsubseteq}$ converges to x in X, d, then $(x, \inf_{i \in I} r_i)$ is the least upper bound of $(x_i, r_i)_{i \in I, \sqsubseteq}$ in $\mathbf{B}(X, d)$.

Show that $\prec^{d^+}$ is the way-below relation on $\mathbf{B}(X, d), \leqslant^{d^+}$—even if X, d is not Smyth-complete.

Conclude that $\mathbf{B}(X, d), \leqslant^{d^+}$ is a continuous poset, and that its Scott topology coincides with the open ball topology of d^+. (None of this holds for general quasi-metric spaces.)

Exercise 7.3.15 (Coproducts of spaces of formal balls) For every family $(X_i, d_i)_{i \in I}$ of hemi-metric spaces, show that $\mathbf{B}(\coprod_{i \in I} X_i, \coprod_{i \in I} d_i)$ is naturally isometric (hence homeomorphic, as well as order isomorphic) to $\coprod_{i \in I} \mathbf{B}(X_i, d_i)$, $\coprod_{i \in I} d_i^+$. The isometry maps $((i, x), r)$ with $x \in X_i$ to $(i, (x, r))$. Using the Romaguera–Valero Theorem 7.3.11, deduce another proof that every coproduct of Smyth-complete quasi-metric spaces is Smyth-complete.

Lemma 7.3.16 *Let X, d and Y, d' be two hemi-metric spaces, and $f: X \to Y$ be a c-Lipschitz map, $c \in \mathbb{R}^+$. Then $\mathbf{B}^c(f): \mathbf{B}(X, d) \to \mathbf{B}(Y, d')$, defined by $\mathbf{B}^c(f)(x, r) = (f(x), c.r)$, is c-Lipschitz.*

In particular, $\mathbf{B}^1$ defines a functor from **HMet** *(resp.,* **QMet***, resp.* **Met***) to itself.*

Proof Indeed, $d^+(\mathbf{B}^c(f)(x, r), \mathbf{B}^c(f)(y, s)) = d^+((f(x), c.r), (f(y), c.s)) = \max(d(f(x), f(y)) - c.r + c.s, 0) \leqslant \max(c.d(x, y) - c.r + c.s, 0) = c.d^+((x, r), (y, s))$. □

7.3.2 Free complete metric spaces, and extension by continuity

Every metric space can be embedded into a complete one, in a canonical way. The idea is that if $(x_n)_{n \in \mathbb{N}}$ is a Cauchy sequence that does not have a limit, we add one. To do so, we declare that the missing limit is defined as the Cauchy sequence itself, up to the fact that we must equate (through $\equiv$ below) all Cauchy sequences that intuitively converge to the same limit – i.e., we take a T_0 quotient, in the sense of Exercise 6.4.4. This classical construction on metric (not quasi-metric) spaces is the subject of Exercise 7.3.17.

One can rephrase this in a categorical way as follows. Let X, d be a metric space. Its Cauchy completion $\mathcal{C}hy(X, d)$ is the *free complete metric space* over the metric space X, d. That is, for every complete metric space Y, d' and every 1-Lipschitz map $f: X \to Y$, f extends to a unique 1-Lipschitz map $g: \mathcal{C}hy(X, d) \to Y$, in the sense that $g \circ \eta_X^{\mathcal{C}hy} = f$.

Exercise 7.3.17 (Cauchy completion) Let X, d be a metric space. The *Cauchy completion* $\mathcal{C}hy(X, d)$ of X, d is built as follows. Given two sequences $(x_n)_{n \in \mathbb{N}}$ and $(y_n)_{n \in \mathbb{N}}$, let $\widehat{d}((x_n)_{n \in \mathbb{N}}, (y_n)_{n \in \mathbb{N}}) = \limsup_{n \in \mathbb{N}} d(x_n, y_n)$ and $(x_n)_{n \in \mathbb{N}} \equiv (y_n)_{n \in \mathbb{N}}$ if and only if $\widehat{d}((x_n)_{n \in \mathbb{N}}, (y_n)_{n \in \mathbb{N}}) = 0$. Let $\mathcal{C}hy(X, d)$ be the set of $\equiv$-equivalence classes $q_{\equiv}((x_n)_{n \in \mathbb{N}})$ of Cauchy sequences $(x_n)_{n \in \mathbb{N}}$ in X, and define $\widehat{d}$ on $\mathcal{C}hy(X, d)$ by quotienting, i.e., $\widehat{d}(q_{\equiv}((x_n)_{n \in \mathbb{N}}), q_{\equiv}((y_n)_{n \in \mathbb{N}})) = \widehat{d}((x_n)_{n \in \mathbb{N}}, (y_n)_{n \in \mathbb{N}})$.

Show that $Chy(X, d), \widehat{d}$ is a complete metric space, and that X, d embeds isometrically into $Chy(X, d), \widehat{d}$, through η_X^{Chy}, defined so that $\eta_X^{Chy}(x)$ is the $\equiv$-equivalence class of the constant sequence whose terms are all equal to x.

Moreover, show that the image $\eta_X^{Chy}[X]$ is dense in $Chy(X, d)$.

Exercise 7.3.18 Let X, d be a metric space. Show that the definition of $\widehat{d}$ can be simplified, for Cauchy sequences $(x_n)_{n \in \mathbb{N}}$ and $(y_n)_{n \in \mathbb{N}}$, to: $\widehat{d}((x_n)_{n \in \mathbb{N}}, (y_n)_{n \in \mathbb{N}})$ is the limit of $(d(x_n, y_n))_{n \in \mathbb{N}, \leqslant}$ in $\overline{\mathbb{R}^+}, \mathsf{d}_{\mathbb{R}}^{sym}$.

Deduce that the Cauchy completion $Chy(X)$ of an ultra-metric space X, d is again ultra-metric.

Every metric space is dense in its Cauchy completion. This is important, because it allows us to extend any uniformly continuous map.

Lemma 7.3.19 (Extension by continuity) *Let A be a dense subset of a topological space X, and f be a continuous map from A to a T_2 space Y. Then f has at most one continuous extension $g: X \to Y$, i.e., one such that f and g coincide on A.*

This continuous extension g exists if X, d is metric, Y, d' is complete metric, and f is uniformly continuous from A, d to Y, d'. Then g is not only continuous but even uniformly continuous, and, if f is c-Lipschitz, then so is g.

Proof By assumption, X is the closure of A, i.e., the set of all limits of nets whose elements are in A, by Proposition 4.7.9. If g extends f and is continuous, then, for every element $x \in X$, which is a limit of a net $(a_i)_{i \in I, \sqsubseteq}$ inside A, necessarily $g(x)$ is a limit of $(f(a_i))_{i \in I, \sqsubseteq}$, since continuous maps preserve limits (Proposition 4.7.18). Since Y is T_2, limits are unique (Proposition 4.7.4), so that the extension g is unique, if it exists.

Conversely, assume X, d metric, Y, d' complete metric, and f uniformly continuous from X, d to Y, d'. Surely the above uniqueness result holds, for Y is T_2 by Lemma 6.1.10. For any $x \in X$, consider any net $(a_i)_{i \in I, \sqsubseteq}$ converging to x. For every $\epsilon > 0$, since f is uniformly continuous, one can find $\eta > 0$ such that, for all $a, a' \in A$ with $d(a, a') < \eta$, $d'(f(a), f(a')) < \epsilon$. For i, i' large enough, $d(a_i, a_{i'}) < \eta$ since every convergent sequence is Cauchy (Lemma 7.1.5, using the fact that d is a metric), so $d'(f(a_i), f(a_{i'})) < \epsilon$. So $(f(a_i))_{i \in I, \sqsubseteq}$ is Cauchy, hence converges to some value since Y, d' is complete. We name this value $g(x)$. Really making a function g out of this requires the Axiom of Choice, and we must also check that this value of $g(x)$ does not depend on $(a_i)_{i \in I, \sqsubseteq}$, only on its limit x. So let $(b_j)_{j \in J, \preceq}$ be any other net converging to x. For any $\epsilon > 0$, let η be as above. For i large enough, $d(x, a_i) < \eta/2$, and for j large enough $d(x, b_j) < \eta/2$. Since d is a metric,

$d(a_i, x) < \eta/2$, so $d(a_i, b_j) < \eta$, whence $d'(f(a_i), f(b_j)) < \epsilon$ for i, j large enough. Since $f(a_i)$ converges to $g(x)$, we also have $d'(g(x), f(a_i)) < \epsilon$ for i large enough, so $d'(g(x), f(b_j)) < 2\epsilon$ for j large enough. Since 2ϵ can be made arbitrarily small, $(f(b_j))_{j \in J, \preceq}$ also converges to $g(x)$.

Let us check that g is uniformly continuous. For every $\epsilon > 0$, let $\eta > 0$ be such that, for all $a, a' \in A$ with $d(a, a') < 3\eta$, $d'(f(a), f(a')) < \epsilon/3$. Let $x, x' \in X$ be any two points with $d(x, x') < \eta$. We claim that $d'(g(x), g(x')) < \epsilon$. Let x be a limit of a net $(a_i)_{i \in I, \sqsubseteq}$ in A, x' be a limit of a net $(b_j)_{j \in J, \preceq}$ in A. For i large enough, $d(x, a_i) < \eta \leqslant 3\eta$, and $d'(g(x), f(a_i)) < \epsilon/3$. For j large enough, similarly $d(x', b_j) < \eta$ and $d'(g(x'), f(b_j)) < \epsilon/3$, so $d'(f(b_j), g(x')) < \epsilon/3$. Since $d(a_i, b'_j) \leqslant d(a_i, x) + d(x, x') + d(x', b_j) = d(x, a_i) + d(x, x') + d(x', b_j) < 3\eta$, $d'(f(a_i), f(b'_j)) < \epsilon/3$. So $d'(g(x), g(x')) \leqslant d'(g(x), f(a_i)) + d'(f(a_i), f(b_j)) + d'(f(b_j), g(x')) < \epsilon$.

Finally, if f is c-Lipschitz, then, for all $x, x' \in X$, let x be a limit of some net $(a_i)_{i \in I, \sqsubseteq}$ in A, and x' be a limit of some net $(b_j)_{j \in J, \preceq}$ in A. For every $\epsilon > 0$, $d'(g(x), f(a_i)) < \epsilon/4$ and $c.d(a_i, x) < \epsilon/4$ for i large enough, and $d'(f(b'_j), g(x')) < \epsilon/4$, $c.d(x', b'_j) < \epsilon/4$ for j large enough, so $d'(g(x), g(x')) < \epsilon/4 + d'(f(a_i), f(b'_j)) + \epsilon/4 \leqslant c.d(a_i, b'_j) + \epsilon/2 \leqslant (\epsilon/4 + c.d(x, x') + \epsilon/4) + \epsilon/2 \leqslant c.d(x, x') + \epsilon$. As ϵ is arbitrary, $d'(g(x), g(x')) \leqslant c.d(x, x')$.

The recourse to nets in the above proof is not necessary. A variant of Exercise 4.7.14 indeed shows that, when X is first-countable, in particular metric, the closure of A in X is the set of all limits of *sequences* of elements of A. □

While Lemma 7.3.19 only applies to T_2, resp. metric spaces, it is easy to extend it to quasi-metric spaces. We only need to make the following observation.

Fact 7.3.20 *Let X, d and Y, d' be quasi-metric spaces, and f be uniformly continuous from X, d to Y, d' (resp., c-Lipschitz). Then f is also uniformly continuous (resp., c-Lipschitz) from X, d^{sym} to Y, d'^{sym}.*

Corollary 7.3.21 *Let X, d and Y, d' be quasi-metric spaces, and A be a dense subset of X, d^{sym}. Assume that Y, d'^{sym} is complete (e.g., if Y, d' is Smyth-complete). Then every uniformly continuous map f from A, d to Y, d' extends to a unique uniformly continuous map g from X, d to Y, d'. If f is c-Lipschitz, then so is g.*

7.4 A weak form of completeness: Yoneda-completeness

Another notion of completeness was proposed by Bonsangue *et al.* (1998). This has come to be known in the literature as *Yoneda-completeness*. We shall see that this notion of completeness also has a natural characterization in terms of formal balls.

Definition 7.4.1 (Yoneda-complete) A hemi-metric space X, d is *Yoneda-complete* if and only if d is T_0 and every Cauchy net in X, d has a d-limit.

The definition of Smyth-complete spaces was the same, except that we required every Cauchy net to have a d^{op}-limit, not a d-limit.

Lemma 7.4.2 (Smyth $\Rightarrow$ Yoneda-complete) *Every Smyth-complete quasi-metric space is Yoneda-complete.*

Proof In a Smyth-complete quasi-metric space, every Cauchy net has a limit in X, d^{sym}, and this is in particular a d-limit by Proposition 7.1.19. □

In metric spaces, Yoneda-completeness won't bring anything new:

Lemma 7.4.3 *For metric spaces, Smyth-completeness, Yoneda-completeness, and completeness are equivalent notions.*

Proof This is a trivial consequence of Corollary 7.1.21. □

However, this will make a difference in quasi-metric spaces. Remember that $\overline{\mathbb{R}^+}, \mathsf{d}_{\mathbb{R}}$ is not Smyth-complete (Exercise 7.2.3). However (see Exercise 7.1.16):

Fact 7.4.4 ($\overline{\mathbb{R}^+}, \mathsf{d}_{\mathbb{R}}$ is Yoneda-Complete) $\mathbb{R} \cup \{-\infty, +\infty\}$, $\mathbb{R} \cup \{+\infty\}$ *and* $\overline{\mathbb{R}^+}$ *are Yoneda-complete in the* $\mathsf{d}_{\mathbb{R}}$ *quasi-metric.*

Example 7.4.5 But $\mathbb{R}, \mathsf{d}_{\mathbb{R}}$, without $+\infty$, is not Yoneda-complete: remember that every monotone net is Cauchy, e.g., $(x_n)_{n \in \mathbb{N}}$ where $x_n = n$. But this does not have any d-limit (i.e., a limsup; see Exercise 7.1.16). On the other hand, $\mathbb{R}, \mathsf{d}_{\mathbb{R}}^{sym}$ is (Smyth, Yoneda) complete.

As for Smyth-completeness, one can replace Cauchy nets by Cauchy-weightable nets. We need the following analogue of Lemma 7.2.9.

Lemma 7.4.6 *Let X, d be a hemi-metric space, and $(x_i)_{i \in I, \sqsubseteq}$ be a Cauchy net in X, d. If some subnet $(x_{\alpha(j)})_{j \in J, \preceq}$ has a d-limit x, then x is also a d-limit of $(x_i)_{i \in I, \sqsubseteq}$.*

Proof For every $y \in X$, $d(x, y) = \inf_{j \in J} \sup_{k \in J, j \preceq k} d(x_{\alpha(k)}, y)$ is less than or equal to $\inf_{j \in J} \sup_{i' \in I, \alpha(j) \sqsubseteq i'} d(x_{i'}, y)$, using the fact that α is monotonic, which is less than or equal to $\inf_{i \in I} \sup_{i' \in I, i \sqsubseteq i'} d(x_{i'}, y)$ since α is cofinal. The latter is just $\limsup_{i \in I, \sqsubseteq} d(x_i, y)$.

So $d(x, y) \leqslant \limsup_{i\in I,\sqsubseteq} d(x_i, y)$. Assume $d(x, y) = \limsup_{j\in J,\preceq} d(x_{\alpha(j)}, y)$ is strictly less than $\limsup_{i\in I,\sqsubseteq} d(x_i, y)$. Fix $a \in \mathbb{R}^+$ and $\epsilon > 0$ such that $\limsup_{j\in J,\preceq} d(x_{\alpha(j)}, y) < a < a + \epsilon \leqslant \limsup_{i\in I,\sqsubseteq} d(x_i, y)$. By the left-hand inequality, there is a $j \in J$ such that $d(x_{\alpha(k)}, y) < a$ for every $k \in J$ above j. Let $i_0 \in I$ be above $\alpha(j)$, and large enough that $d(x_i, x_{i'}) < \epsilon$ for all $i \sqsubseteq i'$ above i_0. For every such $i' \in I$, since α is cofinal, there is a $k \in J$ such that $i' \sqsubseteq \alpha(k)$. Since J is directed, we may assume that $j \preceq k$. So $d(x_{\alpha(k)}, y) < a$. Also, $d(x_i, x_{\alpha(k)}) < \epsilon$, so $d(x_i, y) < a + \epsilon$. This holds whenever $i_0 \sqsubseteq i$, so $\limsup_{i\in I,\sqsubseteq} d(x_i, y) < a + \epsilon$, a contradiction.

So $d(x, y) = \limsup_{i\in I,\sqsubseteq} d(x_i, y)$: x is a d-limit of $(x_i)_{i\in I,\sqsubseteq}$. □

Exercise 7.4.7 Let X, d be a hemi-metric space. Show the following converse to Lemma 7.4.6. Let $(x_i)_{i\in I,\sqsubseteq}$ be a Cauchy net in X, d, with d-limit x. Show that every subnet $(x_{\alpha(j)})_{j\in J,\preceq}$ of $(x_i)_{i\in I,\sqsubseteq}$ is Cauchy and has x as d-limit.

Proposition 7.4.8 *Let X, d be a quasi-metric space. Then X, d is Yoneda-complete if and only if every Cauchy-weightable net in X, d has a d-limit.*

Proof The only if direction is obvious. Conversely, let $(x_i)_{i\in I,\sqsubseteq}$ be a Cauchy net in X, d; it has a Cauchy-weightable subnet by Lemma 7.2.8, which has a d-limit x by assumption; then x is a d-limit of $(x_i)_{i\in I,\sqsubseteq}$ by Lemma 7.4.6. □

And the definition of d-limits takes a simpler form for Cauchy-weightable nets.

Lemma 7.4.9 *Let X, d be a hemi-metric space, and $(x_i, r_i)_{i\in I,\sqsubseteq}$ be a Cauchy-weighted net in X, d. Then x is a d-limit of $(x_i)_{i\in I,\sqsubseteq}$ if and only if, for every $y \in X$, $d(x, y) = \sup_{i\in I}(d(x_i, y) - r_i)$. Moreover, $(d(x_i, y) - r_i)_{i\in I,\sqsubseteq}$ is a monotone net, so the latter sup is the least upper bound of a directed family.*

Proof If $i \sqsubseteq j$, then $d(x_i, x_j) \leqslant r_i - r_j$ since $(x_i, r_i)_{i\in I,\sqsubseteq}$ is Cauchy-weighted. So, for any $y \in X$, $d(x_i, y) - r_i \leqslant d(x_i, x_j) + d(x_j, y) - r_i \leqslant d(x_j, y) - r_j$. Therefore $(d(x_i, y) - r_i)_{i\in I,\sqsubseteq}$ is a monotone net.

It follows that $\sup_{i\in I}(d(x_i, y) - r_i) = \limsup_{i\in I,\sqsubseteq}(d(x_i, y) - r_i)$. The $\geqslant$ direction is trivial. Conversely, we must show that, for every $i \in I$, $d(x_i, y) - r_i \leqslant \inf_{j\in I} \sup_{k\in I, j\sqsubseteq k}(d(x_k, y) - r_k)$. This follows from the fact that, for every $j \in I$, since I is directed there is a $k \in I$ above i and j, and then $d(x_i, y) - r_i \leqslant d(x_k, y) - r_k$.

For every $\epsilon > 0$, $r_i < \epsilon$ for i large enough, i.e., there is an $i \in I$ such that $r_j < \epsilon$ whenever $i \sqsubseteq j$. It follows easily that $\limsup_{i\in I,\sqsubseteq} d(x_i, y) - \epsilon \leqslant \limsup_{i\in I,\sqsubseteq}(d(x_i, y) - r_i) \leqslant \limsup_{i\in I,\sqsubseteq} d(x_i, y)$. Since $\epsilon > 0$ is arbitrary,

$\limsup_{i \in I, \sqsubseteq} d(x_i, y) = \limsup_{i \in I, \sqsubseteq}(d(x_i, y) - r_i) = \sup_{i \in I} d(x_i, y)$. The claim follows. □

Exercise 7.4.10 Let X be $\mathbb{R} \cup \{-\infty, +\infty\}$, resp., $\mathbb{R} \cup \{+\infty\}$, resp. $\overline{\mathbb{R}^+}$. Let $(x_i, r_i)_{i \in I, \sqsubseteq}$ be a Cauchy-weighted net in $X, \mathrm{d}_{\mathbb{R}}$. Show that the $\mathrm{d}_{\mathbb{R}}$-limit (i.e., the limsup) of $(x_i)_{i \in I, \sqsubseteq}$ is $\sup_{i \in I}(x_i - r_i)$, this being the least upper bound of the monotone net $(x_i - r_i)_{i \in I, \sqsubseteq}$.

7.4.1 Building Yoneda-complete spaces

Yoneda-completeness is preserved by many constructions – more than Smyth-completeness.

Lemma 7.4.11 *Let X, d be a quasi-metric space, and define d' as the quasi-metric $d'(x, y) = \min(d(x, y), 1)$. Then X, d is Yoneda-complete if and only if X, d' is, the notions of Cauchy net are the same in each space, and every d-limit of a Cauchy net is also a d'-limit of the net.*

Proof If $(x_i)_{i \in I, \sqsubseteq}$ is Cauchy in X, d', then, for every $\epsilon > 0$, $d'(x_i, x_j) < \min(\epsilon, 1)$ for $i \sqsubseteq j$ large enough. Since $d'(x_i, x_j) < 1$, $d'(x_i, x_j) = d(x_i, x_j)$, so $(x_i)_{i \in I, \sqsubseteq}$ is Cauchy in X, d. Since $d' \leqslant d$, the converse, that every Cauchy net in X, d is Cauchy in X, d', is clear. So Cauchy nets coincide in X, d and in X, d'.

Assume now that x is a d-limit of $(x_i)_{i \in I, \sqsubseteq}$. For every $y \in X$, $d(x, y) = \limsup_{i \in I, \sqsubseteq} d(x_i, y)$. So $d'(x, y) = \limsup_{i \in I, \sqsubseteq} \min(d(x_i, y), 1)$ – although min is not lim sup-continuous (see Exercise 7.1.13, first part), $\min(\_, 1)$ is (Exercise 7.1.13, second part, where the second net is the constant net 1). That is, y is a d'-limit of $(x_i)_{i \in I, \sqsubseteq}$.

Now let $(x_i)_{i \in I, \sqsubseteq}$ be any Cauchy net in X, d'. This is Cauchy in X, d, so it has a d-limit x. This must be a d'-limit of the net. So X, d' is Yoneda-complete. □

Lemma 7.4.12 *Every coproduct of Yoneda-complete quasi-metric spaces is Yoneda-complete.*

Proof We use Proposition 7.4.8. Let X_k, d_k be Yoneda-complete quasi-metric spaces, $k \in K$. Write d for $\coprod_{k \in K} d_k$. Any Cauchy-weighted net $(z_i, r_i)_{i \in I, \sqsubseteq}$ in $\coprod_{k \in K} X_k, \coprod_{k \in K} d_k$ must lie entirely in the same X_k, as in Lemma 7.2.24. In other words, writing z_i as (k_i, x_i), every k_i must be equal to some k. Let x be the d_k-limit of $(x_i)_{i \in I, \sqsubseteq}$, and $z = (k, x)$. For every $y' \in \coprod_{k \in K} X_k$, either y' is of the form (k, y) (with the same k), then $d(z, y') = d_k(x, y) =$

$\limsup_{i \in I, \sqsubseteq} d_k(x_i, y) = \limsup_{i \in I, \sqsubseteq} d(z_i, y')$; or y' is of the form (k', y) with $k' \neq k$, then $d(z, y') = +\infty = \limsup_{i \in I, \sqsubseteq} +\infty = \limsup_{i \in I, \sqsubseteq} d(z_i, y')$. In any case, z is the d-limit of $(z_i)_{i \in I, \sqsubseteq}$. □

While Smyth-complete spaces are stable under finite products, Yoneda-complete spaces are stable under arbitrary (categorical) products.

Lemma 7.4.13 *Let X_k, d_k be hemi-metric spaces, $k \in K$. Equip $X = \prod_{k \in K} X_k$ with the sup quasi-metric (namely $d(\vec{x}, \vec{y}) = \sup_{k \in K} d_k(x_k, y_k)$; see Exercise 6.5.5).*

In X, d-limits are computed componentwise: for every Cauchy net $(\vec{x}_i)_{i \in I, \sqsubseteq}$ in X, d, if $(x_{ik})_{i \in I, \sqsubseteq}$ has a d_k-limit x_k for each $k \in K$, then $\vec{x} = (x_k)_{k \in K}$ is a d-limit of $(\vec{x}_i)_{i \in I, \sqsubseteq}$.

In particular, if X_k, d_k is Yoneda-complete for each $k \in K$, then so is X, d.

Proof Let $(\vec{x}_i, r_i)_{i \in I, \sqsubseteq}$ be a Cauchy-weighted net in X, d. For every $k \in K$, and for all $i \sqsubseteq j$, $d(x_{ik}, x_{jk}) \leqslant d(\vec{x}_i, \vec{x}_j) < r_i - r_j$, so $(x_{ik}, r_i)_{i \in I, \sqsubseteq}$ is Cauchy-weighted. Assume x_k is a d_k-limit of $(x_{ik})_{i \in I, \sqsubseteq}$ for each $k \in K$, and let $\vec{x} = (x_k)_{k \in K}$. For every $\vec{y} \in X$, $d(\vec{x}, \vec{y}) = \sup_{k \in K} d_k(x_k, y_k) = \sup_{k \in K} \sup_{i \in I}(d_k(x_{ik}, y_k) - r_i)$ by Lemma 7.4.9. This is equal to $\sup_{i \in I} \sup_{k \in K}(d_k(x_{ik}, y_k) - r_i)$, that is, to $\sup_{i \in I}(d(\vec{x}_i, \vec{y}) - r_i)$. By Lemma 7.4.9 again, $\vec{y}$ is a d-limit of $(\vec{x}_i)_{i \in I, \sqsubseteq}$.

Now let $(\vec{x}_i)_{i \in I, \sqsubseteq}$ be a Cauchy net in X, d. By Lemma 7.2.8, it has a Cauchy-weightable subnet $(\vec{x}_{\alpha(j)})_{j \in J, \preceq}$. If x_k is a d_k-limit of $(x_{ik})_{i \in I, \sqsubseteq}$ for each $k \in K$, then x_k is also a d_k-limit of the subnet $(x_{\alpha(j)k})_{j \in J, \preceq}$ (Exercise 7.4.7). By the above, $(\vec{x}_{\alpha(j)})_{j \in J, \preceq}$ has a d-limit $\vec{x}$ with $\vec{x} = (x_k)_{k \in K}$. By Lemma 7.4.6, x_k is also a d_k-limit of the Cauchy net $(x_{ik})_{i \in I, \sqsubseteq}$. So d-limits are computed componentwise for general Cauchy nets.

In particular, if X_k, d_k is Yoneda-complete for each $k \in K$, then X, d is Yoneda-complete, and d-limits of Cauchy-weightable nets are computed componentwise. We must additionally check that d is T_0, but this is clear. □

Example 7.4.14 Remember that $(-\infty, 0]^{\mathbb{N}}$ is a countable product of Smyth-complete spaces that is not Smyth-complete. Lemma 7.4.13 (together with the fact that $(-\infty, 0]$, being Smyth-complete, is Yoneda-complete) allows us to conclude that it is Yoneda-complete.

Lemma 7.4.13 has the following converse, confirming that d-limits in categorical products are computed componentwise.

Lemma 7.4.15 *Let X_k, d_k be hemi-metric spaces, $k \in K$. Equip $X = \prod_{k \in K} X_k$ with the sup quasi-metric (call it d). For every Cauchy net $(\vec{x}_i)_{i \in I, \sqsubseteq}$*

in X, d, if $\vec{x} = (x_k)_{k\in K}$ is a d-limit of $(\vec{x}_i)_{i\in I,\sqsubseteq}$, then x_k is a d_k-limit of $(x_{ik})_{i\in I,\sqsubseteq}$ for each $k \in K$.

Proof By Lemma 7.2.8, $(\vec{x}_i)_{i\in I,\sqsubseteq}$ has a Cauchy-weightable subnet. Assume therefore that $(\vec{x}_{\alpha(j)}, r_j)_{j\in J,\preceq}$ is Cauchy-weighted, where α is a sampler. For every $\vec{y} \in X$, $d(\vec{x}, \vec{y})$ equals $\sup_{j\in J}(d(\vec{x}_{\alpha(j)}, \vec{y}) - r_j) = \sup_{j\in J}(\sup_{k\in K} d_k(x_{\alpha(j)k}, y_k) - r_j) = \sup_{k\in K} \sup_{j\in J}(d_k(x_{\alpha(j)k}, y_k) - r_j)$. Fix $k \in K$ and $y \in X_k$. By taking $\vec{y} = \vec{x}$, one obtains $0 = \sup_{k'\in K} \sup_{j\in J}(d_{k'}(x_{\alpha(j)k'}, x_{k'}) - r_j)$, in particular $d_{k'}(x_{\alpha(j)k'}, x_{k'}) \leqslant r_j$ for all $j \in J$ and $k' \in K$.

Now define $\vec{y}$ as the tuple whose kth component is y and whose k'th component is $x_{k'}$ for each $k' \neq k$. Then $d(\vec{x}, \vec{y}) = d_k(x_k, y)$ and, as above, $d(\vec{x}, \vec{y}) = \sup_{k'\in K} \sup_{j\in J}(d_{k'}(x_{\alpha(j)k'}, y_{k'}) - r_j)$, which is equal to $\max(\sup_{j\in J}(d_k(x_{\alpha(j)k}, y) - r_j), \sup_{k'\neq k} \sup_{j\in J}(d_{k'}(x_{\alpha(j)k'}, y_{k'}) - r_j) = \sup_{j\in J}(d_k(x_{\alpha(j)k}, y) - r_j)$. It follows that $d_k(x_k, y) = \sup_{j\in J}(d_k(x_{\alpha(j)k}, y_k) - r_j)$, i.e., that y is a d_k-limit of $(x_{\alpha(j)k})_{j\in J,\preceq}$. By Lemma 7.4.6, and since $(x_{ik})_{i\in I,\sqsubseteq}$ is Cauchy, y is also a d_k-limit of $(x_{ik})_{i\in I,\sqsubseteq}$. □

Exercise 7.4.16 Let X_k, d_k be finitely many Yoneda-complete quasi-metric spaces, $1 \leqslant k \leqslant n$. Show that the tensor product $X = \prod_{k=1}^n X_k$, with the quasi-metric $d = d_1 \otimes \cdots \otimes d_n$ (Exercise 6.5.3) is Yoneda-complete, and that d-limits are componentwise.

Lemma 7.4.17 *Let X, d be a Yoneda-complete quasi-metric space.*

Then $\mathbf{B}(X, d), d^+$ is also a Yoneda-complete quasi-metric space. The d^+-limit of a Cauchy net $(x_i, r_i)_{i\in I,\sqsubseteq}$ of formal balls is (x, r), where x is the d-limit of $(x_i)_{i\in I,\sqsubseteq}$ and $r = \inf_{i\in I} r_i$.

Proof Let $((x_i, r_i), s_i)_{i\in I,\sqsubseteq}$ be a Cauchy-weighted net in $\mathbf{B}(X, d), d^+$. Let $r = \inf_{i\in I} r_i$, and $r'_i = r_i - r$. Whenever $i \sqsubseteq j$, $d^+((x_i, r_i), (x_j, r_j)) \leqslant s_i - s_j$, so $d(x_i, x_j) \leqslant r_i + s_i - r_j - s_j = r'_i + s_i - r'_j - s_j$. Now $\inf_{i\in I}(r'_i + s_i) = \inf_{i\in I} r'_i + \inf_{i\in I} s_i$: indeed, $\inf_{i\in I}(r'_i + s_i) = -\sup_{i\in I}((-r'_i) + (-s_i)) = -\sup_{i\in I}(-r'_i) - \sup_{i\in I}(-s_i)$, since $+$ is Scott-continuous, and $(-r'_i)_{i\in I,\sqsubseteq}$ and $(-s_i)_{i\in I,\sqsubseteq}$ are monotone nets. So $(x_i, r'_i + s_i)_{i\in I,\sqsubseteq}$ is Cauchy-weighted in X, d. Let x be the d-limit of $(x_i)_{i\in I,\sqsubseteq}$. By Lemma 7.4.9, $d(x, y) = \sup_{i\in I}(d(x_i, y) - r'_i - s_i)$ for every $y \in X$. So, for every formal ball (y, s), $d^+((x, r), (y, s)) = \max(\sup_{i\in I}(d(x, y) - r'_i - s_i) - r + s, 0) = \max(\sup_{i\in I}(d(x_i, y) - r'_i - s_i) - r + s, \sup_{i\in I}(-s_i)) = \sup_{i\in I}(\max(d(x_i, y) - r_i - s_i + s, -s_i))$ (since max is Scott-continuous) $= \sup_{i\in I}(d^+((x_i, r_i), (y, s)) - s_i)$, hence (x, r) is a d^+-limit of $(x_i, r_i)_{i\in I,\sqsubseteq}$. This generalizes to Cauchy nets, not just Cauchy-weightable nets, by Lemma 7.2.8 and Lemma 7.4.6. □

The complete subsets of a complete metric space are its closed subsets (Exercise 3.4.7 (b)). This has the following quasi-metric analogue.

Lemma 7.4.18 *Let X, d be a Yoneda-complete quasi-metric space. For any subset A of X, d, the following are equivalent:*

1. *A, d is Yoneda-complete.*
2. *A is d-*closed*, i.e., the d-limit of any Cauchy net $(x_i)_{i \in I, \sqsubseteq}$ in A is in A.*
3. *the d-limit of any Cauchy-weightable net $(x_i)_{i \in I, \sqsubseteq}$ in A is in A.*

Proof $1 \Rightarrow 2$. Let x be the d-limit of $(x_i)_{i \in I, \sqsubseteq}$ in A, and x' be its d-limit in X. For every $y \in X$, $d(x', y) = \limsup_{i \in I, \sqsubseteq} d(x_i, y)$. This holds in particular for every $y \in A$, so x' is the d-limit of $(x_i)_{i \in I, \sqsubseteq}$ in A. A, d is quasi-metric, since d is T_0, as X, d is quasi-metric. So d-limits are unique (Lemma 7.1.17) in A, hence $x = x'$. It follows that x' is in A. Therefore A is d-closed.

$2 \Rightarrow 3$ is obvious.

$3 \Rightarrow 1$. Let $(x_i)_{i \in I, \sqsubseteq}$ be any Cauchy-weightable net in A, d. Then it is also Cauchy-weightable in X, d. Since the latter is Yoneda-complete, $(x_i)_{i \in I, \sqsubseteq}$ has a d-limit x in X. For every $y \in X$, $d(x, y) = \limsup_{i \in I, \sqsubseteq} d(x_i, y)$, hence also for every $y \in A$. So x is the d-limit of $(x_i)_{i \in I, \sqsubseteq}$ in A, and therefore A, d is Yoneda-complete. □

Lemma 7.4.19 *Let X, d and Y, d' be hemi-metric spaces. Let $(f_i)_{i \in I, \sqsubseteq}$ be a Cauchy net of c-Lipschitz maps from X to Y that has a $\sup_X^{\cdot} d'$-limit f in Y^X. Then f is also c-Lipschitz.*

Proof By Lemma 7.2.8, $(f_i)_{i \in I, \sqsubseteq}$ has a Cauchy-weightable subnet, which has f as $\sup_X d'$-limit by Lemma 7.4.6. Without loss of generality, we may therefore assume that $(f_i)_{i \in I, \sqsubseteq}$ is Cauchy-weightable. Therefore let $(f_i, r_i)_{i \in I, \sqsubseteq}$ be Cauchy-weighted in Y^X, $\sup_X d'$.

For every pair of points x, x' in X, and every $i \in I$, $d'(f_i(x), f(x')) \leqslant d'(f_i(x), f_i(x')) + d'(f_i(x'), f(x'))$. By Lemma 7.4.9, $\sup_X d'(f, g)$ is equal to $\sup_{i \in I}(\sup_X d'(f_i, g) - r_i)$, and by taking $g = f$ we obtain $\sup_{i \in I}(\sup_X d'(f_i, f) - r_i) = 0$; in particular $\sup_X d'(f_i, f) \leqslant r_i$. So $d'(f_i(x), f(x')) \leqslant d'(f_i(x), f_i(x')) + r_i$, whence $d'(f_i(x), f(x')) - r_i \leqslant d'(f_i(x), f_i(x'))$. By Lemma 7.4.15, $f(x)$ is a d'-limit of $(f_i(x))_{i \in I, \sqsubseteq}$, and $(f_i(x), r_i)_{i \in I, \sqsubseteq}$ is clearly Cauchy-weighted, so $d'(f(x), f(x')) = \sup_{i \in I}(d'(f_i(x), f(x')) - r_i)$.

From this and $d'(f_i(x), f(x')) - r_i \leqslant d'(f_i(x), f_i(x')) \leqslant c.d(x, x')$, we obtain $d'(f(x), f(x')) \leqslant c.d(x, x')$. □

Exercise 7.4.20 Let f_n be the map from $\mathbb{R}$ to $\mathbb{R}$ defined by: $f_n(r) = 0$ when $r \leqslant 0$, $f_n(r) = 2^n.r$ when $0 < r \leqslant 1/2^n$, and $f_n(r) = 1$ for $r > 1$. Show that

f_n is uniformly continuous, when $\mathbb{R}$ is equipped with the quasi-metric $\mathrm{d}_{\mathbb{R}}$. Show that $(f_n)_{n\in\mathbb{N}}$ is Cauchy, and that its $\sup_{\mathbb{R}} \mathrm{d}_{\mathbb{R}}$-limit is *not* uniformly continuous, although it is continuous. Deduce that the space of uniformly continuous maps from a quasi-metric space to a Yoneda-complete quasi-metric space may fail to be Yoneda-complete.

We sum up these results in the following theorem. Compared to the situation with Smyth-completeness (Theorem 7.4.21), Yoneda-completeness is preserved by arbitrary products, and for spaces of c-Lipschitz maps without assumptions. However, we have no Yoneda-completeness result for spaces of continuous, resp. uniformly continuous maps (see Exercise 7.4.20).

Theorem 7.4.21 (Constructing Yoneda-complete spaces)

1. *Every d-closed subset of a Yoneda-complete quasi-metric space is Yoneda-complete.*
2. *Every coproduct of Yoneda-complete quasi-metric spaces is Yoneda-complete.*
3. *Every finite product, every tensor product, every topological product, and every categorical product of Yoneda-complete quasi-metric spaces is Yoneda-complete.*
4. *Given any set X and any Yoneda-complete quasi-metric space Y, d', the set of all maps Y^X from X to Y with the sup quasi-metric $\sup_X d'$ is Yoneda-complete.*
5. *Given any hemi-metric space X, d and any Yoneda-complete quasi-metric space Y, d', for every $c \in \mathbb{R}^+$ the space of $[X \to Y]_c$ of all c-Lipschitz maps from X to Y with the sup quasi-metric $\sup_X d'$ is Yoneda-complete.*
6. *The space of formal balls of a Yoneda-complete quasi-metric space is Yoneda-complete.*

Proof 1 is true by Lemma 7.4.18. 2 is true by Lemma 7.4.12. 3. For finite and categorical products, this is Lemma 7.4.13. For tensor products, see Exercise 7.4.16. As far as countable topological products $\prod_{k\in\mathbb{N}} X_k, \widetilde{\prod}_{k\in K} d_k$ are concerned, we notice that they are just the categorical products of $X_k, \frac{1}{2^k}\min(d_k, 1)$. Now $X_k, \min(d_k, 1)$ is Yoneda-complete by Lemma 7.4.11, from which we easily deduce that $X_k, \frac{1}{2^k}\min(d_k, 1)$ is Yoneda-complete as well, and we again conclude by Lemma 7.4.13.

4. This is again true by Lemma 7.4.13.

5. Then every d-closed subspace of Y^X is also Yoneda-complete by Item 1 above. Precisely, Lemma 7.4.19 states that $[X \to Y]_c$ is d-closed in Y^X, and we conclude since Y^X is Yoneda-complete by Item 4.

Finally, 6 is true by Lemma 7.4.17. ☐

Exercise 7.4.22 Show that the categories of Yoneda-complete ultra-quasi-metric (resp., ultra-metric) spaces and 1-Lipschitz maps is Cartesian-closed. The finite products and exponentials are defined as in **UQMet** (resp., **UMet**).

Exercise 7.4.23 Show that the category of Yoneda-complete quasi-metric (resp., complete metric) spaces and 1-Lipschitz maps is SMCC.

Lemma 7.2.33, that open subsets of Smyth-complete quasi-metric spaces are Smyth-completely quasi-metrizable, does not seem to have an equivalent for Yoneda-complete spaces. But we have the following closely related result, which also deals not just with one open subset, but with countable intersections of opens.

Lemma 7.4.24 *Let X, d be a Smyth-complete quasi-metric space, and let $(U_n)_{n\in\mathbb{N}}$ be a countable family of open subsets of X, d. Then $\bigcap_{n\in\mathbb{N}} U_n$ is Yoneda-completely quasi-metrizable, i.e., there is a quasi-metric on $\bigcap_{n\in\mathbb{N}} U_n$ that makes it Yoneda-complete, and whose open ball topology is the subspace topology from X, d.*

Proof By Lemma 7.2.33, each U_n has a quasi-metric d_n that makes it Smyth-complete, and whose open ball topology is the subspace topology. Let $d' = \bigvee_{n\in\mathbb{N}} d_n$ on $\bigcap_{n\in\mathbb{N}} U_n$: its open ball topology is again the subspace topology, by Lemma 6.3.11. We claim that $\bigcap_{n\in\mathbb{N}} U_n, d'$ is Yoneda-complete.

Let Z be the subset of $\prod_{n\in\mathbb{N}} U_n$ of all tuples $(x_n)_{n\in\mathbb{N}}$ where $x_m = x_n$ for all $m, n \in \mathbb{N}$. The map sending $x \in \bigcap_{n\in\mathbb{N}} U_n$ to the vector whose components are all equal to x is a bijection onto Z, and an isometry of $\bigcap_{n\in\mathbb{N}} U_n, d'$ onto $Z, \widetilde{\prod}_{n\in\mathbb{N}} d_n$.

$\prod_{n\in\mathbb{N}} U_n, \widetilde{\prod}_{n\in\mathbb{N}} d_n$ is Yoneda-complete by Theorem 7.4.21, Item 3. And Z is $\widetilde{\prod}_{n\in\mathbb{N}} d_n$-closed in $\prod_{n\in\mathbb{N}} U_n$: given any Cauchy net $(\vec{x}_i)_{i\in I,\sqsubseteq}$ in Z, i.e., such that, for each $i \in I$, all the components of $\vec{x}_i$ are the same, its $\widetilde{\prod}_{n\in\mathbb{N}} d_n$-limit $\vec{x}$ is computed componentwise (Lemma 7.4.15), hence all the components of $\vec{x}$ are the same, so $\vec{x}$ is in Z. By Theorem 7.4.21, Item 1, $Z, \widetilde{\prod}_{n\in\mathbb{N}} d_n$ is Yoneda-complete. It follows that $\bigcap_{n\in\mathbb{N}} U_n, d'$ is, too. □

7.4.2 Formal balls and the Kostanek–Waszkiewicz Theorem

We now relate Yoneda-completeness and formal balls. The following elucidates d-limits: they arise from directed sups of formal balls.

Lemma 7.4.25 *Let X, d be a hemi-metric space, and $(x_i, r_i)_{i\in I,\sqsubseteq}$ be a monotone net of formal balls in $\mathbf{B}(X, d)$. If $(x_i)_{i\in I,\sqsubseteq}$ has a d-limit x, and $r = \inf_{i\in I} r_i$, then (x, r) is a least upper bound of $(x_i, r_i)_{i\in I,\sqsubseteq}$ in $\mathbf{B}(X, d)$.*

Proof We first check that (x, r) is an upper bound, i.e., that $d(x_i, x) \leqslant r_i - r$ for every $i \in I$. For every j with $i \sqsubseteq j$, $d(x_i, x) \leqslant d(x_i, x_j) + d(x_j, x) \leqslant r_i - r_j + d(x_j, x) \leqslant r_i - r + d(x_j, x)$. Since x is a d-limit, $\limsup_{j \in I, \sqsubseteq} d(x_j, x) = d(x, x) = 0$, so taking limsups over j, $d(x_i, x) \leqslant r_i - r$, as claimed.

Let (y, s) be any upper bound of $(x_i, r_i)_{i \in I, \sqsubseteq}$ in $\mathbf{B}(X, d)$. For every $i \in I$, $d(x_i, y) \leqslant r_i - s$, and $r_i \geqslant s$. From the latter, we deduce that $r \geqslant s$, and from the former that $\limsup_{i \in I, \sqsubseteq} d(x_i, y) \leqslant \limsup_{i \in I, \sqsubseteq} r_i - s = \inf_{i \in I} r_i - s = r - s$. Since x is the d-limit, $d(x, y) \leqslant r - s$, so $(x, r) \leqslant^{d^+} (y, s)$. So (x, r) is the least upper bound. $\square$

We obtain a form of converse under the assumption that $\mathbf{B}(X, d)$ is a dcpo.

Lemma 7.4.26 *Let X, d be a hemi-metric space, and assume that $\mathbf{B}(X, d), \leqslant^{d^+}$ is a dcpo. Let $(x_i, r_i)_{i \in I, \sqsubseteq}$ be a monotone net of formal balls in $\mathbf{B}(X, d)$, with least upper bound (x, r). Then $r = \inf_{i \in I} r_i$ and x is a d-limit of $(x_i)_{i \in I, \sqsubseteq}$.*

Proof We first observe that $r = \inf_{i \in I} r_i$. We argue as in the Romaguera-Valero Theorem 7.3.11. Let $r_\infty = \inf_{i \in I} r_i$. Since $r_i \geqslant r$ for every $i \in I$, $r_\infty \geqslant r$. Consider the net $(x_i, r_i - r_\infty)_{i \in I, \sqsubseteq}$. This has a least upper bound (x', r'), and $\inf_{i \in I}(r_i - r_\infty) \geqslant r'$. So $r' = 0$. Since (x', r') is an upper bound of $(x_i, r_i - r_\infty)_{i \in I, \sqsubseteq}$, $(x', r' + r_\infty)$ is an upper bound of $(x_i, r_i)_{i \in I, \sqsubseteq}$, i.e., (x', r_∞) is an upper bound of the latter. So $(x, r) \leqslant^{d^+} (x', r_\infty)$, in particular $r \geqslant r_\infty$. So $r = r_\infty$.

Next, we observe that the map $\_ + a : (y, s) \mapsto (y, s + a)$ is Scott-continuous, for every $a \in \mathbb{R}^+$. It is monotonic: if $(y, s) \leqslant^{d^+} (y', s')$, then $d(y, y') \leqslant s - s' = (s + a) - (s' + a)$, so $(y, s + a) \leqslant^{d^+} (y', s' + a)$. And given any directed family $(y_j, s_j)_{j \in J}$ with least upper bound (y, s), $(y, s + a)$ is above every $(y_j, s_j + a)$ by monotonicity, hence above their least upper bound (z, t). By our first observation, $t = \inf_{j \in J}(s_j + a) = s + a$. Since $(y_j, s_j + a) \leqslant^{d^+} (z, t) = (z, s + a)$, for every $j \in J$, $d(y_j, z) \leqslant (s_j + a) - (s + a) = s_j - s$, so (z, s) is above every (y_j, s_j), hence above their least upper bound (y, s). From $(y, s) \leqslant^{d^+} (z, s)$, we deduce $d(y, z) = 0$, so $(y, s + a) \leqslant^{d^+} (z, t)$. So $(y, s + a)$ is equal to the least upper bound (z, t) of $(y_j, s_j + a)_{j \in J}$.

Knowing this, we observe that since $(x_i, r_i) \leqslant^{d^+} (x, r)$—with $r = \inf_{i \in I} r_i$ by the first observation—$d(x_i, x) \leqslant r_i - r$. So, for every $y \in X$, $d(x_i, y) \leqslant d(x_i, x) + d(x, y) \leqslant r_i - r + d(x, y)$. Taking least upper bounds over $i \in I$, $\sup_{i \in I}(d(x_i, y) - r_i) \leqslant d(x, y) - r$.

Assume, for the sake of contradiction, that the inequality is strict. Let $s = \sup_{i \in I}(d(x_i, y) - r_i) < d(x, y) - r$. In particular, $s < +\infty$. For every $i \in I$,

$d(x_i, y) - r_i \leqslant s$, so $d(x_i, y) \leqslant r_i + s$, i.e., $(x_i, r_i + s) \leqslant^{d^+} (y, 0)$. This implies that the least upper bound of $(x_i, r_i + s)_{i \in I}$, which happens to be $(x, r + s)$ since $\_ + s$ is Scott-continuous, is below $(y, 0)$. In other words, $d(x, y) \leqslant r + s$, a contradiction.

So $d(x, y) - r = s = \sup_{i \in I}(d(x_i, y) - r_i)$; i.e., $d(x, y) = \sup_{i \in I}(d(x_i, y) - (r_i - r))$. Observe that $(x_i, r_i - r)_{i \in I, \sqsubseteq}$ is Cauchy-weighted. By Lemma 7.4.9, since y is arbitrary, x is a d-limit of $(x_i)_{i \in I, \sqsubseteq}$. □

The following is the *Kostanek–Waszkiewicz Theorem*, established in slightly more general form by Kostanek and Waszkiewicz (2011, theorem 7.1).

Theorem 7.4.27 (Kostanek–Waszkiewicz) *Let X, d be a quasi-metric space. The following are equivalent:*

1. *X, d is Yoneda-complete.*
2. *Every Cauchy-weightable net in X, d has a d-limit.*
3. *The space $\mathbf{B}(X, d)$ of formal balls is a dcpo in the ordering $\leqslant^{d^+}$.*

In this case, the least upper bound (x, r) of any monotone net $(x_i, r_i)_{i \in I, \sqsubseteq}$ in $\mathbf{B}(X, d)$ is such that $r = \inf_{i \in I} r_i$ and x is the d-limit of $(x_i)_{i \in I, \sqsubseteq}$.

Proof That 1 and 2 are equivalent is true by Proposition 7.4.8.

$2 \Rightarrow 3$. Let $(x_i, r_i)_{i \in I, \sqsubseteq}$ be a monotone net of formal balls. Let $r = \inf_{i \in I} r_i$. For all $i, j \in I$ with $i \sqsubseteq j$, $d(x_i, x_j) \leqslant r_i - r_j$, so $d(x_i, x_j) \leqslant (r_i - r) - (r_j - r)$, whence $(x_i, r_i - r)_{i \in I, \sqsubseteq}$ is a Cauchy-weighted net. So $(x_i)_{i \in I, \sqsubseteq}$ has a d-limit x, by assumption. By Lemma 7.4.25, (x, r) is the least upper bound of $(x_i, r_i)_{i \in I, \sqsubseteq}$. So $\mathbf{B}(X, d)$ is a dcpo.

$3 \Rightarrow 2$. Let $(x_i, r_i)_{i \in I, \sqsubseteq}$ be a Cauchy-weighted net. By 3, Lemma 7.4.26 applies, so D has a least upper bound (x, r), with $r = \inf_{i \in I} r_i = 0$, and x is a d-limit of $(x_i)_{i \in I, \sqsubseteq}$. □

Example 7.4.28 Fact 7.4.4 and Exercise 7.2.3 together show that there are Yoneda-complete spaces that are not Smyth-complete, e.g., $\overline{\mathbb{R}^+}, \mathsf{d}_{\mathbb{R}}$.

Exercise 7.4.29 ($\mathbb{R}_\ell$ is Yoneda-complete) Show that the Sorgenfrey line $\mathbb{R}_\ell, \mathsf{d}_\ell$ is another example of a Yoneda-complete, not Smyth-complete space.

Exercise 7.4.30 Let X be a set with a quasi-ordering $\leqslant$. Show that $X, d_{\leqslant}$ is Yoneda-complete if and only if $\leqslant$ is an ordering, and X is a dcpo with this ordering. First check that the Cauchy nets are exactly the eventually monotone nets, i.e., the nets $(x_i)_{i \in I, \sqsubseteq}$ such that $(x_i)_{i \in \uparrow i_0, \sqsubseteq}$ is a monotone net for some $i_0 \in I$; and that a $d_{\leqslant}$-limit of a monotone net of X is the same thing as a least upper bound of the elements of this net.

Exercise 7.4.31 Lemma 7.2.5 states that Cauchy sequences suffice to characterize completeness in metric spaces (equivalently, Yoneda-completeness, by Lemma 7.4.3). Using Exercise 7.4.30, show that they do *not* suffice to characterize Yoneda-completeness in quasi-metric spaces. That is, find a quasi-metric space X, d where every Cauchy sequence has a d-limit, but is nonetheless not Yoneda-complete.

7.4.3 Yoneda-continuity

Yoneda-complete quasi-metric spaces also enjoy a fixed point theorem, but only for maps that are a bit more than c-Lipschitz. The additional condition was simply called continuity by Bonsangue *et al.* (1998).

Definition 7.4.32 (Yoneda-continuity) Let X, d and Y, d' be hemi-metric spaces. A map f from X to Y is *Yoneda-continuous* if and only if it is uniformly continuous, and, for every Cauchy net $(x_i)_{i \in I, \sqsubseteq}$ in X, d with d-limit x, $f(x)$ is a d'-limit of $(f(x_i))_{i \in I, \sqsubseteq}$.

Note that $(f(x_i))_{i \in I, \sqsubseteq}$ is Cauchy in Y, d': for every $\epsilon > 0$, let $\eta > 0$ be such that $d'(f(x), f(x')) < \epsilon$ whenever $d(x, x') < \eta$; then $d(x_i, x_j) < \eta$ for $i \sqsubseteq j$ large enough, hence $d(f(x_i), f(x_j)) < \epsilon$.

Example 7.4.33 There are uniformly continuous maps that are not Yoneda-continuous. E.g., given two posets X and Y, the uniformly continuous maps from $X, d_\leqslant$ to $Y, d_\leqslant$ are the monotone maps, while the Yoneda-continuous maps are the Scott-continuous maps, as we shall see in Proposition 7.4.38. A simpler example is given in the following exercise.

Exercise 7.4.34 Let $f\colon \overline{\mathbb{R}^+} \to \overline{\mathbb{R}^+}$ map every $r \in \mathbb{R}^+$ to 0, and $+\infty$ to $+\infty$. Show that f is uniformly continuous, and in fact c-Lipschitz for every $c > 0$, but that f is not Yoneda-continuous.

Exercise 7.4.35 (Metric $\Rightarrow$ Yoneda-continuous = uniformly continuous) Show that the Yoneda-continuous maps from a metric space X, d to a metric space Y, d' are exactly the uniformly continuous maps.

Exercise 7.4.36 Let X, d be a hemi-metric space. Show that, for every $x \in X$, $d(\_, x)$ is (1-Lipschitz and) Yoneda-continuous from X, d to $\overline{\mathbb{R}^+}, \mathsf{d}_{\mathbb{R}}$.

As usual, one can trade Cauchy nets for Cauchy-weightable nets.

Lemma 7.4.37 *Let X, d and Y, d' be hemi-metric spaces. A map f from X to Y is Yoneda-continuous if and only if it is uniformly continuous, and, for every*

Cauchy-weightable net $(x_i)_{i \in I, \sqsubseteq}$ in X, d with d-limit x, $f(x)$ is a d'-limit of $(f(x_i))_{i \in I, \sqsubseteq}$.

Proof The only if direction is obvious. Conversely, let $(x_i)_{i \in I, \sqsubseteq}$ be a Cauchy net in X, d with d-limit x. By Lemma 7.2.8, it has a Cauchy-weightable subnet $(x_{\alpha(j)})_{j \in J, \preceq}$, and the latter has x as d-limit (Exercise 7.4.7). By assumption, $f(x)$ is a d'-limit of $(f(x_{\alpha(j)}))_{j \in J, \preceq}$. As we noted earlier, $(f(x_i))_{i \in I, \sqsubseteq}$ is Cauchy, so Lemma 7.4.6 applies: $f(x)$ is a d'-limit of $(f(x_i))_{i \in I, \sqsubseteq}$. □

It follows that c-Lipschitz Yoneda-continuous maps can be characterized using formal balls. Recall from Lemma 7.3.16 that $\mathbf{B}^c(f)(x, r) = (f(x), c.r)$, for every c-Lipschitz map f.

Proposition 7.4.38 (Yoneda-continuity and Scott-continuity) *Let X, d and Y, d' be Yoneda-complete quasi-metric spaces, and f be a c-Lipschitz map from X, d to Y, d', $c \in \mathbb{R}^+$. Then f is Yoneda-continuous if and only if $\mathbf{B}^c(f)$ is Scott-continuous from $\mathbf{B}(X, d)$ to $\mathbf{B}(Y, d')$.*

Proof Assume f Yoneda-continuous. By the Kostanek–Waszkiewicz Theorem 7.4.27, every directed set $(x_i, r_i)_{i \in I}$ in $\mathbf{B}(X, d)$ has a least upper bound (x, r), where $r = \inf_{i \in I} r_i$, x is a d-limit of $(x_i)_{i \in I, \sqsubseteq}$, and $i \sqsubseteq j$ iff $(x_i, r_i) \leq^{d^+} (x_j, r_j)$. Now $(x_i, r_i - r)_{i \in I, \sqsubseteq}$ is Cauchy-weighted, so $f(x)$ is a d'-limit of the net $(f(x_i))_{i \in I, \sqsubseteq}$, since f is Yoneda-continuous. Also, $c.r = \inf_{i \in I} c.r_i$. By Theorem 7.4.27 again, $\mathbf{B}^c(f)(x, r) = (f(x), c.r)$ is the least upper bound of the family $(f(x_i), c.r_i)_{i \in I, \sqsubseteq}$, i.e., of $(\mathbf{B}^c(f)(x_i, r_i))_{i \in I, \sqsubseteq}$. So $\mathbf{B}^c(f)$ is Scott-continuous.

Conversely, assume $\mathbf{B}^c(f)$ is Scott-continuous, and let $(x_i, r_i)_{i \in I, \sqsubseteq}$ be any Cauchy-weighted net. By Scott-continuity, the least upper bound of the monotone net $(f(x_i), c.r_i)_{i \in I, \sqsubseteq}$ must be $(f(x), c.r)$, where (x, r) is the least upper bound of $(x_i, r_i)_{i \in I}$. By Theorem 7.4.27 again, $f(x)$ must be the d-limit of $(f(x_i)_{i \in I, \sqsubseteq}$. By Lemma 7.4.37, f is Yoneda-continuous. □

This allows us to replace arguments on Yoneda-continuity by mere Scott-continuity on formal balls. Proposition 7.4.52 below is a good illustration of this.

7.4.4 The Banach–Rutten Fixed Point Theorem

The following is then a natural analogue of the Banach Fixed Point Theorem, and is due to Rutten (1996, theorem 3.7, item 2).

Theorem 7.4.39 (Banach–Rutten Fixed Point Theorem) *Let X, d be a Yoneda-complete quasi-metric space, and $f : X \to X$ be a c-Lipschitz Yoneda-continuous map, where $c \in [0, 1)$. Assume that there is a point $x_0 \in X$ such that $d(x_0, f(x_0)) < +\infty$.*

Then f has a fixed point x, obtained as the d-limit of the sequence of iterates $(f^n(x_0))_{n \in \mathbb{N}}$. Any two fixed points x and y of f are at infinite distance from each other: if $d(y, x) < +\infty$ then $y \leqslant^{d^+} x$, and if $y \neq x$ then $d^{sym}(x, y) = +\infty$.

Proof Let $r_0 \in \mathbb{R}^+$ be any number above $\frac{d(x_0, f(x_0))}{1-c}$, so that $d(x_0, f(x_0)) \leqslant (1 - c)r_0$, i.e., $(x_0, r_0) \leqslant^{d^+} (f(x_0), c.r_0)$. The latter is equivalent to $(x_0, r_0) \leqslant^{d^+} \mathbf{B}^c(f)(x_0, r_0)$. Since $\mathbf{B}^c(f)$ is c-Lipschitz (Lemma 7.3.16), it is monotonic (using Proposition 6.1.8). So $(\mathbf{B}^c(f)^n(x_0, r_0))_{n \in \mathbb{N}}$ is a chain. Since X, d is Yoneda-complete, by the Kostanek–Waszkiewicz Theorem 7.4.27 this sequence has a least upper bound (x, r) where x is the d-limit of $(f^n(x_0))_{n \in \mathbb{N}}$. By Proposition 7.4.38, $\mathbf{B}^c(f)$ is Scott-continuous, so (as in the Kleene Fixed Point Theorem, Proposition 4.3.6), $\mathbf{B}^c(f)(x, r) = (x, r)$. In particular, $f(x) = x$, so x is a fixed point. The claim that all fixed points are at infinite distance from each other is as in Theorem 7.2.14. □

Corollary 7.4.40 *Let X, d be a Yoneda-complete quasi-metric space such that $d(x, y) < +\infty$ for all $x, y \in X$. Let $f : X \to X$ be any c-Lipschitz Yoneda-continuous map, where $c \in [0, 1)$. Then f has a unique fixed point x. Given any point $x_0 \in X$, x is the d-limit of the sequence $(f^n(x_0))_{n \in \mathbb{N}}$.*

Exercise 7.4.41 (Kleene–Rutten Fixed Point Theorem) Show Rutten's other fixed point theorem, which is an analogue of the Kleene Fixed Point Theorem (Rutten, 1996, theorem 3.7, item 1): let X, d be a Yoneda-complete quasi-metric space, f be a Yoneda-continuous map from X, d to X, d, and $x_0 \in X$ be a point such that $x_0 \leqslant^d f(x_0)$. Show that f has a least fixed point above x_0, obtained as the d-limit of $(f^n(x_0))_{n \in \mathbb{N}}$.

Exercise 7.4.42 (Caristi–Waszkiewicz Fixed Point Theorem) Let X, d be a Yoneda-complete quasi-metric space. A *potential* map on X is a map $\varphi : X \to \mathbb{R}^+$ such that, for every Cauchy net $(x_i)_{i \in I, \sqsubseteq}$ in X with d-limit x, $\varphi(x) \leqslant \liminf_{i \in I, \sqsubseteq} \varphi(x_i)$. (As we shall argue below, this is a form of lower semi-continuity.)

Apply the Bourbaki–Witt Theorem 2.3.1 and the Kostanek–Waszkiewicz Theorem 7.4.27 to show the Caristi–Waszkiewicz Theorem (Waszkiewicz, 2010, theorem 4.1): let X, d be a non-empty Yoneda-complete quasi-metric space, and

$f: X \to X$ be any map (not necessarily continuous), and assume there is a potential map on X such that, for every $x \in X$, $\varphi(f(x)) + d(x, f(x)) \leqslant \varphi(x)$; then f has a fixed point. More precisely, for every $x \in X$, f has a fixed point x_0 that is close to x, in the sense that $d(x, x_0) \leqslant \varphi(x)$. As a hint, show that $\Phi = \{(x, r) \in \mathbf{B}(X, d) \mid r \geqslant \varphi(x)\}$ is a dcpo in the ordering $\leqslant^{d^+}$, where sups are computed as in $\mathbf{B}(X, d)$.

As a corollary, show Caristi's Fixed Point Theorem: let X, d be a non-empty complete metric space, $f: X \to X$ be any map, and assume there is an lsc map $\varphi: X \to \mathbb{R}^+$ such that, for every $x \in X$, $\varphi(f(x)) + d(x, f(x)) \leqslant \varphi(x)$; then f has a fixed point. (Notice that any lsc map $\varphi: X \to \mathbb{R}^+$ is a potential.)

7.4.5 The d-Scott topology

The pervasiveness of the notion of formal balls in everything we have done so far leads us to ask whether the open ball topology is the right one. One argument against the open ball topology is that it is merely a souped-up version of the Alexandroff topology: see Exercise 6.1.15, for example, where we have seen that the open ball topology of $d_\leqslant$ on a set X with a quasi-ordering $\leqslant$ is just the Alexandroff topology of $\leqslant$. This is disappointing: the Scott topology, for one, is a more interesting topology. Let us propose the following.

Definition 7.4.43 (d-Scott topology) Let X, d be a hemi-metric space. The *d-Scott topology* on X is the topology induced from the Scott topology on $\mathbf{B}(X, d)$, quasi-ordered by $\leqslant^{d^+}$, by the map η_X^{EH}.

In other words, look at X as though it were a subspace of $\mathbf{B}(X, d)$, where the latter is equipped with the Scott topology of its quasi-ordering $\leqslant^{d^+}$. Explicitly, the d-Scott-opens of X are the subsets $\{x \in X \mid (x, 0) \in V\}$, where V ranges over the Scott-open subsets of $\mathbf{B}(X, d)$.

Exercise 7.4.44 Let X be a set with a quasi-ordering $\leqslant$. Show that the $d_\leqslant$-Scott topology on X is exactly the Scott topology of $\leqslant$.

Exercise 7.4.45 Let X, d be a Yoneda-complete quasi-metric space. Show that every closed subset of X in its d-Scott topology is d-closed (see Lemma 7.4.18). Conclude that every closed subspace F of X in its d-Scott topology, equipped with the restriction of d to F, is Yoneda-complete.

Is the d-Scott topology really a different topology than the open ball topology? It must be so in general: for $d = d_\leqslant$, it will be different whenever the Scott and the Alexandroff topologies of $\leqslant$ differ. But they are the same when d is *metric*.

Proposition 7.4.46 *Let X, d be a hemi-metric space. The open ball topology is finer than the d-Scott topology on X.*

If d is a metric, then the d-Scott and open ball topologies coincide.

Proof For the first part, we must show that every d-Scott-open U in X is open in the open ball topology. So let $U = \{x \in X \mid (x, 0) \in V\}$ for some Scott-open subset V of $\mathbf{B}(X, d)$, and use Trick 6.1.6: fix $x \in U$, and find an open ball $B^d_{x,<\epsilon}$ that is included in U. To this end, realize that $(x, 0)$ is a least upper bound of the directed family (x, r), $r > 0$, so that $(x, r) \in V$ for some $r > 0$. Now take $\epsilon = r$: for every $y \in B^d_{x,<r}$, $d(x, y) < r$, so $(x, r) \leqslant^{d^+} (y, 0)$, and, as (x, r) is in V, so is $(y, 0)$, whence y is in U. So $B^d_{x,<\epsilon}$ is indeed included in U.

For the second part, assume that d is a metric; then it suffices to show that the open ball $B^d_{x,<\epsilon}$ is d-Scott-open for every $x \in X$ and every $\epsilon > 0$. Let $V = \{(y, s) \mid d(x, y) < \epsilon - s\}$. Using Exercise 7.3.14, which applies since d is metric, V is just $\Uparrow(x, \epsilon)$ in $\mathbf{B}(X, d)$, hence is Scott-open since $\mathbf{B}(X, d)$ is a continuous poset in this case (Proposition 5.1.16). But $\eta_X^{\mathrm{EH}^{-1}}(V) = \{y \in X \mid (y, 0) \in V\}$ is exactly $B^d_{x,<\epsilon}$, so $B^d_{x,<\epsilon}$ is d-Scott-open. ☐

They are also the same in Smyth-complete spaces.

Proposition 7.4.47 *Let X, d be a quasi-metric space. If X, d is Smyth-complete, then the d-Scott and open ball topologies coincide.*

Proof By the first part of Proposition 7.4.46, every d-Scott-open is open in the open ball topology. Conversely, it remains to show that every open ball $B^d_{x,<\epsilon}$ is d-Scott-open. Clearly, $B^d_{x,<\epsilon} = \{y \in X \mid (y, 0) \in V\}$, where $V = \{(y, s) \mid d(x, y) < \epsilon - s\}$. Now $V = \{(y, s) \mid (x, \epsilon) \prec^{d^+} (y, s)\}$, which is Scott-open since $\prec^{d^+}$ is the way-below relation on $\mathbf{B}(X, d)$, by the Romaguera–Valero Theorem 7.3.11 (and using Proposition 5.1.16). ☐

Recall that the Scott topology of a quasi-ordering was sandwiched between the upper and the Alexandroff topologies (Proposition 4.2.18). What plays the role of the upper topology here is the open hole topology of Exercise 6.2.11, whose subbasic open sets are the open holes $T^d_{x,>\epsilon} = \{y \in X \mid d(y, x) > \epsilon\}$. We obtain that the d-Scott topology is sandwiched between the open hole and open ball topologies – at least on Yoneda-complete spaces.

Lemma 7.4.48 *Let X, d be a Yoneda-complete quasi-metric space. The d-Scott topology is finer than the open hole topology.*

Proof We must show that every open hole $T^d_{x,>\epsilon}$ is d-Scott-open. Let $V = T^{d^+}_{(x,0),>\epsilon}$: V is open in the open hole topology of $\mathbf{B}(X, d)$, and is in particular upward-closed in the specialization ordering of the latter, which happens to be $\leqslant^{d^+}$ (Exercise 6.2.11). Let us show that V is Scott-open. Let $(y_i, s_i)_{i \in I, \sqsubseteq}$ be a monotone net of formal balls whose least upper bound (y, s) is in V. The latter means that $d(y, x) > s + \epsilon$. Let $\eta > 0$ be such that $\eta < d(y, x) - s - \epsilon$, so that $d(y, x) > s + \epsilon + \eta$. By the Kostanek–Waszkiewicz Theorem 7.4.27, $s = \inf_{i \in I} s_i$ and y is the d-limit of $(y_i)_{i \in I, \sqsubseteq}$. In particular, $d(y, x) = \limsup_{i \in I, \sqsubseteq} d(y_i, x)$. Since $d(y, x) > s + \epsilon + \eta$, $d(y_i, x) > s + \epsilon + \eta/2$ for i large enough. Since $s = \inf_{i \in I} s_i$, $s + \eta/2 > s_i$ for i large enough. So $d(y_i, x) > s_i + \epsilon$ for some $i \in I$, i.e., $(y_i, s_i) \in V$. ☐

Corollary 7.4.49 *Let X, d be a hemi-metric space. The specialization quasi-ordering of X in its d-Scott topology (as well as in its open ball and open hole topologies) is $\leqslant^d$.*

Exercise 7.4.50 Show that the conclusion of Lemma 7.4.48 also holds when X, d is no longer assumed to be Yoneda-complete quasi-metric, but when (*a*) X, d is a metric space (not necessarily complete), or (*b*) $d = d_{\leqslant}$ for some quasi-ordering $\leqslant$.

Exercise 7.4.51 Let X, d be a hemi-metric space. Bonsangue *et al.* (1998) define the so-called *generalized Scott topology* on X as follows: the generalized Scott-opens of X, d are the subsets U of X such that, for every $x \in U$ and for every Cauchy net $(x_i)_{i \in I, \sqsubseteq}$ in X that has x as d-limit, there is an $\epsilon > 0$ such that $B^d_{x_i, <\epsilon} \subseteq U$ for i large enough. So, in particular, x_i is in U for i large enough, which implies that the complement of U is d-closed. But the required property $B^d_{x_i, <\epsilon} \subseteq U$, with $\epsilon > 0$ independent of i, is stronger.

Show that the generalized Scott-opens form a topology.

Show that the generalized Scott topology can be defined in terms of Cauchy-weightable nets, i.e., U is generalized Scott-open iff, for every $x \in U$ and for every Cauchy-weightable net $(x_i)_{i \in I, \sqsubseteq}$ in X that has x as d-limit, there is an $\epsilon > 0$ such that $B^d_{x_i, <\epsilon} \subseteq U$ for i large enough.

From this, deduce that the generalized Scott topology is finer than the d-Scott topology on any Yoneda-complete quasi-metric space X, d.

Yoneda-continuity then arises as a form of continuity, provided we restrict ourselves to Lipschitz maps, and consider d-Scott topologies.

Proposition 7.4.52 *Let X, d and Y, d' be Yoneda-complete quasi-metric spaces. For every Lipschitz map $f : X \to Y$, f is Yoneda-continuous if and only if it is continuous with respect to the d-Scott and d'-Scott topologies on X and Y, respectively.*

Proof Let f be c-Lipschitz, for some $c \in \mathbb{R}^+$. If f is Yoneda-continuous, then $\mathbf{B}^c(f)$ is Scott-continuous by Proposition 7.4.38. One checks easily that $\mathbf{B}^c(f) \circ \eta_X^{\mathrm{EH}} = \eta_Y^{\mathrm{EH}} \circ f$. So, given any d'-Scott-open subset V of Y, writing V as $\eta_X^{\mathrm{EH}^{-1}}(W)$ for some Scott-open subset W of $\mathbf{B}(Y, d')$, we have $f^{-1}(V) = (\eta_Y^{\mathrm{EH}} \circ f)^{-1}(W) = (\mathbf{B}^c(f) \circ \eta_X^{\mathrm{EH}})^{-1}(W) = \eta_X^{\mathrm{EH}^{-1}}(\mathbf{B}^c(f)^{-1}(W))$, which is d-Scott-open.

Conversely, assume that f is continuous from X with its d-Scott topology to Y with its d'-Scott topology. Let $(x_i, r_i)_{i \in I, \sqsubseteq}$ be a Cauchy-weighted net in X, d such that x is a d-limit of $(x_i)_{i \in I, \sqsubseteq}$. Note that, by Lemma 7.4.25, $(x, 0)$ is the least upper bound of the directed family $(x_i, r_i)_{i \in I, \sqsubseteq}$ in $\mathbf{B}(X, d)$. Assume by way of contradiction that $f(x)$ is not a d'-limit of $(f(x_i))_{i \in I, \sqsubseteq}$. By Lemma 7.4.9, and since $(f(x_i), c.r_i)_{i \in I, \sqsubseteq}$ is Cauchy-weighted, there would be a point $y \in Y$ such that $d'(f(x), y) \neq \sup_{i \in I}(d'(f(x_i), y) - c.r_i)$. Since $\mathbf{B}^c(f)$ is monotonic, $\mathbf{B}^c(f)(x_i, r_i) \leqslant^{d'^+} \mathbf{B}^c(f)(x, 0)$, so $d'(f(x_i), f(x)) \leqslant c.r_i$. Hence $d'(f(x_i), y) \leqslant d'(f(x_i), f(x)) + d'(f(x), y) \leqslant c.r_i + d'(f(x), y)$, for every $i \in I$. Since $d'(f(x), y) \neq \sup_{i \in I}(d'(f(x_i), y) - c.r_i)$, we obtain $d'(f(x), y) > \sup_{i \in I}(d'(f(x_i), y) - c.r_i)$. Pick a real number $\eta > 0$ such that $d'(f(x), y) > \eta > \sup_{i \in I}(d'(f(x_i), y) - c.r_i)$. Then $f(x)$ is in the open hole $T^{d'}_{y, > \eta}$, which is d'-Scott-open by Lemma 7.4.48. Since f is continuous, $f^{-1}(T^{d'}_{y, > \eta})$ is d-Scott-open. Let U be a Scott-open subset of $\mathbf{B}(X, d)$ such that $f^{-1}(T^{d'}_{y, > \eta}) = \eta_X^{\mathrm{EH}^{-1}}(U)$. Since $f(x) \in T^{d'}_{y, > \eta}$, $(x, 0)$ is in U. So (x_i, r_i) is in U for i large enough, and, as U is upward closed, $(x_i, 0)$ is also in U. So $f(x_i) \in T^{d'}_{y, > \eta}$, meaning $d'(f(x_i), y) > \eta$, for i large enough. But this contradicts the fact that $\eta > \sup_{i \in I}(d'(f(x_i), y) - c.r_i)$. □

Exercise 7.4.53 Let X, d be a hemi-metric space. Show that $\eta^{\mathrm{EH}}_{\mathbf{B}(X,d)}$, which maps (x, r) to $((x, r), 0)$, is Scott-continuous. Deduce that the Scott topology on $\mathbf{B}(X, d)$ is finer than the d^+-Scott topology.

Exercise 7.4.54 (Scott = d^+-Scott on formal balls) Let X, d be a hemi-metric space. Define $\mu_X^{\mathrm{EH}} : \mathbf{B}(\mathbf{B}(X, d), d^+) \to \mathbf{B}(X, d)$ as the map that sends $((x, r), s)$ to $(x, r + s)$.

Show that μ_X^{EH} is Scott-continuous. Whenever $((x_i, r_i), s_i)_{i \in I}$ is a directed family in $\mathbf{B}(\mathbf{B}(X, d), d^+)$ with a least upper bound $((x, r), s)$, you will need to show that $(x, r + s)$ is a least upper bound of $(x_i, r_i + s_i)_{i \in I}$. To do so, if (y, t) is another upper bound of the latter family, then you must show that $((y, a), b)$ is an upper bound of $((x_i, r_i), s_i)_{i \in I}$, where $a = \max(t - s, 0)$ and $b = t - a = \min(s, t)$.

Show that $\mu_X^{\mathrm{EH}} \circ \eta_{\mathbf{B}(X,d)}^{\mathrm{EH}}$ is the identity map; i.e., $\mathbf{B}(X, d)$ is a Scott-retract of $\mathbf{B}(\mathbf{B}(X, d), d^+)$ through the retraction μ_X^{EH} and $\eta_{\mathbf{B}(X,d)}^{\mathrm{EH}}$, and the section $\eta_{\mathbf{B}(X,d)}^{\mathrm{EH}}$.

With Exercise 7.4.53, conclude that the d^+-Scott topology on any space $\mathbf{B}(X, d)$ of formal balls always coincides with the Scott topology.

The fact that the d-Scott topology appears as a generalization of the Scott topology allows us to proceed by analogy, and generalize the notions of finite elements and algebraic and continuous posets to the quasi-metric case. We start with finite elements. By using Exercise 5.1.11 (c), one can show that an element x of a poset X is finite if and only if the map $d_{\leqslant}(x, \_)$ is Scott-continuous from X to $\overline{\mathbb{R}^+}^{op}$. Analogously:

Definition 7.4.55 (d-finite) Let X, d be a hemi-metric space. A point $x \in X$ is *d-finite* if and only if the map $d(x, \_)$ is Yoneda-continuous from X, d to $\overline{\mathbb{R}^+}, \mathrm{d}_{\mathbb{R}}^{op}$.

Lemma 7.4.56 *Let X, d be a hemi-metric space. For $x \in X$, the following are equivalent:*

1. *x is d-finite.*
2. *For every Cauchy net $(y_i)_{i \in I, \sqsubseteq}$ in X, d that has a d-limit y, the equation $d(x, y) = \liminf_{i \in I, \sqsubseteq} d(x, y_i)$ holds.*
3. *For every Cauchy-weighted net $(y_i, s_i)_{i \in I, \sqsubseteq}$ in X, d such that $(y_i)_{i \in I, \sqsubseteq}$ has a d-limit y, $d(x, y) = \inf_{i \in I}(d(x, y_i) + s_i)$.*

Moreover, in case 3, $(d(x, y_i) + s_i)_{i \in I, \sqsubseteq}$ is a monotone net in $\overline{\mathbb{R}^+}^{op}$ (not $\overline{\mathbb{R}^+}$).

Compare 2 with the definition of d-limit: for every point x, not just the d-finite ones, $d(y, x)$ is the limsup of $d(y_i, x)$; when x is d-finite, 2 states that we can say the same thing with d replaced by d^{op} and limsups replaced by liminfs.

Proof 1 $\Rightarrow$ 2. If x is d-finite, then, for every Cauchy net $(y_i)_{i \in I, \sqsubseteq}$ in X, d with d-limit y, $d(x, y)$ is a $\mathrm{d}_{\mathbb{R}}^{op}$-limit of $(d(x, y_i))_{i \in I, \sqsubseteq}$. Note that the latter is Cauchy: for every $\epsilon > 0$, $d(y_i, y_j) < \epsilon$ for $i \sqsubseteq j$ large enough, so $d(x, y_j) < d(x, y_i) + \epsilon$, i.e., $\mathrm{d}_{\mathbb{R}}^{op}(d(x, y_i), d(x, y_j)) < \epsilon$ for $i \sqsubseteq j$ large enough. Then the fact that $d(x, y)$ is a $\mathrm{d}_{\mathbb{R}}^{op}$-limit means that, for every $z \in \overline{\mathbb{R}^+}$, $\mathrm{d}_{\mathbb{R}}^{op}(d(x, y), z) = \limsup_{i \in I, \sqsubseteq} \mathrm{d}_{\mathbb{R}}^{op}(d(x, y_i), z)$.

If $d(x, y) \neq +\infty$, then take any $z \in \mathbb{R}^+$ above $d(x, y)$. So $z - d(x, y) = \limsup_{i \in I, \sqsubseteq} \max(z - d(x, y_i), 0) = \max(\limsup_{i \in I, \sqsubseteq}(z - d(x, y_i)), 0)$ (by Exercise 7.1.12) is equal to $\max(z - \liminf_{i \in I, \sqsubseteq} d(x, y_i), 0)$. For $z = d(x, y)$, we obtain that $d(x, y) - \liminf_{i \in I, \sqsubseteq} d(x, y_i) \leqslant 0$, i.e., $d(x, y) \leqslant$

$\liminf_{i \in I, \sqsubseteq} d(x, y_i)$. If the inequality were strict, then there would be a $z \in \mathbb{R}^+$ such that $d(x, y) < z < \liminf_{i \in I, \sqsubseteq} d(x, y_i)$. So $z - d(x, y) > 0$, while $\max(z - \liminf_{i \in I, \sqsubseteq} d(x, y_i), 0) = 0$, a contradiction. So $d(x, y) = \liminf_{i \in I, \sqsubseteq} d(x, y_i)$.

If $d(x, y) = +\infty$, then, for every $z \in \mathbb{R}^+$, $\mathsf{d}_{\mathbb{R}}^{op}(d(x, y), z) = 0$. Since $\mathsf{d}_{\mathbb{R}}^{op}(d(x, y), z) = \limsup_{i \in I, \sqsubseteq} \mathsf{d}_{\mathbb{R}}^{op}(d(x, y_i), z) = 0$, for every $\epsilon > 0$, $\mathsf{d}_{\mathbb{R}}^{op}(d(x, y_i), z) < \epsilon$ for i large enough, i.e., $z - \epsilon < d(x, y_i)$. It follows that $\liminf_{i \in I, \sqsubseteq} d(x, y_i) = +\infty = d(x, y)$.

$2 \Rightarrow 3$. Let $(y_i, s_i)_{i \in I, \sqsubseteq}$ be Cauchy-weighted, and y be a d-limit of $(y_i)_{i \in I, \sqsubseteq}$. Whenever $i \sqsubseteq j$, $d(y_i, y_j) \leqslant s_i - s_j$, so $d(x, y_j) \leqslant d(x, y_i) + s_i - s_j$, hence $d(x, y_j) + s_j \leqslant d(x, y_i) + s_i$. So $(d(x, y_i) + s_i)_{i \in I, \sqsubseteq}$ is a monotone net in $\overline{\mathbb{R}^+}^{op}$. The fact that $\liminf_{i \in I, \sqsubseteq} d(x, y_i) = \inf_{i \in I}(d(x, y_i) + s_i)$ is proved as in Lemma 7.4.9. Alternatively, $(-d(x, y_i) - s_i)_{i \in I, \sqsubseteq}$ is a monotone net in $\mathbb{R} \cup \{-\infty, +\infty\}$, and $\inf_{i \in I} s_i = 0$, so, by Exercise 7.4.10, $\limsup_{i \in I, \sqsubseteq} -d(x, y_i) = \sup_{i \in I}(-d(x, y_i) - s_i)$, from which the result follows by taking opposites. Then 3 follows immediately, using $\liminf_{i \in I, \sqsubseteq} d(x, y_i) = \inf_{i \in I}(d(x, y_i) + s_i)$.

$3 \Rightarrow 1$. The map $d(x, \_)$ is 1-Lipschitz from X, d to $\overline{\mathbb{R}^+}, \mathsf{d}_{\mathbb{R}}^{op}$, by Exercise 6.2.10, for example, or by applying the triangular inequality directly. Let $(y_i, s_i)_{i \in I, \sqsubseteq}$ be Cauchy-weighted, and y be a d-limit of $(y_i)_{i \in I, \sqsubseteq}$. By 3, $d(x, y) = \inf_{i \in I}(d(x, y_i) + s_i)$. So, for every $z \in \mathbb{R}^+$, $\mathsf{d}_{\mathbb{R}}^{op}(d(x, y), z) = \max(z - d(x, y), 0) = \max(\sup_{i \in I}(z - d(x, y_i) - s_i), 0) = \max(\limsup_{i \in I, \sqsubseteq}(z - d(x, y_i)), 0)$ (Exercise 7.4.10) $= \limsup_{i \in I, \sqsubseteq} \max(z - d(x, y_i), 0) = \limsup_{i \in I, \sqsubseteq} \mathsf{d}_{\mathbb{R}}^{op}(d(x, y_i), z)$. We obtain the same equality $\mathsf{d}_{\mathbb{R}}^{op}(d(x, y), z) = \limsup_{i \in I, \sqsubseteq} \mathsf{d}_{\mathbb{R}}^{op}(d(x, y_i), z)$ when $z = +\infty$: if $d(x, y) = \inf_{i \in I}(d(x, y_i) + s_i) = +\infty$, then $d(x, y_i) = +\infty$ for every $i \in I$, so we obtain $\limsup_{i \in I, \sqsubseteq} \mathsf{d}_{\mathbb{R}}^{op}(d(x, y_i), z) = 0 = \mathsf{d}_{\mathbb{R}}^{op}(d(x, y), z)$, and if $d(x, y) = \inf_{i \in I}(d(x, y_i) + s_i)$ is different from $+\infty$, then $\mathsf{d}_{\mathbb{R}}^{op}(d(x, y), z) = +\infty$, and $d(x, y_i) < +\infty$ for i large enough, so $\limsup_{i \in I, \sqsubseteq} \mathsf{d}_{\mathbb{R}}^{op}(d(x, y_i), z) = +\infty$.

Since $\mathsf{d}_{\mathbb{R}}^{op}(d(x, y), z) = \limsup_{i \in I, \sqsubseteq} \mathsf{d}_{\mathbb{R}}^{op}(d(x, y_i), z)$ for every $z \in \overline{\mathbb{R}^+}$, $d(x, y)$ is a $\mathsf{d}_{\mathbb{R}}^{op}$-limit of $(d(x, y_i))_{i \in I, \sqsubseteq}$. It follows that $d(x, \_)$ is Yoneda-continuous by Lemma 7.4.37. □

Exercise 7.4.57 Show that the $\mathsf{d}_{\mathbb{R}}$-finite elements of $\overline{\mathbb{R}^+}$ are the elements of $\mathbb{R}^+$; i.e., r is $\mathsf{d}_{\mathbb{R}}$-finite if and only if $r \neq +\infty$.

Exercise 7.4.58 Let X be a quasi-ordered set, with quasi-ordering $\leqslant$. Show that its $d_{\leqslant}$-finite elements are exactly its finite elements. (Recall Exercise 7.4.30.)

The following lemma shows that metric spaces are very special with respect to d-finiteness.

Proposition 7.4.59 (Metric $\Rightarrow$ every element is d-finite) *Every element of a metric space X, d is d-finite.*

Proof Let X, d be metric, and $x \in X$. Since $d(\_, x)$ is Yoneda-continuous (Exercise 7.4.36), $d(x, \_)$ is, too. So x is finite, by definition. □

Here is the other notable case where this happens.

Lemma 7.4.60 *Every element of a Smyth-complete quasi-metric space X, d is d-finite.*

Proof Let $x \in X$. Every Cauchy net $(y_i)_{i \in I, \sqsubseteq}$ in X, d has a limit y in X, d^{sym}. For every $\epsilon > 0$, $d^{sym}(y, y_i) < \epsilon$ for i large enough, by Lemma 7.1.1. So $d(x, y) \leqslant d(x, y_i) + d(y_i, y) < d(x, y_i) + \epsilon$ and $d(x, y_i) \leqslant d(x, y) + d(y, y_i) < d(x, y) + \epsilon$ for i large enough, i.e., $d(x, y) - \epsilon < d(x, y_i) < d(x, y) + \epsilon$ for i large enough. It follows that $d(x, y) \leqslant \liminf_{i \in I, \sqsubseteq} d(x, y_i)$ and $\limsup_{i \in I, \sqsubseteq} d^{op}(y_i, x) \leqslant d(x, y)$. Since the liminf is always below the limsup (Exercise 7.1.11), here they coincide and equal $d(x, y)$. By Lemma 7.4.56 2, x is d-finite. □

This can be refined into the following theorem, due to Ali-Akbari *et al.* (2009).

Theorem 7.4.61 (Smyth-complete = all points d-finite) *A Yoneda-complete quasi-metric space X, d is Smyth-complete if and only if every point of X is d-finite.*

Proof One direction is true by Lemma 7.4.60. Conversely, assume that every element in X is d-finite. Let $(x_i)_{i \in I, \sqsubseteq}$ be a Cauchy net in X, d, and x be its d-limit. Since $0 = d(x, x) = \limsup_{i \in I, \sqsubseteq} d(x_i, x)$, for every $\epsilon > 0$, $d(x_i, x) < \epsilon$ for i large enough, say i above i_0. Since $(x_i)_{i \in I, \sqsubseteq}$ is Cauchy, we have $d(x_i, x_j) < \epsilon/2$ for $i \sqsubseteq j$ large enough, say above i_1. Since x is d-finite, by Lemma 7.4.56 part 2, $0 = d(x, x) = \liminf_{i \in I, \sqsubseteq} d(x, x_i)$, so there is an $i \in I$ above i_1 such that $d(x, x_i) < \epsilon/2$. For every j above i, $d(x, x_j) \leqslant d(x, x_i) + d(x_i, x_j) < \epsilon$. So, for every j above both i_0 and i, $d^{sym}(x, x_i) < \epsilon$. Using Lemma 7.1.1, x is the limit of $(x_i)_{i \in I, \sqsubseteq}$ in X, d^{sym}. So X, d is Smyth-complete. □

Definition 7.4.62 (Algebraic hemi-metric space) A hemi-metric space X, d is *algebraic* if and only if every element of X is a d-limit of a Cauchy net of d-finite elements of X.

Example 7.4.63 By Proposition 7.4.59 and Lemma 7.4.60, every metric space and every Smyth-complete quasi-metric space X, d is algebraic, in a trivial way: the Cauchy net we take is a constant net.

Example 7.4.64 There are algebraic hemi-metric spaces that are neither metric nor Smyth-complete, such as $\overline{\mathbb{R}^+}, \mathrm{d}_{\mathbb{R}}$. That the latter is not Smyth-complete is true by Exercise 7.2.3. It is algebraic: by Exercise 7.4.57 all its elements except $+\infty$ are d-finite, and $+\infty$ is obtained as the d-limit of the sequence $(n)_{n \in \mathbb{N}}$, for example.

Exercise 7.4.65 Let X be a quasi-ordered set, with quasi-ordering $\leqslant$. Recalling Exercise 7.4.58, show that $X, d_{\leqslant}$ is algebraic (as a hemi-metric space) if and only if X is algebraic (as a quasi-ordered set).

Definition 7.4.66 (Strong basis) Call a *strong basis* B of a hemi-metric space X, d any set of d-finite elements such that every element of X is a d-limit of a Cauchy net of B.

So X has a strong basis iff X, d is algebraic. We may call basis B any subset B such that every element of X is a d-limit of a Cauchy net of elements of B. In a strong basis, we require that all elements be d-finite.

We claimed earlier that dense subsets were too weak a concept in non-T_2 spaces. But density was essential in T_2 spaces. For example, any metric space is dense in its Cauchy completion (Exercise 7.3.17), and this is what allowed us to show the existence of extensions by continuity (Lemma 7.3.19). We claim that dense subsets should be replaced by strong bases in the larger realm of quasi-metric spaces. The following exercise should serve to reassure us that dense subsets and strong bases are close notions.

Exercise 7.4.67 (Strong basis = dense in metric spaces) Let X, d be a metric space. Show that the strong bases of X are exactly its dense subsets (with respect to the open ball topology, or, equivalently, the d-Scott topology, since the two topologies coincide).

Although the d-Scott topology and the open ball topology differ in general, we should also be reassured that the d-Scott topology is described in terms of certain open balls, namely those with d-finite centers, in algebraic Yoneda-complete spaces.

Proposition 7.4.68 *Let X, d be an algebraic Yoneda-complete quasi-metric space, with strong basis B. The family of all open balls $B^d_{x,<\epsilon}$ with $x \in B$ (not $x \in X$), $\epsilon > 0$, is a base of the d-Scott topology on X.*

Proof We first show that $B^d_{x,<\epsilon}$ is d-Scott-open whenever $x \in B$ and $\epsilon > 0$. Clearly, $B^d_{x,<\epsilon} = \{y \in X \mid (y, 0) \in V\}$ where $V = \{(y, s) \mid d(x, y) < \epsilon - s\}$. It is easy to see that V is upward closed. Let $(y_i, s_i)_{i \in I, \sqsubseteq}$ be a directed family of formal balls with least upper bound (y, s) in V. By the Kostanek–Waszkiewicz

Theorem 7.4.27, y is the d-limit of $(y_i)_{i \in I, \sqsubseteq}$, and $s = \inf_{i \in I} s_i$. It follows that $(y_i, s_i - s)_{i \in I, \sqsubseteq}$ is a Cauchy-weighted net, so, by Lemma 7.4.56 part 3, $d(x, y) = \inf_{i \in I}(d(x, y_i) + s_i - s)$. Since $d(x, y) < \epsilon - s$, $d(x, y_i) + s_i - s < \epsilon - s$ for i large enough, so $(y_i, s_i) \in V$. It follows that V is Scott-open, so $B^d_{x,<\epsilon}$ is d-Scott-open.

Conversely, let U be any d-Scott-open subset of X, and write U as $\{y \in X \mid (y, 0) \in V\}$ for some Scott-open subset V of $\mathbf{B}(X, d)$. Let y be any point of U. By assumption, y is the d-limit of some Cauchy net $(y_i)_{i \in I, \sqsubseteq}$ of elements of B. Replacing the latter by a subnet if necessary, we can assume that it is Cauchy-weightable (Lemma 7.2.8). So let $(y_i, r_i)_{i \in I, \sqsubseteq}$ be Cauchy-weighted. As a monotone net of formal balls, its least upper bound is $(y, 0)$, by the Kostanek–Waszkiewicz Theorem 7.4.27. Since $(y, 0)$ is in V, (y_i, r_i) is in V for some $i \in I$. We may assume $r_i > 0$, as this is what the proof of Lemma 7.2.8 ensures. Alternatively, if $r_i = 0$, then $(y_i, 0)$ is the least upper bound of the chain $(y_i, \epsilon)_{\epsilon > 0}$, so $(y_i, \epsilon) \in V$ for some $\epsilon > 0$.

Since $(y_i, r_i) \leqslant^{d^+} (y, 0)$, $d(y_i, y) \leqslant r_i$, so y is in the open ball $B^d_{y_i,<\epsilon}$, where $\epsilon > 0$ is chosen such that $\epsilon < r_i$ (this is where we need $r_i > 0$). And every element z of $B^d_{y_i,<\epsilon}$ is such that $d(y_i, z) \leqslant r_i$, so $(z, 0)$ is above (y_i, r_i), hence is in V: so z is in U. So $y \in B^d_{y_i,<\epsilon} \subseteq U$. Since y_i is in B, the open balls centered at points in B form a base of the d-Scott topology, using Lemma 4.1.10. □

In particular, when all points are d-finite, the d-Scott topology coincides with the open ball topology. This is in particular the case for metric spaces (Proposition 7.4.59), so that we retrieve the result of Proposition 7.4.46, second part; and for Smyth-complete quasi-metric spaces (Lemma 7.4.60), so that we retrieve Proposition 7.4.47.

Exercise 7.4.69 Let X, d be an algebraic Yoneda-complete quasi-metric space, and U be an open subset of X in the generalized Scott topology of Exercise 7.4.51. Show that every point x of U is in some open ball $B^d_{y,<\epsilon}$ contained in U, with y d-finite. Conclude that, in this case, the generalized Scott topology coincides with the d-Scott topology.

Exercise 7.4.70 (Finite products of algebraic spaces) Remember from Exercise 5.1.57 that the product of a finite family of algebraic posets is algebraic, and that the finite elements in the product are the tuples of finite elements.

Show the analogue result for hemi-metric spaces. Let $(X_i, d_i)_{i \in I}$ be a family of hemi-metric spaces, where I is finite, and let $d = \prod_{i \in I} d_i$. Show that the d-finite elements of $\prod_{i \in I} X_i$ are the tuples $\vec{x}$ where x_i is d_i-finite for each $i \in I$, and deduce that if X_i, d_i is algebraic for every $i \in I$, then $\prod_{i \in I} X_i, d$ is algebraic.

Exercise 7.4.71 (Products of pointed algebraic spaces) Remember (Exercise 5.1.57) that the product of a family of algebraic pointed posets is again algebraic, and that the finite elements in the product are tuples of finite elements, all of which but finitely many are least in their respective spaces.

Show the analogue for hemi-metric spaces. Let $(X_i, d_i)_{i \in I}$ be a family of hemi-metric spaces, and d be the sup hemi-metric $d(\vec{x}, \vec{y}) = \sup_{i \in I} d_i(x_i, y_i)$. Assume that each X_i has a least element $\bot_i$ in $\leqslant^{d_i}$. Show that the d-finite elements of $\prod_{i \in I} X_i$ are the tuples $\vec{x}$ such that x_i is d_i-finite for every $i \in I$, and that are *almost bottom*, i.e., such that, for every $\epsilon > 0$, there is a finite subset J of I such that, for every $i \in I \smallsetminus J$, $d_i(x_i, \bot_i) < \epsilon$.

Conclude that, if X_i, d_i is algebraic for every $i \in I$, then $\prod_{i \in I} X_i, d$ is algebraic. A piece of advice: answer Exercise 7.4.70 first.

Having defined algebraicity for hemi-metric spaces, our next logical move should be to define a notion of continuity. This is done by Kostanek and Waszkiewicz (2011), where it is proved that a Yoneda-complete quasi-metric space X, d is continuous iff $\mathbf{B}(X, d), \leqslant^{d^+}$ is a continuous dcpo. Since the actual definition of continuity for hemi-metric spaces is more complicated, we prefer to state the following directly.

Definition 7.4.72 (Continuous hemi-metric space) A Yoneda-complete quasi-metric space X, d is *continuous* if and only if $\mathbf{B}(X, d), \leqslant^{d^+}$ is a continuous dcpo.

It is not true that X, d is algebraic if and only if $\mathbf{B}(X, d), \leqslant^{d^+}$ is an algebraic dcpo. In fact, unless X is empty, no element (x, r) of $\mathbf{B}(X, d)$ is finite (consider the sequence $(x, r + \epsilon)$, $\epsilon > 0$, whose least upper bound is (x, r)), so $\mathbf{B}(X, d)$ is never algebraic.

Exercise 7.4.73 ($\mathbb{R}_\ell$ is continuous, not algebraic) Show that the Sorgenfrey line $\mathbb{R}_\ell, \mathsf{d}_\ell$ is a continuous Yoneda-complete quasi-metric space which is not Smyth-complete. Show that it is not even algebraic: no element of $\mathbb{R}_\ell$ is d_ℓ-finite.

7.5 The formal ball completion

Formal balls allow us to build a natural notion of completion for any hemi-metric space. Instead of building a space of Cauchy nets, or of Cauchy-weighted nets, and then taking a quotient, as in Exercise 7.3.17, we shall only consider certain canonical representatives of the equivalence classes. Because $\mathbf{B}(X, d)$ is an abstract basis (Lemma 7.3.10), one is naturally led to consider (certain) rounded ideals in $\mathbf{B}(X, d), \prec^{d^+}$ (see Proposition 5.1.33). Note that rounded ideals are particular kinds of (downward closed) Cauchy nets. We need to define a quasi-metric on these.

Lemma 7.5.1 (Hausdorff–Hoare quasi-metric) *Let Y, d be a hemi-metric space, understood with its open ball topology, and Z be any space of non-empty subsets of Y. The* Hausdorff–Hoare hemi-metric *on Z is $d_{\mathcal{H}}$, defined by*

$$d_{\mathcal{H}}(D, D') = \sup_{y \in D} \inf_{y' \in D'} d(y, y').$$

Its specialization quasi-ordering $\leqslant^{d_{\mathcal{H}}}$ is given by $D \leqslant^{d_{\mathcal{H}}} D'$ iff $D \subseteq cl(D')$, iff $cl(D) \subseteq cl(D')$.

Proof First, we check that $d_{\mathcal{H}}$ is a hemi-metric. The equality $d_{\mathcal{H}}(D, D) = 0$ is clear. For the triangular inequality, consider three elements D, D', D'' of Z. For all $y \in D$, $y' \in D'$, and $y'' \in D''$, $d(y, y'') \leqslant d(y, y') + d(y', y'')$. So $\inf_{y'' \in D''} d(y, y'') \leqslant d(y, y') + \inf_{y'' \in D''} d(y', y'') \leqslant d(y, y') + d_{\mathcal{H}}(D', D'')$. Since y' is arbitrary, $\inf_{y'' \in D''} d(y, y'') \leqslant \inf_{y' \in D'} d(y, y') + d_{\mathcal{H}}(D', D'')$, and the triangular inequality follows by taking least upper bounds over $y \in D$.

Note that $d_{\mathcal{H}}(D, D') = \sup_{y \in D} d(y, D')$, where the notation $d(y, D')$ was defined in Lemma 6.1.11 as $\inf_{y' \in D'} d(y, y')$. The specialization quasi-ordering of $d_{\mathcal{H}}$ is given by: D is below D' iff $d_{\mathcal{H}}(D, D') = 0$ (Proposition 6.1.8), iff, for every $y \in D$, $d(y, D') = 0$. By Lemma 6.1.11, this is equivalent to: for every $y \in D$, y is in $cl(D')$; equivalently, $cl(D) \subseteq cl(D')$. □

Definition 7.5.2 (Formal ball completion) Let X, d be a hemi-metric space. For every rounded ideal D of $\mathbf{B}(X, d), \prec^{d^+}$, let the *aperture* $\alpha(D)$ of D be $\inf_{(x,r) \in D} r$.

The *formal ball completion* of X, d is the space $\mathbf{S}(X, d)$ of all rounded ideals D of $\mathbf{B}(X, d), \prec^{d^+}$ with zero aperture, with quasi-metric $d^+{}_{\mathcal{H}}$.

The specialization ordering of $d^+{}_{\mathcal{H}}$ simplifies, on rounded ideals, to mere inclusion, instead of inclusion of closures:

Lemma 7.5.3 *Let X, d be a hemi-metric space. The specialization ordering $\leqslant^{d^+{}_{\mathcal{H}}}$ of the rounded ideal completion $\mathbf{RI}(\mathbf{B}(X, d), d^+), d^+{}_{\mathcal{H}}$ and of the formal ball completion $\mathbf{S}(X, d), d^+{}_{\mathcal{H}}$ is inclusion: for all rounded ideals D and D' of formal balls, $d^+{}_{\mathcal{H}}(D, D') = 0$ if and only if $D \subseteq D'$.*

Proof If $D \subseteq D'$, then $D \subseteq cl(D')$, so $d^+{}_{\mathcal{H}}(D, D') = 0$ by Lemma 7.5.1. Conversely, assume $d^+{}_{\mathcal{H}}(D, D') = 0$. So $D \subseteq cl(D')$. For every $b \in D$, since

D is directed with respect to $\prec^{d^+}$ (in the sense of Proposition 5.1.33), there is a $b_1 \in D$ such that $b \prec^{d^+} b_1$, and then also a $b_2 \in D$ such that $b_1 \prec^{d^+} b_2$. Since $b_2 \in D$, b_2 is in $cl(D')$. By Lemma 7.3.10, b_2 is in the interior of $\uparrow b_1$. Using Lemma 4.1.27, $\uparrow b_1$ must then meet D', say at b'. Then $b_1 \leqslant^{d^+} b'$, so $b \prec^{d^+} b'$. Since D' is downward closed, b is in D'. So $D \subseteq D'$. □

Given any set D of formal balls, and any $s \in \mathbb{R}^+$, write $D+s$ for $\{(x, r+s) \mid (x,r) \in D\}$. Whenever $s \leqslant \alpha(D)$, conversely, write $D - s$ for $\{(x, r-s) \mid (x,r) \in D\}$.

Proposition 7.5.4 *Let X, d be a hemi-metric space. Let σ_X be the map from $\mathbf{B}(\mathbf{S}(X,d), d^+{}_{\mathcal{H}})$ to the rounded ideal completion $\mathbf{RI}(\mathbf{B}(X,d), \prec^{d^+})$ defined by $\sigma_X(D,s) = D+s$. Then:*

1. *σ_X is a bijective and 1-Lipschitz map from $\mathbf{B}(\mathbf{S}(X,d), d^+{}_{\mathcal{H}}), d^+{}_{\mathcal{H}}{}^+$ onto $\mathbf{RI}(\mathbf{B}(X,d), \prec^{d^+}), d^+{}_{\mathcal{H}}$; its inverse σ_X^{-1} maps D to $(D-\alpha(D), \alpha(D))$;*
2. *σ_X is an order isomorphism of $\mathbf{B}(\mathbf{S}(X,d), d^+{}_{\mathcal{H}}), \leqslant^{d^+{}_{\mathcal{H}}{}^+}$ onto $\mathbf{RI}(\mathbf{B}(X,d), \prec^{d^+}), \leqslant^{d^+{}_{\mathcal{H}}}$;*
3. *The space of formal balls $\mathbf{B}(\mathbf{S}(X,d), d^+{}_{\mathcal{H}})$ is a continuous dcpo in the $\leqslant^{d^+{}_{\mathcal{H}}{}^+}$ ordering.*

Proof Clearly, the aperture of $D-\alpha(D)$ is zero. Moreover, for every rounded ideal D, for every $s \in \mathbb{R}^+$, it is easy to see that $D+s$ is also a rounded ideal, and that $D-s$ is a rounded ideal if $s \leqslant \alpha(D)$. This follows from the definition of $\prec^{d^+}$, which implies that $(x,r) \prec^{d^+} (x',r')$ iff $(x, r+s) \prec^{d^+} (x', r'+s)$. So σ_X and σ_X^{-1} are well-defined maps, and are inverse to each other.

1. We now check that σ_X is 1-Lipschitz. Let (D,s) and (D',s') be two formal balls on $\mathbf{S}(X,d), d^+{}_{\mathcal{H}}$. We have:

$$\begin{aligned} d^+{}_{\mathcal{H}}(D+s, D'+s') &= \sup_{(x,r)\in D+s} \inf_{(x',r')\in D'+s'} d^+((x,r),(x',r')) \\ &= \sup_{(x,r)\in D} \inf_{(x',r')\in D'} d^+((x,r+s),(x',r'+s')). \end{aligned}$$

Now $d^+((x,r+s),(x',r'+s')) = \max(d(x,x') - r - s + r' + s', 0) \leqslant \max(d(x,x') - r - s + r' + s', -s+s', 0) = \max(\max(d(x,x') - r + r'), 0) - s + s', 0) = \max(d^+((x,r),(x',r')) - s + s', 0)$. (For later use, notice that this inequality is an equality whenever $-s+s' \leqslant 0$, i.e., when $s \geqslant s'$.) So,

$$\begin{aligned} d^+{}_{\mathcal{H}}(D+s, D'+s') &\leqslant \sup_{(x,r)\in D} \inf_{(x',r')\in D'} \max(d^+((x,r),(x',r')) - s + s', 0) \\ &= \max(\sup_{(x,r)\in D} \inf_{(x',r')\in D'} (d^+((x,r),(x',r')) - s + s'), 0) \\ &= d^+{}_{\mathcal{H}}{}^+((D,s),(D',s')). \end{aligned}$$

(And this is an equality if $s \geqslant s'$.)

2. Since σ_X is 1-Lipschitz, it is in particular monotonic, as the ordering $\leqslant^d$ is uniquely determined from d for any hemi-metric d (Fact 7.3.2). Conversely, assume that $D+s \leqslant^{d^+\mathcal{H}} D'+s'$, i.e., that $D+s \subseteq D'+s'$, by Lemma 7.5.3. For every $\epsilon > 0$, pick an element $(x, r) \in D$ with $r < \epsilon$. Since $(x, r+s)$ is in $D+s$, it is in $D'+s'$, so $r+s \geqslant s'$, hence $s+\epsilon \geqslant s'$. Since ϵ is arbitrary, $s \geqslant s'$. We noticed above that this entailed $d^+\mathcal{H}(D+s, D'+s') = d^+\mathcal{H}^+((D, s), (D', s'))$. Since $D+s \leqslant^{d^+\mathcal{H}} D'+s'$, the left-hand side equals 0, so $(D, s) \leqslant^{d^+\mathcal{H}^+} (D', s')$. So σ_X^{-1} is monotonic, hence σ_X is an order isomorphism.

3. By Proposition 5.1.33, and since $\mathbf{B}(X, d), \prec^{d^+}$ is an abstract basis (using Lemma 7.3.10), $\mathbf{RI}(\mathbf{B}(X, d), \prec^{d^+})$ is a continuous dcpo in the inclusion ordering, which is the specialization ordering of $\mathbf{RI}(\mathbf{B}(X, d), \prec^{d^+}), d^+\mathcal{H}^+$ by Lemma 7.5.3. So $\mathbf{B}(\mathbf{S}(X, d), d^+\mathcal{H})$ is also a continuous dcpo in its specialization ordering. □

Note in passing that $d^+\mathcal{H}$-limits in the formal ball completion $\mathbf{S}(X, d)$ have a simple expression.

Lemma 7.5.5 *Let X, d be a hemi-metric space, and $(D_i, r_i)_{i \in I, \sqsubseteq}$ be a Cauchy-weighted net in $\mathbf{S}(X, d), d^+\mathcal{H}$. The $d^+\mathcal{H}$-limit of the Cauchy-weightable net $(D_i)_{i \in I, \sqsubseteq}$ is $\bigcup_{i \in I}(D_i + r_i)$, the union of the monotone net $(D_i + r_i)_{i \in I, \sqsubseteq}$.*

Proof First, Proposition 7.5.4 states that σ_X is an order isomorphism. Since $(D_i, r_i)_{i \in I, \sqsubseteq}$ is a monotone net in $\mathbf{B}(\mathbf{S}(X, d), d^+\mathcal{H})$, $(D_i + r_i)_{i \in I, \sqsubseteq}$ is a monotone net in $\mathbf{RI}(\mathbf{B}(X, d), \prec^{d^+})$. The latter is a dcpo, by Proposition 5.1.33, so this net has a least upper bound $D = \bigcup_{i \in I}(D_i + r_i)$. Again using σ_X, $(D, 0)$ is the least upper bound of the monotone net $(D_i, r_i)_{i \in I, \sqsubseteq}$. Since $\inf_{i \in I} r_i = 0$, we use the Kostanek–Waszkiewicz Theorem 7.4.27 (which applies since $\mathbf{B}(\mathbf{S}(X, d), d^+\mathcal{H}), \leqslant^{d^+\mathcal{H}^+}$ is a dcpo, by Proposition 7.5.4), and obtain that D is the $d^+\mathcal{H}$-limit of $(D_i)_{i \in I, \sqsubseteq}$ in $\mathbf{S}(X, d)$. □

Combining the last part of Proposition 7.5.4 and the Kostanek–Waszkiewicz Theorem 7.4.27, we obtain:

Corollary 7.5.6 (The formal ball completion is Yoneda-complete) *Let X, d be a hemi-metric space. Its formal ball completion $\mathbf{S}(X, d), d^+\mathcal{H}$ is Yoneda-complete.*

The latter means that the space of formal balls of the completion is a dcpo. Proposition 7.5.4 tells us more: the space of formal balls of $\mathbf{S}(X, d)$ is a *continuous* dcpo, i.e., $\mathbf{S}(X, d), d^+\mathcal{H}$ is continuous Yoneda-complete. (It is even algebraic, as we shall see in Exercise 7.5.11.) If additionally the way-below relation on the latter coincided with $\prec^{d^+\mathcal{H}^+}$, then, by the Romaguera–Valero Theorem 7.3.11, $\mathbf{S}(X, d)$ would be Smyth-complete.

In general, $\mathbf{S}(X, d), d^+\mathcal{H}$ is not Smyth-complete. We shall see an example in Exercise 7.5.9: in general $\prec^{d^+\mathcal{H}^+}$ is not the way-below relation on $\mathbf{S}(X, d), d^+\mathcal{H}$.

Remember from Proposition 5.1.33 that, in an abstract basis $B, \prec$, $\Downarrow b$ denotes the set of elements $b' \in B$ such that $b' \prec b$. We use the same notation $\Downarrow(x, r)$ in $\mathbf{RI}(\mathbf{B}(X, d), \prec^{d^+})$. Note that $\Downarrow(x, 0)$ is then in $\mathbf{S}(X, d)$. Remember also (Lemma 6.1.11) that $d^+((x, r), D')$ is by definition $\inf_{(x',r')\in D'} d^+((x, r), (x', r'))$.

Lemma 7.5.7 *Let X, d be a hemi-metric space. For every formal ball (x, r) and every rounded ideal D' of $\mathbf{B}(X, d), \prec^{d^+}$, $d^+\mathcal{H}(\Downarrow(x, r), D') = d^+((x, r), D')$.*

Proof By definition, $d^+\mathcal{H}(\Downarrow(x, r), D') = \sup_{(y,s)\prec^{d^+}(x,r)} d^+((y, s), D')$. Whenever $(y, s) \prec^{d^+} (x, r)$, $d^+((y, s), (x, r)) = \max(d(y, x) - s + r, 0) = 0$, so $d^+((y, s), D') \leqslant d^+((y, s), (x, r)) + d^+((x, r), D') = d^+((x, r), D')$. So $d^+\mathcal{H}(\Downarrow(x, r), D') \leqslant d^+((x, r), D')$. Conversely, among those (y, s) such that $(y, s) \prec^{d^+} (x, r)$, we find $(x, r + \epsilon)$ for each $\epsilon > 0$, so $d^+\mathcal{H}(\Downarrow(x, r), D') \geqslant d^+((x, r+\epsilon), D') \geqslant d^+((x, r), D') - \epsilon$. As ϵ is arbitrary, equality follows. □

It follows that any quasi-metric space embeds into its formal ball completion.

Proposition 7.5.8 (Embedding into $\mathbf{S}(X, d)$) *Let X, d be a hemi-metric space. The map $\eta^{\mathbf{S}}_X : X \to \mathbf{S}(X, d)$ that sends x to $\Downarrow(x, 0)$ is an isometric embedding.*

Proof Clearly, $\Downarrow(x, 0)$ has aperture 0, since it contains all elements (x, ϵ) with $\epsilon > 0$.

By Lemma 7.5.7, $d^+\mathcal{H}(\eta^{\mathbf{S}}_X(x), \eta^{\mathbf{S}}_X(y)) = d^+((x, 0), \Downarrow(y, 0))$. For every element (z, t) of $\Downarrow(y, 0)$, $d(z, y) < t$, so $d(x, y) \leqslant d(x, z) + d(z, y) < d(x, z) + t = d^+((x, 0), (z, t))$. So $d^+((x, 0), \Downarrow(y, 0)) = \inf_{(z,t)\in\Downarrow(y,0)} d^+((x, 0), (z, t)) \geqslant d(x, y)$.

Conversely, $\Downarrow(y, 0)$ contains all the elements (y, ϵ), $\epsilon > 0$, so $d^+((x, 0), \Downarrow(y, 0)) \leqslant d^+((x, 0), (y, \epsilon)) = d(x, y) + \epsilon$. As $\epsilon > 0$ is arbitrary, $d^+((x, 0), \Downarrow(y, 0)) \leqslant d(x, y)$, whence the equality holds. □

Exercise 7.5.9 (The completion of $\mathbb{R}^+$ is $\overline{\mathbb{R}^+}$) Let D be an element of $\mathbf{S}(\mathbb{R}^+, \mathsf{d}_\mathbb{R})$ (resp., $\mathbf{S}(\mathbb{R}, \mathsf{d}_\mathbb{R})$). Let $a = \sup_{(x,r)\in D}(x - r) \in \overline{\mathbb{R}^+}$. Show that, if

$a < +\infty$, then $D = \Downarrow(a, 0)$, and that otherwise D is the set of all formal balls. Deduce that $\mathbf{S}(\mathbb{R}^+, \mathsf{d}_\mathbb{R})$ (resp., $\mathbf{S}(\mathbb{R}, \mathsf{d}_\mathbb{R})$) is isometric to $\overline{\mathbb{R}^+}, \mathsf{d}_\mathbb{R}$ (resp., to $\mathbb{R} \cup \{+\infty\}, \mathsf{d}_\mathbb{R}$).

Conclude that the formal ball completion of a space may fail to be Smyth-complete.

And the completion essentially adds all missing d-limits: while the following Lemma talks about $d^+{}_\mathcal{H}$-limits instead, remember that $d^+{}_\mathcal{H}$ is merely the extension of d to the larger, formal ball completion.

Lemma 7.5.10 (X is d-dense in its completion) *Let X, d be a hemi-metric space. We say that a subset A of X is* d-dense *in X if and only if X is the smallest d-closed subset of X containing A.*

The image $\eta^\mathbf{S}_X[X]$ is $d^+{}_\mathcal{H}$-dense in the formal ball completion $\mathbf{S}(X, d)$. More precisely, every element D of $\mathbf{S}(X, d)$ is the $d^+{}_\mathcal{H}$-limit of some Cauchy-weightable net in $\eta^\mathbf{S}_X[X]$. The canonical choice is the one underlying the Cauchy-weighted net $(\eta^\mathbf{S}_X(x), r)_{(x,r)\in D, \leqslant^{d^+}}$.

Proof Let D be any element of $\mathbf{S}(X, d)$. D is a Cauchy-weighted net. Consider the net $(\eta^\mathbf{S}_X(x), r)_{(x,r)\in D, \leqslant^{d^+}}$. This is a net since D is directed in $\prec^{d^+}$, hence in $\leqslant^{d^+}$. Whenever $(x, r) \leqslant^{d^+} (y, s)$, $d(x, y) \leqslant r - s$, so $d^+{}_\mathcal{H}(\eta^\mathbf{S}_X(x), \eta^\mathbf{S}_X(y)) \leqslant r - s$ since $\eta^\mathbf{S}_X$ is 1-Lipschitz (Proposition 7.5.8). Since its aperture is clearly zero, $(\eta^\mathbf{S}_X(x), r)_{(x,r)\in D, \leqslant^{d^+}}$ is Cauchy-weighted. For every $D' \in \mathbf{S}(X, d)$, we realize that $\sup_{(x,r)\in D}(d^+{}_\mathcal{H}(\eta^\mathbf{S}_X(x), D') - r) = \sup_{(x,r)\in D} d^+((x, r), D')$ (by Lemma 7.5.7) $= d^+{}_\mathcal{H}(D, D')$, by the definition of $d^+{}_\mathcal{H}$. This states exactly that D is the $d^+{}_\mathcal{H}$-limit of the net $(\eta^\mathbf{S}_X(x), r)_{(x,r)\in D, \leqslant^{d^+}}$, using Lemma 7.4.9. □

Exercise 7.5.11 ($\mathbf{S}(X, d)$ is algebraic) Let X, d be a hemi-metric space, and $x \in X$. Show that $\eta^\mathbf{S}_X(x)$ is $d^+{}_\mathcal{H}$-finite in $\mathbf{S}(X, d)$. Conclude that $\mathbf{S}(X, d), d^+{}_\mathcal{H}$ is an algebraic Yoneda-complete quasi-metric space, with strong basis given by $\eta^\mathbf{S}_X[X]$. (In other words, identifying X with its image by $\eta^\mathbf{S}_X$, X itself is a strong basis of its formal ball completion.)

Exercise 7.5.12 (Completion of products $\neq$ product of completions) Deduce from Exercise 7.4.71 that, although $\overline{\mathbb{R}^+}$ is the formal ball completion of $\mathbb{R}^+$ (Exercise 7.5.9), $\overline{\mathbb{R}^+}^\mathbb{N}$ is *not* the formal ball completion of $\mathbb{R}^{+\mathbb{N}}$ (even up to isometry) – and despite the fact that $\overline{\mathbb{R}^+}^\mathbb{N}, \sup_\mathbb{N} \mathsf{d}_\mathbb{R}$ is Yoneda-complete and algebraic. Hint: use Exercise 7.5.11.

Exercise 7.5.13 Let X, d be a hemi-metric space. Assume that $(x_i, r_i)_{i\in I, \sqsubseteq}$ is a Cauchy-weighted net, and that $(x_i)_{i\in I, \sqsubseteq}$ has x as a d-limit. Assume also that, for each $i \in I$, x_i is a d-limit of some net $(x_{ij})_{j \in J_i, \preceq_i}$, where $(x_{ij}, r_{ij})_{j \in J_i, \preceq_i}$ is Cauchy-weighted. Define $(i, j) \leqslant (i', j')$ if and only if $r_i + r_{ij} \geqslant r_{i'} + r_{i'j'}$. Show that $(x_{ij})_{(i,j) \in \coprod_{i \in I} J_i, \leqslant}$ is a Cauchy net, and has x as a d-limit.

Exercise 7.5.14 (d-closure) Let X, d be a hemi-metric space. Call the *d-closure* of a subset A of X the smallest d-closed subset of X, d containing A.

Using Exercise 7.5.13, show that the d-closure of A is exactly the set of d-limits of Cauchy-weightable nets, or, equivalently, of Cauchy nets, of elements of A.

7.5.1 Characterizing Smyth-completeness

We now realize that the formal ball completion also characterizes Smyth-completeness.

Proposition 7.5.15 *Let X, d be a hemi-metric space. The following are equivalent:*

1. *X, d is Smyth-complete.*
2. *$\eta^{\mathbf{S}}_X$ is a bijection.*
3. *$\eta^{\mathbf{S}}_X$ is an isometry of X, d onto $\mathbf{S}(X, d), d^+{}_{\mathcal{H}}$.*

Proof $1 \Rightarrow 2$. Assume X, d is Smyth-complete. In particular, X, d is T_0, and since $\eta^{\mathbf{S}}_X$ is an isometric embedding (Proposition 7.5.8), it is injective. Let us show that it is surjective. By the Romaguera–Valero Theorem 7.3.11, $\mathbf{B}(X, d)$ is a continuous dcpo, and $\prec^{d^+}$ is its way-below relation. Let D be any element of $\mathbf{S}(X, d)$. Since D is directed and $\mathbf{B}(X, d)$ is a dcpo, $\sup D$ exists. Since D has aperture 0, $\sup D$ must be of the form $(x, 0)$ for some $x \in X$. We claim that $D = \Downarrow(x, 0)$.

Since D is directed with respect to $\prec^{d^+}$, any element (y, s) in D is such that $(y, s) \prec^{d^+} (z, t)$ for some $(z, t) \in D$, and $(z, t) \leqslant^{d^+} (x, 0)$. So $(y, s) \in \Downarrow(x, 0)$. Conversely, let (y, s) be any element of $\Downarrow(x, 0)$. So $(y, s) \prec^{d^+} (x, 0) = \sup D$. Since $\prec^{d^+}$ is the way-below relation, it is interpolative, hence $(y, s) \prec^{d^+} (z, t) \prec^{d^+} \sup D$ for some (z, t). By definition of the way-below relation, $(z, t) \leqslant^{d^+} (z', t')$ for some element (z', t') of D, hence $(y, s) \prec^{d^+} (z', t')$. Since D is downward closed in the sense of Proposition 5.1.33, (y, s) is in D. So $D = \Downarrow(x, 0) = \eta^{\mathbf{S}}_X(x)$.

$2 \Rightarrow 3$ is easy, once we recall that $\eta^{\mathbf{S}}_X$ is an isometric embedding (Proposition 7.5.8).

$3 \Rightarrow 1$. Assume $\eta^{\mathbf{S}}_X$ is an isometry. Then, using Lemma 7.3.16, $\mathbf{B}^1(\eta^{\mathbf{S}}_X)$ is an isometry from $\mathbf{B}(X, d), d^+$ onto $\mathbf{B}(\mathbf{S}(X, d), d^+{}_{\mathcal{H}}), d^+{}_{\mathcal{H}}{}^+$, hence an order isomorphism, since the hemi-metrics determine the specialization quasi-ordering, by Fact 7.3.2. Composing this with σ_X (Proposition 7.5.4), we obtain an order isomorphism from $\mathbf{B}(X, d), \leqslant^{d^+}$ onto $\mathbf{RI}(\mathbf{B}(X, d), \prec^{d^+}), \leqslant^{d^+\mathcal{H}}$. The latter is a continuous dcpo by Proposition 5.1.33, so the former is one as well.

The order isomorphism $\sigma_X \circ \mathbf{B}^1(\eta_X^{\mathbf{S}})$ must also preserve and reflect the way-below relation, so $(x, r) \ll (y, s)$ in $\mathbf{B}(X, d), \leqslant^{d^+}$ if and only if $\Downarrow(x, 0) + r$ is way-below $\Downarrow(y, 0) + s$ in $\mathbf{RI}(\mathbf{B}(X, d), \prec^{d^+})$. We note that $\Downarrow(x, 0) + r = \Downarrow(x, r)$: an element (z, t) belongs to the left-hand side iff t can be written $t' + r$, $t' \in \mathbb{R}^+$, so that $(z, t') \prec^{d^+} (x, 0)$, i.e., $d(z, x) < t'$, i.e., $(z, t) \prec^{d^+} (x, r)$. So $(x, r) \ll (y, s)$ iff $\Downarrow(x, r)$ is way-below $\Downarrow(y, s)$. By Proposition 5.1.33, this is equivalent to the existence of $(z, t) \in \Downarrow(y, s)$ such that $\Downarrow(x, r) \subseteq \Downarrow(z, t)$, i.e., to the existence of $(z, t) \prec^{d^+} (y, s)$ such that $(x, r) \leqslant^{d^+} (z, t)$ (the latter coming from the fact that (x, r) is the least upper bound of $\Downarrow(x, r)$, and similarly for (z, t)). Equivalently, $(x, r) \prec^{d^+} (y, s)$. It follows that the way-below relation $\ll$ on $\mathbf{B}(X, d), \leqslant^{d^+}$ coincides with $\prec^{d^+}$. By the Romaguera-Valero Theorem 7.3.11, X, d is Smyth-complete. ☐

Exercise 7.5.16 (Completion of coproducts) Given a family $(X_i, d_i)_{i \in I}$ of hemi-metric spaces, show that $\mathbf{S}(\coprod_{i \in I} X_i, \coprod_{i \in I} d_i)$ is isometric (hence homeomorphic, as well as order isomorphic) to $\coprod_{i \in I} \mathbf{S}(X_i, d_i), \coprod_{i \in I} d_i^+$.

The completion of a product is in general different from the product of the completions (see Exercise 7.5.12), except in the metric case, or for finite products (see Exercise 7.5.31).

7.5.2 Free Yoneda-complete spaces, and extension by continuity

If d is a metric, then $d^+\mathcal{H}$ is also a metric on the formal ball completion, despite the apparent asymmetry of its definition.

Lemma 7.5.17 *Let X, d be a metric space. Then the formal ball completion $\mathbf{S}(X, d), d^+\mathcal{H}$ is also a metric space.*

Proof We must show that $d^+\mathcal{H}(D, D') = d^+\mathcal{H}(D', D)$ whenever D, D' are rounded ideals of formal balls with aperture 0. By symmetry, it is enough to show that $d^+\mathcal{H}(D, D') \leqslant d^+\mathcal{H}(D', D)$, i.e., that, for every formal ball $(x, r) \in D$, $d^+((x, r), D') \leqslant d^+\mathcal{H}(D', D)$. If $a = d^+\mathcal{H}(D', D) = +\infty$, this is clear. Otherwise, fix $\epsilon > 0$, and let us show that $d^+((x, r), D') < a + \epsilon$: since ϵ is arbitrary, this will prove that $d^+((x, r), D') \leqslant a$, hence the lemma.

For every $(x', r') \in D'$, $d^+((x', r'), D) \leqslant a$, so there is a $(y, s) \in D$ such that $d^+((x', r'), (y, s)) < a + \epsilon/3$. Since $\alpha(D') = 0$, we can pick (x', r') so that $r' < \epsilon/3$. Since D is directed, we can assume that (y, s) is above (x, r): replacing (y, s) by a larger formal ball can only decrease $d^+((x', r'), (y, s))$.

So $d(x', y) < r' - s + a + \epsilon/3$, in particular $d(x', y) < a - s + 2\epsilon/3$, with $(x, r) \leqslant^{d^+} (y, s)$. Since d is a metric, $d(y, x') < a - s + 2\epsilon/3$. Since $(x, r) \leqslant^{d^+} (y, s)$, $d(x, y) \leqslant r - s$. So $d(x, x') \leqslant d(x, y) + d(y, x') < a + r - 2s + 2\epsilon/3$. So $d^+((x, r), (x', r')) \leqslant \max(a + r' - 2s + 2\epsilon/3, 0) < a + \epsilon$. It follows that $d^+((x, r), D') < a + \epsilon$, as claimed. □

Proposition 7.5.18 *Let X, d be a metric space. Its formal ball completion $\mathbf{S}(X, d), d^+{}_{\mathcal{H}}$ is a complete metric space, and the image $\eta^{\mathbf{S}}_X[X]$ is dense in it.*

Proof By Lemma 7.5.17, $\mathbf{S}(X, d), d^+{}_{\mathcal{H}}$ is a metric space. It is also Yoneda-complete by Corollary 7.5.6, hence complete by Lemma 7.4.3. Since $d^+{}_{\mathcal{H}}$-limits coincide with ordinary limits in the open ball topology of $\mathbf{S}(X, d), d^+{}_{\mathcal{H}}$ (by Corollary 7.1.21, using the fact that $d^+{}_{\mathcal{H}}$ is a metric), Lemma 7.5.10 states that $\eta^{\mathbf{S}}_X[X]$ is dense in the latter. □

By the extension by continuity Lemma 7.3.19, we obtain immediately:

Proposition 7.5.19 *Let X, d be a metric space, and Y, d' be a complete metric space. Every uniformly continuous map f from X, d to Y, d' has a unique continuous extension g from $\mathbf{S}(X, d), d^+{}_{\mathcal{H}}$ to Y, d', i.e., $g \circ \eta^{\mathbf{S}}_X = f$. Moreover, g is uniformly continuous, and if f is c-Lipschitz, then so is g.*

Corollary 7.5.20 (Free complete metric space) *Let X, d be a metric space. The formal ball completion $\mathbf{S}(X, d), d^+{}_{\mathcal{H}}$ is the* free complete metric space *over X, d: for every complete metric space Y, d' and every 1-Lipschitz map $f : X \to Y$, f extends to a unique 1-Lipschitz map $g : \mathbf{S}(X, d) \to Y$.*

We have already observed that the Cauchy completion $Chy(X, d)$ is the free complete metric space over X, d (see remark before Exercise 7.3.17). So we have two apparently competing definitions of the free complete metric space. The following exercise shows that the two notions match.

> **Exercise 7.5.21** (Metric ⇒ formal ball completion = Cauchy completion) Show that, when X, d is metric, $Chy(X, d)$ and $\mathbf{S}(X, d)$ are isometric, and even naturally in X, d in **Met**. (Reason categorically.)

While Proposition 7.5.19 is only valid for metric spaces, we can extend it to hemi-metric spaces, by replacing uniformly continuous maps by Yoneda-continuous maps.

Proposition 7.5.22 (Extension by continuity, formal ball completion) *Let X, d be a hemi-metric space, and Y, d' be a Yoneda-complete quasi-metric space. Every uniformly continuous map f from X, d to Y, d' has a unique Yoneda-continuous extension g from $\mathbf{S}(X, d), d^+{}_{\mathcal{H}}$ to Y, d', i.e., $g \circ \eta^{\mathbf{S}}_X = f$. Moreover, if f is c-Lipschitz, then so is g.*

Proof For every $D \in \mathbf{S}(X, d)$, $(\eta_X^{\mathbf{S}}(x), r)_{(x,r)\in D, \leqslant^{d+}}$ is a Cauchy-weighted net, so $(\eta_X^{\mathbf{S}}(x))_{(x,r)\in D, \leqslant^{d+}}$ is Cauchy-weightable. By Lemma 7.5.10, D is its $d^+{}_{\mathcal{H}}$-limit. So, if g exists, then $g(D)$ should be the d'-limit of $(g(\eta_X^{\mathbf{S}}(x)))_{(x,r)\in D, \leqslant^{d+}}$, i.e., of $(f(x))_{(x,r)\in D, \leqslant^{d+}}$. Note that $(f(x))_{(x,r)\in D, \leqslant^{d+}}$ is Cauchy, because f is uniformly continuous: for every $\epsilon > 0$, let $\eta > 0$ be such that $d(x, x') < \eta$ implies $d'(f(x), f(x')) < \epsilon$, then, for all (x, r), (x', r') in D such that $(x, r) \leqslant^{d^+} (x', r')$ and $r < \eta$, then $d(x, x') \leqslant r - r' < \eta$, so $d(f(x), f(x')) < \epsilon$.

So g is unique, if it exists. We show existence by defining $g(D)$ as the d'-limit of the Cauchy net $(f(x))_{(x,r)\in D, \leqslant^{d+}}$. We first check the last claim of the Proposition, which is easiest: if f is c-Lipschitz, then so is g. If f is c-Lipschitz, then $\mathbf{B}^c(f)$ is c-Lipschitz as well, by Lemma 7.3.16. Let $D, D' \in \mathbf{S}(X, d)$. By definition of $g(D)$ as a d'-limit, and using Lemma 7.4.9, $d'(g(D), g(D')) = \sup_{(x,r)\in D}(d'(f(x), g(D')) - c.r)$. The number $d'(f(x), g(D'))$ is at most $d'(f(x), f(x')) + d'(f(x'), g(D'))$ for any $(x', r') \in D'$. By Lemma 7.4.9 again, $0 = d'(g(D'), g(D')) = \sup_{(x',r')\in D'} (d'(f(x'), g(D')) - c.r')$, so $d'(f(x'), g(D')) \leqslant c.r'$ for every $(x', r') \in D'$. Taking greatest lower bounds, $d'(f(x), g(D')) \leqslant \inf_{(x',r')\in D'}(d'(f(x), f(x')) + c.r') \leqslant \inf_{(x',r')\in D'}(c.d(x, x') + c.r')$. Therefore, $d'(g(D), g(D')) \leqslant \sup_{(x,r)\in D} \inf_{(x',r')\in D'}(c.d(x, x') + c.r' - c.r) \leqslant c.d^+{}_{\mathcal{H}}(D, D')$, whence g is c-Lipschitz.

In the general case, where f is only uniformly continuous, we show the weaker claim that g is uniformly continuous. For every $\epsilon > 0$, let $\eta > 0$ be such that, whenever $d(x, x') < \eta$, $d'(f(x), f(x')) < \epsilon/3$. Assume that $d^+{}_{\mathcal{H}}(D, D') < \eta$. Let $\eta_0 < \eta$ be such that $d^+{}_{\mathcal{H}}(D, D') < \eta_0$. By definition of $d^+{}_{\mathcal{H}}$, for every $(x, r) \in D$, there is an $(x', r') \in D'$ such that $d^+((x, r), (x', r')) \leqslant \eta_0$. For every $(x'', r'') \in D'$ above (x', r') (with respect to $\leqslant^{d^+}$), $d^+((x, r), (x'', r'')) \leqslant d^+((x, r), (x', r')) + d^+((x', r'), (x'', r'')) \leqslant \eta_0$. In other words, for every $(x, r) \in D$: $(*)$ for (x', r') large enough, $d^+((x, r), (x', r')) \leqslant \eta_0$. On the other hand, we have $r < \eta - \eta_0$ for (x, r) large enough, since D has aperture 0. For any such fixed $(x, r) \in D$ large enough, by $(*)$, $d(x, x') \leqslant r - r' + \eta_0 < \eta$ for (x', r') large enough in D'. So $d'(f(x), f(x')) < \epsilon/3$ for (x', r') large enough in D'. Using the definition of $g(D')$ as a d'-limit, $0 = d'(g(D'), g(D')) = \limsup_{(x',r')\in D', \leqslant^{d+}} d'(f(x'), g(D'))$, so $d'(f(x'), g(D')) < \epsilon/3$ for $(x', r') \in D'$ large enough. Picking one of these (x', r'), we obtain $d'(f(x), g(D')) \leqslant d'(f(x), f(x')) + d'(f(x'), g(D')) < 2\epsilon/3$. Since this holds for $(x, r) \in D$ large enough, $\limsup_{(x,r)\in D, d^+} d'(f(x), g(D')) \leqslant 2\epsilon/3 < \epsilon$, i.e., $d'(g(D), g(D')) < \epsilon$, using the definition of $g(D)$ as a d'-limit. So g is uniformly continuous.

Finally, we show that g is Yoneda-continuous. Let $(D_i, r_i)_{i\in I,\sqsubseteq}$ be a Cauchy-weighted net in $\mathbf{S}(X, d), d^+{}_{\mathcal{H}}$. By Lemma 7.5.5, the $d^+{}_{\mathcal{H}}$-limit D of $(D_i)_{i\in I,\sqsubseteq}$ is $\bigcup_{i\in I} D_i + r_i$, and the family $(D_i + r_i)_{i\in I,\sqsubseteq}$ is directed.

Fix $y \in Y$, and let $c = \limsup_{i\in I,\sqsubseteq} d'(g(D_i), y)$. For every $a \in \mathbb{R}^+$ with $a < c$, we claim that $d'(g(D), y) \geqslant a$. We show this by showing that, for every $(x, r) \in D$, there is an (x', r') above (x, r) in D such that $d'(f(x'), y) > a$, using the fact that $d'(g(D), y) = \limsup_{(x,r)\in D,\leqslant^{d+}} d'(f(x), y)$, by the definition of $g(D)$ as a d'-limit. Given $(x, r) \in D$, there is an $i \in I$ such that $(x, r) \in D_i + r_i$. Since $a < c \leqslant \sup_{j\in I, i\sqsubseteq j} d'(g(D_j), y)$, there is a $j \in I$ above i such that $a < d'(g(D_j), y)$. Since $(x, r) \in D_i + r_i \subseteq D_j + r_j$, $r \geqslant r_j$ and $(x, r - r_j)$ is in D_j. Since $g(D_j)$ is defined as a d'-limit, $d'(g(D_j), y) = \limsup_{(x',r')\in D_j,\leqslant^{d+}} d'(f(x'), y) \leqslant \sup_{(x',r')\in D_j,(x,r-r_j)\leqslant^{d+}(x',r')} d'(f(x'), y)$, so there is an (x', r') in D_j above $(x, r - r_j)$ such that $a < d'(f(x'), y)$. Since (x', r') is above $(x, r - r_j)$, $d(x, x') \leqslant r - r_j - r' \leqslant r - r'$, so (x', r') is also above (x, r), and the claim is proved.

Since a is arbitrary such that $a < c$, we obtain $d'(g(D), y) \geqslant c$. We claim that $d'(g(D), y) = c$. Otherwise, $d'(g(D), y) > c$. Let $b > c$ be such that $d'(g(D), y) > b$. Let $\epsilon = (b - c)/2 > 0$, and let $\eta > 0$ be such that, whenever $d(x, x') < \eta$, $d'(f(x), f(x')) < \epsilon$. For $(x, r) \in D$ large enough, we can require $r < \eta$. Also, (x, r) is in $D_i + r_i$ for some i, hence for i large enough (say $i_0 \sqsubseteq i$), since $(D_i + r_i)_{i\in I,\sqsubseteq}$ is directed. Note that $r \geqslant r_i$ and $(x, r - r_i)$ is in D_i. By the definition of c, for i large enough, $d'(g(D_i), y) < b$. Without loss of generality, we may assume that this is the case whenever $i_0 \sqsubseteq i$. By the definition of $g(D_i)$ as a d'-limit, for $(x', r') \in D_i$ large enough, $d'(f(x'), y) < b$. In particular, this holds for $(x', r') \in D_i$ large enough above $(x, r - r_i)$. So $d(x, x') \leqslant r - r_i - r' < \eta$, so $d'(f(x), y) \leqslant d'(f(x), f(x')) + d'(f(x'), y) < \epsilon + b$. That is, for $(x, r) \in D$ large enough, $d'(f(x), y) < \epsilon + b$. So $d'(g(D), y) \leqslant \epsilon + b < c$, a contradiction.

So, for every $y \in Y$, $d'(g(D), y) = c = \limsup_{i\in I,\sqsubseteq} d'(g(D_i), y)$. This means that $g(D)$ is the d'-limit of $(g(D_i))_{i\in I,\sqsubseteq}$. Therefore g is Yoneda-continuous. □

Exercise 7.5.29 will be used to obtain a refined view of the above Proposition. In Lemma 7.3.19, we have seen that, in metric spaces, one could extend any uniformly continuous map from a dense subset A of X to a complete metric space, to the whole space X. We shall see that a similar result is obtained by replacing the dense subset A by a strong basis, and using Yoneda-continuity. In Proposition 7.5.22, the strong basis is the image of X by $\eta_X^{\mathbf{S}}$; see Exercise 7.5.11.

In categorical words, for every hemi-metric space X, d, $\mathbf{S}(X, d), d^{+}{}_{\mathcal{H}}$ is the free Yoneda-complete quasi-metric space over X, d. But there is a subtle point: while X, d is an object in the category of hemi-metric spaces and uniformly continuous maps, $\mathbf{S}(X, d), d^{+}{}_{\mathcal{H}}$ is in the category of Yoneda-complete quasi-metric spaces and *Yoneda-continuous* maps. In other words:

Fact 7.5.23 (Free Yoneda-complete quasi-metric space) *Let* $\mathbf{HMet}_u$ *be the category of hemi-metric spaces and uniformly continuous maps, and let* $\mathbf{YCQMet}_{u,\mathbf{Y}}$ *be the category of Yoneda-complete quasi-metric spaces and Yoneda-continuous maps. Similarly, let* $\mathbf{YCQMet}_{\mathbf{Y}}$ *be the category of Yoneda-complete quasi-metric spaces and* 1*-Lipschitz Yoneda-continuous maps.*

The underlying functor from $\mathbf{YCQMet}_{u,\mathbf{Y}}$ *to* $\mathbf{HMet}_u$ *(resp., from* $\mathbf{YCQMet}_{\mathbf{Y}}$ *to* $\mathbf{HMet}$*) has a left adjoint* $\mathbf{S}$*, with unit* $\eta^{\mathbf{S}}$*.*

Exercise 7.5.24 Let X and Y be two posets, and write their orderings $\leqslant$. Show that f is uniformly continuous (resp., Yoneda-continuous) from $X, d_{\leqslant}$ to $Y, d_{\leqslant}$ if and only if f is monotonic (resp., Scott-continuous).

Exercise 7.5.25 (Ideal completion as formal ball completion) Recall from Exercise 7.4.30 that given any poset X, with ordering $\leqslant$, the quasi-metric space $X, d_{\leqslant}$ is Yoneda-complete if and only if X is a dcpo. We now have two ways of building a completion for X: as a quasi-metric space, obtaining $\mathbf{S}(X, d_{\leqslant}), d^{+}_{\leqslant\mathcal{H}}$, or as a poset, building the ideal completion $\mathbf{I}(X)$. Show that the two constructions produce isomorphic objects: show that $d^{+}_{\leqslant\mathcal{H}}$ is the quasi-metric $d_{\subseteq}$ induced from the inclusion ordering $\subseteq$, and that $\mathbf{S}(X, d_{\leqslant}), \subseteq$ and $\mathbf{I}(X), \subseteq$ are naturally order isomorphic. (The categorical way of proving this is shorter.)

The purpose of the following exercises is to give another construction of the completion of a hemi-metric space, introduced by Bonsangue *et al.* (1998), and inspired by a fundamental result in category theory known as the Yoneda Lemma.

Exercise 7.5.26 Let $\eta_X^{\mathbf{Y}}$ map every $x \in X$ to the function $d(\_, x)$. Show that $\sup_X \mathsf{d}_{\mathbb{R}}^{op}(\eta_X^{\mathbf{Y}}(x), f) = f(x)$ for every $f \in [X \to \overline{\mathbb{R}^+}]_1$. (This is the Yoneda Lemma, in its so-called $\mathsf{d}_{\mathbb{R}}^{op}$-enriched form.) Deduce that $\eta_X^{\mathbf{Y}}$ is an isometric embedding of X, d into $[X \to \overline{\mathbb{R}^+}]_1, \sup_X \mathsf{d}_{\mathbb{R}}^{op}$.

Exercise 7.5.27 (Yoneda completion) Define the *Yoneda completion* $\mathbf{Y}(X, d)$ of the hemi-metric space X, d as the $\sup_X \mathsf{d}_{\mathbb{R}}^{op}$-closure of $\eta_X^{\mathbf{Y}}[X]$ in $[X \to \overline{\mathbb{R}^+}]_1$. Show that $\mathbf{Y}(X, d), \sup_X \mathsf{d}_{\mathbb{R}}^{op}$ is Yoneda-complete, and that $\sup_X \mathsf{d}_{\mathbb{R}}^{op}$-limits of Cauchy nets are computed componentwise.

Exercise 7.5.28 Let X, d be a hemi-metric space. Show that, for every $x \in X$, $\eta_X^{\mathbf{Y}}(x)$ is $\sup_X \mathsf{d}_{\mathbb{R}}^{op}$-finite in $\mathbf{Y}(X, d)$.

Exercise 7.5.29 (Extension by continuity, algebraic case) In Lemma 7.3.19, we showed that any uniformly continuous map from a dense subset of a metric space to a complete metric space had a unique extension. The purpose of this exercise is to prove an analogous statement for hemi-metric spaces; algebraicity will turn out to be important here.

Let X, d be an algebraic hemi-metric space, and B be a strong basis of X. Let Y, d' be a Yoneda-complete quasi-metric space, and f be a uniformly continuous map from B, d to Y, d'. Show that f has a unique Yoneda-continuous extension g from X, d to Y, d'. Moreover, if f is c-Lipschitz, then so is g.

Exercise 7.5.30 (Yoneda completion = formal ball completion) Show the following result, similar to Proposition 7.5.22, but with the Yoneda completion instead of the formal ball completion. Let X, d be a hemi-metric space, and Y, d' be a Yoneda-complete quasi-metric space. Prove that every uniformly continuous map f from X, d to Y, d' has a unique Yoneda-continuous extension g from $\mathbf{Y}(X, d), \sup_X \mathsf{d}_{\mathbb{R}}^{op}$ to Y, d', i.e., $g \circ \eta_X^{\mathbf{Y}} = f$. Moreover, if f is c-Lipschitz, then so is g.

Conclude that $\mathbf{S}(X, d), d^+{}_{\mathcal{H}}$ is naturally isometric to $\mathbf{Y}(X, d), \sup_X \mathsf{d}_{\mathbb{R}}^{op}$.

Exercise 7.5.31 (Completion of products) Let X_i, d_i be hemi-metric spaces, $i \in I$, and d be the sup hemi-metric $d(\vec{x}, \vec{y}) = \sup_{i \in I} d_i(x_i, y_i)$ on $\prod_{i \in I} X_i$. Let d' be the sup hemi-metric $d'(\vec{D}, \vec{D}') = \sup_{i \in I} d_i^+{}_{\mathcal{H}}(D_i, D_i')$ on $\prod_{i \in I} \mathbf{S}(X_i, d_i)$. Using extension by continuity results, and suitable strong bases, show that:

1. If each X_i has a least element $\bot_i$ in $\leqslant^{d_i}$, then the product $\prod_{i \in I} \mathbf{S}(X_i, d_i), d'$ is isometric to the $d^+{}_{\mathcal{H}}$-closure Z of $\{\eta^{\mathbf{S}}_{\prod_{i \in I} X_i}(\vec{x}) \mid \vec{x} \text{ almost bottom in } \prod_{i \in I} X_i\}$ in $\mathbf{S}(\prod_{i \in I} X_i, d), d^+{}_{\mathcal{H}}$. (This cannot be isometric to the whole completion $\mathbf{S}(\prod_{i \in I} X_i, d), d^+{}_{\mathcal{H}}$, by Exercise 7.5.12.)
2. If each hemi-metric space X_i, d_i is metric, then $\mathbf{S}(\prod_{i \in I} X_i, d), d^+{}_{\mathcal{H}}$ *is* isometric to $\prod_{i \in I} \mathbf{S}(X_i, d_i), d'$.
3. If I is finite, similarly, $\mathbf{S}(\prod_{i \in I} X_i, d), d^+{}_{\mathcal{H}}$ *is* isometric to $\prod_{i \in I} \mathbf{S}(X_i, d_i), d'$.

7.6 Choquet-completeness

One can extend the notion of completeness to even larger classes of spaces in many ways. These include Čech-completeness, Rudin-completeness, and others; see, for example, Bennett *et al.* (2008) (and Exercise 7.6.21). We shall focus on Choquet-completeness. Be aware, though, that although Choquet-completeness will coincide with completeness on metric spaces, it will be a weaker notion than either Smyth- or Yoneda-completeness on quasi-metric spaces.

7.6.1 The strong Choquet game

The definition rests on the so-called *strong Choquet game*. This is a game played on a general topological space X by two players, α and β. Player β starts the play, by picking a point x_0 and an open neighborhood V_0 of x_0. Then α must produce a smaller open neighborhood U_0 of x_0, i.e., one such that $U_0 \subseteq V_0$. Player β must then produce a new point x_1 in U_0, and a new open neighborhood V_1 of x_1, included in U_0. Playing the game defines a sequence of points $(x_n)_{n \in \mathbb{N}}$, and two sequences of opens $(U_n)_{n \in \mathbb{N}}$ and $(V_n)_{n \in \mathbb{N}}$, such that $V_0 \supseteq U_0 \supseteq V_1 \supseteq U_1 \supseteq V_2 \supseteq \cdots$, and $x_n \in V_n$ for every $n \in \mathbb{N}$. We let the two players play for infinitely long. Then we declare the winner of the game to be α if and only if $\bigcap_{n=0}^{+\infty} U_n$ is non-empty (equivalently, $\bigcap_{n=0}^{+\infty} V_n$ is non-empty), and β otherwise.

So β tries to change the open V_n and the point x_n so that eventually no point remains in the intersection $\bigcap_{n=0}^{+\infty} V_n$, while α tries to force the existence of at least one point in the intersection. The game is asymmetric: α cannot change the point x_n, and can only find an open U_n included in V_n, in the hope of preventing β from choosing certain points x_{n+1} at the next step. Moreover, α cannot prevent β from choosing x_n again, as U_n must be an open neighborhood of x_n.

It is profitable to imagine that players play according to some *strategy*. Intuitively, a strategy for one of the players is a map from histories to their next move, where the histories are the finite sequences of moves played in the past. Formally, a *strategy for player* α, a.k.a. an α*-strategy*, is a map σ from sequences of the form $x_0, V_0, U_0, x_1, V_1, U_1, x_2, V_2, \ldots, x_n, V_n$ where $V_0 \supseteq U_0 \supseteq V_1 \supseteq U_1 \supseteq V_2 \supseteq \cdots \supseteq V_n$ is a decreasing sequence of opens and $x_0 \in U_0$, $x_1 \in U_1$, $x_2 \in U_2$, $\ldots$, $x_{n-1} \in U_{n-1}$, $x_n \in V_n$, $n \in \mathbb{N}$ (we call such sequences the α*-histories*) to open subsets U_n with $x_n \in U_n \subseteq V_n$. Similarly, a β*-history* is any sequence of the form $x_0, V_0, U_0, x_1, V_1, U_1, x_2, V_2, \ldots, x_{n-1}, V_{n-1}, U_{n-1}$ where $V_0 \supseteq U_0 \supseteq V_1 \supseteq U_1 \supseteq V_2 \supseteq \cdots \supseteq V_{n-1} \supseteq U_{n-1}$ is a decreasing sequence of opens and $x_0 \in U_0$, $x_1 \in U_1$, $x_2 \in U_2$, $\ldots$, $x_{n-1} \in U_{n-1}$, $n \in \mathbb{N}$, and we define the β*-strategies* as maps τ from such β-histories to pairs (x_n, V_n) where V_n is an open subset of U_{n-1} (if $n \geqslant 1$; otherwise V_n is an arbitrary open subset) and x_n is an element of V_n.

Given an α-strategy σ and a β-strategy τ, the *value* of the strong Choquet game is $\bigcap_{n=0}^{+\infty} V_n = \bigcap_{n=0}^{+\infty} U_n$, where U_n and V_n are defined in the expected way, namely, by induction on n, using the formulae $U_n = \sigma(x_0, V_0, U_0, x_1, V_1, U_1, x_2, V_2, \ldots, x_n, V_n)$ and $(x_n, V_n) = \tau(x_0, V_0, U_0, x_1, V_1, U_1, x_2, V_2, \ldots, x_{n-1}, V_{n-1}, U_{n-1})$. We say that σ is a *winning strategy* for α if and only if α wins the game whatever the β-strategy τ, i.e., provided the value of the game is non-empty for every β-strategy τ.

Definition 7.6.1 (Choquet-completeness) A topological space X is *Choquet-complete* if and only if player α has a winning strategy in the strong Choquet game on X.

Proposition 7.6.2 *Every Yoneda-complete quasi-metric space, in particular every complete metric space, is Choquet-complete in its open ball topology.*

Proof Let X, d be a Yoneda-complete quasi-metric space. By the Kostanek-Waszkiewicz Theorem 7.4.27, $\mathbf{B}(X, d)$ is a dcpo in the ordering $\leqslant^{d^+}$.

The easiest way to explain the proof is to imagine for now that α plays radii r_n of formal balls centered at x_n, in such a way that, whatever β plays, the sequence of formal balls (x_n, r_n) is increasing in $\prec^{d^+}$. Then the opens U_n will be the actual balls $B^d_{x_n, <r_n}$. At step n, assume an α-history $x_0, V_0, U_0 = B^d_{x_0, <r_0}, x_1, V_1, U_1 = B^d_{x_1, <r_1}, \ldots, x_{n-1}, V_{n-1}, U_{n-1} = B^d_{x_{n-1}, <r_{n-1}}, x_n, V_n$. Since $x_n \in V_n$, by Lemma 6.1.5 there is an open ball centered at x_n and included in V_n. Let us choose its radius $r_n > 0$ so that not only $U_n = B^d_{x_n, <r_n} \subseteq V_n$ but also $r_n < r_{n-1} - d(x_{n-1}, x_n)$ whenever $n \geqslant 1$. This is possible since β must have picked x_n from $U_{n-1} = B^d_{x_{n-1}, <r_{n-1}}$, so $d(x_{n-1}, x_n) < r_{n-1}$, i.e., $r_{n-1} - d(x_{n-1}, x_n) > 0$.

By construction, $(x_0, r_0) \prec^{d^+} (x_1, r_1) \prec^{d^+} \cdots \prec^{d^+} (x_n, r_n) \prec^{d^+} \cdots$, so the sequence has a least upper bound (x, r), since $\mathbf{B}(X, d)$ is a dcpo. For every $n \in \mathbb{N}$, $(x_n, r_n) \prec^{d^+} (x, r)$, so $d(x_n, x) < r_n - r \leqslant r_n$, that is, $x \in B^d_{x_n, <r_n} = U_n$. So $\bigcap_{n \in \mathbb{N}} U_n$ is non-empty.

This argument is slightly incorrect, as r_n is computed in terms of r_{n-1}, notably, and this much information is not part of the α-history. However this can be recomputed. In other words, define $U_n = \sigma(x_0, V_0, U_0, x_1, V_1, U_1, x_2, V_2, \ldots, x_n, V_n)$ as follows. If $n = 0$, then U_0 is $B^d_{x_0, <r_0}$ where $r_0 > 0$ is any radius such that $B^d_{x_0, <r_0} \subseteq V_0$. If $n \geqslant 1$ and U_{n-1} is of the form $B^d_{x_{n-1}, <r_{n-1}}$ for some $r_{n-1} > 0$, then let U_n be $B^d_{x_n, <r_n}$ where $r_n > 0$ is any radius such that $B^d_{x_n, <r_n} \subseteq V_n$ and $r_n < r_{n-1} - d(x_{n-1}, x_n)$. (Use the Axiom of Choice to resolve the two "any"s and the "some" above into actual functions.) $\square$

We shall also see in Proposition 8.3.24 that all locally compact sober spaces are Choquet-complete. Although the following is a special case of this, we give an elementary proof.

Lemma 7.6.3 *Every continuous dcpo is Choquet-complete in its Scott topology.*

Moreover, we can require that α's winning strategy σ is stationary, *i.e., that $U_n = \sigma(x_0, V_0, U_0, x_1, V_1, U_1, \dots, x_n, V_n)$ only depends on the last move x_n, V_n. We then write $U_n = \sigma(x_n, V_n)$.*

Proof Let X be a continuous dcpo. Given an α-history $x_0, V_0, U_0, x_1, V_1, U_1, \dots, x_n, V_n$, since x_n is in the Scott-open subset V_n, and X is continuous, there is a point y_n in V_n such that $y_n \ll x_n$. We let α play $U_n = \twoheaduparrow y_n$, and note that (using the Axiom of Choice) this defines a stationary strategy. We realize that, for every $n \in \mathbb{N}$, $y_n \ll y_{n+1}$, since $y_{n+1} \in V_{n+1} \subseteq U_n = \twoheaduparrow y_n$. In particular, $(y_n)_{n\in\mathbb{N}}$ is an increasing sequence, hence has a least upper bound y. Since every Scott-open subset is upward closed, y is in $\bigcap_{n\in\mathbb{N}} V_n$. □

Proposition 7.6.2 has a kind of converse: every Choquet-complete metric space X, d is completely metrizable (see Theorem 7.6.15 below). We cannot require d itself to be a complete metric, though, because of the following.

Exercise 7.6.4 Let $X = (0, 1]$, and consider the following two metrics on X: first, $d(x, y) = |x - y|$; second, $d'(x, y) = |1/x - 1/y|$. Show that X, d is not complete, that X, d' is complete, and that they are topologically equivalent. Conclude that there are Choquet-complete metric spaces whose metric is not itself complete.

Exercise 7.6.5 (Convergence Choquet-complete) We say that player α has a *convergent winning strategy* in the strong Choquet game on the space X (Dorais and Mummert, 2010) iff α can force β to pick the sequence $(x_n)_{n\in\mathbb{N}}$ so that $(V_n)_{n\in\mathbb{N}}$ (or, equivalently, $(U_n)_{n\in\mathbb{N}}$) is a neighborhood base of some point x—using the same notation as above. In particular, x is in $\bigcap_{n\in\mathbb{N}} U_n = \bigcap_{n\in\mathbb{N}} V_n$, so such a strategy is winning. A space X is *convergence Choquet-complete* if and only if α has a convergent winning strategy.

Show that x is then a limit of $(x_n)_{n\in\mathbb{N}}$, justifying the name "convergent." Show that if X, d is Smyth-complete, then X is convergence Choquet-complete, when equipped with the open ball topology of d.

Exercise 7.6.6 Show that, if X is a continuous dcpo, then X_σ is not only Choquet-complete but even convergence Choquet-complete. To be precise, α has a stationary, convergent winning strategy.

Exercise 7.6.7 The winning strategy for α built in the proof of Proposition 7.6.2 is not stationary. On the other hand, if X, d is a complete metric space, show that α has a stationary winning strategy, which we can also take to be convergent. Use Cantor's Theorem; see Exercise 7.3.6.

Theorem 7.6.8 (Baire) *A topological space X is* Baire *if and only if the intersection of countably many dense open subsets of X is still dense in X.*

Every Choquet-complete space is Baire.

Proof Let W_n be dense open subsets in X, $n \in \mathbb{N}$. Given any non-empty open subset V of X, define the following strong Choquet game: initially, β plays $V_0 = V$ and x_0 is any point of V; at steps $n \geqslant 1$, given the last non-empty open subset U_{n-1} played by α, U_{n-1} meets W_{n-1} since W_{n-1} is dense (using Lemma 4.1.30), and β plays $V_n = U_{n-1} \cap W_{n-1}$ and picks any point for x_n from V_n. Since X is Choquet-complete, there is a strategy for α such that the value $\bigcap_{n \in \mathbb{N}} V_n$ of the game is non-empty. That is, $V \cap \bigcap_{n \geqslant 1}(U_{n-1} \cap W_{n-1})$ is non-empty. In particular, the larger set $V \cap \bigcap_{n \in \mathbb{N}} W_n$ is non-empty. Since V is arbitrary, and using Lemma 4.1.30 again, $\bigcap_{n \in \mathbb{N}} W_n$ is dense. □

Combining this with Lemma 7.6.3, every continuous dcpo is Baire in its Scott topology. Combining this with Proposition 7.6.2 instead, every Yoneda-completely quasi-metrizable space is Baire.

Example 7.6.9 (Baire space is Baire) $\mathbb{N}, d_=$ is a trivially complete metric, and countable products of complete metric spaces are complete metric (Theorem 7.2.31, Item 3). So the Baire space $\mathcal{B} = \mathbb{N}^{\mathbb{N}}$ is Choquet-complete, hence Baire.

Exercise 7.6.10 ($\mathbb{Q}$ is not Baire) Show that $\mathbb{Q}$ with its ordinary metric topology is not Baire.

Exercise 7.6.11 ($\mathbb{R}_\ell$ is Baire) Show that the Sorgenfrey line $\mathbb{R}_\ell$ is Choquet-complete, hence Baire.

Exercise 7.6.12 (Banach–Mazur game) A *Banach–Mazur game* on a topological space is played exactly as the strong Choquet game, except that β only provides a non-empty open set V_0 initially (and no point x_0 of V_0), and only a non-empty open subset V_n of U_{n-1} at step n (and no point x_n of V_n); α is only required to provide a non-empty open subset U_n of V_n for each $n \in \mathbb{N}$. Winning is defined as for the strong Choquet game: a winning strategy for α is one such that, for every β-strategy, the value of the game is non-empty. Symmetrically, a strategy for β is winning if, for every α-strategy, the value of the game is empty.

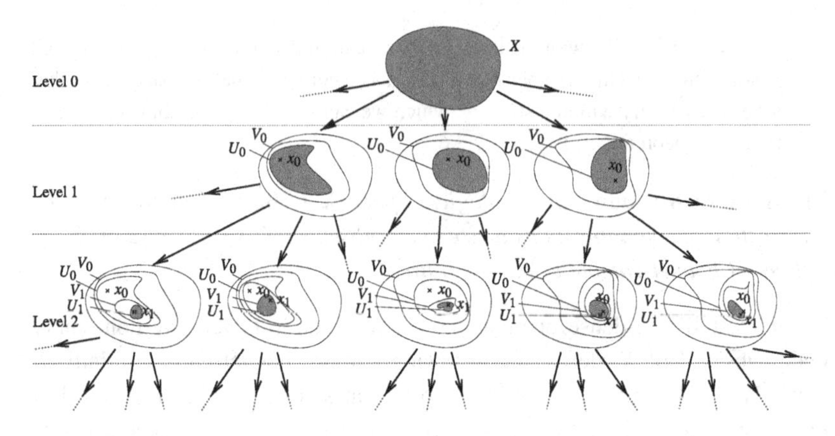

Figure 7.4 An excerpt from the tree $\mathcal{T}(\sigma)$.

Consider the following claims: (*a*) X is Choquet-complete; (*b*) player α has a winning strategy in the Banach–Mazur game on X; (*c*) player β does not have a winning strategy in the Banach–Mazur game on X; (*d*) X is Baire. Show that (*a*) implies (*b*), that (*b*) implies (*c*), and that (*c*) implies (*d*). (Claim (*c*) is in fact equivalent to (*d*); see Exercise 7.6.14.)

7.6.2 Game trees, G_δ subsets, and complete metrizability

Each α-strategy σ gives rise to an infinite tree $\mathcal{T}(\sigma)$ as follows (see Figure 7.4). Its nodes will be in one-to-one correspondence with β-histories $x_0, V_0, U_0, x_1, V_1, U_1, x_2, V_2, \ldots, x_{n-1}, V_{n-1}, U_{n-1}$ that are σ-*compatible*, i.e., such that $\sigma(x_0, V_0, U_0, x_1, V_1, U_1, x_2, V_2, \ldots, x_i, V_i) = U_i$ for each i, $0 \leqslant i \leqslant n-1$. First, at level 0, there is a root node, labeled with the empty sequence. We build as many nodes at level 1, i.e., successors of the root node, as there are possible initial moves x_0, V_0 for β: these nodes are labeled with the β-history x_0, V_0, U_0, where $U_0 = \sigma(x_0, V_0)$. Any node at level $n \geqslant 1$, labeled by a β-history $x_0, V_0, U_0, x_1, V_1, U_1, x_2, V_2, \ldots, x_{n-1}, V_{n-1}, U_{n-1}$, has as many successors as there are pairs x_n, V_n of a possible next move by β, i.e., of pairs of a point $x_n \in U_{n-1}$ and an open neighborhood V_n of x_n included in U_n. The corresponding successor node is then labeled with the β-history $x_0, V_0, U_0, x_1, V_1, U_1, x_2, V_2, \ldots, x_{n-1}, V_{n-1}, U_{n-1}, x_n, V_n, U_n$ where $U_n = \sigma(x_0, V_0, U_0, x_1, V_1, U_1, x_2, V_2, \ldots, x_{n-1}, V_{n-1}, U_{n-1}, x_n, V_n)$.

A *branch* of the tree is any maximal path starting from the root. The branches B are in close correspondence with the strategies τ for β on Z. Indeed, given any β-strategy τ, one defines a branch B by starting from the root node, going to the node x_0, V_0, U_0 obtained by letting β play to obtain x_0, V_0 and $U_0 = \sigma(x_0, V_0)$, then going to the node $x_0, V_0, U_0, x_1, V_1, U_1$ where x_1, V_1 is obtained by letting β play and $U_1 = \sigma(x_0, V_0, U_0, x_1, V_1)$, etc. Conversely, given any branch B, one defines a strategy τ by letting $\tau(x_0, V_0, U_0, x_1, V_1, U_1, x_2, V_2, \ldots, x_{n-1}, V_{n-1}, U_{n-1})$ be x_n, V_n iff there is a (necessary unique) node labeled $x_0, V_0, U_0, x_1, V_1, U_1, x_2, V_2, \ldots, x_{n-1}, V_{n-1}, U_{n-1}, x_n, V_n, U_n$ on the branch B; otherwise, $\tau(x_0, V_0, U_0, x_1, V_1, U_1, x_2, V_2, \ldots, x_{n-1}, V_{n-1}, U_{n-1})$ is arbitrary.

Each branch B defines a unique infinite sequence $x_0, V_0, U_0, x_1, V_1, U_1, x_2, V_2, \ldots, x_{n-1}, V_{n-1}, U_{n-1}, \ldots$, which is the complete play obtained by letting β play using τ. If σ is winning for α, then the corresponding value of the game, i.e., the intersection $val(B)$ of all U_ns, or equivalently all V_ns, is non-empty.

We can now show the following. We have already encountered the G_δ subsets, i.e., the intersections of countably many open subsets, in Exercise 6.3.20.

Proposition 7.6.13 *Every Choquet-complete subspace of a metric space X, d is G_δ in X.*

Proof Let Z be a Choquet-complete subspace of X, d. We shall need to pay special attention to whether we work in Z or in X. We write $B^d_{x,<\epsilon}$ for the open ball of radius ϵ centered at x in X, as usual. The corresponding open ball *in* Z is $B^d_{x,<\epsilon} \cap Z$.

Fix a winning strategy σ for player α on Z, and consider the tree $\mathcal{T}(\sigma)$. Each node N at level $n \geqslant 1$ is labeled by some partial play $z_0, V_0, U_0, z_1, V_1, U_1, z_2, V_2, \ldots, z_{n-1}, V_{n-1}, U_{n-1}$, and we let $z(N) = z_{n-1}$, $V(N) = V_{n-1}$, $U(N) = U_{n-1}$. We say that such a node is *good* iff $V(N) = V_{n-1}$ is some open ball $B^d_{z_{n-1},<\frac{1}{2^m}} \cap Z = B^d_{z(N),<\frac{1}{2^m}} \cap Z$ in Z, for some $m \geqslant n$. Since $U(N)$ is open in Z, there is an open subset $U'(N)$ of X such that $U(N) = U'(N) \cap Z$. For N good, we shall require that $U'(N)$ be included in $B^d_{z(N),<\frac{1}{2^n}}$, replacing $U'(N)$ by $U'(N) \cap B^d_{z(N),<\frac{1}{2^n}}$ if necessary: note that since $U(N) \subseteq V(N) \subseteq B^d_{z(N),<\frac{1}{2^n}} \cap Z$, $(U'(N) \cap B^d_{z(N),<\frac{1}{2^n}}) \cap Z = U'(N) \cap Z = U(N)$. We pick one such $U'(N)$ for each node N. (The Axiom of Choice is not needed here, e.g., one can pick the largest $U'(N)$ for each N.) For each $n \geqslant 1$, let W_n be the union of the open subsets $U'(N)$ (in X) when N ranges over the good nodes at level n.

For every $x \in \bigcap_{n \geqslant 1} W_n$, we claim that x is in Z. This requires us to define a particular strategy for β.

Since $x \in W_1$, there is a good node N_1 at level 1 such that $x \in U'(N_1)$. Write $V(N_1)$ as $B^d_{z_0, < \frac{1}{2^{m_1}}} \cap Z$, with $z_0 \in Z$ and $m_1 \geqslant 1$. Since x is in $U'(N_1)$, it is outside its closed complement $X \smallsetminus U'(N_1)$, so $d(x, X \smallsetminus U'(N_1)) > 0$, by Lemma 6.1.11. Let $n_1 > 1$ be such that $\frac{1}{2^{n_1}} < \frac{1}{2} d(x, X \smallsetminus U'(N_1))$. Since x is in W_{n_1}, there is a good node N'_1, at some possibly unrelated place in $\mathcal{T}(\sigma)$, such that $x \in U'(N'_1)$. Write $V(N'_1)$ as $B^d_{z(N'_1), < \frac{1}{2^{p_1}}} \cap Z$, where $p_1 \geqslant n_1$. We let β start by playing the point $z(N'_1)$, and the open subset $V(N'_1)$. We claim that $V(N'_1) \subseteq U(N_1)$: for every $y \in V(N'_1)$, $d(z(N'_1), y) < \frac{1}{2^{p_1}} \leqslant \frac{1}{2^{n_1}}$; also $d(z(N'_1), x) < \frac{1}{2^{n_1}}$ since $x \in U'(N'_1) \subseteq B^d_{z(N'_1), < \frac{1}{2^{n_1}}}$; so $d(x, y) \leqslant d(x, z(N'_1)) + d(z(N'_1), y) = d(z(N'_1), x) + d(z(N'_1), y) < \frac{2}{2^{n_1}} < d(x, X \smallsetminus U'(N_1))$, and this would be impossible if y were in $X \smallsetminus U'(N_1)$; so $y \in U'(N_1)$, and as $y \in V(N'_1) \subseteq Z$, y is in $U'(N_1) \cap Z = U(N_1)$.

So there is a successor N_2 of N_1 in $\mathcal{T}(\sigma)$ such that $z(N_2) = z(N'_1)$ and $V(N_2) = V(N'_1)$. Moreover, N_2 is a good node. We let β play along in the same way. That is, we write $V(N_2)$ as $B^d_{z_1, < \frac{1}{2^{m_2}}}$ (i.e., $z_1 = z(N'_1)$, $m_2 = p_1$), let $n_2 > 2$ be such that $\frac{1}{2^{n_2}} < \frac{1}{2} d(x, X \smallsetminus U'(N_2))$, and, using the fact that $x \in W_{n_2}$, find a good node N'_2 such that $x \in U'(N'_2)$. As above, $V(N'_2) \subseteq U(N_2)$, so we now let β play the point $z(N'_2)$ and the open subset $V(N'_2)$. There is a successor N_3 of N_2 with $z(N_3) = z(N'_2)$, $V(N_3) = V(N'_2)$, and this is a good node. We write $V(N_3)$ as $B^d_{z_2, < \frac{1}{2^{m_3}}}$, let $n_3 > 3$ be such that $\frac{1}{2^{n_3}} < \frac{1}{2} d(x, X \smallsetminus U'(N_3))$, and, using the fact that $x \in W_{n_3}$, find a good node N'_3 such that $x \in U'(N'_3)$. Since $V(N'_3) \subseteq U(N_3)$, we then let β play $z(N'_3)$ and $V(N'_3)$, and so on.

This play defines a branch N_0 (the root node of $\mathcal{T}(\sigma)$), $N_1, N_2, N_3, \ldots$, in $\mathcal{T}(\sigma)$, in such a way that x is in $U'(N_k)$ for every $k \geqslant 1$. Also, $U'(N_k)$ is included in an open ball of radius $\frac{1}{2^k}$, so the only point of $\bigcap_{k \geqslant 1} U'(N_k)$ is x: any other point y would be such that $d(x, y) < \frac{1}{2^k}$ for every $k \geqslant 1$, so $d(x, y) = 0$, whence $x = y$ since d is a metric. On the other hand, $\bigcap_{k \geqslant 1} U'(N_k)$ contains $\bigcap_{k \geqslant 1} U(N_k)$, and the latter is non-empty since Z is Choquet-complete. Let z be an element of $\bigcap_{k \geqslant 1} U(N_k)$, in particular $z \in Z$. Since $z \in \bigcap_{k \geqslant 1} U'(N_k)$, $z = x$. So x is in Z.

So $\bigcap_{n \geqslant 1} W_n \subseteq Z$. Conversely, we claim that every $z \in Z$ is in every W_n. Let β start by playing z and $B^d_{z, < \frac{1}{2^{m_1}}} \cap Z$ with $m_1 = 1$. There is a good node N_1, successor of the root node N_0 of $\mathcal{T}(\sigma)$, such that $z(N_1) = z$ and $V(N_1) = B^d_{z, < \frac{1}{2^{m_1}}} \cap Z$. In particular, z is in W_1. Since α's reply $U(N_1)$ contains z, β can then play z again, and some open ball $B^d_{z, < \frac{1}{2^{m_2}}} \cap Z$ included in $U(N_1)$

with $m_2 \geqslant 2$, by Fact 6.1.7. There is a good node N_2, successor of N_1, such that $z(N_2) = z$ and $V(N_2) = B^d_{z,<\frac{1}{2^{m_2}}} \cap Z$. In particular, z is in W_2, and we show that z is in W_n for every $n \geqslant 1$ in the same way, by induction on n. (Note that the ability of β to pick a point, and in particular to keep the same point z at each step, is crucial here, and distinguishes Choquet games from Banach–Mazur games; see Exercise 7.6.12.)

So $Z = \bigcap_{n \geqslant 1} W_n$. As a union of open subsets $U'(N)$, W_n is open in X, so Z is G_δ in X. □

The use of trees is again apparent in the following exercise.

Exercise 7.6.14 (Banach–Mazur–Oxtoby Theorem) Recall the Banach–Mazur games from Exercise 7.6.12. Show the *Banach–Mazur–Oxtoby Theorem* (Oxtoby, 1957, theorem 1): a space X is Baire iff player β does not have a winning strategy in the Banach–Mazur game on X. That is, prove the equivalence $(c) \Leftrightarrow (d)$ where (c) and (d) are given in Exercise 7.6.12. This can be summed up by saying that Banach–Mazur games entirely characterize the spaces that are Baire.

In the hard direction $(d) \Rightarrow (c)$, assume β has a winning strategy τ. We say that a finite or infinite sequence of open subsets V_0, U_0, V_1, U_1, $\ldots$, V_n, U_n, $\ldots$ is τ-*compatible* iff each of these open subsets is non-empty, $V_n = \tau(V_0, U_0, V_1, U_1, \ldots, V_{n-1}, U_{n-1})$, and $V_n \supseteq U_n$ for every $n \in \mathbb{N}$ such that this makes sense. Given any τ-compatible finite sequence of odd length V_0, U_0, V_1, U_1, $\ldots$, V_n, consider the set $\mathcal{I}$ of all families $(U_{ni})_{i \in I}$ of non-empty open subsets U_{ni} of V_n such that $(V_{ni})_{i \in I}$ is a family of pairwise disjoint subsets, where $V_{ni} = \tau(V_0, U_0, V_1, U_1, \ldots, V_n, U_{ni})$ for each $i \in I$. Order $\mathcal{I}$ by inclusion. Show that $\mathcal{I}$ has a maximal element $(U_{ni})_{i \in I}$, and that it is such that $\bigcup_{i \in I} V_{ni}$ is dense in V_n, where V_{ni} is defined as above.

Use this to build an infinite tree of open subsets, where paths from the root to a node at level n are labeled X (root), U_0, U_1, $\ldots$, U_{n-1}. For each such node, define $V_i = \tau(V_0, U_0, V_1, U_1, \ldots, V_{i-1}, U_{i-1})$ by induction on $i \leqslant n$, and let the successors of the node be labeled U_{ni}, $i \in I$, where $(U_{ni})_{i \in I}$ is defined from $V_0, U_0, V_1, U_1, \ldots, V_{n-1}, U_{n-1}, V_n$ as above. Let W_{n-1} be the union of all opens that label nodes at level $n \geqslant 1$. Show that each W_n is open and dense, but that their intersection is empty, by noticing that each branch of the tree defines an α-strategy in a canonical way.

Alexandroff showed that the completely metrizable subspaces of a complete metric space are exactly its G_δ subsets. We state this in the following form, due to Choquet (1969, theorem 8.7).

Theorem 7.6.15 (Metric $\Rightarrow$ complete = G_δ = Choquet-complete) *Let X, d be a complete metric space, and Z be a subspace of X. Then the following are equivalent:*

1. *Z is Choquet-complete.*
2. *Z is a G_δ subset of X.*
3. *Z is completely metrizable.*

Proof That 1 implies 2 is true by Proposition 7.6.13. That 2 implies 3 is true by Lemma 7.4.24, since in this case Smyth-completeness, Yoneda-completeness, and completeness agree, by Lemma 7.4.3. Finally, 3 implies 1 by Proposition 7.6.2. □

Corollary 7.6.16 *Let X be a metrizable space. X is completely metrizable if and only if it is Choquet-complete.*

Proof Let d be a metric defining the topology of X. Embed X into its completion $\mathbf{S}(X, d)$ (equivalently, $\mathcal{C}hy(X)$), and apply the equivalence between parts 1 and 3 of Theorem 7.6.15. □

Exercise 7.6.17 Let X, d be a complete metric space. As a special case of Theorem 7.2.31, Item 1, every closed subset F of X, d is complete as well. By Theorem 7.6.15, F must be a G_δ subset of X, d. Show this directly. More generally, using fattenings (Definition 6.1.12), show that any closed subset F of a hemi-metric space X, d is a G_δ subset of X, d^{op}.

Exercise 7.6.18 Let X be a metrizable space. Show that X is Choquet-complete if and only if it is convergence Choquet-complete (see Exercise 7.6.5).

Exercise 7.6.19 (Hausdorff's Theorem) Show that the continuous open images of Choquet-complete spaces are again Choquet-complete. That is, if f is a continuous open surjection from a Choquet-complete space X to a topological space Y, then Y is Choquet-complete.

Deduce *Hausdorff's Theorem*: if X is a completely metrizable space, and Y is a metrizable space that arises as the image of X by an open continuous surjection f, then Y is again completely metrizable.

Exercise 7.6.20 (Dorais–Mummert Theorem) Recall the notion of a convergence Choquet-complete space from Exercise 7.6.5. Show that the continuous open image of any convergence Choquet-complete space is again convergence Choquet-complete. Deduce that any continuous open image of a complete metric space is convergence Choquet-complete.

Show that, for every T_0 space X, X is the continuous open image of a complete metric space Y if and only if X is first-countable and convergence Choquet-complete. This is a variant on the *Dorais–Mummert Theorem*, which you should show as well: for every T_1 space X, X is the continuous open image of a complete metric space Y if and only if it is convergence Choquet-complete (Dorais and Mummert, 2010, theorem 3.3). The same proof, with minor variants, will establish both. Show also that one can take Y ultra-metric, hence zero-dimensional: take Y to be the set of all branches in the tree $\mathcal{T}(\sigma)$, for some convergent winning strategy σ for α, and imitate the proof of Ponomarev's Theorem (Exercise 6.3.14).

Deduce that every T_1 convergence Choquet-complete space is first-countable.

Exercise 7.6.21 (Čech-complete spaces) Let X be a topological space. We say that X is *Čech-complete* if and only if there are countably many open covers $(U_{ni})_{i\in I_n}$, $n \in \mathbb{N}$, such that every filtered family of non-empty closed subsets $(F_j)_{j\in J}$ with the property that, for every $n \in \mathbb{N}$, there are $i \in I_n$ and $j \in J$ such that $F_j \subseteq U_{ni}$, has non-empty intersection.

Show that every complete metric space is Čech-complete (use Cantor's Theorem, Exercise 7.3.6). Show that every regular Čech-complete space is Choquet-complete; moreover, α has a stationary winning strategy. Conclude that a metrizable space is completely metrizable if and only if it is Čech-complete.

Exercise 7.6.22 (Frolík's Theorem) Show that every G_δ subspace X of a compact T_2 space Y is Čech-complete. If X is the intersection of a descending sequence $(V_n)_{n\in\mathbb{N}}$ of open subsets of Y (i.e., $V_n \supseteq V_{n+1}$), let the nth open cover consist of the sets $int(K_{nx}) \cap X$, where K_{nx} is a compact neighborhood x of X included in V_n, one per point $x \in X$. Then use Exercise 4.4.18.

Conversely, let X be $T_{3\frac{1}{2}}$ and Čech-complete. Up to the embedding $\eta_X^{\mathrm{S\check{C}}}$ of Exercise 6.7.23, X is a dense subset of its Stone–Čech compactification $Y = \beta X$. Show that X is a G_δ subset of Y. You can proceed as follows. Given a countable family of open covers $(U_{ni})_{i\in I_n}$, $n \in \mathbb{N}$, of X, write U_{ni} as $V_{ni} \cap X$ for some open subsets V_{ni} of Y, then let $V_n = \bigcup_{i\in I_n} V_{ni}$, and show that $X = \bigcap_{n\in\mathbb{N}} V_n$. In the hard direction, let $(K_j)_{j\in J}$ be the filtered family of compact neighborhoods of $x \in \bigcap_{n\in\mathbb{N}} V_n$, define $F_j = K_j \cap X$, and use Čech-completeness to show that x is in X. You will need Exercise 4.4.18 again.

Deduce *Frolík's Theorem* (Frolík, 1960). This says that the following claims are equivalent: (a) X is a Čech-complete $T_{3\frac{1}{2}}$ space; (b) X is homeomorphic to a G_δ subset of a compact T_2 space; (c) X is homeomorphic to a dense G_δ subset of a compact T_2 space; (d) X is $T_{3\frac{1}{2}}$ and its image by $\eta_X^{\mathrm{S\check{C}}}$ in its Stone–Čech compactification βX is a G_δ subset of βX.

7.7 Polish spaces

Corollary 6.7.17, that separable metric spaces embed into the Hilbert cube, can be refined by additionally assuming completeness. The complete separable metric spaces play an important role in topological measure theory, and deserve a name.

Definition 7.7.1 (Polish space) A *Polish metric space* is a complete, separable metric space. A *Polish space* is a topological space X such that there is a metric d on X that makes X, d complete and separable.

Example 7.7.2 $\mathbb{R}$ is Polish: it is complete metric by Fact 7.2.17, and $\mathbb{Q}$ is a countable dense subset (Example 4.1.31).

It is important in the theory of Polish spaces that one is able to switch the underlying metric d at will, leaving only the topology unchanged. A trivial consequence is:

Fact 7.7.3 *Every topological space homeomorphic to a Polish space is Polish.*

Lemma 7.7.4 *Every closed subset of a Polish space is Polish. The topological product and coproduct of countably many Polish spaces are Polish.*

Proof Recall that a metric space is complete iff it is Smyth-complete (see Lemma 7.4.3). Given Theorem 7.2.31, one needs to check that (sym-)closed subsets of complete separable metric spaces are separable—all subspaces are—and that countable topological products and coproducts of separable metric spaces are separable. Assume X_k, d_k are separable metric spaces, $k \in K$, where $K \subseteq \mathbb{N}$. Let A_k be a countable dense subset of X_k. Using Lemma 4.1.30, it is clear that $\coprod_{k \in K} A_k$ is dense in $\coprod_{k \in K} X_k$; as a countable union of countable sets, it is also countable. The product $\prod_{k \in K} A_k$ is also dense in $\prod_{k \in K} X_k$: every non-empty subset of the latter contains a non-empty basic open subset $\prod_{k \in K} U_k$, where each U_k is open, non-empty, and equal to X_k except for finitely many values of k; since A_k is dense in X_k, pick x_k in $A_k \cap U_k$, using the Axiom of Choice, so $\vec{x} = (x_k)_{k \in K}$ is in $\prod_{k \in K} A_k \cap \prod_{k \in K} U_k$. Finally, as a countable product of countable sets, $\prod_{k \in K} A_k$ is countable. □

Example 7.7.5 ($[0, 1]^{\mathbb{N}}$ is Polish) The Hilbert cube $[0, 1]^{\mathbb{N}}$ is Polish: $[0, 1]$ is closed in the Polish space $\mathbb{R}$, hence Polish by Lemma 7.7.4, first part. By the second part, the Hilbert cube $[0, 1]^{\mathbb{N}}$ is Polish.

Exercise 7.7.6 ($\mathcal{B}$ is Polish) Show that Baire space $\mathcal{B} = \mathbb{N}^{\mathbb{N}}$ is Polish. So there are Polish spaces that are not locally compact.

7.7.1 G_δ subsets of the Hilbert cube

Closed subsets of Polish spaces are Polish. So, for example, $[0, 1]$, $\mathbb{R}^+$, and $[1, +\infty)$ are Polish with their ordinary metric topology. But there are more Polish subspaces; in particular, by Lemma 7.2.33 every open subset of a Polish space is completely quasi-metrizable, and clearly second-countable, hence Polish. All G_δ subsets are:

Proposition 7.7.7 (Polish subspace $= G_\delta$) *The subspaces of a Polish space that are themselves Polish are exactly its G_δ subsets.*

Proof Use the equivalence between parts 2 and 3 in Theorem 7.6.15; realize that any subset of a countably-based space is countably-based, and that separability is equivalent to having a countable base (Fact 6.3.45). □

Exercise 7.7.8 Show that $\mathbb{R} \smallsetminus \mathbb{Q}$, the space of irrational numbers, is Polish.

Here is a purely topological characterization of Polish spaces, expanding on the Urysohn-Tychonoff Theorem 6.3.44.

Proposition 7.7.9 *The Polish spaces are exactly the countably-based, T_3, Choquet-complete spaces.*

Proof If X is Polish, then it is countably-based and T_3 by the Urysohn-Tychonoff Theorem 6.3.44, and Choquet-complete by the $3 \Rightarrow 1$ implication of Theorem 7.6.15. Conversely, if X is countably-based and T_3, then it is (separable) metrizable by Theorem 6.3.44. By Corollary 7.6.16, X is completely metrizable. Since it is also countably-based, it is Polish (Fact 6.3.45). □

Exercise 7.7.10 Show that the Polish spaces are also exactly the countably-based T_3 Čech-complete spaces.

Another characterization is as follows.

Theorem 7.7.11 (Alexandroff) *The Polish spaces are exactly the G_δ subsets of the Hilbert cube $[0, 1]^{\mathbb{N}}$, up to homeomorphism.*

Proof If X, d is Polish metric, then it is in particular separable metric, and therefore embeds through $\eta_d^{\mathbb{N}}$ into $[0, 1]^{\mathbb{N}}$, by Corollary 6.7.17. Now $\eta_d^{\mathbb{N}}[X]$ is homeomorphic to X, hence Polish (Fact 7.7.3), hence G_δ by Proposition 7.7.7.

Conversely, the Hilbert cube $[0, 1]^{\mathbb{N}}$ is Polish (Example 7.7.5). Any G_δ subset of it is Polish, by Proposition 7.7.7, and then any space homeomorphic to a G_δ subset of $[0, 1]^{\mathbb{N}}$ is Polish, too, by Fact 7.7.3. □

The name "G_δ" is certainly not intuitive; "G" is for the German "Gebiet" (domain, area) and A_δ used to denote the class of sets obtained as countable intersections of sets in A (δ being for German "Durchschnitt," i.e., intersection). Dually, the so-called F_σ sets are the unions of countably many closed sets (see Exercise 6.3.20); "F" is for the French "fermé" (closed) and σ for the French "somme" (sum, union) – i.e., the complements of G_δ subsets. One can iterate the construction, and define $G_{\delta\sigma}$ sets, $F_{\sigma\delta}$ sets, and so on. The smallest class Λ of sets containing the open sets, the complements of elements of Λ, and the countable intersections of elements of Λ (hence their countable unions) is called the *Borel σ-algebra* of the space, and its members are the Borel subsets. G_δ and F_σ are particular Borel subsets, and therefore Polish subspaces are Borel. In general, there are many more. See Srivastava (1998) for an in-depth treatment of these, as well as Souslin spaces.

7.7.2 Continuous and ω-continuous models of spaces

Polish spaces have other remarkable properties. We shall see that the space of formal balls of any Polish metric space is ω-continuous, for example.

Definition 7.7.12 (ω-continuous) A poset is *ω-continuous* if and only if it has a countable basis.

The following is due to Norberg (1989, proposition 3.1).

Lemma 7.7.13 (Norberg) *A continuous poset X is ω-continuous if and only if X_σ is countably-based.*

Proof If X is ω-continuous, then it is continuous, by Lemma 5.1.23. Moreover, if B is a countable basis, then the family of all open subsets $\Uparrow b$, $b \in B$, is a countable base of the Scott topology, by Theorem 5.1.27.

Conversely, assume X is continuous and X_σ is countably-based. Let $(U_n)_{n\in\mathbb{N}}$ be a countable base of the Scott topology (see Exercise 6.3.8). For each pair of natural numbers m, n such that $U_m \subseteq \uparrow x \subseteq U_n$ for some $x \in X$, pick one such x and call it b_{mn}. (We use the Axiom of Choice.) Also let I be the set of all pairs (m, n) such that this is possible. We claim that $B = \{b_{mn} \mid (m, n) \in I\}$ is a basis of X.

Whenever $x \ll y$ in X, by interpolation (Proposition 5.1.15) used twice there are elements z, z' such that $x \ll z \ll z' \ll y$. Since $x \ll z$, z is in the open subset $\Uparrow x$, hence there is an $n \in \mathbb{N}$ such that $z \in U_n$ and $U_n \subseteq \Uparrow x$. Similarly, there is an $m \in \mathbb{N}$ such that $z' \in U_m$ and $U_m \subseteq \Uparrow z$. In particular, $U_m \subseteq \uparrow z \subseteq U_n$, so $(m, n) \in I$. Since $U_m \subseteq \uparrow b_{mn} \subseteq U_n$, in particular $b_{mn} \leqslant z'$, so $b_{mn} \ll y$; and $b_{mn} \in U_n \subseteq \Uparrow x$, so $x \ll b_{mn}$. We have proved a refined form of interpolation: if $x \ll y$ then $x \ll b \ll y$ for some $b \in B$. This is enough to ensure that B is a basis (Exercise 5.1.30), and B is clearly countable. □

Exercise 7.7.14 Using Exercises 7.6.6 and 7.6.20, show that every ω-continuous dcpo (when equipped with its Scott topology) is the continuous open image of a complete (ultra-)metric space.

Proposition 7.7.15 *Let X, d be a separable hemi-metric space. $\mathbf{B}(X, d)$ is countably-based in the open ball topology of d^+.*

If X, d is Smyth-complete and separable, in particular if X, d is Polish metric, then $\mathbf{B}(X, d)$ is an ω-continuous dcpo in the $\leqslant^{d^+}$ ordering.

Proof Let $(x_n)_{n \in \mathbb{N}}$ be a countable dense subset of X, d^{sym}. (This assumes X non-empty. The case where X is empty is obvious.) We claim that $B = (B^{d^+}_{(x_n,r),<1/2^p})_{n,p \in \mathbb{N}, r \in \mathbb{Q}}$ is a base of the topology of $\mathbf{B}(X, d)$. Let (y, s) be any formal ball, and U an open neighborhood of (y, s). By Lemma 6.1.5, U contains an open ball $B^{d^+}_{(y,s),<\epsilon}$ for some $\epsilon > 0$. Find $p \in \mathbb{N}$ such that $1/2^p < \epsilon/2$, a point x_n such that $d^{sym}(x_n, y) < 1/2^{p+1}$, and a rational number r such that $|r - s| < 1/2^{p+1}$. Then (y, s) is in $B^{d^+}_{(x_n,r),<1/2^p}$, and $B^{d^+}_{(x_n,r),<1/2^p} \subseteq B^{d^+}_{(y,s),<\epsilon}$: for every $(z, t) \in B^{d^+}_{(x_n,r),<1/2^p}$, $d(x_n, z) - r + t < 1/2^p$, so $d(y, z) - s + t \leqslant d(y, x_n) - s + r + d(x_n, z) - r + t < 1/2^{p+1} + 1/2^{p+1} + 1/2^p < \epsilon$. By Lemma 4.1.10, B is a base. Moreover, B is countable.

If X, d is Smyth-complete and separable, then $\mathbf{B}(X, d)$ is a continuous dcpo by the Romaguera–Valero Theorem 7.3.11. By Lemma 7.3.13, its Scott topology is its open ball topology. Since the latter is countably-based, the former is, too. So $\mathbf{B}(X, d)$ is ω-continuous by Lemma 7.7.13. □

So any Polish space X embeds both as a G_δ subset of the Hilbert cube (Theorem 7.7.11) and as a subset $\eta^{\mathrm{EH}}_X[X]$ of the ω-continuous dcpo $\mathbf{B}(X, d)$, by Lemma 7.3.3. This subset is none other than the set of maximal elements of $\mathbf{B}(X, d)$, which are exactly the formal balls of the form $(x, 0)$.

Proposition 7.7.16 (Model) *Let X be a topological space. A* model *for X is any dcpo Y such that X is homeomorphic to the subspace $\mathrm{Max}\, Y$ of Y_σ consisting of the maximal elements of Y.*

Every completely metrizable space X has a continuous model, and every Polish space X has an ω-continuous model: the space $\mathbf{B}(X, d)$, where d is any complete metric on X.

Proof By the Romaguera–Valero Theorem 7.3.11 for completely metrizable spaces; by Proposition 7.7.15 for Polish spaces. □

Models are sometimes called *computational* models, although no notion of computation is involved. A model Y of a space X is a dcpo that describes approximation schemes for elements of X, in the form of particular elements of Y. The notion of approximation is particularly relevant when Y is a continuous dcpo.

Example 7.7.17 (The model $\mathbf{I}\mathbb{R}$) We have already seen an example of a model, in Exercise 4.2.28: the dcpo $\mathbf{I}\mathbb{R}$ of all non-empty closed intervals $[a, b]$ of reals, ordered by reverse inclusion, is a model of $\mathbb{R}$, and the embedding $\eta_{\mathbf{I}\mathbb{R}}$ maps x to the maximal element $[x, x]$. $\mathbf{I}\mathbb{R}$ is ω-continuous, since the intervals with rational endpoints form a basis (it suffices to check that $[a, b] \ll [c, d]$ iff $a < c$ and $b > d$). One can also check that $\mathbf{I}\mathbb{R}$ is order isomorphic to $\mathbf{B}(\mathbb{R}, \mathsf{d}_{\mathbb{R}}^{sym})$: every interval $[a, b]$ is mapped to the formal ball $(\frac{a+b}{2}, \frac{b-a}{2})$, and, conversely, every formal ball (x, r) is mapped to the interval $[x-r, x+r]$. The order isomorphism is a consequence of Exercise 7.3.4, first part, with $n = 1$.

It is natural to ask which spaces have models, and in particular continuous or ω-continuous models.

Lemma 7.7.18 *For any dcpo Y, $\mathrm{Max}\, Y$ is T_1 in the subspace topology from Y_σ. In particular, any space with a model is T_1.*

Proof Assume x, y are distinct points in $\mathrm{Max}\, Y$. So x is in $Y \smallsetminus \downarrow y$: otherwise $x \leqslant y$, whence $x = y$ since x is maximal. Since the Scott topology is finer than the upper topology (Proposition 4.2.18), $U = Y \smallsetminus \downarrow y$ is Scott-open, so $U \cap \mathrm{Max}\, Y$ is an open neighborhood of x in $\mathrm{Max}\, Y$ that does not contain y. □

The following is due to Martin (2004, theorem 5.1). The idea that Choquet-completeness is fundamental in the study of those spaces that have continuous models is his.

Proposition 7.7.19 *For any continuous dcpo Y, $\mathrm{Max}\, Y$ is Choquet-complete in the subspace topology from Y_σ – α even has a stationary winning strategy on $\mathrm{Max}\, Y$. In particular, any space with a continuous model is Choquet-complete.*

Proof By Lemma 7.6.3, Y is Choquet-complete. There is even a stationary winning strategy σ for α on Y. We let α play on $\mathrm{Max}\, Y$ as follows. Assume β has just played x_n, V_n. By definition, $V_n = W_n \cap \mathrm{Max}\, Y$ for some open neighborhood W_n of x_n in Y_σ. We let α play $U_n \cap \mathrm{Max}\, Y$, where $U_n = \sigma(x_n, W_n)$, and note that this is a stationary strategy. Since σ is winning, in any play $\bigcap_{n \in \mathbb{N}} U_n$ is non-empty. Since Y is a dcpo, i.e., is inductive, we can apply Zorn's Lemma (Theorem 2.4.2): there is an element in $(\bigcap_{n \in \mathbb{N}} U_n) \cap \mathrm{Max}\, Y = \bigcap_{n \in \mathbb{N}}(U_n \cap \mathrm{Max}\, Y)$. □

Exercise 7.7.20 Show that we can improve Proposition 7.7.19, and require α's strategy to be not only stationary but also convergent. That is, any space with a continuous model must be convergence Choquet-complete. Conclude that any space with a continuous model must be the continuous open image of a complete metric space.

The following is *Martin's Second Theorem* (Martin, 2004, corollary 5.3).

Theorem 7.7.21 (Martin) *Let X be a metrizable space. Then X has a continuous model if and only if X is completely metrizable. For such a model, we can take $\mathbf{B}(X, d)$, where d is any complete metric defining the topology of X.*

Proof If X is completely metrizable, then the space of its formal balls is a continuous model, by Proposition 7.7.16. Conversely, if X has a continuous model, then it is Choquet-complete by Proposition 7.7.19. It is therefore completely metrizable, by Corollary 7.6.16. □

Exercise 7.7.22 Show that $\mathbb{Q}$, with the subspace topology from $\mathbb{R}$, $\mathsf{d}_{\mathbb{R}}$, has no continuous model.

And this is *Martin's First Theorem* (Martin, 2003).

Theorem 7.7.23 (Martin) *Let X be a T_3 space. Then X has an ω-continuous model if and only if X is Polish. For such a model, we can take $\mathbf{B}(X, d)$, where d is any complete metric defining the topology of X.*

Proof If X is Polish, then the space of formal balls is an ω-continuous model, by Proposition 7.7.16. Conversely, if X has an ω-continuous model Y, then X is Choquet-complete by Proposition 7.7.19. By Lemma 7.7.13, Y_σ is countably-based. Its subspace Max Y is then clearly countably-based as well. So X is countably-based. Since we assumed X regular, by Proposition 7.7.9 X is Polish. □

The following lemma gives a simple criterion for Max Y to be regular, where Y is a continuous dcpo. This is again due to Martin.

Lemma 7.7.24 (Lawson at the top) *Let Y be a dcpo. We say that Y is* Lawson at the top *if and only if, for every $y \in Y$, $\uparrow y \cap \mathrm{Max}\, Y$ is closed in $\mathrm{Max}\, Y$, equipped with the subspace topology from Y_σ.*

If Y is continuous and Lawson at the top, then $\mathrm{Max}\, Y$ is T_3.

Proof Assume that Y is Lawson at the top. Let $x \in \mathrm{Max}\, Y$, and consider any open neighborhood of x in Max Y. This must be of the form $U \cap \mathrm{Max}\, Y$, where U is Scott-open in Y. By Theorem 5.1.17, there is a point $y \in U$ such that $x \in \twoheaduparrow y$. Since Y is Lawson at the top, $F = \uparrow y \cap \mathrm{Max}\, Y$ is closed in

$\mathrm{Max}\, Y$. We have $x \in (\Uparrow y \cap \mathrm{Max}\, F) \subseteq F \subseteq (U \cap \mathrm{Max}\, Y)$. So $\mathrm{Max}\, Y$ is locally closed, hence regular (Exercise 4.1.21). □

Being Lawson at the top is a property of the so-called Lawson topology; see Exercise 9.1.22.

We now obtain *Lawson's Theorem* (Lawson, 1997b).

Theorem 7.7.25 (Lawson) *The Polish spaces X are exactly those spaces that have an ω-continuous model that is Lawson at the top. For such a model, we can take $\mathbf{B}(X, d)$ where d is any complete metric defining the topology of X.*

Proof If X has an ω-continuous model Y that is Lawson at the top, then $\mathrm{Max}\, Y$ is T_3 by Lemma 7.7.24. So X is Polish, by Martin's First Theorem 7.7.23. Conversely, if X, d is Polish metric, then $\mathbf{B}(X, d)$ is an ω-continuous model by Proposition 7.7.16.

We must show that $\mathbf{B}(X, d)$ is Lawson at the top. Consider any formal ball (x, r). Then $\uparrow(x, r) \cap \mathrm{Max}\, \mathbf{B}(X, d)$ is the set of open balls $(y, 0)$ such that $d(x, y) \leqslant r$. Let $(y, 0)$ be in the complement of $\uparrow(x, r) \cap \mathrm{Max}\, \mathbf{B}(X, d)$, i.e., $d(x, y) > r$. Let $\epsilon > 0$ be such that $r + \epsilon < d(x, y)$. Let $U = B^{d^+}_{(y,0),<\epsilon}$. This is Scott-open in $\mathbf{B}(X, d)$, since the Scott topology coincides with the open ball topology on the latter, using Lemma 7.3.13. We note that $U \cap \mathrm{Max}\, \mathbf{B}(X, d)$ does not meet $\uparrow(x, r) \cap \mathrm{Max}\, \mathbf{B}(X, d)$: if $(z, 0)$ were a point in the intersection, then we would have $d^+((y, 0), (z, 0)) < \epsilon$ (i.e., $d(y, z) < \epsilon$) and $d(x, z) \leqslant r$, hence $d(x, y) \leqslant d(x, z) + d(z, y) = d(x, z) + d(y, z) < \epsilon + r$. This is impossible since $r + \epsilon < d(x, y)$. We have shown that any point of $\mathrm{Max}\, \mathbf{B}(X, d)$ outside $\uparrow(x, r) \cap \mathrm{Max}\, \mathbf{B}(X, d)$ is contained in some open subset of $\mathrm{Max}\, \mathbf{B}(X, d)$ that does not meet $\uparrow(x, r) \cap \mathrm{Max}\, \mathbf{B}(X, d)$. By Trick 4.1.11, the complement of $\uparrow(x, r) \cap \mathrm{Max}\, \mathbf{B}(X, d)$ is open. So $B(X, d)$ is Lawson at the top. □

The space of formal balls is far from being the only possible ω-continuous model of a Polish space. We shall see another class of models, the Smyth models, in Corollary 8.3.27. As an application, we shall see that every locally compact separable metrizable space (i.e., every countably-based locally compact T_3 space) is Polish.

8

Sober spaces

In Chapter 4, we mentioned a connection between topology and order, centered around the notion of specialization quasi-ordering. The connection between topology and order goes much beyond this. For any topological space X, $\mathcal{O}(X)$ is a poset, of a rather special kind, and we start by examining when a poset L is of this form for some space X. The posets that have this property are the spatial lattices, and Stone duality is a canonical way of retrieving the space X of points from the lattice L alone; see Section 8.1. Conversely, the spaces X of points obtained from spatial lattices are exactly the *sober* topological spaces, and any space can be completed to a sober space. This is explored in Section 8.2. The sober spaces have wonderful properties, and the single most important result about sober spaces is the Hofmann–Mislove Theorem, which we establish and whose consequences we state in Section 8.3. This is the starting point of a theory of correspondences between certain spaces and certain lattices: sober spaces and spatial lattices, but also sober locally compact spaces and continuous distributive complete lattices, or continuous dcpos with their Scott topology and completely distributive lattices notably. We then look at limits and colimits in **Top**, and how they are preserved or not by the sobrification functor in Section 8.4.

8.1 Frames and Stone duality

For every topological space X, $\mathcal{O}(X)$ is a complete lattice (see Exercise 5.2.2). Every continuous map $f\colon X \to Y$ defines a function $\mathcal{O}(f)\colon \mathcal{O}(Y) \to \mathcal{O}(X)$, which maps every open subset V of Y to $\mathcal{O}(f)(V) = f^{-1}(V)$. The map $\mathcal{O}(f)$ preserves all least upper bounds (unions) and all finite greatest lower bounds (finite intersections), i.e., it is a *frame homomorphism.* Letting **CLat** be the category of complete lattices and frame homomorphisms, $\mathcal{O}$ then defines a functor from **Top** to $\mathbf{CLat}^{op}$, the opposite category from **CLat**.

A *frame* is any complete lattice that obeys the infinite distributivity law,

$$u \wedge \sup_{i \in I} v_i = \sup_{i \in I}(u \wedge v_i), \tag{8.1}$$

where we now write $\wedge$ for the infimum of two elements. In $\mathcal{O}(X)$, $\wedge$ is intersection, and the infinite distributivity law is clear. (Beware that while greatest lower bounds of two opens are computed as intersections, greatest lower bounds of infinitely many opens are not: see Exercise 5.2.2.)

Let **Frm** be the category of frames and frame homomorphisms. Its opposite category $\mathbf{Loc} = \mathbf{Frm}^{op}$ is the category of *locales*. Then $\mathcal{O}$ defines a functor from **Top** to **Loc**. Indeed, for every $f : X \to Y$ in **Top**, $\mathcal{O}(f)$ is a frame homomorphism from $\mathcal{O}(Y)$ to $\mathcal{O}(X)$, that is, a morphism from $\mathcal{O}(X)$ to $\mathcal{O}(Y)$ in $\mathbf{Loc} = \mathbf{Frm}^{op}$.

Why frames? We might have been content with complete lattices instead: $\mathcal{O}$ also defines a functor from **Top** to $\mathbf{CLat}^{op}$, for example. The historical reason is that free frames over sets exist, while free complete lattices do not. An intriguing property of frames is the following relation with complete Heyting algebras, which arise in logic as the domains of truth-values for intuitionistic logics: read $\wedge$ as "and," $\leqslant$ as "entails," and realize that logical implication $\Rightarrow$ has the distinguishing property that, for all formulae a, b, c, $a \wedge b$ entails c if and only if a entails $b \Rightarrow c$. That is, if you want to prove $b \Rightarrow c$ under the assumption a, then introduce the additional assumption b, and prove c under the two assumptions a and b.

Proposition 8.1.1 (Complete Heyting algebras) *A complete Heyting algebra is a complete lattice in which there is a binary operation* $\Rightarrow$ (residuation, *also known as logical* implication) *such that* $a \wedge b \leqslant c$ *iff* $a \leqslant (b \Rightarrow c)$.

A complete lattice is a frame if and only if it is a complete Heyting algebra.

Proof Let L be a frame. Define $b \Rightarrow c$ as the least upper bound of all elements a such that $a \wedge b \leqslant c$. It is clear that if $a \wedge b \leqslant c$, then $a \leqslant (b \Rightarrow c)$. Conversely, assume $a \leqslant (b \Rightarrow c)$. Then $a \wedge b \leqslant (b \Rightarrow c) \wedge b$. Using the infinite distributivity law, $(b \Rightarrow c) \wedge b = \sup_{a'/a' \wedge b \leqslant c}(a' \wedge b) \leqslant c$, so $a \wedge b \leqslant c$.

Conversely, let L be a complete Heyting algebra. For every element c, $u \wedge \sup_{i \in I} v_i \leqslant c$ iff $\sup_{i \in I} v_i \leqslant (u \Rightarrow c)$, iff $v_i \leqslant (u \Rightarrow c)$ for every $i \in I$, iff $u \wedge v_i \leqslant c$ for every $i \in I$, iff $\sup_{i \in I}(u \wedge v_i) \leqslant c$. Taking $c = u \wedge \sup_{i \in I} v_i$ first, then $c = \sup_{i \in I}(u \wedge v_i)$ yields that $u \wedge \sup_{i \in I} v_i$ is both below and above $\sup_{i \in I}(u \wedge v_i)$. □

Exercise 8.1.2 By Proposition 8.1.1, $\mathcal{O}(X)$ is a complete Heyting algebra for every topological space X. Let U, V be two open subsets of X. What is $U \Rightarrow V$?

Exercise 8.1.3 Let L be a frame. Show that $\wedge$ is Scott-continuous, and reflects $\ll$. Generalize Proposition 5.2.19 from $\mathcal{O}(X)$ to general frames: show that if L is a frame in which $\ll$ is multiplicative, then $\wedge$ preserves and reflects $\ll$.

Exercise 8.1.4 (Posets as thin categories) Every poset L can be considered as a *thin category*, i.e., where there is at most one morphism from A to B, for any pair of objects A, B. Indeed, the objects of L are its elements u, and there is a (unique) morphism from u to v if and only if $u \leqslant v$. Identities are obtained by reflexivity, composition by transitivity.

Show that least upper bounds in posets coincide with coproducts in the corresponding thin category, and that greatest lower bounds coincide with products. Show that residuation $b \Rightarrow c$ is exactly the exponential object c^b, if it exists, and that the complete Heyting algebras are exactly the Cartesian-closed, small, thin categories.

Recall that left adjoints preserve all existing colimits. Why does this imply trivially that every complete Heyting algebra is a frame? Show that in a CCC $\mathbf{C}$, all colimits must distribute over binary products, in the sense that, for every functor F from a small category $\mathbf{J}$ to $\mathbf{C}$ such that $\operatorname{colim} F$ exists, then, for every object B of $\mathbf{C}$, $\operatorname{colim}(F \times B)$ exists, where $F \times B$ is the functor mapping each object $j \in J$ to $F(j) \times B$, and every morphism f to $F(f) \times \mathrm{id}_B$; moreover, that $\operatorname{colim}(F \times B)$ is naturally isomorphic to $(\operatorname{colim} F) \times B$.

Exercise 8.1.5 (Distributive lattice) A lattice L is *distributive* if and only if $u \wedge \sup(v, w) = \sup(u \wedge v, u \wedge w)$ for all $u, v, w \in L$, i.e., iff sup distributes over inf $(= \wedge)$. Show that L is distributive iff inf distributes over $\wedge$, i.e., iff $\sup(u, v \wedge w) = \sup(u, v) \wedge \sup(u, w)$ for all $u, v, w \in L$. (This is called the *dual* distributivity law.)

Exercise 8.1.6 Let L be a complete lattice, and write $\bot$ for its least element, $\top$ for its greatest element. If L is a complete Heyting algebra, then *negation* $\neg u$ is defined as $u \Rightarrow \bot$. A *complete Boolean algebra* is a complete Heyting algebra L in which negation is a *complement*, i.e., $\sup(u, \neg u) = \top$ for every $u \in L$.

Given any T_0 topological space X, show that $\mathcal{O}(X)$ is a complete Boolean algebra iff X has the discrete topology. In particular, there are complete Heyting algebras that are not Boolean.

Given any complete Heyting algebra L, show that, for every $u, v \in L$: (i) $u \wedge \neg u = \bot$, (ii) $u \leqslant \neg\neg u$, (iii) if $u \leqslant v$ then $\neg v \leqslant \neg u$, (iv) $\neg u = \neg\neg\neg u$, (v) $\neg \sup(u, v) \leqslant \neg u \wedge \neg v$.

Given any complete Heyting algebra L, show that $\neg\neg u \leqslant u$ (hence $u = \neg\neg u$) for every $u \in L$ iff L is Boolean. For the latter, you may find it useful to compute $\sup(u, \neg u) \wedge \neg\neg u$ in two different ways, assuming $\sup(u, \neg u) = \top$, and to deduce $\neg\neg u = u \wedge \neg\neg u$. Conversely, show that $\top \leqslant \neg\neg \sup(u, \neg u)$ in any complete Heyting algebra.

Exercise 8.1.7 (The complete Boolean algebra of regular elements of a frame) Given any complete Heyting algebra L, show that, for every $u \in L$, $u = \neg\neg u$ iff u is of the form $\neg v$ for some $v \in L$. Such elements u are called *regular*. Show that the set of regular elements of L, with the ordering $\leqslant$ from L, is a complete Boolean algebra, where the greatest lower bound of regular elements u, v is just $u \wedge v$ as in L, and least upper bounds of families of regular elements $(u_i)_{i \in I}$ are computed as $\neg\neg \sup_{i \in I} u_i$, where the latter sup is computed in L.

Exercise 8.1.8 (Regular opens) A *regular open* subset U of a topological space X is a subset that is equal to the interior of its closure: $U = int(cl(U))$. Show that the regular opens are exactly the regular elements of $\mathcal{O}(X)$. Conclude that the set of regular opens of any space is a complete Boolean algebra.

$\mathcal{O}$ defines a functor from **Top** to **Loc**. We shall see that $\mathcal{O}$ has a right adjoint, pt: **Loc** $\to$ **Top**. This is the starting point of *Stone duality*. That is, Stone duality tries to recover a topological space X from just its lattice of open sets. We won't be able to do this for any topological space, only for the so-called sober spaces, which we shall define below.

Our first task is, given a complete lattice L that we think is the lattice of open subsets of a topological space X, to recover the points of X.

Definition 8.1.9 (Filter) A *filter* F on a lattice L is a non-empty upward closed family of elements of L such that, whenever $u, v \in F$, the greatest lower bound $u \wedge v$ is also in F.

If L is a complete lattice, the filter F is *completely prime* iff, for every family $(u_i)_{i \in I}$ of elements of L whose least upper bound $\sup_{i \in I} u_i$ is in F, u_i is already in F for some $i \in I$.

Example 8.1.10 If $L = \mathcal{O}(X)$, then the filter of all open neighborhoods of $x \in X$ is completely prime.

Being completely prime is akin to Scott-openness, except we do not require $(u_i)_{i \in I}$ to be directed. In particular, every completely prime filter is Scott-open.

Exercise 8.1.11 Show that a filter in L is exactly an upward closed subset of L such that the greatest lower bound of any finite subset of F is again in F. (Observe that the greatest lower bound of an empty subset is $\top$, the largest element of L.)

Example 8.1.10 is the starting point of our reconstruction of the space X from $L = \mathcal{O}(X)$:

Definition 8.1.12 (Point, $\mathsf{pt}(L)$) Let L be a complete lattice. A *point* of L is a completely prime filter of L. Let $\mathsf{pt}(L)$ be the set of points of L, topologized by declaring opens the sets $\mathcal{O}_u = \{x \in \mathsf{pt}(L) \mid u \in x\}$, for every $u \in L$.

Since completely prime filters are now the points of a space $\mathsf{pt}(L)$, we use letters such as x, y instead of F to denote them. Notice the reversal of roles between x and u: $x \in \mathcal{O}_u$ iff $u \in x$.

Proposition 8.1.13 (Hull-kernel topology) *Let L be a complete lattice. The collection of all sets $\mathcal{O}_u$, where $u \in L$, forms a topology on $\mathsf{pt}(L)$, called the* hull-kernel topology *on $\mathsf{pt}(L)$.*

Proof Consider finite intersections: for every completely prime filter x, x is in $\bigcap_{i=1}^n \mathcal{O}_{u_i}$ iff $u_1, \ldots, u_n$ are in x. This is equivalent to requiring $\inf(u_1, \ldots, u_n) \in x$. Indeed, if $\inf(u_1, \ldots, u_n)$ is in x, then all larger elements, in particular $u_1, \ldots, u_n$, are in x since x is upward closed, and the converse direction is true by Exercise 8.1.11. So $\bigcap_{i=1}^n \mathcal{O}_{u_i} = \mathcal{O}_{\inf(u_1,\ldots,u_n)}$.

Unions. For every completely prime filter x, x is in $\bigcup_{i \in I} \mathcal{O}_{u_i}$ iff $u_i \in x$ for some $i \in I$. If this is so, then $\sup_{i \in I} u_i$ is in x, since x is upward closed. Conversely, if $\sup_{i \in I} u_i$ is in x, then some u_i is in x already, because x is completely prime. So $\bigcup_{i \in I} \mathcal{O}_{u_i} = \mathcal{O}_{\sup_{i \in I} u_i}$. □

Exercise 8.1.14 Let L be a complete lattice. Show that the specialization ordering of $\mathsf{pt}(L)$ is inclusion $\subseteq$.

Proposition 8.1.15 (pt) *There is a functor pt from $\mathbf{CLat}^{op}$ to $\mathbf{Top}$, defined on objects L by Definition 8.1.12, and on morphisms $g \colon L \to L'$ in $\mathbf{CLat}^{op}$ (i.e., g is a frame homomorphism from L' to L) by letting $\mathsf{pt}(g) \colon \mathsf{pt}(L) \to \mathsf{pt}(L')$ map every completely prime filter x of L to $g^{-1}(x)$.*

Proof We first have to check that $\mathsf{pt}(g)(x)$ is a point whenever x is a point. This follows from the fact that g is a frame homomorphism. E.g., $\mathsf{pt}(g)(x)$ is completely prime because, if $\sup_{i \in I} u_i \in \mathsf{pt}(g)(x) = g^{-1}(x)$, then $g(\sup_{i \in I} u_i) = \sup_{i \in I} g(u_i) \in x$, so some $g(u_i)$ is in x; i.e., $u_i \in g^{-1}(x)$. Second, we must show that $\mathsf{pt}(g)$ is continuous: $\mathsf{pt}(g)^{-1}(\mathcal{O}_{u'}) = \{x \in \mathsf{pt}(L) \mid g^{-1}(x) \in \mathcal{O}_{u'}\} = \{x \in \mathsf{pt}(L) \mid u' \in g^{-1}(x)\} = \{x \in \mathsf{pt}(L) \mid g(u') \in x\} = \mathcal{O}_{g(u')}$.

That pt preserves identities and composition is clear. □

By restriction, pt defines a functor from $\mathbf{Loc}$ to $\mathbf{Top}$.

The functors $\mathcal{O}$ and pt are not quite inverses of each other. Given any complete lattice L, the elements of $\mathcal{O}(\mathsf{pt}(L))$ are the sets $\mathcal{O}_u$, and it seems natural to equate $u \in L$ with $\mathcal{O}_u \in \mathcal{O}(\mathsf{pt}(L))$. One can do this exactly when L is a so-called spatial lattice.

Definition 8.1.16 (Spatial lattice) A *spatial lattice* is a complete lattice L that is order isomorphic to the lattice $\mathcal{O}(X)$ of some topological space X.

Proposition 8.1.17 *For each complete lattice L, let $\epsilon_L^{Stone} : L \to \mathcal{O}(\mathsf{pt}(L))$ map $u \in L$ to $\mathcal{O}_u$. The map ϵ_L^{Stone} is monotonic and surjective.*

Moreover, the following are equivalent:

1. *L is spatial.*
2. *ϵ_L^{Stone} is injective.*
3. *L is a frame and its points separates its elements, i.e., for any two elements u, v of L such that $u \not\leqslant v$, there is a point x such that $u \in x$ but $v \notin x$.*
4. *ϵ_L^{Stone} is an order isomorphism between L and $\mathcal{O}(X)$, with $X = \mathsf{pt}(L)$.*

Proof First, $\epsilon_L^{\mathrm{Stone}}$ is monotonic: if $u \leqslant u'$, then every element x of $\mathcal{O}_u$ is such that $u \in x$, so $u' \in x$ since x is upward closed, hence $x \in \mathcal{O}_{u'}$. It is surjective by definition of the hull-kernel topology.

1 ⇒ 2. Assume that L is spatial. Without loss of generality, let $L = \mathcal{O}(X)$. If U and U' are two elements of L (opens of X) such that $\mathcal{O}_U \subseteq \mathcal{O}_{U'}$, then, for every element x of U, the filter of all open neighborhoods of x is an element of $\mathsf{pt}(L)$ (see Example 8.1.10), contains U, hence is in $\mathcal{O}_U$, and therefore is also in $\mathcal{O}_{U'}$. So U' is in the filter, i.e., contains x. So $U \subseteq U'$, whenever $\mathcal{O}_U \subseteq \mathcal{O}_{U'}$. In particular, if $\mathcal{O}_U = \mathcal{O}_{U'}$, then $U \subseteq U'$ and $U' \subseteq U$, so $U = U'$: $\epsilon_L^{\mathrm{Stone}}$ is injective.

2 ⇒ 3. Assume that $\epsilon_L^{\mathrm{Stone}}$ is injective. We have seen in the proof of Proposition 8.1.13 that $\bigcup_{i\in I} \mathcal{O}_{v_i} = \mathcal{O}_{\sup_{i\in I} v_i}$, and that $\mathcal{O}_u \cap \mathcal{O}_v = \mathcal{O}_{u\wedge v}$. So $\mathcal{O}_{u\wedge \sup_{i\in I} v_i} = \mathcal{O}_u \cap \bigcup_{i\in I} \mathcal{O}_{v_i} = \bigcup_{i\in I}(\mathcal{O}_u \cap \mathcal{O}_{v_i}) = \mathcal{O}_{\sup_{i\in I}(u\wedge v_i)}$. In other words, $\epsilon_L^{\mathrm{Stone}}(u \wedge \sup_{i\in I} v_i) = \epsilon_L^{\mathrm{Stone}}(\sup_{i\in I}(u \wedge v_i))$. Since $\epsilon_L^{\mathrm{Stone}}$ is injective, the frame distributivity law (8.1) follows.

Moreover, we claim that points separate elements. Assume $u \not\leqslant v$. So $u \wedge v \neq u$, therefore $\mathcal{O}_{u\wedge v} \neq \mathcal{O}_u$, since $\epsilon_L^{\mathrm{Stone}}$ is injective. That is, $\mathcal{O}_u \cap \mathcal{O}_v \neq \mathcal{O}_u$; that is, $\mathcal{O}_u \not\subseteq \mathcal{O}_v$. So there is a point $x \in \mathcal{O}_u$ that is not in $\mathcal{O}_v$, i.e., such that $u \in x$ but $v \notin x$. So x separates u from v.

3 ⇒ 4. To show that $\epsilon_L^{\mathrm{Stone}}$ is injective, i.e., that $\mathcal{O}_u = \mathcal{O}_v$ implies $u = v$, we show the more general claim that $\mathcal{O}_u \subseteq \mathcal{O}_v$ implies $u \leqslant v$. Indeed, otherwise $u \not\leqslant v$, so there is a point x that separates u from v, namely, $x \in \mathcal{O}_u$ but $x \notin \mathcal{O}_v$, and therefore $\mathcal{O}_u \not\subseteq \mathcal{O}_v$. So $\epsilon_L^{\mathrm{Stone}}$ is injective.

The argument above in fact shows that $\epsilon_L^{\mathrm{Stone}}$ is an order embedding whenever it is injective. Since it is also surjective, it is an order isomorphism.
$4 \Rightarrow 1$ is obvious. □

The spatial lattice $\mathcal{O}(\mathsf{pt}(L))$ is the *spatialization* of the complete lattice L. Proposition 8.1.17 shows that a lattice is spatial iff it is order isomorphic to its spatialization.

Example 8.1.18 It is easy to find complete lattices that are not spatial. The complete lattice N_5 on the left of Figure 8.1 is not even distributive, as $b \wedge \sup(a, c) = b \wedge \top = b$, but $\sup(b \wedge a, b \wedge c) = \sup(\bot, c) = c$. The complete lattice M_3 in the middle is not distributive either, as $a \wedge \sup(b, c) = a \wedge \top = a$, but $\sup(a \wedge b, a \wedge c) = \sup(\bot, \bot) = \bot$.

Example 8.1.19 It is only slightly harder to find complete *distributive* lattices that are not spatial. The complete lattice on the right of Figure 8.1 is distributive, but is not a frame, hence is not spatial. Its elements are certain subsets of $\{a\} \cup \mathbb{N}$, where a is some element outside $\mathbb{N}$: namely the whole set (on top), plus the subsets A such that $A \cap \mathbb{N}$ is downward-closed and finite. Sup is union and inf is intersection, from which distributivity is clear. Notice that the supremum of the elements $\downarrow n$, $n \in \mathbb{N}$, is $\{a\} \cup \mathbb{N}$, not their union $\mathbb{N}$. So the frame law fails: $\{a\} \wedge \sup_{n\in\mathbb{N}} \downarrow n = \{a\} \cap (\{a\} \cup \mathbb{N}) = \{a\}$, but $\sup_{n\in\mathbb{N}}(\{a\} \wedge \downarrow n) = \sup_{n\in\mathbb{N}} \emptyset = \emptyset$.
There are even frames that are non-spatial; see Exercise 8.1.25.

One can simplify the description of $\mathsf{pt}(L)$ by replacing the higher-order concept of completely prime filter by *prime elements*:

Proposition 8.1.20 (Prime elements) *A* prime element *of an inf-semi-lattice L with greatest element $\top$ is an element $p \neq \top$ such that, whenever $u \wedge v \leqslant p$, then $u \leqslant p$ or $v \leqslant p$.*

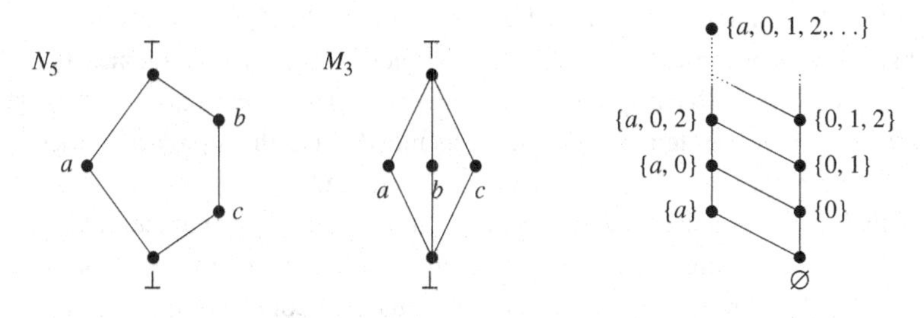

Figure 8.1 Non-spatial lattices.

In any complete lattice L, the completely prime filters of L are exactly the complements of sets of the form $\downarrow p$, p a prime element of L.

Proof Let x be the complement of $\downarrow p$. First, $\top \in x$ since $p \neq \top$. Second, if u and v are in x, then $u \not\leqslant p$ and $v \not\leqslant p$, so $u \wedge v \not\leqslant p$, i.e., x is a filter. Given any family $(u_i)_{i \in I}$, if $\sup_{i \in I} u_i \in x$ then we claim that some u_i is in x. Otherwise $u_i \leqslant p$ for every $i \in I$, so $\sup_{i \in I} u_i \leqslant p$. So x is a completely prime filter.

Conversely, let x be any completely prime filter. Let $(u_i)_{i \in I}$ be the family of all elements outside x, and $p = \sup_{i \in I} u_i$. If p were in x, then some u_i would be in x, a contradiction. So p is not in x. In particular, p is the largest element not in x. So for every $u \in \downarrow p$, u is not in x: if $u \in x$, since $u \leqslant p$ we would have $p \in x$. So $\downarrow p$ does not intersect x, i.e., x is included in the complement of $\downarrow p$. Conversely, if u is not in x, i.e., is equal to some u_i, then by definition $u \leqslant p$, so $u \in \downarrow p$. So x is equal to the complement of $\downarrow p$. □

Given an open subset $\mathcal{O}_u$ in the hull-kernel topology, and a completely prime filter $x = \mathsf{pt}(L) \setminus \downarrow p$, $x \in \mathcal{O}_u$ iff $u \in x$, iff $u \not\leqslant p$. So:

Corollary 8.1.21 *For every complete lattice L, $\mathsf{pt}(L)$ is order isomorphic to the subset of L consisting of its prime elements, with the topology whose open subsets are $\{p$ prime $\mid u \not\leqslant p\}$, where u ranges over L. The order isomorphism maps each completely prime filter x to the prime element p such that x is the complement of $\downarrow p$ in $\mathsf{pt}(L)$.*

Remark 8.1.22 The topology on the set of prime elements given in Corollary 8.1.21 of L is familiar. Consider the upper topology on L^{op}. It induces a topology on the subspace of prime elements: this is our topology.

Example 8.1.23 Look at Figure 8.2 for a few (finite) complete lattices and their sets of prime elements, shown as circled elements. The top two complete lattices are spatial.

The leftmost one is $\mathbb{P}_{\text{fin}}\{0, 1, 2\}$, the lattices of open sets of $\{0, 1, 2\}$ with the discrete topology. (We write 02 instead of $\{0, 2\}$, and similarly for other subsets.) The prime elements are the complements of one-element sets: we indeed retrieve the base space $\{0, 1, 2\}$ this way. This should also give us the intuition that prime elements are elements mostly near the top of the lattice.

The top right lattice is also spatial. One can read off the topological space X which it is the space of opens of: look at the (circled) prime elements, and reverse their ordering. This lattice is the open set lattice of a five-element poset $\{\bot, 1, 2, 12, \top\}$ with $\bot$ least, $\top$ largest, 1 and 2 incomparable, and 12 right above 1 and 2 and below $\top$.

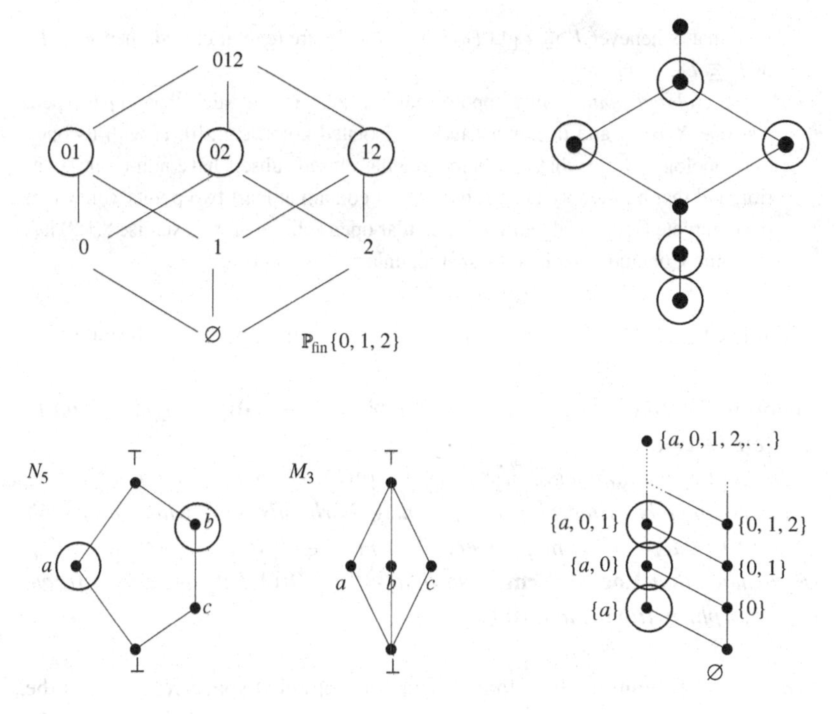

Figure 8.2 Some complete lattices and their prime elements

We have already seen the three lattices at the bottom: they are the non-spatial lattices of Figure 8.1. The middle one, M_3, has no prime element at all. If it were spatial, it would be $\mathcal{O}(\emptyset)$, a one-element set: so it is not spatial. The leftmost one, N_5, has two incomparable elements, so $\mathcal{O}(\mathsf{pt}(K_5))$ would be a four-element lattice, not a five-element one such as N_5. The rightmost one has a space X of points that is homeomorphic to the space $\mathbb{Z}^-$ of non-positive integers with the upper (equivalently, Scott or Alexandroff) topology of $\leqslant$. Then $\mathcal{O}(X)$ is a totally ordered chain, hardly the case of the original lattice.

Exercise 8.1.24 Show that a frame L is spatial if and only if every element $v \in L$ is the greatest lower bound of all prime elements p such that $v \leqslant p$.

Exercise 8.1.25 (Non-spatial frames) Given a topological space X, a *regular closed* subset of X is the complement of a regular open subset of X (Exercise 8.1.8). Show that the closed subsets F of X that arise as complements of prime elements of the complete Boolean algebra of regular open sets are the *irreducible regular closed* subsets, namely the non-empty regular closed subsets F

such that, whenever $F \subseteq F_1 \cup F_2$, where F_1, F_2 are regular closed, then $F \subseteq F_1$ or $F \subseteq F_2$.

An *isolated point* x in a topological space X is one such that $\{x\}$ is open. Assume X is T_2 and does not have any isolated point, e.g., $[0, 1]$ with its open ball topology. Notice that irreducible regular closed subsets have non-empty interior, and that non-empty open subsets must contain at least two points. Show that the complete Boolean algebra of all regular open subsets of X (Exercise 8.1.8) has no point. Conclude that it is not spatial, unless X is empty.

While $\mathcal{O}$ and pt are not exactly inverse to each other, they are adjoints:

Theorem 8.1.26 ($\mathcal{O} \dashv \mathsf{pt}$) *The functor* $\mathsf{pt}\colon \mathbf{Loc} \to \mathbf{Top}$ *is right adjoint to* $\mathcal{O}\colon \mathbf{Top} \to \mathbf{Loc}$.

The unit of the adjunction $\eta_X^{Stone}\colon X \to \mathsf{pt}(\mathcal{O}(X))$ *maps each point* $x \in X$ *to the completely prime filter of open neighborhoods of* x*, while the counit* $\epsilon_L^{Stone}\colon \mathcal{O}(\mathsf{pt}(L)) \to L$ *maps every element* $u \in L$ *to the open subset* $\mathcal{O}_u$. *(Remember that* $\mathbf{Loc} = \mathbf{Frm}^{op}$*, so that* $\epsilon_L^{Stone}\colon \mathcal{O}(\mathsf{pt}(L)) \to L$ *is a frame homomorphism from* L *to* $\mathcal{O}(\mathsf{pt}(L))$*.)*

Proof It is enough to show that, for each topological space X, $\mathcal{O}(X)$ is the free object in **Loc** (for short, the *free locale*) over the topological space X, with respect to $U = \mathsf{pt}$, in the sense of Section 5.5.2. Define $\eta_X^{\mathrm{Stone}}(x)$ as the completely prime filter of open neighborhoods of x.

Given any open subset of $\mathsf{pt}(\mathcal{O}(X))$, which necessarily is of the form $\mathcal{O}_U$ for some open subset U of X, the inverse image $\eta_X^{\mathrm{Stone}^{-1}}(\mathcal{O}_U)$ is the set of points $x \in X$ such that the set of all open neighborhoods of x is in $\mathcal{O}_U$, i.e., such that U is an open neighborhood of x. So $\eta_X^{\mathrm{Stone}^{-1}}(\mathcal{O}_U) = U$. It follows that η_X^{Stone} is continuous.

We now have to check Diagram (5.2). That is, let f be any continuous map from X to $\mathsf{pt}(L)$, for some frame L, and show that there is a unique morphism $h\colon \mathcal{O}(X) \to L$ in **Loc**, i.e., a unique frame homomorphism $h\colon L \to \mathcal{O}(X)$ such that $\mathsf{pt}(h) \circ \eta_X^{\mathrm{Stone}} = f$. If h exists, then, for every $x \in X$, $f(x) = \mathsf{pt}(h)(\eta_X^{\mathrm{Stone}}(x)) = h^{-1}(\eta_X^{\mathrm{Stone}}(x))$. In particular, for every $u \in L$, $h(u)$ is an open neighborhood of a point $x \in X$ iff $u \in f(x)$. So $h(u)$ must be $\{x \in X \mid u \in f(x)\}$, whence h is unique if it exists at all. Since $u \in f(x)$ iff $f(x) \in \mathcal{O}_u$, $h(u)$ is in fact $f^{-1}(\mathcal{O}_u)$. Since $\bigcap_{i=1}^n \mathcal{O}_{u_i} = \mathcal{O}_{\inf(u_1,\ldots,u_n)}$ and $\bigcup_{i\in I} \mathcal{O}_{u_i} = \mathcal{O}_{\sup_{i\in I} u_i}$, h is a frame homomorphism.

The unique h so obtained is, by definition, $\mathrm{ran}_{X,L}(f)$.

Now $\epsilon_L^{\mathrm{Stone}} = \mathrm{ran}_{\mathsf{pt}(L),L}(\mathrm{id}_{\mathsf{pt}(L)})$, so $\epsilon_L^{\mathrm{Stone}}(u) = \mathcal{O}_u$. ☐

Finally, the construction of Diagram (5.2) yields a left adjoint to pt, and we must check that this is $\mathcal{O}$. This left adjoint is defined on morphisms $f: X \to Y$ as $\mathrm{ran}_{X,\mathcal{O}(Y)}(\eta_Y \circ f)$, the unique morphism $h: \mathcal{O}(X) \to \mathcal{O}(Y)$ in **Loc** such that $\mathsf{pt}(h) \circ \eta_X^{\mathrm{Stone}} = \eta_Y^{\mathrm{Stone}} \circ f$. We must check that this is exactly $\mathcal{O}(f)$, the function that maps every $V \in \mathcal{O}(Y)$ to $f^{-1}(V)$. That is, we must check that $\mathsf{pt}(\mathcal{O}(f)) \circ \eta_X^{\mathrm{Stone}} = \eta_Y^{\mathrm{Stone}} \circ f$. For every $x \in X$, $\mathsf{pt}(\mathcal{O}(f))$ maps $\eta_X^{\mathrm{Stone}}(x)$, the set of open neighborhoods of x, to the set of opens V of Y such that $\mathcal{O}(f)(V)$ is an open neighborhood of x, i.e., such that $x \in f^{-1}(V)$: this is the set of open neighborhoods of $f(x)$, i.e., $(\eta_Y^{\mathrm{Stone}} \circ f)(x)$.

8.2 Sober spaces and sobrification

Given any topological space X, one may want to equate each $x \in X$ with the completely prime filter of all open neighborhoods of x, as suggested in Theorem 8.1.26.

Proposition 8.2.1 *For each topological space X, let $\eta_X^{Stone}: X \to \mathsf{pt}(\mathcal{O}(X))$ map $x \in X$ to the completely prime filter of all open neighborhoods of x in X. Then η_X^{Stone} is continuous and almost open.*

Moreover, η_X^{Stone} is injective iff X is T_0. If this is so, then η_X^{Stone} is an embedding.

If η_X^{Stone} is bijective, then it is a homeomorphism.

Proof We have already seen in Theorem 8.1.26 that, for each open subset U of X, $\eta_X^{\mathrm{Stone}^{-1}}(\mathcal{O}_U) = U$, so η_X^{Stone} is continuous. It also follows that η_X^{Stone} is almost open: every open subset U of X arises as $(\eta_X^{\mathrm{Stone}})^{-1}(\mathcal{O}_U)$.

Next, η_X^{Stone} is injective iff any two points that have the same open neighborhoods are equal. This is equivalent to saying that X is T_0. In this case, η_X^{Stone} is an embedding by Proposition 4.10.9.

Assume η_X^{Stone} bijective. Since it is surjective and almost open, it is open (Exercise 4.10.11), and we conclude by Proposition 4.10.4. □

So we can recover X from its lattice $\mathcal{O}(X)$ of open sets, up to homeomorphism, iff X is T_0 and η_X^{Stone} is surjective.

8.2.1 Irreducible closed subsets, and sober spaces

The property that η_X^{Stone} is surjective can be characterized more simply using the notion of an *irreducible* closed subset.

Proposition 8.2.2 (Irreducible closed subsets) *Let X be a topological space. A closed subset F of X is* irreducible *iff it is non-empty and, whenever F is contained in the union of two closed sets F_1 and F_2, $F \subseteq F_1$ or $F \subseteq F_2$. Alternatively, a closed set F is irreducible iff, whenever F is contained in a finite union of closed sets $F_1, \ldots, F_n$, it is contained in one of them.*

Then η_X^{Stone} is surjective iff every irreducible closed subset F of X is the closure $\downarrow x$ of some point $x \in X$.

Proof Assume η_X^{Stone} is surjective, and let F be some irreducible closed subset of X. Consider the set z of all open subsets U that intersect F. This is upward closed. For any two elements U_1, U_2 of z, let F_1 be the complement of U_1, F_2 that of U_2. Since U_1 intersects F, $F \not\subseteq F_1$, and similarly $F \not\subseteq F_2$. Since F is irreducible, F cannot be included in $F_1 \cup F_2$, i.e., $U_1 \cap U_2$ must intersect F. So z is a filter. It is clearly completely prime. Since η_X^{Stone} is surjective, $z = \eta_X^{\mathrm{Stone}}(x)$ for some $x \in X$. By definition, an open subset U of X intersects F iff $U \in z$, iff $x \in U$. In particular, taking U to be the complement of $\downarrow x$ (the closure of x; see Lemma 4.2.7), U cannot intersect F, i.e., $F \subseteq \downarrow x$. Conversely, taking U to be the complement of F, one obtains $x \notin U$, i.e., $x \in F$; in particular, $\downarrow x \subseteq F$. So $F = \downarrow x$.

Conversely, assume that every irreducible closed subset of X is the closure of some point. Let z be any completely prime filter of $\mathcal{O}(X)$. Let U be the union of the open subsets that are not in z. Note that U is not in z either, since z is completely prime. So U is the largest open that is not in z. So the following holds: $(*)$ an open set V is included in U iff V is not in z.

Let F be the complement of U. F is closed. We claim that it is irreducible. We show this by contraposition, i.e., we assume that there are two closed subsets F_1, F_2, such that F is included in neither, and we show that it is not included in $F_1 \cup F_2$. Let U_1, U_2 be the complements of F_1, F_2. Since $F \not\subseteq F_1$, $U_1 \not\subseteq U$ so U_1 is in z. This is by $(*)$, taking $V = U_1$. Similarly, U_2 is in z. Since z is a filter, $U_1 \cap U_2$ is in z. So $U_1 \cap U_2$ is not included in U by $(*)$ again, i.e., F is not included in $F_1 \cup F_2$.

Since F is irreducible, by assumption $F = \downarrow x$ for some $x \in X$. We claim that $z = \eta_X^{\mathrm{Stone}}(x)$. Indeed, for every open subset V of X, V is in $\eta_X^{\mathrm{Stone}}(x)$ iff $x \in V$, while $V \in z$ iff $V \not\subseteq U$ by $(*)$, iff V intersects $F = \downarrow x$, iff $x \in V$. So z and $\eta_X^{\mathrm{Stone}}(x)$ are completely prime filters containing the same elements, and are therefore equal. □

Beware that the irreducible regular closed subsets of Exercise 8.1.25 are not those irreducible closed subsets that are regular: in the definition of the former, we require F_1 and F_2 to be regular closed, while they are only required to be closed in Proposition 8.2.2.

Trick 8.2.3 Let X be a topological space. A closed subset C of X is irreducible if and only if it is non-empty and, whenever C meets an open subset U_1 and an open subset U_2, then C meets their intersection $U_1 \cap U_2$.

Indeed, consider the complements $F_1 = X \setminus U_1$ and $F_2 = X \setminus U_2$, and apply the definition. Note that C meets an open subset U iff it is not included in its complement $F = X \setminus U$.

The T_0 spaces X such that η_X^{Stone} is surjective deserve a name.

Definition 8.2.4 (Sober spaces) A topological space X is *sober* iff it is T_0 and every irreducible closed subset of X is the closure of a (unique) point.

Using Proposition 8.2.2 and Proposition 8.2.1, we obtain:

Fact 8.2.5 *The following are equivalent: X is sober; η_X^{Stone} is bijective; η_X^{Stone} is a homeomorphism.*

So we can retrieve the points of X from just its open sets if and only if X is sober. Note that while a T_0 space is a space that has enough opens to separate each pair of distinct points, a sober space also has enough points, in the sense that any irreducible closed subset is of the form $\downarrow x$, i.e., has a largest point.

Example 8.2.6 $\mathbb{R}$, with its usual, open ball topology, is sober. In general, every T_2 space is sober, as we shall see in Proposition 8.2.12.

Example 8.2.7 $\mathbb{N}_\sigma$ is not sober. Indeed, $\mathbb{N}$ itself is an irreducible closed subset, but not of the form $\downarrow n$ for any $n \in \mathbb{N}$: if $\mathbb{N}$ intersects two open subsets $\uparrow m$ and $\uparrow n$, then it contains $\max(m, n)$, hence one of them, and we conclude by Trick 8.2.3. On the other hand, $\mathbb{N}_\omega$ is sober in its Scott topology. In general, every continuous dcpo is sober, as we shall see in Proposition 8.2.12.

Example 8.2.8 Recall that the cofinite topology on $\mathbb{N}$ has all finite subsets as closed sets, plus $\mathbb{N}$ itself. This is not sober either. Indeed, $\mathbb{N}$ itself, as a closed subset, is irreducible: if $\mathbb{N}$ is included in the union $F_1 \cup F_2$ of two closed subsets, then F_1 or F_2 must be infinite, hence equal to the whole of $\mathbb{N}$ already. This is the only case of an irreducible closed subset that has more than one element. Adding a new element above all the pairwise incomparable elements of $\mathbb{N}$ yields a sober space.

Exercise 8.2.9 Remember the lifting $X_\perp$ of a space X (Exercise 4.4.25). Show that X is sober if and only if $X_\perp$ is sober. (First show that X is T_0 if and only if $X_\perp$ is T_0.) What are the irreducible closed subsets of $X_\perp$?

Exercise 8.2.10 Let X be a topological space. Show that X, as a closed set, is irreducible if and only if any two non-empty open subsets of X have a non-empty intersection.

Dually to the fact that, for every topological space X, $\mathcal{O}(X)$ is spatial (by definition), we have:

Proposition 8.2.11 *For every complete lattice L, $\mathsf{pt}(L)$ is sober.*

Proof Let $X = \mathsf{pt}(L)$. If any two completely prime filters x, y have the same open neighborhoods $\mathcal{O}_u$, by definition they have the same elements u, so $x = y$. So X is T_0.

Let C be an irreducible closed subset of X. Let x be the union of all the elements of C: x is upward closed, and is non-empty since C is non-empty and each element of C is non-empty. We claim that x is a filter. Let $u, v \in x$. Since $u \in x$, u is in some element y of C, so C intersects $\mathcal{O}_u$ (at y). Similarly, C intersects $\mathcal{O}_v$. By Trick 8.2.3, since C is irreducible, C intersects $\mathcal{O}_u \cap \mathcal{O}_v = \mathcal{O}_{u \wedge v}$: there is a $y \in C$ such that $u \wedge v \in y$. Since $y \subseteq x$, $u \wedge v$ is in x. So x is a filter. It is also completely prime: if $\sup_{i \in I} u_i \in x$ then $\sup_{i \in I} u_i \in y$ for some $y \in C$, so u_i is in y, hence in x, for some $i \in I$. So x is an element of X.

We now claim that C is the downward closure of x. For every element y of C, y is included in x, so it only remains to show that x is in C. Otherwise, x is in the complement of C, which is open, hence of the form $\mathcal{O}_u$ for some $u \in L$, using Proposition 8.1.13. By definition of $\mathcal{O}_u$, $u \in x$. By definition of x, there is a $y \in C$ that contains u. So $\mathcal{O}_u$ meets C at y, which is impossible. □

8.2.2 Families of sober spaces

There are (at least) three standard families of sober spaces:

Proposition 8.2.12 (Families of sober spaces) *The following are sober: (a) the T_2 spaces, in particular the metric spaces; (b) the continuous dcpos in their Scott topology; (c) the Smyth-complete quasi-metric spaces in their open ball topology.*

Proof (a) Let X be T_2. In particular, X is T_0. Let C be irreducible closed. If C contained two distinct points x, y, then there would be two disjoint open neighborhoods U_x of x, and U_y of y. Since C meets both U_x and U_y, it would meet $U_x \cap U_y = \emptyset$, by Trick 8.2.3, and this is impossible. So C contains exactly one point.

(b) Let X be a continuous dcpo. Let C be irreducible closed, and consider the set D of all points x such that $x \ll y$ for some $y \in C$. D is non-empty since C is non-empty, and for every $y \in C$, $\Downarrow y$ is directed hence non-empty. Consider two points x_1, x_2 of D. So $\Uparrow x_1$ intersects C, and $\Uparrow x_2$ intersects C. Since X is continuous, by Proposition 5.1.16 $\Uparrow x_1$ and $\Uparrow x_2$ are open, and Trick 8.2.3 implies that $\Uparrow x_1 \cap \Uparrow x_2$ also intersects C. In other words, there is an element y of C such that $x_1, x_2 \ll y$. Since $\ll$ is interpolative on X (Lemma 5.1.32), there is an $x \in X$ such that $x_1, x_2 \ll x \ll y$. So x is in D. This shows that D is directed. Since X is a dcpo, $\sup D$ exists, and is in C since C is Scott-closed. For every $y \in C$, since X is continuous, $y = \sup_{x \ll y} x \leqslant \sup D$. So $\sup D$ is the largest element of C, in particular $C = \downarrow \sup D$.

(c) Finally, let X, d be a Smyth-complete quasi-metric space. Let C be an irreducible closed set of X, d. Let D be the family of formal balls (x, r) where $x \in C$ and $r \in \mathbb{R}^+ \smallsetminus \{0\}$. We claim that D is directed in $\mathbf{B}(X, d), \leqslant^{d^+}$. Indeed, let $(x, r), (x', r') \in D$. Then $B^d_{x,<r}$ and $B^d_{x',<r'}$ both intersect C, so their intersection does, too (Trick 8.2.3). Let $x'' \in B^d_{x,<r} \cap B^d_{x',<r'} \cap C$. So $d(x, x'') < r$ and $d(x', x'') < r'$. Let $r'' = \min(r - d(x, x''), r' - d(x', x''))$. Then (x'', r'') is in D, and above both (x, r) and (x', r'). So D is directed. By the Romaguera–Valero Theorem 7.3.11, D has a least upper bound (y, s), and y is the limit of the net $(x_i)_{i \in I, \leqslant^{d^+}}$ in X, d^{sym}, where $I = D$ and $x_i = x$ whenever $i = (x, r)$. By Lemma 7.1.9, y is also a limit of the net in X, d. Since C is closed, y is in C by Corollary 4.7.10. By definition, for every $x \in C$, for every $r > 0$, $(x, r) \leqslant^{d^+} (y, s)$, i.e., $d(x, y) \leqslant r - s$. So $d(x, y) = 0$, i.e., $x \leqslant^d y$ by Proposition 6.1.8. So C is the downward closure of $y \in C$. □

Exercise 8.2.13 Sobriety is between T_0 and T_2: every sober space is T_0, and every T_2 space is sober by Proposition 8.2.12 (a). Being T_1 is also a property between T_0 and T_2. Show that being T_1 and being sober are incomparable properties; i.e., exhibit a T_1 space that is not sober, and a sober space that is not T_1.

Exercise 8.2.14 Proposition 8.2.12 (b) states that if X is a continuous dcpo, then X_σ is sober. Give an example of a continuous poset X such that X_σ is not sober. (Consider the simplest possible poset that is not a dcpo.)

Show that the Johnstone space $\mathbb{J}$ (Exercise 5.2.15) is a dcpo such that $\mathbb{J}_\sigma$ is not sober. Indication: show that any non-empty open subset of $\mathbb{J}_\sigma$ must contain all points of the form (n, ω), $n \in \mathbb{N}$, except for finitely many, and use Exercise 8.2.10. (Bonus: show that the topology on $\mathbb{N}$ induced by the mapping $n \mapsto (n, \omega)$ is the cofinite topology.)

So both continuity and being a dcpo are needed to make X_σ sober. Conversely, exhibit a dcpo X such that X_σ is sober but X is not continuous. (We shall see later that if X_σ is sober, then X must be a dcpo. See Exercise 8.2.35.)

Exercise 8.2.15 (Quasi-continuous dcpos are sober) Show that every quasi-continuous dcpo is sober in its Scott topology. This extends Proposition 8.2.12 (b).

Exercise 8.2.16 (Continuous Yoneda-complete $\Rightarrow$ sober) Let X, d be a continuous Yoneda-complete quasi-metric space (Definition 7.4.72). Show that X is sober in its d-Scott topology (not its open ball topology). As a hint, imitate the proof of Proposition 8.2.12 (c), and think of using Exercise 7.4.45.

Show how Proposition 8.2.12 (c) arises as a special case.

8.2.3 Sobrification

When a T_0 space X fails to be sober, this is because some points are lacking from X: namely, for any irreducible closed subset F that is not the closure of a point, one point is missing. One may then complete X to add all missing points:

Definition 8.2.17 (Sobrification) For every topological space X, the *sobrification* $\mathcal{S}(X)$ is the set of all irreducible closed subsets of X, with the so-called *lower Vietoris topology*, whose (subbasic) open sets are the sets $\diamond U$ for each open subset U of X; we define $\diamond U$ as the set of all irreducible closed subsets F of X that intersect U.

The reason why we put parentheses around "subbasic" is that the subsets $\diamond U$ do not form just a subbase or even a base, but actually exhaust all the opens of $\mathcal{S}(X)$.

Lemma 8.2.18 *Let X be a topological space. The $\diamond$ operator commutes with all finite intersections and all unions. In particular, the open subsets of X are* exactly *those of the form $\diamond U$, U open in X.*

Proof For any family of opens $(U_i)_{i\in I}$, $\diamond(\bigcup_{i\in I} U_i) = \bigcup_{i\in I} \diamond U_i$. Indeed, F intersects $\bigcup_{i\in I} U_i$ iff F intersects some U_i, $i \in I$. Next, we claim that, for any finite family of opens $U_1, \ldots, U_n$, $\diamond(\bigcap_{i=1}^n U_i) = \bigcap_{i=1}^n \diamond U_i$. Indeed, the left-to-right inclusion is clear. Conversely, any irreducible closed subset C in $\bigcap_{i=1}^n \diamond U_i$ must intersect each U_i, $1 \leqslant i \leqslant n$. Let F_i be the (closed) complement of U_i. If C were not in $\diamond(\bigcap_{i=1}^n U_i)$, i.e., if C did not intersect $U_1 \cap \cdots \cap U_n$, then C would be included in $F_1 \cup \cdots \cup F_n$. Since C is irreducible, C would be included in some F_i, $1 \leqslant i \leqslant n$, contradicting the fact that it intersects U_i. $\square$

Example 8.2.19 Recall from Example 8.2.7 that $\mathbb{N}_\sigma$ is not sober, since the whole of $\mathbb{N}$ is irreducible. The other closed subsets of $\mathbb{N}$ are of the form $\downarrow n$. So $\mathcal{S}(\mathbb{N}_\sigma)$ is homeomorphic to $\mathbb{N}$, plus a new top element added. It is easy to see that the lower Vietoris topology is the Scott topology (also, the upper topology): up to homeomorphism, $\mathcal{S}(\mathbb{N}_\sigma)$ is $\mathbb{N}_\omega$ with its Scott topology.

Example 8.2.20 Recall from Example 8.2.8 that $\mathbb{N}$ with the cofinite topology is not sober either. Its sobrification is $\mathbb{N}$ (whose elements are pairwise incomparable) with a fresh top element $\top$ added, in the topology whose non-empty opens are the complements of finite subsets of $\mathbb{N}$, with $\top$ added: this is the upper topology.

Exercise 8.2.21 Remember that X is sober if and only if $X_\perp$ is sober (Exercise 8.2.9). Show that, in general, $\mathcal{S}(X_\perp) = \mathcal{S}(X)_\perp$, at least up to some natural iso.

While Definition 8.2.17 is natural in the view of Definition 8.2.4, Proposition 8.2.1 suggests another definition of the sobrification of X, as $\mathsf{pt}(\mathcal{O}(X))$. The latter space has a more abstruse definition, but the two spaces are the same, up to iso, as we now claim. In fact, this is the need for the map $f_{\mathcal{S}}$, defined below, to be a homeomorphism that dictated the definition of the lower Vietoris topology on $\mathcal{S}(X)$.

Proposition 8.2.22 ($\mathcal{S}(X) \cong \mathsf{pt}(\mathcal{O}(X))$) *Let X be a topological space. The function $f_{\mathcal{S}} \colon \mathcal{S}(X) \to \mathsf{pt}(\mathcal{O}(X))$ that maps each $C \in \mathcal{S}(X)$ to the set of all open subsets of X that intersect C is a homeomorphism. Moreover, $f_{\mathcal{S}} \circ \eta^{\mathcal{S}}_X = \eta^{Stone}_X$, where $\eta^{\mathcal{S}}_X \colon X \to \mathcal{S}(X)$ is defined by $\eta^{\mathcal{S}}_X(x) = \downarrow x$ for each $x \in X$.*

Proof We first show that $f_{\mathcal{S}}(C)$ is a completely prime filter. If U intersects C and $U \subseteq U'$, then clearly U' also intersects C. If U_1, U_2 are two open subsets that intersect C, then $U_1 \cap U_2$ also intersects C since C is irreducible (Trick 8.2.3). If $(U_i)_{i \in I}$ is a family of open subsets whose union intersects C, then some U_i intersects C.

Next, for every open subset $\mathcal{O}_U$ of $\mathsf{pt}(\mathcal{O}(X))$, $f_{\mathcal{S}}^{-1}(\mathcal{O}_U)$ is the set of all $C \in \mathcal{S}(X)$ such that U is among the open subsets that intersect C. So $f_{\mathcal{S}}^{-1}(\mathcal{O}_U) = \Diamond U$, and $f_{\mathcal{S}}$ is therefore continuous.

Conversely, let $g_{\mathcal{S}}$ map every element of $\mathsf{pt}(\mathcal{O}(X))$, i.e., every completely prime filter of opens x, to the complement of the union U_x of all open subsets that are not in x. Because x is completely prime, U_x is not in x either. So U_x is the largest open subset that is not in x. Since U_x is open, $g_{\mathcal{S}}(x)$ is closed, and we claim that $g_{\mathcal{S}}(x)$ is irreducible. We use Trick 8.2.3. If $g_{\mathcal{S}}(x)$ meets two open subsets V_1 and V_2, then neither V_1 nor V_2 is included in U_x, so V_1 and V_2

are both in x. Since x is a filter, $V_1 \cap V_2$ is in x, so $V_1 \cap V_2$ is not included in U_x, whence $g_{\mathcal{S}}(x)$ meets $V_1 \cap V_2$.

We claim that $g_{\mathcal{S}}$ is continuous: the inverse image of $\diamond U$, U open in X, is the set of completely prime filters x such that the complement of U_x intersects U, i.e., such that U is not included in U_x, i.e., such that $U \in x$, by definition of U_x. This is the open $\mathcal{O}_u$.

We now check that $g_{\mathcal{S}}(f_{\mathcal{S}}(C)) = C$ for every $C \in \mathcal{S}(X)$: $U_{f_{\mathcal{S}}(C)}$ is the largest open not in $f_{\mathcal{S}}(C)$, i.e., that does not intersect C, namely the complement of $cl(C) = C$. And we check that $f_{\mathcal{S}}(g_{\mathcal{S}}(x)) = x$ for every completely prime filter x of opens: $f_{\mathcal{S}}(g_{\mathcal{S}}(x))$ is the set of all opens that intersect $g_{\mathcal{S}}(x)$, i.e., that are not contained in U_x, i.e., that are in x. □

Corollary 8.2.23 ($\mathcal{S}(X)$ is sober) *The sobrification $\mathcal{S}(X) \cong \mathsf{pt}(\mathcal{O}(X))$ of any topological space is sober.*

Proof By Proposition 8.2.11. □

It also follows, using Fact 8.2.5, that:

Fact 8.2.24 *The following are equivalent: (a) X is sober; (b) $\eta^{\mathcal{S}}_X : X \to \mathcal{S}(X)$ is bijective; (c) $\eta^{\mathcal{S}}_X : X \to \mathcal{S}(X)$ is a homeomorphism.*

Exercise 8.2.25 Give a direct and elementary proof of Corollary 8.2.23.

We now stress the fact that, while $\mathcal{S}(X)$ has more points than X, it has exactly the same opens, up to iso.

Lemma 8.2.26 ($\mathcal{S}(X)$ has as many opens as X) *Let X be a topological space. The map $U \mapsto \diamond U$ is an order isomorphism from $\mathcal{O}(X)$ to $\mathcal{O}(\mathcal{S}(X))$.*

Proof Clearly, the map $U \mapsto \diamond U$ is monotonic. By Lemma 8.2.18, it is bijective. It only remains to show that if $\diamond U \subseteq \diamond V$, then $U \subseteq V$. Otherwise, there would be a point x in U but outside V. Then $\downarrow x$ would be in $\diamond U$ but outside $\diamond V$, a contradiction. □

8.2.4 Convergence in sober spaces

Sober spaces can also be characterized through notions of convergence. This uses the following funny notion of a self-convergent net.

Lemma 8.2.27 (Irreducible $\cong$ self-convergent net) *Let X be a topological space. A net $(x_i)_{i \in I, \sqsubseteq}$ in X is* self-convergent *if and only if it converges to x_i for every $i \in I$.*

The irreducible closed subsets of X are exactly the closures of self-convergent nets in X.

More precisely, for every self-convergent net $(x_i)_{i \in I, \sqsubseteq}$ in X, $cl\{x_i \mid i \in I\}$ is irreducible closed, and conversely, every irreducible closed subset C is exactly the set of elements of some self-convergent net $(x_i)_{i \in I, \sqsubseteq}$.

Proof Let C be irreducible closed in X, define I as the set of pairs (U, x) with U open in X and $x \in U \cap C$, and quasi-order I by $(U, x) \sqsubseteq (V, y)$ iff $U \supseteq V$. For any two elements (U_1, x_1) and (U_2, x_2) of I, U_1 meets C and U_2 meets C, so $U_1 \cap U_2$ also meets C, say at x. This is because C is irreducible, and using Trick 8.2.3. Now $(U_1, x_1), (U_2, x_2) \sqsubseteq (U_1 \cap U_2, x)$. So I is directed.

Define a net $(x_i)_{i \in I, \sqsubseteq}$ as follows. For each $i \in I$, i is of the form (U, x), then let $x_i = x$. For every $i \in I$, say $i = (U, x)$, we claim that $x_i = x$ is a limit of the net. Let V be an arbitrary open neighborhood of x. Then $(U \cap V, x)$ is in I, and for every $j \in I$ with $(U \cap V, x) \sqsubseteq j$, j is of the form (W, z) with $W \subseteq U \cap V$, and $x_j = z$ is in W, hence in V. That is, x_j is eventually in V, as desired. Finally, C is (the closure of) $\{x_i \mid i \in I\}$.

Conversely, let $(x_i)_{i \in I, \sqsubseteq}$ be a self-convergent net in X, and let $C = cl\{x_i \mid i \in I\}$. Let U_1, U_2 be two open sets that both meet C. By Corollary 4.1.28, one can find $i_1, i_2 \in I$ such that $x_{i_1} \in U_1$ and $x_{i_2} \in U_2$. By self-convergence, both x_{i_1} and x_{i_2} are limits of the net, so x_i is in U_1 for i large enough, say for every $i \in I$ with $m_1 \sqsubseteq i$, and x_i is in U_2 for i large enough, say with $m_2 \sqsubseteq i$. Since I is directed, find $m \in I$ such that $m_1, m_2 \sqsubseteq m$. Then x_m is in $U_1 \cap U_2$. So C meets $U_1 \cap U_2$. By Trick 8.2.3, C is irreducible. □

Example 8.2.28 The notion of a self-convergent net is trivial in T_2 spaces: in a T_2 space, limits are unique (Proposition 4.7.4), so the self-convergent nets are exactly the constant nets. This agrees with Lemma 8.2.27: by Proposition 8.2.12 (*a*), every T_2 space X is sober, so the irreducible closed subsets of X are exactly the one-element subsets.

Theorem 8.2.29 (Sober = self-convergent nets have unique largest limits) *A space is sober if and only if every self-convergent net $(x_i)_{i \in I, \sqsubseteq}$ has a unique largest limit x. Additionally, $\downarrow x = cl\{x_i \mid i \in I\}$.*

Proof Assume X is sober, and let $(x_i)_{i \in I, \sqsubseteq}$ be an arbitrary self-convergent net. By Lemma 8.2.27, $cl\{x_i \mid i \in I\}$ is irreducible closed, hence of the form $\downarrow x$ for some unique $x \in X$. Every open neighborhood U of x then meets $\{x_i \mid i \in I\}$ (Corollary 4.1.28), hence contains some x_i. Since x_i is a limit of the net, $x_j \in U$ for j large enough. So x is a limit of the net.

Let y be another limit. Since closed subsets contain all limits of their elements (Corollary 4.7.10), $y \in \downarrow x$. So x is the largest limit, and this is unique since X is T_0.

Conversely, assume that every self-convergent net has a unique largest limit. So X is T_0: if $x \leqslant y$ and $y \leqslant x$, then the set of limits of the constant net equal to x is $\downarrow x = \downarrow y$, whose largest elements are x and y; so $x = y$. Moreover, if C is an irreducible closed subset, then by Lemma 8.2.27 C is the set of elements of some self-convergent net $(x_i)_{i \in I, \sqsubseteq}$. Let x be its largest limit. Every element x_i is also a limit, hence $x_i \leqslant x$. So $C \subseteq \downarrow x$. Since C is closed, by Corollary 4.7.10, x is in C. So $C = \downarrow x$. ☐

Exercise 8.2.30 Let X be a topological space, $(x_i)_{i \in I, \sqsubseteq}$ be a self-convergent net in X, and $C = cl\{x_i \mid i \in I\}$. Remember that C is then in $\mathcal{S}(X)$, and that all elements of $\mathcal{S}(X)$ can be obtained this way (Lemma 8.2.27). Show that $(\eta^{\mathcal{S}}_X(x_i))_{i \in I, \sqsubseteq}$ converges to C, and C is even the largest limit of the net $(\eta^{\mathcal{S}}_X(x_i))_{i \in I, \sqsubseteq}$ in $\mathcal{S}(X)$.

In non-T_2 spaces, there is an ample supply of self-convergent nets:

Fact 8.2.31 *Let X be a topological space. Every directed family $(x_i)_{i \in I}$ defines a net $(x_i)_{i \in I, \sqsubseteq}$, where $i \sqsubseteq j$ iff $x_i \leqslant x_j$. This net is always self-convergent.*

Indeed, every open neighborhood U of x_i must contain x_j for every $j \in I$ above i, since U is upward closed (Lemma 4.2.5).

Lemma 8.2.32 *Let X be a topological space, and $(x_i)_{i \in I}$ be a directed family in X. If the net defined by it has a largest limit x, then x is the least upper bound of $(x_i)_{i \in I}$.*

Proof Define $i \sqsubseteq j$ iff $x_i \leqslant x_j$. Assume $(x_i)_{i \in I, \sqsubseteq}$ has a largest limit x. Since $(x_i)_{i \in I, \sqsubseteq}$ is self-convergent (Fact 8.2.31), $x_i \leqslant x$ for every $i \in I$. For every upper bound y of $\{x_i \mid i \in I\}$, we check that $x \leqslant y$ by verifying that every open neighborhood U of x also contains y. Indeed, U contains x_i for i large enough, $x_i \leqslant y$, and U is upward closed by Lemma 4.2.5. ☐

8.2.5 Monotone convergence spaces

From this, we deduce that if X is sober, then it contains all least upper bounds of directed subsets. That is, X is a dcpo in its specialization ordering. We make this more precise.

Definition 8.2.33 (Monotone convergence space) For each T_0 space X, we write X_σ for the set X with the Scott topology of its specialization ordering, extending the notation X_σ for posets X. A *monotone convergence space* is a T_0 space X such that X is a dcpo in its specialization ordering, and whose topology is coarser than that of X_σ, i.e., such that every open is Scott-open.

The monotone convergence spaces were called d-spaces by Wyler (1979).

Proposition 8.2.34 (Sober $\Rightarrow$ monotone convergence) *Every sober space is a monotone convergence space.*

The least upper bound of any directed family D of elements of X is the unique element x such that $cl(D) = \downarrow x$.

Proof By Fact 8.2.31, D defines a self-convergent net $(x_i)_{i \in I, \sqsubseteq}$. Theorem 8.2.29 states that it has a largest limit x. This must be the least upper bound of $(x_i)_{i \in I}$, by Lemma 8.2.32.

So X is a dcpo in its specialization ordering. It remains to show that every open subset U of X is Scott-open. Let D be any directed family of X such that $x \in U$, where $x = \sup D$. Theorem 8.2.29 also states that $\downarrow x = cl(D)$. Since $x \in U$, $cl(D)$ intersects U so D intersects U (Corollary 4.1.28). So U is Scott-open, i.e., open in X_σ. □

Exercise 8.2.35 Let X be a poset. Show that if X_σ is sober, then X is a dcpo, although not necessarily a continuous one. (Give an explicit counterexample.)

Exercise 8.2.36 Give an example of a T_0 space X whose topology is strictly coarser than its Scott topology. (What about taking X T_2?)

Exercise 8.2.37 (Yoneda-complete = monotone convergence space of formal balls) Using Exercise 7.4.53 and the Kostanek–Waszkiewicz Theorem 7.4.27, show that, given a hemi-metric space X, d, its space of formal balls $\mathbf{B}(X, d)$, equipped with its d^+-Scott topology, is a monotone convergence space if and only if X, d is a Yoneda-complete quasi-metric space.

Exercise 8.2.38 (Monotone convergence = directed families converge to their sup) The name "monotone convergence space" is probably best explained through a notion of convergence. Show that a T_0 topological space X is a monotone convergence space if and only if every directed family $(x_i)_{i \in I}$ has a least upper bound, and converges to this least upper bound. (We silently equate the directed family with the net $(x_i)_{i \in I, \sqsubseteq}$, where $i \sqsubseteq j$ iff $x_i \leqslant x_j$.) Moreover, the least upper bound of $(x_i)_{i \in I}$ is its largest limit.

Exercise 8.2.39 (Products of monotone convergence spaces) Using Exercise 8.2.38, for example, show that any product of monotone convergence spaces is a monotone convergence space.

Exercise 8.2.40 (Retracts of monotone convergence spaces) Let X be a retract of a monotone convergence space Y. Show that X is also a monotone convergence space.

Exercise 8.2.41 Show that a quasi-metric space X, d is a monotone convergence space if and only if X with the $\leqslant^d$ ordering is a dcpo, and for every $x \in X$, $d(x, \_)$ is Scott-continuous from X_σ to $\overline{\mathbb{R}^+}^{op}$.

8.2.6 Free sober spaces

Lemma 8.2.42 *There is a functor $\mathcal{S}$ from* **Top** *to the category of sober spaces and continuous maps, defined on morphisms $f: X \to Y$ by $\mathcal{S}(f): \mathcal{S}(X) \to \mathcal{S}(Y)$ maps every $C \in \mathcal{S}(X)$ to $cl(f[C])$.*

In particular, $\mathcal{S}(f)(C)$ is irreducible closed for every $C \in \mathcal{S}(X)$, and $\mathcal{S}(f)$ is continuous: $\mathcal{S}(f)^{-1}(\diamond V) = \diamond f^{-1}(V)$ for every open subset V of Y.

Proof We must first show that, whenever C is irreducible in X, $cl(f[C])$ is irreducible in Y. Indeed, it is non-empty, and if $cl(f[C])$ is included in the union of two closed subsets F_1, F_2 of Y, then $f[C] \subseteq F_1 \cup F_2$, so $C \subseteq f^{-1}(F_1 \cup F_2) = f^{-1}(F_1) \cup f^{-1}(F_2)$. Since C is irreducible, C is included in $f^{-1}(F_1)$ or in $f^{-1}(F_2)$, say the first one. Then $f[C] \subseteq F_1$, so $cl(f[C]) \subseteq F_1$ by the definition of closure.

We check that $\mathcal{S}(f)$ is continuous. We claim that the inverse image of $\diamond V$, V open in Y, is $\diamond f^{-1}(V)$. Indeed, for every $C \in \mathcal{S}(X)$, $\mathcal{S}(f)(C) \in \diamond V$ iff $cl(f[C])$ intersects V, iff $f[C]$ intersects V (Corollary 4.1.28), iff C intersects $f^{-1}(V)$.

We now check that it is a functor. That $\mathcal{S}(\mathrm{id}_X) = \mathrm{id}_{\mathcal{S}(X)}$ is clear. Given $f: X \to Y$ and $g: Y \to Z$ in **Top**, we must show that $\mathcal{S}(g \circ f) = \mathcal{S}(g) \circ \mathcal{S}(f)$, i.e., that, for every $C \in \mathcal{S}(X)$, $cl(g[cl(f[C])]) = cl(g[f[C]])$. Equivalently, we must show that, for every open subset W of Z, W intersects the left-hand side iff it intersects the right-hand side. (The claim will then follow by taking W to be the complement of each side in turn.) Now W intersects the left-hand side iff W intersects $g[cl(f[C])]$, by Corollary 4.1.28, iff $g^{-1}(W)$ intersects $cl(f[C])$, iff $g^{-1}(W)$ intersects $f[C]$ by Corollary 4.1.28 again, iff $f^{-1}(g^{-1}(W))$ intersects C. This is equivalent to the fact that W intersects $cl(g[f[C]])$, still by Corollary 4.1.28. □

Exercise 8.2.43 (Retracts of sober spaces) Show that every retract of a sober space is sober.

Theorem 8.2.44 (Sobrification = free sober space) *For every topological space, $\mathcal{S}(X)$ is a sober space, and its specialization ordering is inclusion $\subseteq$. When X is T_0, the map $\eta^{\mathcal{S}}_X \colon X \to \mathcal{S}(X)$ defined by $\eta^{\mathcal{S}}_X(x) = \downarrow x$, is an embedding, i.e., X appears as a subspace of the sober space $\mathcal{S}(X)$ up to homeomorphism.*

Moreover, $\mathcal{S}(X)$ has the universal property that, for every sober space Y, every continuous map $f \colon X \to Y$ extends to a unique continuous map $f^\star$ from $\mathcal{S}(X)$ to Y, i.e., $f^\star \circ \eta^{\mathcal{S}}_X = f$.

In other words, $\mathcal{S}(X)$ is the free sober space *over the topological space X, i.e., $\mathcal{S} \dashv U$, where U is the forgetful functor from the category of sober spaces to that of topological spaces. It follows that the category of sober spaces and continuous maps is reflective in* **Top**.

Proof We have already seen that $\mathcal{S}(X)$ is sober, in Corollary 8.2.23. Compute the specialization quasi-ordering $\preceq$ of $\mathcal{S}(X)$. Let $C, C' \in \mathcal{S}(X)$. If $C \subseteq C'$, then certainly $C \in \diamond U$ implies $C' \in \diamond U$, from which it easily follows that $C \preceq C'$. Conversely, assume $C \preceq C'$. Take U to be the complement of C' in X. If C were in $\diamond U$, then C' would be in $\diamond U$ since $C \preceq C'$. So $C \notin \diamond U$, i.e., $C \subseteq C'$. In particular, since $\preceq$ is the inclusion ordering, $\mathcal{S}(X)$ is T_0.

We check that $\eta^{\mathcal{S}}_X$ is continuous. More precisely, ${\eta^{\mathcal{S}}_X}^{-1}(\diamond U) = U$ for every open subset U of X, which also shows that $\eta^{\mathcal{S}}_X$ is almost open. Next, $\eta^{\mathcal{S}}_X$ is injective whenever X is T_0, hence is an embedding by Proposition 4.10.9.

Finally, given Y sober and $f \colon X \to Y$ continuous, one can define $f^\star$ as mapping each $C \in \mathcal{S}(X)$ to the unique $y \in Y$ such that $\downarrow y = cl(f[C]) = \mathcal{S}(f)(C)$. We let the reader check the details, using Lemma 8.2.42. However, there is a more direct argument: by Proposition 8.2.22, it suffices to show that there is a unique map $h \colon \mathsf{pt}(\mathcal{O}(X)) \to Y$ such that $h \circ \eta^{\mathsf{Stone}}_X = f$. Since Y is sober, we may consider that Y is of the form $\mathsf{pt}(L)$ for some frame L: indeed $Y \cong \mathsf{pt}(\mathcal{O}(Y))$ (Fact 8.2.5). Then the claim is just Diagram (5.2), i.e., the fact that $\mathcal{O}(X)$ is the free locale over X, which follows from Theorem 8.1.26. □

Exercise 8.2.45 Let X be a topological space, Y be a sober space, and f be a continuous map from X to Y. Check the details of the first argument proposed in the proof of Theorem 8.2.44. That is, show by direct argument that if $f^\star$ exists, then it must map $C \in \mathcal{S}(X)$ to the unique $y \in Y$ such that $\downarrow y = cl(f[C]) = \mathcal{S}(f)(C)$. Show that, for every open subset V of Y, $(f^\star)^{-1}(V) = \diamond(f^{-1}(V))$.

On the other hand, the arguments we used at the end of the proof of Theorem 8.2.44 show that the adjunction $\mathcal{S} \dashv U$ is essentially a reformulation of the adjunction $\mathcal{O} \dashv \mathsf{pt}$. Since η^{Stone}_X is a homeomorphism when X is a sober space, and since similarly $\epsilon_L^{\text{Stone}}$ is an order isomorphism when L is spatial (Proposition 8.1.17), we obtain:

Theorem 8.2.46 (Sober spaces ≡ spatial lattices) *The adjunction $\mathcal{O} \dashv \mathsf{pt}$ restricts to an equivalence between the category of sober spaces and continuous maps on the one hand, and the opposite of the category of spatial lattices and frame homomorphisms on the other hand.*

This is practical. For example:

Trick 8.2.47 To build a continuous map f from a sober space X to a sober space Y, it is enough to find a frame homomorphism g from $\mathcal{O}(Y)$ to $\mathcal{O}(X)$: then there is a unique continuous map f from X to Y such that $g(V) = f^{-1}(V)$ for every open subset V of Y.

Exercise 8.2.48 (Rounded ideal completions are sobrifications) Let $B, \prec$ be an abstract basis (see Lemma 5.1.32). We say that a subset U of B is upward closed iff, whenever $b \in U$ and $b \prec b'$, then $b' \in U$. U is *rounded open* iff U is upward closed and, whenever $b' \in U$, there is a $b \prec b'$ such that $b \in U$. We equip B with the *rounded topology*, whose opens are the rounded opens. Show that this is indeed a topology. Let $\mathcal{R}(B, \prec)$ be B with the rounded topology.

Show that $\mathcal{S}(\mathcal{R}(B, \prec))$ is homeomorphic to the rounded ideal completion $\mathbf{RI}(B, \prec)_\sigma$ with its Scott topology, through the map that sends each rounded ideal to its closure in $\mathcal{R}(B, \prec)$. This result is due to Lawson (1997a), where the rounded topology was called the quasiScott topology.

Deduce from this another proof that every continuous dcpo is sober in its Scott topology.

In the very particular case where $\prec$ is some ordering $\leqslant$, and B is a poset X with ordering $\leqslant$, the rounded ideals are the Alexandroff opens, and the rounded ideal completion $\mathbf{RI}(B, \prec)$ is the ideal completion $\mathbf{I}(X)$. The map sending each (rounded) ideal D to its closure is the identity map. We obtain the following (Hoffmann, 1979b).

Fact 8.2.49 *For every poset X, $\mathbf{I}(X) = \mathcal{S}(X_a)$, meaning not only that the irreducible Alexandroff closed subsets of X are exactly the ideals, but also that the Scott topology coincides with the lower Vietoris topology.*

8.3 The Hofmann–Mislove Theorem

We have noticed that every completely prime filter is, in particular, Scott-open. In general, there are many Scott-open filters in any complete lattice

L, many more than there are completely prime filters. While elements of L correspond to opens of topological spaces, and completely prime filters are points, the Hofmann–Mislove Theorem states the Scott-open filters correspond to compact saturated subsets.

Here is the easy direction.

Lemma 8.3.1 *Let X be a topological space, and K be a compact subset of X. The set of open neighborhoods of K, i.e., of all open subsets U containing K, in X is a Scott-open filter of $\mathcal{O}(X)$.*

Proof This is clearly a filter. That it is Scott-open follows from Proposition 4.4.7. □

Theorem 8.3.2 (Hofmann–Mislove) *Let X be a sober space. For each Scott-open filter $\mathcal{U}$ of $\mathcal{O}(X)$, $Q = \bigcap_{U \in \mathcal{U}} U$ is compact saturated, and the elements of $\mathcal{U}$ are exactly the open neighborhoods of Q.*

Proof Q is clearly saturated. We show the second part of the theorem, which will imply that Q is compact. Assume there is an open neighborhood U of Q that is not in $\mathcal{U}$. The family of open neighborhoods of Q not in $\mathcal{U}$ is inductive: given any chain of open neighborhoods of Q not in $\mathcal{U}$, their union is also an open neighborhood of Q, and cannot be in $\mathcal{U}$; otherwise some member of the chain would already be in $\mathcal{U}$, which is Scott-open. So use Zorn's Lemma (Theorem 2.4.2): there is a maximal open neighborhood U containing Q that is not in $\mathcal{U}$. We claim that its complement F is irreducible. Since $\mathcal{U}$ is a filter, it contains X, so $U \neq X$, hence F is non-empty. If F meets two open subsets U_1 and U_2, neither U_1 nor U_2 can be included in U. Then $U \cup U_1$ and $U \cup U_2$ strictly contain U. By maximality, $U \cup U_1$ and $U \cup U_2$ are both in $\mathcal{U}$, so their intersection is in $\mathcal{U}$, which is a filter. So $(U \cup U_1) \cap (U \cup U_2) = U \cup (U_1 \cap U_2)$ is in $\mathcal{U}$, hence $U \neq U \cup (U_1 \cap U_2)$. It follows that $U_1 \cap U_2$ is not included in U, hence meets F. By Trick 8.2.3, F is irreducible.

Since X is sober, F is the closure $\downarrow x$ of some point $x \in X$. Its complement U is a neighborhood of Q, so x is not in Q. By the definition of Q, there is an element U' of $\mathcal{U}$ that does not contain x, hence that does not intersect F, i.e., that is included in U. But since $\mathcal{U}$ is upward closed, this would imply that U is in $\mathcal{U}$, a contradiction. So $\mathcal{U}$ is exactly the set of open neighborhoods of Q.

It follows that Q is compact: for any directed family $(U_i)_{i \in I}$ of open subsets whose union contains Q, i.e., whose union is in $\mathcal{U}$, some U_i is in $\mathcal{U}$ since $\mathcal{U}$ is Scott-open, i.e., $Q \subseteq U_i$ for some $i \in I$. We conclude by Proposition 4.4.7. □

Corollary 8.3.3 *Let X be a sober space, and K be a subset of X. Then K is compact if and only if the collection of all open neighborhoods of K in X is Scott-open in $\mathcal{O}(X)$.*

Proof Note that this collection is always a filter. The only if direction is true by Lemma 8.3.1. In the if direction, Theorem 8.3.2 tells us that the intersection Q of all open neighborhoods of K is compact. By Proposition 4.2.9, $Q = \uparrow K$. Now use Proposition 4.4.14 to conclude that K is compact. □

Exercise 8.3.4 (The Dolecki–Greco–Lechicki Theorem) Let X be a Čech-complete space, and let $(U_{ni})_{i \in I_n}$ be the nth open cover in the definition of Čech-completeness (see Exercise 7.6.21). Let $\mathcal{U}$ be a Scott-open subset of $\mathcal{O}(X)$ (note: not a Scott-open *filter*). If X is regular and $U \in \mathcal{U}$, show that one can find open subsets $U_0, U_1, \ldots, U_n, \ldots$, and closed subsets $F_0, F_1, \ldots, F_n, \ldots$, such that $U = U_0 \supseteq F_0 \supseteq U_1 \supseteq F_1 \supseteq U_2 \supseteq \cdots$, and such that, for each $n \in \mathbb{N}$, U_n is in $\mathcal{U}$ and is included in a finite union of opens from $(U_{ni})_{i \in I_n}$. Show that the family $\mathcal{V}$ of all open subsets of X that contain some F_n, $n \in \mathbb{N}$, is a Scott-open filter. Using the Hofmann–Mislove Theorem, conclude that if X is T_3, then there is a compact set Q such that $Q \subseteq U$, and such that every open neighborhood of Q is in $\mathcal{U}$.

Recall that a topological space X is consonant (Exercise 5.4.12) if and only if every Scott-open subset of $\mathcal{O}(X)$ is of the form $\bigcup_{i \in I} \blacksquare Q_i$, where each Q_i is compact saturated in X, and the notation $\blacksquare Q$ denotes the collection of open subsets of X that contain Q.

Prove the *Dolecki–Greco–Lechicki Theorem* (Dolecki *et al.*, 1995, theorem 4.1): every T_3 Čech-complete space is consonant.

So we know at least two classes of consonant spaces: the locally compact spaces (Exercise 5.4.12) and the T_3 Čech-complete spaces. Show that Baire space $\mathcal{B} = \mathbb{N}^{\mathbb{N}}$, while not locally compact (Example 4.8.12), is consonant. Now finish Exercise 5.4.25, and show that the compact-open topology on $[\mathcal{B} \to \mathbb{N}]$ coincides with the Isbell topology, and that the Isbell topology on $[\mathcal{B} \to \mathbb{N}]$ is strictly coarser than the natural topology.

8.3.1 Well-filtered spaces

A prominent consequence of the Hofmann–Mislove Theorem is the following.

Proposition 8.3.5 (Well-filteredness) *A topological space X is* well-filtered *iff, for every filtered family $(Q_i)_{i \in I}$ of compact saturated subsets of X, for every open subset U of X, if $\bigcap_{i \in I} Q_i \subseteq U$ then $Q_i \subseteq U$ for some $i \in I$ already. Every sober space is well-filtered.*

Proof Let $\mathcal{U}$ be the set of all open neighborhoods of some Q_i, $i \in I$, i.e., $\{U \in \mathcal{O}(X) \mid \exists i \in I \cdot Q_i \subseteq U\}$. This is a filter because $(Q_i)_{i \in I}$ is a filtered family. As a union of Scott-open filters (by Lemma 8.3.1), it is Scott-open. So $Q = \bigcap_{U \in \mathcal{U}} U$ is compact saturated, and for every open subset U, $Q \subseteq U$ iff $U \in \mathcal{U}$, by the Hofmann–Mislove Theorem 8.3.2. However, $Q = \bigcap_{\substack{U \text{ open} \\ \exists i \in I \cdot Q_i \subseteq U}} U = \bigcap_{i \in I} \bigcap_{\substack{U \text{ open} \\ Q_i \subseteq U}} U = \bigcap_{i \in I} Q_i$, the latter inequality being because Q_i is saturated. We have therefore proved that $\bigcap_{i \in I} Q_i \subseteq U$ iff $Q \subseteq U$, iff $U \in \mathcal{U}$, iff $Q_i \subseteq U$ for some $i \in I$. □

Notice the similarity between well-filteredness and Proposition 5.2.28: in a dcpo, even one that is not sober (such as $\mathbb{J}$; see Exercise 8.2.14), a property similar to well-filteredness holds, where the compact saturated subsets Q_i are restricted to be finitary compacts.

In the proof of Proposition 8.3.5, we have showed that $\bigcap_{i \in I} Q_i$ is compact saturated, but we haven't said so in the statement of the proposition. This is the case in every well-filtered space.

Proposition 8.3.6 (Filtered intersections of saturated compacts) *In a well-filtered space X (in particular a sober space), the intersection $\bigcap_{i \in I} Q_i$ of every filtered family $(Q_i)_{i \in I}$ of compact saturated subsets of X is compact saturated.*

Proof Let $(U_j)_{j \in J}$ be any open cover of $\bigcap_{i \in I} Q_i$, and let $U = \bigcup_{j \in J} U_j$. By Proposition 8.3.5, $Q_i \subseteq U$ for some $i \in I$. Since Q_i is compact, it has a finite subcover $(U_j)_{j \in K}$ ($K \subseteq J$ finite). This is also a finite subcover of $\bigcap_{i \in I} Q_i$. □

An alternative characterization of well-filtered spaces is as follows. This is obtained by considering the closed complement F of the open subset U of the definition.

Fact 8.3.7 *A space X is well-filtered iff, for every closed subset F of X, for every filtered family $(Q_i)_{i \in I}$ of compact saturated subsets such that Q_i intersects F for each $i \in I$, the intersection $\bigcap_{i \in I} Q_i$ also intersects F (and is in particular non-empty).*

Well-filtered spaces are not that different from sober spaces. The two properties are equivalent for locally compact T_0 spaces.

Proposition 8.3.8 (Locally compact well-filtered $T_0 \Rightarrow$ sober) *Every locally compact well-filtered T_0 space is sober.*

Proof Let X be locally compact, T_0, and well-filtered, and let F be an irreducible closed subset of X. Consider the set $\mathcal{U}$ of all opens that intersect F.

This is a filter because F is irreducible (using Trick 8.2.3), and is clearly completely prime. For each $U \in \mathcal{U}$, for each $x \in U \cap F$, let $Q_{x,U}$ be a compact saturated neighborhood of x included in U (and collect them using the Axiom of Choice). This exists since X is locally compact. We claim that the family of all saturated compacts $Q_{x,U}$ is filtered. It is non-empty: take x in F and $U = X$. And given $Q_{x,U}$ and $Q_{y,V}$ in the family, $int(Q_{x,U})$ and $int(Q_{y,V})$ are in $\mathcal{U}$, so their intersection W is also in $\mathcal{U}$. Pick z in $W \cap F$. Then $Q_{z,W}$ is in the family, and included in W, hence both in $Q_{x,U}$ and $Q_{y,V}$.

By Fact 8.3.7, since X is well-filtered, the intersection of all saturated compacts $Q_{x,U}$ intersects F. Let z be a point in F that is in every $Q_{x,U}$. We claim that $F = \downarrow z$. Otherwise there is a point x in F such that $x \not\leq z$, i.e., such that x has an open neighborhood U that does not contain z. Since $x \in U \cap F$, $Q_{x,U}$ is defined. Since $z \notin U$ and $Q_{x,U} \subseteq U$, z is not in $Q_{x,U}$: but this is impossible. □

Exercise 8.3.9 The Johnstone space $\mathbb{J}$ (Exercise 5.2.15) is not core-compact, and not sober (Exercise 8.2.14). Show that it is not well-filtered either. You may first show that, for every subset A of $\mathbb{N}$, $Q_A = \{(m, \omega) \mid m \in A\}$ is compact saturated in $\mathbb{J}$.

Theorem 8.3.10 (Core-compact sober $\Rightarrow$ locally compact) *Every core-compact sober space is locally compact.*

Proof Let X be core-compact and sober. Let x be any point in X, and $U = U_0$ be any open neighborhood of x. By Proposition 5.2.6, there is an open neighborhood U_ω of x such that $x \in U_\omega \Subset U_0$. By interpolation in $\mathcal{O}(X)$, since this is a continuous dcpo, there is an open subset U_1 such that $U_\omega \Subset U_1 \Subset U_0$. Iterating the process, one obtains a sequence of open neighborhoods U_n, $n \in \mathbb{N}$, of x such that $U_\omega \Subset U_n \Subset \cdots \Subset U_1 \Subset U_0$. Let $\mathcal{U}$ be the set of all open subsets V such that $U_n \subseteq V$ for some $n \in \mathbb{N}$. This is clearly a filter.

The key point is that $\mathcal{U}$ is also the set of all open subsets V such that $U_n \Subset V$ for some $n \in \mathbb{N}$. Indeed, if $U_n \Subset V$ then $U_n \subseteq V$, and, conversely, if $U_n \subseteq V$ then $U_{n+1} \Subset V$. It follows that $\mathcal{U}$ is Scott-open: for any directed family of open subsets $(V_i)_{i \in I}$ of X whose union is in $\mathcal{U}$, i.e., such that $U_n \Subset \bigcup_{i \in I} V_i$ for some $n \in \mathbb{N}$, $U_n \subseteq V_i$ for some $i \in I$, whence $V_i \in \mathcal{U}$.

By the Hofmann–Mislove Theorem 8.3.2, the intersection $Q = \bigcap_{n \in \mathbb{N}} U_n$ of all elements of $\mathcal{U}$ is compact saturated. However $U_\omega \subseteq Q$, so $x \in U_\omega \subseteq Q \subseteq U$. Using Trick 4.8.2, X is locally compact. □

So we have a number of pleasing coincidences:

- In the presence of sobriety, core-compactness *is* local compactness (Theorem 5.2.9, Theorem 8.3.10).
- In the presence of local compactness, sobriety *is* well-filteredness (plus being T_0; see Proposition 8.3.5, Proposition 8.3.8).

We also get yet another characterization of core-compact spaces.

Proposition 8.3.11 *A topological space X is core-compact if and only if its sobrification $\mathcal{S}(X)$ is locally compact.*

Proof By Lemma 8.2.26, X and $\mathcal{S}(X)$ have isomorphic lattices of open subsets. So $\mathcal{O}(X)$ is continuous (i.e., X is core-compact) iff $\mathcal{O}(\mathcal{S}(X))$ is continuous, i.e., iff $\mathcal{S}(X)$ is core-compact. But, since $\mathcal{S}(X)$ is sober (Corollary 8.2.23), an equivalent statement is that $\mathcal{S}(X)$ is locally compact, using Theorem 8.3.10 (and Theorem 5.2.9). □

8.3.2 Sober locally compact spaces and the Hofmann–Lawson Theorem

Categorically, remember that the adjunction $\mathcal{O} \dashv \mathsf{pt}$ restricts to an equivalence between the category of sober spaces and continuous maps on the one hand, and the category of spatial lattices and frame homomorphisms on the other; see Theorem 8.2.46. This further specializes to sober *locally compact* spaces and *continuous* spatial lattices:

Theorem 8.3.12 (Sober locally compact spaces ≡ continuous spatial lattices) *The adjunction $\mathcal{O} \dashv \mathsf{pt}$ restricts to an equivalence between the category of sober locally compact spaces (equivalently, well-filtered locally compact T_0 spaces) and continuous maps on the one hand, and the opposite of the category of continuous spatial lattices and frame homomorphisms on the other.*

Proof If X is sober and locally compact, then $\mathcal{O}(X)$ is both a spatial lattice and a continuous dcpo (the latter by Theorem 5.2.9), hence a continuous frame. Conversely, if L is a continuous spatial lattice, then $L \cong \mathcal{O}(X)$ where $X = \mathsf{pt}(L)$ is core-compact, and sober by Proposition 8.2.11, hence locally compact by Theorem 8.3.10. □

While spatial lattices are a somewhat elusive concept, the *continuous* spatial lattices will turn out to be exactly the continuous distributive complete lattices. This starts with the following characterization of continuity in complete lattices.

Proposition 8.3.13 *Let L be a complete lattice. L is continuous if and only if the following doubly infinite distributivity law is satisfied: for all directed families $(u^i_j)_{j \in J_i}$, one for each $i \in I$, $\inf_{i \in I} \sup_{j \in J_i} u^i_j = \sup_{f \in \prod_{i \in I} J_i} \inf_{i \in I} u^i_{f(i)}$.*

Proof Compared with Proposition 5.7.19, the only difference is that we do not require I to be non-empty. When I is empty, however, the equality is the trivial equality $\top = \top$, where $\top$ is the largest element of L. □

In particular, taking $J_1 = \{*\}$, $J_2 = J$, $(u^1_j)_{j \in J_1}$ to be the family consisting of a single element u, and $u^2_j = v_j$, then $u \wedge \sup_{j \in J} v_j = \sup_{j \in J}(u \wedge v_j)$. This is not quite the frame distributivity law, since $(v_j)_{j \in J}$ is required to be directed in the latter equality. But we are close.

Corollary 8.3.14 *Every continuous distributive complete lattice is a frame.*

Proof Let L be a continuous distributive complete lattice, u be an element of L, and $(v_i)_{i \in I}$ be any family of elements of L. Using a trick similar to Trick 4.4.6, $\sup_{i \in I} v_i$ is the least upper bound of the directed family $(v_J)_{J \in \mathbb{P}_{\text{fin}}(I)}$, where v_J is defined as $\sup_{j \in J} v_j$. We note that, for every $J \in \mathbb{P}_{\text{fin}}(I)$, $u \wedge v_J = \sup_{j \in J}(u \wedge v_j)$. This is by induction on the cardinality of J. When J is empty, $u \wedge v_J = u \wedge \bot = \bot = \sup_{j \in J}(u \wedge v_j)$. Otherwise, we are finished, since L is distributive, using the induction hypothesis.

Now $u \wedge \sup_{i \in I} v_i = u \wedge \sup_{J \in \mathbb{P}_{\text{fin}}(I)} v_J = \sup_{J \in \mathbb{P}_{\text{fin}}(I)}(u \wedge v_J)$ (since $(v_J)_{J \in \mathbb{P}_{\text{fin}}(I)}$ is directed, applying the remark after Proposition 8.3.13) $=$ $\sup_{J \in \mathbb{P}_{\text{fin}}(I)} \sup_{j \in J}(u \wedge v_j)$ (since L is distributive) $= \sup_{i \in I}(u \wedge v_i)$. □

Exercise 8.3.15 Let X be a topological space. Show that X is core-compact if and only if, for all directed families $(U^i_j)_{j \in J_i}$, one for each $i \in I$, the union $\bigcup_{f \in \prod_{i \in I} J_i} int(\bigcap_{i \in I} U^i_{f(i)})$ is the interior of $\bigcap_{i \in I} \bigcup_{j \in J_i} U^i_j$.

Exercise 8.3.16 (Raney's Theorem) Historically, the first theorem of the kind given in Proposition 8.3.13 or Proposition 5.7.19 is the following. A *completely distributive* lattice is a complete lattice L in which the following, so-called *complete distributivity* law, is satisfied: for all (not necessarily directed) families $(u^i_j)_{j \in J_i}$, one for each $i \in I$, $\inf_{i \in I} \sup_{j \in J_i} u^i_j = \sup_{f \in \prod_{i \in I} J_i} \inf_{i \in I} u^i_{f(i)}$.

Let L be a complete lattice. We say that u is *way-way-below* v, in notation $u \lll v$, if and only if, for every (not necessarily directed) family $(w_i)_{i \in I}$ of elements of L such that $v \leqslant \sup_{i \in I} w_i$, $u \leqslant w_i$ for some $i \in I$. L is *prime-continuous* iff every element of L is the least upper bound of all elements way-way-below u. Show that L is prime-continuous if and only if it is completely distributive.

For an example of a prime-continuous complete lattice, the lattice of Scott-open subsets of a continuous poset, or more generally the open subsets of any c-space, will do, as we shall see in Lemma 8.3.41. More concretely, $[0, 1]$ is a prime-continuous lattice, and is easily seen to be completely distributive by elementary arguments.

Exercise 8.3.17 Show that in a prime-continuous lattice L, the following interpolation property holds: if $u \lll v$ then $u \lll w \lll v$ for some $w \in L$.

Exercise 8.3.18 Show that a complete lattice L is completely distributive if and only if the *dual* complete distributivity law is satisfied: $\sup_{i \in I} \inf_{j \in J_i} u^i_j = \inf_{f \in \prod_{i \in I} J_i} \sup_{i \in I} u^i_{f(i)}$. In the $\geqslant$ direction, you may want to show that the dual complete distributivity law is a consequence of prime-continuity, following the argument of Proposition 5.7.19.

Conclude that a complete lattice L is prime-continuous if and only if L^{op} is prime-continuous.

A further step towards simplifying Theorem 8.3.12 is the following.

Proposition 8.3.19 *Every continuous distributive complete lattice is a spatial lattice.*

Proof Let L be a continuous distributive complete lattice. L is a frame by Corollary 8.3.14. Let $u, v \in L$ such that $u \not\leqslant v$. Since the specialization ordering of L_σ is $\leqslant$ (Proposition 4.2.18), and the Scott-open filtered subsets form a base (Proposition 5.1.19), there is a Scott-open filtered subset U containing u but not v, using Proposition 4.2.4. Since L is a lattice, U is even a Scott-open filter.

The complement F of U contains v but not u. Moreover, F is Scott-closed, and in particular inductive. So, by Zorn's Lemma (Theorem 2.4.2), there is a maximal element w above v in F. We claim that w is prime. Clearly, $w \neq \top$ since w is not above u. We must then show that if $u_1 \wedge u_2 \leqslant w$, then $u_1 \leqslant w$ or $u_2 \leqslant w$. Since $u_1 \wedge u_2 \leqslant w$, $\sup(u_1 \wedge u_2, w) = w$. By Proposition 8.3.13, L satisfies the doubly infinite distributivity law, and is in particular distributive. So $\sup(u_1 \wedge u_2, w) = \sup(u_1, w) \wedge \sup(u_2, w)$.

Assume now by way of contradiction that $u_1 \not\leqslant w$ and $u_2 \not\leqslant w$. Since $u_1 \not\leqslant w$, $\sup(u_1, w)$ is different from w, and necessarily above w. By the maximality of w, $\sup(u_1, w)$ cannot be in F, so it is in U. Similarly, $\sup(u_2, w)$ is in U. Since U is a filter, their greatest lower bound, namely $\sup(u_1 \wedge u_2, w) = w$, is in U, a contradiction.

We have just shown that if $u \not\leqslant v$, then there is a prime element w above v that is not above u. By Proposition 8.1.20, there is a completely prime filter, namely the complement x of $\downarrow w$, that contains u but not v. It follows that L is spatial, using Proposition 8.1.17. Concretely, we must show that $\epsilon_L^{\mathsf{Stone}}$ is injective, and we show the more general claim that $\mathcal{O}_u \subseteq \mathcal{O}_v$ implies $u \leqslant v$: if u were not below v, then there would be a point x containing u but not v, that is, such that $x \in \mathcal{O}_u$ but $x \notin \mathcal{O}_v$. □

Theorem 8.3.20 (Continuous spatial lattice = continuous distributive complete lattice) *Let L be a continuous complete lattice L. L is spatial if and only if L is distributive.*

Proof Each (continuous) spatial lattice is a frame, hence a distributive complete lattice. The converse direction is by Corollary 8.3.14. □

Putting together Theorem 8.3.12 and Theorem 8.3.20, we obtain the following theorem, due to Hofmann and Lawson (1978).

Theorem 8.3.21 (Hofmann–Lawson) *The adjunction $\mathcal{O} \dashv \mathsf{pt}$ restricts to an equivalence between the category of sober locally compact spaces (equivalently, well-filtered T_0 locally compact spaces) and continuous maps on the one hand, and the opposite of the category of continuous distributive complete lattices and frame homomorphisms on the other hand.*

Exercise 8.3.22 (Strongly locally compact spaces ≡ algebraic distributive complete lattices) Recall that a compact-open subset of a space X is a subset that is both compact and open (Exercise 5.7.18). Call a topological space X *strongly locally compact* if and only if its topology has a base of compact-open subsets, i.e., if and only if, for every $x \in X$, every open neighborhood U of x contains a compact-open set K containing x. (The name "strongly locally compact" is non-standard.)

Let X be a topological space. Show that the finite elements of $\mathcal{O}(X)$ are the compact-open subsets of X. Deduce that X is strongly locally compact if and only if $\mathcal{O}(X)$ is an algebraic complete lattice.

Analogously to Theorem 8.3.20, show that the algebraic spatial lattices are exactly the algebraic distributive complete lattices. Conclude that the adjunction $\mathcal{O} \dashv \mathsf{pt}$ restricts to an equivalence between the category of sober strongly locally compact spaces and continuous maps on the one hand, and the opposite of the category of algebraic distributive complete lattices and frame homomorphisms on the other.

Exercise 8.3.23 Show that every b-space (Exercise 5.1.39) is strongly locally compact.

Sober locally compact spaces have many other properties. Here is one.

Proposition 8.3.24 (Sober locally compact $\Rightarrow$ Choquet-complete, Baire) *Every sober locally compact space (equivalently, well-filtered T_0 locally compact space) is Choquet-complete, and in particular Baire: the intersection of countably many dense open subsets is dense.*

Proof Given an α-history $x_0, V_0, U_0, x_1, V_1, U_1, x_2, V_2, \ldots, x_n, V_n$ where $V_0 \supseteq U_0 \supseteq V_1 \supseteq U_1 \supseteq V_2 \supseteq \cdots \supseteq V_n$ is a decreasing sequence of opens and $x_0 \in U_0, x_1 \in U_1, x_2 \in U_2, \ldots, x_{n-1} \in U_{n-1}, x_n \in V_n, n \in \mathbb{N}$, the α player picks an open neighborhood U_n of x_n of the form $int(Q_n)$ such that Q_n is compact saturated and included in V_n. This is possible because the space is locally compact. Then the value $\bigcap_{n\in\mathbb{N}} V_n$ of the game contains $\bigcap_{n\in\mathbb{N}} Q_n$, which is non-empty since the space is well-filtered: use Fact 8.3.7 with F the whole space. So this space is Choquet-complete, hence Baire by Theorem 7.6.8. $\square$

The Hofmann–Mislove Theorem also provides us with another kind of model, different from the formal ball model (recall Proposition 7.7.16).

Proposition 8.3.25 (Smyth powerdomain) *The* Smyth powerdomain $\mathcal{Q}(X)$ *of a topological space is its space of non-empty compact saturated subsets, ordered by reverse inclusion $\supseteq$.*

$\mathcal{Q}(X)$ is a dcpo if X is well-filtered. In this case, the least upper bound of a directed family $(Q_i)_{i\in I}$ is its intersection $\bigcap_{i\in I} Q_i$.

If X is well-filtered and locally compact, then $\mathcal{Q}(X)$ is a continuous dcpo, where $Q \ll Q'$ iff $Q' \subseteq int(Q)$.

Proof Let $(Q_i)_{i\in I}$ be a directed family in $\mathcal{Q}(X)$, and assume X is well-filtered. By Fact 8.3.7 with $F = X$, $Q = \bigcap_{i\in I} Q_i$ is non-empty. By Proposition 8.3.6, it is compact saturated. It is then clear that Q is the largest element of $\mathcal{Q}(X)$ above (i.e., included in) every Q_i, so $Q = \sup_{i\in I} Q_i$. In particular, $\mathcal{Q}(X)$ is a dcpo.

Assume X is not only well-filtered but also locally compact. If $Q' \subseteq int(Q)$, and Q' is below the least upper bound of the directed family $(Q_i)_{i\in I}$, i.e., if $\bigcap_{i\in I} Q_i \subseteq Q' \subseteq int(Q)$, then $Q_i \subseteq int(Q)$ for some $i \in I$, by Proposition 8.3.5. So $Q_i \subseteq Q$, hence Q is below Q_i. So Q is way-below Q'.

Conversely, assume that Q is way-below Q'. Since Q' is saturated, Q' is the intersection of the filtered family $(U_i)_{i\in I}$ of all the open subsets that contain Q'. For each $i \in I$, there is a compact saturated subset Q_i such that $Q' \subseteq int(Q_i)$ and $Q_i \subseteq U_i$, by interpolation (Proposition 4.8.14). We pick one Q_i for each $i \in I$ by the Axiom of Choice. The family $(Q_i)_{i\in I}$ is filtered (hence directed in $\supseteq$): since $int(Q_i) \cap int(Q_j)$ is an open subset containing Q', it is equal to U_k for some k, and then $Q_k \subseteq U_k \subseteq Q_i \cap Q_j$. Clearly $\bigcap_{i \in I} Q_i \subseteq Q'$, i.e., the least upper bound of the directed family $(Q_i)_{i \in I}$ is above Q'. Since Q is way-below Q', Q is below Q_i for some $i \in I$. That is, $Q_i \subseteq Q$. Since $Q' \subseteq int(Q_i)$, $Q' \subseteq int(Q)$. □

Lemma 8.3.26 *Let X be a well-filtered and locally compact space.*

The subsets $\Box U = \{Q \in \mathcal{Q}(X) \mid Q \subseteq U\}$, where U is open in X, are open in $\mathcal{Q}(X)_\sigma$. Given any base $\mathcal{B}$ of the topology of X, the sets of the form $\Box(U_1 \cup \cdots \cup U_n)$, $n \geqslant 1$, where $U_1, \ldots, U_n \in \mathcal{B}$, are a base of the topology of $\mathcal{Q}(X)_\sigma$.

If X is also T_0, then the map $\eta_X^{\mathcal{Q}} \colon X \to \mathcal{Q}(X)_\sigma$ that sends x to $\uparrow x$ is an embedding.

Proof Let U be open in X. Then $\Box U$ is Scott-open: it is upward closed since, if $Q \in \Box U$ and Q' is above Q, then $Q' \subseteq Q \subseteq U$, hence $Q' \in \Box U$; and if $(Q_i)_{i \in I}$ is a filtered family of compact saturated subsets whose intersection is in $\Box U$, then some Q_i is in $\Box U$ by Proposition 8.3.5.

The sets $\Box U$, U open, form a basis. Indeed, by Proposition 8.3.25 $\mathcal{Q}(X)$ is a continuous dcpo, and by Theorem 5.1.17 the subsets $\twoheaduparrow Q$ form a basis. But $\twoheaduparrow Q = \Box int(Q)$, using Proposition 8.3.25 again.

If $\mathcal{B}$ is a base of the topology of X, then every open subset U is a union of elements U_i of $\mathcal{B}$, $i \in I$. Using Trick 4.4.6, U is the union of the directed family $(U_J)_{J \in \mathbb{P}_{\text{fin}}(I)}$, where U_J is the finite union $\bigcup_{j \in J} U_j$. We now realize that $\Box$ commutes with directed unions: if $Q \in \Box \bigcup_{J \in \mathbb{P}_{\text{fin}}(I)} U_J$, then, since Q is compact, $Q \subseteq U_J$ for some $J \in \mathbb{P}_{\text{fin}}(I)$, using Proposition 4.4.7. Conversely, if $Q \in \Box U_J$, then clearly $Q \in \Box \bigcup_{J \in \mathbb{P}_{\text{fin}}(I)} U_J$. So $\Box U = \bigcup_{J \in \mathbb{P}_{\text{fin}}(I)} \Box U_J$. It follows that the sets of the form $\Box(U_1 \cup \cdots \cup U_n)$ where $U_1, \ldots, U_n \in \mathcal{B}$ already form a base of the topology of $\mathcal{Q}(X)_\sigma$. We can also require $n \geqslant 1$, since $\Box\emptyset$ is empty.

Finally, we realize that $(\eta_X^{\mathcal{Q}})^{-1}(\Box U) = U$ for every basic open $\Box U$; so $\eta_X^{\mathcal{Q}}$ is continuous; and since U is an arbitrary open subset of X, the same formula shows that $\eta_X^{\mathcal{Q}}$ is almost open. When X is T_0, $\eta_X^{\mathcal{Q}}$ is injective, by Proposition 4.2.3. So $\eta_X^{\mathcal{Q}}$ is an embedding by Proposition 4.10.9. □

Corollary 8.3.27 (The Smyth model) *Let X be a well-filtered locally compact T_1 space – e.g., a locally compact T_2 space. Then $\mathcal{Q}(X)_\sigma$ is a continuous model of X, with embedding $\eta_X^{\mathcal{Q}}$. If X is also countably-based, then it is an ω-continuous model.*

Proof First recall that any T_2 space is both T_1 and well-filtered, using Proposition 8.2.12. Now assume X is well-filtered, locally compact, and T_1.

Since X is T_1, $\eta_X^{\mathcal{Q}}(x) = \{x\}$ for every $x \in X$, since the specialization ordering of X is equality, by Proposition 4.2.3. In particular, $\eta_X^{\mathcal{Q}}(x)$ is maximal in $\mathcal{Q}(X)$. Conversely, any maximal element of $\mathcal{Q}(X)$ is a non-empty compact subset that is minimal for inclusion, hence is a one-element set. It only remains to show that $\mathcal{Q}(X)$ is ω-continuous when X is countably-based. Let $\mathcal{B}$ be a countable base of the topology of X. By Lemma 8.3.26, a base of $\mathcal{Q}(X)_\sigma$ is given by the sets $\Box(U_1 \cup \cdots \cup U_n)$, $n \geqslant 1$, where $U_1, \ldots, U_n$ are in $\mathcal{B}$. There are only countably many such sets, so $\mathcal{Q}(X)_\sigma$ is countably-based. We conclude by Norberg's Lemma 7.7.13. □

Together with Martin's Second Theorem 7.7.21 (and Fact 6.3.45 as well as the Urysohn–Tychonoff Theorem 6.3.44 for the second part), we obtain:

Corollary 8.3.28 (Locally compact separable metrizable ⇒ Polish) *Every locally compact and metrizable space is completely metrizable.*

Every locally compact separable metrizable space, i.e., every countably-based locally compact metrizable space, i.e., every countably-based locally compact T_3 space, is Polish.

Exercise 8.3.29 Give an alternate proof of Corollary 8.3.28, using the theory of Choquet-completeness, and in particular Proposition 8.3.24.

Exercise 8.3.30 Give an example of a Polish space that is not locally compact. So the class of locally compact separable metrizable spaces is properly contained in the class of Polish spaces.

8.3.3 Stably locally compact spaces

Theorem 8.2.46 and Theorem 8.3.21 are the first two examples of theorems establishing that $\mathcal{O} \dashv \mathsf{pt}$ restricts to an equivalence between certain categories of sober topological spaces and certain categories of spatial locales.

Here is a third one.

Definition 8.3.31 (Stably locally compact space) A topological space is *stably locally compact* if and only if it is sober, locally compact, and coherent (equivalently, T_0, well-filtered, locally compact, and coherent).

Recall (Definition 5.2.21) that a space is coherent iff the intersection of any two compact saturated subsets is again compact saturated. Recall also that $\ll$ is multiplicative (Definition 5.2.18) iff, whenever $v \ll u_1$ and $v \ll u_2$, then $v \ll u_1 \wedge u_2$, and that a core-coherent space is a topological space X such that $\Subset$ is multiplicative on $\mathcal{O}(X)$. Recall finally that any locally compact coherent space is core-coherent (Lemma 5.2.24).

Lemma 8.3.32 *Every sober, core-compact (equivalently, well-filtered, T_0, and locally compact), and core-coherent space is coherent.*

Proof Let X be a sober, core-compact, and core-coherent space, and Q_1, Q_2 be two compact saturated subsets of X. By Lemma 8.3.1, the filter $\mathcal{F}_1$ of all open neighborhoods of Q_1 is Scott-open in $\mathcal{O}(X)$, and similarly for the filter $\mathcal{F}_2$ of all open neighborhoods of Q_2. Let $\mathcal{F}$ be the filter of all open sets U of X that contain some open of the form $U_1 \cap U_2$ with $U_1 \in \mathcal{F}_1$, $U_2 \in \mathcal{F}_2$.

We claim that $\mathcal{F}$ is Scott-open. Let U be any element of $\mathcal{F}$, and let $U_1 \in \mathcal{F}_1$, $U_2 \in \mathcal{F}_2$ be such that $U_1 \cap U_2 \subseteq U$. Write $\Uparrow V$ for the set of all open subsets U such that $V \Subset U$; this is just $\twoheaduparrow V$ in the dcpo $\mathcal{O}(X)$. Since $\mathcal{F}_1$ is Scott-open in $\mathcal{O}(X)$, and the latter is continuous (since X is core-compact), U_1 is in $\Uparrow V_1$ for some $V_1 \in \mathcal{F}_1$, i.e., $V_1 \in \mathcal{F}_1$ and $V_1 \Subset U_1$, by Theorem 5.1.17. Similarly, there is an element V_2 of $\mathcal{F}_2$ such that $V_2 \Subset U_2$. By Proposition 5.2.19, which applies since X is core-coherent, $\cap$ preserves $\Subset$, so $V_1 \cap V_2 \Subset U_1 \cap U_2$. It follows that $V_1 \cap V_2 \Subset U$, i.e., $U \in \Uparrow(V_1 \cap V_2)$. Since $V_1 \cap V_2$ is in $\mathcal{F}$, we conclude that $\mathcal{F}$ is Scott-open using Trick 4.1.11.

By the Hofmann–Mislove Theorem 8.3.2, the intersection of all elements of $\mathcal{F}$ is therefore compact saturated. But this intersection is $Q_1 \cap Q_2$. ☐

Exercise 8.3.33 (Jung's order-theoretic characterization of coherence) Let X be a continuous dcpo. Show that if $x \ll y$, then $\twoheaduparrow y \Subset \twoheaduparrow x$. Show that, for any two open subsets U and V of X_σ, $U \Subset V$ if and only if $U \subseteq \uparrow E \subseteq V$ for some finite set E. Deduce that X_σ is coherent if and only if, for all pairs of elements $x_1 \ll y_1$ and $x_2 \ll y_2$, there is a finite set E such that: (1) every element z of E is such that $x_1 \ll z$ and $x_2 \ll z$, and (2) every element z such that $y_1 \ll z$ and $y_2 \ll z$ is above some element of E. (Think of the relationships between coherence and core-coherence, and of Rudin's Lemma and its consequences.)

Theorem 8.3.34 (Stably locally compact spaces ≡ arithmetic lattices) *The* arithmetic lattices *are the continuous complete distributive lattices in which the way-below relation is multiplicative.*

The adjunction $\mathcal{O} \dashv$ pt restricts to an equivalence between the category of stably locally compact spaces and continuous maps on the one hand, and the opposite of the category of arithmetic lattices and frame homomorphisms on the other hand.

Proof Use the Hofmann–Lawson Theorem 8.3.21, and refine: when X is stably locally compact, $\mathcal{O}(X)$ is not only a continuous distributive complete lattice, but its way-below relation is multiplicative too, by Lemma 5.2.24; so $\mathcal{O}(X)$ is arithmetic. Conversely, if L is an arithmetic lattice, then L is order isomorphic to $\mathcal{O}(X)$ for some space X that is not only sober and locally compact, but also core-coherent. So X is coherent by Lemma 8.3.32. ☐

The stably locally compact spaces that are also compact are the stably compact spaces. These have a very rich theory of their own: see Chapter 9.

Exercise 8.3.35 Let X be a strongly locally compact space. Using Exercise 5.1.31, show that the following conditions are equivalent: (i) X is coherent; (ii) for any two compact-open subsets K_1 and K_2 of X, $K_1 \cap K_2$ is compact (hence compact-open).

On the lattice-theoretic side, show that, given an algebraic complete lattice L, the way-below relation on L is multiplicative if and only if the greatest lower bound of any two finite elements is again finite.

Show that the adjunction $\mathcal{O} \dashv \mathsf{pt}$ restricts to an equivalence between the category of strongly locally compact, coherent spaces and continuous maps on the one hand, and the opposite of the category of algebraic distributive complete lattices where the greatest lower bound of any two finite elements is finite, and frame homomorphisms, on the other hand.

8.3.4 Sober c-spaces and continuous dcpos

We finish with the sober c-spaces.

Proposition 8.3.36 (Sober c-space = continuous dcpo) *The sober c-spaces are exactly the continuous dcpos with their Scott topology. The way-below relation is defined by $x \lll y$ iff $y \in int(\uparrow x)$.*

Proof By Proposition 5.1.37, if X is a continuous dcpo, then X_σ is a c-space. X_σ is sober, by Proposition 8.2.12 (*b*).

Conversely, let Y be a sober c-space. Define $\lll$ by $x \lll y$ iff $y \in int(\uparrow x)$. Let $\leqslant$ be the specialization ordering (not just a quasi-ordering, since Y is sober hence T_0) of Y, and to mark the difference, let X be Y viewed as a poset.

We claim that $x \lll y$ implies $x \ll y$, where $\ll$ is the way-below relation of X. Let $(z_i)_{i \in I}$ be any directed family of elements of X such that $y \leqslant \sup_{i \in I} z_i$—the least upper bound $\sup_{i \in I} z_i$ exists because Y is a monotone convergence space, by Proposition 8.2.34. In fact, $\sup_{i \in I} z_i$ is the unique element z such that $cl\{z_i \mid i \in I\} = \downarrow z$. Since $y \leqslant z$, y is in $cl\{z_i \mid i \in I\}$, so every open subset containing y meets $\{z_i \mid i \in I\}$ by Lemma 4.1.27. This is the case of $int(\uparrow x)$ since $x \lll y$. So $z_i \in int(\uparrow x)$ for some $i \in I$, in particular $x \leqslant z_i$.

Fix $y \in Y$. We claim that the family D of all elements $x \lll y$ is directed, and has y as least upper bound. Since Y is a c-space, recall that every open neighborhood U of y contains a point x in D. In particular, taking $U = Y$, D is non-empty. Moreover, given any two elements $x, x' \in D$, then taking

$U = int(\uparrow x) \cap int(\uparrow x')$ yields a further point $x'' \in D$ in U, in particular such that $x, x' \leqslant x''$. Finally, the upper bounds of D are the elements that are in all open sets that contain at least one element of D, by definition of the specialization quasi-ordering. By the definition of c-spaces, the open sets that contain at least one element of D are exactly the open neighborhoods of y, so the upper bounds of D are the elements of $\uparrow y$, whence $y = \sup D$.

Using Trick 5.1.20, it follows that X is a continuous dcpo. It remains to show that $Y = X_\sigma$, i.e., that the topology of Y is the Scott topology of X. Since Y is a monotone convergence space (as a sober space; see Proposition 8.2.34), the topology of Y is coarser than that of X_σ. Conversely, we claim that $\twoheaduparrow x$ is open in Y for every $x \in X$. Since such sets form a base of the Scott topology (Theorem 5.1.17), the Scott topology will be coarser than, hence coincide with, that of Y.

To show this, we in fact show that $x \ll y$ is equivalent to $x \lll y$. Then $\twoheaduparrow x$ will be the open set $int(\uparrow x)$. Assume $x \ll y$. Consider the family D of all elements $z \lll y$. This is directed, as we have already seen. Moreover, $y = \sup D$. Since $x \ll y$, there is an element z of D such that $x \leqslant z$. Since $z \in D$, y is in $int(\uparrow z)$. Since $x \leqslant z$, y is also in $int(\uparrow x)$, i.e., $x \lll y$. ☐

We can now improve on Lemma 5.1.43. This said that if X_σ is a retract of Y_σ, where Y is a continuous poset, then X is also a continuous poset. Up to the fact that we now need Y to be a dcpo, not just a poset, our improvement first says that every (topological) retract X of a continuous dcpo is a continuous dcpo, without having to assume that X is of the form Z_σ for some poset Z. Second, it says that X is indeed of the form Z_σ for some continuous dcpo Z.

Corollary 8.3.37 (Retracts of continuous dcpos) *Let Y be any continuous dcpo. Every retract X of Y_σ is a continuous dcpo, and its topology is its Scott topology.*

Proof Y_σ is a sober c-space by Proposition 8.3.36. By Proposition 5.1.40, X is a c-space. X is also sober, by Exercise 8.2.43. So X is a sober c-space, hence is not only a continuous dcpo, but has the Scott topology of its specialization ordering, by Proposition 8.3.36. ☐

Exercise 8.3.38 (Sober b-space = algebraic dcpo) Show that the sober b-spaces (see Exercise 5.1.39) are exactly the algebraic dcpos with their Scott topology.

Exercise 8.3.39 (Sober locally finitary compact = quasi-continuous dcpo) Similarly, use Exercise 8.2.15 to show that the sober locally finitary compact spaces are exactly the quasi-continuous dcpos with their Scott topology.

We can also simplify the characterization of Smyth-complete quasi-metric spaces given in the Romaguera–Valero Theorem 7.3.11.

Theorem 8.3.40 (Smyth-complete quasi-metric = sober space of formal balls) *Let X, d be a quasi-metric space. Then X, d is Smyth-complete if and only if its space of formal balls $\mathbf{B}(X, d)$ is sober in the open ball topology of d^+.*

Proof $\mathbf{B}(X, d)$ is a c-space by Proposition 7.3.9. If it is sober, then by Proposition 8.3.36 it is a continuous dcpo, and its way-below relation $\ll$ coincides with $\prec^{d^+}$, by Lemma 7.3.10. So X, d is Smyth-complete, by the Romaguera-Valero Theorem 7.3.11.

Conversely, if X, d is complete, then by Theorem 7.3.11 $\mathbf{B}(X, d)$ is a continuous dcpo, and its topology is the Scott topology by Lemma 7.3.13. So it is sober, by Proposition 8.3.36 again, or by Proposition 8.2.12 (*b*). □

Theorem 7.3.11 should also be compared with the result of Exercise 8.2.37. This shows that the distinction between Smyth-complete and Yoneda-complete spaces is reflected at the level of spaces of formal balls as the distinction between sober and monotone convergence spaces.

On the localic side, remember Raney's Theorem (Exercise 8.3.16): a prime-continuous lattice is one where every element is the least upper bound of all elements way-way-below it, and the prime-continuous complete lattices are exactly the completely distributive lattices.

Lemma 8.3.41 *Let X be a c-space. Then $\mathcal{O}(X)$ is prime-continuous, and $U \lll V$ if and only if there is an element $y \in V$ such that $U \subseteq \uparrow y$.*

Proof Assume $U \lll V$. Since X is a c-space, V is the union of all sets of the form $int(\uparrow y)$ with $y \in V$. Indeed, by definition every element x of V is in $int(\uparrow y)$ for some $y \in V$. Since $U \lll V$, U must be included in one of these sets $int(\uparrow y)$, hence in $\uparrow y$. Conversely, assume that $U \subseteq \uparrow y$ for some $y \in V$. For every open cover $(W_i)_{i \in I}$ of V, pick $i \in I$ such that $y \in W_i$. Then $U \subseteq W_i$. So $U \lll V$.

Now every open subset V is the union of all sets $int(\uparrow y)$, $y \in V$, and we have just seen that $int(\uparrow y) \lll V$. So $\mathcal{O}(X)$ is prime-continuous. □

Lemma 8.3.42 *Let Y be a sober space such that $\mathcal{O}(Y)$ is prime-continuous. Then Y is a c-space, and in particular is of the form X_σ for some continuous dcpo X.*

Proof Similar to Theorem 8.3.10. Let x be any point in X, and $U = U_0$ be an open neighborhood of x. Since $\mathcal{O}(Y)$ is prime-continuous, U is the union of all opens way-way-below U. One of them contains x, so there is an open $U_\omega \lll U_0$ such that $x \in U_\omega$.

We now observe that the way-way-below relation $\lll$ obeys interpolation. This is Exercise 8.3.17, and the proof is exactly the same as for Proposition 5.1.15. By repeated interpolation, we find open subsets $U_n, n \in \mathbb{N}$, such that $U_\omega \lll \cdots \lll U_n \lll \cdots \lll U_1 \lll U_0$. Let $\mathcal{U}$ be the filter of all open subsets V such that $U_n \subseteq V$ for some $n \in \mathbb{N}$. We observe that $\mathcal{U}$ is also the set of all open subsets V such that $U_n \lll V$ for some $n \in \mathbb{N}$: if $U_n \lll V$ then $U_n \subseteq V$, and if $U_n \subseteq V$ then $U_{n+1} \lll V$. As such, $\mathcal{U}$ is completely prime: if $(V_k)_{k \in K}$ is any family of open subsets whose union is in $\mathcal{U}$, i.e., is such that $U_n \lll \bigcup_{k \in K} V_k$ for some $n \in \mathbb{N}$, then $U_n \subseteq V_k$ for some $k \in K$, i.e., $V_k \in \mathcal{U}$. Since Y is sober, $\mathcal{U}$ is the filter of all open neighborhoods of some point y. Now $x \in U_\omega$, U_ω is included in every element of $\mathcal{U}$ hence in their intersection $\uparrow y$, so $x \in int(\uparrow y)$. Since $U = U_0$ is in $\mathcal{U}$, finally, y is in U.

So Y is a c-space. Since it is sober, it is of the form X_σ for some continuous dcpo X, by Proposition 8.3.36. □

Theorem 8.3.43 (Continuous dcpos ≡ completely distributive lattices) *The adjunction $\mathcal{O} \dashv \mathsf{pt}$ restricts to an equivalence between the category of continuous dcpos with their Scott topology on the one hand, and the opposite of the category of completely distributive lattices on the other.*

Proof Use the Hofmann–Lawson Theorem 8.3.21. We must check that if X is a continuous dcpo, then $\mathcal{O}(X_\sigma)$ is a completely distributive lattice. X_σ is a c-space by Proposition 5.1.37, so $\mathcal{O}(X_\sigma)$ is prime-continuous by Lemma 8.3.41, hence completely distributive by Raney's Theorem.

Conversely, let L be a completely distributive lattice. So L is prime-continuous, in particular continuous. By Theorem 8.3.21, L is order isomorphic to $\mathcal{O}(X)$ for some sober, locally compact space X. Moreover, $\mathcal{O}(X)$ is completely distributive. By Lemma 8.3.42, X is a sober c-space, i.e., a continuous dcpo with its Scott topology. □

Corollary 8.3.44 (Sobrifications of c-spaces) *The continuous dcpos with their Scott topologies are exactly the sobrifications of c-spaces: for every c-space X, $\mathcal{S}(X)$ is a continuous dcpo and its topology is the Scott topology, and conversely every continuous dcpo arises this way up to homeomorphism.*

Proof If X is a c-space, then $\mathcal{O}(X)$ is prime-continuous by Lemma 8.3.41, so $\mathcal{O}(\mathcal{S}(X))$ is, too, by Lemma 8.2.26. Since $\mathcal{S}(X)$ is sober (Corollary 8.2.23), it is a continuous dcpo with its Scott topology by Proposition 8.3.36. Conversely, if Y is a continuous dcpo, then Y_σ is a sober c-space by Proposition 8.2.12 (*b*) and Proposition 5.1.37. As a sober space, it is homeomorphic to $\mathcal{S}(Y_\sigma)$ by Fact 8.2.24, and we conclude. □

Exercise 8.3.45 (Sobrifications of b-spaces) Show that the algebraic dcpos with their Scott topologies are exactly the sobrifications of b-spaces: for every b-space X, $\mathcal{S}(X)$ is an algebraic dcpo and its topology is the Scott topology. Moreover, every algebraic dcpo arises this way up to homeomorphism.

Exercise 8.3.46 (c-space $\cong$ abstract basis) Show the following tight connection between c-spaces and abstract bases (Lemma 5.1.32). Given any abstract basis $B, \prec$, show that the space $\mathcal{R}(B, \prec)$ obtained as B with the rounded topology (Exercise 8.2.48) is a c-space. Conversely, given any c-space X, remember that $X, \lll$ is an abstract basis, where $x \lll y$ iff $y \in int(\uparrow x)$ (Lemma 5.1.41).

Call an abstract basis $B, \prec$ *complete* if and only if, for all $b_1, b_2, b_3 \in B$ such that $b_2 \prec b_3$ and $b_1 \not\prec b_3$, there is an element $b \in B$ such that $b \prec b_1$ and $b \not\prec b_2$. It may be profitable to realize that, for any abstract basis $B, \prec$, the specialization ordering $\leqslant$ of $\mathcal{R}(B, \prec)$ is given by $b_1 \leqslant b_2$ iff, for every $b \in B$ such that $b \prec b_1$, then $b \prec b_2$, and that completeness is then equivalent to the fact that $b_1 \leqslant b_2 \prec b_3$ implies $b_1 \prec b_3$.

Then show that the two constructions are inverse to each other in the case of complete abstract bases. Precisely, show that: (i) $\mathcal{R}(X, \lll) = X$ for every c-space X, and $X, \lll$ is complete; and that: (ii) for every complete abstract basis $B, \prec$, the abstract basis obtained from the c-space $\mathcal{R}(B, \prec)$ is $B, \prec$ itself.

Deduce another proof of Corollary 8.3.44, using rounded ideal completions.

Exercise 8.3.47 (Completely distributive lattice = bicontinuous distributive lattice) A poset X is *bicontinuous* if and only if both X and X^{op} are continuous posets. (Exercise 8.3.18 in particular establishes that all completely distributive lattices are bicontinuous.) Let $\ll$ be the way-below relation of X, and $\gg$ that of X^{op}. We call the latter the *way-above* relation. To dispel any misunderstanding, show that, in general, $x \ll y$ is not equivalent to $y \gg x$. (Even in completely distributive lattices: consider $X = [0, 1]$.)

Then show that the complete lattices that are both distributive and bicontinuous are exactly the completely distributive lattices.

Here is an indication. Let L be a complete lattice that is both distributive and bicontinuous. L and L^{op} are then spatial (why?). By Exercise 8.1.24, since L^{op} is spatial, every element v of L is the least upper bound of all co-prime elements p of L below v; an element of L is *co-prime* if and only if it is prime in L^{op}, i.e., different from the least element $\bot$ and such that, whenever $p \leqslant \sup(u, v)$, then $p \leqslant u$ or $p \leqslant v$. Then show that, whenever p is co-prime, p is way-below u if and only if p is way-way-below u.

Exercise 8.3.48 Deduce from Exercise 8.3.47 that the following statements are equivalent, for a sober, locally compact space Y: (1) Y is a continuous dcpo, i.e., $Y = X_\sigma$ for some continuous dcpo X; (2) $\mathcal{O}(Y)^{op}$ is a continuous poset; (3) the lattice of all closed subsets of Y, ordered by inclusion, is continuous.

Exercise 8.3.49 Let L be a completely distributive lattice. Show that the elements that are least upper bounds of finitely many co-primes form a basis of the continuous dcpo L.

Topological side (spaces)	$\xrightarrow{\mathcal{O}}$ $\xleftarrow[\text{pt}]{}$	Localic side (complete lattices)	
Sober	$\equiv$	Spatial	Theorem 8.2.46
Sober locally compact	$\equiv$	Continuous spatial	Theorem 8.3.12
$\Vert$		$\Vert$	
Well-filtered T_0 locally compact	$\equiv$	Continuous distributive	Theorem 8.3.21
Stably locally compact	$\equiv$	Arithmetic	Theorem 8.3.34
Stably compact	$\equiv$	Fully arithmetic	Exercise 9.1.8
Locally spectral	$\equiv$	Algebraic arithmetic	Fact 9.5.4
Spectral	$\equiv$	Algebraic fully arithmetic	Exercise 9.5.5
Sober Noetherian	$\equiv$	Distributive with the ACC	Theorem 9.7.11
Continuous dcpo	$\equiv$	Completely distributive	Theorem 8.3.43
$\Vert$			
Sober c-space			

Figure 8.3 Some equivalences of categories obtained by Stone duality.

Figure 8.3 provides a summary of equivalences between categories of topological spaces and categories of locales obtained by Stone duality $\mathcal{O} \dashv \mathrm{pt}$. Some of them will be established in Chapter 9, and are easy consequences of those we have already established.

8.4 Colimits and limits of sober spaces

8.4.1 Coproducts and quotients of sober spaces

By Theorem 8.2.44, $\mathcal{S}$ is a left adjoint. It follows immediately that:

Proposition 8.4.1 *$\mathcal{S}$ preserves all colimits.*

The fact that $\mathcal{S}$ is a reflection implies that the colimits in the category of sober spaces and continuous maps are the sobrifications of the corresponding colimits taken in **Top**. We shall see that they are also the T_0 quotients of the latter colimits. The situation is even simpler for coproducts:

Lemma 8.4.2 *Let $(X_i)_{i \in I}$ be a family of sober spaces. Their topological coproduct $\coprod_{i \in I} X_i$ is sober.*

Proof Let F be an irreducible closed subset of the coproduct. As a closed set, F is the union of subsets $\{i\} \times F_i$, $i \in I$, where each F_i is closed. Fix an element (i, x) of F. F is included in $\{i\} \times F_i$ union $\bigcup_{j \neq i} \{j\} \times F_j$, hence in one of them by irreducibility. This must be $\{i\} \times F_i$, since (i, x) is in F. Since F is irreducible again, F_i must be irreducible in X_i. So F_i must be the downward closure of a unique point x_i in X_i. Using Lemma 4.6.7, F must be the downward closure of (i, x_i) in the coproduct. So the coproduct is sober. □

At the same time, the proof of Lemma 8.4.2 gives a concrete description of the sobrification of the coproduct $\coprod_{i \in I} X_i$ (in **Top**) as the coproduct (in **Top** again) of the spaces $\mathcal{S}(X_i)$, up to the fact that we have not checked that the topologies were indeed the same. But this is a direct consequence of Proposition 8.4.1.

Fact 8.4.3 *Let $(X_i)_{i \in I}$ be a family of topological spaces. Then $\mathcal{S}(\coprod_{i \in I} X_i)$ is naturally homeomorphic to $\coprod_{i \in I} \mathcal{S}(X_i)$, where coproducts are taken in* **Top**.

Exercise 8.4.4 Show that the categorical product $\prod_{i \in I} L_i$ of the complete lattices (resp., frames) L_i, $i \in I$, is just their Cartesian product, with the product ordering (i.e., the latter is product in **CLat**, resp., in **Frm**). What is the categorical coproduct of the locales L_i, $i \in I$? Show that both $\mathcal{O}$ and pt preserve coproducts (of topological spaces, resp., of locales).

Quotients of sober spaces are not in general sober: while $\mathbb{R}$ is sober (since it is T_2, by Proposition 8.2.12), we saw in Exercise 4.10.26 that $\mathbb{R}/\mathbb{Q}$ was not even T_0. However, the double quotients (in a similar sense as in Exercise 6.4.5) are sober. Recall that in a topological space Y, $y =_0 y'$ iff y and y' have the same open neighborhoods, iff y is both below and above y' in the specialization quasi-ordering of Y (Exercise 4.10.25).

Lemma 8.4.5 (Double quotient) *Let X be a topological space, and $\equiv$ be an equivalence relation on X. The* double quotient *of X by $\equiv$ is the T_0 quotient $(X/{\equiv})/{=_0}$ of the quotient $X/{\equiv}$. If X is sober, then $(X/{\equiv})/{=_0}$ is sober.*

Proof Let $q_\equiv \colon X \to X/{\equiv}$ be the quotient map. Then $f = \mathcal{S}(q_\equiv) \circ \eta^{\mathcal{S}}_X$ is a continuous map from X to $\mathcal{S}(X/{\equiv})$. We claim that it is quotient. Since quotients are coequalizers in **Top**, and $\mathcal{S}$ preserves all coequalizers (a consequence of Proposition 8.4.1), $\mathcal{S}(q_\equiv)$ is quotient. Since X is sober, $\eta^{\mathcal{S}}_X$ is a homeomorphism (Fact 8.2.24), so f is indeed quotient.

So $\mathcal{S}(X/{\equiv})$ is homeomorphic to $X/{\equiv'}$, where $x \equiv' x'$ iff $\mathcal{S}(q_\equiv)(\eta^{\mathcal{S}}_X(x)) = \mathcal{S}(q_\equiv)(\eta^{\mathcal{S}}_X(x'))$, using Proposition 4.10.22. Since $\mathcal{S}(X)$ is the free sober space above X, with embedding $\eta^{\mathcal{S}}_X$ (Theorem 8.2.44), $\eta^{\mathcal{S}}_X$ is in particular natural, so $\mathcal{S}(q_\equiv) \circ \eta^{\mathcal{S}}_X = \eta^{\mathcal{S}}_{X/\equiv} \circ q_\equiv$. It follows that $x \equiv' x'$ iff $\eta^{\mathcal{S}}_{X/\equiv}(q_\equiv(x)) = \eta^{\mathcal{S}}_{X/\equiv}(q_\equiv(x'))$, iff $q_\equiv(x) =_0 q_\equiv(x')$. Indeed, in general $\eta^{\mathcal{S}}_Y(y) = \eta^{\mathcal{S}}_Y(y')$ iff $\downarrow y = \downarrow y'$ iff y is both below and above y', i.e., iff $y =_0 y'$. So $\mathcal{S}(X/{\equiv})$ is homeomorphic to $(X/{\equiv})/{=_0}$. By Corollary 8.2.23, $(X/{\equiv})/{=_0}$ is sober. □

Theorem 8.4.6 (Colimits of sober spaces) *Any T_0 quotient of a colimit taken in* **Top** *of sober spaces is sober. The category of sober spaces is cocomplete, and colimits are computed as T_0 quotients of the corresponding limits in* **Top**. *The sobrification functor $\mathcal{S}$ preserves all colimits, up to iso.*

Proof It is enough to deal with coproducts and coequalizers. The former are dealt with by Lemma 8.4.2, the latter by Lemma 8.4.5, since quotients are coequalizers in **Top**. □

8.4.2 Products of sober spaces

Products are preserved by sobrification, too. This is rather exceptional for a left adjoint. However, we shall see that not all limits are preserved.

Proposition 8.4.7 *Let $(X_i)_{i \in I}$ be a family of topological spaces. The irreducible closed subsets of $\prod_{i \in I} X_i$ are the subsets of the form $\prod_{i \in I} C_i$, where each C_i is irreducible closed in X_i, $i \in I$.*

Proof In one direction, let C_i be an irreducible closed subset of X_i, for each $i \in I$. We claim that $C = \prod_{i \in I} C_i$ is irreducible closed in $X = \prod_{i \in I} X_i$. First, C is the intersection $\bigcap_{i \in I} \pi_i^{-1}(C_i)$, and is therefore closed. It is non-empty by the Axiom of Choice (specifically, by Theorem 2.4.3), since no C_i is empty. We claim that C is irreducible. Assume C meets two open subsets W_1 and W_2 of $\prod_{i \in I} X_i$. So C meets two basic opens $\prod_{i \in I} U_{1i}$ and $\prod_{i \in I} U_{2i}$ included respectively in W_1 and W_2, with U_{1i} and U_{2i} open and equal to X_i for all $i \in I$ except finitely many. For each $i \in I$, C_i then meets both U_{1i} and U_{2i}, hence $U_{1i} \cap U_{2i}$ by Trick 8.2.3. Using the Axiom of Choice, pick $x_i \in C_i \cap (U_{1i} \cap U_{2i})$. Then $\vec{x} = (x_i)_{i \in I}$ is both in C and in $\prod_{i \in I}(U_{1i} \cap U_{2i}) \subseteq W_1 \cap W_2$. So C meets $W_1 \cap W_2$. By Trick 8.2.3, C is therefore irreducible.

Conversely, let C be irreducible closed in $\prod_{i \in I} X_i$. For each $i \in I$, let $C_i = cl(\pi_i[C]) = \mathcal{S}(\pi_i)(C)$. Using Lemma 8.2.42, C_i is irreducible closed in X_i. Clearly, $C \subseteq \prod_{i \in I} C_i$. Let us show the converse inclusion. Assume on the contrary that there is a tuple $\vec{x} \in \prod_{i \in I} C_i$ that is not in C. Since C is closed, and by definition of the product topology, there is a product $\prod_{i \in I} U_i$ of open subsets, where $U_i = X_i$ for all but finitely many values of i (say for all $i \in I \setminus J$, where J is finite), that contains $\vec{x}$ but does not intersect C. Then C is included in the complement of $\prod_{i \in I} U_i$, which happens to be the finite union (because J is finite) $\bigcup_{i \in J} \pi_i^{-1}(F_i)$, where one lets F_i be the complement of U_i in X_i. Since C is irreducible, there is an $i \in J$ such that $C \subseteq \pi_i^{-1}(F_i)$. However, this implies that $\pi_i[C] \subseteq F_i$, hence $C_i \subseteq F_i$. But the ith component x_i of $\vec{x}$ is in C_i, and in U_i, a contradiction. □

Theorem 8.4.8 *Any product of sober spaces is sober. The sobrification functor preserves products up to natural iso. The iso is the* product map $\prod$ *that sends every $(C_i)_{i \in I} \in \prod_{i \in I} \mathcal{S}(X_i)$ to $\prod_{i \in I} C_i \in \mathcal{S}(\prod_{i \in I} X_i)$.*

Proof Let X_i be sober spaces, $i \in I$. Any irreducible closed subset of X_i is of the form $\prod_{i \in I} C_i$, where C_i is irreducible closed in X_i for every $i \in I$, by Proposition 8.4.7. Since X_i is sober, $C_i = \downarrow x_i$ for some unique point x_i of X_i. Then $\prod_{i \in I} C_i$ is the downward closure of $(x_i)_{i \in I}$ in $\prod_{i \in I} X_i$, using the fact that the specialization quasi-ordering of the product is the product of the specialization quasi-orderings (Lemma 4.5.17). So $\prod_{i \in I} X_i$ is sober.

Now let X_i be arbitrary spaces, $i \in I$. The product map is well defined by Proposition 8.4.7. Let us show that it is continuous. By Lemma 8.2.18, the subsets of the form $\diamond \prod_{i \in I} U_i$, where U_i is open for every $i \in I$ and equal to X_i for all but finitely many values of i, form a base of the topology of $\mathcal{S}(\prod_{i\in I} X_i)$. Its inverse image by $\prod$ is $\prod_{i\in I} \diamond U_i$, which is open since, for all but finitely many values of i, $\diamond U_i = \diamond X_i = \mathcal{S}(X_i)$. So $\prod$ is continuous, using Trick 4.3.3.

Again by Proposition 8.4.7, $\prod$ is bijective. Its inverse sends $C \in \prod_{i\in I} \mathcal{S}(X_i)$ to $(\pi_i[C])_{i\in I}$: write C as $\prod_{i\in I} C_i$, and realize that $C_i = \pi_i[C]$. This is continuous, using Trick 4.5.4, provided the map sending C to $\pi_i[C]$ is continuous for every $i \in I$. The inverse of $\diamond U_i$, where U_i is open in X_i, is $\diamond\pi_i^{-1}(U_i)$, hence this is indeed the case, by Trick 4.3.3.

It remains to show that the $\prod$ iso is natural, or equivalently that its inverse is natural, or equivalently that, given any continuous maps $f_i : X_i \to Y_i, i \in I$, given any irreducible closed subset $C = \prod_{i\in I} C_i$ of $\prod_{i\in I} X_i$, $\mathcal{S}(f_i)(\pi_i[C]) = \pi_i[\mathcal{S}(\prod_{i\in I} f_i)(C)]$. The left-hand side is $cl(f_i[C_i])$, while the right-hand side is $\pi_i[cl(\prod_{i\in I} f_i[C_i])]$. Now, by Exercise 4.5.3, $cl(\prod_{i\in I} f_i[C_i]) = \prod_{i\in I} cl(f_i[C_i])$, and we conclude. □

Exercise 8.4.9 (Products of core-compact spaces) We generalize Exercise 5.2.12 to the case of infinite products. Let X_i, $i \in I$, be core-compact spaces, and assume that all but finitely many of them are compact. By considering their sobrifications, and using the similar result on products of locally compact spaces (Proposition 4.8.10), show that $\prod_{i\in I} X_i$ is core-compact.

8.4.3 Subspaces and equalizers of sober spaces

While products in the category of sober spaces coincide with product in **Top**, equalizers will also exist in the category of sober spaces, and will also be defined as certain (sober) subspaces. However $\mathcal{S}$ will *not* preserve equalizers.

Lemma 8.4.10 *Let X be a topological space, and A be a subspace of X. For each irreducible closed subset C of A, the closure $cl(C)$ of C in X is irreducible closed in X, and $C = cl(C) \cap A$.*

Proof Assume $cl(C)$ is included in the union of two closed subsets F_1 and F_2 of X. Then C is included in the union of $F_1 \cap A$ and $F_2 \cap A$, which are closed in A by Proposition 4.9.4. Since C is irreducible, C is contained in one of them, say $F_1 \cap A$. In particular, $C \subseteq F_1$, so $cl(C) \subseteq F_1$. Hence $cl(C)$ is irreducible. Clearly, $C \subseteq cl(C) \cap A$. Conversely, one can write C as $F \cap A$ for some closed subset F of X by Proposition 4.9.4. Since $C \subseteq F$, $cl(C) \subseteq F$, so $cl(C) \cap A \subseteq F \cap A = C$. □

Lemma 8.4.11 *Let X be a topological space, and A be a subspace of X. Then $\mathcal{S}(A)$ is a subspace of $\mathcal{S}(X)$, up to iso: precisely, let $i : A \to X$ be the inclusion map, then $\mathcal{S}(i)$ is an embedding of $\mathcal{S}(A)$ into $\mathcal{S}(X)$.*

Proof We first claim that $\mathcal{S}(i)$ is injective. Let C, C' be two irreducible closed subsets of A such that $\mathcal{S}(i)(C) = \mathcal{S}(i)(C')$, i.e., such that $cl(C) = cl(C')$, where cl denotes closure in X. Then $cl(C) \cap A = cl(C') \cap A$, whence $C = C'$ by Lemma 8.4.10. Next, $\mathcal{S}(i)$ is continuous by Lemma 8.2.42, and $\mathcal{S}(i)^{-1}(\diamond V) = \diamond i^{-1}(V) = \diamond(V \cap A)$ for every open subset V of X. Every open subset of $\mathcal{S}(A)$ is of the form $\diamond U$ with U open in A, by Lemma 8.2.18, and U is of the form $V \cap A$ for some open subset V of X by definition. Then $\diamond U = \mathcal{S}(i)^{-1}(\diamond V)$, showing that $\mathcal{S}(i)$ is almost open. It follows that $\mathcal{S}(i)$ is an embedding, by Proposition 4.10.9. □

Since embeddings are just subspace inclusions up to iso, it follows that $\mathcal{S}$ preserves embeddings, i.e., that, whenever i is an embedding of topological spaces, so is $\mathcal{S}(i)$.

Lemma 8.4.12 *Let X be a sober space, and Y be a topological space. The equalizer $[g_1 = g_2]$ of two continuous maps $g_1, g_2 \colon X \to Y$ in* **Top** *is sober.*

In particular, the equalizer of two morphisms g_1, g_2 in the category of sober spaces exists and coincides with their equalizer $[g_1 = g_2]$ in **Top**.

Proof Let C be irreducible closed in $[g_1 = g_2]$. By Lemma 8.4.10, $cl(C)$ is irreducible closed in X. Since X is sober, there is a unique point $x \in X$ such that $cl(C)$ is the downward closure of x in X. We claim that x is in $[g_1 = g_2]$. Otherwise, $g_1(x) \neq g_2(x)$, so $g_1(x) \not\leqslant g_2(x)$ or $g_2(x) \not\leqslant g_1(x)$, where $\leqslant$ is the specialization ordering of X (this is an ordering, not just a quasi-ordering, because X is sober, hence T_0). Assume that $g_1(x) \not\leqslant g_2(x)$, without loss of generality. Let U be the complement of $\downarrow g_2(x)$ in X. U is open in X (Lemma 4.2.7), and contains $g_1(x)$. Since $g_1(x) \in U$, x is in the open subset $g_1^{-1}(U)$. Since x is also in the closure of C in X, C must meet $g_1^{-1}(U)$ by Lemma 4.1.27. Let x' be any point in $C \cap g_1^{-1}(U)$. Since $x' \in C \subseteq [g_1 = g_2]$, $g_1(x') = g_2(x')$, and since $x' \in C \subseteq cl(C) = \downarrow x$, $x' \leqslant x$, whence $g_1(x') = g_2(x') \leqslant g_2(x)$, using the fact that every continuous map is monotonic (Proposition 4.3.9). On the other hand, since $x' \in g_1^{-1}(U)$ and $U = X \setminus \downarrow g_2(x)$, $g_1(x') \not\leqslant g_2(x)$, a contradiction.

So $cl(C)$ is the downward closure of x in X, and x is in $[g_1 = g_2]$. Every element x' of $[g_1 = g_2]$ below x is in $\downarrow x = cl(C)$, hence in $cl(C) \cap [g_1 = g_2] = C$, by Lemma 8.4.10. (Note that the term "below" is unambiguous, as the specialization quasi-ordering of $[g_1 = g_2]$ is the restriction of the specialization ordering of X, by Proposition 4.9.5. Also, this quasi-ordering must then be an ordering, so $[g_1 = g_2]$ is T_0.) Conversely, every element of C is in $cl(C)$, hence below x. So $cl(C)$ is the downward closure of x in $[g_1 = g_2]$, therefore $[g_1 = g_2]$ is sober.

The second part of the Lemma follows immediately. □

However, $\mathcal{S}$ does *not* preserve equalizers. This would mean that, given two continuous maps g_1, g_2 from X to Y, $\mathcal{S}[g_1 = g_2]$, together with the closure map cl sending each $C \in \mathcal{S}[g_1 = g_2]$ to its closure in $\mathcal{S}(X)$ (i.e., cl is $\mathcal{S}(i)$, where i is the canonical inclusion map $i\colon [g_1 = g_2] \to X$), would be an equalizer of $\mathcal{S}(g_1), \mathcal{S}(g_2)\colon \mathcal{S}(X) \to \mathcal{S}(Y)$. While $\mathcal{S}(i) = cl$ is an embedding, that is, $\mathcal{S}[g_1 = g_2]$ is indeed a subspace of $\mathcal{S}(X)$ up to iso, it fails to be the subspace $[\mathcal{S}(g_1) = \mathcal{S}(g_2)]$. Here is a counterexample, due to Hoffmann (1979a, remark 1.5). Let X and Y both be $\mathbb{N}$ with its cofinite topology (Exercise 4.2.21; see also Example 8.2.8), g_1 be the identity map, and g_2 be any permutation without any fixed point, say the map that sends every even number x to $x + 1$ and every odd number x to $x - 1$. Note that g_2 is continuous, because the inverse image of any finite set is finite. Now $[g_1 = g_2]$ is empty, so $\mathcal{S}[g_1 = g_2]$ is empty. However, $\mathcal{S}(X) = \mathcal{S}(Y)$ is $\mathbb{N}$ plus a fresh element $\top$ (denoting $\mathbb{N}$ itself) above all elements of $\mathbb{N}$, $\mathcal{S}(g_1)$ and $\mathcal{S}(g_2)$ both map $\top$ to $\top$, hence $[\mathcal{S}(g_1) = \mathcal{S}(g_2)] = \{\top\}$. Since the latter is non-empty, in particular it is non-isomorphic to $\mathcal{S}[g_1 = g_2]$. □

Theorem 8.4.13 (Limits of sober spaces) *Any limit taken in* **Top** *of sober spaces is sober. The category of sober spaces and continuous maps is complete, and limits are computed as in* **Top**. *The sobrification functor does not preserve all limits, but preserves all products, up to natural iso.*

Proof Recall that any limit can be obtained as an equalizer of two morphisms whose source and target objects are products. Using Theorem 8.4.8 and Lemma 8.4.12, and limits taken in **Top** of sober spaces is sober. The second part of the theorem is then an easy consequence, as is the third part, using the fact that $\mathcal{S}$ preserves products (Theorem 8.4.8) but not equalizers. □

In particular, $\mathcal{S}$ cannot be right adjoint to any functor. This means that, given any sober space X, there is in general no smallest space Z of which X could be the sobrification, up to iso.

8.4.4 Binary products and Galois connections

The case of binary products is particularly interesting. Another way of showing that $\mathcal{S}$ preserves binary products would be to show that the $\mathcal{O}$ functor from **Top** to **Loc** preserves binary products. Since $\mathcal{S}$ is $\mathsf{pt} \circ \mathcal{O}$, and pt, as a right adjoint, preserves products, this would provide a proof. However, although it is true that **Loc** has products, $\mathcal{O}$ does not preserve them; see Johnstone (1982, sections 2.12–2.14).

So we won't reprove the fact that $\mathcal{S}$ preserves binary products this way. However, the binary products in the category of locales have an elegant description, and we devote this section to that.

Exercise 8.4.14 Show that the category of *spatial* locales, i.e., the opposite of the category of spatial lattices and frame homomorphisms, has all limits, and that $\mathcal{O}$, as a functor from the category of sober spaces to the category of spatial locales, preserves them all (including equalizers). Hint: reason categorically. As a side-note, Johnstone's counterexample (Johnstone, 1982, sections 2.12–2.14) is a product of two sober spaces; from this we can show that binary products in the category of spatial locales do not coincide with binary products in the larger category of all locales.

Binary products in the category of spatial locales must be binary coproducts in the category of spatial lattices. Let us look for a convenient way of describing them.

A *poset adjunction* $\alpha \dashv \gamma$ between two posets L and L' is a pair of maps $\alpha \colon L \to L'$ and $\gamma \colon L' \to L$ such that $\alpha(u) \leqslant u'$ if and only if $u \leqslant \gamma(u')$, for all $u \in L$, $u' \in L'$. This has a number of equivalent definitions. Let us say that the former is Definition 1. Definition 2 is: a poset adjunction is a pair of maps α, γ that are monotonic, and such that $u \leqslant \gamma(\alpha(u))$ for every $u \in L$, and $\alpha(\gamma(u')) \leqslant u'$ for every $u' \in L'$. Definition 3 applies when L' is a complete lattice, and requires the following: α is a map that preserves all existing least upper bounds, i.e., whenever $(u_i)_{i \in I}$ is a family of elements of L that has a least upper bound, then $\sup_{i \in I} \alpha(u_i) = \alpha(\sup_{i \in I} u_i)$; in this case, γ is uniquely determined from α by defining $\gamma(u')$ as the largest element $u \in L$ such that $\alpha(u) \leqslant u'$, for each $u' \in L'$ (in particular, the family of all elements u such that $\alpha(u) \leqslant u'$ has a least upper bound). Definition 4 applies when L is a complete lattice, and requires instead that γ preserve all existing greatest lower bounds, in which case α is uniquely determined by letting $\alpha(u)$ be the least element u' such that $u \leqslant \gamma(u')$.

An illustration is given in Figure 8.4, in the simple case where L and L' are both totally ordered. This should give an intuition that a poset adjunction is almost an order isomorphism, except for some flat segments (horizontal or vertical), where α and γ will pick one end of the segment (materialized as a bullet $\bullet$) and not the other (materialized by a small arc).

Exercise 8.4.15 Show that all the above definitions of a poset adjunction are equivalent. (For 3 $\Rightarrow$ 1, show that α must be monotonic, and $\alpha(\gamma(u')) \leqslant u'$ for every $u' \in L'$. Once 3 has been dealt with, you may want to deal with 4 by noticing that $\alpha \dashv \gamma$ is a poset adjunction between L and L' according to Definition 1, or 2, if and only if $\gamma \dashv \alpha$ is a poset adjunction between L'^{op} and L^{op}.)

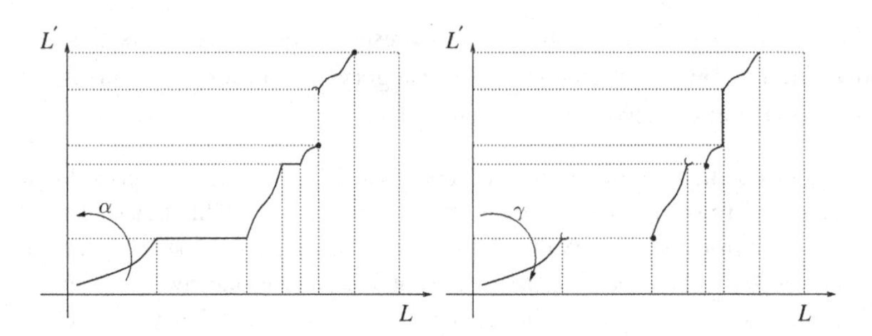

Figure 8.4 A poset adjunction.

Exercise 8.4.16 Show that a poset adjunction L and L' is nothing else than an adjunction between the thin categories L and L'. This is the reason why we chose the notation $\alpha \dashv \gamma$, and the name "poset adjunction."

Adjunctions were invented much later than poset adjunctions, or rather than the equivalent notion of Galois connections. Somehow, confusingly, our notion of poset adjunction is exactly what is now usually called a Galois connection in abstract interpretation (Cousot and Cousot, 1977). Historically, however, a *Galois connection* between two posets L and L' is a poset adjunction between L and L'^{op}. Explicitly:

Definition 8.4.17 (Galois connection) Let L, L' be posets. Call a map from L to L' *antitone* if and only if it reverses order, i.e., an antitone map from L to L' is a monotonic map from L to L'^{op}.

A *Galois connection* between L and L' is a pair (α, γ) of maps from L to L' and conversely such that $u' \leqslant \alpha(u)$ iff $u \leqslant \gamma(u')$ for all $u \in L$, $u' \in L'$ (Definition 1); equivalently, a pair of antitone maps (α, γ) such that $u \leqslant \gamma(\alpha(u))$ for every $u \in L$ and $u' \leqslant \alpha(\gamma(u'))$ for every $u' \in L'$ (Definition 2).

If L and L' are complete lattices, then we also have the following equivalent definitions. First, both α and γ must map least upper bounds to greatest lower bounds, i.e., $\alpha(\sup_{i \in I} u_i) = \inf_{i \in I} \alpha(u_i)$ and similarly for γ. And conversely, if $\alpha : L \to L'$ maps least upper bounds to greatest lower bounds, then there is a unique map $\gamma : L' \to L$ such that (α, γ) is a Galois connection: $\gamma(u')$ is the largest element u such that $u' \leqslant \alpha(u)$ (Definition 3). If $\gamma : L' \to L$ maps least upper bounds to greatest lower bounds, then there is a unique map $\alpha : L \to L'$ such that (α, γ) is a Galois connection: $\alpha(u)$ is the largest element u' such that $u \leqslant \gamma(u')$ (Definition 4).

Proposition 8.4.18 (Opens as galois connections) *Let X, Y be two topological spaces. For every open subset W of $X \times Y$, (α_W, γ_W) defines a Galois connection between $\mathcal{O}(X)$ and $\mathcal{O}(Y)$, where:*

- *for each $U \in \mathcal{O}(X)$, $\alpha_W(U)$ is the largest open subset V of Y such that $U \times V \subseteq W$;*
- *for each $V \in \mathcal{O}(Y)$, $\gamma_W(V)$ is the largest open subset U of X such that $U \times V \subseteq W$.*

Concretely, $\alpha_W(U) = int(\bigcap_{x \in U} W_{|x})$, and $\gamma_W(V) = int(\bigcap_{y \in V} W^{|y})$ (see Exercise 4.5.10).

Proof First, define $\alpha_W(U)$ as the union of all the open subsets V of Y such that $U \times V \subseteq W$. Then $U \times \alpha_W(U)$ is the union of all open rectangles $U \times V$, when V ranges over the open subsets such that $U \times V \subseteq W$, and is therefore included in W. It follows that $\alpha_W(U)$ is the largest open subset V of Y such that $U \times V \subseteq W$, as announced. Similarly, $\gamma_W(V)$ is the largest open subset U of X such that $U \times V \subseteq W$. It remains to show that, for all opens U of X, V of Y, $\alpha_W(U) \supseteq V$ if and only if $U \subseteq \gamma_W(V)$. This is clear, as both conditions are equivalent to $U \times V \subseteq W$.

Concretely, $int(\bigcap_{x \in U} W_{|x})$ is an open subset V such that, for every $x \in U$, for every $y \in V$, y is in $W_{|x}$, i.e., $(x, y) \in W$. So $U \times V \subseteq W$, whence $V \subseteq \alpha_W(U)$. Conversely, for every $y \in \alpha_W(U)$, (x, y) is in W, so $y \in W_{|x}$. It follows that $\alpha_W(U) \subseteq \bigcap_{x \in U} W_{|x}$. Since $\alpha_W(U)$ is open, it must then be contained in the interior V of $\bigcap_{x \in U} W_{|x}$. The formula $\gamma_W(V) = int(\bigcap_{y \in V} W^{|y})$ is proved similarly. □

This construction was already the essential argument in Exercise 5.4.10. Beware, however: Proposition 8.4.18 does not state that open subsets W of $X \times Y$ are in bijection with Galois connections (see Exercise 8.4.23). However, it gives us a hint that such Galois connections should be the elements of the product of two locales.

Lemma 8.4.19 *For all complete lattices L, L', let* $\mathrm{Gal}(L, L')$ *be the set of all Galois connections between L and L'. Order this by $(\alpha, \gamma) \leqslant (\alpha', \gamma')$ if and only if $\alpha(u) \leqslant \alpha'(u)$ for every $u \in L$, i.e., if and only if $\gamma(u') \leqslant \gamma'(u')$ for every $u' \in L'$.*

$\mathrm{Gal}(L, L')$ *is a complete lattice, and greatest lower bounds are computed* pointwise, *i.e., the greatest lower bound of a family $(\alpha_i, \gamma_i)_{i \in I}$ is (α, γ) where $\alpha(u) = \inf_{i \in I} \alpha_i(u)$ for every $u \in L$, and $\gamma(u') = \inf_{i \in I} \gamma_i(u')$ for every $u' \in L'$.*

Proof We must first show that the two definitions of $\leqslant$ on $\mathrm{Gal}(L, L')$ are equivalent. If $\alpha(u) \leqslant \alpha'(u)$ for every $u \in L$, then in particular for every $u' \in L'$, $u' \leqslant \alpha(\gamma(u')) \leqslant \alpha'(\gamma(u'))$, using Definition 2 and the assumption, so $\gamma(u') \leqslant \gamma'(u')$ by Definition 1. The converse implication is proved similarly.

It follows that $\mathrm{Gal}(L, L')$ is order isomorphic, using Definition 3, to the set of all functions $\alpha: L \to L'$ that map least upper bounds to greatest lower bounds, and, using Definition 4, to the set of all functions $\gamma: L' \to L$ that map least upper bounds to greatest lower bounds, both ordered pointwise. Each of these sets is a complete lattice. E.g., for every family $(\alpha_i)_{i\in I}$ of functions from L to L' that map least upper bounds to greatest lower bounds, let $\alpha(u) = \inf_{i\in I} \alpha_i(u)$ for each $u \in L$: if u is a least upper bound of a family $(u_j)_{j\in J}$, then $\alpha(u) = \inf_{i\in I} \inf_{j\in J} \alpha_i(u_j) = \inf_{j\in J} \inf_{i\in I} \alpha_i(u_j) = \inf_{j\in J} \alpha(u_j)$. So α maps least upper bounds to greatest lower bounds, and is therefore the greatest lower bound of the family $(\alpha_i)_{i\in I}$. Since all greatest lower bounds exist, so do all least upper bounds, and the formulae for greatest lower bounds given in the statement of the Proposition hold. □

Beware that least upper bounds are *not* computed pointwise: there is no reason why we should have $\alpha(u) = \sup_{i\in I} \alpha_i(u)$, for example, where (α, γ) is the least upper bound of $(\alpha_i, \gamma_i)_{i\in I}$ in $\mathrm{Gal}(L, L')$.

Showing that $\mathrm{Gal}(L, L')$ is a frame (not just a complete lattice) when L and L' both are is more involved, since Lemma 8.4.19 gives no idea what least upper bounds would look like in $\mathrm{Gal}(L, L')$. Much as opens in $X \times Y$ were unions of open rectangles, we shall show that elements of $\mathrm{Gal}(L, L')$ are least upper bounds of similarly defined rectangles, and that these least upper bounds *are* pointwise.

Definition 8.4.20 Let L, L' be two complete lattices. Write $\top$ for their largest elements, and $\bot$ for their least elements. Given $u \in L$, $u' \in L'$, let the *$u \times u'$-rectangle* be the pair $(\alpha_{u\times u'}, \gamma_{u\times u'})$, where:

$$\alpha_{u\times u'}(v) = \begin{cases} \top & \text{if } v = \bot \\ u' & \text{if } v \neq \bot, v \leqslant u \\ \bot & \text{if } v \not\leqslant u \end{cases} \qquad \gamma_{u\times u'}(v') = \begin{cases} \top & \text{if } v' = \bot \\ u & \text{if } v' \neq \bot, v' \leqslant u' \\ \bot & \text{if } v' \not\leqslant u'. \end{cases}$$

Every $u \times u'$-rectangle is a Galois connection, i.e., $v' \leqslant \alpha_{u\times u'}(v)$ iff $v \leqslant \gamma_{u\times u'}(v')$. Indeed, $v' \leqslant \alpha_{u\times u'}(v)$ is equivalent to the following disjunction: $v = \bot$; or $v \neq \bot$, $v \leqslant u$, and $v' \leqslant u'$; or $v' = \bot$ and $v \not\leqslant u$. And $v \leqslant \gamma_{u\times u'}(v')$ is equivalent to the following disjunction: $v' = \bot$; or $v' \neq \bot$, $v' \leqslant u'$, and $v \leqslant u$; or $v = \bot$ and $v' \not\leqslant u'$. These disjunctions are equivalent, as a tedious case analysis shows. The following exercise should help one grasp the intuition behind the funny definition.

Exercise 8.4.21 Show that, given any two topological spaces X and Y, and any open rectangle $W = U \times U'$ in $X \times Y$, (α_W, γ_W) is the $U \times U'$-rectangle in $\mathrm{Gal}(\mathcal{O}(X), \mathcal{O}(Y))$.

The following is reminiscent of the fact that every element of $\mathcal{O}(X \times Y)$ is a union of open rectangles.

Lemma 8.4.22 *Let L, L' be complete lattices. For all $u \in L$, $u' \in L$, and $(\alpha, \gamma) \in \mathrm{Gal}(L, L')$, $(\alpha_{u\times u'}, \gamma_{u\times u'}) \leqslant (\alpha, \gamma)$ if and only if $u' \leqslant \alpha(u)$, if and only if $u \leqslant \gamma(u')$.*

Every Galois connection (α, γ) between L and L' is the least upper bound of all rectangles $\alpha_{u\times u'}$ below it (i.e., with $u' \leqslant \alpha(u)$ or, equivalently, with $u \leqslant \gamma(u')$).

Moreover, this least upper bound is pointwise*: for every $v \in L$, $\alpha(v)$ is the largest element of the form $\alpha_{u\times u'}(v)$ where u and u' range over the elements of L, resp. L' such that $u' \leqslant \alpha(u)$; and for every $v' \in L'$, $\gamma(v')$ is the largest element of the form $\gamma_{u\times u'}(v)$ where u and u' range over the elements of L, resp. L' such that $u \leqslant \gamma(u')$.*

Proof We first show that $(\alpha_{u\times u'}, \gamma_{u\times u'}) \leqslant (\alpha, \gamma)$ if and only if $u' \leqslant \alpha(u)$, if and only if $u \leqslant \gamma(u')$.

For any element (α, γ) of $\mathrm{Gal}(L, L')$, if (α, γ) is above the $u \times u'$-rectangle, then $\alpha_{u\times u'}(u) \leqslant \alpha(u)$, so $u' \leqslant \alpha(u)$ (even when $u = \bot$, where $\alpha(u) = \top$, since α maps least upper bounds to greatest lower bounds—consider the least upper bound of the empty family). Conversely, if $u' \leqslant \alpha(u)$, then we claim that $\alpha_{u\times u'}(v) \leqslant \alpha(v)$ for every $v \in L$: if $v = \bot$, this is because $\alpha(\bot) = \top$; if $v \neq \bot$ and $v \leqslant u$, $\alpha_{u\times u'}(v) = u' \leqslant \alpha(u) \leqslant \alpha(v)$, since α is antitone; and if $v \not\leqslant u$, then $\alpha_{u\times u'}(v) = \bot \leqslant \alpha(v)$.

This allows us to show that $\alpha(v)$ is the largest element of the form $\alpha_{u\times u'}(v)$ where u and u' range over the elements of L, resp. L' such that $u' \leqslant \alpha(u)$, and similarly for γ. Since $u' \leqslant \alpha(u)$ iff $\alpha_{u\times u'} \leqslant \alpha$, $\alpha_{u\times u'}(v)$ is below $\alpha(v)$ for all such u, u'. Conversely, let $u = v$, $u' = \alpha(v)$. If $v = \bot$, then $\alpha_{u\times u'}(v) = \top$, but then $\alpha(v)$ equals $\top$ as well. If $v \neq \bot$, then $\alpha_{u\times u'}(v) = u' = \alpha(v)$, and we are done.

So α is the pointwise least upper bound of all rectangles below it. It must then be an upper bound in $\mathrm{Gal}(L, L')$ of all these rectangles, and it must be least, since any other upper bound α' must be such that $\alpha_{u\times u'} \leqslant \alpha'$ for every $u \in L$ and every $u' \leqslant \alpha(u)$; in particular for $u' = \alpha(u)$, $\alpha_{u\times u'}(u) \leqslant \alpha'(u)$, i.e., $\alpha(u) = u' \leqslant \alpha'(u)$ (even when $u = \bot$, where $\alpha(u) = \alpha'(u) = \top$).

A similar result holds for γ. □

Exercise 8.4.23 Exercise 8.4.21 may give you the false impression that the $\mathcal{O}$ functor preserves binary products. This is wrong, although an explicit counterexample seems too complicated to study here: see Johnstone (1982, section 2.14). However, this will hold whenever one of the two spaces is core-compact, as we shall see right away.

Let X and Y be topological spaces, and assume X is core-compact. As a first step, let (α, γ) be a Galois connection between $\mathcal{O}(X)$ and $\mathcal{O}(Y)$. Let $W \in \mathcal{O}(X \times Y)$ be defined as $\bigcup_{U \in \mathcal{O}(X)} U \times \alpha(U)$. Show that $\alpha \leqslant \alpha_W$. To show the converse, $\alpha_W \leqslant \alpha$, you will need to show that every open rectangle $U \times V$ included in W is such that $V \leqslant \alpha(U)$. For every point $x \in U$, fix an open neighborhood U_x of x such that $U_x \Subset U$ (why does this exist?). Then, for each $y \in V$, use the definition of W and of $\Subset$ to build an open neighborhood V_{xy} of y such that $V_{xy} \subseteq \alpha(U_x)$, and conclude.

Finally, show that $\mathrm{Gal}(\mathcal{O}(X), \mathcal{O}(Y))$ is isomorphic to $\mathcal{O}(X \times Y)$ whenever X (or Y) is core-compact.

Lemma 8.4.24 *Let L, L' be two complete lattices. The greatest lower bound of the $u \times u'$-rectangle and of the $v \times v'$-rectangle is the $(u \wedge v) \times (u' \wedge v')$-rectangle.*

Proof Greatest lower bounds are computed pointwise, by Lemma 8.4.19. Let (α, γ) be the greatest lower bound of the $u \times u'$-rectangle and of the $v \times v'$-rectangle. For every $w \in L$, α maps w to $\top$ if $w = \bot$, and otherwise to $u' \wedge v'$ if w is both below u and below v, i.e., if $w \leqslant u \wedge v$ and $w \neq \bot$; finally, if $w \not\leqslant u$ or $w \not\leqslant v$, then $\alpha(w) = \bot$: if $w \not\leqslant u$, then $\alpha(w) = \bot \wedge \alpha_{v \times v'}(w)$, and if $w \not\leqslant v$, then $\alpha(w) = \alpha_{u \times u'}(w) \wedge \bot$. This is exactly the definition of $\alpha_{(u \wedge v) \times (u' \wedge v')}$. ☐

Recall from Proposition 8.1.1 that a complete Heyting algebra is the same thing as a frame, except that it is presented in terms of residuation instead of the frame distributivity law.

Proposition 8.4.25 (Gal preserves frames) *Let L, L' be two complete Heyting algebras. Then* $\mathrm{Gal}(L, L')$ *is a complete Heyting algebra.*

Given two elements (α, γ), (α', γ') of $\mathrm{Gal}(L, L')$, *their residuation is $(\alpha \Rightarrow \alpha', \gamma \Rightarrow \gamma')$, where $(\alpha \Rightarrow \alpha')(v) = \inf_{\substack{u \in L, u' \in L' \\ u' \leqslant \alpha(u)}} (u' \Rightarrow \alpha'(u \wedge v))$ for every $v \in L$, and $(\gamma \Rightarrow \gamma')(v') = \inf_{\substack{u \in L, u' \in L' \\ u \leqslant \gamma(u')}} (u \Rightarrow \gamma'(u' \wedge v'))$ for every $v' \in L'$.*

Accordingly, we shall write $(\alpha \Rightarrow \alpha', \gamma \Rightarrow \gamma')$ for (α'', γ'').

Proof Let α'' abbreviate $\alpha \Rightarrow \alpha'$, and γ'' abbreviate $\gamma \Rightarrow \gamma'$. We check that (α'', γ'') is a Galois connection using the first definition of Galois connections (see Definition 8.4.17). The statement $v' \leqslant \alpha''(v)$ is equivalent to: for all $u \in L$, $u' \in L'$, if $u' \leqslant \alpha(u)$ then $v' \leqslant (u' \Rightarrow \alpha'(u \wedge v))$; using the definition of residuation $\Rightarrow$, this is equivalent to: for all $u \in L$, $u' \in L'$, if $u' \leqslant \alpha(u)$ then $u' \wedge v' \leqslant \alpha'(u \wedge v)$; using the fact that (α, γ) and (α', γ') are Galois connections, this is equivalent to: for all $u \in L$, $u' \in L'$, if $u \leqslant \gamma(u')$ then $u \wedge v \leqslant \gamma'(u' \wedge v')$; and by using similar steps in reverse, this is equivalent to $v \leqslant \gamma''(v')$. So (α'', γ'') is a Galois connection.

We now check that (α'', γ'') is residuation $(\alpha, \gamma) \Rightarrow (\alpha', \gamma')$ in $\mathrm{Gal}(L, L')$. It suffices to show that, for every map $\beta: L \to L'$ that sends least upper bounds to greatest lower bounds, $\beta \leqslant \alpha''$ iff $\beta \wedge \alpha \leqslant \alpha'$. Using the characterizations of $\leqslant$ and $\wedge$ given in Lemma 8.4.19, if $\beta \leqslant \alpha''$, then, by definition, for every $v \in L$, for all $u \in L$, $u' \in L'$ with $u' \leqslant \alpha(u)$, $\beta(v) \leqslant (u' \Rightarrow \alpha'(u \wedge v))$, so $\beta(v) \wedge u' \leqslant \alpha'(u \wedge v)$. Taking $u = v$ and $u' = \alpha(v)$, we obtain $\beta(v) \wedge \alpha(v) \leqslant \alpha'(v)$. Conversely, if $\beta \wedge \alpha \leqslant \alpha'$, then, for every $v \in L$, for all $u \in L$, $u' \in L'$ with $u' \leqslant \alpha(u)$, $\beta(v) \wedge u' \leqslant \beta(v) \wedge \alpha(u) \leqslant \beta(u \wedge v) \wedge \alpha(u \wedge v)$ (since β and α are antitone) $\leqslant \alpha'(u \wedge v)$ (by assumption). This means that $\beta(v) \leqslant \alpha''(v)$ for every $v \in L$.

In particular, it follows that $\mathrm{Gal}(L, L')$ is a complete Heyting algebra. □

Exercise 8.4.26 Show that the $\mathcal{O}$ functor preserves residuation. In other words, given two topological spaces X and Y, and two open subsets W and W' of $X \times Y$, show that $\alpha_{W \Rightarrow W'} = \alpha_W \Rightarrow \alpha_{W'}$, and $\gamma_{W \Rightarrow W'} = \gamma_W \Rightarrow \gamma_{W'}$, where $W \Rightarrow W'$ is residuation of W and W' in the frame $\mathcal{O}(X \times Y)$ (see Exercise 8.1.2).

Exercise 8.4.27 Let L, L' be two complete lattices. A *C-ideal* is a downward closed subset $\mathcal{I}$ of rectangles in $\mathrm{Gal}(L, L')$ such that, for every $u \in L$ and every family $(u'_i)_{i \in I}$ in L' such that the $u \times u'_i$-rectangle is in $\mathcal{I}$ for every $i \in I$, the $u \times (\sup_{i \in I} u'_i)$-rectangle is also in $\mathcal{I}$; and, symmetrically, for every family $(u_i)_{i \in I}$ in L and every $u' \in L'$ such that the $u_i \times u'$-rectangle is in $\mathcal{I}$ for every $i \in I$, the $(\sup_{i \in I} u_i) \times u'$-rectangle is also in $\mathcal{I}$. The C-ideals are compared using inclusion.

Show that the space of all C-ideals is order isomorphic to $\mathrm{Gal}(L, L')$.

Show that least upper bounds in the space of all C-ideals are computed as follows. Let $R(L, L')$ be the set of all subsets of rectangles in $\mathrm{Gal}(L, L')$ containing the $\bot \times \bot$ rectangle, ordered by inclusion, and notice that least upper bounds are just unions in $R(L, L')$. Let $F: R(L, L') \to R(L, L')$ map any $\mathcal{I} \in R(L, L')$ to the set of all $u \times u'$-rectangles where $u' \leqslant \sup_{i \in I} u'_i$ for some family $(u'_i)_{i \in I}$ in L' such that $\mathcal{I}$ contains all the $u \times u'_i$-rectangles, $i \in I$, plus all $u \times u'$-rectangles where $u \leqslant \sup_{i \in I} u_i$ for some family $(u_i)_{i \in I}$ in L such that $\mathcal{I}$ contains all the $u_i \times u'$-rectangles, $i \in I$. Show that the least upper bound of a family $(\mathcal{I}_j)_{j \in J}$ in the space of C-ideals is the least fixed point of F above $\bigcup_{j \in J} \mathcal{I}_j$. (Recall the fixed point theorems of Section 2.3.3.)

Let L'' be a third complete lattice. Use fixed point induction (Fact 2.3.3) to show the following: let $h\colon R(L, L') \to L''$ be any map such that $h(\bigcup_{j\in J} \mathcal{I}_j) = \sup_{j\in J} h(\mathcal{I}_j)$ for every family $(\mathcal{I}_j)_{j\in J}$ in $R(L, L')$, and such that $h(F(\mathcal{I})) = h(\mathcal{I})$ for every $\mathcal{I} \in R(L, L')$; then h restricts to a sup-preserving map from the set of all C-ideals to L''. This whole exercise is meant as a hint in solving Exercise 8.4.28.

Exercise 8.4.28 (Frame coproducts, locale products) Show that $\mathrm{Gal}(L, L')$ is the coproduct of L and L' in **Frm** (equivalently, their product in the category of locales). The first injection $\iota_1\colon L \to L + L'$ maps u to the $u \times \top$-rectangle, the second injection $\iota_2\colon L' \to L+L'$ maps u' to the $\top \times u'$-rectangle, and copairing of $f\colon L \to L''$ and $f'\colon L' \to L''$ is $[f, f']\colon \mathrm{Gal}(L, L') \to L''$, which maps (α, γ) to $\sup_{u\in L} f(u) \wedge f'(\alpha(u))$. To show that $[f, f']$ preserves least upper bounds, you are advised to use Exercise 8.4.27.

The category of locales has all products, not just binary products. Given a family $(L_i)_{i\in I}$ of locales, its locale product is built as follows. A (generalized) rectangle is a tuple $(u_i)_{i\in I}$ in $\prod_{i\in I} L_i$ such that either $u_i = \bot$ for every $i \in I$, or $u_i \neq \bot$ for every $i \in I$ and $u_i = \top$ for all but finitely many $i \in I$. Rectangles are ordered pointwise. A (generalized) C-ideal is a downward closed subset of rectangles such that, for every family of rectangles $(\vec{u}_j)_{j\in J}$ that coincide on all components except possibly $i \in I$ (i.e., $u_{jk} = u_{j'k}$ for all $j, j' \in J$ and $k \neq i$), its pointwise least upper bound is still in the C-ideal. It can be shown that the lattice of all such C-ideals is the frame coproduct, hence the locale product, of $(L_i)_{i\in I}$ (Johnstone, 1982, section 2.12).

The following is due to Shmuely (1974, theorem 2.6), and has a very short proof using Stone duality.

Theorem 8.4.29 (Shmuely's Theorem) *For all completely distributive lattices L and L',* $\mathrm{Gal}(L, L')$ *is a completely distributive lattice.*

Proof By Theorem 8.3.43, L and L' are lattices of Scott-open subsets of two continuous dcpos X, X', up to iso. Every continuous dcpo is locally compact, by Corollary 5.1.36. By Exercise 8.4.23, $\mathrm{Gal}(L, L')$ is order isomorphic to $\mathcal{O}(X_\sigma \times X'_\sigma)$. However, $X_\sigma \times X'_\sigma$ equals $(X \times X')_\sigma$, and $X \times X'$ is a continuous dcpo by Proposition 5.1.54. So $\mathcal{O}(X_\sigma \times X'_\sigma)$ is completely distributive, hence also $\mathrm{Gal}(L, L')$ by Theorem 8.3.43 again. □

9

Stably compact spaces and compact pospaces

9.1 Stably locally compact spaces, stably compact spaces

We have already introduced the notion of stably locally compact space in Definition 8.3.31. A stably compact space is only moderately more constrained:

Definition 9.1.1 (Stably compact space) A topological space is *stably compact* if and only if it is stably locally compact and compact.

In other words, a space is stably compact if and only if it is sober, locally compact, compact, and coherent; alternatively, if and only if it is T_0, well-filtered, locally compact, compact, and coherent.

Example 9.1.2 $\mathbb{S}$ is stably compact. More generally, every finite poset is stably compact in its Scott (equivalently, Alexandroff, upper) topology.

Example 9.1.3 $[0, 1]_\sigma$ is stably compact. This can be checked directly. Its non-empty closed subsets are all of the form $[0, t] = \downarrow t$, so $[0, 1]_\sigma$ is sober. If x is in the open $U = (t, 1]$, then x is also in the interior of the compact saturated subset $[\frac{x+t}{2}, 1] = \uparrow \frac{x+t}{2}$, so $[0, 1]_\sigma$ is locally compact. $[0, 1]_\sigma = \uparrow 0$ is compact, and the intersection of two non-empty compact saturated subsets $[t, 1] \cap [u, 1]$ is the compact saturated subset $[\max(t, u), 1]$.

Example 9.1.4 $[0, 1]$, this time with the open ball topology of $\mathsf{d}_{\mathbb{R}}^{sym}$, is stably compact. Indeed, every compact T_2 space is stably compact: sobriety is by Proposition 8.2.12 (*a*), local compactness is by Proposition 4.8.8, compactness is obvious, and coherence is because compact and closed subsets coincide, and the intersection of two closed subsets is closed.

The stably compact spaces have a wonderfully rich theory, and first arose from Leopoldo Nachbin's study of *compact pospaces* (Nachbin, 1948, 1965).

These were put into modern shape by Johnstone (1982). Our presentation owes much to Alvarez-Manilla *et al.* (2004).

As another example, every continuous dcpo is sober by Proposition 8.2.12 (*b*), and locally compact by Corollary 5.1.36. And every pointed poset is compact in its Scott-topology. So:

Fact 9.1.5 *Every coherent continuous (resp., and pointed) dcpo is stably locally compact (resp., stably compact) in its Scott topology.*

In particular:

Fact 9.1.6 *Every continuous complete lattice and every bc-domain is stably compact in its Scott topology.*

Indeed, by Lemma 5.7.15, every bc-domain is compact, locally compact, and coherent.

Example 9.1.7 $[0, 1]_\sigma$ is a continuous complete lattice, so we retrieve the result of Example 9.1.3.

Exercise 9.1.8 (Stably compact spaces $\equiv$ fully arithmetic lattices) Call *fully arithmetic* any arithmetic lattice in which the top element $\top$ is finite (i.e., $\top \ll \top$; the name "fully arithmetic" is non-standard). Show the following Stone-duality type characterization of stably compact spaces: the adjunction $\mathcal{O} \dashv \mathsf{pt}$ restricts to an equivalence between the category of stably compact spaces on the one hand, and the opposite of the category of fully arithmetic lattices and frame homomorphisms on the other.

Exercise 9.1.9 Show that $\mathbb{S}_\sigma^I$ and $[0, 1]_\sigma^I$ are stably compact spaces, for every set I. (A_σ^I denotes the order-theoretic product A^I, with its Scott topology.)

There is not much difference either between stably locally compact and stably compact spaces. Remember from Exercise 4.4.25 that the lifting $X_\perp$ is X plus a fresh element $\perp$ added below all others. The opens of $X_\perp$ are those of X, plus $X_\perp$ itself. The compact saturated subsets of $X_\perp$ are those of X plus $X_\perp$ itself by Exercise 4.4.26. It follows that $X_\perp$ is compact, and that X is locally compact if and only if $X_\perp$ is (Exercise 4.8.6), that X is sober if and only if $X_\perp$ is (Exercise 8.2.9), and that X is coherent if and only if $X_\perp$ is. The following is then immediate.

Proposition 9.1.10 *Let X be a topological space. Then X is stably locally compact if and only if $X_\perp$ is stably compact.*

9.1.1 From compact pospaces to stably compact spaces

A good source of stably compact spaces is provided by compact pospaces in their upward topology, as we shall see in this section. We shall see later that all stably compact spaces are obtained this way.

Definition 9.1.11 (Pospace) A *pospace* (short for *partially ordered space*) is a pair $(Z, \preceq)$ of a topological space Z and a partial ordering $\preceq$ on Z whose graph $(\preceq) = \{(x, y) \in Z \times Z \mid x \preceq y\}$ is closed in $Z \times Z$.

The fact that $(\preceq)$ is closed is better understood through notions of convergence: using Corollary 4.7.10, we insist that, for every net $(x_i, y_i)_{i \in I, \sqsubseteq}$ that converges to (x, y), if $x_i \preceq y_i$ for every $i \in I$, then $x \preceq y$ as well.

This seemingly innocuous property has the following consequence:

Proposition 9.1.12 *Every pospace is T_2.*

Proof If x and y are two distinct points of Z, then $x \not\preceq y$ or $y \not\preceq x$, i.e., (x, y) or (y, x) is not in $(\preceq)$. Assume without loss of generality that (x, y) is in the (open) complement of $(\preceq)$. By definition of the product topology, there is an open rectangle $U \times V$ containing (x, y) and not intersecting $(\preceq)$. Then U and V are disjoint neighborhoods of x and y, respectively. □

Additionally requiring compactness has strong consequences for pospaces. The following lemma will prepare us in showing order normality (Proposition 9.1.14).

Lemma 9.1.13 *Let $(Z, \preceq)$ be a compact pospace. For every closed subset F of Z, its downward closure $\downarrow_\preceq F$ and its upward closure $\uparrow_\preceq F$ with respect to $\preceq$ are both closed in Z.*

Proof $\downarrow_\preceq F$ is the image by the first projection π_1 of $(\preceq) \cap (Z \times F)$. Since $(Z, \preceq)$ is a pospace, $(\preceq) \cap (Z \times F)$ is closed in $Z \times Z$, hence compact (Proposition 4.4.15), since $Z \times Z$ is compact by Tychonoff's Theorem 4.5.12. Since π_1 is continuous, its image $\downarrow_\preceq F$ is compact in Z (Proposition 4.4.13), hence closed since Z is T_2 by Proposition 9.1.12. We reason similarly for $\uparrow_\preceq F = \pi_2[(\preceq) \cap (F \times Z)]$. □

Proposition 9.1.14 (Compact pospaces are order normal) *Every compact pospace $(Z, \preceq)$ is* order normal, *i.e., for every pair of disjoint closed subsets F and F' such that F is downward closed and F' is upward closed in $\preceq$, there are disjoint open subsets U and U' such that U is downward closed in $\preceq$ and contains F, and U' is upward closed in $\preceq$ and contains F' (see Figure 9.1).*

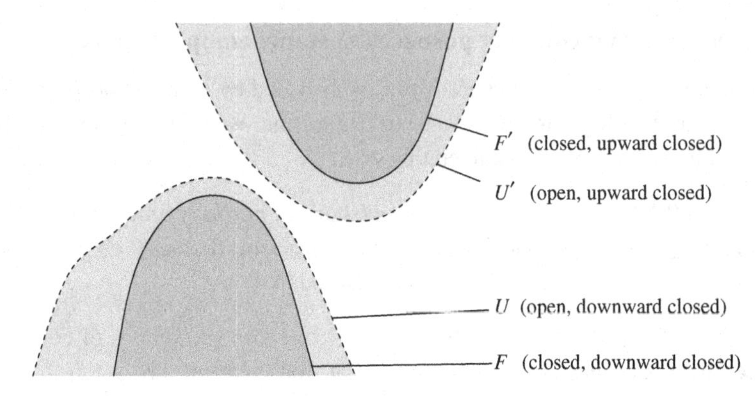

Figure 9.1 Order normality.

Proof Z is compact and T_2 by Proposition 9.1.12. By Proposition 4.4.17, Z is normal, so one can find disjoint open subsets V and V' such that $F \subseteq V$ and $F' \subseteq V'$. Let U be the complement of $\uparrow_{\preceq}(Z \smallsetminus V)$, and U' be the complement of $\downarrow_{\preceq}(Z \smallsetminus V')$. U is downward closed, and U' is upward closed. By Lemma 9.1.13, U and U' are open. Moreover, $U \subseteq V$ and $U' \subseteq V'$, so U and U' are disjoint.

Since $F \subseteq V$, F does not intersect $Z \smallsetminus V$. It follows that F does not intersect $\uparrow_{\preceq}(Z \smallsetminus V)$; otherwise there would be an element $y \in F$ and an element $x \in Z \smallsetminus V$ such that $x \preceq y$: then x would be in F, since F is downward closed, hence x would be in V, a contradiction. So F is included in U. Similarly, $F' \subseteq U'$. □

Lemma 9.1.15 (Compact pospaces are order T_2) *Every compact pospace $(Z, \preceq)$ is order T_2, i.e., for every pair of points x, y such that $x \not\preceq y$, there are disjoint open neighborhoods U of x and V of y where U is additionally upward closed in $\preceq$ and V is downward closed in $\preceq$.*

Proof By Proposition 9.1.12, Z is T_2, hence T_1, so $\{x\}$ and $\{y\}$ are closed in Z (see Exercise 4.1.17). Let $F = \uparrow_{\preceq}\{x\}$, $F' = \downarrow_{\preceq}\{y\}$. By Lemma 9.1.13, F and F' are closed in Z. Moreover, they are disjoint; otherwise there would be an element z such that $x \preceq z$ and $z \preceq y$. By order normality (Proposition 9.1.14), one can find an upward closed open subset U containing F and a downward closed open subset V containing F' such that U and V are disjoint. In particular, $x \in U$ and $y \in V$. □

Exercise 9.1.16 (Pospaces are order T_1) Let $(Z, \preceq)$ be a pospace, not necessarily compact. Show that $(Z, \preceq)$ is *order* T_1, i.e., that, whenever $x \not\preceq y$, there is an upward closed open neighborhood U of x and a downward closed open neighborhood V of y such that U does not contain y and V does not contain x.

Show that this is the best we can ask for without compactness. You may consider $Z = ([0,1]+[0,1])/{\equiv}$, where both copies of $[0,1]$ come with their open ball topology, and $\equiv$ is defined by: $(1,x) \equiv (2,y)$ iff $x = y$ and $x \neq 0$. Define $\preceq$ on Z by: $q_{\equiv}(i,x) \preceq q_{\equiv}(j,y)$ iff $0 < x \leqslant y$, or $i = j$ and $0 = x = y$. (Z is essentially the half-open interval $(0,1]$ plus two copies of 0 below all other elements.) Then look at $q_{\equiv}(1,0)$ and $q_{\equiv}(2,0)$.

The upward closed, open subsets form a topology, and similarly for the downward closed, open subsets. Let us give names to these topologies.

Definition 9.1.17 (Upward, downward topology) Let $(Z, \preceq)$ be a pospace. The *upward topology* of $(Z, \preceq)$ is the family of all open subsets of Z that are upward closed in $\preceq$. The *downward topology* of $(Z, \preceq)$ is the family of all open subsets of Z that are downward closed in $\preceq$.

Lemma 9.1.18 *Let $(Z, \preceq)$ be a pospace. The specialization quasi-ordering of Z with its upward topology is exactly $\preceq$.*

Proof Let X be Z with its upward topology. Let $\leqslant$ be the specialization quasi-ordering of X. If $x \preceq y$ then $x \leqslant y$, since every open subset of X is upward closed in $\preceq$ by definition. Conversely, assume $x \not\preceq y$. Since $(Z, \preceq)$ is order T_1 (see Exercise 9.1.16), there is an upward closed open subset U containing x but not y, i.e., $x \not\leqslant y$. □

Example 9.1.19 $[0,1]$ with the open ball topology of $\mathsf{d}_{\mathbb{R}}^{sym}$ is compact, and the graph $(\leqslant)$ of its usual ordering is closed in $[0,1]^2$. Indeed, $(\leqslant)$ is the inverse image of the closed set $(-\infty, 0]$ of $\mathbb{R}$ by the continuous map $x, y \mapsto x - y$ (see Exercise 4.8.8). The open subsets in the upward topology are exactly the intervals $(t, 1]$ with $0 \leqslant t \leqslant 1$, plus $[0,1]$ and the empty set. This is exactly the Scott topology on $[0,1]$. We have seen that $[0,1]_\sigma$ was stably compact, and this is no accident.

Proposition 9.1.20 (Compact pospace ⇒ stably compact space) *Let $(Z, \preceq)$ be a compact pospace. Then Z with the upward topology is a stably compact space. Its compact saturated subsets are exactly the upward closed, closed subsets of Z.*

Proof Let X be Z with its upward topology. Since all open subsets of X are open in Z, all compact subsets of Z are compact in X: if K is compact in Z, then any cover of K by opens of X is a cover by opens of Z, hence contains a finite subcover. A trivial consequence is that X itself is compact.

We claim that X is locally compact. Let $x \in X$, and U be any upward closed open subset of Z containing x. Let $F = Z \setminus U$: F is closed, and downward closed. As in Proposition 9.1.12, $\uparrow_{\preceq} x$ is closed and upward closed, and does not intersect F. Since $(Z, \preceq)$ is order normal (Proposition 9.1.14), there are disjoint open subsets V and W of Z, where W contains F and is downward closed, and V contains $\uparrow_{\preceq} x$, hence x, and is upward closed. Let $Q = Z \setminus W$. Q is closed in Z, hence compact in Z (Proposition 4.4.15), hence compact in X. Since $V \cap W = \emptyset$, $V \subseteq Q$. Since $F \subseteq W$, $Q \subseteq U$. That is, $x \in V \subseteq Q \subseteq U$. So X is locally compact, by Trick 4.8.2.

We make a parenthesis, and prove the last part of the proposition. Given any downward closed, closed subset F of Z, F is compact in Z, hence in X. Moreover, F is downward closed in $\preceq$, i.e., in the specialization ordering $\leqslant$ of X (Lemma 9.1.18), so it is saturated in X. Conversely, let Q be any saturated compact subset of X. In particular, Q is upward closed in Z. We claim that Q is closed in Z. Let y be any point outside Q. For every element $x \in Q$, $x \not\leqslant y$, otherwise y would be in Q, so $x \not\preceq y$. Since $(Z, \preceq)$ is order T_2, there is in particular an upward closed open subset U_x of Z containing x and a disjoint (downward closed) open subset V_x of Z containing y. (We pick U_x, V_x for each $x \in Q$ using the Axiom of Choice.) Note that U_x is open in X by definition, and that $(U_x)_{x \in Q}$ is an open cover of Q in X. Since Q is compact, it has a finite subcover $(U_x)_{x \in E}$, where E is a finite subset of Q. Let $U = \bigcup_{x \in E} U_x$, $V = \bigcap_{x \in E} V_x$. U and V are open in Z, and disjoint, and V contains x. Since $Q \subseteq U$, V is an open subset of Z containing x that does not meet Q. So the complement of Q is open, by Trick 4.1.11, i.e., Q is closed in Z.

Now consider any family $(Q_i)_{i \in I}$ of compact saturated subsets of X. For every $i \in I$, Q_i is both upward closed and closed in Z, so their intersection is also upward closed and closed in Z, hence compact saturated in X. Coherence is a special case of this observation. For every closed subset F of X, i.e., if F is downward closed and closed in Z, then, using Proposition 4.4.15 twice, F is compact in the compact space Z and Q_i is closed in the T_2 space Z (Proposition 9.1.12). So, if Q_i intersects F for each $i \in I$, then $\bigcap_{i \in I} Q_i$ also intersects the compact subset F (Proposition 4.4.9). By Fact 8.3.7, X is well-filtered. Finally, X is T_0, because its specialization quasi-ordering is $\preceq$, which is an ordering. □

The following exercise gives a domain-theoretic way of producing compact pospaces.

Exercise 9.1.21 (Lawson topology) Let X be a poset, with specialization ordering $\leqslant$. The *Lawson topology* on X is generated by the Scott-open subsets and the complements of finitary compacts $\uparrow E$ (E finite). Write X_λ for X with its Lawson topology. Show that $(X_\lambda, \leqslant)$ is a pospace whenever X is a continuous poset, or more generally when X is a quasi-continuous poset (see Exercise 5.1.34). We have already encountered the Lawson topology in a special case, in the proof of Proposition 5.6.30.

Exercise 9.1.22 Show that a dcpo X is Lawson at the top (Lemma 7.7.24) if and only if its Scott and Lawson topologies induce the same subspace topology on $\operatorname{Max} X$.

9.1.2 From stably compact spaces to compact pospaces

Stably compact spaces exhibit a nice form of duality, called *de Groot duality*. This is instrumental in showing that all stably compact spaces arise from compact pospaces.

Definition 9.1.23 (Cocompact topology, de Groot dual) Given a topological space X, a *cocompact* subset of X is the complement $X \smallsetminus Q$ of a compact saturated subset Q of X.

The *cocompact topology* on X is the topology generated by the cocompact subsets of X. We write X^{d} for X with the cocompact topology: this is the *de Groot dual* of X.

Fact 9.1.24 *The cocompacts form a base of the cocompact topology.*

Indeed, the intersection of finitely many cocompacts is cocompact: taking complements, this just means that the union of finitely many compact saturated subsets of X is compact saturated (Proposition 4.4.12).

When X is stably compact, the cocompact topology has a simpler description:

Lemma 9.1.25 *When X is well-filtered, coherent and compact (in particular when X is stably compact), the elements of the cocompact topology are exactly the cocompacts of X: the intersection of any family of compact saturated subsets of X is compact saturated.*

Proof We show the second part of the Lemma, namely that the intersection of any family $(Q_i)_{i\in I}$ of compact saturated subsets is compact saturated. The first part follows, since then any union of cocompacts will be cocompact. If I is empty, $\bigcap_{i\in I} Q_i = X$ is compact. Otherwise, $\bigcap_{i\in I} Q_i$ is the intersection of the filtered family of all subsets of the form $\bigcap_{i\in J} Q_i$, where J ranges over the non-empty finite subsets of I. For each such J, $\bigcap_{i\in J} Q_i$ is compact saturated since X is coherent. Then any filtered intersection of compact saturated subsets is compact saturated since X is well-filtered, using Proposition 8.3.6. $\square$

Definition 9.1.26 (Patch topology) Let X be a topological space. The *patch topology* on X is the coarsest topology that is finer than the topologies of X and of X^{d}. We write X^{patch} for X with its patch topology.

It is slightly cumbersome to talk about "open subsets of X^{patch}," and we shall often call such sets the *patch-open* subsets of X. Similarly, a subset of X is *patch-closed* iff it is closed in X^{patch}, *patch-compact* iff it is compact in X^{patch}.

It is time we clarified the various relationships between closed, compact, patch-closed, patch-compact subsets of a set X:

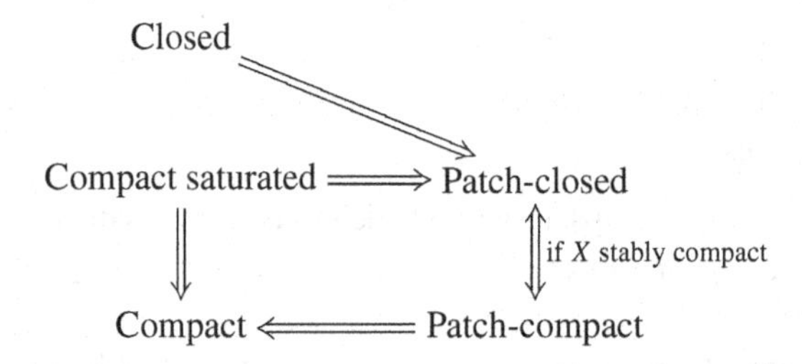

The equivalence between patch-closed and patch-compact subsets when X is stably compact follows from the fact that closed and compact subsets of a compact T_2 space are the same thing (Proposition 4.4.15), and from the following result.

Proposition 9.1.27 (Stably compact ⇒ patch-compact) *Let X be a topological space. Then X is patch-compact if and only if X is well-filtered, coherent, and compact.*

In particular, every stably compact space is patch-compact.

Proof Let X be well-filtered, coherent, and compact. We use Alexander's Subbase Lemma (Theorem 4.4.29). A subbase of the patch topology is given by the subsets of the form U (U open in X) and $X \setminus Q$ (Q compact saturated in X). Consider any open cover of X^{patch} by subbasic patch-open subsets U_i, $i \in I$ (where each U_i is open in X), and $X \setminus Q_j$, $j \in J$ (where each Q_j is compact saturated in X). That is, $X \subseteq \bigcup_{i \in I} U_i \cup (X \setminus \bigcap_{j \in J} Q_j)$. Letting $Q = \bigcap_{j \in J} Q_j$, it must then be that $Q \subseteq \bigcup_{i \in I} U_i$. Q is compact saturated by Lemma 9.1.25, so, by the Heine–Borel Property, there is a finite subset I' of I such that $Q \subseteq \bigcup_{i \in I'} U_i$. By well-filteredness, there is a finite subset J' of J such that $\bigcap_{j \in J'} Q_j \subseteq \bigcup_{i \in I'} U_i$. It follows that the family of subsets U_i, $i \in I'$, and $X \setminus Q_j$, $j \in J'$, is a finite subcover, showing that X^{patch} is indeed compact.

Conversely, assume X patch-compact. Let U be an open subset of X, and $(Q_j)_{j \in J}$ be a non-empty family of compact saturated subsets of X such that $\bigcap_{j \in J} Q_j \subseteq U$. Then $X \subseteq U \cup \bigcup_{j \in J}(X \setminus Q_j)$. Since X is patch-compact, there is a finite subset J_0 of J such that $X \subseteq U \cup \bigcup_{j \in J_0}(X \setminus Q_j)$ or $X \subseteq \bigcup_{j \in J_0}(X \setminus Q_j)$. The latter case is subsumed by the former, so $X \subseteq U \cup \bigcup_{j \in J_0}(X \setminus Q_j)$, i.e., $\bigcap_{j \in J_0} Q_j \subseteq U$. When $(Q_j)_{j \in J}$ is filtered, we obtain that some Q_k, $k \in J$, is included in every Q_j, $j \in J_0$, so $Q_k \subseteq U$. So X is well-filtered.

Now let Q_1 and Q_2 be two compact saturated subsets of X, and $(U_i)_{i \in I}$ be an open cover of $Q_1 \cap Q_2$. So $X \subseteq \bigcup_{i \in I} U_i \cup (X \setminus Q_1) \cup (X \setminus Q_2)$. Since X is patch-compact, there is a finite subset I_0 of I such that X is included in $\bigcup_{i \in I_0} U_i \cup (X \setminus Q_1) \cup (X \setminus Q_2)$, so $Q_1 \cap Q_2 \subseteq \bigcup_{i \in I_0} U_i$. Therefore $Q_1 \cap Q_2$ is compact, whence X is coherent.

Finally, X is compact, because every compact subset of X^{patch} is compact in X, whose topology is coarser. □

Recall that compact T_2 topologies are minimal, in the sense that any compact topology coarser than a T_2 topology coincides with it (Theorem 4.4.27). The following is a similar result, for the case of well-filtered coherent compact spaces.

Lemma 9.1.28 *Let X be a well-filtered coherent compact space, $\mathcal{B}$ be a set of open subsets of X, and $\mathcal{K}$ be a set of cocompact subsets of X such that the following order T_2 kind of property holds: for every pair of points x, y of X with $x \not\leq y$, there is an open subset $U \in \mathcal{B}$ and a cocompact subset $U' \in \mathcal{K}$ such that $x \in U$, $y \in U'$, and $U \cap U' = \emptyset$.*

Then $\mathcal{B}$ is a subbase of the topology of X.

Proof Let U be any open subset of X. Let x be an arbitrary element of U. For every element $y \in X \setminus U$, it must be that $x \not\leq y$, since U is upward closed, so by assumption there is an open subset U_{xy} in $\mathcal{B}$ containing x and a cocompact subset $U'_{xy} = X \setminus Q_{xy}$ in $\mathcal{K}$ containing y such that $U_{xy} \cap U'_{xy} = \emptyset$, i.e., $U_{xy} \subseteq Q_{xy}$. (We pick U_{xy}, U'_{xy}, and Q_{xy} by appealing, as usual, to the Axiom of Choice.) Since $\bigcup_{y \in X \setminus U} U'_{xy}$ contains $X \setminus U$, $\bigcap_{y \in X \setminus U} Q_{xy}$ is included in U. Write $\bigcap_{y \in X \setminus U} Q_{xy}$ as the filtered intersection of $Q_E = \bigcap_{y \in E} Q_{xy}$ over all finite subsets E of $X \setminus U$. Since X is coherent and compact, Q_E is compact saturated. Since X is well-filtered, some Q_E is included in U. So $\bigcup_{y \in E} U'_{xy}$ contains $X \setminus U$. Since $\bigcup_{y \in E} U'_{xy}$ is disjoint from $U_x = \bigcap_{y \in E} U_{xy}$, U_x is in particular disjoint from $X \setminus U$, hence included in U. Note that U_x is a finite intersection of elements of $\mathcal{B}$. Since $x \in U_x \subseteq U$ for each $x \in U$, $U = \bigcup_{x \in U} U_x$ is in the topology generated by $\mathcal{B}$, which allows us to conclude. □

Corollary 9.1.29 *Let X be a set with two topologies on it, $\mathcal{O}_1$ and $\mathcal{O}_2$. Let $\leqslant_1$ be the specialization quasi-ordering of X with topology $\mathcal{O}_1$, and $\leqslant_2$ be that of $\mathcal{O}_2$. Assume that X with the $\mathcal{O}_1$ topology is locally compact; that X with the $\mathcal{O}_2$ topology is well-filtered, coherent, and compact; that, for all $x, y \in X$, $x \leqslant_1 y$ implies $x \leqslant_2 y$; that every $\mathcal{O}_1$-open is $\mathcal{O}_2$-open; and that every $\mathcal{O}_1$-cocompact is $\mathcal{O}_2$-cocompact.*

Then $\mathcal{O}_1$ and $\mathcal{O}_2$ are the same topology.

Proof Let $\mathcal{B} = \mathcal{O}_1$, $\mathcal{K}$ be the set of $\mathcal{O}_1$-cocompact subsets, and $\leqslant$ be $\leqslant_2$. If $x \not\leqslant y$, i.e., $x \not\leqslant_2 y$, then $x \not\leqslant_1 y$, so there is an $\mathcal{O}_1$-open subset V such that $x \in V$ and $y \notin V$. Since X is locally compact in the $\mathcal{O}_1$ topology, there is an $\mathcal{O}_1$-compact saturated subset Q and an $\mathcal{O}_1$-open subset U of X such that $x \in U \subseteq Q \subseteq V$. Let $U' = X \smallsetminus Q$: U' is in $\mathcal{K}$, U is in $\mathcal{B}$, x is in U, y is in U' (since $y \notin V$, y is not in Q), and $U \cap U' = \emptyset$ (because $U \subseteq Q$). Since every $\mathcal{O}_1$-open is $\mathcal{O}_2$-open, $\mathcal{B}$ is a set of open subsets of X in its $\mathcal{O}_2$ topology. Since every $\mathcal{O}_1$-cocompact is $\mathcal{O}_2$-cocompact, $\mathcal{K}$ is a set of cocompact subsets in the $\mathcal{O}_2$ topology. So we can apply Lemma 9.1.28: $\mathcal{B} = \mathcal{O}_1$ is a subbase of $\mathcal{O}_2$, hence $\mathcal{O}_1 = \mathcal{O}_2$. □

Definition 9.1.30 (Nachbin pospace) Given any locally compact T_0 topological space X, its *Nachbin pospace* is $(X^{\mathsf{patch}}, \leqslant)$, where $\leqslant$ is the specialization quasi-ordering of X.

Lemma 9.1.31 *The Nachbin pospace $(X^{\mathsf{patch}}, \leqslant)$ of any locally compact T_0 space X is indeed a pospace.*

Proof Let X be locally compact T_0. First, since X is T_0, $\leqslant$ is an ordering. Next, if $x \not\leqslant y$, then there is an open subset U such that $x \in U$ but $y \notin U$. Since X is locally compact, there is a compact saturated subset Q such that $x \in int(Q)$ and $Q \subseteq U$. Now $int(Q) \times (X \smallsetminus Q)$ is an open neighborhood of (x, y) in $X^{\mathsf{patch}} \times X^{\mathsf{patch}}$, and is included in the complement of the graph $(\leqslant)$; otherwise there would be an $x' \in int(Q)$ and a $y' \notin Q$ with $x' \leqslant y'$, but since Q' is upward closed, this leads to a contradiction. So $(\leqslant)$ is closed, hence $(X^{\mathsf{patch}}, \leqslant)$ is a pospace. □

Moreover, the upward topology of $(X^{\mathsf{patch}}, \leqslant)$ is finer than the topology of X, i.e., every open subset of X is patch-open and upward closed. There is no reason why the two topologies should coincide in general. They do if we assume X stably compact.

Theorem 9.1.32 (Nachbin pospace of a stably compact space) *Let X be stably compact.*

Then $(X^{\mathsf{patch}}, \leqslant)$ is a compact pospace, the upward topology of $(X^{\mathsf{patch}}, \leqslant)$ coincides with the topology of X, and the downward topology coincides with the topology of X^{d}.

Proof $(X^{\mathsf{patch}}, \leqslant)$ is a pospace by Lemma 9.1.31, and is compact by Proposition 9.1.27.

Let us show that the upward topology $\mathcal{O}_2$ of $(X^{\mathsf{patch}}, \leqslant)$ is exactly the topology $\mathcal{O}_1$ of X. $\mathcal{O}_1$ and $\mathcal{O}_2$ induce the same specialization quasi-ordering, namely $\leqslant$; in the case of $\mathcal{O}_2$, this is by Lemma 9.1.18, exploiting the fact that $(X^{\mathsf{patch}}, \leqslant)$ is a pospace.

X with the $\mathcal{O}_1$ topology is locally compact, since it is stably compact. Since $(X^{\mathsf{patch}}, \leqslant)$ is a compact pospace, X with the $\mathcal{O}_2$ topology is stably compact by Proposition 9.1.20, hence well-filtered, coherent, and compact.

The $\mathcal{O}_1$-opens are the open subsets of X, and every open subset U of X is by definition patch-open, and upward closed, so every $\mathcal{O}_1$-open is $\mathcal{O}_2$-open. Given any compact saturated subset Q in X, Q is patch-closed by definition of X^{patch}. Since Q is not only patch-closed but also upward closed, Proposition 9.1.20, last part, tells us that Q is compact saturated in the upward topology of $(X^{\mathsf{patch}}, \leqslant)$. Taking complements, we have just shown that every $\mathcal{O}_1$-cocompact is $\mathcal{O}_2$-cocompact. We can now apply Corollary 9.1.29 and conclude that $\mathcal{O}_1 = \mathcal{O}_2$, i.e., that the original topology of X is exactly the upward topology of its Nachbin pospace.

Finally, the closed subsets in the downward topology are exactly the complements of the downward closed open subsets of X^{patch}, i.e., are exactly the upward closed, closed subsets of X^{patch}. By Proposition 9.1.20 , last part, these are the compact saturated subsets of the upward topology, which, as we have seen, are just the compact saturated subsets of X. So the open subsets in the downward topology are exactly the cocompacts of X, i.e., the opens of X^{d}. $\square$

Example 9.1.33 We have seen that $[0, 1]_\sigma$ is stably compact (Example 9.1.3, Example 9.1.7). Its patch topology has a subbasis of open subsets of the form $(t, 1]$ (which are Scott-open) and $[0, t)$ (which are cocompact): this is just the open ball topology of $[0, 1]$. So the Nachbin pospace of $[0, 1]_\sigma$ is $([0, 1], \leqslant)$, a compact pospace indeed.

Proposition 9.1.34 *Let $(Z, \preceq)$ be a compact pospace. The topology of Z coincides with the patch topology of the space Z equipped with its upward topology, i.e., the topology of Z is the coarsest one containing both the upward closed open subsets and the downward closed open subsets of Z. Equivalently, for every $x \in Z$, for every open neighborhood V of x, there is an upward* *closed open subset U of Z and a downward closed open subset U' of Z such that $x \in U \cap U' \subseteq V$.*

Proof The equivalence of the last part with the other claims is by Trick 4.1.11. That the patch topology of Z with its upward topology is the coarsest one containing both the upward closed open subsets and the downward closed open subsets of Z is by Proposition 9.1.20.

Let $x \in Z$, and V be an open neighborhood of x. Let K be the complement of V. Since Z is compact and K is closed, by Proposition 4.4.15 K is compact, too. For each point y of K, y must be different from x, so $x \not\leqslant y$ or $y \not\leqslant x$ since Z is T_0 (even T_2, by Proposition 9.1.12). Let K_1 be the set of all points $y \in K$ such that $x \not\leqslant y$, and $K_2 = K \setminus K_1$. When $y \in K_1$, there is an open, upward closed neighborhood $U_y^\uparrow$ of x and an open, downward closed neighborhood $U_y^\downarrow$ of y such that $U_y^\uparrow \cap U_y^\downarrow = \emptyset$, since Z is order T_2 (Lemma 9.1.15). When $y \notin K_1$, the roles are reversed and there is an open, upward closed neighborhood $V_y^\uparrow$ of y and an open, downward closed neighborhood $V_y^\downarrow$ of x such that $V_y^\uparrow \cap V_y^\downarrow = \emptyset$. (We pick $U_y^\uparrow$, $U_y^\downarrow$, $V_y^\uparrow$, and $V_y^\downarrow$ using the Axiom of Choice.) The family of all open subsets $U_y^\downarrow$, $y \in K_1$, and $V_y^\uparrow$, $y \in K_2$, covers K. Since K is compact, there is a finite subset E_1 of K_1 and a finite subset E_2 of K_2 such that $K \subseteq \bigcup_{y \in E_1} U_y^\downarrow \cup \bigcup_{y \in E_2} V_y^\uparrow$. Let $U = \bigcap_{y \in E_1} U_y^\uparrow$, and $U' = \bigcap_{y \in E_2} V_y^\downarrow$. This is open since E_1 and E_2 are finite. By construction, $x \in U \cap U'$, and it is easy to check that $U \cap U'$ does not meet $\bigcup_{y \in E_1} U_y^\downarrow \cup \bigcup_{y \in E_2} V_y^\uparrow$, hence does not meet K, and therefore is included in V. □

Example 9.1.35 Proposition 9.1.34 is converse to Theorem 9.1.32. That is, compared to Example 9.1.33, take the compact pospace $([0, 1], \leqslant)$ and realize that its upward topology is the Scott topology: we retrieve the stably compact space $[0, 1]_\sigma$.

Exercise 9.1.36 (Lawson = patch for continuous posets, for quasi-continuous dcpos) Show that, if X is a continuous poset, then the upper topology of X^{op} coincides with the de Groot dual of the Scott topology of X, i.e., $(X^{op})_u = (X_\sigma)^{\mathsf{d}}$. (Hint: use Theorem 5.2.30. Also, do not confuse the *upper* topology (see Proposition 4.2.12) with the *upward* topology (see Definition 9.1.17)). Deduce that, in this case, $X_\lambda = (X_\sigma)^{\mathsf{patch}}$. Recall from Exercise 9.1.21 that X_λ is X with its Lawson topology.

Show the same results when X is a quasi-continuous dcpo instead.

Exercise 9.1.37 Deduce from Exercise 9.1.36 (or prove directly) that $\mathbb{R} = (\mathbb{R}_\sigma)^{\mathsf{patch}} = (\mathbb{R}_\sigma)_\lambda$, i.e., that the open ball topology on $\mathbb{R}$ coincides with the patch, as well as with the Lawson topology on the poset of real numbers.

Any well-formed duality should be such that the double dual of X equals X. This is exactly what happens with de Groot duality in the case of stably compact spaces.

Theorem 9.1.38 (de Groot duality, $X^{\mathsf{dd}} = X$) *For every stably compact space X, X^{d} is also stably compact, and $X^{\mathsf{dd}} = X$.*

In other words, the cocompact topology of X^{d} is exactly the original topology of X.

Proof $(X^{\mathsf{patch}}, \leqslant)$ is a compact pospace by Theorem 9.1.32. It is then immediate that $(X^{\mathsf{patch}}, \geqslant)$, where $\geqslant$ is the opposite ordering of $\leqslant$, is also a compact pospace. We obtain X^{d} as X with the downward topology of $(X^{\mathsf{patch}}, \leqslant)$ or equivalently as X with the upward topology of $(X^{\mathsf{patch}}, \geqslant)$, by definition of these topologies. Since the latter topology turns the space into a stably compact space, by Proposition 9.1.20 X^{d} is stably compact. Theorem 9.1.32 also tells us that X with the downward topology of $(X^{\mathsf{patch}}, \geqslant)$ is X^{dd}. But this is exactly the upward topology of $(X^{\mathsf{patch}}, \leqslant)$, so $X^{\mathsf{dd}} = X$. □

We have seen another duality for hemi-metric spaces, between a hemi-metric d and its opposite d^{op}. For symcompact spaces, this duality *coincides* with de Groot duality.

Theorem 9.1.39 (Duality on symcompact spaces) *Let X, d be a quasi-metric space. The following are equivalent:*

1. *X, d is symcompact, i.e., X, d^{sym} is compact.*
2. *With its open ball topology, X, d and X, d^{op} are stably compact and are de Groot duals.*

In this case, X, d^{sym}, with its open ball topology and the ordering $\leqslant^d$, coincides with the Nachbin pospace of X, d.

Proof Write X for X, d with its open ball topology, X^{op} for X, d^{op} with its open ball topology, and X^{sym} for X, d^{sym} with its open ball topology.

$1 \Rightarrow 2$. We first show that $(X^{sym}, \leqslant^d)$ is a compact pospace. It is enough to show that $(\leqslant^d)$ is closed in $X^{sym} \times X^{sym}$. Let $(x, y) \notin (\leqslant^d)$, i.e., $x \not\leqslant^d y$, or equivalently $d(x, y) \neq 0$, by Proposition 6.1.8. Let $\epsilon < d(x, y)/2$, $\epsilon > 0$. Then $B^{d^{sym}}_{x,<\epsilon} \times B^{d^{sym}}_{y,<\epsilon}$ is an open neighborhood of (x, y) that does not intersect $(\leqslant^d)$: if it intersected it, say at (x', y'), then we would have $d(x, x') \leqslant d^{sym}(x, x') < \epsilon$, $d(x', y') = 0$, and $d(y', y) \leqslant d^{sym}(y, y') < \epsilon$, so $d(x, y) < 2\epsilon$ by the triangular inequality, contradicting $2\epsilon > d(x, y)$. So the complement of $(\leqslant^d)$ is open by Trick 4.1.11.

We have just shown more: letting $\mathcal{B}$ be the set of open balls of X, d, and $\mathcal{K}$ be the set of open balls of X, d^{op}, for every pair of points x, y of X with $x \not\leqslant^d y$ there is an open subset $U \in \mathcal{B}$ and a subset $U' \in \mathcal{K}$ such that $x \in U$, $y \in U'$, and $U \cap U' = \emptyset$. This is part of the assumptions of Lemma 9.1.28. We use the latter on the space X^o defined as X with the upward topology of $(X^{sym}, \leqslant^d)$. By Proposition 9.1.20, X^o is stably compact, in particular well-filtered, coherent, and compact. It remains to check that $\mathcal{B}$ is a collection of open subsets of X^o, and that $\mathcal{K}$ consists of cocompact subsets of X^o. Every element of $\mathcal{B}$ is open in X^{sym}, as a consequence of Proposition 6.1.19, and upward closed in $\leqslant^d$, since the latter is the specialization quasi-ordering of X, d (Proposition 6.1.8), so is open in X^o. Every element $B^{d^{op}}_{x,<\epsilon}$ of $\mathcal{K}$ is also open in X^{sym}, by the same reasoning, and downward closed: for any element y of $B^{d^{op}}_{x,<\epsilon}$ and $y' \leqslant^d y$, $d(y, x) < \epsilon$ and $d(y', y) = 0$, so $d(y', x) < \epsilon$ by the triangular inequality. The complement Q of $B^{d^{op}}_{x,<\epsilon}$ is then closed and upward closed in X^{sym}. By Proposition 9.1.20, Q is compact saturated in X^o. So every element of $\mathcal{K}$ is indeed cocompact.

We can therefore apply Lemma 9.1.28: $\mathcal{B}$ is a subbase of the topology of X^o. In other words, the topology of X^o is the open ball topology of X, d, i.e., $X^o = X$. In particular, X is stably compact.

Reasoning similarly, by exchanging d and d^{op} we obtain that X^{op} is also stably compact. Moreover, $X^{op} = (X^{op})^o$, i.e., the topology of X^{op} consists of the open subsets of X^{sym} that are upward closed in $\leqslant^{d^{op}}$, i.e., downward closed in $\leqslant^d$. These are exactly the opens of $(X^o)^{\mathsf{d}} = X^{\mathsf{d}}$, by Theorem 9.1.32. So X and X^{op} are de Groot duals.

$2 \Rightarrow 1$. Since $X^{op} = X^{\mathsf{d}}$, the open subset $B^{d^{op}}_{x,<\epsilon}$ of X^{op} is cocompact in X. So the symmetric open ball $B^{d^{sym}}_{x,<\epsilon} = B^d_{x,<\epsilon} \cap B^{d^{op}}_{x,<\epsilon}$ is open in X^{patch}. It follows that every open of X^{sym} is open in X^{patch}: the patch topology is finer than that of X^{sym}. However, the topology of X^{sym} is T_2, the patch topology is compact, so the two topologies coincide by Theorem 4.4.27. That is, $X^{sym} = X^{\mathsf{patch}}$ is compact. □

Exercise 9.1.40 Let X, d be a symcompact quasi-metric space. Show that the closed ball $B^d_{x,\leqslant\epsilon} = \{y \in X \mid d(x, y) \leqslant \epsilon\}$, $\epsilon > 0$, is compact saturated in X with its open ball topology.

Exercise 9.1.41 (Stably compact $\Rightarrow$ compactly generated) Let X be a stably compact space, and consider an open subset U of $\mathcal{C}X$, where $\mathcal{C}$ is the class of all compact T_2 spaces (see Definition 5.6.5). Imitating one of the arguments we used in the proof of Proposition 5.6.30, show that U is upward closed. By considering $X^{\mathsf{patch}} \in \mathcal{X}$, and the identity $\mathcal{C}$-probe, show that U is patch-open.

Conclude that U is open in X. From this, deduce that every stably compact space is compactly generated.

Exercise 9.1.42 (Coherent ω-continuous dcpo $\Rightarrow$ Polish in Lawson topology) Show that the Lawson topology of an ω-continuous poset X is countably-based. Concretely, if B is a basis of X, then the sets $\Uparrow b$ and $X \setminus \uparrow b$, $b \in B$, form a subbase of X_λ.

Deduce that every coherent ω-continuous dcpo X is Polish in its Lawson topology (equivalently, its patch topology, if X is pointed).

Exercise 9.1.43 Let K be a patch-compact subset of a stably compact space X. Show that $\downarrow K$ is closed, hence that $\downarrow K = cl(K)$.

9.2 Coproducts and retracts of stably compact spaces

We first observe that finite coproducts of stably compact spaces are stably compact. Moreover, coproducts commute with the Nachbin pospace construction.

Proposition 9.2.1 (Finite coproducts of stably compact spaces) *Let $X_1, \ldots, X_n$ be finitely many stably compact spaces. Their topological coproduct $\coprod_{i=1}^n X_i$ is stably compact. Moreover, the Nachbin pospace of $\coprod_{i=1}^n X_i$ is $(\coprod_{i=1}^n X_i^{\mathrm{patch}}, \coprod_{i=1}^n \leqslant_i)$, where $\leqslant_i$ is the specialization ordering of X_i, $1 \leqslant i \leqslant n$.*

Proof For each i, $1 \leqslant i \leqslant n$, $(X_i^{\mathrm{patch}}, \leqslant_i)$ is a compact pospace by Theorem 9.1.32. Then $Z = \coprod_{i=1}^n X_i^{\mathrm{patch}}$ is compact (Exercise 4.6.6). Let $\preceq$ be the coproduct ordering $\coprod_{i=1}^n \leqslant_i$ (see Exercise 4.6.7). For any pair of points (i, x), (j, y) in $\coprod_{i=1}^n X_i$ such that $(i, x) \not\preceq (j, y)$, either $i \neq j$, then $(\{i\} \times X_i) \times (\{j\} \times X_j)$ is an open neighborhood of $((i, x), (j, y))$ that is included in the complement of $(\preceq)$; or $i = j$, $x \not\leqslant_i y$, so there is an open rectangle $U \times V$ in $X_i^{\mathrm{patch}} \times X_i^{\mathrm{patch}}$ containing (x, y) and included in the complement of $(\leqslant_i)$, since $(X_i^{\mathrm{patch}}, \leqslant_i)$ is a pospace (Lemma 9.1.31); then $(\{i\} \times U) \times (\{i\} \times V)$ is an open neighborhood of $((i, x), (j, y))$ included in the complement of $(\preceq)$. This shows that $(\preceq)$ is closed, hence $(Z, \preceq)$ is a compact pospace.

The opens of Z in the upward topology are the upward closed open subsets of Z, i.e., the unions $\bigcup_{i=1}^n \{i\} \times U_i$, where each U_i is open and upward closed in X_i^{patch}. Equivalently, U_i is open in X_i by Theorem 9.1.32. So Z with the upward topology is exactly $\coprod_{i=1}^n X_i$. In particular, the latter is stably compact. Finally, by the definition of $\preceq$, the Nachbin pospace of $\coprod_{i=1}^n X_i$ is exactly $(Z, \preceq)$. □

In general, finite colimits (taken in **Top**) of stably compact spaces fail to be stably compact:

Exercise 9.2.2 Give an example of a stably compact space X and an equivalence relation $\equiv$ on X such that $Z = X/{\equiv}$ is not stably compact. (Hint: remember Exercise 9.1.16.)

Exercise 9.2.2 shows that quotients, i.e., coequalizers of stably compact spaces, may fail to be stably compact. However:

Proposition 9.2.3 (Retracts of stably compact spaces) *Let $r: Y \to Z$ be a retraction. If Y is T_0, resp. locally compact, resp. compact, resp. well-filtered, resp. coherent, then so is Z.*

In particular, every retract of a stably compact (resp., stably locally compact) space is stably compact (resp., stably locally compact).

Remember also that any retract of a sober space is sober (Exercise 8.2.43).

Proof Let $s: Z \to Y$ be the associated section. When Y is T_0, the claim is Lemma 4.10.32. When Y is locally compact, resp., compact, this is Proposition 4.10.33.

When Y is well-filtered, we show that Z is, too, as follows. Consider any filtered family $(Q_i)_{i \in I}$ of compact saturated subsets of Z and an open subset U of Z such that $\bigcap_{i \in I} Q_i \subseteq U$. For every $i \in I$, $\uparrow s[Q_i]$ is compact saturated in Z, by Proposition 4.4.13 and Proposition 4.4.14, and included in $r^{-1}(Q_i)$. Indeed, if $z \in \uparrow s[Q_i]$, then there is an element $y \in Q_i$ such that $s(y) \leqslant z$, so $y = r(s(y)) \leqslant r(z)$, because r, as a continuous map, is monotonic (Proposition 4.3.9); then $r(z) \in Q_i$ since Q_i is saturated, i.e., upward closed. Since $\bigcap_{i \in I} Q_i \subseteq U$, we obtain that $\bigcap_{i \in I} \uparrow s[Q_i] \subseteq \bigcap_{i \in I} r^{-1}(Q_i) = r^{-1}(\bigcap_{i \in I} Q_i) \subseteq r^{-1}(U)$. Since Y is well-filtered, $\uparrow s[Q_i] \subseteq r^{-1}(U)$ for some $i \in I$. It follows that $Q_i \subseteq U$, again using $r \circ s = \mathrm{id}_Y$. So Z is well-filtered.

Similarly, when Y is coherent, let Q_1, Q_2 be two compact saturated subsets of Z. Then $\uparrow s[Q_1]$ and $\uparrow s[Q_2]$ are compact saturated in Y, hence also their intersection $Q = \uparrow s[Q_1] \cap \uparrow s[Q_2]$. Since r is continuous, $r[Q]$ is also compact in Z by Proposition 4.4.13. We claim that $r[Q] = Q_1 \cap Q_2$, and this will show that $Q_1 \cap Q_2$ is compact, hence that Z is coherent. In one direction, every element z of $Q_1 \cap Q_2$ equals $r(s(z))$, which is in $r[Q]$ since $s(z) \in Q$. Conversely, any element of $r[Q]$ is of the form $r(y)$ with $y \in Q$, i.e., such that $s(z_1) \leqslant y$ and $s(z_2) \leqslant y$ for some $z_1 \in Q_1$, $z_2 \in Q_2$. So

$z_1 = r(s(z_1)) \leqslant r(y)$, and similarly $z_2 \leqslant r(y)$, so $r(y)$ is both in Q_1 and in Q_2, using the fact that both are saturated. So Z is coherent. □

9.3 Products and subspaces of stably compact spaces

The same nice situation is obtained with products, finite or infinite.

Proposition 9.3.1 (Products of stably compact spaces) *Let $(X_i)_{i \in I}$ be any family of stably compact spaces. Their topological product $\prod_{i=1}^n X_i$ is stably compact. Moreover, the Nachbin pospace of $\prod_{i=1}^n X_i$ is $(\prod_{i=1}^n X_i^{\mathrm{patch}}, \prod_{i=1}^n \leqslant_i)$, where $\leqslant_i$ is the specialization ordering of X_i, $i \in I$.*

Proof For each $i \in I$, $(X_i^{\mathrm{patch}}, \leqslant_i)$ is a compact pospace by Theorem 9.1.32. Then $Z = \prod_{i \in I} X_i^{\mathrm{patch}}$ is compact by Tychonoff's Theorem 4.5.12. Let $\preceq$ be the product ordering $\prod_{i \in I} \leqslant_i$; this is the specialization quasi-ordering of $\prod_{i \in I} X_i$ by Lemma 4.5.17. For any pair of points $\vec{x}$, $\vec{y}$ in $\prod_{i \in I} X_i$ such that $x \not\preceq y$, there must be an index $i \in I$ such that $x_i \not\leqslant y_i$. Since $(X_i^{\mathrm{patch}}, \leqslant_i)$ is a pospace (Lemma 9.1.31), there is an open rectangle $U \times V$ in $X_i^{\mathrm{patch}} \times X_i^{\mathrm{patch}}$ containing (x_i, y_i) and included in the complement of $(\leqslant_i)$. So $\pi_i^{-1}(U) \times \pi_i^{-1}(V)$ is an open rectangle containing $(\vec{x}, \vec{y})$ that is included in the complement of $(\preceq)$. This shows that $(Z, \preceq)$ is a compact pospace.

We claim that the upward topology of Z coincides with the topology of $\prod_{i \in I} X_i$.

In one direction, for every $i \in I$, for every open subset U of X_i, $\pi_i^{-1}(U)$ is the inverse image by π_i of the patch-open subset U of X_i, and is upward closed. Since the sets of the form $\pi_i^{-1}(U)$, U open in X_i, $i \in I$, generate the topology of $\prod_{i \in I} X_i$, this topology is coarser than the upward topology of Z.

Conversely, we use Lemma 9.1.28. Let X be Z with the upward topology of Z, and $\mathcal{B}$ be the set of all subsets of the form $\pi_i^{-1}(U)$, U open in X_i, $i \in I$. We have just seen that this is a set of open subsets of X. Let $\mathcal{K}$ be the set of complements of products $\prod_{i \in I} Q_i$ of compact saturated subsets Q_i of X_i, $i \in I$. By Tychonoff's Theorem, and since compact saturated subsets in X_i are closed in X_i^{patch}, hence compact in X_i^{patch} (by Proposition 4.4.15, since X_i^{patch} is T_2 by Proposition 9.1.12), for every $i \in I$, such products $\prod_{i \in I} Q_i$ are compact in Z, hence closed in Z. Moreover, they are upward closed in $\preceq$. By Proposition 9.1.20, they are compact saturated in the upward topology. So every element of $\mathcal{K}$ is cocompact in X. Finally, assume $\vec{x}$, $\vec{y}$ are elements of X such that $\vec{x} \not\preceq \vec{y}$. For some $i \in I$, $x_i \not\leqslant_i y_i$. By Lemma 9.1.15 applied to the compact pospace $(X_i^{\mathrm{patch}}, \leqslant_i)$, there is an upward closed open neighborhood U of x_i in X_i^{patch} (i.e., an open subset of X_i, by Theorem 9.1.32) and a downward closed open neighborhood V of y_i in X_i^{patch} (i.e., a cocompact subset of X_i, by Theorem 9.1.32) such that $U \cap V = \emptyset$. Then $\pi_i^{-1}(U)$ is an element of $\mathcal{B}$; writing $Q_i = X_i \smallsetminus V$ and $Q_j = X_j$ for every $j \in I$, $j \neq i$, $\pi_i^{-1}(V)$ is the complement of $\prod_{j \in I} Q_j$, and is therefore in $\mathcal{K}$; and clearly $\pi_i^{-1}(U) \cap \pi_i^{-1}(V) = \emptyset$. Lemma 9.1.28 then applies: $\mathcal{B}$ is a subbase of the topology of X, i.e., the topology of $\prod_{i \in I} X_i$ coincides with the topology of X, which is the upward topology of Z.

Finally, by definition of $\preceq$, the Nachbin pospace of $\prod_{i \in I} X_i$ is exactly $(Z, \preceq)$. □

Proposition 9.3.1 entails that $\mathbb{S}^I$ and $([0,1]_\sigma)^I$ are stably compact, for every set I. This is not quite what we asked for in the second part of Exercise 9.1.9, but close: since $\mathbb{S}$ and $[0,1]$ are pointed continuous dcpos, $\mathbb{S}^I = \mathbb{S}^I_\sigma$ and $([0,1]_\sigma)^I = [0,1]^I_\sigma$, using Proposition 5.1.56.

Exercise 9.3.2 (Stably compact spaces are not CCC) Find a family of continuous maps from $[0,1]$ to $[0,1]$ that is not equicontinuous. Use the Arzelà–Ascoli Theorem 6.6.11 to show that $[[0,1] \to [0,1]]$ is not compact in the core-open topology, nor in any conjoining topology. From this, deduce that the category of stably compact spaces and continuous maps is not Cartesian-closed, and that the category of compact T_2 spaces and continuous maps is not Cartesian-closed either.

General limits of stably compact spaces, taken in **Top**, are not stably compact in general. More specifically, there are equalizers of pairs of maps between stably compact spaces that are not stably compact:

Example 9.3.3 Consider the two maps $f, g : [0,1]_\sigma \to \mathbb{S}$, where $f(t) = 1$ iff $t \in (0,1]$, and $g(t) = 1$ for every $t \in [0,1]$. Their equalizer in **Top** is $(0,1]_\sigma$, which is not even compact.

However, certain important subspaces of stably compact spaces are stably compact.

Proposition 9.3.4 (Patch-closed subspaces) *Every patch-closed subset F of a stably compact space X defines a stably compact subspace of X, still written F, with specialization ordering $\preceq$ obtained as the restriction of that of X to F. Moreover, the patch topology on F, i.e., the topology of F^{patch}, is exactly the topology on F induced from X^{patch}.*

Proof By assumption F is closed in X^{patch}. Since X^{patch} is compact, F is compact in X^{patch}, by Proposition 4.4.15. But compact subsets and compact subspaces are the same thing; see Exercise 4.9.11. So F with the subspace topology induced from X^{patch} is a compact space. To make this clear, write F_p for F with this topology. F itself will denote the subspace F, with topology induced from X.

Let $\preceq$ be the restriction of the specialization ordering $\leqslant$ of X to the subspace F. The graph $(\preceq)$ equals $(\leqslant) \cap (F \times F)$. Since $(X^{\mathsf{patch}}, \leqslant)$ is a pospace (Lemma 9.1.31), $(\leqslant)$ is closed in $X^{\mathsf{patch}} \times X^{\mathsf{patch}}$, so $(\preceq)$ is closed in $F \times F$, where the latter is taken with the subspace topology, induced from $X^{\mathsf{patch}} \times X^{\mathsf{patch}}$. It is easy to see that this is the same as the topology of $F_p \times F_p$, since both topologies are generated by open rectangles $U \times V$, where U and V are open in F_p. So $(F_p, \preceq)$ is a compact pospace. By Proposition 9.1.20, F with the upward topology of $(F_p, \preceq)$ is stably compact.

We claim that this topology is exactly the topology induced from that of X. Any open subset U in the latter topology is of the form $V \cap F$, where V is open in X. Then V is open in X^{patch} and upward closed in $\leqslant$ (by Theorem 9.1.32), so U is open in F_p and upward closed in $\preceq$, hence open in the upward topology of $(F_p, \preceq)$.

Conversely, it is more practical to reason with closed sets, and show that every closed subset F' of F_p that is downward closed in $\preceq$ is closed in F with the topology induced from X. By Proposition 4.9.4, F' is of the form $F'' \cap F$ for some closed subset F'' of X^{patch}. In other words, F' itself is closed in X^{patch}. By Lemma 9.1.13, $\downarrow_{\leqslant} F'$ is also closed in X^{patch}, and downward closed, hence closed in X. We claim that $F' = \downarrow_{\leqslant} F' \cap F$. Indeed, $F' \subseteq \downarrow_{\leqslant} F' \cap F$ is clear, while, for every $x \in \downarrow_{\leqslant} F' \cap F$, there is point $y \in F'$ such that $x \leqslant y$, whence $x \preceq y$, so that $x \in F'$, since F' is downward closed in $\preceq$. But $F' = \downarrow_{\leqslant} F' \cap F$ displays F' as a closed subset of F, where the latter is equipped with the topology induced from X, by Proposition 4.9.4. This finishes the proof that the upward topology of $(F_p, \preceq)$ is exactly the subspace topology on F, induced from X. It follows that $F^{\mathsf{patch}} = F_p$, using Proposition 9.1.34, and we conclude. □

Subspaces of stably compact spaces that are not patch-closed may fail to be stably compact. Actually, they can be any T_0 topological space. This can be said in a more positive way, as follows. Recall from Exercise 9.1.9 that $\mathbb{S}^I$ is stably compact for every set I. Note also that $\mathbb{S}^I$ is order isomorphic to $\mathbb{P}(I)$, with the inclusion ordering: each $J \in \mathbb{P}(I)$ is mapped to the tuple $(\chi_J(i))_{i \in I}$. In particular, $\mathbb{S}^I$ is homeomorphic to $\mathbb{P}(I)_\sigma$. We call either space the *powerset* of I.

Proposition 9.3.5 (Embeddings into powersets) *Let X be a topological space. The map $\eta_{\mathbb{S}}: X \to \mathbb{S}^{\mathcal{O}(X)}$, such that $\eta_{\mathbb{S}}(x) = (\chi_U(x))_{U \in \mathcal{O}(X)}$ for every $x \in X$, is monotonic, and is an order embedding if and only if X is T_0.*

As a map from X to the stably compact space $\mathbb{S}^{\mathcal{O}(X)}$, $\eta_{\mathbb{S}}$ is continuous, almost open, and an embedding if and only if X is T_0.

Proof Recall that $\mathbb{S}^I$ is a continuous dcpo, for every set I, and its specialization ordering is the pointwise ordering. Since χ_U is monotonic for every open subset U of X, $\eta_{\mathbb{S}}$ is monotonic. If $\eta_{\mathbb{S}}(x) \leqslant \eta_{\mathbb{S}}(y)$, then, for every $U \in \mathcal{O}(X)$, if $\chi_U(x) = 1$ then $\chi_U(y) = 1$, i.e., $x \leqslant y$, by definition of the specialization quasi-ordering $\leqslant$. So $\eta_{\mathbb{S}}$ is an order embedding iff $\leqslant$ is an ordering, i.e., X is T_0.

The only non-trivial open subset of $\mathbb{S}$ is $\{1\}$, so the subsets of the form $\pi_U^{-1}\{1\}$, U open in X, form a subbase of $\mathbb{S}^{\mathcal{O}(X)}$. Their inverse image by $\eta_{\mathbb{S}}$ is the open subset U, so $\eta_{\mathbb{S}}$ is continuous by Trick 4.3.3. It also follows that, for every open subset U of X, there is an open subset V of $\mathbb{S}^{\mathcal{O}(X)}$ such that $U = \eta_{\mathbb{S}}^{-1}(V)$, namely $V = \pi_U^{-1}\{1\}$. So $\eta_{\mathbb{S}}$ is almost open. By Proposition 4.10.9, $\eta_{\mathbb{S}}$ is an embedding iff it is injective, iff any two points that have the same open neighborhoods are equal, iff X is T_0. □

Exercise 9.3.6 (Powersets and sobrification) Seeing $\mathbb{S}^{\mathcal{O}(X)}$ as the powerset $\mathbb{P}(\mathcal{O}(X))$, $\eta_{\mathbb{S}}(x)$ is just the set of all open neighborhoods of x. All of them are completely prime filters of $\mathcal{O}(X)$. So $\eta_{\mathbb{S}}$ also defines a continuous and almost open map η from X to the subspace Y of $\mathbb{P}(\mathcal{O}(X))_\sigma$ consisting of completely prime filters of $\mathcal{O}(X)$. Show that Y is exactly $\mathsf{pt}(\mathcal{O}(X))$ (i.e., the topology induced from the Scott topology on $\mathbb{P}(\mathcal{O}(X))$ is the hull-kernel topology), and η is exactly the map $\eta_X^{\mathsf{Stone}}: X \to \mathsf{pt}(\mathcal{O}(X))$ of Theorem 8.1.26.

A remarkable point of the construction of Proposition 9.3.5 is that every continuous map from X to a continuous complete lattice Y extends to one on the whole powerset $\mathbb{S}^{\mathcal{O}(X)}$:

Proposition 9.3.7 *Let X be a topological space, and L be a continuous complete lattice. For every continuous map f from X to L_σ, there is a map $\widehat{f}$ from the powerset $\mathbb{S}^{\mathcal{O}(X)}$ to L_σ such that $\widehat{f}$ preserves all least upper bounds (and is in particular continuous), and $\widehat{f}(\eta_{\mathbb{S}}(x)) = f(x)$ for every $x \in X$ (extension).*

Proof Equate $\mathbb{S}^{\mathcal{O}(X)}$ with $\mathbb{P}(\mathcal{O}(X))$, and define $\widehat{f}(\vec{x})$ for every family $\vec{x}$ of open subsets of X as the least upper bound of all elements $u \in L$ such that $f^{-1}(\twoheaduparrow u)$ is in $\vec{x}$.

We claim that $\widehat{f}$ preserves all least upper bounds. For any family $(\vec{x}_i)_{i\in I}$ of elements of $\mathbb{P}(\mathcal{O}(X))$, $\widehat{f}(\bigcup_{i\in I}\vec{x}_i)$ is the least upper bound of all elements $u \in L$ such that $f^{-1}(\Uparrow u)$ is in some $\vec{x}_i$, $i \in I$. This is the least upper bound of all elements $\widehat{f}(\vec{x}_i)$. Since $\widehat{f}$ preserves all least upper bounds, it is in particular Scott-continuous, hence continuous by Proposition 4.3.5.

Finally, $\widehat{f}(\eta_{\mathbb{S}}(x))$ is the least upper bound of all elements $u \in L$ such that $f^{-1}(\Uparrow u)$ is an open neighborhood of x, i.e., such that $u \ll f(x)$. This is just $f(x)$, since L is continuous. □

This is particularly important when L is a closed interval $[a, b]$ of $\mathbb{R}$. A map $f: X \to \mathbb{R}$ is *bounded* if and only if there is a closed interval $[a, b]$ of $\mathbb{R}$ such that $\mathrm{Im}\, f \subseteq [a, b]$. Extending slightly the result of Exercise 4.9.10, the Scott topology on $[a, b]$ coincides with the subspace topology, induced from $\mathbb{R}_\sigma$. It follows immediately that:

Corollary 9.3.8 *Let X be a topological space. For every bounded continuous map f from X to $\mathbb{R}_\sigma$, there is a bounded map $\widehat{f}$ from the powerset $\mathbb{S}^{\mathcal{O}(X)}$ to $\mathbb{R}_\sigma$ that preserves all least upper bounds and is in particular continuous, and such that:*

$$\widehat{f}(\eta_{\mathbb{S}}(x)) = f(x) \quad \textit{for every } x \in X \textit{ (extension)}$$

$$\sup_{\vec{x}\in\mathbb{S}^{\mathcal{O}(X)}} \widehat{f}(\vec{x}) = \sup_{x\in X} f(x)$$

$$\inf_{\vec{x}\in\mathbb{S}^{\mathcal{O}(X)}} \widehat{f}(\vec{x}) = \inf_{x\in X} f(x).$$

Proof This follows from Proposition 9.3.7 by taking $L = [a, b]$, $a = \inf_{x\in X} f(x)$, $b = \sup_{x\in X} f(x)$, except that we only obtain $\inf_{\vec{y}\in\mathbb{S}^{\mathcal{O}(X)}} \widehat{f}(\vec{y}) \geqslant a$ this way. However, $\inf_{\vec{y}\in\mathbb{S}^{\mathcal{O}(X)}} \widehat{f}(\vec{y}) \leqslant \inf_{x\in X} \widehat{f}(\eta_{\mathbb{S}}(x)) = \inf_{x\in X} f(x) = a$; and similarly for $\sup_{\vec{x}\in\mathbb{S}^{\mathcal{O}(X)}} \widehat{f}(\vec{x})$. □

Exercise 9.3.9 (Scott's Theorem, and injective spaces) Let X, Y be two topological spaces, and $i: X \to Y$ be a continuous map. Show that the map $\mathbb{P}(\mathcal{O}(i))$ from $\mathbb{P}(\mathcal{O}(Y))$ to $\mathbb{P}(\mathcal{O}(X))$ that maps each family $\vec{y}$ of open subsets of Y to the family of all open subsets $i^{-1}(V)$, $V \in \vec{y}$, preserves all least upper bounds.

Using this observation and Proposition 9.3.7, show that, if i is an embedding, then, for every continuous complete lattice L, any continuous map $f: X \to L_\sigma$ gives rise to a map $g: Y \to L_\sigma$ that preserves all least upper bounds and such that $g \circ i = f$.

A topological space Z is *injective* if and only if, for every topological space Y, for every subspace X of Y, every continuous map $f: X \to Z$ extends to a continuous map from the whole space Y to Z. Observe that we have just proved that every continuous complete lattice with its Scott topology is injective. Show that every injective space Z is a retract of $\mathbb{S}^{\mathcal{O}(Z)}$, and is therefore a continuous complete lattice, and comes with its Scott topology.

From all this, deduce *Scott's Theorem*: the injective topological spaces are exactly the continuous complete lattices with their Scott topology.

Exercise 9.3.10 Intermediate between the powerset $\mathbb{P}(\mathcal{O}(X))$ of Proposition 9.3.5 and the space of completely prime filters of $\mathcal{O}(X)$ (Exercise 9.3.6), let us consider the space $\mathbb{F}(\mathcal{O}(X))$ of all non-trivial filters of open subsets of X, ordered by inclusion. Recall from Definition 8.1.9 that a filter of open subsets is a non-empty family $\mathcal{F}$ of open subsets such that, for all opens U, V such that $U \subseteq V$ and $U \in \mathcal{F}$, $V \in \mathcal{F}$, and such that the intersection of any two elements of $\mathcal{F}$ is in $\mathcal{F}$. $\mathcal{F}$ is *non-trivial* if and only if $\mathcal{F} \neq \mathcal{O}(X)$, iff $\emptyset \notin \mathcal{F}$.

Show that $\mathbb{F}(\mathcal{O}(X))$ is a Scott domain (an algebraic bc-domain; see Exercise 5.7.18) whenever X is non-empty, that least upper bounds of directed families are just unions, and that the finite elements of $\mathbb{F}(\mathcal{O}(X))$ are the filters $\blacksquare U = \{V \in \mathcal{O}(X) \mid U \subseteq V\}$, U non-empty open in X.

Recall that $\eta_{\mathbb{S}}(x)$ is the set of all open neighborhoods of $x \in X$. Show that $\eta_{\mathbb{S}}$ is a continuous almost open map from X to $\mathbb{F}(\mathcal{O}(X))_\sigma$, and is an embedding iff X is T_0. Moreover, show that $\eta_{\mathbb{S}}[X]$ is dense in $\mathbb{F}(\mathcal{O}(X))_\sigma$.

Exercise 9.3.11 Let X be a non-empty topological space, and Z be a bc-domain. Show that every continuous map $f: X \to Z_\sigma$ extends to a Scott-continuous map $\check{f}$ from $\mathbb{F}(\mathcal{O}(X))$ to Z, defined by $\check{f}(\mathcal{F}) = \sup_{U \in \mathcal{F}} \inf f[U]$—where the outer sup is directed. By extension, we again mean that $\check{f} \circ \eta_{\mathbb{S}} = f$. Show that $\check{f}$ is the largest continuous such extension (use Scott's Formula, and specifically Corollary 5.1.61, on the function g that maps each finite element $\blacksquare U$ of $\mathbb{F}(\mathcal{O}(X))$ to $\inf f[U]$).

Exercise 9.3.12 (Densely injective = bc-domain) A non-empty topological space Z is *densely injective* if and only if, for every topological space Y, for every dense subspace X of Y, every continuous map $f: X \to Z$ extends to a continuous map from Y to Z.

Similarly to Scott's Theorem (Exercise 9.3.9), show that the densely injective spaces Z are exactly the bc-domains, and that among all continuous extensions of $f: X \to Z$ to Y, where X is dense in Y and Y is non-empty, there is a largest one, which maps every $y \in Y$ to $\sup_{V \text{ open neighborhood of } y \text{ in } Y} \inf f[V \cap X]$. (For the second part, use Exercise 9.3.11. To show that every densely injective space Z is a bc-domain, observe that Z is, up to homeomorphism, a dense subspace of $\mathbb{F}(\mathcal{O}(Z))$, so that the identity map on Z extends to a map r from $\mathbb{F}(\mathcal{O}(Z))$ to Z; show that r is a retraction, hence that Z is a bc-domain.)

9.4 Patch-continuous, perfect maps

Definition 9.4.1 (Patch-continuous, perfect) Let X and Y be two topological spaces. The map $f : X \to Y$ is *patch-continuous* if and only f is continuous from X^{patch} to Y^{patch}.

We say that $f : X \to Y$ is *perfect* if and only if f is continuous and $f^{-1}(Q)$ is compact in X for every compact saturated subset Q of Y.

An equivalent definition of perfectness requires $f^{-1}(Q)$ to be saturated as well.

Lemma 9.4.2 *Let X and Y be two topological spaces. The map $f : X \to Y$ is perfect if and only if f is continuous and $f^{-1}(Q)$ is compact saturated in X for every compact saturated subset Q of Y.*

Proof It suffices to show that $f^{-1}(Q)$ is saturated whenever Q is. Since f is continuous, f is monotonic (Proposition 4.3.9). Then, given any element $x \in f^{-1}(Q)$ and any point x' above x in X, $f(x) \leqslant f(x')$. Since $f(x) \in Q$, $f(x')$ is in Q, too, so $x' \in f^{-1}(Q)$. □

There are a number of inequivalent definitions of perfectness. Some authors require in addition that $f[F]$ be closed for every closed subset F (see Exercise 9.4.7 below), and others require f to be surjective. We have followed Alvarez-Manilla *et al.* (2004) here. The same observation holds for "proper" (Exercise 9.4.7).

By Lemma 9.4.2, given any perfect map, the inverse images of opens are open, and the inverse images of cocompacts are cocompact. It follows immediately that:

Lemma 9.4.3 *Every perfect map is patch-continuous.*

The converse fails:

Example 9.4.4 The map $- : \mathbb{R}_\sigma \to \mathbb{R}_\sigma$ is not perfect, in fact not even continuous, as it is not monotonic. However, it is patch-continuous. Recall indeed that $-$ is continuous from $\mathbb{R}$ to $\mathbb{R}$, where $\mathbb{R}$ is equipped with its open ball topology (Example 4.3.8), and $\mathbb{R}$ coincides with $(\mathbb{R}_\sigma)^{\mathsf{patch}}$, by Exercise 9.1.37.

Monotonic patch-continuous maps are always perfect, assuming stable compactness:

Proposition 9.4.5 *Let X and Y be two topological spaces, and assume that X is stably compact. The perfect maps from X to Y are exactly the patch-continuous, monotonic maps from X to Y.*

Proof One direction is true by Lemma 9.4.3. In the other direction, assume $f: X \to Y$ is patch-continuous and monotonic. For every open subset V of Y, $f^{-1}(V)$ is then both patch-open (by patch-continuity) and upward closed (by monotonicity). Since X is stably compact, the patch-open upward closed subsets of X are exactly its open subsets, by Theorem 9.1.32. So f is continuous. For every cocompact subset $Y \setminus Q$ of Y, where Q is compact saturated, $Y \setminus Q$ is patch-open and downward closed, so $f^{-1}(Y \setminus Q)$ is also patch-open and downward closed. By Theorem 9.1.32 again, $f^{-1}(Y \setminus Q)$ is cocompact, so its complement $f^{-1}(Q)$ is compact saturated. □

The assumption that X is stably compact cannot be weakened, say to X stably locally compact:

Example 9.4.6 Consider $X = (0, 1]_\sigma$, $Y = [0, 1]_\sigma$, and the canonical inclusion i from X into Y. $X^{\mathrm{patch}} = (0, 1]_\lambda$ equals $(0, 1]$ with its open ball topology, and $Y^{\mathrm{patch}} = [0, 1]_\lambda$ is $[0, 1]$ with its open ball topology, so i is easily seen to be patch-continuous. It is monotonic, too, but it is not perfect, since the inverse image of the compact saturated set $[0, 1]$ is the non-compact set $(0, 1]$.

Exercise 9.4.7 (Proper maps) Let X, Y be two topological spaces. A *proper* map $f: X \to Y$ is a perfect map such that $\downarrow f[F]$ is closed for every closed subset F of X. Show that if X is well-filtered and Y is locally compact, then this is not a new notion: every perfect map is already proper. You may proceed as follows. Given any point $y \notin \downarrow f[F]$, consider the family of all compact saturated neighborhoods of y. Show that this is filtered, and that one of these must be disjoint from $\downarrow f[F]$. (Reason by contradiction.)

Exercise 9.4.8 Show that a map $f: X \to Y$ is proper if and only if it is continuous, $\downarrow f[F]$ is closed for every closed subset F of X, and $f^{-1}(\uparrow y)$ is compact for every $y \in Y$.

Exercise 9.4.9 (Proper images of stably compact spaces) Let X, Y be two topological spaces, and f be a surjective map from X to Y. Show that: (*a*) if X is compact and f is continuous, then Y is compact; (*b*) if X is well-filtered and f is perfect, then Y is well-filtered; (*c*) if X is coherent and f is perfect, then Y is coherent; (*d*) if X is locally compact and f is proper, then Y is locally compact. (Hint: given $y \in Y$ and an open neighborhood V of y in Y, find a compact saturated subset Q in X, included in $f^{-1}(V)$, and whose interior contains $f^{-1}(\uparrow y)$.)

Conclude that, for any stably compact space X, if $f: X \to Y$ is a proper surjective map and Y is T_0, then Y is stably compact.

Pospaces form a category, in which the morphisms $f: (X, \preceq) \to (Y, \leqslant)$ are the maps that are both continuous from X to Y and *order-preserving*, i.e., such that $x \preceq x'$ implies $f(x) \leqslant f(x')$. (While one would naturally call the order-preserving maps monotonic, we wish to avoid any confusion that may arise from the fact that any continuous map is already monotonic – with respect to specialization quasi-orderings, not to $\preceq$ and $\leqslant$.)

Proposition 9.4.10 (Compact pospaces $\equiv$ stably compact spaces) *There is an equivalence between the category of compact pospaces and continuous, order-preserving maps on the one hand, and the category of stably compact spaces and perfect maps on the other hand.*

The two functors defining the equivalence are the identity on morphisms, and map the compact pospace $(Z, \preceq)$ to the stably compact space Z with its upward topology, and conversely the stably compact space X to $(X^{\mathsf{patch}}, \leqslant)$ where $\leqslant$ is the specialization ordering of X.

Proof If X is a stably compact space, then $F(X) = (X^{\mathsf{patch}}, \leqslant)$ is a compact pospace by Theorem 9.1.32. Conversely, define $G(Z, \preceq)$ as Z with its upward topology. When $(Z, \preceq)$ is a compact pospace, $G(Z, \preceq)$ is stably compact by Proposition 9.1.20. For every stably compact space X, $G(F(X))$ is X with the upward topology of $(X^{\mathsf{patch}}, \leqslant)$: this is just X, by Theorem 9.1.32.

Conversely, for every compact pospace $(Z, \preceq)$, $F(G(Z, \preceq))$ is the pospace $(X^{\mathsf{patch}}, \leqslant)$, where X is Z with its upward topology, and $\leqslant$ is the specialization quasi-ordering of X. Using Lemma 9.1.18, $\leqslant$ coincides with $\preceq$. By Proposition 9.1.34, the topologies of X^{patch} and of Z are the same. So $F(G(Z, \preceq)) = (Z, \preceq)$.

Finally, F and G are defined as identities on morphisms. We conclude by Lemma 9.4.3 and Proposition 9.4.5. ☐

Perfect maps are more constrained than continuous maps. While Scott-continuous maps from $\mathbb{R}$ to $\mathbb{R}$ may have jumps, perfect maps must be patch-continuous, i.e., continuous from $\mathbb{R}$ to $\mathbb{R}$ with the ordinary topology, and in particular they cannot have any jumps. Moreover, they must be monotonic. The following separation property is therefore a rather strong one.

Theorem 9.4.11 (Urysohn–Nachbin Theorem) *Let X be a well-filtered, locally compact, and compact space, e.g., a stably compact space. For every compact saturated subset Q and for every closed subset F such that $Q \cap F = \emptyset$, there is a perfect map $f: X \to [0, 1]_\sigma$ that is identically 0 on F and identically 1 on Q.*

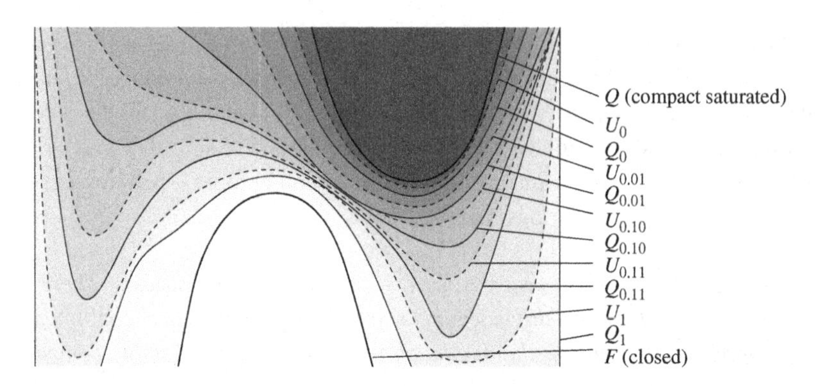

Figure 9.2 The Urysohn–Nachbin construction.

Proof We use Proposition 6.3.17. (Instead, we could use repeated interpolation between opens and compact saturated subsets, as in Urysohn's Theorem 6.3.15. The case $Q \subseteq U_0 = int(Q_0) \subseteq Q_0 \subseteq U_{0.01} = int(Q_{0.01}) \subseteq Q_{0.01} \subseteq U_{0.10} = int(Q_{0.10}) \subseteq Q_{0.10} \subseteq U_{0.11} = int(Q_{0.11}) \subseteq Q_{0.11} \subseteq U_1 = int(Q_1) \subseteq Q_1 \subseteq X \setminus F$ is illustrated on Figure 9.2.)

The family $\mathcal{U}$ of all open subsets of X is an upward scale, and since X is well-filtered and compact, the family $\mathcal{F}$ of all compact saturated subsets of X is a downward scale. Since X is locally compact, Proposition 4.8.14 tells us that interpolation holds, i.e., $(\mathcal{U}, \mathcal{F})$ is a dual scale. So, given $Q \cap F = \emptyset$ with $Q \in \mathcal{F}$ and $X \setminus F \in \mathcal{U}$, there is map $f : X \to [0, 1]$ that is identically 1 on Q, identically 0 on F, and such that $f^{-1}(t, 1]$ is open for every $t \in [0, 1]$ (so f is open from X to $[0, 1]_\sigma$), and $f^{-1}[u, 1]$ is compact saturated for every $u \in [0, 1]$. The cocompact subsets of $[0, 1]_\sigma$ are exactly the downward closed open subsets of $[0, 1]$, by Theorem 9.1.32, and are therefore of the form $[0, u)$ with $0 \leqslant t \leqslant 1$, $\emptyset$, or $[0, 1]$. It follows that f is perfect. □

Exercise 9.4.12 (Perfect insertion) Let X be a stably compact space, h be a continuous map from X to $(\mathbb{R} \cup \{-\infty, +\infty\})_\sigma$, and g be a map from X to $\mathbb{R} \cup \{-\infty, +\infty\}$ that is continuous from X^{d} to $(\mathbb{R} \cup \{-\infty, +\infty\})_\sigma^{\mathsf{d}}$, i.e., such that $g^{-1}[t, +\infty]$ is compact (saturated) in X for every $t \in \mathbb{R} \cup \{-\infty, +\infty\}$. Using the appropriate modification of the Katětov–Tong Insertion Theorem, show that if $g \leqslant h$ then there is a perfect map f from X to $(\mathbb{R} \cup \{-\infty, +\infty\})_\sigma$ (in particular, a continuous map from X^{patch} to $\mathbb{R} \cup \{-\infty, +\infty\}$) such that $g \leqslant f \leqslant h$.

Exercise 9.4.13 (Approximation by perfect maps) Let X be a stably compact space, and g be a perfect map from X to $\overline{\mathbb{R}^+}_\sigma$. For every $\epsilon > 0$, show that the map $g_\epsilon : x \mapsto \max(g(x) - \epsilon, 0)$ is perfect and $g_\epsilon \ll g$ in $[X \to \overline{\mathbb{R}^+}_\sigma]$: consider a directed family $(f_i)_{i \in I}$ in $[X \to \overline{\mathbb{R}^+}_\sigma]$ whose pointwise least upper bound is above g; for each $x \in X$ such that $g(x) > \epsilon$, pick $i_x \in I$ such that $f_{i_x}(x) > g(x) - \epsilon/2$, show that $U_x = \{z \in X \mid f_{i_x}(z) > g(z) - \epsilon/2\}$ is patch-open and contains x, and consider the patch-compact subset $Q = g^{-1}[\epsilon, +\infty)$.

Using Exercise 9.4.12 (and recalling Exercise 5.1.13), conclude that if X is a stably compact space, then $[X \to \overline{\mathbb{R}^+}_\sigma]$ is a continuous dcpo, with a basis of perfect maps, and similarly for $[X \to [0, 1]_\sigma]$.

9.5 Spectral spaces

Recall that a compact-open subset of a space X is a subset that is both compact and open. We have already encountered the notion in defining Scott domains (Exercise 5.7.18), and in defining strongly locally compact spaces (Exercise 8.3.22). The following notion of spectral space is due to Hochster (1969). This arose in the study of so-called spectra of rings, of which we shall give a glimpse in Exercise 9.5.38.

Definition 9.5.1 (Spectral space) A *locally spectral space* is a sober, strongly locally compact, and coherent space. A *spectral space* is a locally spectral space that is also compact.

(Note that Hochster's notion of locally spectral space differs from ours.)

Using Exercise 8.3.35, an equivalent definition is:

Fact 9.5.2 *A locally spectral space is a sober space X in which the family $\mathcal{KO}(X)$ of all compact-open subsets is not just a base of the topology, but is also closed under binary intersections. In a spectral space, $\mathcal{KO}(X)$ is stable under all finite intersections, including the empty one.*

Since every strongly locally compact space is locally compact, it is clear that:

Fact 9.5.3 (Spectral $\Rightarrow$ stably compact) *Every (locally) spectral space is stably (locally) compact.*

We can also reformulate Exercise 8.3.22 and Exercise 8.3.35 as follows.

Fact 9.5.4 (Locally spectral spaces $\equiv$ algebraic arithmetic lattices) *The adjunction $\mathcal{O} \dashv \mathsf{pt}$ restricts to an equivalence between the category of sober strongly locally compact spaces on the one hand, and the opposite of the category of algebraic distributive complete lattices and frame homomorphisms on the other.*

It also restricts to an equivalence between the category of locally spectral spaces on the one hand, and the opposite of the category of algebraic arithmetic lattices and frame homomorphisms on the other. The algebraic arithmetic lattices are exactly the algebraic distributive complete lattices in which the greatest lower bound of any two finite elements is finite.

Exercise 9.5.5 (Spectral spaces ≡ algebraic fully arithmetic lattices) Show that the adjunction $\mathcal{O} \dashv \mathsf{pt}$ restricts to an equivalence between the category of spectral spaces on the one hand, and the opposite of the category of algebraic fully arithmetic lattices and frame homomorphisms on the other.

Exercise 9.5.6 Show that the algebraic fully arithmetic lattices are exactly the algebraic distributive complete lattices where the top element $\top$ is finite, and where the greatest lower bound of any two finite elements is finite.

Example 9.5.7 Every finite T_0 space is spectral. (Exercise: prove this.)

Example 9.5.8 $[0, 1]_\sigma$ is not spectral: its only compact-open subsets are $\emptyset$ and $[0, 1]$, and are far from generating the Scott topology.

Example 9.5.9 Let $\mathbb{I}$ be the following poset. Its elements are those of $[0, 1]$, plus fresh, pairwise distinct elements t^+, $t \in [0, 1)$, inserted right after t. That is, we have $0 < 0^+ < \cdots < s < s^+ < \cdots < t < t^+ < \cdots < 1$, for all $s < t$ in $[0, 1)$. Equip $\mathbb{I}$ with its Scott topology. The sup of any family of elements of $\mathbb{I}$ that are strictly below t^+ is below t, so every element of the form t^+ is finite: the only directed families whose sup is above t^+ must contain t^+ or a larger element. On the other hand, since t is the sup of all elements s^+ with $s < t$ whenever $t \neq 0$, and 0 is finite, $\mathbb{I}$ is an algebraic dcpo with basis of finite elements consisting of 0 and t^+, $t \in [0, 1)$. We obtain a base of the Scott topology on $\mathbb{I}$ by considering the open subsets $\uparrow 0$ and $\uparrow t^+$, $t \in [0, 1)$. They are all compact. Conversely, since $\uparrow t$ is not open unless $t = 0$, they enumerate $\mathcal{KO}(\mathbb{I})$ exactly. Noting that $\uparrow s^+ \cap \uparrow t^+ = \uparrow \max(s, t)^+$, $\mathcal{KO}(\mathbb{I})$ is closed under finite intersections. So $\mathbb{I}_\sigma$ is spectral.

Example 9.5.10 Exercise 9.5.5 shows that $\mathsf{pt}(L)$ is a spectral space whenever L is an algebraic fully arithmetic lattice. By Exercise 9.5.11 below, it follows that $\mathsf{pt}(\mathbf{I}(L))$ is a spectral space for every distributive complete lattice L. E.g., $\mathsf{pt}(\mathbf{I}(\mathbb{P}(E)))$ is spectral for any set E.

We shall see that there are many spectral spaces: Johnstone's Theorem 9.5.14 below states that every stably compact space is the retract of one. Example 9.5.10 is the key idea that leads to this result.

We shall need the results of the following two exercises to establish Johnstone's Theorem. They are not difficult, and are a good way to get familiar with the required notions.

Exercise 9.5.11 (Ideal completion of distributive pointed lattices) Remember from Exercise 5.1.49 that the ideal completion $\mathbf{I}(L)$ of a pointed lattice L is a complete lattice. Show that if L is also distributive, then the least upper bound of any two elements D_1, D_2 of $\mathbf{I}(L)$ is the family of all elements $\sup(d_1, d_2)$ with $d_1 \in D_1$ and $d_2 \in D_2$. From this, deduce that if L is a distributive pointed lattice, then $\mathbf{I}(L)$ is a distributive complete lattice.

Expanding on this, show that if L is a distributive pointed lattice, then the least upper bound of any family $(D_i)_{i \in I}$ of elements of $\mathbf{I}(L)$ is the family of all elements of the form $\sup_{i \in J} d_i$, where each d_i is taken from D_i and J is a finite subset of I. (When J is empty, this sup denotes $\bot$, the least element of L.) Conclude that if L is a distributive pointed lattice, then $\mathbf{I}(L)$ is an algebraic fully arithmetic lattice.

Exercise 9.5.12 ($\mathbf{I} \dashv \mathcal{K}$, algebraic fully arithmetic $\equiv$ bounded distributive lattice) By Exercise 5.5.3, $\mathbf{I}(X)$ is the free dcpo over the poset X, so $\mathbf{I}$ is in particular a functor from the category of posets and monotone maps to the category of dcpos and Scott-continuous maps.

Recall that a bounded lattice is a lattice with a least element and a largest element. A *bounded lattice homomorphism* is a monotone map that preserves all finite greatest lower bounds and all finite least upper bounds—i.e., binary sups and infs, and also least and largest elements.

Define a functor $\mathcal{K}$ as follows. On objects, for each algebraic fully arithmetic lattice L, $\mathcal{K}(L)$ is the lattice of finite elements of L. Show that $\mathcal{K}(L)$ is a bounded distributive lattice. On morphisms, for each frame homomorphism from the algebraic fully arithmetic lattice L to the algebraic fully arithmetic lattice L' that maps finite elements of L to finite elements of L', define $\mathcal{K}(f)$ as the restriction of f to $\mathcal{K}(L)$.

For every bounded distributive lattice L, let $\eta_L \colon L \to \mathcal{K}(\mathbf{I}(L))$ map each $x \in L$ to $\downarrow x$. Show that this defines a bounded lattice homomorphism. Show that, given any algebraic fully arithmetic lattice L', and any bounded lattice homomorphism f from L to $\mathcal{K}(L')$, there is a unique frame homomorphism h from $\mathbf{I}(L)$ to L' that preserves finite elements, such that $\mathcal{K}(h) \circ \eta_L = f$.

From this, deduce that $\mathcal{K}$ is a functor that is right adjoint to the restriction of the functor $\mathbf{I}$, from the category of bounded distributive lattices and bounded lattice homomorphisms, to the category of algebraic fully arithmetic lattices and frame homomorphisms that map finite elements to finite elements.

Then show that this adjunction $\mathbf{I} \dashv \mathcal{K}$ is an equivalence of categories.

Proposition 9.5.13 *Given any continuous dcpo L, let $r_L \colon \mathbf{I}(L) \to L$ be the retraction defined by $r_L(D) = \sup D$, with section $s_L \colon L \to \mathbf{I}(L)$ defined by*

$s_L(u) = \Downarrow u$, as given in Theorem 5.1.48. If L is a fully arithmetic lattice, then $\mathsf{pt}(s_L)$ is a retraction of $\mathsf{pt}(\mathbf{I}(L))$ onto $\mathsf{pt}(L)$, with section $\mathsf{pt}(r_L)$.

Proof Since L is a fully arithmetic lattice, it is complete, hence pointed. It is also distributive by definition. Using Exercise 9.5.11, $\mathbf{I}(L)$ is a complete lattice. Least upper bounds are also described in this exercise, while greatest lower bounds are intersections (Exercise 5.1.49).

The fact that $\mathsf{pt}(s_L)$ is a retraction, with section $\mathsf{pt}(r_L)$, is clear, up to the fact that pt is a functor from the *opposite* of the category of frames to **Top**, hence reverses composition. In particular, $\mathsf{pt}(s_L) \circ \mathsf{pt}(r_L) = \mathsf{pt}(r_L \circ s_L) = \mathsf{pt}(\mathrm{id}_L) = \mathrm{id}_{\mathsf{pt}(L)}$. The only tricky point is that we need to show that $\mathsf{pt}(s_L)$ and $\mathsf{pt}(r_L)$ are well defined at all: using Proposition 8.1.15, we need to show that s_L is a frame homomorphism from L to $\mathbf{I}(L)$, and r_L is a frame homomorphism from $\mathbf{I}(L)$ to L. This is where we need L to be arithmetic, which allows us to use the explicit formulae for least upper and greatest lower bounds recalled above.

Let $(D_i)_{i\in I}$ be any family of elements of $\mathbf{I}(L)$, i.e., of ideals in L. Their least upper bound D is the set of elements $\sup_{i\in J} d_i$, J a finite subset of I, $d_i \in D_i$. So $r_L(D)$ is the least upper bound of these quantities, which is easily seen to coincide with $\sup_{i\in I} \sup D_i = \sup_{i\in I} r_L(D_i)$. So r_L commutes with sups. If I is finite, we must now show that $\sup(\bigcap_{i\in I} D_i)$ is the greatest lower bound of $(\sup D_i)_{i\in I}$. This is clearly a lower bound. In the converse direction, since L is continuous, it is enough to show that every element u that is way-below $\inf_{i\in I} \sup D_i$ is below $\sup(\bigcap_{i\in I} D_i)$. So $u \ll \sup D_i$ for every $i \in I$, whence there is an element $d_i \in D_i$ such that $u \leqslant d_i$, for each $i \in I$. Let $d = \inf_{i\in I} d_i$. Then $u \leqslant d$, and d is in $\bigcap_{i\in D_i}$, which allows us to conclude that r_L is a frame homomorphism.

Let $(u_i)_{i\in I}$ be a family of elements of L. Then $s_L(\sup_{i\in I} u_i)$ is the set of elements u such that $u \ll \sup_{i\in I} u_i$. If $u \ll \sup_{i\in I} u_i$, then by interpolation (Proposition 5.1.15, which applies since L is continuous) there is a $v \in L$ such that $u \ll v \ll \sup_{i\in I} u_i$. Writing $\sup_{i\in I} u_i$ as the directed least upper bound of sups of finite subfamilies, there is a finite subset J of I such that $v \leqslant \sup_{i\in J} u_i$, whence $u \ll \sup_{i\in J} u_i$. Since each u_i is the least upper bound of the directed family $\Downarrow u_i$ (L is continuous), $\sup_{i\in J} u_i$ is the least upper bound of the directed family of elements $\sup_{i\in I} d_i$ where each d_i is way-below u_i. (This is directed because $\prod_{i\in I} \Downarrow d_i$ is itself directed.) So $u \leqslant \sup_{i\in J} d_i$ for some collection of elements $d_i \in \Downarrow u_i = s_L(u_i)$, $i \in J$. Then $\sup_{i\in J} d_i$ is in $\sup_{i\in I} s_L(u_i)$, by definition of least upper bounds in $\mathbf{I}(L)$ (Exercise 9.5.11). By downward closure, u is in $\sup_{i\in I} s_L(u_i)$. Conversely, if $u \in \sup_{i\in I} s_L(u_i)$, then $u = \sup_{i\in J} d_i$ for some finite subset J of I and $d_i \ll u_i$ for each $i \in J$.

It is easy to see that $u \ll \sup_{i \in J} u_i$, using the fact that J is finite. In particular, $u \ll \sup_{i \in I} u_i$, so $u \in s_L(\sup_{i \in I} u_i)$. So s_L preserves least upper bounds.

We need to show that s_L also preserves all finite greatest lower bounds. It is enough to show that it preserves top elements, and binary greatest lower bounds. For the latter, the elements u of $s_L(v \wedge w)$ are those such that $u \ll v \wedge w$. Since $\ll$ is multiplicative, these are exactly the elements u such that $u \ll v$ and $u \ll w$, i.e., the elements of $\twoheaddownarrow v \cap \twoheaddownarrow w = s_L(v) \cap s_L(w)$. For the former, the top element $\top$ of L is finite since L is fully arithmetic, so $s(\top) = L = \downarrow \top$ is the largest ideal. So s_L is a frame homomorphism, and we conclude. □

We can now prove *Johnstone's Theorem* (Johnstone, 1982, theorem VII 4.6):

Theorem 9.5.14 (Stably compact = retract of spectral) *The stably compact spaces are exactly the retracts of spectral spaces.*

In particular, if X is a stably compact space, then it is a retract of the spectral space $\mathsf{pt}(\mathbf{I}(\mathcal{O}(X)))$, the retraction maps each completely prime filter $\mathcal{F}$ of ideals of open subsets of X to the unique point $x \in X$ whose open neighborhoods are exactly the open subsets U such that $\{V \in \mathcal{O}(X) \mid V \Subset U\} \in \mathcal{F}$, and the section maps $x \in X$ to the prime filter of all ideals D of open subsets of X that contain some open neighborhood of x.

Proof If Y is spectral, then it is stably compact by Fact 9.5.3. So any retract of Y is again stably compact, using Proposition 9.2.3.

Conversely, take $Y = \mathsf{pt}(\mathbf{I}(\mathcal{O}(X)))$, assuming X stably compact. So $L = \mathcal{O}(X)$ is a fully arithmetic lattice, by Exercise 9.1.8, whence Proposition 9.5.13 applies: $\mathsf{pt}(s_{\mathcal{O}(X)})$ is a retraction of Y onto $\mathsf{pt}(\mathcal{O}(X))$. Since X is sober, the latter is homeomorphic to X through the η_X^{Stone} map (Fact 8.2.5). Also, $\mathbf{I}(L)$ is an algebraic fully arithmetic lattice (Exercise 9.5.11, second part), so $Y = \mathsf{pt}(\mathbf{I}(L))$ is spectral (Fact 9.5.4, second part). The given definition of the retraction and section arise from their characterizations as $\mathsf{pt}(s_{\mathcal{O}(X)})$ and $\mathsf{pt}(r_{\mathcal{O}(X)})$, respectively. □

Example 9.5.15 Let $X = [0, 1]_\sigma$. This is stably compact, but not spectral. Apply the construction of Theorem 9.5.14. $\mathcal{O}(X)$ consists of the subsets $(t, 1]$, forming an ascending chain as t ranges from 1 to 0, plus the top element $[0, 1]$. For any ideal D in $\mathbf{I}(\mathcal{O}(X))$ not containing $[0, 1]$, write D as $\{(t_i, 1] \mid i \in I\}$. Their union is $(t, 1]$ with $t = \inf_{i \in I} t_i$, so that D is either $\downarrow(t, 1]$ (which we write D_t) or $\downarrow(t, 1] \setminus \{(t, 1]\}$ (which we write D_t^-, and is defined, i.e., non-empty, iff $t \neq 1$). If $s < t$, we have $D_t^- \subset D_t \subset D_s^- \subset D_s$. In particular, $\mathbf{I}(\mathcal{O}(X))$ is totally ordered. So every element of $\mathbf{I}(\mathcal{O}(X))$ (except the top element $\downarrow[0, 1]$) is prime. Using Proposition 8.1.20, the elements of $\mathsf{pt}(\mathbf{I}(\mathcal{O}(X)))$

can be equated with the prime elements of $\mathbf{I}(\mathcal{O}(X))$. Up to this identification, the hull-kernel opens are $\mathcal{O}_u = \{p \text{ prime} \mid u \not\subseteq p\}$, where u ranges over all elements of $\mathbf{I}(\mathcal{O}(X))$.

The map that sends D_t to t ($t \in [0, 1]$) and D_t^- to t^+ ($t \in [0, 1)$) is an order isomorphism with the poset $\mathbb{I}$ of Example 9.5.9. Up to this isomorphism, the hull-kernel opens are mapped to the subsets $\mathbb{I} \setminus \downarrow u$, $u \in \mathbb{I}$, plus the whole of $\mathbb{I}$: this is the upper topology on $\mathbb{I}$. Since $\uparrow t^+ = \mathbb{I} \setminus \downarrow t$ is open in the upper topology, it is finer than the Scott topology, hence coincides with it. The spectral space $\mathsf{pt}(\mathbf{I}(\mathcal{O}([0, 1]_\sigma)))$ is therefore the space $\mathbb{I}_\sigma$ of Exercise 9.5.9, up to homeomorphism.

The retraction is the obvious one: it maps both t and t^+ in $\mathbb{I}$ to $t \in [0, 1]$, while the section maps $t \in [0, 1]$ to $t \in \mathbb{I}$.

De Groot duality is simplified in spectral spaces, as Proposition 9.5.17 below shows.

Lemma 9.5.16 *Let X be a strongly locally compact space. For every compact saturated subset Q of X, for every open subset U containing Q, there is a compact-open subset K such that $Q \subseteq K \subseteq U$.*

In particular, the compact saturated subsets of X are exactly the intersections of filtered families $(K_i)_{i \in I}$ of compact-open subsets.

Proof For each $x \in Q$, let K_x be a compact-open neighborhood of x included in U. There is a family $(K_x)_{x \in Q}$, using the Axiom of Choice. This is an open cover of Q, so it has a finite subcover $(K_x)_{x \in E}$. Then let $K = \bigcup_{x \in E} K_x$.

For the second part, any intersection of a family of compact-open subsets is compact saturated, since X is sober (Proposition 8.3.6). Conversely, let Q be a compact saturated subset of X. The family $(K_i)_{i \in I}$ of compact-open neighborhoods of Q is non-empty, since X itself is compact. It is also filtered: if K_i and K_j are two compact-open neighborhoods of Q, then $K_i \cap K_j$ is an open neighborhood of Q, and by the first part of the Lemma, there is a compact-open neighborhood of Q included in $K_i \cap K_j$.

Clearly, $Q \subseteq \bigcap_{i \in I} K_i$. Conversely, since Q is saturated, it suffices to show that $\bigcap_{i \in I} K_i \subseteq U$ for every open neighborhood U of Q. The first part of the Lemma shows that there is even an $i \in I$ such that $K_i \subseteq U$. □

Proposition 9.5.17 *Let X be a spectral space. The complement map $K \mapsto X \setminus K$ is an order isomorphism from $\mathcal{KO}(X^{\mathsf{d}})$ onto $\mathcal{KO}(X)^{op}$. X^{d} is spectral, and the cocompact topology on X has a base of complements of compact-open subsets of X.*

Proof If K is compact-open in X^{d}, then $X \setminus K$ is cocompact in X^{d}, hence open in X, and closed in X^{d}, hence compact saturated in X, since $X^{\mathsf{dd}} = X$ (Theorem 9.1.38). So $X \setminus K$ is compact-open in X. The complement map is then a well-defined, monotone map from $\mathcal{KO}(X^{\mathsf{d}})$ to $\mathcal{KO}(X)^{op}$. It is clearly its own inverse, hence an order isomorphism.

Let $X \setminus Q$ be any open subset of X^{d}, i.e., Q is compact saturated in X. By Lemma 9.5.16, Q is the intersection of a filtered family $(K_i)_{i \in I}$ of compact-open subsets of X. So $X \setminus Q$ is a union of a (directed) family of complements of elements of $\mathcal{KO}(X)$, and these are elements of $\mathcal{KO}(X^{\mathsf{d}})$ by the first part of the Proposition. □

Exercise 9.5.18 (Spectral map) Let X be a well-filtered space (e.g., a stably compact space), and Y be a spectral space. Show that the perfect maps $f: X \to Y$ are exactly the *spectral maps*, i.e., the maps f such that $f^{-1}(K)$ is compact-open in X, for every compact-open subset K of Y.

Exercise 9.5.19 (Spectral space ≡ bounded distributive lattice) Given a diagram of functors $\mathbf{C}_1 \underset{U}{\overset{F}{\rightleftarrows}} \mathbf{C}_2 \underset{U'}{\overset{F'}{\rightleftarrows}} \mathbf{C}_3$ where $F \dashv U$ and $F' \dashv U'$ are adjunctions, show that $F'F \dashv UU'$ is also an adjunction. (Up to questions of sizes of categories, this is composition in the category of adjunctions.) Show that if $F \dashv U$ and $F' \dashv U'$ are equivalences of categories, then so is $F'F \dashv UU'$.

Deduce from Exercise 9.5.12 that there is an equivalence of categories $\mathcal{KO} \dashv$ $\mathbf{ptI}$, where $\mathcal{KO}$ is a functor from the category of spectral spaces and perfect maps to the opposite of the category of bounded distributive lattices and bounded lattice homomorphisms, which maps each spectral space to its lattice of compact-open subsets.

Show that the de Groot dual construction takes a particularly simple form through this equivalence of categories: for any spectral space X, $\mathcal{KO}(X^{\mathsf{d}})$ and $\mathcal{KO}(X)^{op}$ are order isomorphic. Conversely, for every bounded distributive lattice L, $\mathbf{ptI}(L^{op})$ and $\mathbf{ptI}(L)^{\mathsf{d}}$ are homeomorphic.

Exercise 9.5.20 (Stone spaces) The original form of Stone duality is the equivalence $\mathcal{KO} \dashv \mathbf{ptI}$ of Exercise 9.5.19, in the following special case. A *Boolean algebra* is a bounded distributive lattice L with a *complement* (or *negation*) operation $\neg: L \to L$ such that $\sup(u, \neg u) = \top$ and $u \wedge \neg u = \bot$ for every $u \in L$.

First compare this with Exercise 8.1.6, and show that every complete Boolean algebra is indeed a Boolean algebra.

Then show that $\mathcal{KO} \dashv \mathbf{ptI}$ restricts to an equivalence between the category of Stone spaces and continuous maps and the opposite of the category of Boolean algebras and bounded lattice homomorphisms. A *Stone space* is a zero-dimensional compact T_0 space. Recall that a space is zero-dimensional (Exercise 4.1.34) iff its topology has a base consisting of clopen subsets. You should observe that every Stone space is a particular form of spectral space (one that is T_2, in particular), and that every continuous map between Stone spaces is actually perfect.

Exercise 9.5.21 (Stone Representation Theorem) For any set X, $\mathbb{P}(X)$, ordered by inclusion, is a Boolean algebra, with complement as negation. More generally, call a *field of sets* any subset L of a Boolean algebra of the form $\mathbb{P}(X)$ such that L contains $\emptyset$ and X, and is closed under binary intersections, binary unions, and complements. Clearly, every field of sets is again a Boolean algebra.

Using Exercise 9.5.20, prove *Stone's Representation Theorem*: up to order isomorphism, the Boolean algebras are exactly the fields of sets.

Similarly, prove that, up to order isomorphism, the bounded distributive lattices are exactly the *rings of sets*, i.e., the subsets of lattices of the form $\mathbb{P}(X)$ that contain $\emptyset$ and X, and are closed under binary intersections and binary unions (but not necessarily under complements).

Exercise 9.5.22 (Ultrafilters, spectrum of a Boolean algebra) Let L be a Boolean algebra. Show that $b \Rightarrow c$, defined as $\sup(\neg b, c)$, is a residuation operation (see Proposition 8.1.1: a bounded distributive (not necessarily complete) lattice with a residuation is called a *Heyting algebra*). You may need to use the dual distributivity law of Exercise 8.1.5. Show that $\neg\neg a = a$ for every $a \in L$, and that $\neg a = (a \Rightarrow \bot)$. Conversely, show that a Heyting algebra L such that $\neg\neg a \leqslant a$ for every $a \in L$, where $\neg a$ is defined from the residuation $\Rightarrow$ of L as $a \Rightarrow \bot$, is a Boolean algebra. This is very close to Exercise 8.1.6. Show also that, in a Boolean algebra L, $\neg \sup(a, b) = \neg a \wedge \neg b$ for all $a, b \in L$.

An *ultrafilter* on L is a filter (a non-empty upward closed subset that is closed under the formation of binary greatest lower bounds; see Definition 8.1.9) such that, for every $u \in L$, u or $\neg u$ is in L, but not both. In particular, an ultrafilter of subsets (Definition 4.7.34) of X is just an ultrafilter on $\mathbb{P}(X)$. Show that D is a prime element (see Proposition 8.1.20) of $\mathbf{I}(L)$ if and only if its complement $F = L \setminus D$ is an ultrafilter.

The *spectrum* $\mathrm{spec}(L)$ of L is the set of all ultrafilters on L, and comes with the *Stone topology*, whose basic open subsets are $\mathcal{O}'_u = \{F \in \mathrm{spec}(L) \mid u \in F\}$, $u \in L$. Show that $\mathrm{spec}(L)$ is homeomorphic to $\mathbf{ptI}(L)$. So there is also an equivalence $\mathcal{KO} \dashv \mathrm{spec}$ between the category of Stone spaces and continuous maps on the one hand, and the opposite of the category of Boolean algebras and bounded lattice homomorphisms on the other.

The Nachbin pospace of a spectral space is also special.

Definition 9.5.23 (Priestley space) A *Priestley space* is a pair $(Z, \preceq)$ of a compact space Z and a partial ordering $\preceq$ satisfying the following *Priestley order separation* property: for every pair of points x, y such that $x \not\preceq y$, there is an upward closed clopen subset U containing x but not y.

The Priestley order separation property is exactly the order T_2 property of Lemma 9.1.15, with the added requirement that the open subset V there should be equal to the complement of U. So $U = Z \smallsetminus V$ is both open and closed, i.e., clopen.

Theorem 9.5.24 *For every spectral space X, the Nachbin pospace $(X^{\mathrm{patch}}, \leqslant)$, where $\leqslant$ is the specialization ordering of X, is a Priestley space. X^{patch} is zero-dimensional: a base of the patch topology consisting of clopen subsets is given by the differences $K \smallsetminus K'$ of two compact-open subsets K, K' of X.*

Proof X is stably compact by Fact 9.5.3, so $(X^{\mathrm{patch}}, \leqslant)$ is a compact pospace, by Theorem 9.1.32. The open subsets of X^{patch} are unions of intersections $U \cap (X \smallsetminus Q) = U \smallsetminus Q$ of an open subset U of X and a cocompact subset $X \smallsetminus Q$ of X. Let $x \in X$, and $U \smallsetminus Q$ be such a patch-open neighborhood of x. Since X is spectral, there is a compact-open subset K of U that contains x. By Lemma 9.5.16, $K \smallsetminus Q$ is also a union of sets of the form $K \smallsetminus K_i$, $i \in I$, where each K_i is a compact-open neighborhood of Q in X. Let K' be some K_i such that $x \in K \smallsetminus K_i$. So x is in a subset of the form $K \smallsetminus K'$ included in $U \smallsetminus Q$, with K and K' compact-open. Therefore, such subsets form a base of the patch topology, by Lemma 4.1.10.

Since K is compact saturated in X, it is closed in X^{d} hence also in X^{patch}; since K' is open in X, it is also open in X^{patch}, so $K \smallsetminus K'$ is closed in X^{patch}. Also, K is open in X hence in X^{patch}, and K' is compact saturated in X hence closed in X^{d} hence closed in X^{patch}; so $K \smallsetminus K'$ is also open in X^{patch}. So $K \smallsetminus K'$ is clopen. □

Exercise 9.5.25 The Sorgenfrey line $\mathbb{R}_\ell$ is zero-dimensional (Exercise 4.1.34). Show that it is not spectral, not locally spectral, and not the Nachbin pospace of any spectral space.

Theorem 9.5.26 *Let $(Z, \preceq)$ be a Priestley space. Then Z with the upward topology is a spectral space.*

Proof Let X be Z with the upward topology. By Proposition 9.1.20, X is stably compact. Let $x \in X$, and U be an open neighborhood of x in X. So U is open in Z and upward closed in $\preceq$. Let K be the complement of U in Z. So K is closed, hence compact in Z, by Proposition 4.4.15. For each $y \in K$, x is not below y; otherwise y would be in U. So, by the Priestley order separation property, there is an upward closed clopen subset U_y of Z that contains x but not y. (Use the Axiom of Choice to pick one U_y for each $y \in K$.) Using Theorem 9.1.32, U_y is open in X since it is upward closed and open in Z; also, U_y is compact saturated in X since it is closed hence compact in Z, and upward closed. So U_y is compact-open in X.

Since $Z \setminus U_y$ is also open in Z, $(Z \setminus U_y)_{y \in K}$ is an open cover of K in Z. Since K is compact in Z, there is a finite subset E of K such that $(Z \setminus U_y)_{y \in E}$ is an open cover of K in Z. Then $\bigcap_{y \in E} U_y$ is compact-open in X, contains x, and does not meet K, hence is included in U. By Lemma 4.1.10, the compact-open subsets of X form a base of its topology. Since X is coherent, $\mathcal{KO}(X)$ is also closed under binary intersections, so X is spectral. $\square$

Exercise 9.5.27 Let $(Z, \preceq)$ be a Priestley space. Show that Z is zero-dimensional, by a direct argument.

Similarly to Proposition 9.3.1, we have:

Proposition 9.5.28 (Products of spectral spaces) *The product of any family of spectral spaces is spectral. The product $(\prod_{i \in I} Z_i, \prod_{i \in I} \preceq_i)$ of any family $(Z_i, \preceq_i)$ of Priestley spaces, $i \in I$, is a Priestley space.*

Proof We prove the second part first. By Tychonoff's Theorem 4.5.12, $Z = \prod_{i \in I} Z_i$ is compact. Let $\preceq$ be the product ordering $\prod_{i \in I} \preceq_i$, and let $\vec{x}, \vec{y} \in Z$ be such that $\vec{x} \not\preceq \vec{y}$. For some $i \in I$, $x_i \not\preceq_i y_i$, so there is a clopen subset U_i of Z_i containing x_i but not y_i, and such that U_i is upward closed in $\preceq_i$. Let $U = \pi_i^{-1}(U_i)$, where π_i is the ith projection. U is open, and since $Z \setminus U = \pi_i^{-1}(Z_i \setminus U_i)$, U is also closed. So U is clopen, and contains $\vec{x}$ but not $\vec{y}$. So $(Z, \preceq)$ is a Priestley space.

For the first part, let $(X_i)_{i \in I}$ be a family of spectral spaces, and let $\leqslant_i$ be the specialization ordering of X_i. Let $X = \prod_{i \in I} X_i$, and $\leqslant$ be the specialization ordering of X. For each $i \in I$, $(X_i^{\mathsf{patch}}, \leqslant_i)$ is a Priestley space by Theorem 9.5.24, and by Proposition 9.3.1 $(X^{\mathsf{patch}}, \leqslant)$ is their product. We have shown above that $(X^{\mathsf{patch}}, \leqslant)$ was a Priestley space, too. So X^{patch} with the upward topology of $\leqslant$, which is just X by Theorem 9.1.32, is spectral by Theorem 9.5.26. $\square$

General limits of spectral spaces (in **Top**) are not spectral in general. Since equalizers are subspaces up to homeomorphism, all subspaces of spectral spaces are limits of spectral spaces. But Proposition 9.3.5 shows that the class of such spaces is exactly the class of all T_0 spaces: every limit of T_0 spaces is T_0, and every T_0 space X arises as a subspace of the space $\mathbb{S}^{\mathcal{O}(X)}$ (which is spectral by Proposition 9.5.28), up to homeomorphism. Still, we have an analogue of Proposition 9.3.4:

Proposition 9.5.29 (Patch-closed subspaces of spectral spaces) *Every patch-closed subset F of a spectral space X defines a spectral subspace of X.*

Proof Write $\leqslant$ for the specialization ordering of X. By Proposition 9.3.4, the specialization ordering of F coincides with the restriction of $\leqslant$ to F. Also, F^{patch} has the subspace topology of X^{patch}. Let $x, y \in F$ with $x \not\leqslant y$. Since X is spectral, there is an upward closed clopen subset U of X such that $x \in U$ and $y \notin U$. Then $U \cap F$ is open in F by definition, closed in F by Proposition 4.9.4, and clearly upward closed in the restriction of $\leqslant$ to F. So F is spectral. □

As far as coproducts are concerned, we have the following analogue of Proposition 9.2.1.

Exercise 9.5.30 (Finite coproducts of spectral spaces) Let $X_1, \ldots, X_n$ be finitely many spectral spaces. Show that their topological coproduct $\coprod_{i=1}^{n} X_i$ is spectral.

Contrarily to the case of stably compact spaces (Proposition 9.2.3), however, retracts of spectral spaces are in general not spectral: consider any stably compact space that is not spectral (e.g., $[0, 1]_\sigma$), and use Johnstone's Theorem 9.5.14.

We shall now show that every spectral space arises as a limit of *finite* T_0 spaces (i.e., of finite posets, with any of the usual topologies, Scott, Alexandroff, or upper). Moreover, these limits are of a particular kind themselves, and can be restricted to be projective limits. We begin by explaining the dual notion of an inductive limit.

Remember (Section 4.12.3) that a colimit colim F in a category $\mathbf{C}$ is a universal cocone for some functor F from some shape category $\mathbf{J}$ to $\mathbf{C}$. When $\mathbf{J}$ is a directed poset (remember that every poset is a category; see Exercise 8.1.4), colim F is called an *inductive limit*.

More concretely, let I be a set, with a quasi-ordering $\sqsubseteq$ that makes I directed, let $(A_i)_{i \in I}$ be a family of objects in $\mathbf{C}$, indexed by I, and let $s_{ij} : A_i \to A_j$ be morphisms, for each pair $i \sqsubseteq j$ in I, such that $s_{ii} = \mathrm{id}_{A_i}$ and $s_{jk} \circ s_{ij} = s_{ik}$ whenever $i \sqsubseteq j \sqsubseteq k$. Such a family $(s_{ij} : A_i \to A_j)_{i,j \in I, i \sqsubseteq j}$ is called an *inductive system* of objects of $\mathbf{C}$.

Such an inductive system is the description of a functor F, which maps the object $i \in I$ to A_i, and the unique morphism from i to j in I to s_{ij}. The colimit of this functor, if it exists, is an object A, together with morphisms $s_i : A_i \to A$, $i \in I$, such that $s_j \circ s_{ij} = s_i$ whenever $i \sqsubseteq j$—this defines a cocone; also, A is a *universal* (smallest) cocone: if there is another one A' with morphisms $s'_i : A_i \to A'$ such that $s'_j \circ s_{ij} = s'_i$, then there is a unique map $f : A \to A'$ such that $s'_i = f \circ s_i$ for every $i \in I$.

For example, if $\mathbf{C}$ is the category of sets, and $A_i \subseteq A_j$ whenever $i \sqsubseteq j$, assume that s_{ij} is the inclusion map. Then $A = \bigcup_{i \in I} A_i$ is the corresponding colimit. More generally, taking the inductive limit A of an inductive system $(s_{ij} : A_i \to A_j)_{i,j \in I, i \sqsubseteq j}$ of spaces means *gluing* A_i together, by taking the coproduct of the spaces A_i, then quotienting, i.e., equating $x_i \in A_i$ with $s_{ij}(x_i) \in A_j$, whenever $i \sqsubseteq j$.

The following states that the algebraic fully arithmetic complete lattices are exactly the colimits of finite distributive lattices in **Frm**. (Note that every finite distributive lattice is trivially fully arithmetic and complete.)

Proposition 9.5.31 *Let L be an algebraic fully arithmetic complete lattice. We say that $A \subseteq L$ is an* approximating sublattice *of L if and only if A is finite, all elements of A are finite in L, $\bot$ and $\top$ are in A, and the greatest lower bound and least upper bound in L of any two elements of A is in A.*

Let $\mathbf{J}_L$ be the family of all approximating sublattices of L, ordered by inclusion. Together with the canonical inclusion maps, $\mathbf{J}_L$ defines an inductive system of finite distributive lattices, whose colimit in **Frm** *is L.*

By $\mathbf{J}_L$ defining an inductive system, we mean that there is a functor that maps each object A of $\mathbf{J}_L$ to A itself, as an object of **Frm**, and each inclusion $A \subseteq B$ of approximating sublattices to the corresponding inclusion map.

Proof It is clear that every approximating sublattice A is distributive, because L itself is distributive, and infs and sups in A are computed as in L. To show that $\mathbf{J}_L$ is an inductive system, we must first show that $\mathbf{J}_L$ is directed. $\mathbf{J}_L$ is non-empty, since $\{\bot, \top\}$ is one approximating sublattice. If A and B are two approximating sublattices, let C be the set of all finite sups of finite infs of elements from $A \cup B$. C is finite, since it has no more elements than $\mathbb{P}(\mathbb{P}(A \cup B))$. C is clearly closed under finite sups, and is closed under finite infs, because L is distributive. It remains to show that the elements of C are finite. This follows from the fact that all finite infs of finite elements are finite in a fully arithmetic lattice ($\top$ is finite, and the inf of two finite elements is finite; see Exercise 8.3.35), and that all finite sups of finite elements are finite (by Exercise 5.1.12). So C is an approximating sublattice, hence $\mathbf{J}_L$ is directed.

We must now check that the inclusion maps are frame homomorphisms. This is clear, since infs and sups are computed in each approximating sublattice as in L. So $\mathbf{J}_L$ indeed defines an inductive system of frames. For clarity, let s_{AB} be the inclusion map of the approximating sublattice A into the approximating sublattice B, and s_A be the inclusion map of A into L. Clearly $s_B \circ s_{AB} = s_A$, so L defines a cocone.

Let us show that L is universal. Let L' be any frame, and assume that L' defines a cocone, in the sense that there are frame homomorphisms $s'_A : A \to L'$ such that $s'_B \circ s_{AB} = s'_A$ whenever $A \subseteq B$ are approximating sublattices. For every finite element u of L, $A_u = \{\bot, u, \top\}$ is an approximating sublattice. Let $f(u) = s'_{A_u}(u)$, for each finite element u of L. Since L' defines a cocone, and A_u is the smallest approximating sublattice containing u, $f(u) = s'_A(u)$ for any approximating sublattice containing u.

We observe that f defines a monotonic map from the set of finite elements of L to L'. Monotonicity comes from the fact that if $u \leqslant v$, then $A = \{\bot, u, v, \top\}$ is an approximating sublattice, and $s'_A(u) \leqslant s'_A(v)$. So $f(u) \leqslant f(v)$. Using Exercise 5.1.62, f has a unique Scott-continuous extension $\mathfrak{r}(f)$ from L to L'.

The map $u, v \mapsto \sup(u, v)$ preserves and reflects $\ll$ on L. Reflection is because sup is Scott-continuous, and by Proposition 5.1.66. Preservation is because if $u \ll u'$ and $v \ll v'$, then every directed family $(w_i)_{i \in I}$ whose sup is above $\sup(u', v')$ contains elements above u and v, respectively; by directedness, some w_i is above both, hence above $\sup(u, v)$. By Proposition 5.1.67, $\mathfrak{r}(f)(\sup(u, v)) = \sup(\mathfrak{r}(f)(u), \mathfrak{r}(f)(v))$ for all $u, v \in L$. We also have $\mathfrak{r}(f)(\bot) = f(\bot)$ since $\bot$ is finite, and $f(\bot) = s'_A(\bot) = \bot$ since any map s'_A is a frame homomorphism. It follows that $\mathfrak{r}(f)$ commutes with finite sups. Since $\mathfrak{r}(f)$ is also Scott-continuous, i.e., commutes with sups of directed families, and sups of arbitrary families $(u_i)_{i \in I}$ are sups of the directed family $(\sup_{i \in J} u_i)_{J \in \mathbb{P}_{\text{fin}}(I)}$, $\mathfrak{r}(f)$ commutes with all least upper bounds.

The map $u, v \mapsto u \wedge v$ also preserves and reflects $\ll$ on L (Exercise 8.1.3). By Proposition 5.1.67 again, $\mathfrak{r}(f)(u \wedge v) = \mathfrak{r}(f)(u) \wedge \mathfrak{r}(f)(v)$ for all $u, v \in L$. Since $\top$ is finite, $\mathfrak{r}(f)(\top) = f(\top) = s'_A(\top) = \top$ since any map s'_A is a frame homomorphism. So $\mathfrak{r}(f)$ commutes with all finite greatest lower bounds. Hence $\mathfrak{r}(f)$ is a frame homomorphism from L to L'.

Given any approximating sublattice A of L, for every $u \in A$, $\mathfrak{r}(f)(s_A(u)) = \mathfrak{r}(f)(u) = f(u)$ (since u is finite) $= s'_A(u)$. So $\mathfrak{r}(f) \circ s_A = s'_A$. It is easy to see that $\mathfrak{r}(f)$ is the unique frame homomorphism g such that $g \circ s_A = s'_A$ for all A: f is the unique monotonic map such that $f \circ s_A = s'_A$ for all A, and $\mathfrak{r}(f)$ is the unique Scott-continuous extension of f to the whole of L. So L is indeed the colimit of $\mathbf{J}_L$. $\square$

Dually, the limits in a category $\mathbf{C}$ are exactly the colimits in the opposite category $\mathbf{C}^{op}$. Similarly, a *projective limit* in $\mathbf{C}$ is any limit of a functor $F: \mathbf{J}^{op} \to \mathbf{C}$, where $\mathbf{J}$ is a directed quasi-ordered set. (That is, this is a colimit in $\mathbf{C}^{op}$ of the functor $F: \mathbf{J} \to \mathbf{C}^{op}$.)

Concretely, let I be a set, with a quasi-ordering $\sqsubseteq$ that makes I directed, let $(A_i)_{i \in I}$ be a family of objects in $\mathbf{C}$, indexed by I, and let $r_{ij}: A_j \to A_i$ be morphisms, for each pair $i \sqsubseteq j$ in I, such that $r_{ii} = \mathrm{id}_{A_i}$ and $r_{ij} \circ r_{jk} = r_{ik}$ whenever $i \sqsubseteq j \sqsubseteq k$. Such a family $(r_{ij}: A_j \to A_i)_{i,j \in I, i \sqsubseteq j}$ is called a *projective system* of objects of $\mathbf{C}$.

The limit of the functor defined by such a projective system, if it exists, is an object A, together with morphisms $r_i: A \to A_i$, $i \in I$, such that $r_{ij} \circ r_j = r_i$ whenever $i \sqsubseteq j$—this defines a cone; and A is a *universal* (smallest) cone: if there is another one A' with morphisms $r'_i: A' \to A_i$ such that $r_{ij} \circ r'_j = r'_i$, then there is unique map $f: A' \to A$ such that $r'_i = r_i \circ f$ for every $i \in I$.

Recalling the general shape of limits in **Top**, the projective limit A of a projective system $(r_{ij}: A_j \to A_i)_{i,j \in I, i \sqsubseteq j}$ is built, up to a unique homeomorphism, as the subspace of $\prod_{i \in I} A_i$ consisting of those tuples $\vec{x}$ such that $r_{ij}(x_j) = x_i$ whenever $i \sqsubseteq j$. One can see A as an ideal space of which each A_i is a rough picture: each element $\vec{x} \in A$ is projected to $r_i(\vec{x}) = x_i$ in A_i.

When each A_i is finite, one can see each A_i as a *discretization* of A, a kind of finite raster image that approximates the actual landscape A. If j is above i, one can think of A_j as another discretization of the same landscape A, only with better resolution. The following theorem then says that the landscapes that can be approximated by such raster images are exactly the spectral spaces. Similar ideas have been used in the field of digital topology; see, e.g., Smyth (1995). The following theorem is due to Hochster (1969, proposition 10).

Theorem 9.5.32 (Spectral = projective limit of finite spaces) *The spectral spaces are exactly the projective limits of finite T_0 spaces in* **Top**.

Proof Let $(r_{ij}: A_j \to A_i)_{i,j \in I, i \sqsubseteq j}$ be a projective system of finite T_0 spaces. Since each A_i is finite, it is spectral (Example 9.5.7), so $Y = \prod_{i \in I} A_i$ is spectral by Proposition 9.5.28. The limit of the projective system is the subspace X of $Y = \prod_{i \in I} A_i$ consisting of all tuples $\vec{x}$ such that $r_{ij}(x_j) = x_i$ whenever $i \sqsubseteq j$.

Since r_{ij} is continuous from A_j to A_i, and A_j is finite, r_{ij} is perfect: for every compact saturated subset Q of A_i, $r_{ij}^{-1}(Q)$ is upward closed in A_j since

r_{ij} is continuous hence monotonic (Proposition 4.3.9); as A_j is finite, $r_{ij}^{-1}(Q)$ is then (finitary) compact. By Lemma 9.4.3, r_{ij} is patch-continuous. Since π_j is continuous from $\prod_{i \in I} A_i^{\text{patch}}$ to A_j^{patch}, and using Proposition 9.3.1, π_j is patch-continuous from Y to A_j. So $r_{ij} \circ \pi_j$ is patch-continuous.

The subset $\Delta_i = \{(x, x) \mid x \in A_i\}$ of the stably compact space $A_i \times A_i$ is patch-closed (indeed, finite). Since $r_{ij} \circ \pi_j$ and π_i are patch-continuous, $(r_{ij} \circ \pi_j) \times \pi_i$ is, too (see Exercise 4.5.5, working on the patch topologies). So $[(r_{ij} \circ \pi_j) \times \pi_i]^{-1}(\Delta_i)$ is patch-closed in Y. Now $X = \bigcap_{\substack{i,j \in I \\ i \sqsubseteq j}} [(r_{ij} \circ \pi_j) \times \pi_i]^{-1}(\Delta_i)$ is also patch-closed in Y. So X is a spectral subspace, by Proposition 9.5.29.

Conversely, let X be a spectral space, and apply Proposition 9.5.31 to $L = \mathcal{O}(X)$. Indeed L is an algebraic fully arithmetic complete lattice (Exercise 9.5.5). By Stone duality, the composition of pt with the functor defined by the inductive system $\mathbf{J}_L$ (i.e., the spaces are $\mathsf{pt}(A)$ where A ranges over approximating sublattices of $L = \mathcal{O}(X)$, and, whenever $A \subseteq B$, we let $r_{AB} : \mathsf{pt}(B) \to \mathsf{pt}(A)$ be $\mathsf{pt}(s_{AB})$, where s_{AB} is the inclusion map of A into B), yields a projective system of spectral spaces, of which $\mathsf{pt}(L)$ is a limit since pt is a right adjoint (see Theorem 8.1.26). However, when A is finite, $\mathsf{pt}(A)$ is finite as well. Also, $\mathsf{pt}(L)$ is homeomorphic to X. So X is a projective limit of finite T_0 spaces. □

Exercise 9.5.33 The above proof goes through Stone duality. Putting all the arguments together, one sees that the spectral space X is obtained from finite spaces as follows. Given any finite family $K_1, \ldots, K_n$ of compact-open subsets of X, the finite unions of finite intersections of elements from this family form an approximating sublattice A of $\mathcal{O}(X)$, and all of them are obtained this way. Define $\equiv_A$ on X by $x \equiv_A y$ if and only if x and y belong to the same K_is, i.e., for every i, $1 \leqslant i \leqslant n$, $x \in K_i$ iff $y \in K_i$.

Show that the finite space $\mathsf{pt}(A)$ is obtained as the quotient space $X/{\equiv_A}$, up to homeomorphism. In other words, the rasterized version $\mathsf{pt}(A)$ looks like Figure 9.3 (with $n = 4$ here).

Together with Johnstone's Theorem 9.5.14, Theorem 9.5.32 implies:

Corollary 9.5.34 *The stably compact spaces are exactly the retracts of projective limits of finite T_0 spaces in* **Top**.

Given a stably compact space X, one rasterizes it as in Figure 9.3, except now $K_1, \ldots, K_n$ are not restricted to compact-open subsets, but are general open subsets. Putting all the finite pieces together, through the computation of a projective limit, yields $\mathsf{pt}(\mathbf{I}(\mathcal{O}(X)))$, of which X is a retract (in particular, a subspace), by Johnstone's Theorem 9.5.14.

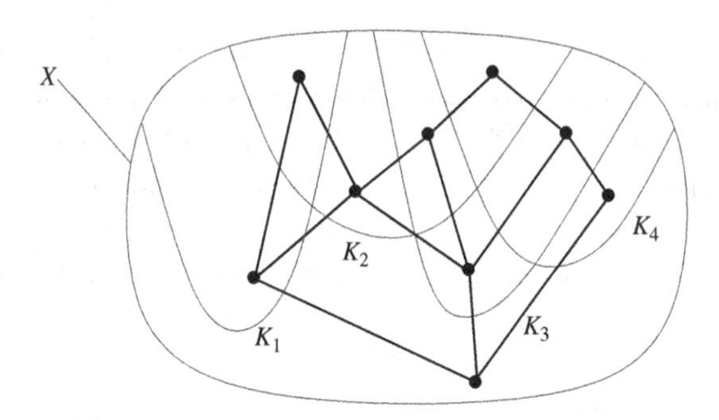

Figure 9.3 Rasterizing a spectral space.

Exercise 9.5.35 (Spectral = patch-closed subset of a powerset) For every set I, show that every patch-closed subset of the powerset $\mathbb{S}_\sigma^I$ is spectral (in the subspace topology).

Conversely, let X be a spectral space. Consider the following variant of $\eta_{\mathbb{S}}: X \to \mathbb{S}^{\mathcal{O}(X)}$ (Proposition 9.3.5): $\eta'_{\mathbb{S}}: X \to \mathbb{S}^{\mathcal{KO}(X)}$ is the map that sends $x \in X$ to $(\chi_U(x))_{U \in \mathcal{KO}(X)}$.

Show that $\eta'_{\mathbb{S}}$ is a perfect embedding. To show that the inverse images of compact saturated subsets are compact, consider finitary compacts first, and recall that any compact saturated subset of a continuous poset is the intersection of a filtered family of finitary compacts (Theorem 5.2.30).

From this, deduce that $\eta'_{\mathbb{S}}[X]$ is patch-closed in $\mathbb{S}_\sigma^{\mathcal{KO}(X)}$.

Conclude that the spectral spaces are exactly the patch-closed subsets of powersets $\mathbb{S}_\sigma^I$, up to homeomorphism (Hochster, 1969, proposition 9).

Together with Johnstone's Theorem 9.5.14, Exercise 9.5.35 entails:

Fact 9.5.36 *The stably compact spaces are exactly the retracts of patch-closed subsets of powersets.*

Exercise 9.5.37 (Spectral spaces are not CCC) Using Exercise 5.5.5, show that the category of spectral spaces and continuous maps is not Cartesian-closed.

The spectral spaces were so named because they are exactly the spectra of commutative rings in their Zariski topology: this is *Hochster's Theorem* (Hochster, 1969, theorem 6). Proving this is difficult, and would lead us too far astray. Exercise 9.5.38 provides the easy direction of this theorem. This already requires some algebra.

Exercise 9.5.38 (Hochster's Theorem, easy part) A *ring* R (with unit) is a set with the following structure. First, R is an Abelian group, i.e., there is an addition operation $+\colon R \times R \to R$, a constant $0 \in R$, and a map $-\colon R \to R$ satisfying $(a+b)+c = a+(b+c), a+0 = 0+a = a, a+b = b+a, a+(-a) = (-a)+a = 0$ for all $a, b, c \in R$. We also write $a-b$ for $a+(-b)$. Second, R has a multiplication $.\colon R \times R \to R$ (and we write ab for $a.b$), with unit $1 \in R$, such that $(ab)c = a(bc)$, $1a = a1 = a$ for all $a, b, c \in R$. Also, the additive structure distributes over multiplication: $a(b+c) = ab + ac$, and $(b+c)a = ba + ca$, $0a = a0 = 0$, and $a(-b) = -ab$, $(-a)b = -ab$. A *commutative* ring additionally satisfies $ab = ba$ for all $a, b \in R$. An important example is the commutative ring $\mathbb{Z}$ of integers.

Fix a commutative ring R. Given a subset A of R, and $b \in R$, write $bA = \{ba \mid a \in A\}$. Given two subsets A and B of R, write $AB = \{ab \mid a \in A, b \in B\}$ and $A+B = \{a+b \mid a \in A, b \in B\}$. An *ideal* I of R is a subgroup of R (i.e., a subset such that $0 \in I$, $-a \in I$ for every $a \in I$, and $a+b \in I$ for all $a, b \in I$) that also satisfies $RI = I$, that is, $ab \in I$ whenever $a \in R$ and $b \in I$. For instance, the ideals of $\mathbb{Z}$ are the subsets $n\mathbb{Z}$, where $n \in \mathbb{N}$.

Order ideals by inclusion. Show that the poset $\mathrm{Id}(R)$ of all ideals is a complete lattice, that the greatest lower bound of a family of ideals $(I_j)_{j \in J}$ is their intersection; its least element is $\{0\}$; its greatest element is R itself; the least upper bound of a pair of ideals I_1, I_2 is $I_1 + I_2$; and the least upper bound of a directed family of ideals is their union.

$\mathrm{Id}(R)$ may fail to be a frame. For every ideal I, its *radical* $\sqrt{I}$ is the set of elements $a \in R$ such that $a^k \in I$ for some $k \in \mathbb{N}$ with $k \geqslant 1$; a^k is defined by induction on k by $a^0 = 1, a^{k+1} = a.a^k$. Show that $\sqrt{I}$ is an ideal. (You can use the binomial formula $(a+b)^n = \sum_{i=0}^{n} \binom{n}{i} a^i b^{n-i}$, where $\binom{n}{i} = \frac{n(n-1)\dots(n-i+1)}{i(i-1)\dots 2\,1}$, and kc is the sum of k copies of $c \in R$, where $k \in \mathbb{N}$.) An ideal I such that $I = \sqrt{I}$ is called a *radical ideal*; for instance, the radical ideals of $\mathbb{Z}$ are those $n\mathbb{Z}$ where n is a product of pairwise distinct prime numbers. Show that the radical $\sqrt{I}$ of the ideal I is always radical, and is the smallest radical ideal containing I. E.g., in $\mathbb{Z}$, $\sqrt{n\mathbb{Z}} = p_1 p_2 \dots p_k \mathbb{Z}$, where $p_1, p_2, \dots, p_k$ are the distinct prime factors of n.

Let $\mathrm{Rad}(R)$ be the poset of all radical ideals of R, ordered by inclusion. Show that $\mathrm{Rad}(R)$ is also a complete lattice. Show that greatest lower bounds are intersections, $\sqrt{\{0\}}$ is the least element, R the largest, least upper bounds of directed families are unions, and the least upper bound of $I_1, I_2 \in \mathrm{Rad}(R)$ is $\sqrt{I_1 + I_2}$.

Additionally, show that $\mathrm{Rad}(R)$ is an algebraic fully arithmetic lattice, and that its finite elements are the ideals of the form $\sqrt{(A)}$, with $A \subseteq R$ finite. (A) denotes the smallest ideal containing A. This is the ideal *generated by* A. When $A = \{a_1, \dots, a_n\}$, we also write $(A) = (a_1, \dots, a_n)$, and this is equal to the set of all sums $\sum_{a \in A} r_a a$, where each $r_a \in R$. You will need to show that $\sqrt{(A)} \cap \sqrt{(B)} = \sqrt{(AB)}$ for all finite subsets A and B of R.

Conclude that $\mathsf{pt}(\mathrm{Rad}(R))$ is a spectral space, whose lattice of open subsets is order isomorphic to $\mathrm{Rad}(R)$, for every commutative ring R.

An ideal I of R is *prime* if and only if $1 \notin I$ and, for any two elements $a, b \in R$ such that $ab \in I$, $a \in I$ or $b \in I$. (The prime ideals of $\mathbb{Z}$ are those of the form $p\mathbb{Z}$, p a prime number.) Show that every prime ideal is radical, and that the prime elements of $\mathrm{Rad}(R)$ are exactly the prime ideals. The *spectrum* of the commutative ring R is the set of prime ideals of R, with the *Zariski topology*, whose closed subsets are the subsets $F_I = \{p \text{ prime ideal of } R \mid I \subseteq p\}$, for each (radical) ideal I of R. Using Proposition 8.1.20, show that the spectrum of a commutative ring R is homeomorphic to $\mathsf{pt}(\mathrm{Rad}(R))$. In particular, the spectrum of a commutative ring is always a spectral space. Hochster's Theorem states that the spectral spaces are exactly the spectra of commutative rings; in particular, every spectral space arises as the spectrum of some commutative ring. See Hochster's original paper (Hochster, 1969), or Banaschewski (1996), which is closer to our presentation.

9.6 Bifinite domains

A simple modification of the characterization of spectral spaces as projective limits of finite T_0 spaces (i.e., finite posets) yields the so-called *bifinite domains*, a.k.a. Plotkin's *SFP-domains*. Just like bc-domains, bifinite domains form one of the important categories used in domain theory (Abramsky and Jung, 1994). In particular, this category will turn out to be Cartesian-closed.

We require the maps $r_{ij} : A_j \to A_i$ to be retractions, and a bit more.

Definition 9.6.1 (ep-pair, projection) An *ep-pair* from a topological space X to a topological space Y is a pair of continuous maps $X \underset{r}{\overset{s}{\rightleftarrows}} Y$ such that $r \circ s = \mathrm{id}_X$ and $s \circ r \leqslant \mathrm{id}_Y$. Then r is called a *projection* of Y onto X.

The inequality $s \circ r \leqslant \mathrm{id}_Y$ is with respect to the specialization quasi-ordering of Y. ep-pairs are usually called *embedding-projection pairs*. We will not use the latter phrase, and in particular will not call s an embedding (it is more than just an embedding), since it induces some confusion with (ordinary) embeddings.

There are several equivalent definitions of ep-pairs. One is to say that r is a retraction and s a matching section, such that additionally $s \circ r \leqslant \mathrm{id}_Y$. Another is to say that $s \dashv r$ is a poset adjunction (see Section 8.4.4) that is also a section-retraction pair. In this case, each of s, r determines the other one: $r(y)$ is the largest $x \in X$ such that $s(x) \leqslant y$, and $s(x)$ is the smallest $y \in Y$ such that $x \leqslant r(y)$. So a projection is merely a retraction that is also a right adjoint.

Definition 9.6.2 (Bifinite domain) A *bifinite domain* is a limit in **Top** of a projective system $(r_{ij} : A_j \to A_i)_{i,j \in I, i \sqsubseteq j}$ where each A_i is a finite poset, and each r_{ij} is a projection.

Beware that bifinite domains sometimes refer to *pointed* bifinite domains in the literature. We do not assume a least element.

Each finite poset A_i comes with a unique topology whose specialization ordering is the one given on A_i (Exercise 4.2.15). So a bifinite domain is a projective limit of finite T_0 spaces. It follows immediately from Theorem 9.5.32 (and Fact 9.5.3) that:

Fact 9.6.3 (Bifinite ⇒ spectral) *Every bifinite domain is spectral, in particular stably compact.*

Definition 9.6.4 (Retractive system, ep-system) A family $\left(A_i \underset{r_{ij}}{\overset{s_{ij}}{\rightleftarrows}} A_j\right)_{i,j \in I, i \sqsubseteq j}$, with I directed in $\sqsubseteq$, is a *retractive system* of topological spaces if and only if r_{ij}, s_{ij} is a retraction-section pair for all $i, j \in I$ with $i \sqsubseteq j$, such that: $r_{ii} = s_{ii} = \mathrm{id}_{A_i}$ for every $i \in I$, $r_{ij} \circ r_{jk} = r_{ik}$ and $s_{jk} \circ s_{ij} = s_{ik}$ whenever $i \sqsubseteq j \sqsubseteq k$.

An *ep-system* additionally is such that $s_{ij} \dashv r_{ij}$ is an ep-pair.

In particular, not only is $(r_{ij} : A_j \to A_i)_{i,j \in I, i \sqsubseteq j}$ a projective system, but it is also true that $(s_{ij} : A_i \to A_j)_{i,j \in I, i \sqsubseteq j}$ is an inductive system of topological spaces.

Lemma 9.6.5 *Let $(r_{ij} : A_j \to A_i)_{i,j \in I, i \sqsubseteq j}$ be a projective system of topological spaces, where each r_{ij} is a projection. Let s_{ij} be the unique associated section. Then the family $\left(A_i \underset{r_{ij}}{\overset{s_{ij}}{\rightleftarrows}} A_j\right)_{i,j \in I, i \sqsubseteq j}$ is an ep-system.*

Proof There is a unique left adjoint to r_{ii}; since $r_{ii} = \mathrm{id}_{A_i}$, id_{A_i} is one; s_{ii} is another, so $s_{ii} = \mathrm{id}_{A_i}$. When $i \sqsubseteq j \sqsubseteq k$, we claim that $s_{jk} \circ s_{ij}$ is left adjoint to $r_{ik} = r_{ij} \circ r_{ik}$. Indeed, $(s_{jk} \circ s_{ij})(x) \leqslant y$ iff $s_{ij}(x) \leqslant r_{jk}(y)$ iff $x \leqslant (r_{ij} \circ r_{jk})(y)$. But s_{ik} is the unique left adjoint of r_{ik}. So $s_{jk} \circ s_{ij} = s_{ik}$. □

A simpler categorical definition of ep-systems is as follows. Let **C** be **Top**, or any subcategory. Then let $\mathbf{C}^{\mathrm{ep}}$ be the category whose objects are the same as

$\mathbf{C}$, but whose morphisms from A to B are ep-pairs $A \underset{r}{\overset{s}{\rightleftarrows}} B$. Composition is given by $B \underset{r'}{\overset{s'}{\rightleftarrows}} C \circ A \underset{r}{\overset{s}{\rightleftarrows}} B = A \underset{r \circ r'}{\overset{s' \circ s}{\rightleftarrows}} C$. Then the ep-systems in $\mathbf{C}$ are just the inductive systems in $\mathbf{C}^{ep}$, and also the projective systems in $(\mathbf{C}^{ep})^{op}$.

Let us now compare the inductive limit A of $(s_{ij} : A_i \to A_j)_{i,j \in I, i \sqsubseteq j}$ and the projective limit X of $(r_{ij} : A_j \to A_i)_{i,j \in I, i \sqsubseteq j}$. Recall that X is the set of tuples $\vec{x} \in \prod_{i \in I} A_i$ such that $r_{ij}(x_j) = x_i$ whenever $i \sqsubseteq j$. The following exercise states that A is, up to homeomorphism, just the union of all A_i, $i \in I$. X is in general larger. So the situation is particularly simple here: in general, an inductive limit would fold the spaces A_i up, identifying several elements, and this does not happen with retractive systems.

Exercise 9.6.6 Let $\left(A_i \underset{r_{ij}}{\overset{s_{ij}}{\rightleftarrows}} A_j \right)_{i,j \in I, i \sqsubseteq j}$ be a retractive system of topological spaces. Let X be the projective limit of $(r_{ij} : A_j \to A_i)_{i,j \in I, i \sqsubseteq j}$.

For all $i, j \in I$, show that there is a continuous map $rs_{ij} : A_i \to A_j$ such that $rs_{ij} = r_{jk} \circ s_{ik}$ whenever $k \in I$ is above both i and j. If $i \sqsubseteq j$, then $rs_{ij} = s_{ij}$. If $j \sqsubseteq i$, then $rs_{ij} = r_{ji}$.

Let $s_i : A_i \to A$ map $x_i \in A_i$ to $(rs_{ij}(x_i))_{j \in I}$. Show that s_i is an embedding, i.e., a homeomorphism onto its image $X_i = s_i[A_i]$. Show that $(X_i)_{i \in I, \sqsubseteq}$ is a monotone net of subsets of X. Let $A = \bigcup_{i \in I} X_i$. Show that A is the inductive limit of $(s_{ij} : A_i \to A_j)_{i,j \in I, i \sqsubseteq j}$, up to homeomorphism: show that A, together with $s_i : A_i \to A$, $i \in I$, forms a universal cocone for $(s_{ij} : A_i \to A_j)_{i,j \in I, i \sqsubseteq j}$.

Lemma 9.6.7 *Let* $\left(A_i \underset{r_{ij}}{\overset{s_{ij}}{\rightleftarrows}} A_j \right)_{i,j \in I, i \sqsubseteq j}$ *be an ep-system with projective limit* X *and inductive limit* A. *Let* $s_i : A_i \to A$ *be the maps given in Exercise 9.6.6. The family* $(s_i \circ \pi_i)_{i \in I, \sqsubseteq}$ *is a monotone net of continuous maps from* X *to* A, *whose least upper bound in* $[X \to X]$ *is* id_X.

Proof We must first show that if $i \sqsubseteq i'$, then $s_i \circ \pi_i \leqslant s_{i'} \circ \pi_{i'}$, i.e., that, for every $\vec{x} \in X$, for every $j \in I$, $rs_{ij}(x_i) \leqslant rs_{i'j}(x_{i'})$. Since $\vec{x} \in X$, $x_i = r_{ii'}(x_{i'})$, and since $s_{ii'}$ is left adjoint to $r_{ii'}$, it follows that $s_{ii'}(x_i) \leqslant x_{i'}$. Now, picking some $k \in I$ above i, i', and j, $rs_{ij}(x_i) = r_{jk}(s_{ik}(x_i)) = r_{jk}(s_{i'k}(s_{ii'}(x_i))) \leqslant$

$r_{jk}(s_{i'k}(x_{i'}))$ (since continuous maps are monotone, by Proposition 4.3.9) $= rs_{i'j}(x_{i'})$. So $(s_i \circ \pi_i)_{i \in I, \sqsubseteq}$ is a monotone net.

We then claim that, for every $i \in I$, $s_i \circ \pi_i \leqslant \text{id}_X$, i.e., that, for every $\vec{x} \in X$, $s_i(x_i) \leqslant \vec{x}$. Equivalently, we must show that, for every $j \in I$, $rs_{ij}(x_i) \leqslant x_j$. Let $k \in I$ be above i and j. Then $rs_{ij}(x_i) = r_{jk}(s_{ik}(x_i)) \leqslant r_{jk}(x_k)$, by the same arguments as above, and $r_{jk}(x_k) = x_j$.

Finally, let $f : X \to X$ be any continuous map above every $s_i \circ \pi_i$. For every $\vec{x} \in X$, for every $i \in I$, we have $s_i(x_i) \leqslant f(\vec{x})$, hence $rs_{ii}(x_i) = x_i$ is below the ith component of $f(\vec{x})$. So $\vec{x} \leqslant f(\vec{x})$. It follows that $\sup_{i \in I}(s_i \circ \pi_i) = \text{id}_X$. □

When each A_i is a finite poset, the maps $s_i \circ \pi_i$ form a directed family of *idempotent deflations*:

Definition 9.6.8 (Deflation) Let X be a topological space. A *deflation* on X is a continuous map $f : X \to X$ such that $f \leqslant \text{id}_{[X \to X]}$ and whose image $f[X]$ is finite. It is *idempotent* if and only if $f \circ f = f$.

We obtain:

Theorem 9.6.9 *The bifinite domains are exactly the dcpos X with a* generating system of idempotent deflations, *i.e., a directed family of idempotent deflations $(f_i)_{i \in I}$ whose least upper bound is* id_X. *The topology of X is the Scott topology, X is algebraic, and the finite elements of X are exactly the elements of $\bigcup_{i \in I} f_i[X]$.*

Proof Assume X is a bifinite domain, obtained as the projective limit of some ep-system $\left(A_i \underset{r_{ij}}{\overset{s_{ij}}{\rightleftarrows}} A_j \right)_{\substack{i,j \in I \\ i \sqsubseteq j}}$, where each A_i is a finite poset. We have seen in Lemma 9.6.7 that $f_i = s_i \circ \pi_i$ defined a directed family of idempotent deflations on X, whose least upper bound is id_X.

We now observe that, for every $x \in A_i$, $s_i(x)$ is the smallest element $\vec{y}$ of X such that $x \leqslant y_i$. In one direction, the ith component of $\vec{y} = s_i(x)$ is $y_i = r_{ii}(x)$, which is above x indeed. In the other direction, assume that $\vec{y} \in X$ is such that $x \leqslant y_i$. For every $j \in I$, pick a $k \in I$ above both i and j; then $x \leqslant y_i = r_{ik}(y_k)$, and since s_{ik} is left adjoint to r_{ik}, $s_{ik}(x) \leqslant y_k$; since r_{jk} is continuous hence monotonic (Proposition 4.3.9), $r_{jk}(s_{ik}(x)) \leqslant r_{jk}(y_k) = y_j$, i.e., $rs_{ij}(x) \leqslant y_j$. Since this holds for every $j \in I$, $s_i(x) \leqslant \vec{y}$.

For every $i \in I$, and $\vec{x} \in X$, we claim that $\uparrow_X f_i(\vec{x}) = \uparrow_X s_i(x_i)$ is open in X. We write $\uparrow_X$ instead of $\uparrow$ to make it clear in which space the upward closure is taken. Since $s_i(x_i)$ is the least $\vec{y} \in X$ such that $y_i \in \uparrow_{A_i} x_i$, $\uparrow_X f_i(\vec{x}) =$

$\uparrow_X s_i(x_i) = \pi_i^{-1}(\uparrow_{A_i} x_i)$, where we use π_i to denote the ith projection from X to A_i. $\uparrow_{A_i} x_i$ is open in A_i (with, e.g., its Scott topology), since A_i is finite. Since π_i is continuous, $\uparrow_X f_i(\vec{x})$ is open in X.

We claim that the subsets of the form $\uparrow_X f_i(\vec{x})$ are a base of the topology of X. Let U be an open neighborhood of some point $\vec{x}$ of X. Using Lemma 4.1.10, it is enough to show that $f_i(\vec{x}) \in U$ for some $i \in I$, since then $\uparrow_X f_i(\vec{x})$ will be an open neighborhood of $\vec{x}$ included in U. Since the topology of X is the subspace topology from $\prod_{j\in I} A_j$, U contains an open neighborhood of $\vec{x}$ of the form $\pi_{i_1}^{-1}(U_1) \cap \dots \cap \pi_{i_m}^{-1}(U_m)$, where $i_1, \dots, i_m \in I$, U_1 is open in A_1, ..., and U_m is open in A_m. Since each A_j is finite, we can replace each U_j by the even smaller open subset $\uparrow_{A_{i_j}} x_{i_j}$, and we still get an open neighborhood of $\vec{x}$. We have seen that $\pi_{i_j}^{-1}(\uparrow_{A_{i_j}} x_{i_j}) = \uparrow_X f_{i_j}(\vec{x})$, so $x \in \uparrow_X f_{i_1}(\vec{x}) \cap \dots \cap \uparrow_X f_{i_m}(\vec{x}) \subseteq U$. Let $i \in I$ be above $i_1, \dots, i_m$. Then f_i is above $f_{i_1}, \dots, f_{i_m}$, so $f_i(\vec{x}) \in \uparrow_X f_{i_1}(\vec{x}) \cap \dots \cap \uparrow_X f_{i_m}(\vec{x}) \subseteq U$.

In particular, X is a b-space (see Exercise 5.1.39). As a limit of finite, hence sober spaces, in **Top**, X is sober, by Theorem 8.4.13. By Exercise 8.3.38, X is therefore an algebraic dcpo with its Scott topology. Each element of the form $f_i(\vec{x})$ is such that $\uparrow f_i(\vec{x})$ is open, hence is finite (Exercise 5.1.11). Since $\vec{x}$ is the least upper bound of $(f_i(\vec{x}))_{i\in I}$, the elements in the ranges of f_i, $i \in I$, form a basis of finite elements of X. But the minimal basis contains all finite elements (Exercise 5.1.25), so the finite elements are exactly those of $\bigcup_{i\in I} f_i[X]$.

For future reference, we note that $\bigcup_{i\in I} f_i[X] = \bigcup_{i\in I} s_i[A_i]$ is also the inductive limit of the given ep-system (Exercise 9.6.6).

Conversely, let X be a dcpo with a generating system $(f_i)_{i\in I}$ of idempotent deflations. We claim that X is algebraic and that $\bigcup_{i\in I} f_i[X]$ is a basis of finite elements, hence exactly the set of all finite elements of X. Each element $f_i(x)$ is finite: if $f_i(x) \leqslant \sup_{j\in J} y_j$ for some directed family $(y_j)_{j\in J}$, then $f_i(x) = f_i(f_i(x)) \leqslant \sup_{j\in J} f_i(y_j)$, so $f_i(x) \leqslant f_i(y_i)$ for some $j \in J$ (since $\{f_i(y_j) | j \in J\} \subseteq f_i[X]$ is a finite set), whence $f_i(x) \leqslant y_j$. And each $x \in X$ is the sup of the directed family of finite elements $f_i(x)$, $i \in I$.

We build another generating system of idempotent deflations $(f_J)_{J\in\mathbb{P}_{\text{fin}}(I), J\neq\emptyset}$. We build f_J by induction on the cardinality of J. It will be an invariant that, for every $J \in \mathbb{P}_{\text{fin}}(I)$, there is an $i \in I$ such that $f_J = f_i$. We let $f_{\{i\}} = f_i$ for each $i \in I$. For every $J \in \mathbb{P}_{\text{fin}}(I)$ of cardinality at least 2, we let f_J be any f_i, $i \in I$, that is above every f_j, $j \in J$, and above every f_K with $K \subseteq J$, $K \neq J$, and $K \neq \emptyset$. This exists because $(f_i)_{i\in I}$ is directed, there are only finitely such maps f_i and f_K, and each f_K equals f_k for some $k \in I$.

For every $J \in \mathbb{P}_{\text{fin}}(I)$, $J \neq \emptyset$, let $A_J = \bigcup_{K \subseteq J, K \neq \emptyset} f_K[X]$. This is a finite poset. When $J \subseteq J'$, let $s_{JJ'}: A_J \to A_{J'}$ be the inclusion map, and let $r_{JJ'}$ be the restriction of f_J to $A_{J'}$.

For every $a \in A_J$, write a as $f_K(x)$ for some $K \subseteq J$, $K \neq \emptyset$, and some $x \in X$. Then $(r_{JJ'} \circ s_{JJ'})(a) = f_J(f_K(x))$. Since $f_J \leqslant \mathrm{id}_X$, this is below $f_K(x) = a$. Since f_J is above f_K, this is also above $f_K(f_K(x)) = f_K(x) = a$. So $r_{JJ'} \circ s_{JJ'} = \mathrm{id}_{A_J}$.

Since $f_J \leqslant \mathrm{id}_X$, $s_{JJ'} \circ r_{JJ'} \leqslant \mathrm{id}_{A_{J'}}$. So $s_{JJ'} \dashv r_{JJ'}$ is an ep-pair. Clearly, $s_{JJ} = \mathrm{id}_{A_J}$ and $s_{J'J''} \circ s_{JJ'} = s_{JJ''}$ whenever $J \subseteq J' \subseteq J''$. Since $r_{JJ'}$ is the unique right adjoint of $s_{JJ'}$, the equalities $r_{JJ} = \mathrm{id}_{A_J}$ and $r_{JJ'} \circ r_{J'J''} = r_{JJ''}$ are immediate; the argument is symmetric to that of Lemma 9.6.5. So

$$\left(A_J \underset{r_{JJ'}}{\overset{s_{JJ'}}{\rightleftarrows}} A_{J'} \right)_{\substack{J, J' \in \mathbb{P}_{\text{fin}}(I) \\ J \subseteq J'}} \quad \text{is an ep-system.}$$

Let X' be the projective limit of this ep-system. Using the first part of the proof, X' is an algebraic dcpo, and its finite elements are the inductive limit of the ep-system. Since each $s_{JJ'}$ is an inclusion, another inductive limit is $\bigcup_{J \in \mathbb{P}_{\text{fin}}(I)} A_J = \bigcup_{J \in \mathbb{P}_{\text{fin}}(I)} f_J[X] = \bigcup_{i \in I} f_i[X]$. These inductive limits must be homeomorphic, hence order isomorphic, and the second one is the basis of finite elements of X. So X and X' have order isomorphic bases of finite elements. From Proposition 5.1.33, any two continuous dcpos with order isomorphic bases are order isomorphic. So X_σ and X' are homeomorphic. □

A by-product of the above proof is a better understanding of the relation between inductive and projective limits of ep-systems in **Top**:

Proposition 9.6.10 *Let* $\left(A_i \underset{r_{ij}}{\overset{s_{ij}}{\rightleftarrows}} A_j \right)_{\substack{i,j \in I \\ i \sqsubseteq j}}$ *be an ep-system of finite posets, and X be its projective limit in* **Top**. *Then X is a bifinite domain, and its basis A of finite elements is also the inductive limit of the ep-system, up to order isomorphism.*

Exercise 9.6.11 (Limit–Colimit Coincidence) The colimit of an ep-system of topological spaces is in general smaller than its limit. One can however redo all the above in the category **Cpo** of dcpos and Scott-continuous maps. The projective limit of a projective system $(r_{ij}: A_j \to A_i)_{\substack{i,j \in I \\ i \sqsubseteq j}}$ of dcpos is the sub-poset X of the poset product $\prod_{i \in I} A_i$ (beware: this may be different from the topological product) of those tuples $\vec{x}$ such that $r_{ij}(x_j) = x_i$ for all $i, j \in I$ with $i \sqsubseteq j$.

Now let $\left(A_i \underset{r_{ij}}{\overset{s_{ij}}{\rightleftarrows}} A_j \right)_{\substack{i,j \in I \\ i \sqsubseteq j}}$ be an ep-system of dcpos. Defining rs_{ij}, s_i, and X_i as in Exercise 9.6.6, and proving the analogue of Lemma 9.6.7 in **Cpo**, show that X is also the inductive limit of $(s_{ij} : A_i \to A_j)_{\substack{i,j \in I \\ i \sqsubseteq j}}$ in **Cpo**.

Such an object, which is both a colimit of the inductive system and a limit of the projective system underlying the given ep-system, is called a *bilimit*.

Exercise 9.6.12 Let $\left(A_i \underset{r_{ij}}{\overset{s_{ij}}{\rightleftarrows}} A_j \right)_{\substack{i,j \in I \\ i \sqsubseteq j}}$ be an ep-system of finite posets. Show that the projective limits of the retractive systems $(r_{ij} : A_j \to A_i)_{\substack{i,j \in I \\ i \sqsubseteq j}}$ are the same in **Cpo** and in **Top**. More precisely, let X be the projective limit in **Cpo**, and X' be the projective limit in **Top**. Show that $X' = X_\sigma$.

So the bifinite domains are also the bilimits in **Cpo** of ep-systems of finite posets. (This is the origin of the term *bifinite*.)

Proposition 9.6.13 (Products of bifinite domains) *Let $(X_i)_{i \in I}$ be a family of bifinite domains, and assume that all but finitely many of them are pointed. Then the (topological, equivalently poset) product $\prod_{i \in I} X_i$ is bifinite.*

Proof Let I_0 be the finite subset of those $i \in I$ such that X_i is not pointed. Otherwise, write $\perp$ the least element of those that are pointed. By Theorem 9.6.9, each X_i is an algebraic, hence continuous (Corollary 5.1.24) dcpo. Their topological product then coincides with their poset product, by Proposition 5.1.56. Also, each X_i has a generating system of idempotent deflations $(f_{ij})_{j \in J_i}$. Without loss of generality, we can assume that the constant map equal to $\perp$ is among this system, whenever $i \in I \setminus I_0$; write ϵ for some $j \in J_i$ such that f_{ij} is this constant map. Let J be the set of all tuples $\vec{j} \in \prod_{i \in I} J_i$ such that $j_i = \epsilon$ for every $i \in I \setminus I_0$ except finitely many. Then define $g_{\vec{j}} : \prod_{i \in J} X_i \to \prod_{i \in J} X_i$ by $g_{\vec{j}}(\vec{x}) = (f_{ij_i}(x_i))_{i \in I}$. One easily checks that $g_{\vec{j}}$ is an idempotent deflation, and forms a directed family whose least upper bound is the identity on $\prod_{i \in J} X_i$. We then conclude by Theorem 9.6.9. □

Theorem 9.6.14 (**B** is CCC) *The category* **B** *of bifinite domains and continuous maps is Cartesian-closed. The exponential of X and Y is $[X \to Y]$ with the Scott topology of the pointwise ordering.*

Proof Using Theorem 9.6.9, let $(f_i)_{i \in I}$ and $(g_j)_{j \in J}$ be generating systems of idempotent deflations on X, resp. Y. Also, X and Y are dcpos, so continuity coincides with Scott-continuity, by Proposition 4.3.5. Let $fg_{ij} : [X \to Y] \to [X \to Y]$ map $h \in [X \to Y]$ to $g_j \circ h \circ f_i$. It is easy to see that fg_{ij} is

Scott-continuous, and idempotent. Given any $h, h' \in [X \to Y]$, if $h(x) = h'(x)$ for every $x \in f_i[X]$, then $fg_{ij}(h) = fg_{ij}(h')$. So, if m is the number of elements in B_j and n the number of elements in A_i, then there are at most m^n elements $fg_{ij}(h)$, $h \in [X \to Y]$. It follows that fg_{ij} is an idempotent deflation. Also, $(fg_{ij})_{(i,j)\in I\times J}$ is directed, since $(f_i)_{i\in I}$ and $(g_j)_{j\in J}$ are both directed. And $(\sup_{(i,j)\in I\times J} fg_{ij}(h))(x) = \sup_{i\in I} \sup_{j\in J} g_j(h(f_i(x))) = \sup_{i\in I} h(f_i(x)) = h(\sup_{i\in I} f_i(x))$ (since h is Scott-continuous) $= h(x)$. So $(fg_{ij})_{(i,j)\in I\times J}$ is a generating system of idempotent deflations on $[X \to Y]$. So $[X \to Y]$ is a bifinite domain.

The topological products of finitely many bifinite domains coincide with their poset products, and are the products in **B**, using Proposition 9.6.13. Next, we show that $[X \to Y]$ is an exponential object in **B** as in the case of **Cpo** (Proposition 5.5.4). □

Exercise 9.6.15 (Scott domain ⇒ pointed bifinite) Let X be a Scott domain. For each finite subset B of finite elements of X, define $f_B(x) = \sup(\downarrow x \cap B)$. Show that f_B is an idempotent deflation. Use this to show that every Scott domain is a pointed bifinite domain.

Also, give an example of a pointed bifinite domain that is not a Scott domain.

Exercise 9.6.16 Given any subset B of a poset X, let $mub(B)$ be the set of minimal upper bounds of B in X. That is, if $B = \{a_1, \dots, a_n\}$, then $mub(B)$ is the set of minimal elements of $ub(B) = \uparrow a_1 \cap \dots \cap \uparrow a_n$. The set $mub(B)$ may be empty, finite, or infinite in general. Show that if X is an algebraic dcpo such that X_σ is a spectral space (e.g., a bifinite domain), then, for every finite set B of finite elements of X, $mub(B)$ is again a finite set of finite elements of X, and $ub(B) = \uparrow mub(B)$. (We say that X has *Jung's Property m*.)

Exercise 9.6.17 (Pointed bifinite = property m + finite mub-closure of finite elements) We can improve our understanding of the relation between Scott domains and bifinite domains (Exercise 9.6.15), and at the same time give an internal description of bifinite domains. This is easier in the case of *pointed* bifinite domains. The general case was dealt with by Jung (1988).

The *mub-closure* of a subset B of X is the smallest subset $mub^\infty(B)$ of X containing B and such that, for every finite subset C of B, $mub(C) \subseteq mub^\infty(B)$. Show that if X is a bifinite domain, then, for every finite set B of finite elements, $mub^\infty(B)$ is again a finite set of finite elements. Probably the easiest way to show this is to consider an ep-system $\left(A_i \underset{r_{ij}}{\overset{s_{ij}}{\rightleftarrows}} A_j \right)_{\substack{i,j\in I\\ i\sqsubseteq j}}$ whose projective limit is X (up to homeomorphism), and to show that its inductive limit A (in **Top**) is such that $mub^\infty(B)$ is a finite subset of A for every finite subset B of A.

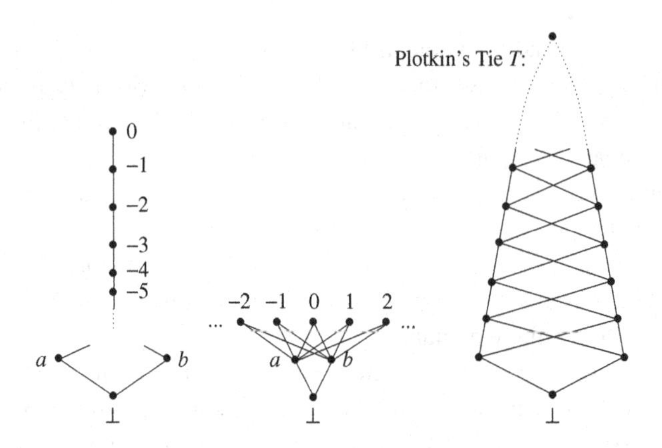

Figure 9.4 Some pointed algebraic posets that are not bifinite.

Conversely, assume that X is a pointed algebraic dcpo satisfying property m and such that, for every finite set B of finite elements of X, $mub^{\infty}(B)$ is again a finite set of finite elements of X. Imitating Exercise 9.6.15, for every finite subset B of finite elements of X containing the bottom element $\perp$, let $f_B(x) = \sup(\downarrow x \cap mub^{\infty}(B))$, and conclude that X is a bifinite domain.

So the pointed bifinite domains are exactly the pointed algebraic dcpos satisfying property m and such that, for every finite set B of finite elements of X, $mub^{\infty}(B)$ is again a finite set of finite elements of X.

Example 9.6.18 Figure 9.4 shows the three standard examples of pointed algebraic dcpos that are not bifinite. All elements of each are finite, except the top element of the rightmost one. The leftmost one is such that $mub\{a, b\}$ is empty, and the middle one has an infinite $mub\{a, b\}$, so these two are not spectral in their Scott topology, by Exercise 9.6.16. The rightmost one, *Plotkin's tie*, is the dcpo T obtained as the set of pairs $(i, m) \in \{0, 1\} \times \mathbb{N}$ with $(i, m) \leqslant (j, n)$ iff $i < j$ or $(i, m) = (j, n)$, with a bottom element $\perp$ and a top element added. This one is spectral, but not bifinite by Exercise 9.6.17, as $mub^{\infty}\{(0, 0), (1, 0)\}$ is infinite – this is the whole set except $\perp$ and the top element.

Definition 9.6.19 (RB-domain) An **RB**-*domain* is any retract of a bifinite domain.

Exercise 9.6.20 (bc-domain ⇒ **RB**-domain) Show that every bc-domain is an **RB**-domain.

By Johnstone's Theorem 9.5.14:

Fact 9.6.21 *Every* **RB**-*domain is stably compact.*

Using Proposition 9.6.13, finite products are computed in **B** as in **Top**, and by Theorem 9.6.14, **B** is Cartesian-closed. So Exercise 5.5.5 applies: the category **RB** of **RB**-domains is Cartesian-closed. We shall describe the exponential objects precisely in Theorem 9.6.24.

Theorem 9.6.22 *The* **RB**-*domains are exactly the dcpos X with a* generating system of deflations, *i.e., a directed family of deflations $(g_i)_{i\in I}$ whose least upper bound is* id_X. *The topology of X is the Scott topology; X is a continuous dcpo, $g_i(x) \ll x$ for all $i \in I$ and $x \in X$, and $\bigcup_{i\in I} g_i[X]$ is a basis.*

So the difference between this and bifinite domains is that here we do not require the deflations to be idempotent.

Proof If X is an **RB**-domain, say the retract of a bifinite domain Y, then let $(f_i)_{i\in I}$ be a generating system of idempotent deflations on Y, using Theorem 9.6.9. Also let $r: Y \to X$ be the retraction, and $s: X \to Y$ be the section. Define g_i as $r \circ f_i \circ s$. It is easy to see that $(g_i)_{i\in I}$ is a generating system of deflations on X. Y is also a algebraic dcpo in its Scott topology (in particular, continuous, by Corollary 5.1.24), so X is a continuous dcpo in its Scott topology by Corollary 8.3.37. Also, remember that the elements $f_i(y)$, $i \in I$, $y \in Y$, form the basis of finite elements of Y. In particular, $f_i(s(x))$ is way-below $s(x)$ in Y. So, by Exercise 5.1.44, $g_i(x) = r(f_i(s(x)))$ is way-below $r(s(x)) = x$ in X. Since $(g_i(x))_{i\in I}$ is a directed family of elements way-below x whose least upper bound is x, we conclude that the elements $g_i(x)$, $i \in I$, $x \in X$, form a basis of X, using Trick 5.1.28.

Conversely, assume that X is a dcpo with a generating system of deflations $(g_i)_{i\in I}$. We proceed somewhat similarly to the proof of Theorem 9.6.9 and build another generating system of deflations $(g_J)_{J\in\mathbb{P}_{\mathrm{fin}}(I)}$, by letting g_J be any $g_i, i \in I$, above every g_j, $j \in J$, and every g_K with $K \subseteq J$, $K \neq J$. Whenever $J \subseteq J'$, $g_J \leqslant g_{J'}$, and, for every $J \in \mathbb{P}_{\mathrm{fin}}(I)$, there is an $i \in I$ such that $g_J = g_i$. Let Y be the set of tuples $\vec{y} = (y_J)_{J\in\mathbb{P}_{\mathrm{fin}}(I)}$ where, for each J, y_J is in the finite poset $A_J = \bigcup_{K\subseteq J} g_K[X]$, and such that $y_J \leqslant y_{J'}$ whenever $J \subseteq J'$. In particular, $\vec{y}$ defines a directed family in X, so one can define $r: Y \to X$ by $r(\vec{y}) = \sup_{J\in\mathbb{P}_{\mathrm{fin}}(I)} y_J$. We also define $s: X \to Y$ by $s(x) = (g_J(x))_{J\in\mathbb{P}_{\mathrm{fin}}(I)}$. Since each g_J equals g_i for some $i \in I$, and every g_i is below some g_J (namely, $J = \{i\}$), $\sup_{J\in\mathbb{P}_{\mathrm{fin}}(I)} g_J(x) = x$ for every $x \in X$. So $r \circ s = \mathrm{id}_X$.

We equip Y with the Scott topology of the pointwise ordering. Then Y is a dcpo. Moreover, the map $f_J: Y \to Y$ defined by $f_J(\vec{y}) = (y_{J\cap J'})_{J'\in\mathbb{P}_{\mathrm{fin}}(I)}$

is an idempotent deflation. So Y_σ is a bifinite domain, by Theorem 9.6.9. It is easy to see that r and s are Scott-continuous. So X is a Scott-retract of Y. ☐

Since products of retracts can be rewritten as retracts of the relevant products, Proposition 9.6.13 entails:

Proposition 9.6.23 (Products of **RB**-domains) *Let $(X_i)_{i \in I}$ be a family of* **RB***-domains, and assume that all but finitely many of them are pointed. Then the (topological, equivalently poset) product $\prod_{i \in I} X_i$ is an* **RB***-domain.*

Theorem 9.6.24 (**RB** is CCC) *The category* **RB** *of* **RB***-domains and continuous maps is Cartesian-closed. The exponential of X and Y is $[X \to Y]$ with the Scott topology of the pointwise ordering.*

Proof The proof is as for Theorem 9.6.14, using deflations instead of idempotent deflations, thanks to Theorem 9.6.22. ☐

The following exercises are concerned with the study of an even larger Cartesian-closed category of stably compact, continuous dcpos, the **FS**-domains (Jung, 1990).

Exercise 9.6.25 Let X be a dcpo, and f and g be two Scott-continuous maps from X to X. Given a subset M of X, we say that M *separates* f from g, in notation $f \sqsubseteq_M g$, iff, for every $x \in X$, there is an $m \in M$ such that $f(x) \leqslant m \leqslant g(x)$. We say that f is *finitely separated* from g, in notation $f \sqsubseteq g$, iff $f \sqsubseteq_M g$ for some finite subset M of X. Observe that any deflation f is finitely separated from id_X. But there are continuous maps that are finitely separated from id_X which are not deflations: show that the map $x \mapsto \max(x - \frac{1}{n}, 0)$ from $[0, 1]$ to $[0, 1]$, for any $n \in \mathbb{N}$, $n \geqslant 1$, is an example.

Show that if $f \sqsubseteq_M \mathrm{id}_X$, where M is finite, then, for every $x \in X$, $f(x) \ll x$. Deduce that there is an $m \in M$ such that $f^2(x) \leqslant m \ll x$, where f^2 denotes $f \circ f$.

Exercise 9.6.26 (**FS**-domain ⇒ stably compact) An **FS**-*domain* is a dcpo X such that the identity map id_X is the least upper bound of a directed family $(f_i)_{i \in I}$ of continuous maps $f_i \sqsubseteq \mathrm{id}_X$. So every **RB**-domain is an **FS**-domain. (The converse is not known.)

Show that $(f_i^2)_{i \in I}$ is another directed family of continuous maps such that $f_i^2 \sqsubseteq \mathrm{id}_X$ and whose least upper bound is id_X. Using this, show that every **FS**-domain is a continuous dcpo that is stably compact in its Scott topology (use Exercise 8.3.33 to show coherence).

Exercise 9.6.27 (Retracts of **FS**-domains) Show that the Scott-retracts of **FS**-domains are **FS**-domains.

Exercise 9.6.28 (Products of **FS**-domains) Let $(X_i)_{i \in I}$ be a family of **FS**-domains, and assume that all but finitely many are pointed. Show that the (topological, poset) product $\prod_{i \in I} X_i$ is an **FS**-domain.

Exercise 9.6.29 (**FS** is CCC) Let X and Y be dcpos, f be a Scott-continuous map from X to X such that $f \sqsubseteq_M \mathrm{id}_X$, and g be a Scott-continuous map from Y to Y such that $g \sqsubseteq_N \mathrm{id}_Y$. Let F_{fg} be the map from $[X \to Y]$ to $[X \to Y]$ defined by $F_{fg}(h) = g^2 \circ h \circ f^2$. Define an equivalence relation $\sim$ on $[X \to Y]$ by $h_1 \sim h_2$ iff $\uparrow g(h_1(m)) \cap N = \uparrow g(h_2(m)) \cap N$ for every $m \in M$. Pick an element $\overline{h}$ from each $\sim$-equivalence class, so that $h \sim \overline{h}$, and let P be the set of elements $g \circ \overline{h} \circ f$ thus obtained. Show that $F_{fg} \sqsubseteq_P \mathrm{id}_{[X \to Y]}$, and that P is finite whenever M and N are. (As a hint, first show that $g \circ h \circ f \leqslant \overline{h}$ and $g \circ \overline{h} \circ f \leqslant h$.)

Deduce that if X and Y are **FS**-domains, then the dcpo $[X \to Y]$ of Scott-continuous maps from X to Y is again an **FS**-domain.

Conclude that the category **FS** of **FS**-domains and Scott-continuous maps is Cartesian-closed. This was established by Jung (1990, theorem 3) for pointed **FS**-domains, but pointedness is not required.

Exercise 9.6.30 (Algebraic **FS** = algebraic **RB** = bifinite) Show that every algebraic **FS**-domain is bifinite. To this end, assume that $(f_i)_{i \in I}$ is a directed family of continuous maps such that $f_i \sqsubseteq_{M_i} \mathrm{id}_X$, M_i finite, and let A_i be the set of fixed points of f_i. Show that $A_i \subseteq M_i$ is finite. Using the fact that X is algebraic, and conversely that $f_i(x) \ll x$, show that the finite elements of X are exactly the elements of $\bigcup_{i \in I} A_i$. Next, define $g_i : X \to X$ as $f_i^{2n_i - 2}$, where n_i is the cardinality of M_i, and show that $(g_i)_{i \in I}$ is a generating system of idempotent deflations.

Conclude that the notions of algebraic **FS** domains, algebraic **RB**-domains, and bifinite domains all coincide.

It follows from Exercise 9.6.30 that none of the pointed algebraic dcpos of Figure 9.4 are **RB**-domains, or even **FS**-domains.

Exercise 9.6.31 (Uniform approximation) Let X be a topological space. We say that a continuous map $f : X \to X$ is *uniformly approximating* iff, for every open subset U of X, there is a compact saturated subset Q such that $f^{-1}(U) \subseteq Q \subseteq U$, and, for every compact saturated subset Q of X, there is an open subset U such that $Q \subseteq U \subseteq \uparrow f[Q]$. In other words, $f^{-1}(U)$ provides an approximant (an open subset way-below) U in $\mathcal{O}(X)$, and $\uparrow f[Q]$ provides a compact saturated subset way-below Q in $\mathcal{Q}(X)$ (see Proposition 8.3.25), under mild conditions.

Show that if X is a dcpo in its Scott topology, and $f \sqsubseteq \mathrm{id}_X$, then f is uniformly approximating. This yields many examples of uniformly approximating functions. Conversely, show that if X is a stably compact space, and f is uniformly approximating, then $f^2 \sqsubseteq \mathrm{id}_X$. (This is tricky. For each $x \in X$, let $U_x = int(\uparrow f(x))$, $V_x = f^{-1}(X \smallsetminus \downarrow f(x))$, and let Q_x be a compact saturated subset such that

$f^{-1}(V_x) \subseteq Q_x \subseteq V_x$. Show that $U_x \setminus Q_x$ is patch-open, and contains x, and use the fact that X is patch-compact to obtain a finite subset E. Then check that $f^2 \sqsubseteq_M \mathrm{id}_X$ where $M = f[E]$.)

If X is a topological space, with specialization ordering $\leqslant$, write $(\leqslant) = \{(x, y) \in X \times X \mid x \leqslant y\}$. If f is uniformly approximating on X, write $(\uparrow f) = \{(x, y) \in X \times X \mid f(x) \leqslant y\}$. Call a topological space X *uniformly approximated* if and only if it is sober and there is a family of uniformly approximating maps $f_i : X \to X$ such that $((\uparrow f_i))_{i \in I}$ is a filtered family and $\bigcap_{i \in I}(\uparrow f_i) = (\leqslant)$.

Show that every **FS**-domain is uniformly approximated in its Scott topology.

Conversely, show that every uniformly approximated space X is stably compact and has the Scott topology of its specialization ordering. (The hardest part is showing coherence. You should prove it last. Given a family $(f_i)_{i \in I}$ of uniformly approximating maps as above, first show that $(f_i)_{i \in I}$ is directed, and that $\sup_{i \in I} f_i(x) = x$ for every $x \in X$. Once you have shown that the topology of X is the Scott topology, show that every open subset U is the union of the directed family $(f_i^{-1}(U))_{i \in I}$. Using this, show that the filter of all intersections $U_1 \cap U_2$ of open neighborhoods U_1, U_2 of two given compact saturated subsets Q_1, Q_2 is Scott-open.) Conclude that X is an **FS**-domain in its Scott topology.

So uniformly approximated spaces are a topological reformulation of the notion of **FS**-domains. This is due to Jung and Sünderhauf (1998).

Exercise 9.6.32 Let X be a pointed continuous dcpo such that $[X \to X]$ is a coherent continuous dcpo. Since $[X \to X]$ is also pointed, $[X \to X]_\sigma$ is stably compact by Fact 9.1.5. Recall that the elementary step function $U \searrow y$ maps every element of the open subset U to y, and all other elements to $\bot$ (Definition 5.7.9).

Show that: (1) if $g \ll \mathrm{id}_X$, then there are finitely many elements $d'_1 \ll d_1$, $d'_2 \ll d_2, \ldots, d'_n \ll d_n$ in X such that $\bigcap_{i=1}^n \uparrow(\twoheaduparrow d_i \searrow d'_i) \subseteq \twoheaduparrow g$, in particular $\bigcap_{i=1}^n \uparrow(\twoheaduparrow d_i \searrow d'_i) \subseteq \uparrow g$; (2) one can find elements e_i, e'_i, $1 \leqslant i \leqslant n$, in X such that $d'_i \ll e'_i \ll e_i \ll d_i$, and then $(\twoheaduparrow d_i \searrow d'_i) \ll (\twoheaduparrow e_i \searrow e'_i)$; (3) for every $a \in X$, let r_a be the function that maps every $x \leqslant a$ to x and every $x \not\leqslant a$ to a; show that r_a is continuous and that, letting $I_a = \{i \in \{1, \ldots, n\} \mid e_i \ll a\}$, r_a is in $\bigcap_{i \in I_a} \uparrow(\twoheaduparrow e_i \searrow e'_i)$; (4) for every subset I of $\{1, \ldots, n\}$, there is a finite set N_I of $[X \to X]$ such that $\bigcap_{i \in I} \uparrow(\twoheaduparrow e_i \searrow e'_i) \subseteq \uparrow N_I \subseteq \bigcap_{i \in I} \twoheaduparrow(\twoheaduparrow d_i \searrow d'_i)$ (use Theorem 5.2.30); (5) for every $a \in X$, there is a map $h_a \in N_{I_a}$ such that $h_a \leqslant r_a$; (6) for every $h \in [X \to X]$ such that $h \leqslant r_a$ and $h \in \bigcap_{i \in \{1,\ldots,n\}, d_i \ll a} \uparrow(\twoheaduparrow d_i \searrow d'_i)$, for every $x \leqslant a$, $g(x) \leqslant h(x)$ (use the auxiliary function h' defined by $h'(x) = h(x)$ if $x \leqslant a$, $h'(x) = x$ otherwise, and use (1) to show that $g \leqslant h'$); (7) show that the set H of all maps $\{h_a \mid a \in X\}$ is finite and, for each $h \in H$, pick some element $a_h \in X$ such that $h = h_{a_h}$: if $h = h_a$, show that $h(a) \leqslant a_h$ and $h(a_h) \leqslant a$ (use

$h_a \leqslant r_a, r_{a_h}$); (8) given any $f \ll \mathrm{id}_X$, find $g \in [X \to X]$ such that $f \leqslant g^2$ and $g \ll \mathrm{id}_X$, and apply the previous results (in particular, (6) both to r_a and to r_{a_h}) to show that, for every $a \in X$, $f(a) \leqslant h(a_h) \leqslant a$, where $h = h_a$. From this, deduce that if $f \ll \mathrm{id}_X$, then $f \sqsubseteq \mathrm{id}_X$.

Conclude that the pointed **FS**-domains are exactly the pointed continuous dcpos X such that $[X \to X]$ is a coherent continuous dcpo (Jung, 1990, corollary 9). So the category of pointed **FS**-domains is the largest full Cartesian-closed subcategory of **Cpo** whose objects consist of pointed coherent continuous dcpos.

Exercise 9.6.32 is the main step in proving Jung's Theorem (Jung, 1990, corollary 10): every Cartesian-closed category of pointed continuous dcpos and continuous maps is either a category of **FS**-domains or a category of so-called **L**-domains. (An **L**-domain X is a pointed continuous dcpo such that $\downarrow x$ is a complete lattice for every $x \in X$. Beware that sups do not need to be computed in $\downarrow x$ as in X.) Also, the categories of pointed **FS**-domains and of **L**-domains are Cartesian-closed. So there are exactly two maximal Cartesian-closed categories of pointed continuous dcpos.

9.7 Noetherian spaces

We finish with the following class of spaces, which plays an important role in algebraic geometry. More recently, they have started playing a role in computer science, and particularly in the verification of certain infinite-state systems (Goubault-Larrecq, 2010).

Definition 9.7.1 (Noetherian space) A *Noetherian space* is a topological space where every open subset is compact.

Example 9.7.2 Every finite topological space is Noetherian.

Example 9.7.3 $\mathbb{N}_\sigma$ is Noetherian. Indeed, every element is finite, so $\uparrow n$ is open for every $n \in \mathbb{N}$. There cannot be any other open, except the empty set. So they are all (finitary) compact.

Example 9.7.4 $[0, 1]_\sigma$ is not Noetherian, e.g., the open subsets $(t, 1]$ with $t > \frac{1}{2}$ form a directed open cover of $(\frac{1}{2}, 1]$, but no $(t, 1]$ with $t > \frac{1}{2}$ contains $(\frac{1}{2}, 1]$.

Example 9.7.5 The spectral space $\mathbb{I}$ of Example 9.5.9 is not Noetherian either: the union of the open subsets $\uparrow t^+$, $t > \frac{1}{2}$, is open, but not compact, as no subset $\uparrow t^+$, $t > \frac{1}{2}$, contains it all.

Proposition 9.7.6 *A set Y with a quasi-ordering $\sqsubseteq$ has the* ascending chain condition (ACC) *iff every infinite ascending chain $y_0 \sqsubseteq y_1 \sqsubseteq \cdots \sqsubseteq y_k \sqsubseteq \cdots$ stabilizes, i.e., there is an $N \in \mathbb{N}$ such that $y_k \sqsubseteq y_N$ for every $k \geqslant N$.*

Let X be a topological space. Then X is Noetherian iff $\mathcal{O}(X)$ has the ascending chain condition.

Proof Assume X is Noetherian, and let $U_0 \subseteq U_1 \subseteq \cdots \subseteq U_k \subseteq \cdots$ be an infinite ascending chain in $\mathcal{O}(X)$. $U = \bigcup_{n \in \mathbb{N}} U_n$ is open, hence compact. Notice that the family $(U_n)_{n \in \mathbb{N}}$ is directed, so $U \subseteq U_N$ for some $N \in \mathbb{N}$. For every $k \geqslant N$, then, $U_k \subseteq \bigcup_{n \in \mathbb{N}} U_n \subseteq U_N \subseteq U_k$, so $U_k = U_N$.

Conversely, assume X is not Noetherian, and construct an infinite non-stabilizing ascending chain. Let U be a non-compact open subset of X. There is an open cover $(V_i)_{i \in I}$ of U that has no finite subcover. By induction on $k \in \mathbb{N}$, build an infinite sequence of opens W_k, as follows. At each step W_k will be a finite union of opens of the form V_i, $i \in I$. Moreover, the sequence $W_1, W_2, \ldots, W_k, \ldots$ will be strictly ascending. Assume $W_1, W_2, \ldots, W_{k-1}$ have been built. By assumption, W_{k-1} does not contain U; otherwise W_{k-1} would induce a finite subcover of U. So there is an element x_k of U outside W_{k-1}. This x_k must belong to some V_{i_k}, $i_k \in I$. Let $W_k = W_{k-1} \cup V_{i_k}$. Since $x_k \in W_k$ but $x_k \notin W_{k-1}$, W_{k-1} is strictly contained in W_k. $\square$

Proposition 9.7.7 (Noetherian = hereditarily compact) *Every subspace of a Noetherian space is Noetherian.*

The Noetherian spaces are exactly the hereditarily compact *spaces, i.e., the spaces X whose subspaces are all compact.*

Proof Let X be a Noetherian space, and $Y \subseteq X$. Let $U_0 \subseteq U_1 \subseteq \cdots \subseteq U_k \subseteq \cdots$ be an infinite ascending chain of open subsets of Y. Write $U_k = V_k \cap Y$ for some open subset V_k of X. Let $W_k = V_0 \cup V_1 \cup \cdots \cup V_k$. Since $U_k = U_0 \cup U_1 \cup \cdots \cup U_k$, $U_k = W_k \cap Y$. $W_0 \subseteq W_1 \subseteq \cdots \subseteq W_k \subseteq \cdots$ is an infinite ascending chain of open subsets of X, so, by Proposition 9.7.6, there is an $N \in \mathbb{N}$ such that $W_k = W_N$ for every $k \geqslant N$. Then $U_k = W_k \cap Y = W_N \cap Y = U_N$. So $U_0 \subseteq U_1 \subseteq \cdots \subseteq U_k \subseteq \cdots$ stabilizes. We conclude by Proposition 9.7.6.

If X is Noetherian, then we have just seen that every subspace of X is Noetherian, in particular compact. Conversely, if every subspace of X is compact, then every subset of X is (Exercise 4.9.11), in particular every open subset. So X is Noetherian, by definition. $\square$

Noetherian spaces are also characterized in terms of convergence. Recall (Lemma 8.2.27) that a self-convergent net is one that converges to each of its elements.

Theorem 9.7.8 (Noetherian = every net has a self-convergent subnet) *Let X be a topological space. The following are equivalent: (i) X is Noetherian; (ii) every sequence $(x_n)_{n \in \mathbb{N}}$ contains a cluster point, i.e., some x_n, $n \in \mathbb{N}$, is a cluster point of the sequence; (iii) every net $(x_i)_{i \in I, \sqsubseteq}$ contains a cluster point, i.e., some x_i, $i \in I$, is a cluster point of the sequence; (iv) every net in X has a self-convergent subnet.*

Proof (i) $\Rightarrow$ (iii). Assume X is Noetherian, and let $(x_i)_{i \in I, \sqsubseteq}$ be a net in X. Consider the subspace $Y = \{x_i \mid i \in I\}$ of X. By Proposition 9.7.7, Y is compact. By Proposition 4.7.22, the net $(x_i)_{i \in I, \sqsubseteq}$ has a cluster point in Y, say x_i. For every open neighborhood U of x_i in X, $U \cap Y$ is an open neighborhood of x_i in Y. So for every $j \in I$, there is a $j' \in I$ above j such that $x_{j'}$ is in $U \cap Y$, hence in U. So x_i is also a cluster point of the net in X.

(iii) $\Rightarrow$ (iv). Let $(x_i)_{i \in I, \sqsubseteq}$ be a net in X. Let I' be the subset of all indices $i \in I$ such that x_i is a cluster point of the net. By (iii), I' is non-empty. Given any two elements $i_1, i_2 \in I'$, consider the net $(x_i)_{i \in I_{i_1 i_2}, \sqsubseteq}$, where $I_{i_1 i_2}$ is the subset of all those $i \in I$ such that $i_1, i_2 \sqsubseteq i$. By (iii) again, there is an $i \in I_{i_1 i_2}$ such that x_i is a cluster point of $(x_i)_{i \in I_{i_1 i_2}, \sqsubseteq}$, hence of the original net $(x_i)_{i \in I, \sqsubseteq}$. So $i_1, i_2 \sqsubseteq i$ and $i \in I'$. That is, I' is directed. The same argument, using $i_2 = i_1$, shows that I' is also cofinal in I, i.e., that, for every $i \in I$, there is an $i' \in I'$ such that $i \sqsubseteq i'$.

Now assume $(x_i)_{i \in I, \sqsubseteq}$ is an ultranet. The cluster points of an ultranet are exactly its limits. So $(x_i)_{i \in I, \sqsubseteq}$ converges to each x_i, $i \in I'$. It is easy to see that each limit of $(x_i)_{i \in I, \sqsubseteq}$ is also a limit of $(x_i)_{i \in I', \sqsubseteq}$, because I' is cofinal. So $(x_i)_{i \in I', \sqsubseteq}$ is self-convergent.

In the general case where $(x_i)_{i \in I, \sqsubseteq}$ is an arbitrary net in X, we can form a subnet that is an ultranet by Kelley's Theorem 4.7.35, and further extract a self-convergent subnet of the latter by the above construction. So $(x_i)_{i \in I, \sqsubseteq}$ has a self-convergent subnet.

(iv) $\Rightarrow$ (ii). Assume that every net has a self-convergent subnet. Then, in particular, any element of the subnet is a cluster point of the net, and this certainly applies to nets that are sequences.

(ii) $\Rightarrow$ (i). We show the contrapositive, namely that if X is not Noetherian, then it contains a sequence $(x_n)_{n \in \mathbb{N}}$ of which no x_m is a cluster point. By Proposition 9.7.6, there must be a strictly ascending chain $U_0 \subset U_1 \subset \cdots \subset U_n \subset \cdots$ of open subsets of X ($\subset$ is strict inclusion). For each $n \geqslant 1$, pick an element x_n in U_{n+1} that is not in U_n. For every $m \in \mathbb{N}$, we claim that x_m is not a cluster point of $(x_n)_{n \in \mathbb{N}}$. Indeed, otherwise every open neighborhood U of x_m would contain infinitely many x_ns. In particular, this would hold of

$U = U_{m+1}$. However the only x_ns that are in U_{m+1} are those such that $n \leqslant m$, and there are only finitely many of them. □

The really interesting Noetherian spaces are those that are sober, and one can easily complete any Noetherian space to one that is sober as well:

Lemma 9.7.9 *A space X is Noetherian iff its sobrification $\mathcal{S}(X)$ is Noetherian.*

Proof For any space X, being Noetherian is a property of the lattice $\mathcal{O}(X)$, independently of the actual set X of points, by Proposition 9.7.6. Then apply Lemma 8.2.26. □

Since $\mathcal{KO}(X) = \mathcal{O}(X)$ in a Noetherian space, the following is trivial:

Fact 9.7.10 (Noetherian $\Rightarrow$ spectral) *Every sober Noetherian space is spectral, in particular stably compact.*

Sober Noetherian spaces have a simple characterization through Stone duality:

Theorem 9.7.11 (Sober Noetherian spaces $\equiv$ distributive complete lattices with the ACC.) *The adjunction $\mathcal{O} \dashv$ pt restricts to an equivalence between the category of sober Noetherian spaces and continuous maps on the one hand, and the opposite of the category of distributive complete lattices with the ascending chain condition and frame homomorphisms on the other hand.*

Proof Use the Hofmann–Lawson Theorem 8.3.21. We must check that if X is sober Noetherian, then $\mathcal{O}(X)$ is a distributive complete lattice with the ascending chain condition: this follows from Proposition 9.7.6.

Conversely, let L be a distributive complete lattice with the ascending chain condition. Every element u of L is finite. Indeed, otherwise there would be a directed family $(u_i)_{i \in I}$ of elements below u, where no u_i equals u, and whose least upper bound is u. Using a similar argument as in Proposition 9.7.6, we build an ascending chain $(w_n)_{n \in \mathbb{N}}$ of elements of the family that does not stabilize: once $w_0 < w_1 < \cdots < w_{n-1}$ have been built, we build w_n as some u_i strictly above w_{n-1}: this must exist; otherwise the least upper bound of $(u_i)_{i \in I}$ would be w_{n-1}, which is impossible.

So L is in particular algebraic, hence continuous. Also, a lattice where every element is finite has a multiplicative way-below relation, since $u \ll v$ is equivalent to $u \leqslant v$. (In the non-trivial direction, when $u \leqslant v$, then $u \ll u \leqslant v$, so $u \ll v$ by Proposition 5.1.4.) So L is order isomorphic to $\mathcal{O}(X)$ for some stably locally compact space X, by Theorem 8.3.34. Moreover, $\mathcal{O}(X)$ has the ascending chain condition, so X is Noetherian by Proposition 9.7.6. □

And sober Noetherian spaces are entirely characterized as specific posets, as we show below, with a unique topology. Recall that $\leqslant$ is well-founded on X if and only if there is no infinite strictly descending chain $x_1 > x_2 > \cdots > x_n > \cdots$ in X.

Theorem 9.7.12 (Fundamental theorem of sober Noetherian spaces) *We say that a poset X has* property T *if and only if $X = \downarrow E$ for some finite set E. It has* property W *if and only if, for all $x, y \in X$, the set $\uparrow x \cap \uparrow y$ of common upper bounds of x and y is of the form $\downarrow E$ for some finite set E.*

The sober Noetherian spaces are exactly the posets X whose ordering $\leqslant$ is well-founded and that satisfy properties T and W, in the upper topology. The closed subsets of X are exactly the finitary closed subsets $\downarrow E$. The topology of X^{d} is the Alexandroff topology of $\geqslant$.

Proof Assume X is sober and Noetherian. For every $x \in X$, $\downarrow_X x$ is closed by Lemma 4.2.7, so its complement is open, and compact since X is Noetherian. (We write $\downarrow_X x$ instead of $\downarrow x$ to make clear in which space the downward closure is taken.) X is spectral (Fact 9.7.10), so Proposition 9.5.17 applies: $\downarrow_X x$ is compact-open in X^{d}. As $\downarrow_X x = \uparrow_{X^{\mathsf{d}}} x$, the upward closure of any point in X^{d} is open in X^{d}. It follows that every upward closed A subset of X^{d}, which is equal to $\bigcup_{x \in X} \uparrow_{X^{\mathsf{d}}} x$, is open in X^{d}. So the topology of X^{d} is the Alexandroff topology of the specialization ordering $\geqslant$ of X^{d} (Proposition 4.2.11).

Also, every closed subset F of X is compact saturated in X^{d}, by Theorem 9.1.38; and $(\uparrow_{X^{\mathsf{d}}} x)_{x \in F}$ is an open cover of F in X^{d}, so there is a finite subset E of F such that $F \subseteq \bigcup_{x \in E} \uparrow_{X^{\mathsf{d}}} x$, whence $F \subseteq \uparrow_{X^{\mathsf{d}}} E = \downarrow_X E$, and therefore $F = \downarrow E$ is finitary closed. Applying this to $F = X$ yields property T, and applying it to $\uparrow x \cap \uparrow y$ yields property W. Since every closed subset is finitary closed, the topology of X is in particular the upper topology of $\leqslant$ (Proposition 4.2.12). Finally, any infinite strictly descending sequence $x_0 > x_1 > \cdots > x_n > \cdots$ yields an infinite strictly ascending sequence of open subsets $(X \smallsetminus \downarrow x_0) \subset (X \smallsetminus \downarrow x_1) \subset \cdots \subset (X \smallsetminus \downarrow x_n) \subset \cdots$, contradicting Proposition 9.7.6. So $\leqslant$ is well-founded.

Conversely, assume X is a poset whose ordering $\leqslant$ is well-founded and satisfies properties T and W. Write $\supset$ for strict containment, i.e., $A \supset B$ iff $A \supseteq B$ and $A \neq B$. A strictly descending sequence $\downarrow E_0 \supset \downarrow E_1 \supset \cdots \supset \downarrow E_n \supset \cdots$, where each E_n is finite, is a *bad sequence* if and only if it is infinite. A finite subset E of X is *bad* if and only if $\downarrow E$ starts some bad sequence, i.e., iff $E = E_0$ for some bad sequence as above. Otherwise, E is *good*.

If there is a bad subset, then there is one of minimal cardinality, E_0. Given E_0, there is a bad sequence as above starting with $\downarrow E_0$, so there is a finite subset E_1 such that $\downarrow E_0 \supset \downarrow E_1$, and E_1 is itself bad. Pick one such E_1 of minimal cardinality. Similarly, pick a minimal E_2 such that $\downarrow E_1 \supset \downarrow E_2$ and E_2 is bad, and repeat. We obtain a *minimal bad sequence* $\downarrow E_0 \supset \downarrow E_1 \supset \cdots \supset \downarrow E_n \supset \cdots$, where minimality means that, for every $n \in \mathbb{N}$, every E'_n such that $\downarrow E_{n-1} \supset \downarrow E'_n$ and that contains strictly fewer elements than E_n is good.

Since the sequence is bad, $E_n \neq \emptyset$ for every $n \neq \emptyset$. Also, since E_n is finite, $\downarrow E_n = \downarrow \text{Max}\, E_n$, where $\text{Max}\, E_n$ is the set of maximal elements of E_n. By minimality, $E_n = \text{Max}\, E_n$, i.e., the elements of E_n are pairwise incomparable.

We claim that each E_n contains exactly one element. Assume the contrary: let n be minimal such that E_n contains at least two elements. For each $x \in E_n$, since the elements of E_n are incomparable, $\downarrow x \cap E_n$ has strictly fewer elements than E_n, hence is good. So the sequence $\downarrow(\downarrow x \cap E_n) \supseteq \downarrow(\downarrow x \cap E_{n+1}) \supseteq \cdots$ can only contain finitely many distinct subsets. It follows that there is an $N_x \geqslant n$ such that $\downarrow(\downarrow x \cap E_m) = \downarrow(\downarrow x \cap E_{N_x})$ for every $m \geqslant N_x$. Since E_n is finite, there is a natural number N above every N_x, $x \in E_n$. For every $m \geqslant N$, $\downarrow E_m = \downarrow(\downarrow E_n \cap E_m)$ (since $\downarrow E_n \supseteq \downarrow E_m$) $= \bigcup_{x \in E_n} \downarrow(\downarrow x \cap E_m) = \bigcup_{x \in E_n} \downarrow(\downarrow x \cap E_N)$ (since N and m are both above each N_x) $= \downarrow(\downarrow E_n \cap E_N) = \downarrow E_N$ (since $\downarrow E_n \supseteq \downarrow E_N$). This contradicts $\downarrow E_N \supset \downarrow E_{N+1}$ when $m = N + 1$.

So every E_n contains exactly one element. Write $E_n = \{x_n\}$. Since $\downarrow E_0 \supset \downarrow E_1 \supset \cdots \supset \downarrow E_n \supset \cdots$, we obtain an infinite strictly descending chain $x_0 > x_1 > \cdots > x_n > \cdots$, a contradiction. So every finite subset E of X is good.

Let F be a closed subset of X with the upper topology. By definition, F is an intersection of finitary closed subsets, i.e., of subsets of the form $\downarrow E_i$ with E_i finite, $i \in I$. Write F as $\bigcap_{J \in \mathbb{P}_{\text{fin}}(I)} \bigcap_{i \in J} \downarrow E_j$. Property W implies that the intersection of any two finitary closed subsets $\downarrow E$ and $\downarrow E'$ is again finitary closed, since $\downarrow E \cap \downarrow E' = \bigcup_{x \in E, x' \in E'} (\downarrow x \cap \downarrow x')$ is a finite union of finitary closed subsets. So, together with property T (when J is empty), $\bigcap_{i \in J} \downarrow E_j$ is of the form $\downarrow E_J$ for some finite subset E_J of X. The family $(\downarrow E_J)_{J \in \mathbb{P}_{\text{fin}}(I)}$ is filtered. We claim that $F = \downarrow E_J$ for some $J \in \mathbb{P}_{\text{fin}}(I)$. Otherwise, pick $J_0 \in \mathbb{P}_{\text{fin}}(I)$. Since $F \neq \downarrow E_{J_0}$, there is an element in $\downarrow E_{J_0} \setminus F$, hence in $\downarrow E_{J_0} \setminus E_{J_1}$ for some $J_1 \in \mathbb{P}_{\text{fin}}(I)$. By filteredness, i.e., by replacing J_1 by $J_0 \cup J_1$ if necessary, $\downarrow E_{J_0} \supseteq \downarrow E_{J_1}$, and the inclusion must be strict. Since $F \neq E_{J_1}$, similarly we find $J_2 \in \mathbb{P}_{\text{fin}}(I)$ such that $\downarrow E_{J_1} \supset \downarrow E_{J_2}$. Continuing this way, we obtain a bad sequence, a contradiction. So $F = \downarrow E_J$ for some $J \in \mathbb{P}_{\text{fin}}(E)$. So every closed subset F is finitary closed.

Now let $U_0 \subseteq U_1 \subseteq \cdots \subseteq U_n \subseteq \cdots$ be an infinite ascending chain of open subsets of X. For each $n \in \mathbb{N}$, $X \setminus U_n$ is a finitary closed subset $\downarrow E_n$. Since every finite subset is good, E_0 is good, so the chain $\downarrow E_0 \supseteq$

$\downarrow E_1 \supseteq \cdots \supseteq \downarrow E_n \supseteq \cdots$ contains only finitely many distinct closed subsets. So $U_0 \subseteq U_1 \subseteq \cdots \subseteq U_n \subseteq \cdots$ stabilizes: by Proposition 9.7.6, X is Noetherian.

Finally, let us show that X_u is sober. Since $\leqslant$ is an ordering, X_u is T_0. Let F be any irreducible closed subset. Since F is finitary closed, write F as $\downarrow\{x_1, \ldots, x_n\}$. As F is irreducible (Proposition 8.2.2) and F is included in the union of the finitely many closed subsets $\downarrow x_1, \ldots, \downarrow x_n$, $F \subseteq \downarrow x_i$ for some i, $1 \leqslant i \leqslant n$. Hence $F = \downarrow x_i$. So X_u is sober. □

Exercise 9.7.13 Show that Plotkin's tie T (Figure 9.4) is sober and Noetherian in its Scott topology. Find an example of a bifinite domain that is not Noetherian. (One can even find totally ordered, bounded Scott domains that are not Noetherian.) So the classes of sober Noetherian and of bifinite domains are incomparable classes of spectral spaces.

Exercise 9.7.14 For every topological space X, the *Hoare powerspace* $\mathcal{H}_{\mathsf{V}}(X)$ of X is the space of all non-empty closed subsets of X, with a topology that we again call the *lower Vietoris topology* (as in Definition 8.2.17): its subbasic opens are those of the form $\Diamond U = \{F \in \mathcal{H}_{\mathsf{V}}(X) \mid F \cap U \neq \emptyset\}$, U open in X. So $\mathcal{S}(X)$ is a subspace of $\mathcal{H}_{\mathsf{V}}(X)$, for example. One can identify $\mathcal{H}_{\mathsf{V}}(X)_\perp$ with the space of all closed subsets of X, including the empty set, with a similarly defined lower Vietoris topology.

Show that $\mathcal{H}_{\mathsf{V}}(X)$ and $\mathcal{H}_{\mathsf{V}}(X)_\perp$ are sober Noetherian for every Noetherian space X; conversely, if $\mathcal{H}_{\mathsf{V}}(X)$ is Noetherian, or if $\mathcal{H}_{\mathsf{V}}(X)_\perp$ is Noetherian, then so is X.

Here is another characterization of Noetherian spaces, this time relying on subbases. This is a variant of Alexander's Subbase Lemma.

Lemma 9.7.15 (Bad sequence lemma) *Let Y be a topological space, and $\mathcal{B}$ be a subbase of the topology of Y. If Y is not Noetherian, then there is a* bad sequence $U_0, U_1, \ldots, U_n, \ldots$, *i.e., a sequence of elements of $\mathcal{B}$ such that U_n is not included in $U_0 \cup U_1 \cup \cdots \cup U_{n-1}$ for any $n \in \mathbb{N}$.*

Proof Since Y is not Noetherian, there is a non-compact open subset U. By Alexander's Subbase Lemma (Theorem 4.4.29), there is an open cover $(V_i)_{i \in I}$ of U consisting of elements of $\mathcal{B}$ such that no finite subfamily covers U. We then define U_n by induction on $n \in \mathbb{N}$, i.e., by assuming that $U_0, U_1, \ldots, U_{n-1}$ have already been chosen among $(V_i)_{i \in I}$ so that $U_k \not\subseteq U_0 \cup U_1 \cup \cdots \cup U_{k-1}$ for every k, $0 \leqslant k \leqslant n-1$. Since $U_0, U_1, \ldots, U_{n-1}$ do not cover U, there must be a point $x \in U \setminus \bigcup_{k=0}^{n-1} U_k$. By definition, $x \in V_i$ for some $i \in I$: let $U_n = V_i$. □

Exercise 9.7.16 (R.-E. Hoffmann's Theorem) Let X be a T_0 topological space. The *Skula topology* on X is the topology generated by the open subsets of X and the downward closed subsets. Show that X with the Skula topology is always T_2.

Show that X is sober and Noetherian if and only if X is compact in its Skula topology (Hoffmann, 1979a, theorem 3.1); then the Skula topology coincides with the patch topology on X.

There is another remarkable relation between Noetherian spaces and particular quasi-orderings, which have turned out to be important in computer science, and particularly in verification (Finkel and Schnoebelen, 2001).

Proposition 9.7.17 (Well quasi-ordering) *Let X be a set, equipped with a quasi-ordering $\leqslant$. The following are equivalent:*

1. *X_a, i.e., X with the Alexandroff topology of $\leqslant$, is Noetherian.*
2. *Every sequence $(x_n)_{n \in \mathbb{N}}$ of elements of X has an ascending subsequence, i.e., a subsequence $(x_{n_k})_{k \in \mathbb{N}}$ such that $x_{n_0} \leqslant x_{n_1} \leqslant \cdots \leqslant x_{n_k} \leqslant \cdots$ (and $n_0 < n_1 < \cdots < n_k < \cdots$).*
3. *For every sequence $(x_n)_{n \in \mathbb{N}}$ of elements of X, there is a pair $i < j$ such that $x_i \leqslant x_j$.*
4. *$\leqslant$ is well-founded and has no infinite antichain; an* antichain *is a set of pairwise incomparable elements.*
5. *Every upward closed subset of X is of the form $\uparrow E$ for some finite set E.*

If $\leqslant$ satisfies any of the above conditions, then it is called a well quasi-ordering, *or* wqo.

Proof $1 \Rightarrow 2$. Assume X_a is Noetherian. By Theorem 9.7.8, $(x_n)_{n \in \mathbb{N}}$ has a self-convergent subnet $(x_{\alpha(j)})_{j \in J, \preceq}$. Pick $j_0 \in J$, and let $n_0 = \alpha(j_0)$. Since $(x_{\alpha(j)})_{j \in J, \preceq}$ converges to x_{n_0}, and $\uparrow x_{n_0}$ is open in X_a, $x_{\alpha(j)}$ is in $\uparrow x_{n_0}$, i.e., above x_{n_0}, for j large enough. Pick $j_1 \in J$ so large that not only $x_{n_0} \leqslant x_{\alpha(j)}$ for every j above j_1, but also such that j_1 is above j_0. This is possible since J is directed. Let $n_1 = \alpha(j_1)$. In particular $x_{n_0} \leqslant x_{n_1}$. Since $(x_{\alpha(j)})_{j \in J, \preceq}$ converges to x_{n_1}, by a similar argument $x_{\alpha(j)}$ is above x_{n_1} for j large enough above j_1. Pick such a j, call it j_2, and let $n_2 = \alpha(j_2)$, so in particular $x_{n_1} \leqslant x_{n_2}$. Continuing in the same way, we obtain an ascending subsequence $x_{n_0} \leqslant x_{n_1} \leqslant \cdots \leqslant x_{n_k} \leqslant \cdots$.

$2 \Rightarrow 3$. Pick $i = n_0$, $j = n_1$.

$3 \Rightarrow 4$. If $x_0 > x_1 > \cdots > x_n > \cdots$ is an infinite strictly descending chain, then the existence of $i < j$ such that $x_i \leqslant x_j$ contradicts $x_i > x_j$. If instead A is an infinite antichain, then one can enumerate a countably infinite subset

$\{x_0, x_1, \ldots, x_n, \ldots\}$ of A, and again the existence of $i < j$ such that $x_i \leqslant x_j$ contradicts the fact that x_i and x_j are incomparable.

$4 \Rightarrow 5$. Let U be an upward closed subset of X. For every $x \in U$, we claim that x is above some minimal element of U. This is proved by well-founded induction (see Section 2.3.4): if x is already minimal in U, then we are done; otherwise one can find $y \in U$ with $y < x$, and the induction hypothesis tells us that y is above some minimal element of U. Let E be the set of minimal elements of U. So $U = \uparrow E$. Clearly, E is an antichain, hence is finite.

$5 \Rightarrow 1$. By 5 the open subsets of X_a are finitary compact, hence compact. □

While Noetherian spaces form a relatively small class of spaces compared to, say, the spectral spaces, several interesting spaces are Noetherian.

Proposition 9.7.18

(*i*) *Every finite space is Noetherian, and in general any space with only finitely many open subsets;*
(*ii*) $\mathbb{N}_a$, *the space of all natural numbers in the Alexandroff (or upper, or Scott) topology of its natural ordering, is Noetherian;*
(*iii*) *every subspace of a Noetherian space is Noetherian;*
(*iv*) *for every Noetherian space X, X with a coarser topology is again Noetherian;*
(*v*) *the image of a Noetherian space by a continuous map is Noetherian; in particular every quotient and every retract of a Noetherian space is Noetherian;*
(*vi*) *every finite coproduct of Noetherian spaces is Noetherian;*
(*vii*) *every finite product of Noetherian spaces is Noetherian.*

Proof (*i*) and (*iv*) are trivial consequences of Proposition 9.7.6.

(*ii*) The open subsets of $\mathbb{N}_a$ (or $\mathbb{N}_u$, $\mathbb{N}_\sigma$) are the upward closed subsets $\uparrow n$, $n \in \mathbb{N}$, plus the empty set. Each of them is compact, as a finitary compact.

(*iii*) is true by Proposition 9.7.7. (*v*) Let $f : X \to Y$ be continuous and X be Noetherian. Without loss of generality, assume $Y = f[X]$, as f is also continuous from X to the subspace $f[X]$. Given any open subset V of Y, $f^{-1}(V)$ is open in X since f is continuous. Since X is Noetherian, $f^{-1}(V)$ is compact, so $f[f^{-1}(V)]$ is compact in Y (Proposition 4.4.13). But f is surjective, so $f[f^{-1}(V)] = V$. (*vi*) Let $X_1, \ldots, X_n$ be Noetherian spaces. Any open subset V of $\coprod_{i=1}^n X_i$ is the union of $\iota_i[U_i]$, where $U_i = \{x \mid (i, x) \in U\}$ is open. Since X_i is Noetherian, U_i is compact, and since ι_i is continuous (Proposition 4.6.2), $\iota_i[U_i]$ is compact. As a finite union of compact subsets, V is compact (Proposition 4.4.12).

(*vii*) Since the one-element space is finite, hence Noetherian, it is enough to show that the product of two Noetherian spaces X_1, X_2 is Noetherian. By Lemma 9.7.9, $\mathcal{S}(X_1)$ and $\mathcal{S}(X_2)$ are Noetherian, hence they have the upper topologies of well-founded orderings with properties T and W by Theorem 9.7.12. The product ordering is clearly again well-founded, and has properties T and W. Moreover, $\mathcal{S}(X_1) \times \mathcal{S}(X_2)$ has the upper topology of the product ordering: the product topology is finer than the upper topology (Exercise 4.5.19), and conversely every open rectangle is of the form $(X_1 \smallsetminus \downarrow E_1) \times (X_2 \times \downarrow E_2)$ with E_1 and E_2 finite (Theorem 9.7.12 again), whose complement is $(\downarrow E_1 \times X_2) \cup (X_1 \times \downarrow E_2)$; since X_1 and X_2 have property T, this complement is open in the upper topology. Using Theorem 9.7.12, $\mathcal{S}(X_1) \times \mathcal{S}(X_2)$ is Noetherian. By Theorem 8.4.8, the latter is homeomorphic to $\mathcal{S}(X_1 \times X_2)$, which must then be Noetherian. So $X_1 \times X_2$ is Noetherian, by Lemma 9.7.9. □

Exercise 9.7.19 Give another proof of the fact that the product of two Noetherian spaces X_1 and X_2 is Noetherian, this time based on self-convergent nets, using Theorem 9.7.8.

Exercise 9.7.20 Show that, whenever $\mathcal{B}$ is a base of the topology of a Noetherian space Y, then every open subset of Y is a finite union of elements of $\mathcal{B}$. Conclude that if X_1 and X_2 are Noetherian, then every open subset of $X_1 \times X_2$ is a *finite* union of open rectangles.

Exercise 9.7.21 Let L and L' be two distributive complete lattices. Show that if L and L' both have the ascending chain condition, then the complete lattice $\mathrm{Gal}(L, L')$ of all Galois connections between L and L' (Lemma 8.4.19) is distributive and has the ascending chain condition. (Use Stone duality.)

Show that both distributivity and the ascending chain condition are needed of L and L' for $\mathrm{Gal}(L, L')$ to have the ascending chain condition. Show indeed that if $\mathrm{Gal}(L, L')$ has the ascending chain condition, and L and L' are non-trivial (i.e., contain at least two elements, i.e., have distinct top and bottom elements), then L and L' must have the ascending chain condition. Also, by letting $L = L'$ be the complete lattice obtained as $\mathbb{N} \cup \{\bot, \top\}$, where any two natural numbers are incomparable, above $\bot$, and below $\top$, show that there are (non-distributive) complete lattices L, L' with the ascending chain condition such that $\mathrm{Gal}(L, L')$ does *not* have the ascending chain condition.

Exercise 9.7.22 Show that infinite coproducts and infinite products of Noetherian spaces are not in general Noetherian.

Exercise 9.7.23 (Noetherian spaces do not form a CCC) Show that the specialization ordering of every Noetherian space X is well-founded. Show that the pointwise ordering on $[\mathbb{N}_a \to \mathbb{S}]$ is not well-founded, where $\mathbb{N}_a$ is $\mathbb{N}$ with the Alexandroff topology of its natural ordering $\leqslant$. Conclude that the category of Noetherian spaces and continuous maps is not Cartesian-closed.

Exercise 9.7.24 (Dickson's Lemma) For every $k \in \mathbb{N}$, show that the pointwise ordering $\mathbb{N}^k$ is wqo. This is known as *Dickson's Lemma.*

Exercise 9.7.25 (The Rado structure) The *Rado structure* X_{Rado} (Rado, 1954) is a close cousin and predecessor of the Johnstone space (Exercise 5.2.15). This is the set $\{(m, n) \in \mathbb{N}^2 \mid m < n\}$, ordered by $\leqslant_{\text{Rado}}$: $(m, n) \leqslant_{\text{Rado}} (m', n')$ iff $m = m'$ and $n \leqslant n'$, or $n < m'$. Show that this is indeed an ordering. Using Dickson's Lemma, show that $\leqslant_{\text{Rado}}$ is a wqo.

Show that the ideals of X_{Rado}, apart from those of the form $\downarrow(m, n)$—where downward closure is relative to $\leqslant_{\text{Rado}}$—are those of the form $\omega_i = \{(i, n) \mid n \geqslant i + 1\} \cup \{(m, n) \in X_{\text{Rado}} \mid n \leqslant i - 1\}$, or $\omega = X_{\text{Rado}}$. See Figure 9.5, where one of the sets ω_i is shown in dark gray.

Conclude that $\mathbf{I}(X_{\text{Rado}}) = \mathcal{S}(X_{\text{Rado}\,a})$, while Noetherian, has a specialization ordering (inclusion) that is not wqo. In particular, the lower Vietoris topology is the Scott topology, and also the upper topology of $\subseteq$, but is strictly coarser than the Alexandroff topology.

Similarly, show that $\mathcal{H}_{\mathsf{V}}(X_{\text{Rado}\,a})$ and $\mathcal{H}_{\mathsf{V}}(X_{\text{Rado}\,a})_\perp$, while Noetherian, are not wqo in their specialization ordering, and that the lower Vietoris topology is strictly coarser than the Alexandroff topology.

A (finite) *word* over a set X is a finite sequence $x_1 x_2 \dots x_n$ of elements of X. Its *letters* are $x_1, x_2, \dots, x_n$, and its *length* is the natural number n. We

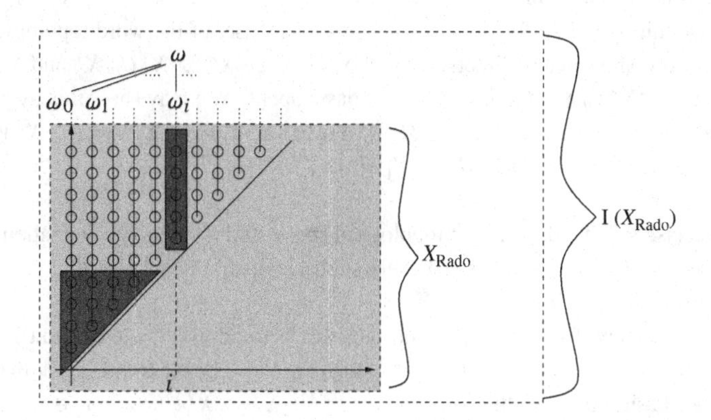

Figure 9.5 Ideals in the Rado structure.

write ε for the empty word (i.e., when $n = 0$), and $w.w'$ for the concatenation of two words w and w'. For each $x \in X$, we also write x for the word of length 1 whose sole letter is x. Given any subsets A, B of X^*, write $A.B = \{ww' \mid w \in A, w' \in B\}$, A^* the set of words over A. We also abbreviate $A.B$ as AB.

Definition 9.7.26 (X^*, word topology) For every topological space X, the *word topology* on X^* is the coarsest one containing the subsets $X^*U_1X^*U_2 X^* \dots X^*U_nX^*$ as opens, where $n \in \mathbb{N}$, and U_1, U_2, $\dots$, U_n are open subsets of X. We write X^* for the space of all finite words over X with this topology.

A *subword* of a word $w = x_1x_2\dots x_n$ is any word $x_{i_1}x_{i_2}\dots x_{i_m}$ where $1 \leqslant i_1 < i_2 < \dots < i_m \leqslant n$, $m \in \mathbb{N}$. Given a word $w = x_1x_2\dots x_n$, the words *embedded* in w are those of the form $x'_{i_1}x'_{i_2}\dots x'_{i_m}$ where $1 \leqslant i_1 < i_2 < \dots < i_m \leqslant n$, $m \in \mathbb{N}$, and $x'_{i_1} \leqslant x_{i_1}, x'_{i_2} \leqslant x_{i_2}, \dots, x'_{i_m} \leqslant x_{i_m}$. In other words, an embedded word is obtained from a subword by decreasing each of its letters. We write $w \leqslant^* w'$ if and only if w is embedded in w', and call $\leqslant^*$ the *word embedding* quasi-ordering. The subbasic opens $X^*U_1X^*U_2X^* \dots X^*U_nX^*$ can therefore be understood as the subsets of words that have a subword in the (fixed) open rectangle $U_1U_2\dots U_n$.

Exercise 9.7.27 Show that the concatenation function $cat\colon X^* \times X^* \to X^*$ is continuous, and that the function $i\colon X \to X^*$ that maps the letter x to x as a word, is continuous. Show that, for every continuous map $f\colon X \to Y$, the map $f^*\colon X^* \to Y^*$ that sends every sequence $x_1x_2\dots x_n$ of X^* to $f(x_1)f(x_2)\dots f(x_n)$ is continuous as well.

Exercise 9.7.28 Show that the subsets $X^*U_1X^*U_2X^* \dots X^*U_nX^*$ as defined in Definition 9.7.26 form a base, not just a subbase, of the word topology. More precisely, show that the intersection of $X^*U_1X^*U_2X^* \dots X^*U_mX^*$ and $X^*V_1X^* V_2X^* \dots X^*V_nX^*$ is a finite union of basic opens. You may show this by proving that $X^* \cap \mathcal{V} = \mathcal{V}$, $\mathcal{U} \cap X^* = \mathcal{U}$, $X^*U_1\mathcal{U} \cap X^*V_1\mathcal{V} = X^*U_1(\mathcal{U} \cap X^*V_1\mathcal{V}) \cup X^*V_1(X^*U_1\mathcal{U} \cap \mathcal{V}) \cup X^*(U_1 \cap V_1)(\mathcal{U} \cap \mathcal{V})$.

Exercise 9.7.29 Let X be a topological space, and $\leqslant$ be its specialization quasi-ordering. It should be clear that every subbasic open $X^*U_1X^*U_2X^* \dots X^*U_nX^*$ is upward closed in $\leqslant^*$.

Conversely, for every open subset $\mathcal{U}$ of X^*, and every open subset U of X, define $\mathcal{U}/U$ as follows. If $\mathcal{U} = X^*$, then $\mathcal{U}/U = \emptyset$; otherwise, $\mathcal{U}$ is the union of all basic opens of the form $X^*U_{i1}X^*U_{i2}X^* \dots X^*U_{in_i}X^*$, $i \in I$, that are included in $\mathcal{U}$, by Exercise 9.7.28, and then $n_i \geqslant 1$ for every $i \in I$: in this case, we let $\mathcal{U}/U$ be the union of all basic opens $X^*(U_{i1} \cap U)X^*U_{i2}X^* \dots X^*U_{in_i}X^*$. Show that $\mathcal{U}/U$ and $X^*U\mathcal{U}$ are open in X^*.

When $A \subseteq X$, let $A^?$ be the set of length 1 words whose (sole) letter is in A, plus ε. For every closed subset F of X, for every closed subset $\mathcal{F}$ of X^*, letting $U = X \setminus F, \mathcal{U} = X^* \setminus \mathcal{F}$, show that the complement of $F^?\mathcal{F}$ in X^* is $X^*X\mathcal{U} \cup \mathcal{U}/U$. Deduce that $F_1^? F_2^? \dots F_n^?$ is closed in X^* for every sequence of closed subsets F_1, $F_2, \dots, F_n$ in X.

The downward closure of a word $w = x_1x_2 \dots x_n$ with respect to $\leqslant^*$ is $(\downarrow x_1)^?(\downarrow x_2)^? \dots (\downarrow x_n)^?$. Conclude that the specialization quasi-ordering of X^* is the word embedding quasi-ordering $\leqslant^*$.

Exercise 9.7.30 Let X be a set equipped with a quasi-ordering $\leqslant$. Show that the word topology on $(X_a)^*$ is the Alexandroff topology of $\leqslant^*$. (Hint: the upward closure of $x_1x_2 \dots x_n$ in X^* is $X^*(\uparrow x_1)X^*(\uparrow x_2)X^* \dots X^*(\uparrow x_n)X^*$.)

We now use a minimal bad sequence argument in X^*, similar to the one we used in Theorem 9.7.12. Recall from Lemma 9.7.15 that a bad sequence of elements of a subbase $\mathcal{B}$ is a sequence of opens in $\mathcal{B}$ such that none is included in the union of those that precede it.

Lemma 9.7.31 (Minimal bad sequence lemma) *Let Y be a topological space, and $\mathcal{B}$ be a subbase of the topology of Y. Assume that there is a well-founded quasi-ordering $\sqsubseteq$ on $\mathcal{B}$. If Y is not Noetherian, then there is a* minimal *bad sequence $U_0, U_1, \dots, U_n, \dots$, i.e., one such that any other sequence $V_0, V_1, \dots, V_n, \dots$, that is lexicographically smaller (i.e., for some $k \in \mathbb{N}$, $V_k \sqsubseteq U_k$, $V_k \neq U_k$, while $V_0 = U_0, V_1 = U_1, \dots, V_{k-1} = U_{k-1}$) is good.*

Proof By Lemma 9.7.15, there is a bad sequence. Take U_0 minimal in $\sqsubseteq$ so that it starts a bad sequence. Then take U_1 minimal so that U_0, U_1 start a bad sequence. Iterate the process, defining U_n minimal given $U_0, U_1, \dots, U_{n-1}$ so that there is a bad sequence starting with $U_0, U_1, \dots, U_{n-1}, U_n$, and diagonalize: the sequence $U_0, U_1, \dots, U_n, \dots$, is itself bad, and clearly minimal in the indicated sense. ☐

Lemma 9.7.32 *Let X be a Noetherian space, $(a_n)_{n\in\mathbb{N}}$ be a sequence of points in X, and U_n be an open neighborhood of a_n in X for each $n \in \mathbb{N}$. Then there is a subsequence $(a_{n_k})_{k\in\mathbb{N}}$ (i.e., $n_0 < n_1 < \dots < n_k < \dots$) such that $a_{n_k} \in \bigcap_{j=0}^{k} U_{n_j}$ for every $k \in \mathbb{N}$.*

Proof By Theorem 9.7.8 (ii), $(a_n)_{n\in\mathbb{N}}$ contains a cluster point a_{n_0}. Since $a_{n_0} \in U_{n_0}$, there is a subsequence $a_{n_0}, a_{n_{01}}, \dots, a_{n_{0k}}, \dots$ of $(a_n)_{n\in\mathbb{N}}$ whose elements are all in U_{n_0}. Apply Theorem 9.7.8 (ii) again on the subsequence

$a_{n01}, \ldots, a_{n0k}, \ldots$: let a_{n_1} be one of its cluster points, yielding a further subsequence $a_{n_1}, a_{n11}, \ldots, a_{n1k}, \ldots$ of $(a_n)_{n\in\mathbb{N}}$ whose elements are all not only in U_{n_0}, but also in U_{n_1}. Then iterate the process. □

The following is a topological generalization of a result known as Higman's Lemma. The latter will be the topic of Exercise 9.7.34.

Theorem 9.7.33 (Topological Higman Lemma) *Let X be a topological space. Then X^* is Noetherian if and only if X is.*

Proof Assume that X is Noetherian but X^* is not. Let $\mathcal{B}$ be the base of all non-empty open subsets of the form given in Definition 9.7.26. Quasi-order the elements of $\mathcal{B}$ by $X^*U_1X^*U_2X^* \ldots X^*U_mX^* \sqsubseteq X^*U'_1X^*U'_2X^* \ldots X^*U'_nX^*$ iff the word $U_1U_2 \ldots U_m$ is a subword of the word $U'_1U'_2 \ldots U'_n$, i.e., iff there is a strictly increasing map $g\colon \{1, 2, \ldots, m\} \to \{1, 2, \ldots, n\}$ such that $U_i = U'_{g(i)}$ for every i, $1 \leqslant i \leqslant m$.

By Lemma 9.7.31, there is a minimal bad sequence $\mathcal{U}_0, \mathcal{U}_1, \ldots, \mathcal{U}_n, \ldots$ of elements of $\mathcal{B}$. Write $\mathcal{U}_n$ as $X^*U_{n1}X^*U_{n2}X^* \ldots X^*U_{nm_n}X^*$. Note that $m_n \neq 0$, since otherwise $\mathcal{U}_n$ would be the whole of X^*, and this would imply $\mathcal{U}_{n+1} \subseteq \mathcal{U}_n$, contradicting the fact that the sequence is bad. Since the sequence is bad, for each $n \in \mathbb{N}$ there is a word $w_n \in \mathcal{U}_n$ that is not in $\mathcal{U}_0 \cup \mathcal{U}_1 \cup \cdots \cup \mathcal{U}_{n-1}$. Using the fact that $m_n \neq 0$, one can write w_n as $w'_n a_n w''_n$, where $w'_n \in X^*$, $a_n \in U_{n1}$, and $w''_n \in X^*U_{n2}X^* \ldots X^*U_{nm_n}X^*$.

Now use Lemma 9.7.32 to obtain a subsequence $(a_{n_k})_{k\in\mathbb{N}}$ (i.e., $n_0 < n_1 < \cdots < n_k < \cdots$) such that $a_{n_k} \in \bigcap_{j=0}^{k} U_{n_j1}$ for every $k \in \mathbb{N}$. Consider the sequence $\mathcal{U}_0, \mathcal{U}_1, \ldots, \mathcal{U}_{n_0-1}, \mathcal{U}'_0, \mathcal{U}'_1, \ldots, \mathcal{U}'_k, \ldots$, where, for each $k \in \mathbb{N}$, $\mathcal{U}'_k = X^*U_{n_k2}X^* \ldots X^*U_{n_km_{n_k}}X^*$ is obtained from $\mathcal{U}_{n_k}$ by removing the leading $X^*U_{n_k1}$. This is lexicographically smaller than $\mathcal{U}_0, \mathcal{U}_1, \ldots, \mathcal{U}_n, \ldots$, hence it must be a good sequence. So one of its elements must be contained in the union of all previous elements. We cannot have $\mathcal{U}_n \subseteq \mathcal{U}_0 \cup \mathcal{U}_1 \cup \cdots \cup \mathcal{U}_{n-1}$ for any $n \leqslant n_0 - 1$ since the sequence we started with was bad, so there must be a $k \in \mathbb{N}$ such that $\mathcal{U}'_k \subseteq \mathcal{U}_0 \cup \mathcal{U}_1 \cup \cdots \cup \mathcal{U}_{n_0-1} \cup \mathcal{U}'_0 \cup \cdots \cup \mathcal{U}'_{k-1}$. In particular, w''_{n_k} is in $\mathcal{U}_0 \cup \mathcal{U}_1 \cup \cdots \cup \mathcal{U}_{n_0-1} \cup \mathcal{U}'_0 \cup \cdots \cup \mathcal{U}'_{k-1}$. However, w''_{n_k} cannot be in any $\mathcal{U}_i$ with $i \leqslant n_0 - 1$; otherwise the larger (in $\leqslant^*$) word $w_{n_k} = w'_{n_k} a_{n_k} w''_{n_k}$ would also be in $\mathcal{U}_i$; this is because $\mathcal{U}_i$ is upward closed in $\leqslant^*$. So w''_{n_k} must be in some $\mathcal{U}'_j = X^*U_{n_j2}X^* \ldots X^*U_{n_jm_{n_j}}X^*$, $j \leqslant k-1$. Since $a_{n_k} \in \bigcap_{j=0}^{n_k} U_{n_j1}$, a_{n_k} is in U_{n_j1}, so $w_{n_k} = w'_{n_k} a_{n_k} w''_{n_k}$ is in $X^*U_{n_j1}X^*U_{n_j2}X^* \ldots X^*U_{n_jm_{n_j}}X^* = \mathcal{U}_{n_j}$, a contradiction. So X^* is Noetherian.

Conversely, recall that a space is Noetherian if and only if it has no infinite ascending chain of opens (Proposition 9.7.6). If X^* is Noetherian, then any infinite strictly ascending chain $U_0 \subset U_1 \subset \cdots \subset U_k \subset \cdots$ of opens of

X induces an infinite strictly ascending chain $X^*U_0X^* \subset X^*U_1X^* \subset \cdots \subset X^*U_kX^* \subset \cdots$ of opens in X^*, a contradiction. So X is Noetherian. □

Exercise 9.7.34 (Higman's Lemma) Let $\leqslant$ be a quasi-ordering on a set X. Show that $\leqslant^*$ is wqo on X^* if and only if $\leqslant$ is wqo on X. This is *Higman's lemma* (Higman, 1952).

Exercise 9.7.35 (Multisets) Let X be a topological space. Define an equivalence relation $\equiv$ on X^* by $w \equiv w'$ if and only if w is a permutation of w', i.e., iff $w = a_1a_2\dots a_n$ for some $a_1, a_2, \dots, a_n \in X$ and $w' = a_{\pi(1)}a_{\pi(2)}\dots a_{\pi(n)}$ for some bijective map π from $\{1, 2, \dots, n\}$ to $\{1, 2, \dots, n\}$. The elements of $X^{\circledast} = X^*/\equiv$ are called *multisets*. We write $\{\!|a_1, a_2, \dots, a_n|\!\}$ for the equivalence class of $a_1a_2\dots a_n$. Multisets can be thought of as finite sets, except that each element can occur several times; e.g., $\{\!|a, a, b|\!\}$ is distinct from $\{\!|a, b|\!\}$.

Topologize $X^{\circledast}$ with the quotient topology. Then show that $X^{\circledast}$ is Noetherian if and only if X is.

Exercise 9.7.36 (Words, in the prefix topology) Let $X_1, X_2, \dots, X_n, \dots$ be any countable sequence of topological spaces. A *heterogeneous word* w on the latter is any sequence $x_1x_2\dots x_n$ of elements from $X_1, X_2, \dots, X_n$, respectively, $n \in \mathbb{N}$.

A *telescope* on $(X_n)_{n\geqslant 1}$ is a sequence $\mathcal{U} = U_0, U_1, \dots, U_n, \dots$ of opens, where $U_n \in \prod_{i=1}^n X_i$ for each $n \in \mathbb{N}$, and such that $U_nX_{n+1} \subseteq U_{n+1}$ for every $n \in \mathbb{N}$. (We write AB for $A \times B$, for simplicity.) A telescope is *wide* iff $U_n = \prod_{i=1}^n X_i$ for some $n \in \mathbb{N}$, i.e., for all sufficiently large $n \in \mathbb{N}$.

Given any telescope $\mathcal{U} = U_0, U_1, \dots, U_n, \dots$ on $(X_n)_{n\geqslant 1}$, let $\lfloor \mathcal{U} \rangle$ be the set of heterogeneous words w over $X_1, X_2, \dots, X_n, \dots$ such that $w \in U_{|w|}$, where we write $|w|$ for the length of w.

$\rhd_{n=1}^{+\infty} X_n$ is the space of all *heterogeneous words* over $(X_n)_{n\geqslant 1}$, i.e., the disjoint union of all spaces $\prod_{i=1}^n X_i$, $n \in \mathbb{N}$, with the *prefix topology*, which is given by the trivial open $\emptyset$, plus all subsets of the form $\lfloor \mathcal{U} \rangle$, $\mathcal{U}$ a wide telescope on $(X_n)_{n\geqslant 1}$. Show that this is indeed a topology.

Show that its specialization quasi-ordering is the *prefix quasi-ordering* $\leqslant^{\rhd}$ (also known as the *short-lex* quasi-ordering), defined by: $a_1a_2\dots a_m \leqslant^{\rhd} b_1b_2\dots b_n$, where $a_i, b_i \in X_i$ for all i, iff $m \leqslant n$, $a_1 \leqslant b_1$, $a_2 \leqslant b_2, \dots,$ and $a_m \leqslant b_m$. (The prefix ordering $\leqslant^{\mathrm{pref}}$ of Exercise 6.1.21 is a special case when $\leqslant$ is equality on each X_i.)

Show that the map $i_n \colon X_1 \times \cdots \times X_n \to \rhd_{n=1}^{+\infty} X_n$ that sends each n-tuple $(a_1, a_2, \dots, a_n)$ to the word $a_1a_2\dots a_n$ is continuous. Show also that the map $cons \colon X_1 \times \rhd_{n=2}^{+\infty} X_n \to \rhd_{n=1}^{+\infty} X_n$ that sends $a_1, a_2a_3\dots a_n$ to $a_1a_2a_3\dots a_n$ is continuous.

Finally, show that if $X_1, X_2, \ldots, X_n, \ldots$ are all Noetherian, then $\rhd_{n=1}^{+\infty} X_n$ is Noetherian.

Exercise 9.7.37 Letting $X_1 = X_2 = \cdots = X_n = \cdots = \{0, 1\}$ with equality as ordering, show that the prefix ordering $\leqslant^{\rhd}$ on $\rhd_{n=1}^{+\infty} X_n$ is not in general a wqo. Conclude that even when $X_1, X_2, \ldots, X_n$ are all Alexandroff-discrete, the prefix topology on $\rhd_{n=1}^{+\infty} X_n$ is in general strictly coarser than the Alexandroff topology of $\leqslant^{\rhd}$.

Let us finally deal with finite *trees*, also known as *terms*. For every set X, let $\mathcal{T}(X)$ denote the set of all terms built using function symbols from X. They are inductively defined by: whenever $t_1, t_2, \ldots, t_n$ are terms, and for every $f \in X$, the tree $f(t_1, t_2, \ldots, t_n)$, whose root is labeled f and whose ordered list of successors is $t_1, t_2, \ldots, t_n$, is a term. The base case is implicit, and occurs when $n = 0$, where we write f instead of $f()$.

Example 9.7.38 Figure 9.6 displays a finite tree, with nodes labeled by elements of a set containing f, g, a, b, and c. As a term, this would be written $g(f(f(c, g(a, b))), f(b), g(f(a, b), f(a, b), c))$.

Our terms are usually called *ground* terms, i.e., terms without variables. In general, given any set $\mathcal{V}$ of so-called variables, one defines the set $\mathcal{T}(X, \mathcal{V})$ of terms with variables from $\mathcal{V}$ as the smallest one that contains $\mathcal{V}$, and such that, whenever $t_1, t_2, \ldots, t_n \in \mathcal{T}(X, \mathcal{V})$ and $f \in X$, $f(t_1, t_2, \ldots, t_n) \in \mathcal{T}(X, \mathcal{V})$. We shall not deal with this seemingly more general notion: indeed let $X + \mathcal{V}$ be the disjoint sum of X with $\mathcal{V}$, where $\mathcal{V}$ is equipped with the discrete topology, so that $\mathcal{T}(X, \mathcal{V})$ can be seen as the subspace of $\mathcal{T}(X + \mathcal{V})$ of those terms where variables are applied to no argument at all (i.e., each subterm $v(t_1, t_2, \ldots, t_n)$ with $v \in \mathcal{V}$ is such that $n = 0$). We shall equip $\mathcal{T}(X, \mathcal{V})$ with the subspace topology, so that all our results transfer immediately.

When $\vec{t}$ is a word $t_1 t_2 \ldots t_n$ over $\mathcal{T}(X)$, we write $f(\vec{t})$ instead of $f(t_1, t_2, \ldots, t_n)$. It is sometimes convenient to be able to talk about subterms

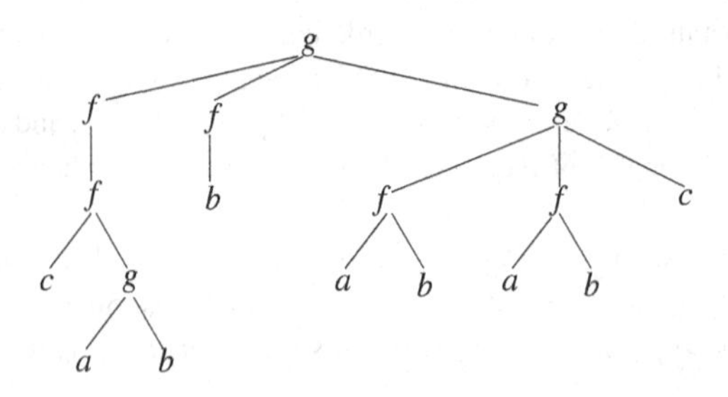

Figure 9.6 A term.

and positions p, together with the *subterm* $t_{|p}$ *of* t *at position* p. A *position* is a finite word over $\mathbb{N}$. The empty word ε is always a position in any term t, and $t_{|\varepsilon} = t$. Whenever $t_{|p}$ is defined, and is not a variable, i.e., $t_{|p}$ is of the form $f(t_1, \ldots, t_n)$, then pi is a position in t for every i, $1 \leqslant i \leqslant n$, and $t_{|pi} = t_i$. The *size* of a term is the number of its positions. We write $t[s]_p$ for the term t, except that the subterm at position p has been replaced by s.

Definition 9.7.39 (Tree topology) Let X be any topological space, and $Y = \mathcal{T}(X)$. The *simple tree expressions* are inductively defined by: if U is open in X, and $\pi_1, \ldots, \pi_n$ are simple tree expressions, then $\Diamond U(\pi_1 \mid \cdots \mid \pi_n)$ is a simple tree expression. It denotes the set of terms t that have a subterm $f(\vec{t})$ with $f \in U$ and $\vec{t} \in Y^*\pi_1 Y^* \ldots Y^*\pi_n Y^*$.

The *tree topology* on $Y = \mathcal{T}(X)$ is the coarsest one containing the (denotations of) all simple tree expressions.

The base case of the inductive definition above is the case $n = 0$: in this case, $\Diamond U()$ is the set of terms that contain a subterm $f(\vec{t})$ with $f \in U$, and $\vec{t}$ arbitrary.

Generalize the notation $\Diamond U(\pi_1 \mid \cdots \mid \pi_n)$ as follows. We define $\Diamond U \cdot \mathcal{U}$, where U is any open subset of X, and $\mathcal{U}$ is any open subset of $\mathcal{T}(X)^*$, as the set of all terms that have a subterm of the form $f(\vec{t})$ with $f \in U$ and $\vec{t} \in \mathcal{U}$. We shall see that these subsets are always open in $Y = \mathcal{T}(X)$. For now, we know that they are open whenever $\mathcal{U}$ is an *elementary open*, i.e., of the form $Y^*\pi_1 Y^* \ldots Y^*\pi_n Y^*$.

Exercise 9.7.40 Show that every finite intersection of simple tree expressions can be rewritten as a finite union of simple tree expressions. (For clarity, find an explicit formula for the intersection of $\Diamond U \cdot \mathcal{U}$ and $\Diamond U' \cdot \mathcal{U}'$. When $\mathcal{U}$ and $\mathcal{U}'$ are elementary opens, you may use the formulae given in Exercise 9.7.28.) In particular, the (denotations of) simple tree expressions form a base of the tree topology.

Exercise 9.7.41 Let X be any topological space. Show that, for every open subset U of X, for every open subset $\mathcal{U}$ of $\mathcal{T}(X)^*$, $\Diamond U \cdot \mathcal{U}$ is open in $\mathcal{T}(X)$.

Definition 9.7.42 (Tree embedding) Let $\leqslant$ be a quasi-ordering on a set X. The *tree embedding* quasi-ordering $\preceq_\leq$ on $\mathcal{T}(X)$ is defined by induction on the sum of the sizes of the terms s, t, by $s \preceq_\leq t$ iff:

- either $t = g(t_1, \ldots, t_n)$ and $s \preceq_\leq t_j$ for some j, $1 \leqslant j \leqslant n$;
- or $s = f(\vec{s})$, $t = g(\vec{t})$, $f \leqslant g$ and $\vec{s} \preceq^*_\leq \vec{t}$.

We recall here that $\preceq^*_\leq$ is the word embedding quasi-ordering on words, understanding that the letters (which are terms themselves here) are quasi-ordered by $\preceq_\leq$. The purpose of the following exercises is to show that $\preceq_\leq$ is the specialization quasi-ordering of $\mathcal{T}(X)$.

Exercise 9.7.43 Let X be a topological space, and $\leqslant$ be its specialization quasi-ordering. Show that every open subset of $\mathcal{T}(X)$ is upward closed with respect to $\preceq_\leq$. So every closed subset of $\mathcal{T}(X)$ is downward closed with respect to $\preceq_\leq$.

Exercise 9.7.44 Let X be a topological space, and $Y = \mathcal{T}(X)$. For every open subset U of X, every open subset $\mathcal{U}$ of Y^*, and every open subset V of Y, let $\Diamond U \cdot \mathcal{U} \,//\, V$ be the set of all terms containing a subterm $f(\vec{t}) \in V$ with $f \in U$ and $\vec{t} \in \mathcal{U}$. Show that $\Diamond U \cdot \mathcal{U} \,//\, V$ is open in $\mathcal{T}(X)$.

For every closed subset F of X, and all closed subsets $\mathcal{F}_1, \mathcal{F}_2, \ldots, \mathcal{F}_n$ of Y, let $F^?(\mathcal{F}_1^? \mathcal{F}_2^? \ldots \mathcal{F}_n^?)$ denote the union of $\mathcal{F}_1, \mathcal{F}_2, \ldots, \mathcal{F}_n$ with the set of those terms $f(\vec{t})$ such that $f \in F$ and $\vec{t} \in \mathcal{F}_1^? \mathcal{F}_2^? \ldots \mathcal{F}_n^?$. (See Exercise 9.7.29 for the notation $\mathcal{F}_1^? \mathcal{F}_2^? \ldots \mathcal{F}_n^?$.) Show that $F^?(\mathcal{F}_1^? \mathcal{F}_2^? \ldots \mathcal{F}_n^?)$ is closed in Y: let $U = X \smallsetminus F$, V be the complement of $\mathcal{F}_1 \cup \mathcal{F}_2 \cup \cdots \cup \mathcal{F}_n$ in Y, and $\mathcal{U}$ be the complement of $\mathcal{F}_1^? \mathcal{F}_2^? \ldots \mathcal{F}_n^?$ in Y^*, then show that $F^?(\mathcal{F}_1^? \mathcal{F}_2^? \ldots \mathcal{F}_n^?)$ is the complement of $(\Diamond X \cdot \mathcal{U} \,//\, V) \cup (\Diamond U \cdot Y^* \,//\, V)$.

Conclude that if $\leqslant$ is the specialization quasi-ordering of X, then the tree embedding quasi-ordering $\preceq_\leq$ is the specialization quasi-ordering of $\mathcal{T}(X)$.

Let $\mathcal{B}$ be the subbase of all non-empty simple tree expressions. We plan to use Lemma 9.7.31, and quasi-order $\mathcal{B}$ by the following relation $\sqsubseteq$. For any simple tree expressions $\pi = \Diamond U(\pi_1 \mid \cdots \mid \pi_m)$ and $\pi' = \Diamond U'(\pi'_1 \mid \cdots \mid \pi'_n)$, we let $\pi \sqsubseteq \pi'$ iff either $\pi = \pi'$ or $\pi \sqsubseteq \pi'_j$ for some j, $1 \leqslant j \leqslant n$. In other words, if one reads simple tree expressions $\pi = \Diamond U(\pi_1 \mid \cdots \mid \pi_n)$ as terms $U(\pi_1, \ldots, \pi_n)$, then $\sqsubseteq$ is the subterm ordering.

Lemma 9.7.45 *Let X be a topological space, $Y = \mathcal{T}(X)$, and $\mathcal{B}$ be the basis of all non-empty simple tree expressions on Y, quasi-ordered by the above relation $\sqsubseteq$. Assume a minimal bad sequence $\pi_0, \pi_1, \ldots, \pi_n, \ldots$ of elements of $\mathcal{B}$, and write π_n as $\Diamond U_n(\pi_{n1} \mid \cdots \mid \pi_{nm_n})$ for each $n \in \mathbb{N}$.*

Let $V_{ni} = \pi_{ni} \setminus \bigcup_{p=0}^{n-1} \pi_p$, $n \in \mathbb{N}$, $1 \leqslant i \leqslant m_n$, and let Y' be the space Y, with the topology generated by the subsets V_{ni}. Then Y' is Noetherian.

Proof Assume Y' is not Noetherian. By Lemma 9.7.15, there would be a bad sequence $V_{n_0 i_0}, V_{n_1 i_1}, \ldots, V_{n_q i_q}, \ldots$ Among the indices $n_0, n_1, \ldots, n_q, \ldots$, there is a least one, say n_q. Then the suffix $V_{n_q i_q}, V_{n_{q+1} i_{q+1}}, \ldots$ is again bad. Re-indexing if necessary, we can therefore assume that n_0 is least among n_0, $n_1, \ldots, n_q, \ldots$

The sequence $\pi_0, \pi_1, \ldots, \pi_{n_0-1}, \pi_{n_0 i_0}, \pi_{n_1 i_1}, \ldots, \pi_{n_q i_q}, \ldots$ is lexicographically smaller than the minimal bad sequence $\pi_0, \pi_1, \ldots, \pi_n, \ldots$, and is therefore good. So, for some $q \in \mathbb{N}$, $\pi_{n_q i_q}$ is included in $\pi_0 \cup \pi_1 \cup \cdots \cup \pi_{n_0-1} \cup \pi_{n_0 i_0} \cup \pi_{n_1 i_1} \cup \cdots \cup \pi_{n_{q-1} i_{q-1}}$. (It cannot be the case that π_n is included in $\pi_0 \cup \pi_1 \cup \cdots \cup \pi_{n-1}$ for some $n \leqslant n_0$, since $\pi_0, \pi_1, \ldots, \pi_n, \ldots$ is bad.) Take any element t of $V_{n_q i_q} = \pi_{n_q i_q} \setminus (\pi_0 \cup \pi_1 \cup \cdots \cup \pi_{n_0-1} \cup \pi_{n_0} \cup \cdots \cup \pi_{n_q-1})$, in particular t is in $\pi_{n_q i_q}$ hence in $\pi_0 \cup \pi_1 \cup \cdots \cup \pi_{n_0-1} \cup \pi_{n_0 i_0} \cup \pi_{n_1 i_1} \cup \cdots \cup \pi_{n_{q-1} i_{q-1}}$. Since t is not in $\pi_0 \cup \pi_1 \cup \cdots \cup \pi_{n_0-1}$, t must be in some $\pi_{n_r i_r} \setminus (\pi_0 \cup \pi_1 \cup \cdots \cup \pi_{n_0-1} \cup \pi_{n_0} \cup \cdots \cup \pi_{n_q-1})$ with $r < q$, hence in $V_{n_r i_r}$. It follows that $V_{n_q i_q}$ is included in $V_{n_0 i_0} \cup V_{n_1 i_1} \cup \cdots \cup V_{n_{q-1} i_{q-1}}$, contradicting the fact that $V_{n_0 i_0}, V_{n_1 i_1}, \ldots, V_{n_q i_q}, \ldots$ is bad. □

The following is a topological generalization of a result known as Kruskal's Theorem, which we shall see in Exercise 9.7.48.

Theorem 9.7.46 (Topological Kruskal Theorem) *Let X be a topological space. Then X is Noetherian if and only if $\mathcal{T}(X)$ is.*

Proof Assume that X is Noetherian but $Y = \mathcal{T}(X)$ is not.

By Lemma 9.7.31, there is a minimal bad sequence $\pi_0, \pi_1, \ldots, \pi_n, \ldots$ of elements of $\mathcal{B}$. Write π_n as $\Diamond U_n(\pi_{n1} \mid \cdots \mid \pi_{nm_n})$. Since the sequence is bad, for each $n \in \mathbb{N}$ there is a term $t_n \in \pi_n$ that is not in $\pi_0 \cup \pi_1 \cup \cdots \cup \pi_{n-1}$. By definition, there is a position p_n such that $t_{n|p_n}$ is a subterm of the form $f_n(\vec{t_n})$ with $f_n \in U_n$ and $\vec{t_n} \in Y^* \pi_{n1} Y^* \ldots Y^* \pi_{nm_n} Y^*$ (abbreviating $\mathcal{T}(X)$ as Y). In particular, $\vec{t_n}$ must contain a subword $t_{n1} \ldots t_{nm_n}$, where $t_{n1} \in \pi_{n1}, \ldots, t_{nm_n} \in \pi_{nm_n}$. Now replacing $f_n(\vec{t_n})$ by $f_n(t_{n1}, \ldots, t_{nm_n})$ in t_n yields a term that is still in $\pi_n = \Diamond U_n(\pi_{n1} \mid \cdots \mid \pi_{nm_n})$, and which is smaller that t_n in $\preceq_\leq$, so cannot be in $\pi_0 \cup \pi_1 \cup \cdots \cup \pi_{n-1}$, using Exercise 9.7.43. Without loss of generality, we can therefore assume that $\vec{t_n}$ is exactly the subword $t_{n1} \ldots t_{nm_n}$, where $t_{n1} \in \pi_{n1}, \ldots, t_{nm_n} \in \pi_{nm_n}$.

Let V_{ni}, $n \in \mathbb{N}$, $1 \leqslant i \leqslant m_n$, and Y' be given as in Lemma 9.7.45. We claim that the word $\vec{t_n} = t_{n1} \ldots t_{nm_n}$ is in $Y'^* V_{n1} Y'^* \ldots Y'^* V_{nm_n} Y'^*$. To prove this, it is enough to observe that t_{ni} is in V_{ni} for every i, $1 \leqslant i \leqslant m_n$, i.e., that it is in π_{ni} (which is clear), but not in any π_p, $p < n$. Indeed, if t_{ni} were in π_p, then the larger term t_n (with respect to $\preceq_\leq$) would also be in π_p by Exercise 9.7.43, and this is impossible.

By Lemma 9.7.45, Y' is Noetherian. By Theorem 9.7.33, Y'^* is then Noetherian, too. Using Proposition 9.7.18 (vii), $X \times Y'^*$ is Noetherian. So there is an index $k \in \mathbb{N}$ such that $(f_k, \vec{t_k})$ is a cluster point of $(f_n, \vec{t_n})_{n \in \mathbb{N}}$, by Theorem 9.7.8 (ii). In particular, there are infinitely many indices $n \in \mathbb{N}$ such that $(f_n, \vec{t_n}) \in U_k \times Y'^* V_{k1} Y'^* \ldots Y'^* V_{km_k} Y'^*$, in particular such that

$f_n \in U_k$ and $\vec{t_n} \in Y^* \pi_{k1} Y^* \dots Y^* \pi_{km_k} Y^*$. Take one with $n > k$. Then $t_n \in \Diamond U_k(\pi_{k1} \mid \cdots \mid \pi_{km_k}) = \pi_k$, which is impossible.

So $\mathcal{T}(X)$ is Noetherian.

Conversely, assume $\mathcal{T}(X)$ is Noetherian. Given any ascending chain of opens $U_0 \subseteq U_1 \subseteq \cdots \subseteq U_k \subseteq \cdots$ in X, the corresponding ascending chain $\Diamond U_0() \subseteq \Diamond U_1() \subseteq \cdots \subseteq \Diamond U_k() \subseteq \cdots$ stabilizes, using Proposition 9.7.6. But $\Diamond U_k() \subseteq \Diamond U_n()$ implies $U_k \subseteq U_n$: if there were an element $f \in U_k \setminus U_n$, $f()$ would be in $\Diamond U_k() \setminus \Diamond U_n()$. So $U_0 \subseteq U_1 \subseteq \cdots \subseteq U_k \subseteq \cdots$ also stabilizes. By Proposition 9.7.6, X is therefore Noetherian. □

Exercise 9.7.47 Recall that $\mathcal{T}(X, \mathcal{V})$ is the space of all terms $t \in \mathcal{T}(X + \mathcal{V})$ such that the variables that occur in t are applied to no argument at all. Write v (instead of $v()$) for the variable v seen as a term. $\mathcal{T}(X, \mathcal{V})$ comes with the topology induced from $\mathcal{T}(X + \mathcal{V})$.

Let the *application map* $@ : X \times \mathcal{T}(X, \mathcal{V})^* \to \mathcal{T}(X, \mathcal{V})$ send $(f, \vec{t})$ to $f(\vec{t})$. Show that @ is continuous.

Show that the map $\xi : X \to \mathcal{T}(X, \mathcal{V})$ that sends f to the term $f()$ is continuous, and that the map $v : \mathcal{V} \to \mathcal{T}(X, \mathcal{V})$ that sends every variable v to v, seen as a tree, is continuous, where $\mathcal{V}$ is equipped with the discrete topology.

Given two topological spaces X and X', a set $\mathcal{V}$, a continuous map $h : X \to X'$, and any function $g : \mathcal{V} \to \mathcal{T}(X')$, show that the map $\mathcal{T}(h, g) : \mathcal{T}(X, \mathcal{V}) \to \mathcal{T}(X')$ defined by $\mathcal{T}(h, g)(v) = g(v)$ for every $v \in \mathcal{V}$, and $\mathcal{T}(h, g)(f(t_1, \dots, t_n)) = h(f)(\mathcal{T}(h, g)(t_1), \dots, \mathcal{T}(h, g)(t_n))$, is continuous.

Given any $v \in \mathcal{V}$, let $u[v := t]$ be the *substitution* of t for v in u, i.e., the term obtained from u by replacing every occurrence of v by t. Using the above, show that the map $u, t \mapsto u[v := t]$ is continuous from $\mathcal{T}(X, \mathcal{V}) \times \mathcal{T}(X)$ to $\mathcal{T}(X)$.

Exercise 9.7.48 (Kruskal's Theorem) Let X be a set equipped with a quasi-ordering $\leqslant$. Show that the tree topology on $\mathcal{T}(X_a)$ is the Alexandroff topology of $\preceq_\leq$. (Hint: let $Y = \mathcal{T}(X)$, $s = f(s_1, \dots, s_m)$, then the upward closure $\uparrow_Y s$ of s in Y is $\Diamond(\uparrow_X f)(\uparrow_Y s_1 \mid \cdots \mid \uparrow_Y s_m)$.)

Deduce *Kruskal's Theorem* (Kruskal, 1960): $\preceq_\leq$ is wqo on $\mathcal{T}(X)$ if and only if $\leqslant$ is wqo on X.

Exercise 9.7.49 (Noetherian rings) The notion of Noetherian space arose from the study of those (commutative) rings that are *Noetherian*, in the sense that every infinite ascending chain of ideals stabilizes. Show that the spectrum of a Noetherian ring is a sober Noetherian space. (The converse fails: there are non-Noetherian rings whose spectrum is a Noetherian space.)

Exercise 9.7.50 Show that a commutative ring R is Noetherian if and only if every ideal of R is finitely generated, i.e., of the form (A) for some finite set A.

Exercise 9.7.51 (Rings of polynomials) Given a commutative ring R, a *polynomial* P (in one variable X, and with coefficients in R) is any expression of the form $a_k X^k + a_{k-1} X^{k-1} + \cdots + a_1 X + a_0$ with $a_k, a_{k-1}, \ldots, a_1, a_0 \in R$. Formally, a polynomial is just an infinite sequence $a_0 a_1 \ldots a_k \ldots$ of elements of R that are equal to 0 except for finitely many. We also write it $\sum_{i=0}^{k} a_i X^i$, or $\sum_{i \in \mathbb{N}} a_i X^i$. The terms $a_i X^i$ are the *monomials*, with *coefficient* a_i. If $a_i = 0$ for every $i \in \mathbb{N}$, then the polynomial is the *zero polynomial*, written 0. Otherwise, one can assume $a_k \neq 0$; then k is the *degree* of the polynomial, $a_k X^k$ is its *leading monomial*, and a_k is its *leading coefficient*. (We agree that the degree is extended to the case of the zero polynomial, by letting it be $-\infty$ in this case.) The set $R[X]$ of all polynomials is a commutative ring, with addition defined coefficient-wise, i.e., $\sum_{i \in \mathbb{N}} a_i X^i + \sum_{i \in \mathbb{N}} b_i X^i = \sum_{i \in \mathbb{N}} (a_i + b_i) X^i$, and product defined by $(\sum_{i \in \mathbb{N}} a_i X^i).(\sum_{j \in \mathbb{N}} b_j X^j) = \sum_{m \in \mathbb{N}} (\sum_{\substack{i,j \in \mathbb{N} \\ i+j=m}} a_i b_j) X^m$.

Observe that every non-zero polynomial can be written in a unique way as $a_k X^k + P$, where $a_k X^k$ is the leading monomial and P is of degree at most $k-1$. Call this its *standard form*. Given any ideal I in $R[X]$, observe that the set L of leading coefficients of non-zero polynomials in I, plus the zero of R, is an ideal in R. If R is Noetherian, let $L = (a_1, \ldots, a_n)$. By definition, there are polynomials in standard form $a_i X^{k_i} + P_i$, $1 \leqslant i \leqslant n$ in I. Let J be the ideal they generate. Show that $I = J + I_{<k}$, where $I_{<k}$ denotes the set of polynomials of degree strictly less than k in I. Then, considering the ideal of all coefficients of X^{k-1} in polynomials in $I_{<k}$, show that there are finitely many polynomials $P'_1, \ldots, P'_m$ in $I_{<k}$ such that every element of I of degree at most $d-1$ is a linear combination $\sum_{j=1}^{m} r_j P'_j$, with each $r_j \in R$. Conclude that I is finitely generated.

From this, deduce the *Hilbert Basis Theorem*: for every Noetherian ring R, $R[X]$ is also a Noetherian ring.

The commutative ring of polynomials in m variables, $R[X_1, \ldots, X_m]$, is defined as R if $m = 0$, and as $R[X_1, \ldots, X_{m-1}][X_m]$ otherwise. Show that $R[X_1, \ldots, X_m]$ is Noetherian whenever X is a Noetherian ring. Conclude that $\mathbb{Z}[X_1, \ldots, X_m]$, $\mathbb{Q}[X_1, \ldots, X_m]$, $\mathbb{R}[X_1, \ldots, X_m]$, and $\mathbb{C}[X_1, \ldots, X_m]$ are Noetherian rings. ($\mathbb{C} = \mathbb{R}^2$ is the set of complex numbers. The pair (a, b) is written $a + ib$, with addition $(a + ib) + (a' + ib') = (a + a') + i(b + b')$, $-(a + ib) = -a + i(-b)$, and product $(a + ib).(a' + ib') = (aa' - bb') + i(ba' + ab')$.)

Exercise 9.7.52 (Kaplansky's Theorem) Let R be a commutative ring. Show that, for any two ideals I and I' of R, $F_I \subseteq F_{I'}$ iff $\sqrt{I'} \subseteq \sqrt{I}$. (Use the homeomorphism from the last part of Exercise 9.5.38, and realize that $F_I = F_{\sqrt{I}}$.) From this and Theorem 9.7.12, deduce *Kaplansky's Theorem*: every radical ideal I of a Noetherian ring R is the intersection of finitely many prime ideals.

Exercise 9.7.53 (The Zariski topology on $\mathbb{C}^m$) Let $m \in \mathbb{N}$. For every $\vec{x} \in \mathbb{C}^m$, and every polynomial $P \in \mathbb{Z}[X_1, \ldots, X_m]$, one can evaluate P at $\vec{x}$: informally, $P(\vec{x})$ is obtained by replacing X_1 by $x_1, \ldots, X_m$ by x_m, and simplifying. Let $\varphi(\vec{x}) = \{P \in \mathbb{Z}[X_1, \ldots, X_m] \mid P(\vec{x}) = 0\}$. Show that $\varphi(\vec{x})$ is a prime ideal of $R = \mathbb{Z}[X_1, \ldots, X_m]$.

The *Zariski topology* on $\mathbb{C}^m$ is the coarsest topology that makes φ continuous as a map from $\mathbb{C}^m$ to the spectrum of $R = \mathbb{Z}[X_1, \ldots, X_m]$. That is, its closed subsets are $\varphi^{-1}(F_I) = \{\vec{x} \in \mathbb{C}^m \mid \forall P \in I \cdot P(\vec{x}) = 0\}$. (The latter, i.e., the sets of common zeroes of sets of polynomials with coefficients in $\mathbb{Z}$, are called the *algebraic sets*. This has nothing to do with algebraic dcpos.) Show that $\mathbb{C}^m$ with the Zariski topology is a Noetherian space, whose topology is coarser than the usual, product topology on $\mathbb{C}^m$. You will need the following deep theorem, called *Hilbert's Nullstellensatz* (Zariski, 1947): for every ideal I of R, $\sqrt{I} = \bigcap_{\vec{x} \in \varphi^{-1}(F_I)} \varphi(\vec{x})$, i.e., the polynomials that evaluate to 0 on every $\vec{x}$ that is a common zero to all elements of I is the radical $\sqrt{I}$ of I.

Using Kaplansky's Theorem, show that, up to homeomorphism, the spectrum of $\mathbb{Z}[X_1, \ldots, X_m]$ is the sobrification of $\mathbb{C}^m$ in its Zariski topology.

References

Abramsky, S. and Jung, A. 1994. Domain Theory. Pages 1–168 of Abramsky, S., Gabbay, D. M., and Maibaum, T. S. E. (eds.), *Handbook of Logic in Computer Science*, vol. 3. Oxford University Press.

Adámek, J., Herrlich, H., and Strecker, G. E. 2009. *Abstract and Concrete Categories: The Joy of Cats*. Dover.

Ali-Akbari, M., Honari, B., Pourmahdian, M., and Rezaii, M. M. 2009. The Space of Formal Balls and Models of Quasi-Metric Spaces. *Mathematical Structures in Computer Science*, **19**, 337–355.

Alvarez-Manilla, M., Jung, A., and Keimel, K. 2004. The Probabilistic Powerdomain for Stably Compact Spaces. *Theoretical Computer Science*, **328**(3), 221–244.

Arens, R. F. 1946. A Topology for Spaces of Transformations. *Annals of Mathematics*, **47**(3), 480–495.

Banaschewski, B. 1996. Radical Ideals and Coherent Frames. *Commentationes Mathematicae Universitas Carolinae*, **37**(2), 349–370.

Bennett, H. R., Lutzer, D. J., and Reed, G. M. 2008. Domain Representability and the Choquet Game in Moore and BCO-Spaces. *Topology and Its Applications*, **155**, 445–458.

Bing, R. H. 1951. Metrization of Topological Spaces. *Canadian Journal of Mathematics*, **3**, 175–186.

Bonsangue, M. M., van Breugel, F., and Rutten, J. J. M. M. 1998. Generalized Metric Spaces: Completion, Topology, and Powerdomains via the Yoneda Embedding. *Theoretical Computer Science*, **193**, 1–51.

Bourbaki, N. 1971. *Eléments de Mathématique: Topologie Générale, Chapitres 1–4*. Diffusion C.C.L.S.

Choquet, G. 1969. *Lectures on Analysis*. Mathematics Lecture Note Series, vol. I: Integration and Topological Vector Spaces. W. A. Benjamin.

Cohen, P. J. 2008. *Set Theory and the Continuum Hypothesis*. Dover. First published by W. A. Benjamin, 1966.

Cousot, P. and Cousot, R. 1977. Abstract Interpretation: A Unified Lattice Model for Static Analysis of Programs by Construction or Approximation of Fixpoints. Pages 238–252 of *Conference Record of the Fourth Annual ACM SIGPLAN-SIGACT Symposium on Principles of Programming Languages*. Los Angeles, CA. ACM Press.

Day, B. J. 1972. A Reflection Theorem for Closed Categories. *Journal of Pure and Applied Algebra*, **2**(1), 1–11.

Dieudonné, J. 1944. Une Généralisation des Espaces Compacts. *Journal de Mathématiques Pures et Appliquées*, **23**, 65–76.

Dolecki, S., Greco, G. H., and Lechicki, A. 1995. When Do the Upper Kuratowski Topology (Homeomorphically, Scott Topology) and the Co-Compact Topology Coincide? *Transactions of the American Mathematical Society*, **347**(8), 2869–2884.

Dorais, F. G. and Mummert, C. 2010. Stationary and Convergent Strategies in Choquet Games. *Fundamenta Mathematicae*, **209**, 59–79.

Edalat, A. and Heckmann, R. 1998. A Computational Model for Metric Spaces. *Theoretical Computer Science*, **193**, 53–73.

Erné, M. 1991. The ABC of Order and Topology. Pages 57–83 of *Category Theory at Work, Proceedings of a Workshop*. Research and Exposition in Mathematics, vol. 18. Heldermann.

Ershov, Y. L. 1973. The Theory of A-Spaces. *Algebra and Logic*, **12**(4), 209–232.

Ershov, Y. L. 1997. The Bounded Complete Hull of an α-Space. *Theoretical Computer Science*, **175**, 3–13.

Escardó, M. and Heckmann, R. 2001–2002. Topologies on Spaces of Continuous Functions. *Topology Proceedings*, **26**(2), 545–564.

Escardó, M., Lawson, J. D., and Simpson, A. 2004. Comparing Cartesian Closed Categories of (Core) Compactly Generated Spaces. *Topology and Its Applications*, **143**(1–3), 105–146.

Finkel, A. and Schnoebelen, P. 2001. Well-Structured Transition Systems Everywhere! *Theoretical Computer Science*, **256**(1–2), 63–92.

Frolík, Z. 1960. Generalizations of the G_δ-Property of Complete Metric Spaces. *Czechoslovak Mathematical Journal*, **10**(3), 359–379.

Gierz, G., Hofmann, K. H., Keimel, K., Lawson, J. D., Mislove, M., and Scott, D. S. 2003. *Continuous Lattices and Domains*. Encyclopedia of Mathematics and Its Applications, vol. 93. Cambridge University Press.

Goubault-Larrecq, J. 2010. Noetherian Spaces in Verification. Pages 2–21 of Abramsky, S., Meyer auf der Heide, F., and Spirakis, P. (eds.), *Proceedings of the 37th International Colloquium on Automata, Languages, and Programming (ICALP'10), Part II*. 6199. Lecture Notes in Computer Science. Springer.

Goubault-Larrecq, J. and Mackie, I. 1997. *Proof Theory and Automated Deduction*. Applied Logic Series, vol. 6. Kluwer Academic Publishers.

Higman, G. 1952. Ordering by Divisibility in Abstract Algebras. *Proceedings of the London Mathematical Society*, **2**(7), 326–336.

Hochster, M. 1969. Prime Ideal Structure in Commutative Rings. *Transactions of the American Mathematical Society*, **142**, 43–60.

Hoffmann, R.-E. 1979a. On the Sobrification Remainder ${}^sX - X$. *Pacific Journal of Mathematics*, **83**(1), 145–156.

Hoffmann, R.-E. 1979b. Sobrification of Partially Ordered Sets. *Semigroup Forum*, **17**, 123–138.

Hofmann, K.-H., and Lawson, J. D. 1978. The Spectral Theory of Distributive Continuous Lattices. *Transactions of the American Mathematical Society*, **246**, 285–310.

Iwamura, T. 1944. A Lemma on Directed Sets. *Zenkoku Shijo Sugaku Danwakai*, **262**, 107–111. In Japanese.

Johnstone, P. 1982. *Stone Spaces*. Cambridge Studies in Advanced Mathematics, vol. 3. Cambridge University Press.

Jung, A. 1988. *Cartesian Closed Categories of Domains*. Ph.D. thesis, Technische Hochschule Darmstadt.

Jung, A. 1990. The Classification of Continuous Domains. Pages 35–40 of *Proceedings of the Fifth Annual IEEE Symposium on Logic in Computer Science*. IEEE Computer Society Press.

Jung, A. and Sünderhauf, P. 1998. Uniform Approximation of Topological Spaces. *Topology and Its Applications*, **83**(1), 23–37.

Kelley, J. L. 1950. Convergence in Topology. *Duke Mathematical Journal*, **17**(3), 277–283.

Kelley, J. L. 1955. *General Topology*. University Series in Higher Mathematics. Van Nostrand Reinhold. Reprinted 1975 in Graduate Texts in Mathematics, Springer.

Kostanek, M. and Waszkiewicz, P. 2011. The Formal Ball Model for $\mathcal{Q}$-Categories. *Mathematical Structures in Computer Science*, **21**(1), 1–24.

Kruskal, J. B. 1960. Well-Quasi-Ordering, the Tree Theorem, and Vazsonyi's Conjecture. *Transactions of the American Mathematical Society*, **95**(2), 210–225.

Künzi, H.-P. A. 2009. An Introduction to Quasi-Uniform Spaces. Pages 239–304 of Mynard, F. and Pearl, E. (eds.), *Beyond Topology*. Contemporary Mathematics, vol. 486. American Mathematical Society.

Lang, S. 2005. *Algebra*. 3rd edn. Graduate Texts in Mathematics. Springer.

Lawson, J. D. 1997a. The Round Ideal Completion via Sobrification. *Topology Proceedings*, **22**, 261–274.

Lawson, J. D. 1997b. Spaces of Maximal Points. *Mathematical Structures in Computer Science*, **7**, 543–555.

Lawvere, F. W. 2002. Metric Spaces, Generalized Logic, and Closed Categories. *Reprints in Theory and Applications of Categories*, **1**, 1–27. Originally published in *Rendiconti del Seminario Matematico e Fisico di Milano*, XLIII (1973), 135–166.

Mac Lane, S. 1971. *Categories for the Working Mathematician*. Graduate Texts in Mathematics, vol. 5. Springer.

Markowsky, G. 1976. Chain-Complete Posets and Directed Sets with Applications. *Algebra Universalis*, **6**, 53–68.

Martin, K. 2003. The Regular Spaces with Countably Based Models. *Theoretical Computer Science*, **305**, 299–310.

Martin, K. 2004. *Topological Games in Domain Theory*. Technical report PRG-RR-02-04. Programming Research Group, University of Oxford.

Mendelson, E. 1997. *An Introduction to Mathematical Logic*. 4th edn. Chapman and Hall.

Mislove, M. 1998. Topology, Domain Theory and Theoretical Computer Science. *Topology and Its Applications*, **89**, 3–59.

Moore, E. H. and Smith, H. L. 1922. A General Theory of Limits. *American Journal of Mathematics*, **44**(2), 102–121.

Myhill, J. and Shepherdson, J. C. 1955. Effective Operations on Partial Recursive Functions. *Zeitschrift für Mathematische Logik und Grundlagen der Mathematik*, **1**(4), 310–317.

Nachbin, L. 1948. Sur les Espaces Uniformes Ordonnés. *Comptes Rendus de l'Académie des Sciences*, **MR 9**(455), 774–775. English translation in Nachbin (1965).

Nachbin, L. 1965. *Topology and Order*. Van Nostrand. Translated from the 1950 monograph *Topologia e Ordem* (in Portuguese). Reprinted by Robert E. Kreiger Publishing Co., Huntington, NY, 1967, 1976.

Nagata, J.-I. 1950. On a Necessary and Sufficient Condition of Metrizability. *Journal of the Institute of Polytechnics, Osaka City University, Series A*, **1**(2), 93–100.

Noble, N. 1969. Ascoli Theorems and the Exponential Map. *Transactions of the American Mathematical Society*, **143**, 393–411.

Norberg, T. 1989. Existence Theorems for Measures on Continuous Posets, with Applications to Random Set Theory. *Mathematica Scandinavica*, **64**, 15–51.

Oxtoby, J. 1957. The Banach-Mazur Game and Banach Category Theorem. Chap. 7, pages 159–163 of Dresher, M., Tucker, A. W., and Wolfe, P. (eds.), *Contributions to the Theory of Games*. Annals of Mathematical Studies, vol. III. Princeton University Press.

Ponomarev, V. I. 1960. Axioms of Countability and Continuous Mappings. *Bulletin de l'Académie Polonaise des Sciences, Série Sciences Mathématiques Astronomiques et Physiques*, **8**, 127–134.

Rado, R. 1949. Axiomatic Treatment of Rank in Infinite Sets. *Canadian Journal of Mathematics*, **1**, 337–343.

Rado, R. 1954. Partial Well-Ordering of Sets of Vectors. *Mathematika*, **1**, 89–95.

Romaguera, S. and Valero, O. 2010. Domain Theoretic Characterisations of Quasi-Metric Completeness in Terms of Formal Balls. *Mathematical Structures in Computer Science*, **20**, 453–472.

Rutten, J. J. M. M. 1996. Elements of Generalized Ultrametric Domain Theory. *Theoretical Computer Science*, **170**, 349–381.

Scott, D. S. 1975. Data Types as Lattices. Pages 579–651 of *Proceedings of the International Summer Institute and Logic Colloquium*. Lecture Notes in Mathematics, vol. 499. Springer.

Shmuely, Z. 1974. The Structure of Galois Connections. *Pacific Journal of Mathematics*, **54**(2), 209–225.

Smirnov, Y. 1951. A Necessary and Sufficient Condition for Metrizability of Topological Space. *Doklady Akademia Nauk SSSR N.S.*, **77**, 197–200.

Smyth, M. B. 1988. Quasi-Uniformities: Reconciling Domains with Metric Spaces. Pages 236–253 of Main, M., Melton, A., Mislove, M., and Schmidt, D. (eds.), *Mathematical Foundations of Programming Language Semantics*. Lecture Notes in Computer Science, vol. 298. Springer.

Smyth, M. B. 1995. Inverse Limits of Graphs. Pages 397–409 of *Proceedings of the 2nd Imperial College Department of Computing Workshop on Theory and Formal Methods 1994*. Imperial College Press.

Sorgenfrey, R. H. 1947. On the Topological Product of Paracompact Spaces. *Bulletin of the American Mathematical Society*, **53**, 631–632.

Srivastava, S. M. 1998. *A Course on Borel Sets*. Graduate Texts in Mathematics. Springer.

Steenrod, N. E. 1967. A Convenient Category of Topological Spaces. *Michigan Mathematical Journal*, **14**, 133–152.

Stone, A. H. 1948. Paracompactness and Product Spaces. *Bulletin of the American Mathematical Society*, **54**, 977–982.

Streicher, T. 2006. *Domain-Theoretic Foundations of Functional Programming*. World Scientific.

Strickland, N. P. 2009. *Compactly Generated Spaces*. Course Notes for Cambridge Part III course on homotopy theory. Available at http://neil-strickland.staff.shef.ac.uk/courses/homotopy/cgwh.pdf

Urysohn, P. S. 1925. Zum Metrisationsproblem. *Mathematische Annalen*, **94**, 309–315.

Waszkiewicz, P. 2010. Common Patterns for Metric and Ordered Fixed Point Theorems. Pages 83–87 of Santocanale, L. (ed.), *Proceedings of the 7th Workshop on Fixed Points in Computer Science*. Available at http://hal.archives–ouvertes.fr/docs/00/51/23/77/pdf/proceedings.pdf.

Weihrauch, K. and Schneider, U. 1981. Embedding Metric Spaces into CPO's. *Theoretical Computer Science*, **16**(1), 5–24. Elsevier.

Wilson, W. A. 1931. On Quasi-Metrizable Spaces. *American Journal of Mathematics*, **53**, 675–684.

Winskel, G. 1993. *The Formal Semantics of Programming Languages: An Introduction*. Foundations of Computing. MIT Press.

Wyler, O. 1979. Dedekind Complete Posets and Scott Topologies. Pages 384–389 of Banaschewski, B. and Hoffman, R.-E. (eds.), *Continuous Lattices*. Lecture Notes in Mathematics, Springer.

Zariski, O. 1947. A New Proof of Hilbert's Nullstellensatz. *Bulletin of the American Mathematical Society*, **LIII**, 362–368.

Zhang, H. 1993. *Dualities of Domains*. Ph.D. thesis, Department of Mathematics, Tulane University.

Notation index

$\check{f}$ (extension) 418
$\ulcorner E \urcorner$ 7
$\ulcorner m, n \urcorner$ 6
$(\preceq)$ 399
(A) 439
$(\in)$ 142
$+$ (coproduct) 80, 117
$\dashv$ (adjunction) 173
$\dot{\to}$ (natural transformation) 173
$\blacksquare U$ 418
$\blacksquare Q$ 156
$\Downarrow b$ 127
$\Subset$ 141
@ 472
$\bot$ 10
$\twoheaddownarrow x$ 122
$\downarrow A$ (downward closure) 9
$\downarrow_{\preceq} F$ (downward closure) 399
$\downarrow E$ (downward closure) 56
$\downarrow x$ (downward closure) 9, 55
$\wedge$ 341
$\Rightarrow$ (residuation) 342, 394
$\neg$ (negation) 343, 429
$U \searrow y$ (elementary step function) 196
$\sqsubseteq$ 450
$\sqsubseteq_M$ 450
$\subsetneq$ 61
$\subseteq$ (inclusion map) 96
$\top$ 10
$\uparrow A$ (upward closure) 8
$\uparrow_{\preceq} F$ (upward closure) 399
υ 472
$\uparrow x$ (upward closure) 8
$\twoheaduparrow x$, $\twoheaduparrow E$ 124
$\vee$ 10
$\wedge$ 10
ξ 472
$<$ 8
$\ll$ (way-below) 120, 128
$\lll$ (way-way-below) 370
$-\!\!\!\ll$ 377
$\leqslant$ 8, 54
$\leqslant^*$ 464
$\leqslant^\sharp$ 145
$\preceq_{\leq}$ 469
$\leqslant_{\text{Rado}}$ 463
$\leqslant^{\text{pref}}$ 208
$\Diamond U$ 459
$\diamond U$ 356
$\Diamond U \cdot \mathcal{U}$ 469
$\Diamond U \cdot \mathcal{U} /\!/ V$ 470
$\Diamond U(\pi_1 \mid \ldots \mid \pi_n)$ 469
$\langle f, g \rangle$ (pairing of two maps) 75, 113
$=_0$ 104, 237
$\equiv$ 8
$>$ 8
$\geqslant$ 8
$\gg$ (way-above) 381
$[{}^{-1}(V) \in \mathcal{U}]$ 154
$[0, 1]_\sigma$ 60, 97
$[Q \subseteq U]$ 154
$\lfloor \mathcal{U} \rangle$ (telescope) 467

$[U \Subset^{-1} V]$ 151
$[X \to Y]$ 140, 149, 178
$[X \to Y]^{\mathrm{sup}}$ 243
$[X \to Y]_1$ 247
$[X \to Y]_{\mathcal{C}}$ 181
$[X \to Y]^{\mathrm{I}}$ 154
$[X \to Y]_c$ 247
$[X \to Y]^{\circledcirc}$ 151
$[X \to Y]^{\natural}$ 159
$[X \to Y]_{\mathrm{u}}$ 278
$[f, g]$ (copairing of two maps) 81
$[g_1 = g_2]$ (equalizer) 116

0 (initial object) 117
! (unique morphism) 113, 117
1 (terminal object) 113

$\alpha(D)$ (aperture) 312
$A + B$ (coproduct) 117
$A +_X B$ (pushout) 117
α 324
App (application) 149
αX (Alexandroff one-point compactification) 93
$A \times B$ (product) 113
$A \times_X B$ (pullback) 115

B (bifinite domains) 446
$\mathcal{B}$ (Baire space) 94, 101, 143, 162, 327, 335, 366
$\mathbf{B}(X, d)$ 280
$\mathbf{B}^c$ (functor) 286
BCDom (bc-domains) 200
β 324
$\perp$ 10, 64, 70, 196
$\Box U$ 374
βX (Stone–Čech compactification) 257
$B^d_{x,<\epsilon}$ (open ball) 22, 204
$B^d_{x,\leqslant\epsilon}$ (closed ball) 24, 280

Cat 173
CCLat 200
$\mathbf{C}^{\mathrm{ep}}$ 441
χ_U (characteristic map) 67
$\mathcal{C}hy(X, d)$ (Cauchy completion) 286
$cl(A)$ (closure) 51
CLat 341
coker(g_1, g_2) 118
colim F 117
$\mathbf{C}^{op}$ 117
$\mathcal{C}X$ 182

$\widehat{d}$ 286
$d(\_, A)$ 213
$D + s$ 313
$D - s$ 313
d-closed 294, 302
Cpo (category of dcpos) 112, 174
Δ (diagonal map) 75
∇ (codiagonal map) 81
$\delta(A)$ (diameter) 281
$d_{\mathcal{H}}$ (Hausdorff–Hoare quasi-metric) 312
d_ℓ 216
d^{op} 207
d^{sym} 207

ϵ (counit) 173
ε (empty word) 208, 463
η (unit) 173
$\eta_X^{\mathcal{C}hy}$ 287
η_X^{EH} 280
$\eta_{\mathbb{IR}}$ 62
$\eta_X^{\mathcal{S}}$ 357, 363
$\eta_X^{\mathrm{S\check{C}}}$ 257
η_X^{Stone} 350, 351
$\eta_X^{\mathbf{Y}}$ 322

$f(\_, y)$ 76
$f(x, \_)$ 76
$f^\star$ 363
$f + g$ (coproduct of two maps) 81
$f^{-1}(V)$ (inverse image) 38
$f[A]$ (image) 6, 39, 69
$f_{|B}$ 140
$\mathcal{Q}_{\mathrm{fin}}(X)$ 145
$\mathbb{F}(\mathcal{O}(X))$ 418
Frm 341
$f \times g$ (product of two maps) 75

Gal(L, L') 391, 462

$\mathcal{H}(X)$ 459
HMet 212
$\mathbf{HMet}_{\mathrm{u}}$ 322
H_n (harmonic number) 270
$\mathrm{Hom}_{\mathbf{C}}(X, X')$ (homset) 174

$\mathbb{I}$ 424
I_2 (dyadic numbers) 219
Id(R) 439
id_X (identity) 66, 112
inf A (infimum) 9
$int(A)$ (interior) 51

ι_i (injection) 80, 117
$\mathbb{IR}$ 62, 100, 338
$\mathbf{I}(X)$ (ideal completion) 131

$\mathbb{J}$ 143

$\mathcal{K}(L)$ 132, 425
$\ker(g_1, g_2)$ (pullback) 115
$\mathcal{KO}(X)$ 423

$\Lambda_X(f)$ (currification) 149
$\mathrm{lan}_{X,Y}$ 175
$\lim F$ 115
$\liminf_{i \in I, \sqsubseteq} r_i$ 264
$\limsup_{i \in I, \sqsubseteq} r_i$ 264
$\lim_{n \in \mathbb{N}} x_n$ 22
Loc 342

$\mathbf{Map}_\mathcal{C}$ 180
Max 337
Met 212
Mon 173
$mub(B)$ 447
$mub^\infty(B)$ 447

$\mathbb{N}$ 5
$\mathcal{N}_2$ 122
$\mathbb{N}_\omega$ 121

$\circ$ (composition) 66, 111
$\mathcal{O}^d$ 204
Ord (category of posets) 112, 174
$\mathcal{O}(X)$ 141

$\mathbb{P}(A)$ (powerset) 5, 7
PCpo (category of pointed dcpos) 112
$\mathbb{P}_{\mathrm{fin}}(A)$ 7
$\coprod_{i \in I} X_i$ (coproduct) 80
π_i (projection) 74, 113, 115
$\triangleright_{n=1}^{+\infty} X_n$ 467
$\prod_{i \in I} X_i$ (product) 74
$\prec$ 126
$\prod$ (product map) 385
$\mathsf{pt}(L)$ (space of points) 345

$\mathbb{Q}$ 7, 52, 92
$q_\equiv$ 8
QMet 212
quasi-ordered set 8

$\mathfrak{r}$ (Scott's Formula) 137
$\mathfrak{r}'(f)$ 138

$\mathcal{R}(B, \prec)$ 364, 381
$\overline{\mathbb{R}^+}$ 19
$\mathbb{R}^+_\sigma$ 60, 97
$\mathbb{R}/\mathbb{Q}$ 104, 238
$\mathrm{Rad}(R)$ 439
$\mathrm{ran}_{X,Y}$ 175
RB (retracts of bifinite domains) 450
$\mathbf{RI}(B, \prec)$ 127
$\mathbb{R}_\ell$ 53, 83, 85, 92, 143, 156, 216, 217, 228, 235, 268, 285, 298, 311, 327, 431
$\mathbb{R}_\sigma$ 60

$\mathbb{S}$ (Sierpiński space) 50
$\mathbf{S}(X, d)$ (formal ball completion) 312
Set (category of sets) 112
Σ 208
Σ^* 208
σ_X 313
$\mathrm{spec}(L)$ (spectrum) 430
$\sup A$ (supremum) 9
$\sup_X d$ (sup hemi-metric) 41, 241
$\mathcal{S}(X)$ 356

$\mathcal{T}(X)$ 468
$t[s]_p$ 468
$t_{|p}$ 468
$\otimes$ 239
$\rightarrow^*$ (reflexive-transitive closure) 14
$\rightarrow^+$ (transitive closure) 14
Top (category of topological spaces) 112
$\top$ 10, 52, 196
$\mathbf{Top}_0$ (category of T_0 spaces) 113, 177
$\mathbf{Top}_\mathcal{C}$ 182
$\mathcal{T}(\sigma)$ 328
$T^d_{x, >\epsilon}$ (open hole) 211

$\mathcal{U}/U$ 464

$\bigvee_{i \in I} d_i$ 240

YCQMet 322
$\mathbf{YCQMet_{u,Y}}$ 322
$\mathbf{YCQMet_Y}$ 322
$W_{|x}$ 76
$W^{|y}$ 76

$X^{\circledast}$ 467
X^I 78
X_a 56
$X_\perp$ 72
X^{d} 403
X_λ 192, 402

X^{op} 8, 240
X^{patch} 404
X_{Rado} 463
X_σ 59, 361
$X^\top$ 196
X_u 57
$\mathbf{Y}(X, d)$ (Yoneda completion) 322
Y^X 169, 171, 249
$\mathbb{Z}$ 7
$\mathbb{Z}^-$ 195

Index

A. H. Stone's Theorem 226
above 8
abstract basis 126
AC 15
ACC 454
accessible 50
adjunction 173
admissible 196
Alexander's Subbase Lemma 73
Alexandroff one-point compactification 93
  topology 56
Alexandroff-discrete 57
algebraic hemi-metric space 308
  poset 122
  set 474
almost bottom 311
  open map 100
α-history 324
α-strategy 324
alphabet 208
antichain 460
antisymmetric 8
antitone 390
aperture 312
apex 114, 117
application map 149, 472
applicatively continuous 163
approximating sublattice 434
Arens' Theorem 258
arithmetic lattice 376
Arzelà–Ascoli Theorem 43, 246
  –Kelley Theorem 168
ascending chain condition 454
  sequence 263, 275
associative 112
associator 248
Axiom of Choice 15

b-space 130, 372
bad sequence 457, 459
Baire 327, 373
  space 94, 101, 143, 162, 327, 335, 366
Banach Fixed Point Theorem 34, 271
  –Mazur game 327
  –Mazur–Oxtoby Theorem 331
  –Rutten Fixed Point Theorem 300
base (of a topology) 47
basis 309
  (of a poset) 125
bc-domain 196, 418
bdcpo 136
below 8
β-history 324
β-strategy 324
bicontinuous poset 381
bifinite domain 440
bilimit 446
binary relation 8
Boolean algebra 343, 429
Borel–Lebesgue property 31, 67
  Theorem 28
bottom 10
bounded hemi-metric 215
  lattice homomorphism 425
  map 417
  poset 10, 196
Bourbaki–Witt Fixed Point Theorem 11
branch 328

$\mathcal{C}$-continuous 180
$\mathcal{C}$-generated 182
c-Lipschitz 34, 209
$\mathcal{C}$-probe 180
c-space 129
Cantor's Theorem 7, 281

cardinality 16
Caristi–Waszkiewicz Fixed Point Theorem 301
Cartesian-closed 178
category 111
Cauchy completion 286
  net 262
  sequence 33, 262
  -weightable net 269
  -weighted net 269
cc-topology 157
CCC 178
Čech-complete 333, 366
center 22, 24, 204, 211, 280
chain 10
  -complete poset 10
  -open 62
characteristic map 67
Choquet game 324
  -complete 325
clopen 47, 53, 107, 431
closed ball 24, 280
  subset 22, 47
closure 51
cluster point 86
coarse topology 50
coarser (topology) 47
cocompact 403
  topology 403
cocomplete category 119
cocone 117
codiagonal map 81
coefficient 473
coequalizer 118
cofinal 61, 87
cofinite topology 60
coherence conditions 248
  diagram 173
coherent 145
cokernel pair 118
colimit 117
collection 5
comb 109
commutator 249
compact 27, 67
  convergence 157
  -open subset 200, 372, 423
  -open topology 154
compactly generated 189, 192, 410
compactwise uniform convergence 244
complement 343, 429
complete 289
  abstract basis 381
  category 116
  Heyting algebra 342
  lattice 9
completely distributive 370
  prime filter 344
  regular 213
completion 286, 312
composition 66, 111
cone 114
conjoining (topology) 153
connected 107
  component 109
consistently complete 195
consonant 156, 366
continued fraction 101
continuity at x 66, 209
  in the first (resp., second) argument 76
continuous (function) 38, 63, 209
  convergence 41, 158
  hemi-metric space 311, 356
  poset 122
convergence 20, 22, 83, 326
  Choquet-complete 326
  from the right 83
converging continuously 41, 158
copairing 81
coproduct 117
  hemi-metric 236
  of two maps 81
  quasi-ordering 81
  topology 80
core-coherent 144
core-compact 141
core-compactly generated space 187
core-open topology 151
coreflection 177
coreflective subcategory 177
counit 173
countable 6
countably infinite 6
  -based 216
cover 67
currification 149

Day's Theorem 184
d-convergence 265
dcpo 61

d-dense 316
de Groot dual 403
deflation 443
degree 473
dense subset 52
densely injective 418
descending chain 14
  sequence 85, 333
d-finite 306
diagonal map 75
diameter 281
dichotomy 107
Dickson's Lemma 463
Dieudonné's Theorem 223
directed 59, 127
  -complete 61
  Hilbert cube 253
discrete topology 50
distance 203
distributive lattice 343
d-limit 265
Dolecki–Greco–Lechicki Theorem 366
domain of a function 6
double quotient 238, 384
downward closed 9, 127
  closure 9
  scale 220, 222
  topology 401
d-Scott topology 302
duration 237
dyadic number 219

ϵ-η formula 38
ϵ-fattening 206
ec-topology 157
edge 111
elementary open 464, 469
  step function 196
embedding 100, 464, 469
empty word 208
endofunctor 113
ep-pair 440
ep-system 441
equalizer 116
equicontinuity 245
equicontinuous 245
equivalence class 8
  of categories 177
  relation 8
Escardó–Lawson–Simpson Lemma 181, 188
Euclidean metric 20, 281
even continuity 166
  convergence 156, 157
evenly continuous 166
eventually 83
  ascending 263
exponentiable 169
exponential 169, 171, 249
  topology 153
extension 137
extension by continuity 287

$\mathcal{F}$-usc 222
far below 198
fattening 206
field of sets 430
filter 89, 344, 418
filtered 68
finer (topology) 47
finitary closed 56
  compact 72
finite 6, 122
  category 116
  powerset 7
finitely cocomplete 119
finitely complete 116
first projection 113
  -countable 85
fixed point 11
  induction 13
forgetful functor 113
formal ball 279, 280
forward-Cauchy net 262
frame 341
  homomorphism 341
free category 112
  complete metric space 286, 319
  locale 350
  object 175
  sober space 363
Frolík's Theorem 333
from the right (convergence) 83
F_σ subset 221, 336
FS-domain 450
full subcategory 170
fully arithmetic lattice 398
functor 113

Galois connection 390
gauge 252
G_δ subset 221, 329, 335
generalized Banach Fixed Point Theorem 271

Scott topology 304
generated by (ideal) 439
(topology) 48
generating system of deflations 449
idempotent deflations 443
graph 111
morphism 112
greatest 9
lower bound 9
ground term 468

harmonic number 270
has cardinality smaller than 16
Hausdorff 49
–Hoare quasi-metric 312
Maximality Theorem 15
Hausdorff's Theorem 332
Heine's Theorem 210
Heine–Borel property 31, 67
hemi-metric 203
space 203
hemi-metrizable 215
hereditarily compact 454
heterogeneous word 467
Heyting algebra 342, 430
Higman's Lemma 467
Hilbert Basis Theorem 473
cube 253
Hilbert's Nullstellensatz 474
history 324
Hoare powerspace 459
Hochster's Theorem 438
Hofmann–Lawson Theorem 372
–Mislove Theorem 365
homeomorphism 98
homset 174

ideal 131, 439
completion 131
idempotent deflation 443
identity 66, 112
image 6, 39, 69
implication 342
inclusion functor 177
map 96
incomparable 10
indiscrete topology 50
induced topology 95
induction hypothesis 15
inductive limit 433
poset 16
inf-semi-lattice 10
infimum 9
infinite 6
word 7
inflationary 11
initial object 117
injection 80
interior 51
intermediate value 108
interpolating compact 94
interpolation 94, 123, 126, 218, 220, 371
interval 107
inverse image 38
irrational number 101, 335
irreducible 349, 351
irreflexive 15
Isbell topology 154
iso 112
isolated point 350
isometric embedding 212
isometry 212
isomorphism 98, 112
iterate 34, 271, 301

Johnstone space 143
Johnstone's Theorem 427
joint continuity 76

k-closed 189
k-open 189
k-topology 189
Kaplansky's Theorem 473
Kelley's Theorem 90
kernel pair 118
Kleene Fixed Point Theorem 64
–Rutten Fixed Point Theorem 301
Knaster–Tarski Fixed Point Theorem 13
Kolmogoroff 50
Kostanek–Waszkiewicz Theorem 298, 361
Kruskal's Theorem 472

L^1 metric 20
L^2 metric 20, 281
large enough 83
lattice 10
Lawson at the top 339, 403
topology 192, 402
leading (monomial, coefficient) 473
least 9
upper bound 9
Lebesgue number 30

left adjoint 173
  unit 248
length 208
letter 208
level 328
lifting 72
limit 20, 83, 115
  inferior 264
  superior 264
linear ordering 10
Lipschitz 34, 209
locale 342
locally compact 91
  finitary compact 131
  finite 223
  spectral 423
lower semi-continuous 65
  Vietoris topology 356, 459
L^p metric 20
lsc 65

Martin's First Theorem 339
  Second Theorem 339
maximal 9
metric space 20
  topology 46
metrizable 215
minimal 9
  upper bounds 447
minimality of compact T_2 topologies 72
model 337
modular 145
monoid 173
monoidal 248
  category 249
  closed 249
monomial 473
monotone convergence space 361
  net 84
monotonic 10
morphism 97, 112
mub-closure 447
multifunction 15
multiplicative 144
multiset 467

Nachbin pospace 406
natural isomorphism 177
  number 5
  topology 159
  transformation 173
negation 343, 429
neighborhood 95, 365
net 83
Noetherian induction 14
  part 14
  relation 14
  ring 472
  space 453
non-trivial filter 90, 418
normal space 51

object 112
one-point compactification 93
open ball 22, 204
  topology 48, 204
open cover 67
open hole 211
  topology 211
open map 98
open neighborhood 53
open rectangle 74
open subset 24, 46
opposite 207
  category 117
  ordering 8
order T_1 400
  T_2 400
  embedding 10
  isomorphism 10
  normal 399
  -preserving 421
ordering 8

pairing 75
paracompact 222
partial ordering 8
partially ordered set 8
  space 399
partition of unity 224
patch topology 404
  -closed 404
  -compact 404
  -continuous 419
  -open 404
path 108
  -connected 108
perfect map 419
plane 18
Plotkin's tie T 448
point 46, 345
  at infinity 93

pointed poset 10, 64, 70, 134
pointwise closed 165
  convergence 40, 89, 140, 162
  ordering 79, 140, 178
Polish space 334, 375, 411
polynomial 473
Ponomarev's Theorem 217
poset 8
  adjunction 389
  coproduct 82
  product 79
position 468
pospace 399
post-fixed point 14
potential 301
power 78
powerset 5, 7, 415, 430
  (finite) 7
precompact 36, 251
prefix 208
  quasi-ordering 467
preservation of limits 85
preserving $\ll$ 138
Priestley order separation 431
  space 431
prime element 347
prime ideal 439
  -continuous 370
product 113, 115
  (of sets) 16
  hemi-metric 239
  map 385
  of posets 79
  of two maps 75
  quasi-ordering 78
  topology 74
productive class 181, 188
projection 74, 113, 115, 440
projective limit 436
proper map 420
property m 447
  T 457
  W 457
pullback 115
pushout 117

quasi-continuous 128
  -metric 203
  -metric space 203
  -ordering 8
quotient 8
  hemi-metric 237
  map 103
  topology 102

radially convex 257
radical ideal 439
radius 22, 24, 204, 211, 280
Rado structure 463
Rado's Selection Lemma 78
Raney's Theorem 370
rational number 7
RB-domain 448
real plane 18
rectangle 74, 392
refinement 222
reflecting $\ll$ 138
reflection 177
reflective subcategory 177
reflexive 8
  -transitive closure 14
regular closed 349
  element 344
  open 344
  space 50
relation 8
relatively compact 141, 168
restriction 140
retract 104, 106, 172, 180
  of bifinite domain 448
retraction 104, 172, 180
retractive system 441
right (convergence from) 83
  adjoint 173
  unit 248
ring 438
  of sets 430
Romaguera–Valero Theorem 283, 379
rounded ideal 127
  completion 127
rounded open 364
  topology 364
Rudin's Lemma 145

sampler 87
$\equiv$-saturated 102
saturated 56
saturation 56
scale 220, 222
Scott domain 200
  topology 59
Scott's Formula 137

Theorem 417
Scott-closed 59
-continuous 63
-open 59
-retract 106
second projection 113
second-countable 216
section 104, 172, 180
selection 15
self-convergent 358
separable 233
metrizable 234
separate continuity 76
separating maps 450
sequence 20
sequential space 194
sequentially closed 22, 82
compact 27
continuous 38
open 24
set 5
short-lex 467
Sierpiński space 50
σ-compatible 328
simple tree expression 469
singleton 50
size 468
small 5
SMCC 250
Smyth powerdomain 373
-complete 268, 308, 317, 326, 337, 354, 379
quasi-ordering 145
sober 283, 353
sobrification 356
Sorgenfrey line 53, 83, 85, 92, 143, 156, 216, 217, 228, 235, 268, 285, 298, 311, 327, 431
plane 228
source 111
spatial lattice 346
spatialization 347
specialization
quasi-ordering 54
spectral 423
map 429
spectrum 430, 439
splitting (topology) 153
stabilize 454
stable scale 222
stably compact 397
locally compact 375
stationary strategy 326
step function 196
Stone duality 344, 429
–Čech compactification 257
Representation Theorem 430
space 429
topology 430
strategy 324
strict 215
ordering 15
part 8
strictly descending 14
strong basis 309
strongly locally compact 372
productive class 181
subadditive 215
subbase (of a topology) 48
subcategory 170
subcover 67
subnet 87
subsequence 27, 87
subspace 240
topology 95
substitution 472
subterm 468
subword 464
$\rightarrow$-successor 14
sup hemi-metric 240, 241
metric 41
-semi-lattice 10
supremum 9
sym-closed 277
symcompact 252, 409
symmetric 8
monoidal 249
monoidal closed 250
symmetrization 207
symmetry 20

T_0 50, 203
T_0 quotient 104, 237
T_1 50, 203
T_2 49
T_3 50
$T_{3\frac{1}{2}}$ 213
T_4 51
target 111
Tarski's Fixed Point Theorem 13
telescope 467
tensor product 239, 248
unit 248

term 468
terminal object 113
thin category 343
Tietze–Urysohn Extension Theorem 222
top 10
topological product of hemi-metric spaces 240
topological space 46
topologically equivalent 214
topology 46
  defined by a family of hemi-metrics 216
total ordering 10
totally bounded 252
  ordered 10
transformation 173
transitive 8
  closure 14
tree 328, 468
  embedding 469
  topology 469
triangular inequality 19, 20
trip 237
Tychonoff's Theorem 32, 76

$\mathcal{U}$-lsc 222
$(\mathcal{U}, \mathcal{F})$-continuous 222
ultra-hemi-metric 208
  -metric 208
  -quasi-metric 208
  filter 90, 430
  net 89
uncountable 7
uniform convergence 40, 242
uniformly approximated space 452
  approximating 451
  continuous 42, 210
unit 173
universal cone 114
up to iso 113
upper bound 9
  semi-continuous 65
  topology 57
upward closed 8, 364
  closure 8
  scale 220, 222
  topology 401
Urysohn Lemma 218
  –Carruth Theorem 257
  –Nachbin Theorem 421
  Property 218
usc 65

value of a game 324
vertex 111
Vietoris topology 356, 459

ω-continuous 336
wavy cross 76
way-above 381
  -below 120, 128
  -way-below 370
weakly productive class 188
well quasi-ordering 460
well-filtered 71, 366
well-founded 14
  induction 14
wide telescope 467
winning strategy 324
word 208, 463
  embedding 464
  topology 464
wqo 460

Yoneda completion 322
  -complete 289, 325, 356, 361
  -continuity 299
  Lemma 322

Zariski topology 439, 474
zero-dimensional 53, 429, 431
ZF 4
ZFC 4
Zorn's Lemma 16

Printed in the United States
by Baker & Taylor Publisher Services